# Lecture Notes in Mechanical Engineering

**Lecture Notes in Mechanical Engineering (LNME)** publishes the latest developments in Mechanical Engineering—quickly, informally and with high quality. Original research reported in proceedings and post-proceedings represents the core of LNME. Volumes published in LNME embrace all aspects, subfields and new challenges of mechanical engineering. Topics in the series include:

- Engineering Design
- Machinery and Machine Elements
- Mechanical Structures and Stress Analysis
- Automotive Engineering
- Engine Technology
- Aerospace Technology and Astronautics
- Nanotechnology and Microengineering
- Control, Robotics, Mechatronics
- MEMS
- Theoretical and Applied Mechanics
- Dynamical Systems, Control
- Fluid Mechanics
- Engineering Thermodynamics, Heat and Mass Transfer
- Manufacturing
- Precision Engineering, Instrumentation, Measurement
- Materials Engineering
- Tribology and Surface Technology

To submit a proposal or request further information, please contact the Springer Editor of your location:

**China:** Dr. Mengchu Huang at mengchu.huang@springer.com
**India**: Priya Vyas at priya.vyas@springer.com
**Rest of Asia, Australia, New Zealand:** Swati Meherishi
at swati.meherishi@springer.com
**All other countries:** Dr. Leontina Di Cecco at Leontina.dicecco@springer.com

To submit a proposal for a monograph, please check our Springer Tracts in Mechanical Engineering at http://www.springer.com/series/11693 or contact Leontina.dicecco@springer.com

**Indexed by SCOPUS. All books published in the series are submitted for consideration in Web of Science.**

More information about this series at http://www.springer.com/series/11236

Abdelmjid Saka · Jean-Yves Choley ·
Jamel Louati · Zakaria Chalh ·
Maher Barkallah · Mohammed Alfidi ·
Mounir Ben Amar · Fakher Chaari ·
Mohamed Haddar
Editors

# Advances in Integrated Design and Production

## Proceedings of the 11th International Conference on Integrated Design and Production, CPI 2019, October 14–16, 2019, Fez, Morocco

Springer

*Editors*
Abdelmjid Saka
Ecole Nationale des Sciences Appliquées
Fès, Morocco

Jamel Louati
National School of Engineers of Sfax
Sfax, Tunisia

Maher Barkallah
National School of Engineers of Sfax
Sfax, Tunisia

Mounir Ben Amar
Laboratoire des Sciences des Procédés
Villetaneuse, France

Mohamed Haddar
National School of Engineers of Sfax
Sfax, Tunisia

Jean-Yves Choley
Institut supérieur de mécanique de Paris
Saint-Ouen, France

Zakaria Chalh
Ecole Nationale des Sciences Appliquées
Fès, Morocco

Mohammed Alfidi
Ecole Nationale des Sciences Appliquées
Fès, Morocco

Fakher Chaari
National School of Engineers of Sfax
Sfax, Tunisia

ISSN 2195-4356 ISSN 2195-4364 (electronic)
Lecture Notes in Mechanical Engineering
ISBN 978-3-030-62198-8 ISBN 978-3-030-62199-5 (eBook)
https://doi.org/10.1007/978-3-030-62199-5



This Springer imprint is published by the registered company Springer Nature Switzerland AG
The registered company address is: Gewerbestrasse 11, 6330 Cham, Switzerland

# Preface

Design and manufacturing represent the technical activities that have played a fundamental role in creating and shaping products and systems.

Starting in the mid-90s, the CPI conference attempts to provide answers to current questions in the fields of design and manufacturing.

CPI2019 was the 11th in a series of biennial events in the field of integrated design and manufacturing. The general objective of this conference is to contribute and discuss the fundamental principles, applications and experiences in the field of new technologies such as industry 4.0, cyber-physical systems and the Internet of things.

CPI2019 has been organized in cooperation with the Laboratory of Engineering Systems and Applications, ENSA, Fez, Morocco, the Mechanics, Modeling and Production Research Laboratory, ENIS, Sfax, Tunisia and the QUARTZ, SupMéca, Paris, France, from October 14 to 16, 2019.

CPI2019 offered again an exciting technical program as well as networking opportunities. Outstanding scientists and industry leaders accepted the invitation for keynote speeches:

- Professor Jean Yves CHOLEY, Laboratory QUARTZ, SUPMECA, Paris, France.
- Professor Aiguo MING, Department of Mechanical and Intelligent Systems Engineerings, UEC Tokyo, Japan.
- Professor Nacer HAMZAOUI, Laboratory Vibration and Acoustic, INSA Lyon, France.
- Professor Samir LAMOURI, Ecole Supérieure des Arts et Métiers, Paris, France.
- Professor Ahmed Rahmani, Center de Recherche en Informatique, Signale et Automatique, École centrale de Lille, France.
- Professor Mohamed Najib ICHCHOU, Laboratory of Tribology and System Dynamic, Ecole centrale de Lyon, France.
- Professor Marc ZOLGHADRI, Laboratory QUARTZ, SUPMECA, Paris, France.

- Professor Mounir BEN AMAR, LSPM, CNRS, Université Paris, France.
- Professor Hind Bril El Haouzi, Department of Mechanical Engineering, Université de Lorraine, CRAN.
- Professor Abdelkhalek EL HAMI, Laboratory Optimization and Reliability in Structural Mechanics, INSA-Rouen, France.
- Professor Sid-Ali ADDOUCHE, Laboratory QUARTZ, Université Paris8, France.

The CPI2019 conference takes up the following topics in its variety and discusses the state of the art and future trends under the global theme "Smart Industry":

- BI and big data
- Cloud computing
- Cyber-physical production systems
- Design of complex and multiphysical systems
- Dependability of industrial systems
- Digital factory
- Dynamics of structures and machines
- Embedded systems
- E-learning
- Green economy—recycling, remanufacturing, reuse and recovery of waste
- Industrial instrumentation
- Industrial process engineering—quality, maintenance and logistics
- Internet of things
- Knowledge management systems—collaborative engineering
- Machine learning
- Manufacturing processes—eco-manufacturing
- Materials—behavior and structure
- Mechatronics
- Modeling, identification and control of dynamic systems
- Modeling theory: languages and tools
- Modeling and design of products and production systems—eco-design
- Multimedia processing
- Numerical control and industrial supervision—diagnostics
- Optimization of products and processes
- Planning and scheduling of industrial systems
- Plant 4.0: Additive manufacturing—technologies, materials and industrialization
- PLM
- Robotics
- Wireless sensors
- Industrial applications and experiences on these themes

All contributions were subject to a double-blind review. The review process was very competitive. We had to review near 230 submissions. A team of 95 reviewers did this terrific job. Our special thanks go to all of them.

Due to the time and conference schedule restrictions, we could finally accept only the best 120 submissions for presentation. The conference had again about 160 participants from different countries from all continents.

The following submission types have been accepted:

- Full paper, short paper
- Work in progress, poster
- Special sessions
- Workshops, tutorials

The proceedings essentially consist of full paper submissions by proposed by the moderators of the sessions after oral presentation of the paper.

Finally, the epilogue brings the proceedings to a conclusion. It discusses many of the challenges in designing and manufacturing in the context of new technologies.

We would like to thank the organizing committee, scientific committee and all participants coming from Tunisia, Algeria, France, Saudi Arabia, Spain and Portugal. Thanks to Springer for support of CPI2019.

This conference would not have appeared without the support of several persons and companies. We are grateful to Sidi Mohamed Ben Abdellah University for the support. Special thanks also to the National School of Applied Sciences of Fez mainly to the Director who support greatly the conference. We thank also Association Marocained'Ingénierie et de Développement (AMID). And finally, of course, great thanks to all the others sponsors.

Abdelmjid Saka
Jean-Yves Choley
Jamel Louati
Zakaria Chalh
Maher Barkallah
Mohammed Alfidi
Mounir Ben Amar
Fakher Chaari
Mohamed Haddar

# Contents

# $H_\infty$ Control for Half Vehicle Active Suspension Systems in Finite Frequency Domain

Loubna Mrabah[1(✉)], Jamal Mrazgua[2], Mohamed Ouahi[1], and EL Houssaine Tissir[2]

[1] Engineering, Systems and Applications Laboratory National School of Applied Sciences (ENSA), University of Sidi Mohammed Ben Abdellah, Fes, Morocco
mrabahloubna@gmail.com, mohamed.ouahi@usmba.ac.ma

[2] Faculty of Sciences, LESSI department of physics, University Sidi Mohammed Ben Abdellah, B.P. 1796, Fes-Atlas, Fes, Morocco
{jamal.mrazgua,elhoussaine.tissir}@usmba.ac.ma

**Abstract.** This article is devoted to the study of $H_\infty$ control problem of active suspension systems for the vertical model of half-vehicle in the finite frequency domain, and the design a state-return controller based on linear matrix inequalities (LMIS) such as the resulting closed-loop system is asymptotically stable, by basing on the generalized Kalman_ Yacubovich_ Popov (GKYP) Lemma to isolate the body from road disturbances and improve the ride comfort in the specific frequency band [4–8] hz. The theory is illustrated by simulation results to demonstrate the effectiveness of the proposed method.

**Keywords:** $H_\infty$ control · Half-vehicle · Active suspension systems · Finite frequency · GKYP lemma · Linear matrix inequalities

## 1 Introduction

Car suspension plays a vital rule in handling of a vehicle, it must not only guide the wheels to ensure optimal contact between the wheel and the road, but also it must be able to filter effectively unevenness of the road for the comfort of the occupants of the vehicle. Traditional car suspensions have been the subject of much research to optimize their performance. In recent years, several studies have shown that the most realistic way to improve the performance of modern suspensions is the addition of active suspension systems [11]. Many approaches have been presented such as robust $H_\infty$ approach [5,12], and [18], $H_\infty$ control [4,7,20], and [21], and Fuzzy logic control [8,13,15], and [16]. More attention has been devoted to the $H_\infty$ control of active suspension.

However, active suspension systems may only belong to certain frequency bands, and driving comfort is known to be sensitive to the frequency. From the ISO-2631 standard. The most important objective for vehicle suspension


A. Saka et al. (Eds.): CPI 2019, LNME, pp. 1–12, 2021.
https://doi.org/10.1007/978-3-030-62199-5_1

systems is the improvement of driving comfort. In other words, the main task is to design a controller that stabilizes the vertical movement of the vehicle body and thus isolating the force transmitted to passengers [1], and [14]. These results can actually achieve performance of suspension of the desired vehicle, in particular driving comfort. It's worth it mention that most approaches are considered in the frequency domain [3,6,7,15–17], and [19]. Recently, various control approaches have been proposed for vehicle systems. For example, Control For Vehicle Active Suspension Systems In Finite Frequency Domain [21]. The same work was done here for the half-vehicle model including pitch and heave modes was invented to simulate ride characteristics of a simplified whole vehicle, which leads to significant improvement in ride and handling [2], and [8].

The Kalman–Yakubovich–Popov lemma (KYP) [9,10]. It only applicable for the finite frequency band and it allows us to characterize various properties of dynamical systems in the frequency domain in terms of LMI.

Here, we searched the $H_\infty$ control problem over finite frequency ranges for vehicle active suspension systems. The generalized Kalman–Yakubovich–Popov (KYP) lemma, gives sufficient conditions that are established to ensure that the associated error system is stable and satisfies a prescribed $H_\infty$ performance in finite frequency domain. The theoretical results provided as LMIs can be solved with numerical example to illustrate the effectiveness of the proposed approach. **Notations** : A>0 and A<0 denote positive definiteness and negative definiteness, $*$ denote the symmetric terms in symmetric matrix, $I$ denote the identity matrix with appropriate dimension.

## 2 Problem Statement and Preliminaries

A Half Vehicle Suspension Model is shown in Fig. 1. The differential equations describing such systems can be found in [8].

Let us define the following state variables:
$x_1(t)$=$z_{sf}(t)$-$z_{uf}(t)$ as the front suspension deflection.
$x_2(t)$=$\dot{z}_{sf}(t)$ as the vertical speed of mass suspended before.
$x_3(t)$=$z_{sr}(t)$-$z_{ur}(t)$ as the rear suspension deflection.
$x_4(t)$=$\dot{z}_{sr}(t)$ as the vertical speed of mass suspended back.
$x_5(t)$=$z_{uf}(t)$-$z_{rf}(t)$ as the front tire deflection.
$x_6(t)$=$\dot{z}_{uf}(t)$ as the vertical speed of unsprung mass before.
$x_7(t)$=$z_{ur}(t)$-$z_{rr}(t)$ as the rear tire deflection.
$x_8(t)$=$\dot{z}_{ur}(t)$ as the vertical speed of unsprung mass back.
$w(t)$ as the disturbance input.

$$w(t) = \begin{bmatrix} \dot{z}_{rf}(t) \\ \dot{z}_{rr}(t) \end{bmatrix}$$

Defining : – The state vector

$$x(t) = \begin{bmatrix} x_1(t)\; x_2(t)\; x_3(t)\; x_4(t)\; x_5(t)\; x_6(t)\; x_7(t)\; x_8(t) \end{bmatrix}^\top$$

– The input vector

$$u(t) = \begin{bmatrix} u_f(t) \\ u_r(t) \end{bmatrix}$$

We can write the dynamic equation in the following state-space from:

$$\dot{x}(t) = Rx(t) + Mw(t) + Nu(t) \tag{1}$$

where $x(t) \in \mathbb{R}^8$ is the state vector; $u(t)$ is the input vector; and $w(t)$ is the disturbance input. With

$$R = \begin{bmatrix} 0 & 1 & 0 & 0 & 0 & -1 & 0 & 0 \\ -k_{sf}a_1 & -c_{sf}a_1 & -k_{sr}a_2 & c_{sr}a_2 & 0 & c_{sf}a_1 & 0 & c_{sr}a_2 \\ 0 & 0 & 0 & 1 & 0 & 0 & 0 & -1 \\ -k_{sf}a_2 & -c_{sf}a_2 & -k_{sr}a_3 & -c_{sr}a_3 & 0 & c_{sf}a_2 & 0 & c_{sr}a_3 \\ 0 & 0 & 0 & 0 & 0 & 1 & 0 & 0 \\ \frac{k_{sf}}{m_{uf}} & \frac{c_{sf}}{m_{uf}} & 0 & 0 & \frac{-k_{tf}}{m_{uf}} & \frac{-c_{sf}}{m_{uf}} & 0 & 0 \\ 0 & 0 & 0 & 0 & 0 & 0 & 0 & 1 \\ 0 & 0 & \frac{k_{sr}}{m_{uf}} & \frac{c_{sr}}{m_{uf}} & 0 & 0 & \frac{-k_{tr}}{m_{ur}} & \frac{-c_{sr}}{m_{ur}} \end{bmatrix}$$

$$N = \begin{bmatrix} 0 & 0 \\ a_1 & a_2 \\ 0 & 0 \\ a_2 & a_3 \\ 0 & 0 \\ \frac{-1}{m_{uf}} & 0 \\ 0 & 0 \\ 0 & \frac{-1}{m_{ur}} \end{bmatrix}, \; M = \begin{bmatrix} 0 & 0 \\ 0 & 0 \\ 0 & 0 \\ 0 & 0 \\ -1 & 0 \\ 0 & 0 \\ 0 & -1 \\ 0 & 0 \end{bmatrix} \tag{2}$$

and

$$a_1 = \frac{1}{m_s} + \frac{l_1^2}{I_\theta}; \quad a_2 = \frac{1}{m_s} - \frac{l_1 l_2}{I_\theta}; \quad a_3 = \frac{1}{m_s} + \frac{l_1^2}{I_\theta}.$$

In the longitudinal half-vehicle model, the two pitching movements $\ddot{\theta}(t)$ and the vertical acceleration center-of-mass $\ddot{z}_c(t)$ are chosen as the control output $z_1(t)$. So to improve ride comfort it is mandatory to keep the transfer function as small as possible over the frequency band (4–8) Hz and we should ensure the firm uninterrupted contact of wheels to road and the dynamic tire load should be small. That are

$$\begin{aligned} k_{tf}(z_{uf}(t) - z_{rf}(t)) &\leq F_f \\ k_{tr}(z_{ur}(t) - z_{rr}(t)) &\leq F_r \\ z_{sf}(t) - z_{uf}(t) &\leq z_{fmax} \\ z_{sr}(t) - z_{ur}(t) &\leq z_{rmax} \end{aligned} \tag{3}$$

Where $z_{fmax}$ and $z_{rmax}$ are the maximum suspension deflection.

The static loads of the tires can be calculated by:

$$F_f + F_r = (m_s + m_{uf} + m_{ur})g$$
$$F_r(l_1 + l_2) = m_s g l_1 + m_{ur} g(l_1 + l_2) \tag{4}$$

The other constraint imposed is from the limited power of the actuator, that are

$$|u_f(t)| \leq u_{fmax}$$
$$|u_r(t)| \leq u_{rmax} \tag{5}$$

In order to satisfy the performance requirements, the controlled outputs are:

$$z_1(t) = \begin{bmatrix} \ddot{z}_c(t) \\ \ddot{\theta}(t) \end{bmatrix}$$
$$z_2(t) = \begin{bmatrix} \frac{z_{sf}(t)-z_{uf}(t)}{z_{fmax}} & \frac{z_{sr}(t)-z_{ur}(t)}{z_{rmax}} & \frac{k_{tf}(z_{uf}(t)-z_{rf}(t))}{F_f} & \frac{k_{tr}(z_{ur}(t)-z_{rr}(t))}{F_r} & \frac{u_f(t)}{u_{fmax}} & \frac{u_r(t)}{u_{rmax}} \end{bmatrix}^T \tag{6}$$

Taking account of (6), the vehicle suspension control system is described by:

$$\dot{x}(t) = Rx(t) + Mw(t) + Nu(t)$$
$$z_1(t) = S_1 x(t) + F_1 u(t)$$
$$z_2(t) = S_2 x(t) + F_2 u(t) \tag{7}$$

where $R$, $M$, and $N$ are defined in (2), and

$$S_1 = \begin{bmatrix} \frac{-k_{sf}}{m_s} & \frac{-c_{sf}}{m_s} & \frac{-k_{sr}}{m_s} & \frac{-c_{sr}}{m_s} & 0 & \frac{c_{sf}}{m_s} & 0 & \frac{c_{sr}}{m_s} \\ \frac{l_1 k_{sf}}{I_\theta} & \frac{l_1 c_{sf}}{I_\theta} & \frac{-l_2 k_{sr}}{I_\theta} & \frac{-l_2 c_{sr}}{I_\theta} & 0 & \frac{-l_1 c_{sf}}{I_\theta} & 0 & \frac{-l_2 c_{sr}}{I_\theta} \end{bmatrix}, \quad F_1 = \begin{bmatrix} \frac{1}{m_s} & \frac{1}{m_s} \\ \frac{-l_1}{I_\theta} & \frac{-l_2}{I_\theta} \end{bmatrix}$$
$$S_2 = \begin{bmatrix} \frac{1}{z_{fmax}} & 0 & 0 & 0 & 0 & 0 & 0 & 0 \\ 0 & 0 & \frac{1}{z_{rmax}} & 0 & 0 & 0 & 0 & 0 \\ 0 & 0 & 0 & 0 & \frac{k_{tf}}{F_f} & 0 & 0 & 0 \\ 0 & 0 & 0 & 0 & 0 & 0 & \frac{k_{tr}}{F_r} & 0 \\ 0 & 0 & 0 & 0 & 0 & 0 & 0 & 0 \\ 0 & 0 & 0 & 0 & 0 & 0 & 0 & 0 \end{bmatrix}, \quad F_2 = \begin{bmatrix} 0 & 0 \\ 0 & 0 \\ 0 & 0 \\ 0 & 0 \\ \frac{1}{u_{fmax}} & 0 \\ 0 & \frac{1}{u_{rmax}} \end{bmatrix} \tag{8}$$

Denote $G(j\omega)$ as the transfer function. The finite frequency $H_\infty$ control problem is to design a controller such that the closed-loop system guarantees

$$sup_{\omega_1<\omega<\omega_2} \|G(j\omega)\|_\infty < \gamma \tag{9}$$

where $\gamma > 0$. From the safety and mechanical structure, the following constraints should be guaranteed.

$$|u_f(t)| \leq u_{fmax}$$
$$|u_r(t)| \leq u_{rmax} \tag{10}$$

$$|\{z_2(t)\}_i| \leq 1 \ , \ i = \ 1, \ 2, \ 3, \ 4, \ 5, \ 6 \tag{11}$$

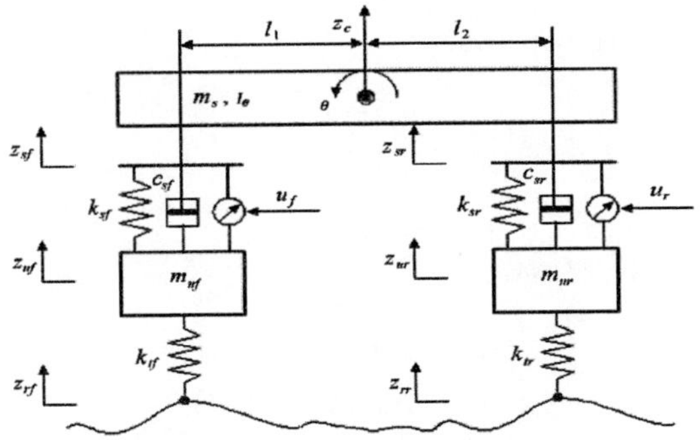

**Fig. 1.** A Half-vehicle Model.

## 3 Controller Design

**Lemma 1** *(Generalized KYP Lemma)[10]: consider the linear system($\bar{A}$, $\bar{B}$, $\bar{C}$, $\bar{D}$). Given a symmetric matrix $\Pi$, the following statements are equivalent:*

1. *The finite frequency inequality*

$$\begin{bmatrix} G(j\omega) \\ I \end{bmatrix}^T \Pi \begin{bmatrix} G(j\omega) \\ I \end{bmatrix} < 0, \quad \omega_1 < \omega < \omega_2 \tag{12}$$

2. *The existed symmetric matrices $P$, and $Q > 0$ satisfying*

$$\begin{bmatrix} \bar{A} & \bar{B} \\ I & 0 \end{bmatrix}^T \Pi \begin{bmatrix} \bar{A} & \bar{B} \\ I & 0 \end{bmatrix} + \begin{bmatrix} \bar{C} & \bar{D} \\ 0 & I \end{bmatrix}^T \Xi \begin{bmatrix} \bar{C} & \bar{D} \\ 0 & I \end{bmatrix} < 0 \tag{13}$$

Where

$$w_c = \frac{w_1 + w_2}{2}, \ \Pi = \begin{bmatrix} -Q & P + jw_cQ \\ P - jw_cQ & -w_1w_2Q \end{bmatrix}, \ \Xi = \begin{bmatrix} I & 0 \\ 0 & -\gamma^2 I \end{bmatrix}$$

**Lemma 2** *(Projection Lemma)[4]: Given a symmetric matrix $\Sigma$, and two matrices $\Gamma$ and $\Lambda$, the problem,*

$$\Sigma + \Gamma^T Z \Lambda + (\Gamma^T Z \Lambda)^T < 0 \tag{14}$$

*is solvable with respect to decision matrix $Z$ if and only if*

$$\Gamma^{\perp^T} \Sigma \Gamma^{\perp} < 0 , \qquad \Lambda^{\perp^T} \Sigma \Lambda^{\perp} < 0 \tag{15}$$

*where $\Gamma^{\perp}$ and $\Lambda^{\perp}$ denote the orthogonal complement of $\Gamma$ and $\Lambda$, respectively.*

We design a state return controller:

$$u(t) = \begin{bmatrix} u_f(t) \\ u_r(t) \end{bmatrix} = Kx(t) \tag{16}$$

where K is the gain matrix to be designed.

Substituting (16) in (7) gives,

$$\begin{aligned} \dot{x}(t) &= \bar{R}x(t) + \bar{M}w(t) \\ z_1(t) &= \bar{S}x(t) + \bar{D}w(t) \\ z_2(t) &= \bar{S}_2 x(t) \end{aligned} \tag{17}$$

where

$$\begin{bmatrix} \bar{R} & \bar{M} \\ \bar{S} & \bar{D} \end{bmatrix} = \begin{bmatrix} R+NK & M \\ S_1 + F_1 K & 0 \end{bmatrix}, \bar{S}_2 = S_2 + F_2 K \tag{18}$$

We state the following theorem.

**Theorem 1:** *Giving positive scalars $\gamma$, and $\rho$. And leaving the state return controller of the form* (16) *exists. The closed–loop system in* (17) *is asymptotically stable and satisfies $sup_{\omega_1<\omega<\omega_2}\|G(jw)\|_\infty < \gamma$, while the constraints in* (10) *and* (11) *are guaranteed, if there are symmetric matrices $P$, $P_1 > 0$, $Q > 0$ and matrix $T$ satisfying*

$$\begin{bmatrix} -T^T - T & P_1 + T\bar{R} - T^T \\ * & \bar{R}^T T^T + T\bar{R} \end{bmatrix} < 0 \tag{19}$$

$$\begin{bmatrix} -Q - T - T^T & T1 & T\bar{M} & 0 \\ * & T2 & T\bar{M} & \bar{S}_1^T \\ * & * & -\gamma^2 I & 0 \\ * & * & * & -I \end{bmatrix} < 0 \tag{20}$$

$$\begin{aligned} T1 &= P + jw_c Q + T\bar{R} - T^T \\ T2 &= -w_1 w_2 Q + \bar{R}^T T^T + T\bar{R} \end{aligned}$$

$$\begin{bmatrix} -P_1 & \sqrt{\rho}K \\ * & -u_{max}^2 I \end{bmatrix} \leq 0 \tag{21}$$

$$\begin{bmatrix} -P_1 & \sqrt{\rho}\{S_2\}_i \\ * & -I \end{bmatrix} < 0, i = 1,2,3,4,5,6. \tag{22}$$

*where $w_c = (w_1 + w_2)/2$ is given scalar, and $\rho = 1$*

**Proof 1:** *By using schur complement inequality* (20) *can be written in the form:*

$$\begin{bmatrix} -Q-T-T^T & P+jw_cQ+T\bar{R}-T^T & T\bar{M} \\ * & -w_1w_2Q+\bar{R}^TT^T+T\bar{R} & T\bar{M} \\ * & * & -\gamma^2 I \end{bmatrix} + \begin{bmatrix} 0 & 0 & 0 \\ 0 & \bar{S_1}^T\bar{S}_1 & 0 \\ 0 & 0 & 0 \end{bmatrix} < 0 \quad (23)$$

*Which can be written as follows,*

$$\begin{bmatrix} -Q & P+jw_cQ & 0 \\ P-jw_cQ & -w_1w_2Q+\bar{S}_1^T\bar{S}_1 & 0 \\ 0 & 0 & -\gamma^2 I \end{bmatrix} + \begin{bmatrix} T \\ T \\ 0 \end{bmatrix} \begin{bmatrix} -I & \bar{R} & \bar{M} \end{bmatrix} + \begin{bmatrix} -I \\ \bar{R}^T \\ \bar{M}^T \end{bmatrix} \begin{bmatrix} T^T & T^T & 0 \end{bmatrix} < 0 \quad (24)$$

*Using Lemma* (2)*, we have*

$$\begin{bmatrix} \bar{R} & \bar{M} \\ I & 0 \\ 0 & I \end{bmatrix}^T \begin{bmatrix} -Q & P+jw_cQ & 0 \\ P-jw_cQ & -w_1w_2Q+\bar{S}_1^T\bar{S}_1 & 0 \\ 0 & 0 & -\gamma^2 I \end{bmatrix} * \begin{bmatrix} \bar{R} & \bar{M} \\ I & 0 \\ 0 & I \end{bmatrix} < 0 \quad (25)$$

*Which is equivalent to*

$$\begin{bmatrix} \bar{R} & \bar{M} \\ I & 0 \end{bmatrix}^T \begin{bmatrix} -Q & P+jw_cQ \\ P-jw_cQ & -w_1w_2Q \end{bmatrix} \begin{bmatrix} \bar{R} & \bar{M} \\ I & 0 \end{bmatrix} + \begin{bmatrix} \bar{S}_1^T\bar{S}_1 & 0 \\ 0 & -\gamma^2 I \end{bmatrix} < 0 \quad (26)$$

*Hence; we obtain the condition* (13)*. By Lemma* (1)*, the finite frequency* (12) *is verified.*

*Now, condition* (19) *can be written as,*

$$\begin{bmatrix} 0 & P_1 \\ P_1 & 0 \end{bmatrix} + \begin{bmatrix} T \\ T \end{bmatrix} \begin{bmatrix} -I & \bar{R} \end{bmatrix} + \begin{bmatrix} -I \\ \bar{R}^T \end{bmatrix} \begin{bmatrix} T^T & T^T \end{bmatrix} < 0 \quad (27)$$

*Which, implies by Lemma*(2) *that*

$$\begin{bmatrix} \bar{R} \\ I \end{bmatrix}^T \begin{bmatrix} 0 & P_1 \\ P_1 & 0 \end{bmatrix} \begin{bmatrix} \bar{R} \\ I \end{bmatrix} < 0 \quad (28)$$

*The condition* (28)*, implies that the system* (17) *is asymptotically stable when* $\omega(t) = 0$*.*

*From* (10) *and* (16) *we have, consider*

$$max_{t\geq 0}|u(t)|^2 = max_{t\geq 0}\|Kx(t)\|_2^2 \leq u_{max}^2$$

$$max_{t\geq 0}|x^T(t)K^TKx(t)| \leq u_{max}^2$$

*Using the transformation* $\bar{x}(t) = P_1^{1/2}x(t)$

$$max_{t\geq 0}|\bar{x}^T(t)P_1^{-1/2}K^TKP_1^{-1/2}\bar{x}(t)|$$

$$\leq \rho.\lambda_{max}(P_1^{-1/2}K^TKP_1^{-1/2}) \leq u_{max}^2 \quad (29)$$

*The condition* (11) *gives*

$$max_{t\geq 0}|\{y_2(t)\}_i|^2 = max_{t\geq 0}\|x^T(t)\{S_2\}_i^T\{S_2\}_i x(t)\|_2 \leq 1 \ , i = 1,2,3,4,5,6;$$

$$max_{t\geq 0}|\{y_2(t)\}_i|^2 \leq \rho.\lambda_{max}(P_1^{-1/2}\{S_2\}_i^T\{S_2\}_i P_1^{-1/2}) \leq 1 \ , i = 1,2,3,4,5,6 \quad (30)$$

*From* (29), *and* (30) *we can write,*

$$\rho P_1^{-1/2} K^T K P_1^{-1/2} \leq u_{max}^2 I \quad (31)$$

$$\rho P_1^{-1/2}\{S_2\}_i^T\{S_2\}_i P_1^{-1/2} \leq I, i = 1,2,3,4,5,6 \quad (32)$$

*By schur complement* (31), *and* (32), *are equivalent to* (21), *and* (22) *respectively. This completes the proof.*

**Remark 1:** *The resulting feasibility problem on the Theorem 1 is nonlinear, it cannot be directly addressed by optimisation LMI. In order to solve the problem of non−linearity, we carry out a transformation for the inequalities* (19)-(22) *with the matrices* $V_1 = diag\{T^{-1}, T^{-1}\}$, $V_2 = diag\{T^{-1}, T^{-1}, I, I\}$, $V_3 = diag\{I, T^{-1}\}$.

**Theorem 2:** *Giving positive scalars* $\gamma$, *and* $\rho$. *And leaving the state return controller of the form* (16) *exists. The closed−loop system in* (17) *is asymptotically stable and satisfies* $sup_{\omega_1<\omega<\omega_2}\|G(jw)\|_\infty < \gamma$, *while the constraints in* (10) *and* (11) *are guaranteed, if there are symmetric matrices* $P$, $P_1 > 0$, $Q > 0$ *and matrices* $T$, $\bar{K}$ *satisfying*

$$\begin{bmatrix} -\bar{T}^T - \bar{T} & \bar{P}_1 - \bar{T} + R\bar{T}^T + N\bar{K}^T \\ * & R\bar{T}^T + N\bar{K} + \bar{T}R^T + \bar{K}^T N^T \end{bmatrix} < 0 \quad (33)$$

$$\begin{bmatrix} T1 & T2 & M & 0 \\ * & T3 & M & T4 \\ * & * & -\gamma^2 I & 0 \\ * & * & * & -I \end{bmatrix} < 0, \quad (34)$$

$$T1 = -\bar{Q} - \bar{T}^T - \bar{T}$$
$$T2 = \bar{P} + jw_c\bar{Q} + R\bar{T}^T + N\bar{K} - \bar{T}$$
$$T3 = -w_1 w_2\bar{Q} + \bar{T}R^T + \bar{K}^T N^T + R\bar{T}^T + N\bar{K}$$
$$T4 = \bar{T}S_1^T + \bar{K}^T F_1^T$$

$$\begin{bmatrix} -\bar{P}_1 & \sqrt{\rho}\bar{K} \\ * & -u_{max}^2 I \end{bmatrix} \leq 0, \quad (35)$$

$$\begin{bmatrix} -\bar{P}_1 & \sqrt{\rho}\{S_2\}_i\bar{T} \\ * & -I \end{bmatrix} < 0, \ i = 1,2,3,4,5,6. \quad (36)$$

The feedback gain matrix is given by

$$K = \bar{K}\bar{T}^{-1}. \tag{37}$$

**Proof 2:** *Taking* $V_1 = diag\{T^{-1}, T^{-1}\}$, $V_2 = diag\{T^{-1}, T^{-1}, I, I\}$, $V_3 = diag\{I, T^{-1}\}$.

*Multiplying the inequalities* (19)–(22) *of the Theorem 1, on the left by* $V_1^T$, $V_2^T$, *and* $V_3^T$, *on the right by* $V_1$, $V_2$, *and* $V_3$ *respectively.*

*If we take* $\bar{Q} = (T^{-1})^T Q T^{-1}$, $\bar{P} = (T^{-1})^T P T^{-1}$, $\bar{P}_1 = (T^{-1})^T P_1 T^{-1}$ $\bar{K} = KT^{-1}$, *and* $\bar{T} = T^{-1}$.
*We get the inequalities* (33)–(36) *of the Theorem 2. This completes the proof.*

**Remark 2:** *Note that the linear matrix inequality* (34) *has complex variables. The LMI* (34) *can be converted into a larger dimension LMI of real variables. The inequality* (34) *can be addressed like the inequality* $S_1 + jS_2 < 0$ *that is equivalent to* $\begin{bmatrix} S_1 & S_2 \\ -S_2 & S_1 \end{bmatrix} < 0$.

**Remark 3:** *If we take Q=0 we can use Theorems 1, and 2. To solve the* $H_\infty$ *control problem in the Entire Frequency* ($EF$) *in the domain of active suspension systems.*

## 4 Numerical Example

Longitudinal half-vehicle vertical model parameters with active suspension system are listed in the Table 1 [2].

**Table 1.** This is the half-vehicle model parameters

| Settings | Values |
|---|---|
| $m_s$ | 575 kg |
| $m_{uf}$, $m_{ur}$ | 60 kg |
| $k_{sf}$, $k_{sr}$ | 16812 N/m |
| $k_{tf}$, $k_{tr}$ | 190000 N/m |
| $c_{sf}$, $c_{sr}$ | 1000 N.s/m |
| $l_1$ | 1.38 m |
| $l_2$ | 1.36 m |
| $I_\theta$ | 769 kg.m$^2$ |
| $z_{fmax}$, $z_{rmax}$ | 0.08 m |
| $u_{fmax}$, $u_{rmax}$ | 1500 N |
| $F_f$ | 4014.5 N |
| $F_r$ | 3580.5 N |

Inequalities (33)–(36) has a feasible solution. In the case of optimal $\gamma$, the minimum guaranteed closed−loop $H_\infty$ performance obtained is $\gamma_{min} = 1.2849$.

According to the Eq. (37). We can get the gain matrix $K_F$ to design in (16) in the finite frequency domain

$$K_F = 10^4 \times [-1.1098 \; -0.0426 \; -0.0819 \; -0.0033 \; 9.6941 \; -0.0435 \; -0.0175 \; 0.0003 \\ -0.0831 \; -0.0034 \; -1.1226 \; -0.0434 \; -0.0186 \; 0.0003 \; 9.7195 \; -0.0437] \quad (38)$$

The $H_\infty$ state feedback controller in the entire frequency is given by:

$$K_E = 10^4 \times [0.3947 \; 0.4729 \; 0.0723 \; 0.0448 \; -4.1567 \; -0.6268 \; 0.9341 \; 0.0147 \\ 0.0729 \; 0.0448 \; 0.4000 \; 0.4763 \; 0.9581 \; 0.0148 \; -4.0944 \; -0.6297] \quad (39)$$

By the simulation, the responses of the open−loop system and the closed−loop system are compared in Fig. 2.

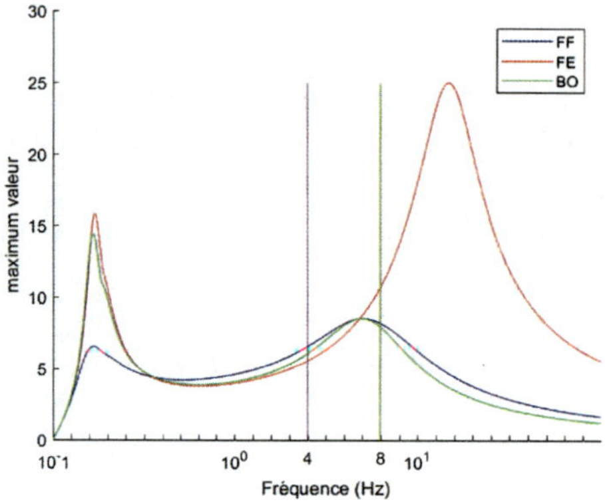


**Fig. 2.** Frequency response of vertical acceleration.

By solving Theorem 2, the minimum $H_\infty$ attenuation level $\gamma_{min}$ are listed in Table 2:

**Table 2.** optimal performance $\gamma_{min}$

| Settings | Values |
|---|---|
| F.F | $\gamma_{min} = 1.2849$ |
| F.E | $\gamma_{min} = 3.1287$ |

## 5 Conclusion

In this article, we have studied problem of $H_\infty$ control of automotive active suspension system for the half−vehicle model of longitudinal type in the Finite Frequency domain, by using the generalized lemma of Kalman−Yakobovich−popov, the mechanical structure performance and safety constraints are guaranteed within their allowable limits and driving comfort has been improved in the specific frequency band [4–8] hz. Simulation results are given to illustrate the results.

## References

1. Cao, J.-T., Liu, H.-H., Li, P., Brown, D.J., Dimirovski, G.: A study of electric vehicle suspension control system based on an improved half-vehicle model. Int. J. Autom. Comput. **04**(3), 236–242 (2007)
2. Lin, J.-S., Huang, C.-J.: Nonlinear backstepping active suspension design applied to a half-car model. Veh. Syst. Dyn. Int. J. Veh. Mech. Mobil. **42**(6), 373–393 (2004)
3. Li, H., Yu, J., Hilton, C., Liu, H.: Adaptive sliding-mode control for nonlinear active suspension vehicle systems using T–S fuzzy approach. IEEE Trans. Ind. Electron. **60**(8), 3328–3338 (2013)
4. Gahinet, P., Apkarian, P.: A linear matrix inequality approach to $H_\infty$ control. Int. J. Robust Nonlinear Control **4**(4), 421–448 (1994)
5. Moradi, S.M., Akbari, A., Mirzaei, M.: An offline LMI-based robust model predictive control of vehicle active suspension system with parameter uncertainty. Trans. Inst. Meas. Control **41**, 1699–1711 (2018)
6. Gao, H., Sun, W., Kaynak, O.: Vibration suppression of vehicle active suspension systems in finte frequency domain. In: Joint 48th IEEE Conference on Decision and Control and 28th Chinese Control Conference Shanghai, P.R.China, 16–18 December 2009
7. Sun, W., Gao, H., Kaynak, K.: Finite frequency $H_\infty$ control for vehicle active suspension systemes. IEEE Trans. Control Syst. Technol. **19**(2), 1–7 (2011)
8. Choi, H.D., Lee, C.J., Lim, M.T.: Fuzzy preview control for half-vehicle electro-hydraulic suspension system. Int. J. Control Autom. Syst. **16**(X), 1–12 (2018)
9. Iwasaki, T., Hara, S.: Generalized KYP lemma: unified frequency domain inequalities with design applications. IEEE Trans. Autom. Control **50**(1), 41–59 (2005)
10. Romao, L.B.R.R., de Oliveira, M.C., Peres, P.L.D., Oliveira, R.C.L.F.: State-feedback and filtering problems using the generalized KYP lemma. In: 2016 IEEE Conference on Computer Aided Control System Design (CACCD), Part of 2016 IEEE Multi-conference on Systems and Control, 19–22 September 2016
11. Yu, Q., Ma, J.: Analysis of automotive acceleration performance improvement due to active suspension. IEEE Trans. Autom. Control J. Veh. Technol. **37**(3), 20–22 (2005)
12. Zhou, S., Feng, G., Lam, J., Xu, S.: Robust $H_\infty$ control for discrete fuzzy systems via basis-dependent Lyapunov functions. Inf. Sci. **174**, 197–217 (2005)
13. Feng, G.: A survey on analysis and design of model-based fuzzy control systems. IEEE Trans. Fuzzy Syst. **14**(5), 676–697 (2006)
14. Els, P.S., Theron, N.J., Uys, P.E., Thoresson, M.J.: The ride comfort vs. handing compromise for off-road vehicles. J. Terramech. **44**, 303–317 (2007)

15. Er Rachid, I., Chaibi, R., Tissir, E.H., Hmamed, A.: Observer based $H_\infty$ control with finite frequency specifications for discrete time T−S fuzzy systems. Int. J. Syst. Sci. (2018). https://doi.org/10.1080/00207721.2018.1536236
16. Chaibi, R., Er Rachid, I., Tissir, E.H., Hmamed, A.: Finite frequency static output feedback $H_\infty$ control of continuous time T−S fuzzy systems. J. Circuits Syst. Comput. **28**(2), 1950023 (2019). https://doi.org/10.1142/S0218126619500233
17. Er Rachid, I., Chaibi, R., El Haiek, B., Tissir, E.H., Hmamed, A.: Robust observer based controller design for uncertain discrete−time systems in finite frequency domain. In: IEEE Conferences, $6^{th}$ International Conferences on Multimedia Computing and Systems (ICMCS 2018) (2018)
18. El-amrani, A., Boukili, B., Hmamed, A., Adel, E.M.E.: Robust $H_\infty$ filters for uncertain systems with finite frequency specifications. J. Control Autom. Electr. Syst. (2017). https://doi.org/10.1007/s40313-017-0336-9
19. El-Amrani, A., Boukili, B., El Hajjaji, A., Hmamed, A.: $H_\infty$ model reduction for T-S fuzzy systems over finite frequency ranges. Optim. Control Appl. Meth. 1–18 (2018). https://doi.org/10.1002/oca.2422
20. Mrazgua, J., Tissir, E.H., Ouahi, M.: Fuzzy fault-tolerant $H_\infty$ control approach for nonlinear active suspension systems with actuator failure. Procedia Comput. Sci. **148**, 465–474 (2019)
21. Mrabah, L., Tissir, E.H., Ouahi, M.: $H_\infty$ control for vehicle active suspension systems in finite frequency domain. In: 2019 5th International Conference on Optimization and Applications (ICOA). https://doi.org/10.1109/ICOA.2019.8727652.

# Double Frequency Filtering in One Dimensional Comb-Like Phononic Structure Containing a Segment Defect

Ilyass El Kadmiri[1(✉)], Youssef Ben-Ali[1], Abdelaziz Ouariach[1], Aissam Khaled[1,2], and Driss Bria[1]

[1] Laboratory of Materials, Waves, Energy and Environment, Team of Waves, Acoustic, Photonic and Materials, Mohamed Premier University, Oujda, Morocco
ilyaskadmiri@hotmail.com
[2] National School of Applied Sciences, ENSAH, Al Hoceima, Morocco

**Abstract.** In this paper, we show theoretically that we can obtain double frequencies filtering using one dimensional comb-like phononic structure containing one defect at the segment level. The proposed structure is built of segment periodicity and grafted in each site by a finite number of resonators. The transmission spectrum and the band structure of this phononic system are theoretically presented using the Green functions approach based on the formalism of the interface response theory for acoustic waves propagating in comb-like structure which present a wide band gap. The presence of a defect in this such structure creates filtered frequencies which strongly depend on the defect length, the number of cellules N and the position of defect J. These characteristics of the considered structure give rise to a good information for the design of ultra-fine frequency phononic filters situated inside the large comb-like band gaps.

**Keywords:** Phononic comb-like · Filtering · Defect modes · Transmission peak

## 1 Introduction

The study of acoustic and elastic wave propagation in periodic band gap materials, known as phononic crystals, was receiving a growing interest during the last decades. Phononic crystals represent periodic artificial structures that have been proposed to control the propagation of acoustic waves. These materials exhibit the pass bands separated by large phononic band gaps in which the acoustic waves cannot propagate. The presence of a defect in this structure gives rise to limited states (defect modes) inside gaps. Potential applications of these defected structures as very selective frequency filters were studied theoretically and experimentally especially in 1D, quasi 1D and 2D structures [1–3]. Studies of one-dimensional periodic layered media [4, 5] are conducted as analogs of 2D and 3D systems. These structures are feasible at any wavelength scale, and they involve only straightforward analytical calculations.

Some researchers are interested in the propagation of elastic waves in the 1D phononic structure; this system is composed of an infinite 1D segment along which


A. Saka et al. (Eds.): CPI 2019, LNME, pp. 13–21, 2021.
https://doi.org/10.1007/978-3-030-62199-5_2

finite sites of resonators are grafted periodically. Abdelkrim Khelif et al, present a novel way of forming phononic crystal waveguides by coupling a series of line defect resonators, the dispersion proprieties of these coupled-resonator acoustic waveguides CRAW (finite length of CRAW structure acts as an efficient selective acoustic filter) can be engineered by using their geometrical parameters [6]. J. M. Escalante et al. studied the dispersion of an acoustic crystal waveguide formed by the evanescent coupling of a defective cavity chain and supported the propagation of slow waves. These acoustic cavity-coupled waveguides (CRAW) are similar to optical cavity-coupled waveguides formed in photonic crystals. The dispersion of the CRAW can be controlled by increasing the distance between the cavities, thereby reducing its coupling and smoothing the dispersion ratio [7]. Y. Pennec et al. demonstrated theoretically the simultaneous existence of phoxonic band gaps in a periodic silicon strip waveguide. The unit-cell of this one-dimensional waveguide contains a hole in the middle and two symmetric stubs on the sides. The insertion of a cavity inside the perfect structure provides simultaneous confinement of phononic and optical waves suitable to enhance the phonon-photon interaction [8]. M. S. Kushwaha et al. present the structure of the band and the transmission coefficient for the propagation of the acoustic wave in a system composed of N′ dangling lateral branches (DSB) periodically grafted into each of the N equidistant sites in a thin tube. A periodic schematic of large gaps for aerating DSB in a thin water tube is obtained. The design of the system with open tubes allows to achieve the lowest space below a threshold frequency and extends to zero, which provides a completely discrete band structure and transmission spectrum [9]. Z. G. Wang et al. present an analytical calculation for the propagation of the sound wave in phononic crystals containing Helmholtz resonators. The band structures, the transmission spectra and defect modes are investigated for different geometries using the interface response theory based on the green function. They showed that the acoustic band structure of the model differs fundamentally from conventional acoustic-elastic cases and richer due to the coexistence of resonance and Bragg gaps. They showed that the presence of a defective resonator in the system can lead to resonance modes within the gaps [10]. Chen Shen et al. reported that in space and time modulated cascaded resonators, acoustic transmission does not reciprocate. An analysis method based on mode expansion was proposed and showed strong irreversibility in two resonator systems with spatially biased modulation phase. The theory has been numerically verified by a finite-difference time-domain simulation (FDTD) and only one resonator has a unidirectional isolation coefficient greater than 25 dB [11]. Arturo Santillan et al. show that the acoustic transparency and slow sound propagation can be carried out with detuned acoustic resonators (DAR), which mimics the effect of electromagnetically induced transparency (EIT) in atomic physics. Sound propagation in a pipe with a series of laterally connected DARs is studied, with neighboring DAR units separated by a distance that is much smaller than the wavelength [12]. They also exhibit the phenomenon of sound transparency, which greatly reduces the propagation of sound from a series of paired DAR coupled laterally to the waveguide. By organizing four identical DAR pairs along the waveguide, the wavelength spacing between adjacent pairs is equal, and they show that this arrangement has unique narrow-band transmission characteristics and strong dispersion. They show that there is a direct coupling between each pair of DARs, which cannot be explained by the interference of waves radiated by these resonators. For smaller detuned values and if the cross-sectional area of the cavity is increased, this destructive coupling will become apparent [13]. A. Khettabi et al. studied a periodic structure formed by Helmholtz resonators (HR). They use two methods of analysis: the Transfer Matrix Method (TMM) and the Interface Response theory [14]. Also, they present a structure composed of a number of identical structural components. Each cell is formed by a simple expansion chamber. the expansion chamber is often used to reduce noise in automotive [15]. The size of the band gap in these composites structures is a critical role for various device applications. Some of the authors reported that the electromagnetic transmission spectrum of 1D comb-like structures exhibit enormous gaps. These structures are composed of an infinite or finite 1D segment of the length $d_1$, along which N′ side resonators of the height $d_2$ are grafted on each site. Y. Ben Ali et al. studied new filters based on the defect modes by using a resonator defect in this structure which containing left-handed materials [16]. They also showed the possibility of filtering two electromagnetic frequencies by the presence of one defective segment in a 1D photonic comb-like structure [17]. Moreover, they studied the effect of the presence of a resonator defect in a one-dimensional photonic comb-like structure which contains a finite number of resonators defect [18].

In this work, we show that there is a double frequency filtering based on two defect modes in an enormous gap by causing a defect at the segment level in the one-dimensional phononic comb-like structure, this is presented in Fig. 1(b). This structure has potential applications as acoustic filters or waveguides, etc.

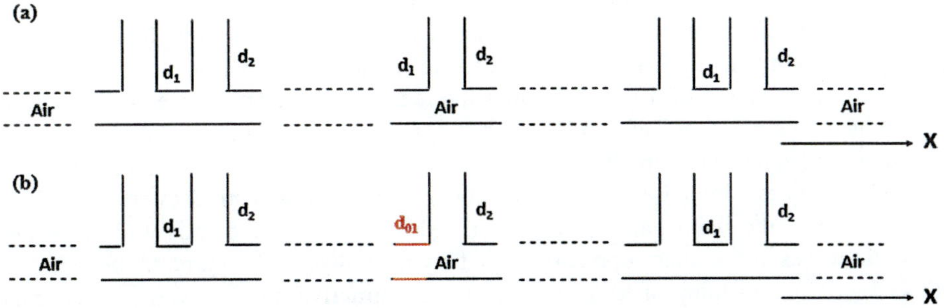


**Fig. 1.** **(a)** One-dimensional comb-like phononic structure containing N periods. **(b)** Same as **(a)** except that there is a segment defect of the length $d_{01}$ located in the middle of the periodic structure.

## 2 Results and Discussions

In this work, we illustrate the acoustic band structure and the transmission coefficient for a phononic comb-like structure containing a defect at the segment level. For the ideal structure, we operate the following geometrical parameters: the length of the segment and the height of the resonator are respectively, $d_1 = 1D$ and $d_2 = 0.5d_1$. While, the length of the defective segment is indicated $d_{01}$, and it is placed between

the sites J and J + 1, where 1 < J < N, and N represents the number of cells. The reduced frequency is given by $\Omega = \frac{\omega d_1}{c}$, with $d_1$ represents the period of the structure, c is the velocity of the phononic waves (c = 348 m/s) and ω is the pulsation ($s^{-1}$). We show that the perfect one-dimensional phononic comb-like exhibits several ranges of frequencies where the phononic waves are forbidden to propagate (gaps). The inclusion of a defective
segment in this perfect comb-like structure creates confined acoustic modes inside the band gap.

## 2.1 Properties of a Perfect Comb-Like Structure

In Fig. 2, we present the acoustic band structure, the transmission coefficient, the phase, the phase time and the group velocity of the ideal comb-like structure. In Fig. 2(a), the variation of the dispersion relation in the region of reduced frequency Ω = [0, 20] versus the real part of reduced Bloch vector K $d_1$, shows clearly the existence of three acoustic pass bands separated by wide phononic band gaps (PBG) where the acoustic waves cannot propagate. Similar behavior is exhibited by the Fig. 2(b) which gives the transmission spectrum of a finite phononic crystal made of N = 8 cells. The band structure and the transmission coefficient exhibit pass bands separated by large gaps, which could be of potential interest for acoustic waveguide structure. The latter originates both from the periodicity of the structure, related to the length $d_1$, and from the resonant behavior of the lateral branches, related to the length $d_2$. We note that a one-dimensional phononic comb-like structure contains a wide gap when compared with photonic crystals [5]. Due to the gap, the acoustic wave can be controlled for a broad range of applications, in particular, the manufacture of filtering or multiplexing devices, noise reduction and improvement of acoustic attenuation performance. Figure 2(c) indicates the variation of the phase versus the reduced frequency Ω; we notice that the phase increases monotonically in the band gap with a jump of π around three values of frequencies (equal to Ω = 6.28, Ω = 12.59 and Ω = 18.85), after these three frequency values, the phase increases. It can be noted that the destructive interferences are responsible of the band gap (these gaps are due to zero of transmission) since two acoustic waves are in phase opposition, so, they neutralize each other and the phase, in this case, is odd (a jump of π), as well as, the constructive interferences (two acoustic waves are in phase and the important vibration) are responsible of the permissible band. Figure 2(d) shows the variation of the phase time (the phase time is the time necessary for the propagation of the acoustic wave inside the structure before its transmission) versus the reduced frequency Ω, we note that the phase time of a few discrete modes is extremely high (63 s). We also observe that there are four gaps; these gaps are seen as zero transmission ranges in the transmission spectrum. The phase time keeps the same value around of 9.61 s within the pass bands. The appearance of enormous gaps makes to guide, manipulate and conduct the acoustic waves. Figure 2(e) presents the variation of the group velocity $V_g/c$ as a function of the reduced frequency Ω, it can be noted that inside the band gaps, the group velocity $V_g/c$ does not exceed 4.5. However, abnormal dispersion occurs in the pass bands and supra-luminal velocities such as $0.20 \leq V_g/c \leq 0.83$ is expected.

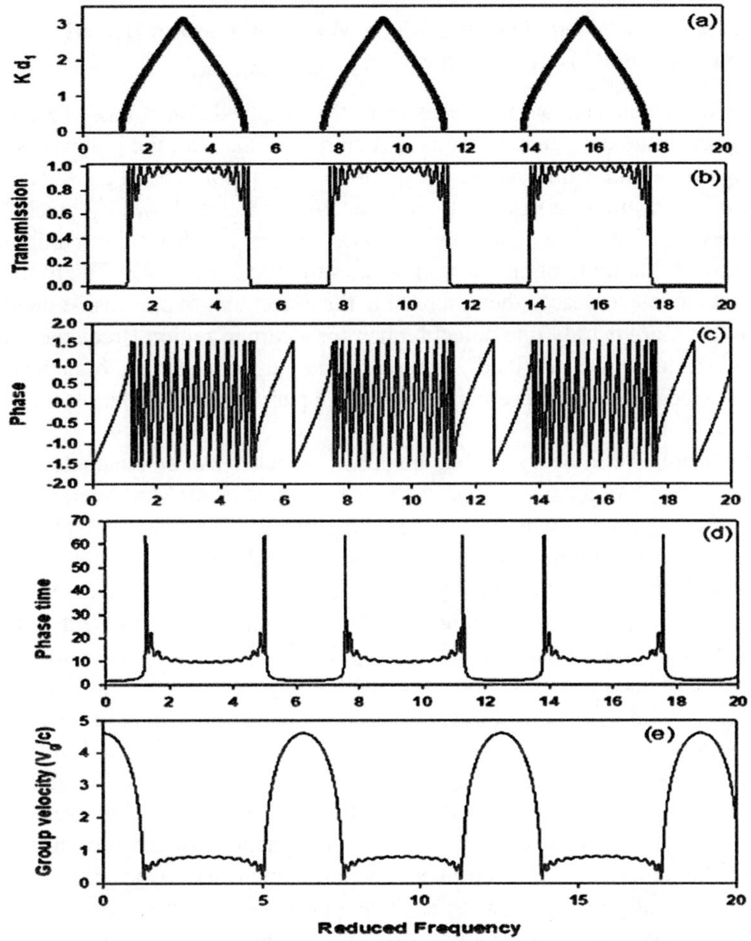


**Fig. 2.** Curve (a) represents the band structure of a perfect infinite structure. Curve (b) represents the transmission spectrum through a perfect finite comb-like structure with $d_1 = 1D$, $d_2 = 0.5\ d_1$ and $N = 8$. (c), (d) and (e) represent respectively the phase, the phase time and the group velocity in the same situation of the curve (b).

It is indicated for a normal incidence, that the characteristics of the PBGs depend on the acoustic properties. Our purpose is to produce one or several defect modes inside these gaps by the creation of a segment defect in the comb-like phononic structure.

Two papers mention that several experimental works have been devoted to periodic stub structures: Fey and Robertson have proved that these systems may exhibit a series of phononic gaps due to the Helmholtz resonance and standing-wave cavity modes within the gaps, the system exhibits narrow bands of negative group delay. Theocharis et al. have shown a limit of slow sound propagation and transparency in loss Helmholtz resonant periodic structures [19, 20]. Also, by engineering acoustic stub waveguides, several works have revealed the possibility to design acoustic metamaterials with negative density and/or modulus [21–23].

## 2.2 Transmission Coefficient and Eigen Modes of a Finite Defective Phononic Comb-Like Structure

In this section, we investigate the behavior of the localized defect modes created inside the band gaps of the defective phononic structure. In Fig. 3(a), the gray areas represent the pass bands and the white areas correspond to the phononic band gaps. The presence of the defective segment gives rise to acoustic eigen modes shown by black branches. The frequencies of these acoustic eigen modes are derived from the maximum transmission through the finite phononic comb-like structure with N = 8. The frequencies of these defect modes decrease when increasing the defect length $d_{01}$. Inside the band gap, we notice that there is only one defect branch for a certain region (frequency range and defect length range well-defined), whereas there are two defect branches in other regions, these localized modes emerge to the high pass bands and merge to a lower pass bands.

In conclusion, it can be noted that the defect branches can be conducted in the wide gap by the variation of a defect length $d_{01}$ at the segment level. Moreover, we can define the order of the band gap where we are looking to have the defect branches.

In addition, we examine the behavior of defect modes in the transmission spectrum as a function of the reduced frequency by setting distinct values of the defect length $d_{01}$. From Fig. 3(b) ($d_{01}$ = 0.2 $d_1$), we remark the appearance of one defect mode with a significant value of transmission in the $2^{nd}$ gap. In Fig. 3(c) ($d_{01}$ = 2 $d_1$), we observe the existence of two defect modes inside gaps with a maximum transmission, while in the Fig. 3(d) ($d_{01}$ = 2.5 $d_1$), we note the similar phenomenon described in Fig. 3(c) but the defect modes have a low transmission rate. The results of this figure show clearly that one or two filtered frequencies are created by the creation of one defect at the segment level, which strongly depends on the defect length $d_{01}$; these characteristics of the considered structure give rise to an excellent information for the design of ultra-fine frequency phononic filters situated inside the large comb-like band gaps.

From a general point of view, to achieve a phononic selective filter inside a wider band gap, it is necessary to design a structure in which the transmission coefficient exhibits well-defined characteristics and sensitive to the propagation of acoustic waves. It can be noted that our structure with the presence of the defect in the segment of length $d_{01}$ is better compared to the presence of a defect at the segment level in the photonic or phononic loop structure which creates a single defect mode in the band gap [5, 24–26].

Then, Fig. 3(e) represents the zoom of the Fig. 3(a), the blue and the red points represent the defect branches (defect modes) located between $d_{01}$ = [0 − 0.5] $d_1$, which the reduced frequency of these two defect modes is located between 4.9 and 7.7. Presently, we examine the variation of the transmission coefficient of these two defect branches versus the length $d_{01}$ which appear in the Fig. 3(e). We note that the transmission of the first branch (the blue branch) which appears inside the $2^{nd}$ gap is decreased with the increase $d_{01}$, while the transmission of the second branch (the red branch) is increased when $d_{01}$ increase. This behavior is always present in each zone of Fig. 3(a) when we have two defect branches. Also, we observe that the two defect branches having the same maximum of transmission (T = 100%) in the range of length $d_{01}$ = [0.1 − 0.4] $d_1$, with the reduced frequency $\Omega$ of these two defect modes are located between $\Omega$ = [4.9 − 7.7]. From Fig. 3(g), we note that the quality factor (the ratio of the central frequency and the full width at half maximum of the transmittance modes) of the red branch is decreased, this behavior is due to the increase of the width at half maximum of the transmission peak, while the quality factor of the blue branch is increasing when increases the defect length $d_{01}$. This result demonstrates that our defect gives rise to two branches acoustic (red and blue points) with two high quality factors (Q = 25488 for the red branch when $d_{01}/d_1 = 0.02$ and Q = 27278 for the blue branch when $d_{01}/d_1 = 0.45$).

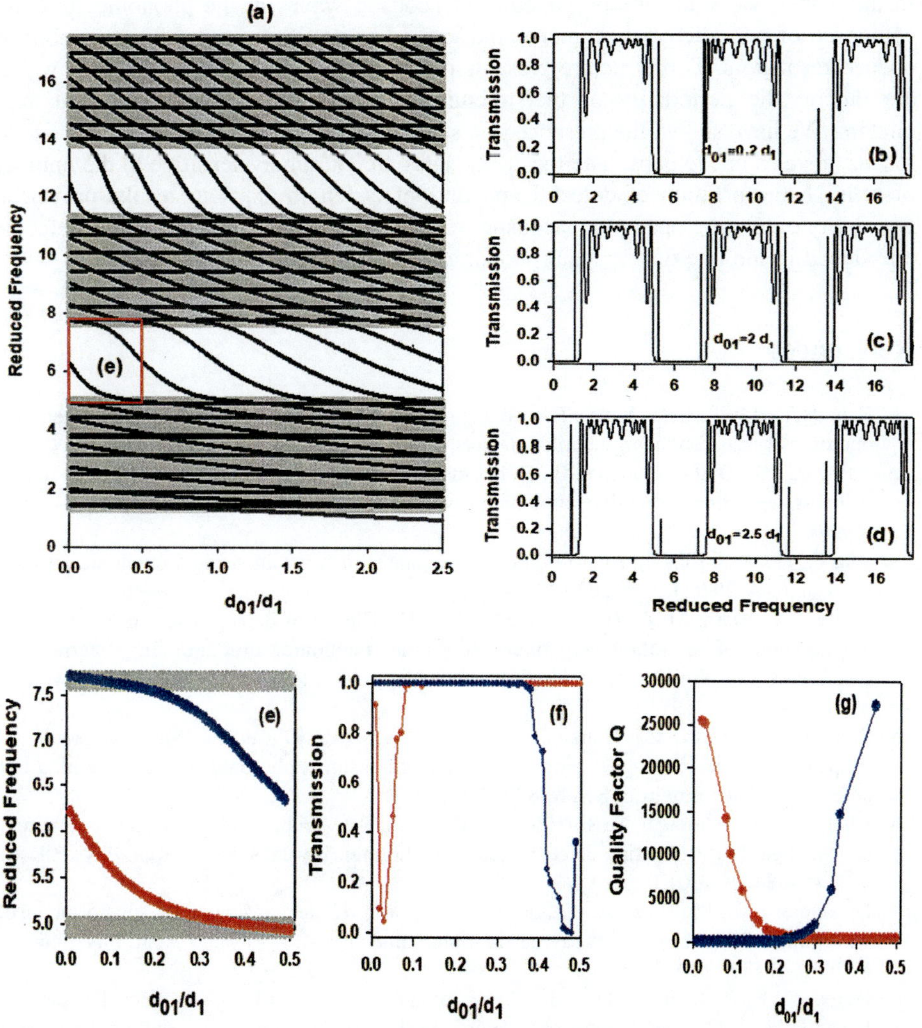


**Fig. 3.** (a) Variation of the reduced frequency versus the length $d_{01}$ with $d_1 = 1D$, $d_2 = 0.5\ d_1$, N = 8 and J = 4. (b), (c) and (d) represent the transmission spectrum versus the reduced frequency for different values of the defect length $d_{01}$, namely $d_{01} = 0.2\ d_1$, $2\ d_1$ and $2.5\ d_1$. Curve (e) represents the zoom of the curve (a). Curve (f) represents the transmission maximum for the two branches appeared in the area (e). Curve (g) represents the quality factor Q of the two branches versus $d_{01}/d_1$.

In practical applications, one generally inclines to design a narrow filter inside large gaps, which can modulate the defect modes in a larger frequency range with a very high transmission rate and high-quality factor by changing the defect length $d_{01}/d_1$.

## 3 Conclusion

In this paper, we study the propagation of acoustic waves in the phononic structure composed of periodicity of segments and grafted in its extremity by finite number of opened resonators. This structure presents pass bands separated by large gaps, which are due to the periodicity of the structure and the resonance modes of the resonators. We have shown the presence of a segment defect in this structure gives rise to localized states inside gaps. The localized states are extremely sensitive to the number of cells N, the position of defect J and the defect length $d_{01}$. Our results prove the possibility of filtering one or two acoustic frequencies when an adequate single defect is introduced in the one-dimensional phononic comb-like structure.

## References

1. Hsu, H.T., Lee, M.H., Yang, T.J., Wang, Y.C., Wu, C.J.: A multichanneled filter in a photonic crystal containing coupled defects. Prog. Electromagn. Res. **117**, 379–392 (2011)
2. Vasseur, J.O., Djafari-Rouhani, B., Dobrzynski, L., Akjouj, A., Zemmouri, J.: Defect modes in one-dimensional comblike photonic waveguides. Phys. Rev. B **59**(20), 13446–13452 (1999)
3. Wu, F., Hou, Z., Liu, Z., Liu, Y.: Point defect states in two-dimensional phononic crystals. Phys. Lett. A **292**(3), 198–202 (2001)
4. Khaled, A., Elamri, F.Z., El Kadmiri, I., Bria, D.: Effects of defect layers insertion on the transmission of a submerged one-dimensional phononic structure. In: International Conference on Wireless Technologies, Embedded and Intelligent Systems, Fez-Morocco, pp. 1–6. IEEE (2019)
5. Ben-Ali, Y., Tahri, Z., Bouzidi, A., Jeffali, F., Bria, D., Azizi, M., Nougaoui, A.: Propagation of electromagnetic waves in a one-dimensional photonic crystal containing two defects. J. Mater. Environ. Sci. **8**, 870–876 (2017)
6. Khelif, A., Mohammadi, S., Eftekhar, A.A., Adibi, A., Aoubiza, B.: Acoustic confinement and waveguiding with a line-defect structure in phononic crystal slabs. J. Appl. Phys. **108**(8), 084515–084520 (2010)
7. Escalante, J.M., Martínez, A., Laude, V.: Dispersion relation of coupled-resonator acoustic waveguides formed by defect cavities in a phononic crystal. J. Phys. D Appl. Phys. **46**(47), 475301–475308 (2013)
8. Pennec, Y., Rouhani, B.D., Li, C., Escalante, J.M., Martínez, A., Benchabane, S., Papanikolaou, N.: Band gaps and cavity modes in dual phononic and photonic strip waveguides. AIP Adv. **1**(4), 041901–041909 (2011)
9. Kushwaha, M.S., Akjouj, A., Djafari-Rouhani, B., Dobrzynski, L., Vasseur, J.O.: Acoustic spectral gaps and discrete transmisson in slender tubes. Solid State Commun. **106**(10), 659–663 (1998)

10. Wang, Z.G., Lee, S.H., Kim, C.K., Park, C.M., Nahm, K., Nikitov, S.A.: Acoustic wave propagation in one-dimensional phononic crystals containing Helmholtz resonators. J. Appl. Phys. **103**(6), 064907–064917 (2008)
11. Shen, C., Li, J., Jia, Z., Xie, Y., Cummer, S.A.: Nonreciprocal acoustic transmission in cascaded resonators via spatiotemporal modulation. Phys. Rev. B **99**(13), 134306–134312 (2019)
12. Santillán, A., Bozhevolnyi, S.I.: Acoustic transparency and slow sound using detuned acoustic resonators. Phys. Rev. B **84**(6), 064304–064309 (2011)
13. Santillán, A., Bozhevolnyi, S.I.: Demonstration of slow sound propagation and acoustic transparency with a series of detuned resonators. Phys. Rev. B **89**(18), 184301–184311 (2014)
14. Khettabi, A., Bria, D., Elmalki, M.: New approach applied to analyzing a periodic Helmholtz resonator. JMES **8**(3), 816–824 (2017)
15. Khettabi, A., Antraoui, I.: Study of a finite network of one-dimensional periodic expansion chambers by the transfer matrix method and Sylvester theorem. AIP Conf. Proc. **2074**(1), 020003–020015 (2019)
16. Ben-Ali, Y., Tahri, Z., Falyouni, F., Bria, D.: Study about a filter using a resonator defect in a one-dimensional photonic comb containing a left-hand material. In: 1st International Conference on Electronic Engineering and Renewable Energy, Saidia-Morocco, pp. 146–156. Springer (2018)
17. Ben-Ali, Y., Tahri, Z., Ouariach, A., Bria, D.: Double frequency filtering by photonic comb-like. In: International Symposium on Advanced Electrical and Communication Technologies, Rabat-Morocco, pp. 1–6. IEEE (2018)
18. Ben-Ali, Y., Tahri, Z., Bria, D.: Electromagnetic filters based on a single negative photonic comb-like structure. Prog. Electromagn. Res. **92**, 41–56 (2019)
19. Munday, J.N., Bennett, C.B., Robertson, W.M.: Band gaps and defect modes in periodically structured waveguides. J. Acoust. Soc. Am. **112**(4), 1353–1358 (2002)
20. Fang, N., Xi, D., Xu, J., Ambati, M., Srituravanich, W., Sun, C., Zhang, X.: Ultrasonic metamaterials with negative modulus. Nat. Mater. **5**(6), 452–456 (2006)
21. Wang, Z.G., Lee, S.H., Kim, C.K., Park, C.M., Nahm, K., Nikitov, S.A.: Effective medium theory of the one-dimensional resonance phononic crystal. J. Phys. Condens. Matter **20**(5), 055209–055214 (2008)
22. Lee, S.H., Park, C.M., Seo, Y.M., Wang, Z.G., Kim, C.K.: Composite acoustic medium with simultaneously negative density and modulus. Phys. Rev. Lett. **104**(5), 054301–054304 (2010)
23. Lee, K.J.B., Jung, M.K., Lee, S.H.: Highly tunable acoustic metamaterials based on a resonant tubular array. Phys. Rev. B **86**(18), 184302–184306 (2012)
24. Akjouj, A., Al-Wahsh, H., Sylla, B., Djafari-Rouhani, B., Dobrzynski, L.: Stopping and filtering waves in phononic circuits. J. Phys. Condens. Matter **16**(1), 37–44 (2003)
25. Mir, A., Akjouj, A., Vasseur, J.O., Djafari-Rouhani, B., Fettouhi, N., El Boudouti, E.H., Zemmouri, J.: Observation of large photonic band gaps and defect modes in one-dimensional networked waveguides. J. Phys. Condens. Matter **15**(10), 1593–1598 (2003)
26. Little, B.E., Foresi, J.S., Steinmeyer, G., Thoen, E.R., Chu, S.T., Haus, H.A., Greene, W.: Ultra-compact Si-SiO2 microring resonator optical channel dropping filters. In: International Conference on Photonics Technology Letters, pp. 549–551. IEEE (1998)

# A New Approach to Model Discrete Event Systems

Mohammed Msaaf(✉) and Fouad Belmajdoub

Laboratory of Industrial Technologies (LTI), Faculty of Sciences and Technologies Fez, University Sidi Mohamed Ben Abdellah, Fez, Morocco
mohammed.msaaf@usmba.ac.ma, fbelmajdoub@yahoo.fr

**Abstract.** During the last decade, a significant research progress has been drawn in relation to the modelling of discrete events systems. This is mainly due to its crucial role that plays in different other research fields in relation to discrete events systems like supervisory control, diagnosis, maintenance… however traditional modelling tools used with this kind of system present some disadvantages especially in front of complex systems. Through this work we establish a new approach to model this kind of system using historical data and some notion of formal language theory.

**Keywords:** Discrete event systems · Modelling · Timed word · Statistical model

## 1 Introduction

The study of discrete event systems (DES) require the description of the behavior of all his component of this system. This description can be made using many types of models, the most used ones are: finite automata [1–4], Petri nets [5, 6], logical expressions [7], grafcet [8] and other tools [9–11]. These tools give a good estimation of the functioning mode of the considered DES. However, the application of these tools became more and sometimes impossible in the case of large and complex DES (such as aeronautical systems, military systems, nuclear power plants…) and they are exposed to many modelling problems such as the states number combinatory explosion that often gives complicate and huge models.

In opposition to these usual tools, statistical models, which are based on historical data, represent a good alternative to these models and more adequate to this kind of system. Moreover, they are easy to implement to any kind of systems and did not require a deep knowledge of the considered DES. The aim of this paper is to develop a new modelling approach based on historical data of the DES and that can be easily implementable to any kind of DES especially when it is about large and complex ones.

This paper will be organized as follows, the second section gives a short presentation of complex discrete event systems and the definition of timed word, the third section presents a new approach to model DES and we end up with a conclusion.


A. Saka et al. (Eds.): CPI 2019, LNME, pp. 22–27, 2021.
https://doi.org/10.1007/978-3-030-62199-5_3

## 2 Preliminary and Definition

### 2.1 Complex Discrete Event Systems

DES are dynamic systems where the state space is discrete and the switching from a state to another one is made according to events occurrence. These events can be of two different natures, they can be observable if they represent a signal coming from a sensor or a controller or they can be unobservable if we cannot detect them with available sensors. Generally, a DES can be described by two sets of events: $\Sigma_o$ the set of observable events and $\Sigma_u$ the set of unobservable events.

A DES is considered as complex if it is the combination of many heterogeneous and independent subsystems that are connected and communicate with each other in order to achieve a well-defined global goal.

### 2.2 Timed Word

A timed word over a set of events (considered as alphabets) $\Sigma$ is an ordered set of events (or a sequence of events) where each event is coupled to its occurrence date. So $\sigma = e(1)^{d_1}e(2)^{d_2}\ldots e(h)^{d_h}$ is a timed word where $e(i) \in \Sigma$ is the $i^{th}$ occurred event and $d_i \in \mathbb{R}^+ (0 \le d_1 \le \ldots \le d_h$ is its occurrence date. For example, $a^{1.0}\ b^{2.7}\ c^{4.1}$ is a timed word where the value 4.1 represent the date occurrence of the event c. the set of timed words over a set of event $\Sigma$ will be noted $TW^*(\Sigma)$ [12]. More information about timed words and formal language can be found in [2].

**Remark:** some authors [13, 14] consider another notation of timed words: Instead of coupling the event $e(i)$ to its occurrence date computed from the record beginning of the current word, they are coupled with the date computed from the occurrence of event $e(i-1)$. In other words timed word will be noted $\sigma_j = e(1)^{t_1}e(2)^{t_2}\ldots e(h)^{t_h}$ where $t_i$ is the elapsed time between the occurrence of events $e(i-1)$ and $e(i)$. The two notations are equivalent and we can pass from one of them to the other one by considering the variable change $t_i = d_i - d_{i-1}$.

## 3 Statistical Model of DES

A statistical model is a data base deduced from historical data of the considered system and containing a description of all possible behavior of the considered DES. Models deduced from historical data recorded during a significant duration are often considered as the most ideal representation of the real system functioning.

In the case of DES, the construction of such a model can be made using historical data composed of generated events and their occurrence dates. This historical data should be recorded in both normal and abnormal functioning modes of the DES.

A DES works generally according to a repetitive cycle in which it performs many tasks in order to achieve a principal goal and repeat the same tasks in the next similar cycles (each task is usually delimited by two events). Events generated in each cycle (observable and unobservable events) can be organized in the form of a finite timed word where each event is coupled with its occurrence date. So $\sigma_j = e(1)^{d_1} e(2)^{d_2} \ldots e(h)^{d_h}$ represents the timed word describing the functioning during the j$^{th}$ cycle.

At the beginning of each cycle, the record of events began and a timer $d$ is set to zero. When an event occurred, it is saved with the current timer value. This data is organised in the form of timed words.

**Remark:** In the case of DES that does not work according to cycles, we can take only one timed word $\sigma$ that contains generated event recorded during a significant period of functioning that cover all possible functioning modes of the DES.

A DES can work according to different cycle types in each one the DES can perform different tasks, this generates different timed words with different events and with different dimensions (number of events). In order to be able to treat these timed words, we propose to decompose them in the form of temporal windows with the same number of observable events. This decomposition can be made following the definition 2, in which we consider the projection $P_o(w) : TW^*(\Sigma) \rightarrow \Sigma_o$ that erase time and unobservable events from a timed word $w$.

**Definition:** (temporal windows from a timed word) let us consider $\sigma_j = e(1)^{d_1} e(2)^{d_2} \ldots e(s)^{d_s} \in \mathrm{TW}^*(\Sigma)$ as the recorded timed word within the $j^{th}$ cycle of the DES with $s = |\sigma_j|$ and $|P_o(\sigma_j)| = h_j$ ($|.|$ represent the number of events). $w_{j,i}^k = w_i^k(\sigma_j) = e(i)^{t_i} e(i+1)^{t_{i+1}} \ldots e(i+p)^{t_{i+p}} \in \mathrm{TW}^*(\Sigma)$ is the i$^{th}$ temporal windows from $\sigma_j$. i.e. $w_{j,i}^k$ is the i$^{th}$ sequence of events of $\sigma_j$ that contains k observable events, such that $k = |P_o(w_i^k(\sigma_j))| (k < |w_{(j,i)}^k|)$. In the rest of this work, temporal windows are used to describe the functioning of DES [15].

All constructed temporal windows coupled with their cycle type will be organized in a set noted $W^k$ that will be considered as the statistical model of the considered DES. The construction of such a model need a study phase of the DES in which the timed word of each cycle is constructed from collected event and their occurrence dates.

**Remark:** according to the parameters k a DES can have several statistical models $W^k$. The choice of this parameter will be the aim of another work.

## 4 Application Case

Let us consider the trolley shown in Fig. 1, which is considered as DES described by the following sets: $\Sigma_o = \{e_{p0}, e_{pA}, e_{pB}, e_{pC}\}$ and $\Sigma_u = \{e_{AB}, e_{BC}\}$ (faulty events are not considered in this work). All elements of these sets are described in the Table 1. It can perform two types of cycle described as follows:

*Cycle 1: the trolley starts from the position P0, go until PC passing by the positions PA and PB and it is subjected to manual operations between each two successive positions and go back to the initial position P0. This give the timed word:*

$$\sigma_1 = e_{p0}^{t1}\, e_{pA}^{t2}\, e_{AB}^{t3}\, e_{pB}^{t4}\, e_{BC}^{t5}\, e_{pC}^{t6}\, e_{pB}^{t7}\, e_{pA}^{t8}\, e_{p0}^{t9}$$

*Cycle 2: the trolley starts from the position P0, go until PB and passing by the position PA and it is subjected to manual operations between the position P0 and PA and go back to the initial position P0. This give the timed word:*

$$\sigma_2 = e_{p0}^{t10}\, e_{pA}^{t11}\, e_{AB}^{t12}\, e_{pB}^{t13}\, e_{pA}^{t14}\, e_{p0}^{t15}$$

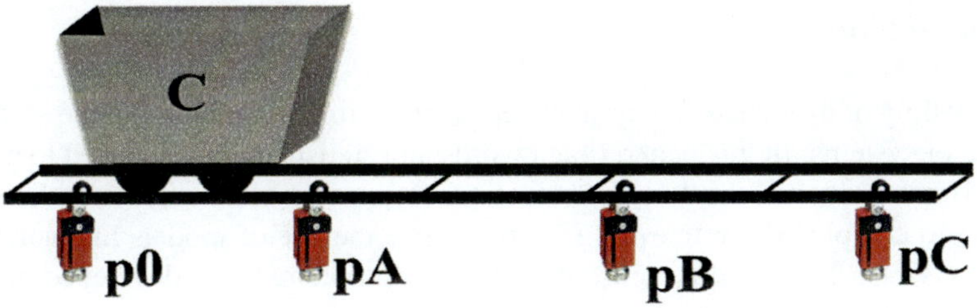


**Fig. 1.** The trolley.

**Table 1.** The event generated by the trolley.

| | Event | Description |
|---|---|---|
| Observable events | $e_{p0}$ | Trolley in the initial position |
| | $e_{pA}$ | Trolley in the position $P_A$ |
| | $e_{pB}$ | Trolley in the position $P_B$ |
| | $e_{pC}$ | Trolley in the position $P_C$ |
| Unobservable events | $e_{AB}$ | Manual actions between $P_A$ and $P_B$ are over |
| | $e_{BC}$ | Manual actions between $P_B$ and $P_C$ are over |

From these two timed words and according to the chosen dimension k we can get different statistical models $W^k$. For $k = 3$ the Fig. 2 show how we can get temporal windows from $\sigma_1$. So as a statistical model for the DES described above we can get:

$$W^3 = \left\{ \left( w_1^3 = e_{p0}^{t1}\, e_{pA}^{t2}\, e_{AB}^{t3}\, e_{pB}^{t4},\, c1 \right), \left( w_2^3 = e_{pA}^{t2}\, e_{AB}^{t3}\, e_{pB}^{t4}\, e_{BC}^{t5}\, e_{pC}^{t6},\, c1 \right), \right.$$
$$\left( w_3^3 = e_{pB}^{t4}\, e_{BC}^{t5}\, e_{pC}^{t6}\, e_{pB}^{t7},\, c1 \right), \left( w_4^3 = e_{pC}^{t6}\, e_{pB}^{t7}\, e_{pA}^{t8},\, c1 \right), \left( w_5^3 = e_{pB}^{t7}\, e_{pA}^{t8}\, e_{p0}^{t9},\, c1 \right),$$
$$\left( w_6^3 = e_{p0}^{t10}\, e_{pA}^{t11}\, e_{AB}^{t12}\, e_{pB}^{t13},\, c2 \right), \left( w_7^3 = e_{pA}^{t11}\, e_{AB}^{t12}\, e_{pB}^{t13}\, e_{pA}^{t14},\, c2 \right),$$
$$\left. \left( w_8^3 = e_{pB}^{t13}\, e_{pA}^{t14}\, e_{p0}^{t15},\, c2 \right) \right\}.$$

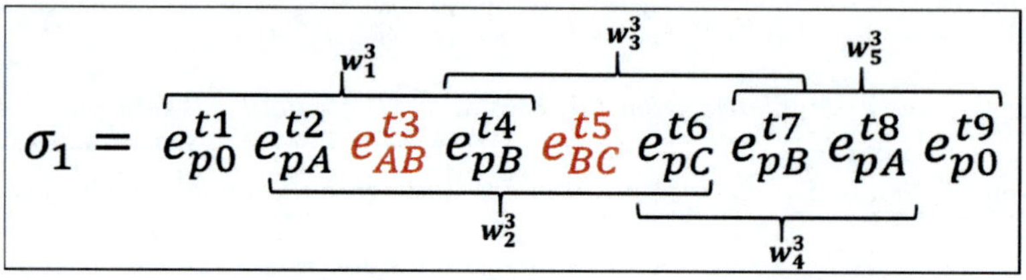


**Fig. 2.** The construction of temporal windows from $\sigma_1$.

## 5 Conclusion

The modelling of discrete event systems represent an important phase in the studying of this type of systems. In this paper, timed words and statistical data are used to construct a model that describes the functioning of this type of systems. This model is more adequate to complex discrete event systems where the use of models like automata or Petri nets is a hard task and is confronted by many problems. This developed modelling method has many advantages:

- *Its implementation can be done only by watching the behaviour of the DES and does not require a deep knowledge of conventional modelling tools;*
- *It integrate time which represents an important parameter in many recent studies of DES like checking the violation of temporal constraint;*
- *It describe the functioning of DES with timed words with similar dimensions which can easily treated by some other tools like neural network*

## References

1. Sampath, M., et al.: Diagnosability of discrete-event systems. IEEE Trans. Autom. Control **40**(9), 1555–1575 (1995)
2. Hopcroft, J.E., Motwani, R., Ullman, J.D.: Introduction to automata theory, languages, and computation. ACM SIGACT News **32**(1), 60–65 (2001)
3. Lamperti, G., Zanella, M.: Flexible diagnosis of discrete-event systems by similarity-based reasoning techniques. Artif. Intell. **170**(3), 232–297 (2006)
4. Rozé, L., Cordier, M.-O.: Diagnosing discrete-event systems: extending the "diagnoser approach" to deal with telecommunication networks. Discrete Event Dyn. Syst. **12**(1), 43–81 (2002)
5. Aghasaryan, A., et al.: Fault detection and diagnosis in distributed systems: an approach by partially stochastic Petri nets. Discrete Event Dyn. Syst. **8**(2), 203–231 (1998)
6. Genç, Ş., Lafortune, S.: A distributed algorithm for on-line diagnosis of place-bordered Petri nets. IFAC Proc. Vol. **38**(1), 68–73 (2005)
7. Wang, Y.: Supervisory control of Boolean discrete-event systems (2000). National Library of Canada = Bibliothèque nationale du Canada
8. Carré-Ménétrier, V., Zaytoon, J.: Grafcet: behavioural issues and control synthesis. Eur. J. Control **8**(4), 375–401 (2002)

9. Harel, D.: Statecharts: a visual formalism for complex systems. Sci. Comput. Program. **8**(3), 231–274 (1987)
10. Silveira, M.d.: Sur la distribution avec redondance partielle de modèles à événements discrets pour la supervision de procédés industriels, Toulouse 3 (2003)
11. Rayhane, H.: Surveillance des systèmes de production automatisés: Détection et Diagnostic. Docteur de l'Institut National Polytechnique de Grenoble et de l'Ecole Mohammedia d'Ingénieurs de Rabat, Année (2004)
12. Alur, R., Henzinger, T.A.: Back to the future: towards a theory of timed regular languages. In: Proceedings of the 33rd Annual Symposium on Foundations of Computer Science. IEEE (1992)
13. Bouyer, P., Chevalier, F., D'Souza, D.: Fault diagnosis using timed automata. In: International Conference on Foundations of Software Science and Computation Structures. Springer (2005)
14. Cassez, F., Grastien, A.: Predictability of event occurrences in timed systems. In: International Conference on Formal Modeling and Analysis of Timed Systems. Springer (2013)
15. Msaaf, M., Belmajdoub, F.: Fault diagnosis and prognosis in discrete event systems using statistical model and neural networks. Int. J. Mechatron. Autom. **6**(4), 173–182 (2018)

# Towards a Rigorous Approach to Coordinate Stakeholders of a Multi-energy Cyber-Physical System

Elmehdi Azzouzi[1,2](✉), Audrey Jardin[1], Daniel Bouskela[1], Faïda Mhenni[2], and Jean-Yves Choley[2]

[1] EDF Lab, Chatou, France
{elmehdi.azzouzi, audrey.jardin, daniel.bouskela}@edf.fr
[2] Quartz Laboratory – Supmeca Paris (ISAE), Saint-Ouen, France
{faida.mhenni, jean-yves.choley}@supmeca.fr

**Abstract.** Large distributed systems such as Multi-Energy Systems are complex systems that can be assimilated to Cyber-Physical Systems of Systems with multiple interactions between different elements and disciplines at different levels. They have the particularity of having dynamic configurations that are subject to multiple factors going from environmental aspects to political and social opinions and decisions. Moreover, Multi-Energy Systems involve a large number of stakeholders such as producers, distributors, suppliers, regulators, etc. with weak coordination. This number is constantly increasing as a result of the energy market deregulation and the high penetration of new decentralized renewable energies where regular consumers become producers. A pressing need has then emerged for new tools and methodologies: (1) to first tackle the lack of coordination while taking into consideration the concerns of each stakeholder that are often contradictory; (2) secondly to assess the system performances under multiple possible configurations and scenarios. This paper introduces our thoughts and progress concerning a rigorous approach using formal contracts. This approach proposes an engineering methodology to coordinate the multiple stakeholders using dedicated tools to predict the behaviour of the different facets of the system, a facet being defined as the way a single stakeholder views the system. The notion of formal contract allows to better structure the relationships between stakeholders and gives the opportunity to verify by simulation whether the system design complies with the requirements. A simple case study of coordination between two stakeholders is presented to demonstrate the methodology.

**Keywords:** Distributed systems · Multi-Energy Systems · Cyber-Physical Systems of Systems · Dynamic configurations · Stakeholders · Decentralized renewable energies · Stakeholders' coordination · Simulation · Scenarios · Formal contract


A. Saka et al. (Eds.): CPI 2019, LNME, pp. 28–41, 2021.
https://doi.org/10.1007/978-3-030-62199-5_4

# 1 Introduction

Multi-Energy Cyber-Physical Systems (ME-CPS) are large distributed Systems of Systems with multiple interactions between them. They gather manifold complex and dynamic systems such as production units working with diverse energy sources (renewable or non-renewable) with possible storage units, electric network, heat network, consumers such as commercial or residential buildings and also industries, etc.

ME-CPS have the particularity of having high-stakes in terms of societal, environmental and financial aspects, etc. They are characterized by being vital systems where people live, they have strong interactions with their environment, and they require high investments for their development, construction, operation, and dismantlement.

The growing awareness concerning the environmental stake has led to an important mutation in the way of thinking about the energy production and management while having an important share of uncertainties concerning the future evolution regulations, production technologies, climate change, etc. which requires considering multiple realistic possible scenarios when developing them.

Nowadays, the energy landscape is becoming more and more complex with new interactions on the power network as a consequence of the high penetration of new decentralized renewable energies, and also the deregulation of the energy market with the entry of new actors and where regular consumers became producers. Thus, they have dynamic configurations and are subject to constant modifications and improvements in order to meet the evolving need of its stakeholders.

ME-CPS are Systems of Systems (SoS) that gather a large number of stakeholders that can enter or leave the system at any phase of the system lifecycle (e.g. an energy producer, a client or an energy supplier that chooses to take part or leave the ME-CPS leading to a considerable impact on the system, thus requiring a reconfiguration). One of the main challenges is coordination between these stakeholders and agreeing on how to interact, what to exchange, with which performances and so on.

In this context, our contribution in this work is to propose a formal and executable approach for more efficient coordination between stakeholders. Our aim is to give stakeholders means to assess their systems' performances by challenging them against formal requirements for a better understanding of the system dynamics and having thus more efficient negotiations between them.

The paper is organized as follows. A state of the art concerning the stakeholders' theory and formal approaches dealing with requirements are introduced in Sect. 2. Section 3 introduces an overview of the methodology that is being developed. A case study where the methodology is applied is presented in Sect. 4. Finally, a summary is given in Sect. 5.

# 2 State of the Art

## 2.1 About Stakeholders and Stakeholders' Coordination

Referring to the INCOSE Handbook (Walden et al. 2015), a stakeholder is:

*"A stakeholder is any entity (individual or organization) with a legitimate interest in the system. When nominating stakeholders, business management will take into account all those who may be affected by or able to influence the system"*

One of the first works that got interested in studying stakeholders management was introduced in (Freeman 1984) called the "Strategic Management: A Stakeholder Approach". Since then, multiple researchers from different communities got interested in the thematic and manifold points of views on the subject can be found in the literature. An exhaustive state of the art is introduced in (Parmar et al. 2010).

The main focus of this theory is identifying all possible stakeholders in an organization in order to ensure the best integration between social and ethics with business and companies.

The existing works and especially the stakeholder theory of Freeman (Freeman 2019; Phillips et al. 2003) focuses on taking into consideration external stakeholders when designing a system or when taking strategic decisions in companies. It aims at ensuring the best integration between social and ethics with business and companies. It is ensured by characterizing stakeholders based on their interests, rights, and how could they affect the future product of the business plan of the company. For that matter, Freeman's theory proposed a classification of stakeholders based on their potential to cooperate or threaten corporate strategy. They can be (i) Supportive (ii) Marginal (iii) Non-supportive (iv) Mixed Blessing. Based on this classification, different strategies can be perused regarding these stakeholders such as (i) Offensive strategy, (ii) Defensive strategy, (iii) Swing strategy, (iv) Hold strategies.

This methodology can have important added value for a specific entity that aims at developing a system or product, however, it is not adapted for the problematic that we intend to tackle. The stakeholder theory considers stakeholders as external elements that influence on one way or another some specific system, which doesn't meet our main goal that is to rather integrate ME-CPS stakeholders into a common co-design framework, and to provide them with appropriate tools and methods allowing them to have a better apprehension of their systems, and as a consequence (i) preventing stakeholders from over-specifying their needs, (ii) having more productive negotiations and (iii) better coordination.

Our methodology focuses on how to coordinate and manage stakeholders taking part of large ME-CPS in order to accomplish objectives at a high level, notably the level of System of Systems (SoS) while ensuring the concerns of each stakeholder.

## 2.2 Formal Verification and Validation

Designing the future energy system requires testing a large number of possible scenarios and configurations while modeling their behavior and dynamics in detail in order to have the right system that is adapted to future needs.

The paper (Azzouzi et al. 2019) has emphasized the lack of formal design approaches suitable for ME-CPS, especially when dealing with formal requirement and thus the automatic Verification and Validation (V&V) of the design.

In order to make automatic V&V of ME-CPS stakeholders' requirements that are integrated into contracts, it is mandatory to have an adapted language and tool enabling to formally specify requirements and temporal constraints on physical systems.

In literature, the most used formal approach for requirement engineering is model checking (Rozier 2011; Roehm et al. 2016) that comes originally from software engineering. This approach allows exploring exhaustively all possible states of a system in order to check some model properties such as reachability, safety, invariance, etc. However, this technique is not adapted for physical systems that have infinite possible states and cannot be resumed to a finite state machine.

In this context, a formal requirement language named FORM-L (FOrmal Requirements Modeling Language) (Nguyen 2017) was developed as part of the ITEA MODRIO European project. FORM-L aims at capturing requirements using close to natural language and well-defined syntax and semantics. With FORM-L, a requirement such as "Any pumps in the system must not cavitate more than once in a sliding time period of a week." can formally be expressed. A new temporal language called ETL (Bouskela and Jardin 2018) was developed in this context to specify the V&V semantics of the requirements temporal aspects and validate, using simulation, whether they are satisfied or not. ETL was implemented in the Modelica language by developing a library called ReqSysPro. A compiler is under development to automatically translate FORM-L models into ReqSysPro models in order to simulate them in Modelica environment.

## 3 Methodology

### 3.1 Preamble

Figure 1 was introduced in (Bouskela et al. 2017). It shows the necessary elements for creating a verification model enabling to check whether the design of a system fulfills its requirements under multiple scenarios. Three elements are needed: requirement model, behavioral model, and architectural model.

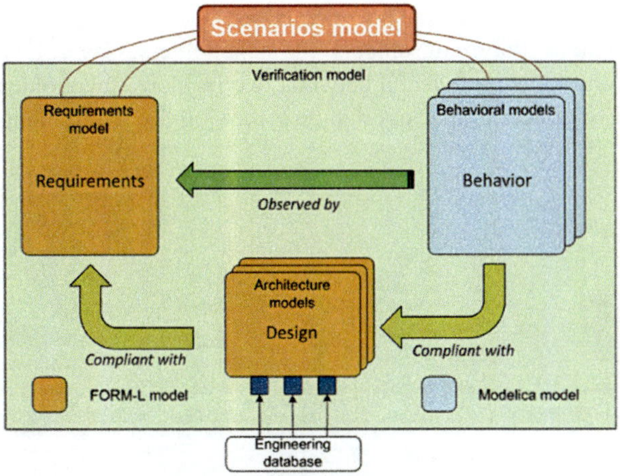


**Fig. 1.** Verification model

The idea behind this figure was defining the nature of the relationships between the three elements that are generally developed by different engineering teams. It has led to creating the concepts "bindings" and "observers" that were discussed in (Bouskela et al. 2017) and that will be used in the current methodology.

The methodology presented in this paper is an extension of the methodology developed as part of the MODRIO European project and is based on the same verification concept with similar elements, but for a more realistic purpose. Indeed, when designing a ME-CPS, we don't have a unique book of specifications gathering all system requirements, however, we have multiple concerns of different stakeholders with leads to multiple books of specifications that could be contradictory with each other. Thus, having consistent requirements on the same system is a challenge by itself.

In order to cope with this problematic, our methodology is based on a Contract Based Design (CBD) paradigm presented in (Sangiovanni-Vincentelli et al. 2012). It offers means to partially tackle the complexity of multi-stakeholders systems. Indeed, working in a framework that is structured using contracts ensures, on the one hand, a high level of design autonomy per discipline as each stakeholder has some precisely defined assumptions and guarantees to be fulfilled. On the other hand, it ensures easy and correct integration of the different parts of the system developed separately provided that all contract requirements are fulfilled. The paradigm of CBD will not be developed in detail in this paper.

### 3.2 Methodology Concepts

In this work, we propose a way to combine existing concepts such as (i) requirements model, (ii) architecture model, (iii) behavioral model, (iv) scenarios model, (v) bindings, (vi) observers, (vii) verification model and (viii) contracts, together with new concepts such as (i) stakeholder model, (ii) contracts model and (iii) stakeholder interaction model, in order to coordinate multiple stakeholders using contracts.

#### 3.2.1 Contract

In order to structure the requirements of stakeholders, several contracts will be established. A contract is established between two or more stakeholders that interact in some way or have concerns that overlap at some point. Requirements in the contract

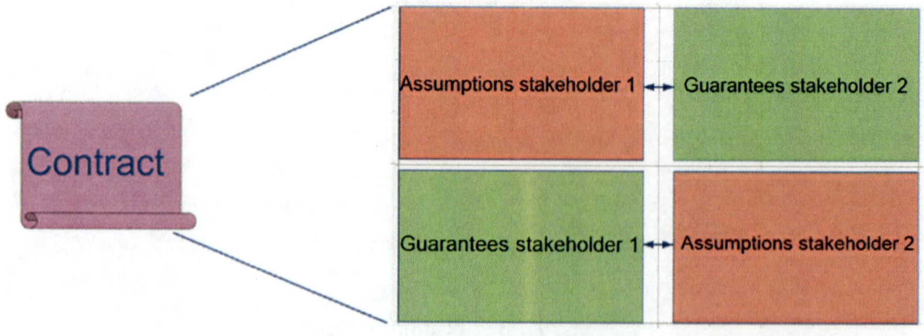


**Fig. 2.** Contract object

will be structured using the concept of Assumptions/Guarantees for each stakeholder. Indeed, the assumptions of each stakeholder refer to what he requires from the other stakeholders, the guarantees on the other side represent what the stakeholder has to ensure to others (Fig. 2).

### 3.2.2 Stakeholder

The object "stakeholder" used in our methodology contains five elements (see Fig. 3):

- An architectural part which represents the perimeter that will be affected to the stakeholder from the architectural model of the overall SoS. In other words, a global architecture encompassing all the interactions between different systems will be defined. Perimeters of stakeholders will be defined and allocated to them (see Sect. 3.2.4).
- A behavioral model describing the behavior of the elements captured in the architectural model. Different models and tools can be used for modeling different facets of ME-CPS.
- The internal requirement model contains all the proper requirements of the stakeholder that will not be shared with other stakeholders. They represent the high-level concerns of the stakeholder.
- Bindings connect models of different nature in order to exchange information for the purpose of verification using simulation. In this case, it assembles the four elements introduced above.
- A set of observers that allows transforming the system physical behavioral state such as mass flow-rates into functional ones that are needed for the internal requirement model.

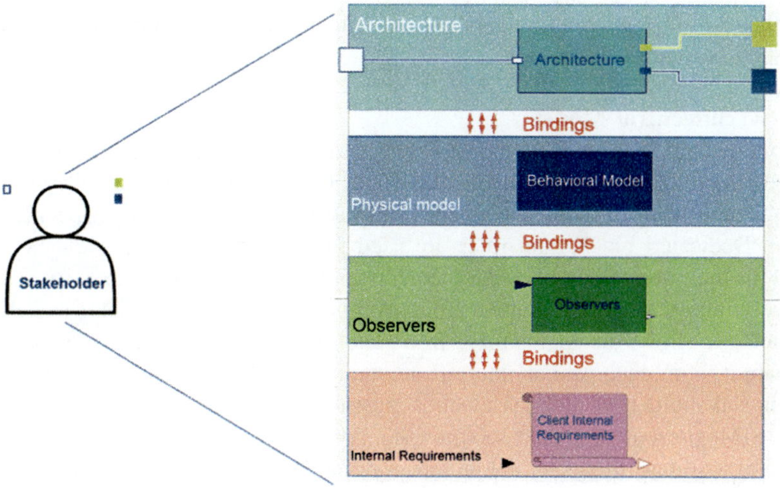


**Fig. 3.** Stakeholder object

### 3.2.3 Contracts Model

A contracts model is a topology of contracts that are established between a set of stakeholders (Fig. 4). This model gathers at the same time all stakeholders' objects as well as all the contracts between them. This topology is not defined nor fixed at the beginning of a project. Indeed, when designing ME-CPS, only a few stakeholders or even only one stakeholder who comes with the idea of a design project may be involved in the beginning. Then, new stakeholders progressively take part of the SoS adding new assumptions and guarantees. It is the case of a town council that first comes up with the idea of creating an eco-friendly neighborhood. Then, other stakeholders such as energy producers, distributors, land developers, etc. will gradually integrate the design project.

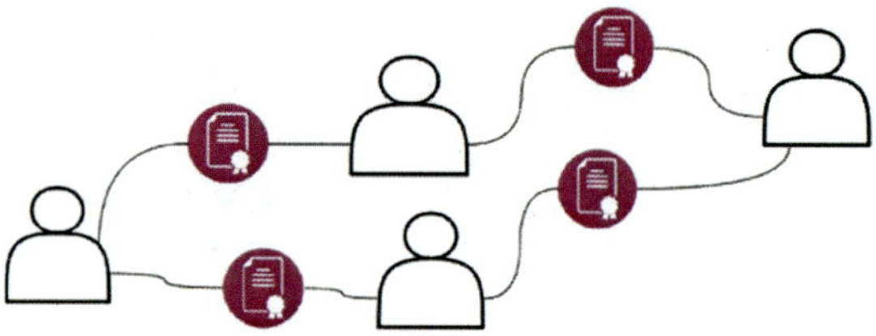

**Fig. 4.** Contracts model

Coordination processes should be proposed in order to identify the topology of contracts with the properties that satisfy the concerns of all stakeholders. The topology of contracts will surely evolve during design phases based on the negotiations that will take place and also during operation phase as stakeholders are subject to withdrawal, changes, etc. For instance, a stakeholder that used to be only responsible for electricity supply can become an electricity producer, which changes the topology of contracts established between stakeholders.

### 3.2.4 Architectural Model

This notion catches up with the notion of "Architecture description" introduced in the standard ISO 42010 (ISO/IEC/IEEE 42010:2011). It describes the architecture of the overall system from different views according to the perspective of a specific system concern. Dedicated blocks and rules are used for representing each view in order to explicit all the interactions between elements and manage their complexity.

A very important notion that we represent in the architecture model is stakeholders' perimeters. It allows us to identify, from one side, which elements are part of the responsibility for each stakeholder, from another side, to identify the interactions between stakeholders which eases the construction of the Stakeholder Interaction Model (SIM) presented in next section (Fig. 5).

During a system lifecycle, the architecture model and stakeholders' perimeters are subject to different modifications. At design phases, the architecture can evolve either by refining its elements or by having concurrent architectures that can be considered and evaluated. During the operational phase, the architecture component can also strongly evolve regarding new elements that can integrate the perimeter of the overall ME-CPS and add new interactions with its elements. Existing elements can also come off the perimeter of the system and leads to a new re-configuration. Stakeholders' perimeters can also evolve regarding their roles. (e.g. A supplier that evolves his electricity supply service into electricity plus gas or heat supply).

### 3.2.5 Stakeholder Interaction Model (SIM)

The stakeholder Interaction Model (SIM) allows going from a vision of architecture description, where we see the interactions between systems, into a description of interactions between stakeholders which allows highlighting what is exchanged and what are relationships between them. This representation allows and facilitates the establishment of contracts between stakeholders as the interactions between them are more explicit. The current representation of interactions is especially helpful when dealing with manifold stakeholders with multiple interactions and dependencies between their systems.

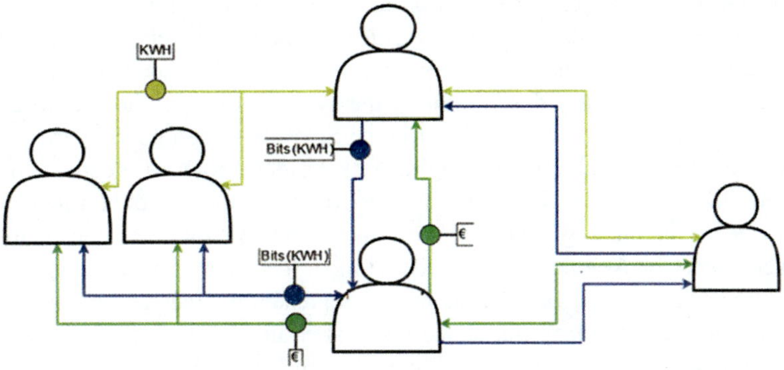


**Fig. 5.** Stakeholder interaction model

### 3.2.6 Verification Model

The verification model gathers at the same time the contract model and the stakeholders' interaction model in order to make verification and validation using simulation. Moreover, it encompasses a scenarios model that will be connected to all elements of the verification model via the bindings. Doing so allows automatically generating parameters representing the scenarios that we want to test, and challenge the requirements with parameters defined for that specific case.

### 3.2.7 Concepts Arrangement

The particularity of our methodology is that there are no frozen chronological processes that dictate what concept to use at what moment of the design. The concepts above can indeed be developed in parallel for the main purpose to simulate models representing the behavior of stakeholders systems as well as their requirements. Nevertheless, refinement steps for requirements, architectures and behavioral models should always follow specific guidelines that will ensure the consistency between different abstraction levels, as well as for the iterations loops during negotiation phases between stakeholders where specific processes should be followed in order to converge to a set of contract that satisfies all stakeholders.

The idea of this section is to introduce our current reflection concerning the arrangement of the concepts presented above with the main ambition of having an executable model that will be simulated in order to challenge stakeholders' requirements with the ME-CPS dynamic behavior.

When dealing with ME-CPS, we generally don't start from scratch as they will always depend on the existing legacy system and will be connected to it (e.g. it is the case for the electric grid where most of the electric systems would be plugged to after development). Thus, we can start by defining the architecture of the existing elements, and what are the interactions between them, that will probably evolve during the design of the new system. Multiple new architectures can be proposed and evaluated during the design. Moreover, and as specified in the standard ISO 42010 (ISO/IEC/IEEE 42010:2011), the architectural description can have one or multiple views describing one or multiple viewpoints or facets in order to answer stakeholder concern, while using dedicated conventions, and without neglecting the dependencies between them.

Having the architectural description of the system, we can allocate different subsystems to stakeholders in order to develop the SIM the stakeholders' perimeters. This perimeter can always evolve depending on stakeholders' roles at a certain moment in the system lifecycle. The SIM will help to develop the contract model between stakeholders by having a global vision on all the interactions between them. At this level, each object 'stakeholder' can be developed. The related architectural perimeter will be allocated to it. A behavioral model representing the architectural elements should be then developed by the concerned stakeholder. Finally, the proper requirements should be introduced into the stakeholder object (see Sect. 3.2.2 for stakeholder object elements).

In parallel with the SIM and using this representation, the contract model can be defined by introducing the assumptions and guarantees of stakeholders regarding each other.

Having the two elements SIM and the contract model, automatic verification and validation can be launched in order to challenge the requirements specified by stakeholders under different scenarios. It is not mandatory to gather all stakeholders' objects and all the defined contracts in order to make testing using simulation, it is even recommended to make analysis using a limited number of stakeholders and contracts in order to have reasonable models with reasonable simulation times.

Simulations will allow identifying, from one side whether each stakeholder can ensure their guarantees under the assumptions that were specified in the contract or not, and from the other side verifying that their proper requirements were satisfied.

The aim of this analysis via simulation is to empower the negotiations between stakeholders with tools that will allow them to have a better apprehension of the impact of their requirements, their systems, as well as their design choices on the overall system.

Our methodology should allow via processes that will be introduced in future works having negotiation loops where each stakeholder can update his requirement parameters in order to converge towards a contract or set of contracts that satisfy at the same time his internal concerns as well as the requirements of all related stakeholders.

Stakeholders will be coordinated once they will find a set of contracts that satisfies each one of them.

## 4 Case Study: Coordination Between an Electricity Supplier and a Client

Our first attempt to introduce the coordination of stakeholders in the methodology started with a simple case study called Heated Room. This example was inspired by the work of Bouissou et al. (2013) and is based on the Modelica model that was developed in that context.

The model represents a room with a heater that is regulated by a controller which attempts to keep the temperature of the room between 18 and 22° while considering an external temperature of 13°.

This case study was adapted to our problematic in which we consider two stakeholders, an electricity supplier and a customer. Each one of them has a different concern and interest. The supplier aims at making business, selling his electricity, and also proposing a service to the clients and thus maximize his profit. The client, on the other hand, will require an efficient system to control the temperature of his room, and also not paying more than a certain amount of money per month for the electricity and the service.

These requirements were formally defined and captured via the Form-L language (Nguyen 2017) for handling requirement as follow:

As a proper requirement of the supplier:

```
requirement profitability is
    when eEndYear
    ensure supplier.revenue > 1.2 * supplier.cost;
end profitability;
```

As a proper requirement of the client:

```
requirement maxBill is
    when eEndMonth
    ensure client.heating.bill<300€;
end maxBill;
```

In the contract between these two stakeholders, we chose a unique guarantee of the supplier, which is at the same time the assumption of the client:

```
requirement thermalConfort is
    always ensure client.room.temperature is in [18 , 22]*_C
end thermalConfort;
```

After specifying the requirement in FormL language, we translated them into Modelica blocks with the ReqSysPro library in order to have an executable requirement model. Requirements were integrated respectively in the internal requirement model of the supplier, client and in the contract between them.

The architecture in Fig. 6 shows the overall system with the interactions between all its elements notably the electricity supply which represent the electric grid to which the heater of the room will be connected, the room of the client that will be heated and the controller that will receive data from the room, and send order to the heater in order to regulate the room temperature.

In this architecture, the perimeter of each stakeholder is defined referring to the role of each one of them and the kind of service that is proposed to the client. In this case study, we have considered that the supplier offers as a service to the client a connected box that controls the heating of his room. Another allocation of architectural elements could have been considered such as integrating the control part in the perimeter of the client if he considers that his knowledge and equipment are able to ensure the same service for lower costs. In this case, other requirements would be considered and the contract would be different.

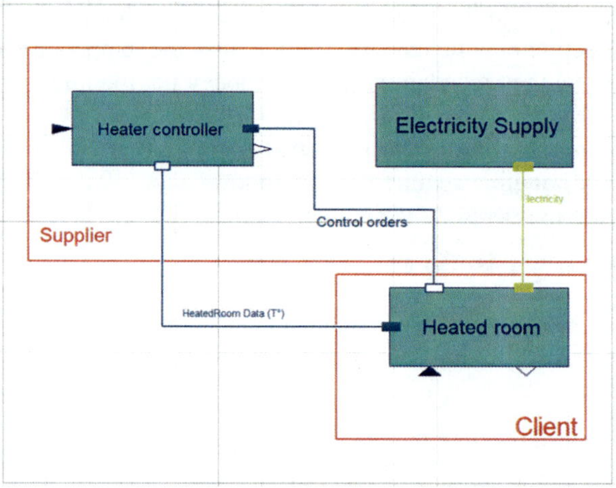


**Fig. 6.** Architecture model

Defining the perimeter in the architectural model allows us to identify the interactions between the systems of each stakeholder. Indeed, we can see in Fig. 6 that there are three connections between the supplier and the client.

The verification model that contains the stakeholder interaction model in Fig. 7 shows us the interactions between the supplier and the client. We can retrieve the three connections from the architectural model.

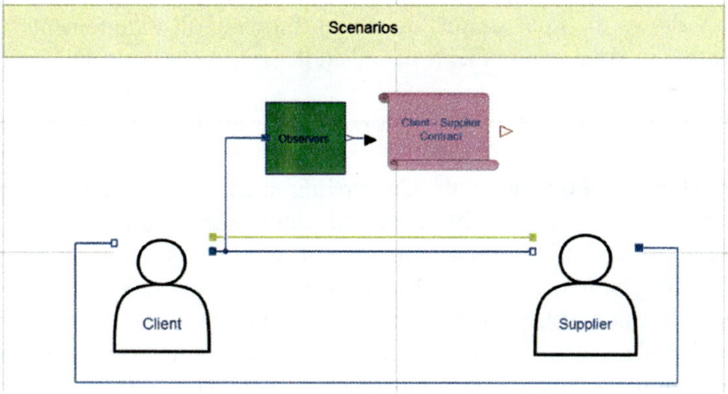


**Fig. 7.** Verification model

Simulation has been launched using the verification model in Fig. 7. It allowed us to analyze and challenge the requirements in the contract as well as in the internal models of each stakeholder. Indeed, after testing 3 sets of requirement parameters, we were able to identify the one that allows satisfying all the requirement from both sides. These parameters concerned especially the price of electricity proposed by the supplier that went a little bit down to meet the client requirement, and at the other hand, the client changed the maximum value fixed initially for his bill in exchange of having an additional service to ensure his thermal comfort.

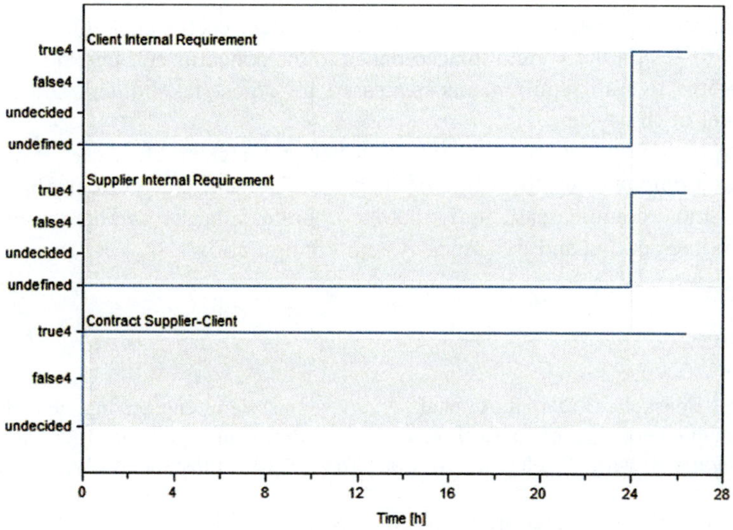


**Fig. 8.** Verification simulation results

Figure 8 shows the final simulation results showing all requirements satisfaction, notably the two internal requirements of stakeholders as well as the requirements in the contract between them.

The simulation was done on a duration of 24 h, and requirement parameters were adapted for that purpose.

In the first graph of the figure, the client requirement was undefined during the 24 h of the day and turned to true at the end of the simulation ensuring that his bill wasn't higher than the specified threshold at the end of a certain period.

The second graph concerning the supplier requirement is similar to the first requirement. It turned to true at the end of the simulation time.

The third graph concerning the requirement on the contract was verified all along the simulation time as the requirement that concerns the heat comfort needed to be true continuously.

## 5 Conclusion

This paper presented an overview of a co-design methodology that aims at coordinating stakeholders taking part in a large Multi-Energy Cyber-Physical-System. The idea behind the methodology is to tackle the complexity of such systems by structuring them using formal contracts between stakeholders. The approach aims at giving stakeholders means to formally and automatically assess their system performances by challenging them against formal requirements, and thus having a better understanding of their dynamics as well as allowing them taking accurate and optimal decisions during negotiations.

Our idea is to create a co-design framework where each stakeholder will have the autonomy to design the system in accordance to his concerns and his know-how while respecting the overall requirements expressed by other stakeholders and also at the global level of the system.

A simple demonstration of coordination between stakeholders was also presented and allowed emphasizing the value of such an approach. Our current objective is to work on a more complex case study that incorporates a larger number of stakeholders with more interactions and dynamics system configuration.

## References

Azzouzi, E., Bouskela, D., Jardin, A., et al.: A survey on systems engineering methodologies for large multi-energy cyber-physical systems. In: Proceedings of the 13th Annual IEEE International Systems Conference SysCon 2019 (2019). https://doi.org/10.1109/SYSCON.2019.8836741

Bouissou, M., Chraïbi, H., Chubarova, I.: Critical comparison of two user friendly tools to study Piecewise Deterministic Markov Processes (PDMP). In: Safety, Reliability and Risk Analysis (2013)

Bouskela, D., Jardin, A.: ETL: a new temporal language for the verification of cyber-physical systems. In: 2018 Annual IEEE International Systems Conference (SysCon), pp. 1–8 (2018)

Bouskela, D., Nguyen, T., Jardin, A.: Toward a rigorous approach for verifying cyber-physical systems against requirements. Can. J. Electr. Comput. Eng. (2017). https://doi.org/10.1109/cjece.2016.2630421

Freeman, E.R.: Strategic Management: A Stakeholder Approach. Pitman Publ. Inc. (1984)

Freeman, R.E.: The Stakeholder Approach Revisited. Zeitschrift für Wirtschafts- und Unternehmensethik (2019). https://doi.org/10.5771/1439-880x-2004-3-228

ISO/IEC/IEEE: Systems and software engineering — Architecture description 42010 (2011)

Parmar, B.L., Freeman, R., Harrison, J., et al.: Stakeholder theory: the state of the art. Acad. Manag. Ann. **3**, 403–445 (2010). https://doi.org/10.1080/19416520.2010.495581

Nguyen, T.: A modelling & simulation based engineering approach for socio-cyber-physical systems. In: Proceedings of the 2017 IEEE 14th International Conference on Networking, Sensing and Control, ICNSC 2017 (2017)

Phillips, R., Freeman, R.E., Wicks, A.C.: What Stakeholder Theory is Not. Bus. Ethics Q. (2003). https://doi.org/10.5840/beq200313434

Roehm, H., Oehlerking, J., Heinz, T., Althoff, M.: STL model checking of continuous and hybrid systems. In: International Symposium on Automated Technology for Verification and Analysis, pp. 412–427. Springer (2016)

Rozier, K.Y.: Linear temporal logic symbolic model checking. Comput. Sci. Rev. (2011). https://doi.org/10.1016/j.cosrev.2010.06.002

Sangiovanni-Vincentelli, A., Damm, W., Passerone, R.: Taming Dr. Frankenstein: Contract-Based Design for Cyber-Physical Systems. Eur. J. Control (2012). https://doi.org/10.3166/ejc.18.217-238

Walden, D.D., Roedler, G.J., Forsberg, K.J., et al.: Systems Engineering Handbook (2015)

Modelica and the Modelica Association, information available on. https://www.modelica.org/

# Road Profile and Vehicle State Estimation Using Unknown Input Observer

Fatima Ezzahra Saber(✉), Mohamed Ouahi, and Abdelmjid Saka

Laboratory of Ingénierie, Systèmes et Applications, National School of Applied Sciences, Sidi Mohammed Ben Abdellah University, Fez, Morocco
{fatimaezzahra.saber,mohamed.ouahi,abdelmjid.saka}@usmba.ac.ma

**Abstract.** Advanced driver assistance systems are mainly aimed at the automotive industry. These systems are dedicated to optimizing the safety of drivers about dangerous situations by informing the vehicle driver or by acting on vehicle dynamics. For this purpose, estimation and observation are major tools to progress the development of ADAS. In this study, we describe a new method to estimate the road profile elevation based on Unknown Input Observers. These observers are designed using a mathematical model of a quarter car model developed in Matlab to analyze the vertical vibration of a vehicle. The necessary measurements are the accelerations of the centers of the wheels. The validation of observes is carried out by means of realistic simulator Callas to demonstrate the robustness of the proposed method.

**Keywords:** Road profile · Vehicle dynamics · Suspension · Unknown input observer · Driver assistance systems

## 1 Introduction

The development of Advanced Driver Assistance Systems (ADAS) has contributed mainly to decreasing the number of accidents while improving comfort and safety in vehicles.

The attributes of the road are one of the most important values that influence the behavior of the vehicle. The dynamics of the control system only works as we hope in all driving maneuvers on all road surfaces, if the vertical acceleration from the center of each wheel of the vehicle is known. Due to economic and Technology reasons, road profiles, roll angle, speeds and heights of the centers of the wheels, the heights of the corners of the chassis, speed and height of the center of gravity of the sprung mass are not measurable in the vehicles. Several studies have been conducted to characterize the different profiles of road used by vehicles [12,23]. [18] offers the estimation of vehicle inertial parameters. other article uses vertical models to control the suspension [4,20,24]. Imine [9] using the sliding-mode observers for systems with unknown inputs for estimated the road profile. In the literature, there was the various synthesis of The design of


A. Saka et al. (Eds.): CPI 2019, LNME, pp. 42–54, 2021.
https://doi.org/10.1007/978-3-030-62199-5_5

observers, such as a sliding mode observer and Kalman-filter based method [5]. Some authors used a full car model or a quarter car model in [1]. The observer-based methods. Imine [10] showed that the profile can be estimated by applying the sliding mode observer. [17] proposed a state feedback control of a complete model of the vehicle to the rejection of disturbance. The control by the fuzzy logic is used by [13] for the control of a quarter of the vehicle ([4,7,11] and [20]), half-vehicle model to four degrees of freedom ([6,14] and [18]) are presented in this article allows to model the movement of the vehicle. Observers are designed using the "unknown input affecting output" framework ([2] and [15]). The gains of the observers are computed by solving linear matrix inequalities [8]. This present document allows the construction of the observers on upright models of the increasing complexity of vehicles going from the quarter of vehicle model (Sect. 2) to go to a half-vehicle model. In a different case, the methodology will be deployed systematically. The outline of this paper is: the first section describes the vehicle modeling. The second section discusses the design of two kinds of unknown input observers. The final Section, the unknown input observers are tested and Data issued from the vehicle simulator Callas are used to evaluate Observers' performance.

## 2 Vehicle Dynamic Model

### 2.1 The Quarter-Car Model

This section describes the quarter of the vehicle model, depicted in (Fig. 1), is commonly used in work on the monitoring and analysis of the handling, passenger comfort and vibration isolation [19,21] and [6].

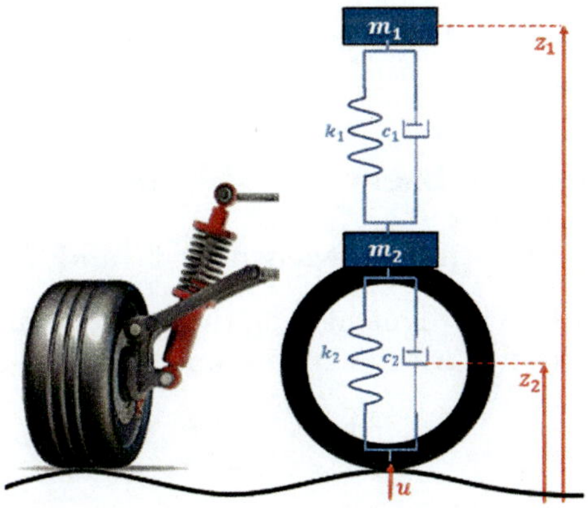


**Fig. 1.** quarter of the vertical model

Adopting the following notation:

$K_1$ : Spring Stiffness of suspension $[N.m^{-1}]$
$K_2$ : Spring stiffness of tire $[N.m^{-1}]$
$C_1$ : Damping coefficient of suspension $[N.s.m^{-1}]$
$C_2$ : Damping coefficient of tire $[N.s.m^{-1}]$
$m_1$ : Sprung mass $[Kg]$
$m_2$ : Mass of the wheel $[Kg]$
$z_1$ : Height of the sprung mass $[m]$
$z_2$ : Height of the center of the wheel $[m]$
$u$ : Road profile $[m]$

Vertical forces of the system are formulated in terms of vertical elongation of the suspension and wheel. After applying equations of the Lagrange in question and the linearization of these equations around a static equilibrium point. The dynamics equations are formulated by:

$$\begin{cases} m_1\ddot{z}_1 = \underbrace{-K_1(z_1 - z_2) - C_1(\dot{z}_1 - \dot{z}_2)}_{\text{some of the forces of the suspension}} \\ m_2\ddot{z}_2 = -[-K_1(z_1 - z_2) - C_1(\dot{z}_1 - \dot{z}_2)] \\ \qquad - \underbrace{K_2(z_2 - u) - C_1\dot{z}_2}_{\text{some of the forces of the wheel}} \end{cases} \tag{1}$$

The aim of the two observers presented in this paper is to calculate the road profile, speed and height of the center of the wheel, speed and height of the corners of the chassis that is not measured. The sensor used measurements of the vertical accelerations from the center of the wheel and chassis. In the first case, the two measures are considered as available, it will be noted OEI1. In the second case, only the vertical acceleration from the center of the wheel will be used and the observer will be OEI2.

## 3 Unknown Input Observers Design

### 3.1 OEI1: Observer with Two Measures $\ddot{z}_1$ and $\ddot{z}_2$

**State-Space Model:** The dynamics of the quarter-car model (1) can be expressed as a linear system with unknown input affecting outputs:

$$\Sigma_1 \begin{cases} \dot{x} = Ax + Ru \\ y = \mathbf{C}x + \mathbf{D}u \end{cases} \tag{2}$$

In this expression, $x = \begin{pmatrix} z_1 \; z_2 \; \dot{z}_1 \; \dot{z}_2 \end{pmatrix}^T$ is the state vector, $u$ is the unknown input and $y = \begin{pmatrix} \ddot{z}_1 \\ \ddot{z}_2 \end{pmatrix}$ is the output vector. $A$ is the state matrix, $R$ is the input matrix associated with the unknown input, $\mathbf{C}$ is the observation matrix and $\mathbf{D}$ represents the influence of unknown input on measures.

$$A = \begin{pmatrix} 0 & 0 & 1 & 0 \\ 0 & 0 & 0 & 1 \\ -\frac{K_1}{m_1} & \frac{K_1}{m_1} & -\frac{C_1}{m_1} & \frac{C_1}{m_1} \\ \frac{K_1}{m_2} & -\frac{K_1+K_2}{m_2} & \frac{C_1}{m_2} & -\frac{C_1+C_2}{m_2} \end{pmatrix}$$

$$R = \begin{pmatrix} 0 \\ 0 \\ 0 \\ \frac{K_2}{m_2} \end{pmatrix}; \mathbf{D} = \begin{pmatrix} \mathbf{D}_1 \\ \mathbf{D}_2 \end{pmatrix} = \begin{pmatrix} 0 \\ \frac{K_2}{m_2} \end{pmatrix}$$

$$\mathbf{C} = \begin{pmatrix} \mathbf{C}_1 \\ \mathbf{C}_2 \end{pmatrix} = \begin{pmatrix} -\frac{K_1}{m_1} & \frac{K_1}{m_1} & -\frac{C_1}{m_1} & \frac{C_1}{m_1} \\ \frac{K_1}{m_2} & -\frac{K_1+K_2}{m_2} & \frac{C_1}{m_2} & -\frac{C_1+C_2}{m_2} \end{pmatrix}$$

Several studies have been done for unknown input observation for linear dynamical systems [16] and [22]. The system (2) respects the following assumptions:

- The unknown input ($u$) is the only input.
- The number of unknown inputs is less or equal to the number of measures.
- The matrix $\mathbf{D}$ is full column rank $Rang(\mathbf{D}) = 1$.

The unknown input $u$ can be expressed from the first observation equation of system (2) because $\mathbf{D}_2 = \frac{K_2}{m_2} \neq 0$ is invertible.

If the state $x$ is known, the estimation of the road profile ($u$) can be calculated from the following equation:

$$u = \mathbf{D}_2^{-1}(y_2 - \mathbf{C}_2 x) \tag{3}$$

By substituting (3) into (2), we obtain:

$$\Sigma_1 \begin{cases} \dot{x} &= (A - R\mathbf{D}_2^{-1}\mathbf{C}_2)x + R\mathbf{D}_2^{-1}y_2 \\ y_1 &= \mathbf{C}_1 x \end{cases} \tag{4}$$

The system with a known input $y_1 = \ddot{z}_1$ and a measure $y_2 = \ddot{z}_2$.

$$\Sigma_1 \begin{cases} \left(\dot{z}_1\ \dot{z}_2\ \ddot{z}_1\ \ddot{z}_2\right)^T &= F\left(z_1\ z_2\ \dot{z}_1\ \dot{z}_2\right)^T \\ & \quad + G\ddot{z}_2 \\ \ddot{z}_1 &= \mathbf{C}_1\left(z_1\ z_2\ \dot{z}_1\ \dot{z}_2\right)^T \end{cases} \tag{5}$$

In order to construct the unknown input observer, we study the detectability of the system. Detectability matrix:

$O_{dt} = \begin{pmatrix} sI_4 - N & -R \\ 0_{2\times 4} & D \end{pmatrix}$, $\forall s \in \mathbb{C}, \mathrm{Re(s)} \geq 0$.

The matrix N is given by the Eq. (6).

The system is detectable because due to the mechanical properties of the vehicle, this matrix is full rank ($Rang(O_{dt}) = 5$). It is possible to estimate the road profile from Eq. (3).

**Observer Design OEI1:** The adequate solution is to rebuild simultaneously the state $x$ and the unknown input $u$ by using the available measurements that constitute the output $y$.

For the linear time-invariant systems. To be able to characterize the robustness performances, the gains of observer OEI1 are obtained by solving a linear matrices inequalities problem.

The full order observer ([2] and [15]) for system (5) can be expressed as:

$$obs\begin{cases} \dot{z} = Nz + My_2 + Ly_1 \\ \hat{x} = z \end{cases} \tag{6}$$

where $\hat{x} = \left(\hat{z}_1\ \hat{z}_2\ \dot{\hat{z}}_1\ \dot{\hat{z}}_2\right)^T$ is the observed state. $N$, $M$ and $L$ are matrices to be choose to have an observation error which asymptotically goes to zero.

The observation error $e = x - \hat{x}$ has the following dynamics

$$\dot{e} = \dot{x} - \dot{z}$$

replacing $x$ by $e + \hat{x}$ and $y_1$ by $\mathbf{C}_1 x$.

$$\dot{e} = (F - L\mathbf{C}_1)e + (F - L\mathbf{C}_1 - N)\hat{x} + (G - M)y_2 \tag{7}$$

The estimation error converges asymptotically to 0 if and only if the matrices $N$, $M$ and $L$ are choose to satisfy the following conditions:

$$\begin{cases} F - L\mathbf{C}_1 \text{ Hurwitz matrix} \\ F - L\mathbf{C}_1 - N = 0 \\ G - M = 0 \end{cases} \tag{8}$$

With these conditions, the estimation error (7) dynamic becomes:

$$\dot{e} = (F - L\mathbf{C}_1)e \tag{9}$$

the determination of matrices $N$, $M$ and $L$ satisfying the constraints (8) is as follows.

*Matrix M*
Using the condition $G - M = 0$, $M$ is determined by $M = G$ where $G$ is the input matrix of system (5).

*Matrix L*
The matrix $L$ is chosen so that the observation error (9) is asymptotically stable.

Using the following Lyapunov function, the convergence of the observer is guaranteed if there exists a symmetric positive matrix $X$ that the Lyapunov function $V(e) = e^T X e$ presents the following properties:

$$\forall e \neq 0 \begin{cases} V(e) > 0 \\ \dot{V}(e) < 0 \end{cases}$$

This can be reformulated using Eq. (9):

$$\begin{cases} X > 0 \\ (F - L\mathbf{C}_2)^T X + X(F - L\mathbf{C}_1) < 0 \end{cases} \tag{10}$$

Applying the schur complement [3,8] and using

$$W = XL$$

The linear matrix inequalities system (10) can be written as:

$$\begin{pmatrix} -X & 0 \\ 0 & F^T X + XF - (\mathbf{C}_1^T W^T - W\mathbf{C}_1) \end{pmatrix} < 0 \tag{11}$$

The gain matrix is obtained by first solving the LMI (11) with respect to $X$ and $W$. In a second time, $L$ is determined by:

$$L = X^{-1} W \tag{12}$$

*Matrix N*
This matrix $N$ is calculated using the second equation of system (8):

$$F - L\mathbf{C}_1 - N = 0 \longrightarrow N = F - L\mathbf{C}_1$$

*In summary*
To summarize, the observer OEI1 allows calculating simultaneously the speed and height of the center of the wheel, the speed and the height of the corners of the chassis and the road profile (unknown input) by using the vertical accelerations from the center of wheel and chassis. The OEI1 is designed as:

$$OEI1 \begin{cases} \begin{pmatrix} \dot{\hat{z}}_1 \\ \dot{\hat{z}}_2 \\ \ddot{\hat{z}}_1 \\ \ddot{\hat{z}}_2 \end{pmatrix} = N \begin{pmatrix} \hat{z}_1 \\ \hat{z}_2 \\ \dot{\hat{z}}_1 \\ \dot{\hat{z}}_2 \end{pmatrix} + M\ddot{z}_2 + L\ddot{z}_1 \\ \hat{u} \quad = \mathbf{D}_2^{-1} \left( \ddot{z}_2 - \mathbf{C}_2 \begin{pmatrix} \hat{z}_1 \\ \hat{z}_2 \\ \dot{\hat{z}}_1 \\ \dot{\hat{z}}_2 \end{pmatrix} \right) \end{cases} \tag{13}$$

### 3.2 OEI2: Observer with One Measure $\ddot{z}_2$

**State-Space Model:** the vertical accelerations is supposed measured on the center of wheel $\ddot{z}_2$. The aim is to calculate the speed and height of the center of the wheel, the speed and the height of the corners of the chassis and the road profile. System $\Sigma_2$ used for this observer is different from $\Sigma_1$ (2) by the observation equation.

$$\Sigma_2 \begin{cases} \dot{x} &= Ax + Ru \\ y_2 &= \mathbf{C}_2 x + \mathbf{D}_2 u \end{cases} \tag{14}$$

This system verifies also the hypotheses that allow designing an observer (3.1).

His observability matrix is:

$$O = \left(\mathbf{C}_2 \; \mathbf{C}_2 A \; \mathbf{C}_2 A^2 \; \mathbf{C}_2 A^3\right)^T$$

The determinant of this matrix is:

$$det(O) = -K_1^4 K_2^2 / m_1^2 m_2^4$$

The system is observable because the determinant is not zero because of the existence of the suspension and the tire.

The matrix $\mathbf{D}_2$ is invertible. It is possible to estimate the road profile from Eq. (3).

**Observer Design OEI2:** The full order observer ([2] and [15]) for system (14). It is expressed as follows:

$$obs \begin{cases} \dot{z} = Nz + Ly_2 \\ \hat{x} = z \end{cases} \tag{15}$$

where $\hat{x} = \left(\hat{z}_1 \; \hat{z}_2 \; \dot{\hat{z}}_1 \; \dot{\hat{z}}_2\right)^T$ is the observed state.

The observation error dynamic $e = x - \hat{x}$ is expressed by:

$$\dot{e} = Ne + (A - L\mathbf{C}_2 - N)x + (R - L\mathbf{D}_2)u \tag{16}$$

The error estimation converges asymptotically to 0 if and only if the matrices $N$ and $L$ are satisfied the following conditions:

$$\begin{cases} N \text{ Hurwitz matrix} \\ A - L\mathbf{C}_2 - N = 0 \\ R - LD\mathbf{D}_2 = 0 \end{cases} \tag{17}$$

The estimation error dynamic (16) becomes:

$$\dot{e} = Ne \tag{18}$$

*Matrix L*

The matrix $D_2$ is invertible and square, we calculate $L$ from (17) :

$$L = R\mathbf{D}_2^{-1}$$

*Matrix N*
Using the matrix $L$, we obtain matrix $N$:

$$N = A - R\mathbf{D}_2^{-1}\mathbf{C}_2$$

*In summary*
To summarize, the observer OEI2 allows calculating simultaneously the speed and height of the center of the wheel, the speed and the height of the corners of the chassis and the road profile (unknown input) by using the vertical accelerations on the center of wheel and chassis. It's written

$$OEI2 \begin{cases} \begin{pmatrix} \dot{\hat{z}}_1 \\ \dot{\hat{z}}_2 \\ \ddot{\hat{z}}_1 \\ \ddot{\hat{z}}_2 \end{pmatrix} = N \begin{pmatrix} \hat{z}_1 \\ \hat{z}_2 \\ \dot{\hat{z}}_1 \\ \dot{\hat{z}}_2 \end{pmatrix} + L\ddot{z}_2 \\ \hat{u} \quad = \mathbf{D}_2^{-1} \left( \ddot{z}_2 - \mathbf{C}_2 \begin{pmatrix} \hat{z}_1 \\ \hat{z}_2 \\ \dot{\hat{z}}_1 \\ \dot{\hat{z}}_2 \end{pmatrix} \right) \end{cases} \tag{19}$$

## 4 Simulations and Discussions

**Vehicle Simulator:** Realistic simulators are useful intermediate between real experiments and simplified models which are used to design controllers or state-observers.

Callas is a vehicle dynamic program distributed worldwide by OKTAL company. This is truly and advantage since the measurement of all those variables is rare and expensive, which is again a motivation of the present study that proposes to estimate the unaffordable measurements.

**Simulation Conditions:** To achieve the qualitative assessment of the two observers, we are going to roll the vehicle on a floor in stairs such as that shown in Fig. 2. On this route, three steps are present. Their respective heights 1, 2 and 3 cm. The plateaus of the steps have a length of 50 m. The pilot was instructed to make the journey at a constant speed of 20 km/h. The quarter vehicle considered in this experiment is the left front quarter of the vehicle.

**Model Validation:** Model validation is the first stage of a software sensor design. parameterization of open-loop models (2) used in the construction of observers is verified using a linear stationary quarter of the vehicle model, whose input is the road profile $u$ measured and considered as known. This model is called bo in the paper.

We took the initial conditions of the observer and the system as well as the software of callas so what are zero.

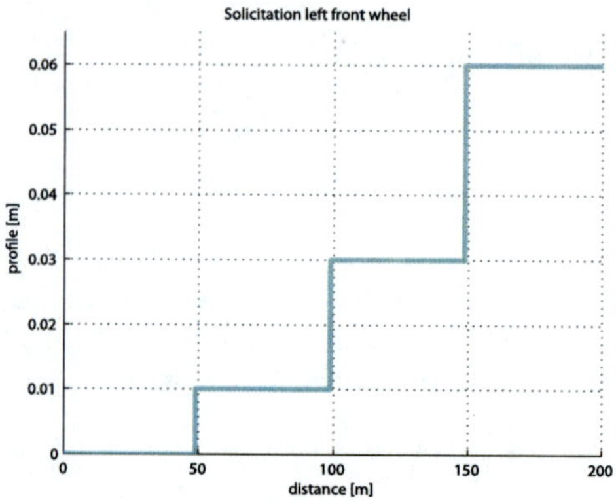


**Fig. 2.** Profile of road used for the reconstruction of profile from a quarter vehicle model

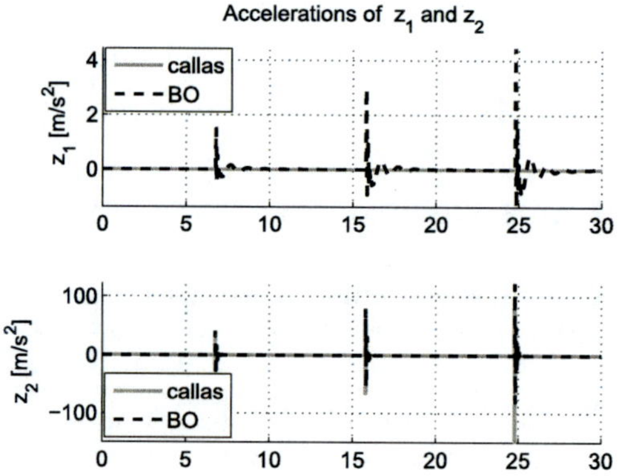


**Fig. 3.** Vertical accelerations of the sprung mass and the center of the wheel. Callas simulation compared with open-loop model (2) with known road profile

Figure 3 represents in its upper part the measurement of vertical acceleration of the chassis $\ddot{z}_1$ and in its lower part that of the unsprung mass (center of the wheel) $\ddot{z}_2$.

This indicates that the model of acceleration of the center of the wheel is correct when the resultant road profile applied to the wheel is known.

**Observer Validation:** Figure 4 presents the results of reconstruction of the heights of the wheel center 4(b) and of the chassis 4(a) for the model (2) ("BO" label) controlled by the profile considered as known and the observers OEI1 (13) and OEI2 (19). Reference Callas is also represented. On the lower of the two figures, one can see the estimation error committed by the three methods.

Regarding the reconstruction of the height of the center of the wheel, for these experimental conditions virtual, an errorless than a millimeter for the

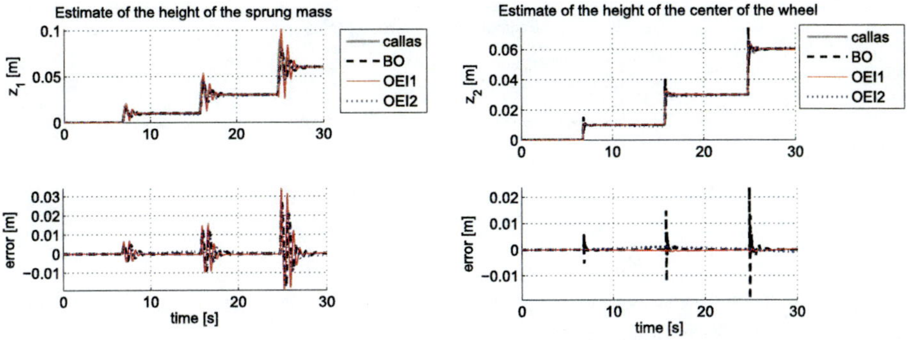


(a) Sprung mass

(b) Center of the wheel

**Fig. 4.** Observers OEI1 and OEI2: estimations of the height of the sprung mass $\hat{z}_1$ and the height of the center of the wheel $\hat{z}_2$, Callas simulation compared with open-loop model (2) and with unknown input observers OEI1 (13) and OEI2 (19). Modelling error and observation error

reconstruction of three steps. Noting, a slight drift of the height for the observer OEI2 using a single measurement of acceleration.

The height of the sprung mass is correctly reconstructed by the two observers. However, the transient response of the observer OEI1 using two measurements of acceleration is much better. The transient error is of the order of magnitude of the step height for the observer using only the measurement of acceleration at the center of the wheel.

Figure 5 presents the reconstruction of the road profile ($u$ unknown input) by the two observers (OEI1 and OEI2) compared to the reference Callas with the estimation error.

Noting that the various trays are correctly reconstructed and that the drift slight observed on the reconstruction of the height of the wheel for the observer (19) is also present in the reconstruction of the profile. In average, the absolute value of the Observation error is $\approx 0$. Finally, in the transitional phases in response to the rise of walking, the maximum errors found are of the same order of magnitude for the two observers, but the damping is better when the two acceleration measurements are used.

In view of these results, it can be concluded that the two observers have offered very good performance for estimating the state and the unknown input.

**Robustness of Observer:** The Fig. 6 shows the response of the observer OEI1 (13) when crossing the stairs described in Fig. 2 when the value of the sprung mass is considered in the design of the observer is changed from 80 to 120[%] of the nominal mass identified. The order of magnitude of the error is small.

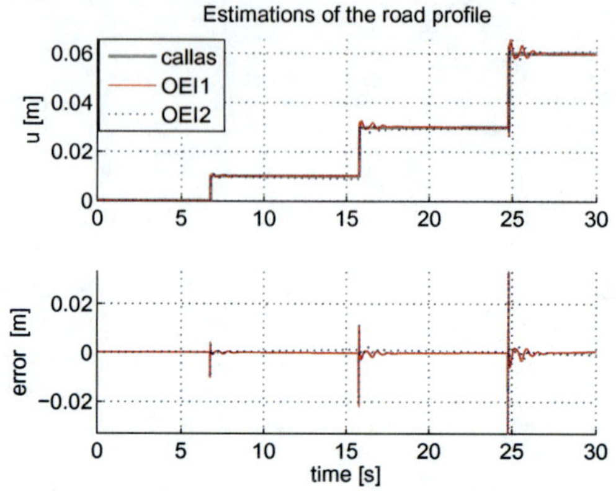


**Fig. 5.** Observers OEI1 and OEI2: estimations of the road profile, Callas simulation compared with unknown input observers OEI1 (13) and OEI2 (19). Observation error

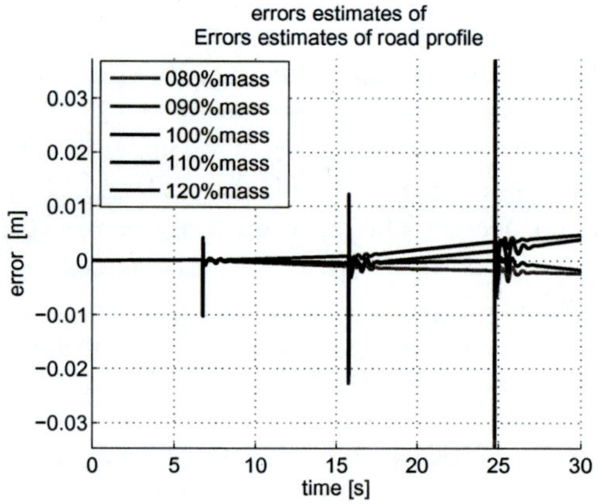


**Fig. 6.** errors of observers OEI1 (13): errors estimates of road profile when the sprung mass is changed of 80 to 120[%] of the nominal mass identified

A study of the average errors committed by the two observers was conducted to changes in model parameters from 80 to 120[%] of their nominal values identified. This study leads to the same qualitative performance recorded for the estimation of the status and profile. The two observers are considered robust to parametric variations.

However, when the additive measurement noise is applied to the measurement signals used by the two observers, problems of divergence of the algorithm are found.

## 5 Conclusion

In this paper, the reconstruction of the road profile and vertical state for embedded systems is achieved. The algorithms have been faced with a realistic vehicle simulator with the advantages of disadvantages of virtual experiments. A study of average errors committed by both observers was conducted to model parameter variations from 80 to 120[These observers differ for small variations in the initial conditions. However, since the measurement noise additive is applied to the measurement signals used by two observers, divergence problems of the algorithm are recognized.

## References

1. Arat, M.A., Taheri, S., Holweg, E.: Road profile estimation for active suspension applications. SAE Int. J. Passeng. Cars Mech. Syst. **8**(2), 492–500 (2015)
2. Boutayeb, M., Darouach, M., Rafaralahy, H.: Generalized state-space observers for chaotic synchronization and secure communication. IEEE Trans. Circ. Syst. **49**, 345–349 (2002)
3. Boyd, S., El Ghaoui, L., Feron, E., Balakrishnan, V.: Linear Matrix Inequalities in System and Control Theory. Studies in Applied Mathematics. SIAM, Philadelphia (1994)
4. Chantranuwathana, S., Huei Peng, H.: Adaptive robust force control for vehicle active suspensions. Int. J. Adapt. Control Sig. Process. **18**, 83–102 (2004)
5. Doumiati, M., Martinez Molina, J., Sename, O., Dugard, L., Lechner, D.: Road profile estimation using an adaptive Youla-Kucera parametric observer: Comparison to real profilers. Control Eng. Pract. **61**, 270–278 (2016)
6. Du, H., Zhang, N., Lam, J.: Parameter-dependent input-delayed control of uncertain vehicle suspensions. J. Sound Vib. **317**, 537–556 (2008)
7. Fischer, D., Isermann, R.: Mechatronic semi-active and active vehicle suspension. Control Eng. Pract. **12**, 1353–1367 (2004)
8. Hassibi, A., How, J., Boyd, S.: A path-following method for solving BMI problems in control. Am. Control Conf. San Diego, Calif. **2**, 1385–1389 (1999)
9. Imine, H., M'Sirdi, N., Delanne, Y.: Sliding-mode observers for systems with unknown inputs: application to estimating the road profile. J. Automot. Eng. **219**, 1–9 (2005)
10. Imine, H.: Observation d'états d'un véhicule pour l'eestimation du profil des les traces de roulement. Ph.D. thesis, Université de Versailles-Saint-Quentin-en-Yvelines (Décembre 2003)
11. Koch, G., Fritsch, O., Lohmann, B.: Potential of low bandwidth active suspension control with continuously variable damper. Control Eng. Pract. **18**, 1251–1262 (2010)
12. Kropac, O., Mucka, P.: Be careful when using the international roughness index as an indicator of road uneveness. J. Sound Vib. **287**, 989–1003 (2005)
13. Liang, L., Ming, Y.A.C.W., Bing, W.: Road DNA based localization for autonomous vehicles. In: IEEE Intelligent Vehicles Symposium, pp. 883–888 (2016)
14. Lin, J.H., Huang, C.J.: Nonlinear backstepping active suspension design applied to a half-car model. Veh. Syst. Dyn. **42**(6), 373–393 (2004)

15. Liu, F.: Synthèse d'observateurs entrées inconnues pour les systèmes non linéaires. doctorat de l'université de basse-normandie, Université de Basse-Normandie (décembre 2007)
16. Maquin, D., Gaddouna, B., Ragot, J.: Estimation of unknown inputs in linear systems. Am. Control Conf. **1**, 1195–1197 (1994)
17. Prattichizzo, D., Marcorelli, P., Bicchi, A., Vicino, A.: Geometric disturbance decoupling control of vehicules with active suspensions. In: IEEE International Conference on Control Applications, Trieste, Italy, pp. 253–257 (1998)
18. Rozyn, M., Zhang, N.: A method for estimation of vehicle inertial paramrters. Veh. Syst. Dyn. **48**(5), 547–565 (2010)
19. Sammier, D.: Sur la modélisation et la commande des suspensions des véhicules automobiles. Ph.D. thesis, Institut National Polytechnique de Grenoble (INPG) (Novembre 2001)
20. Savaresi, S., Poussot Vassal, C., Spelta, C., Sename, O., Dugard, L.: Semi-active suspension control design for vehicles. Elsevier (2010), ISBN 978-0-08-096678-6. http://hal.archives-ouvertes.fr/hal-00537826/en/
21. Smith, M., Wang, F.: Controller parametrisation for disturbance response decoupling: Application to vehicle active suspension control. IEEE Trans. Control Syst. Technol. **10**(3), 393–407 (2002)
22. Stotsky, A., Kolmanovsky, I.: Simple unknown input estimation techniques for automotive applications. Am. Control Conf. **5**, 3312–3317 (2001)
23. Tamboli, J.A., Joshi, S.G.: Optimum design of a passive suspension system of a vehicle subjected to actual random road excitations. J. Sound Vib. **219**, 193–205 (1999)
24. Yagiz, N., Hacioglu, Y.: Backstepping control of a vehicle with active suspensions. Control Eng. Pract. **16**, 1457–1467 (2008)

# Modeling and Control an Inverted Pendulum with Two Arms

Boutaina Elkinany(✉), Mohammed Alfidi, Soukaina Krafes, and Zakaria Chalh

Laboratory of Engineering Systems and Applications, Sidi Mohammed Ben Abdellah University, Fez, Morocco
boutaina.elkinany@gmail.com

**Abstract.** The unicycle robot designates an actuated system that has a single wheel ensuring its stability. Accordingly, unicycle robot has been the subject of researchers interest regarding its high degree of instability which enables this system to move in all directions without falling on the ground. This paper aims at presenting our vision of unicycle mobile robot based on modeling an inverted pendulum with two arms, a mathematical representation is analyzed using the Lagrangian dynamic formulation. Moreover, we will illustrate by Matlab software to apply the virtual reality, that will allow us to check on the motion movement of this system, after that, we will apply the LQR control algorithm in order to reach the good stability of the inverted pendulum with two arms.

**Keywords:** Inverted pendulum with two arms virtual reality · LQR stability

## 1 Introduction

The frequency of using wheeled mobile robots in various domains is growing day after day. The most challenge that we need to confront while using a mobile robot is navigation over different spaces no matter what are the limits of the area. The quantity of the mobile robot conceptions takes into consideration robots form constraints. To underline this mentioned issue, this introduction aims at explaining the development and progression of mobile robots from four wheels to one wheel.

Wheeled mobile robots are the vehicles which have some interaction points with the ground. Consequently, important problems are generated due to this contact, such as the slip phenomenon and this is exactly why minimizing the contact point with the ground is a very challenging and interesting topic for researchers.

To start with four wheeled robots, [1]. Maciej Trojnacki Analyzed the issue of dynamics modeling of this system by describing the robot kinematic structure and presenting the system in three cases.

Systems and Applications Laboratory in National School of Applied Sciences Fez, Morocco.


A. Saka et al. (Eds.): CPI 2019, LNME, pp. 55–64, 2021.
https://doi.org/10.1007/978-3-030-62199-5_6

As for the three wheeled robots [2]. Focused on the modeling and control of the motion of a back wheel drive tricycle mobile robot [3]. Used a Scratchy tricycle model in order to present the development of a multi robot task allocation method by employing the fuzzy inference system.

Regarding the two wheeled robot [4]. Focused on bicycle model using the virtual holonomic constraint method in order to enforce this system to traverse a strictly convex curve with bounded roll angle and bounded speed, also, [5], developed the two wheeled vehicle dynamic model and analyze the system stability by applying some control algorithms. Furthermore, [6]. Azadeh Kheirandish described the behavior of electric bicycle system powered by fuel cell by elaborating a new computing strategy based on fuzzy cognitive map which has the advantages to controlling and stabilizing the system.

At last, the one wheeled robot is a system that can navigate in different areas using only one contact point with the ground, so viewing its particularity; it has attached many researchers to mentor studies about the system [7]. Designed an unicycle model with balance bar, dynamic model was presented basing on an appell equation, as for the simulation results show that the system could attend stability after a transient period of time, additionally, [8]. Proposed a unicycle robot with a reaction wheel, also built a prototype by attaching some electronic components to a body in order to become a self-governing system, similarly, [9]. Used the reaction wheel for obtaining stability on the roll angle by pivoting the disc to engender momentum. While the wheeled robot is used for obtaining balancing on the pitch angle by rotating wheel to move in all directions.

On the other hand, many control algorithms were proposed to stabilize the unicycle robot. Linear controllers; for examples PID controller and LQR were implemented in many research works [9–11]. Some researchers [1,7]. Used nonlinear controllers such as: sliding mode controller and backstepping controller. Likewise, [13]. Designed linear and nonlinear control methods in order to stabilize 5 degrees of freedom Spherical Inverted Pendulum. Intelligent controls were also applied to the robot, such as the tagaki-Sugeno approach using the fuzzy logic [14,15].

In the present paper, We intend to invent a proper unicycle robot model, based on a gradual progress which starts by modeling an inverted pendulum with two arms, then ameliorating this model by adding other components, and at last replacing the arms by a gyroscope.

In our approach of unicycle robot model, we aim at directing the robot in an unusual way that is from a free point related with the two arms.

As we have mentioned, a perturbation $\tau$ is apply on the two arms what is generated a rotation of the arms $\theta_1$ and $\theta_2$, also the oscillation of the inverted pendulum angle $\theta$ around his equilibrium point, so to achieve balance in this system, the LQR control algorithm will be proposed and simulated.

## 2 Mathematical Modeling

The below figure illustrates best our conceptualization of the unicycle robot; the inverted pendulum and two arms rotate once we apply a torque at arms level taking into consideration that the angles between both arms stay fixed in 120° during their rotation.

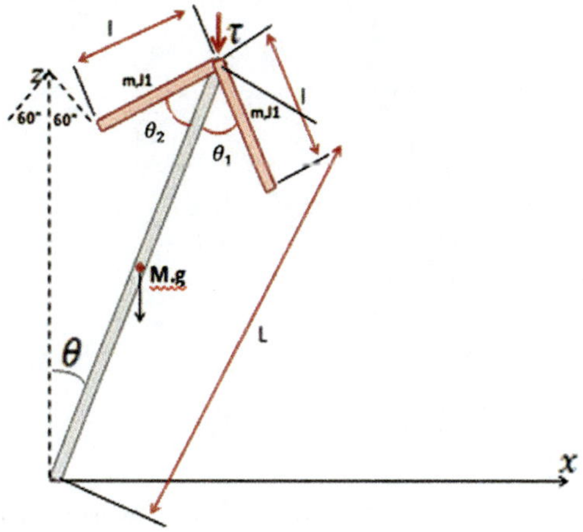


**Fig. 1.** Inverted pendulum with two arms system.

*Remark:* Arms angles do not influence their rotation according to another modeling as it mentionned in Fig. 2.

The below table shows the technical parameters for using this system:

**Table 1.** System parameters.

| Parameters | Description |
|---|---|
| M | Inverted pendulum mass |
| m | Arm mass |
| l | Arm length |
| L | Inverted pendulum length |
| g | Gravity Coefficient 9.81 |
| $\theta$ | Inverted pendulum rotation angle |
| $\theta_1$ | Rotation angle of the first arm |
| $\theta_2$ | Rotation angle of the second arm |
| $\tau$ | Applied perturbation on the two arms |

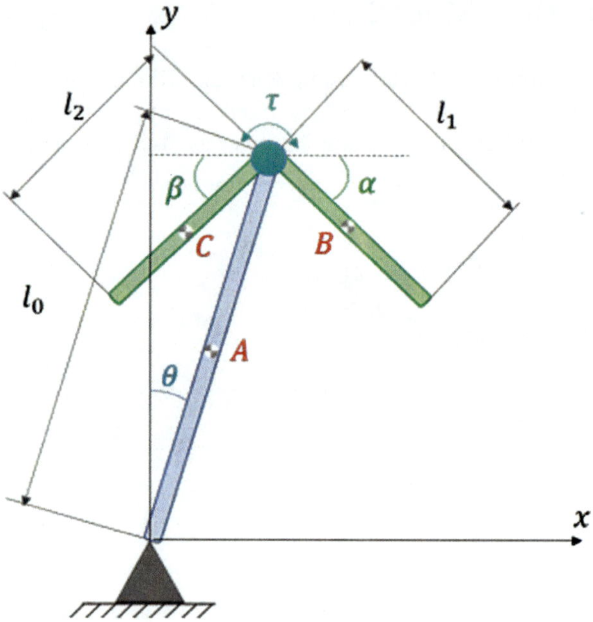


**Fig. 2.** Inverted pendulum with two arms system.

After calculating the time derivate, we can put them into the kinetic and potential energy expressions as follow:
The kinetic energy is given by:

$$E_c = \frac{1}{6}(M+m)L^2\dot{\theta}^2\frac{1}{6}ml^2\dot{\theta}_1^2 + \frac{1}{6}ml^2\dot{\theta_2}^2 + \frac{1}{2}mlL\dot{\theta}\dot{\theta}_1\cos(\theta+\theta_1) \\ -\frac{1}{2}mlL\dot{\theta}_1\dot{\theta}_2\cos(\theta_1+\theta_2) \quad (1)$$

The potential energy is given by:

$$E_p = (\frac{1}{2}.M+2m).L.g.cos(\theta) - \frac{1}{2}m.l.g.cos(\theta_1) - \frac{1}{2}m.l.g.cos(\theta_2) \quad (2)$$

Lagrangian is obtained by:

$$\begin{cases} \frac{d}{dt}(\frac{\partial\mathcal{L}}{\partial\theta'}) - (\frac{\partial\mathcal{L}}{\partial\theta}) = 0 \\ \frac{d}{dt}(\frac{\partial\mathcal{L}}{\partial\theta_1'}) - (\frac{\partial\mathcal{L}}{\partial\theta_1}) = \tau_{\theta_1} \\ \frac{d}{dt}(\frac{\partial\mathcal{L}}{\partial\theta_2'}) - (\frac{\partial\mathcal{L}}{\partial\theta_2}) = \tau_{\theta_2} \end{cases}$$

Equation 1 :

$$(M.l^2 + J_1) + m(4l^2\ddot{\theta} + l^2\ddot{\theta_1}cos(\theta + \theta_1) - l^2\dot{\theta_1}^2 sin(\theta + \theta_1)) + 4\,ml^2\ddot{\theta} - m.l^2\ddot{\theta_2}cos(\theta + \theta_2) - ml^2\dot{\theta_2}^2 sin(\theta - \theta_2) - (M + 4\,m)glsin\theta = 0 \quad (3)$$

Equation 2 :

$$(\frac{l^2}{4} + J_2)\ddot{\theta_1} + ml^2\ddot{\theta}cos(\theta + \theta_1) - ml^2\dot{\theta}^2 sin(\theta - \theta_1) + \frac{m}{2}glsin\theta_1 = \tau_{\theta_1} \quad (4)$$

Equation 3 :

$$(\frac{l^2}{4} + J_3)\ddot{\theta_2} - ml^2\ddot{\theta}cos(\theta - \theta_2) - ml^2\dot{\theta}^2 sin(\theta_2 - \theta) + \frac{m}{2}glsin\theta_2 = \tau_{\theta_2} \quad (5)$$

After linearizing the mathematical modeling, the state-space is given by:

$$\dot{x} = A.x + B.u \Longleftrightarrow \begin{bmatrix} \dot{\theta} \\ \ddot{\theta} \\ \dot{\theta_1} \\ \ddot{\theta_1} \\ \dot{\theta_2} \\ \ddot{\theta_2} \end{bmatrix} = \begin{bmatrix} 0\,1\,0\,0\,0\,0 \\ a\,0\,b\,0\,c\,0 \\ 0\,0\,0\,1\,0\,0 \\ d\,0\,e\,0\,f\,0 \\ 0\,0\,0\,0\,0\,1 \\ g\,0\,h\,0\,i\,0 \end{bmatrix} . \begin{bmatrix} \theta \\ \dot{\theta} \\ \theta_1 \\ \dot{\theta_1} \\ \theta_2 \\ \dot{\theta_2} \end{bmatrix} + \begin{bmatrix} 0 \\ k \\ 0 \\ l \\ 0 \\ p \end{bmatrix} .F \quad (6)$$

With,

$$a = \frac{(M + 4.m).g.l}{\alpha}, \quad b = \frac{3.g.l.m}{2}, \quad c = -b, \quad d = -a \quad e = \frac{-9.g.m.l}{2.\alpha} - \frac{3.g}{2.l},$$
$$f = \frac{9.g.m.l}{2.\alpha} \quad g = 3.a \quad h = f, \quad i = \frac{-3.g}{2}(\frac{1}{l} + \frac{3.m.l}{\alpha}), \quad k = \frac{-6}{\alpha}$$
$$l = \frac{3}{m.l^2} + \frac{1}{l} + \frac{2.g}{\alpha}, \quad p = \frac{-3}{m.l^2} - \frac{6.m.l^2}{\alpha} \quad \alpha = M.l^2 + J_1 + 2.m.l^2,$$

## 3 Modeling Simulation by Matlab Software Virtual Reality

To design and model the dynamic of this inverted pendulum with two arms, it is necessary to visualize the motions of the dynamic model. This animation serves to detect all freedom degrees of this system. It is also helpful in the selection of the suitable controller.

In order to create the virtual reality in Matlab Software, two steps must be followed: the first step is to draw the model using V-Realm builder and the second step is to use matlab simulink(M-Script) to animate the virtual reality by varying the state of the object and extracting motion graphs as seen below. The below figure shows the virtual reality of the studied model:

**Fig. 3.** The motion of the dynamic model in virtual reality

### 3.1 Simulations and Discussions

When a torque is applied on the top of the inverted pendulum, both arms rotate as the inverted pendulum swings to the left and right of its center of mass. Figure 4, 5 and Fig. 6 show the motion of the dynamic model when a torque is applied.

Figures 4 and 5 depict the angle rotation of the two arms which remain 120° regardless of the rotation motions. Figure 6, on the other hand, pictures the oscillations of the inverted pendulum when a servomotor is connected to the unicycle arms.

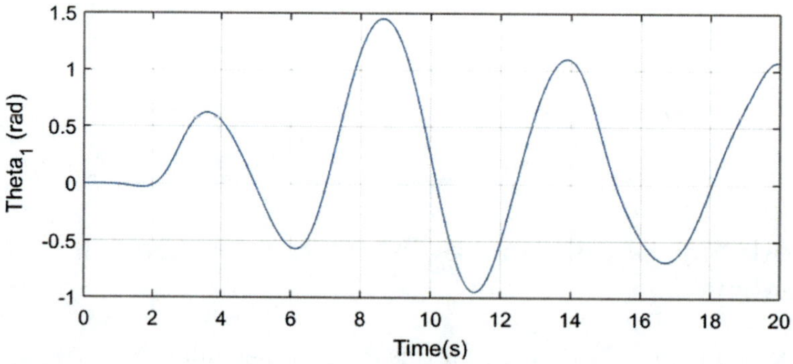


**Fig. 4.** Motion of the first arm

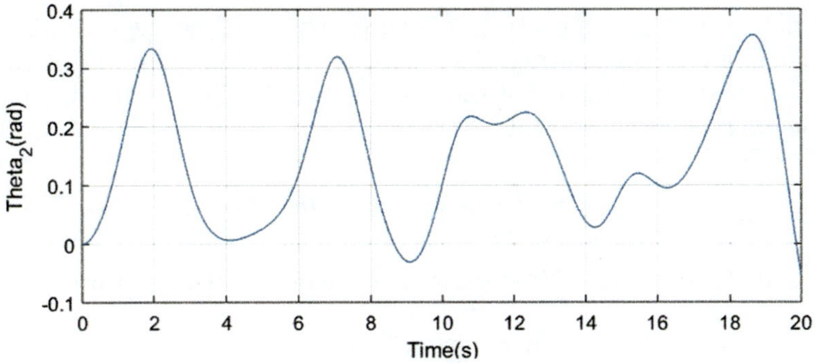


**Fig. 5.** Motion of the second arm

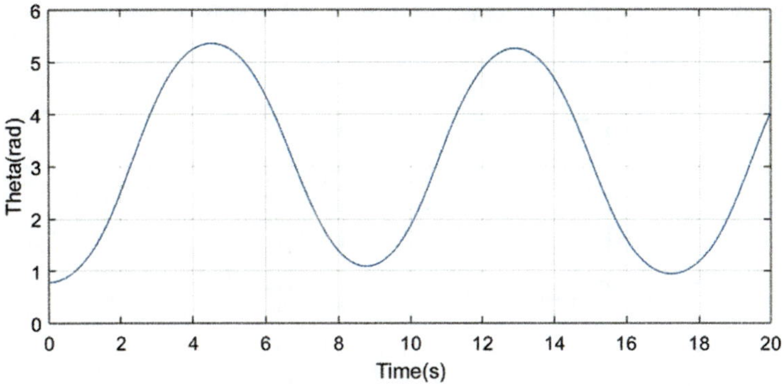


**Fig. 6.** Motion of the inverted pendulum.

## 4 Inverted Pendulum with Two Arms Control Algorithm

In order to evaluate the stability of the proposed model, we will apply the linear quadratic regulator command in the following section, so as to reach the good balancing of the system previously described in (6).

The LQR method consists of finding the gain matrix k, in the manner that the feedback control state equals to:

$$u = -k.x \tag{7}$$

To determine the feedback control, it is necessary to define the weighting matrices Q and R, these two matrices have an enormous impact on LQR controller effectiveness and performance of the system stabilization, and to minimize the quadratic criterion function which is defined by:

$$J = \int_0^\infty (x^T Q x + u^T R u)\ \mathrm{dt} \tag{8}$$

Once Q and R fixed, the LQR state feedback can be calculated by:

$$K = lqr(A, B, Q, R) \tag{9}$$

So,in order to appear the effectiveness of this method algorithm. Figure 7 and 8 depict successively the control diagram of the LQR controller and the experiment results which demonstrate the good stability and the fast response.

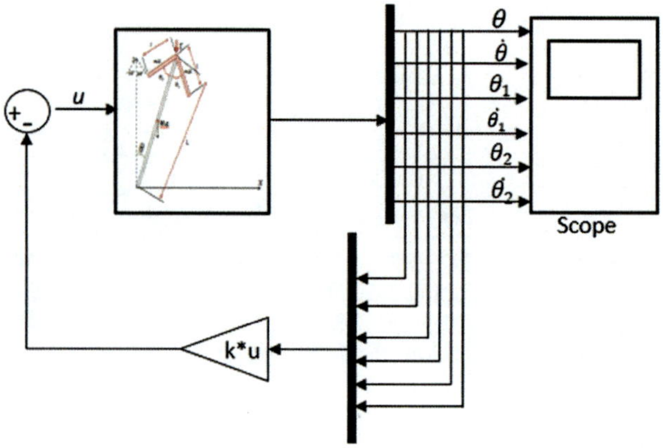


**Fig. 7.** Control effort of the LQR.

As shown, the Fig. 8 represents the behavior of the inverted pendulum with two arms after applying the LQR controller.

It is conspicuous from the simulation results that the LQR controller enforces the model to converge to the upright position within [0; 3] s.

So, it is evident to demonstrate the rapid response and the good stability around the equilibrium point.

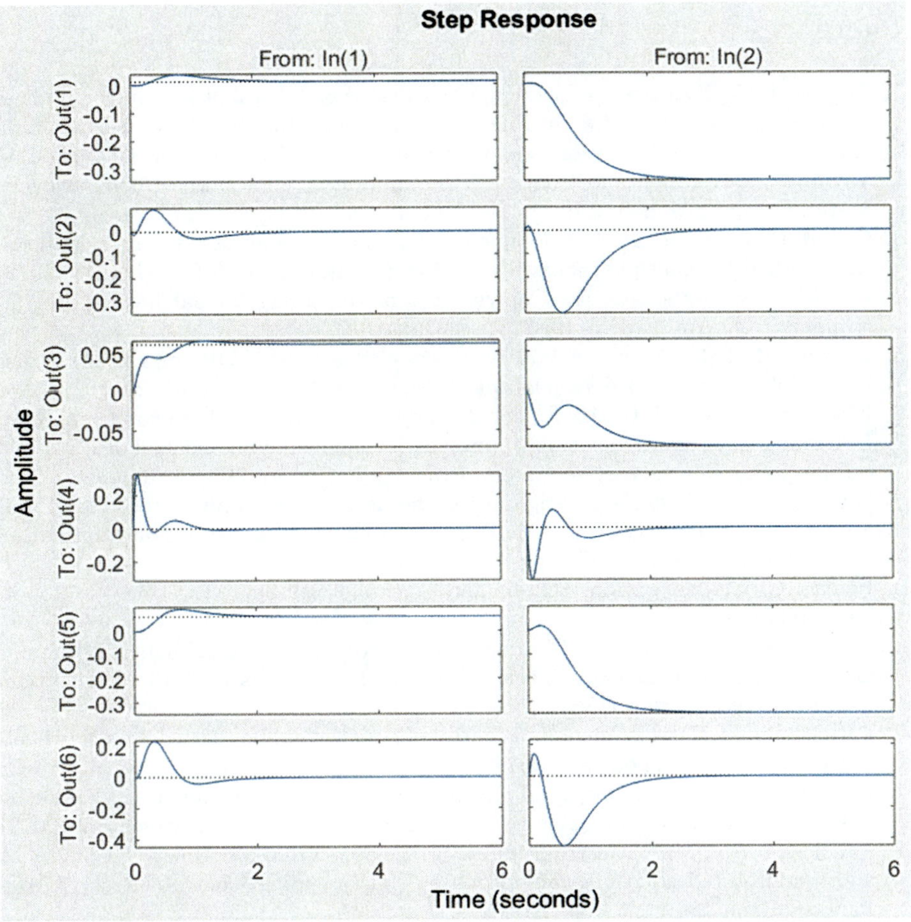


**Fig. 8.** LQR controller for the inverted pendulum with two arms.

## 5 Conclusion

In this present paper, we depicted a nonlinear dynamic model of an inverted pendulum with two arms which rotates on (o,x,z)references. The mathematical model was established by applying the lagrangian approach, the mechanical model was tested by using Matlab simulink software and the virtual reality, on the other hand, the stability of the inverted pendulum with two arms was attended by using the LQR control algorithm, simulation results was shown the good performance.

## References


1. Trojnacki, M.: Dynamics model of a four-wheeled mobile robot for control applications — a three-case study. In: Filev, D. et al. (eds.) Intelligent Systems 2014. 99 Advances in Intelligent Systems and Computing, vol. 323. Springer, Cham (2015)
2. Yavin, Y.: Navigation and Control of the motion of a back wheel drive tricycle. Mathl. Comput. Modell. **25**(11), 23–29 (1997)
3. Chand, P., Carnegie, D.: Development of a reduced human user input task allocation method for multiple robots. Proc. Robot. Auton. Syst. **60**, 1231–1244 (2012)
4. Consolini, L., Maggiore, M.: Control of a bicycle using virtual holonomic constraints. Proc. Automatica **49**, 2831—2839 (2013)
5. Garziad, M., Saka, A.: A comparative assessment between LQR and PID strategies in control of two wheeled vehicle. Int. J. Eng. Res. Afr. **43**, 59–70 (2019)
6. Kheirandish, A., Motlagh, F., Shafiabady, N., Dahari, M., Wahab, A.K.A.: Dynamic fuzzy cognitive network approach for modelling and control of PEM fuel cell for power electric bicycle system. Proc. Appl. Energy **202**, 20–31 (2017)
7. Guo, L., He, K., Song, Y.: Design of the sliding mode controller for a kind of unicycle robot. In: Proceedings of the IEEE International Conference on Information and Automation Ningbo, China, August 2016
8. Pereira, G.: Modeling, Construction and Control of Self-balancing Unicycle Robot, Sao Paulo (2017)
9. Rosyidi, M., Binugroho, E., Charel, E., Dewanto, R.S., Pramadihanto, D.: Speed and balancing control for unicycle robot. In: Proceedings of 2016 International Electronics Symposium (IES)
10. Parnichkun, M.: Development and control of a unicycle robot with double flywheels (2016). www.elsevier.com/locate/mechatronics
11. Daoxiong, G., Qi, P., Guoyu, Z., Wenbo, D.: LQR control for a self-balancing unicycle robot. In: World Congress on Intelligent Control and Automation (2012)
12. Vallejo-Alarcón, M.A., Castro-Linares, R., Velasco-Villa, M.: Unicycle-type robot and quadrotor leader-follower formation backstepping control. Proc. IFAC-PapersOnLine **48**(19), 51–56 (2015)
13. Krafes, S., Chalh, Z., Saka, A.: Backstepping controller design for a 5 DOF spherical inverted pendulum. Int. J. Eng. Res. Afr. **41**, 37–50 (2015)
14. Castillo, O., Aguilar, L., Cardenas, S.: Fuzzy logic tracking control for unicycle mobile robots. Eng. Lett. **13**(2), 93–97 (2006). Advance online publication: 4 August 2006
15. Xua, J.X., Guoa, Z., Leea, T.: Synthesized design of a fuzzy logic controller for an underactuated unicycle. Proc. Fuzzy Sets Syst. **207**, 77–93 (2012)

# Stability and $H_\infty$ Performance for 2-D Discrete Systems with Time-Varying Delays

Mohamed Oubaidi(✉), Zakaria Chalh, and Mohammed Alfidi

Laboratory of Engineering, Systems and Applications, National School of Applied Sciences, Sidi Mohamed Ben Abdellah University, Fez, Morocco
{mohamed.oubaidi,zakaria.chalh}@usmba.ac.ma, alfidi_mohammed@yahoo.fr

**Abstract.** This paper investigates both the stability and $H_\infty$ performance for a class of 2-D discrete systems with time-varying delays described by Fornasini-Marchesini (FM) second model. A new sufficient condition for asymptotic stability with $H_\infty$ performance of these systems is developed based on differences of Lyapunov functionals proposed through introducing free weighting matrices. The findings are later tested via linear matrix inequality (LMI) feasibility. A numerical example is presented to demonstrate the effectiveness and benefits of the result obtained in this study.

**Keywords:** 2-D discrete time-delayed systems · $H_\infty$ performance · FM second model · LMI

## 1 Introduction

The 2-D systems are found in various practical and physical processes where the information propagation occurs in two independent directions such as gas absorption and water stream heating. During the last decade, the research on 2-D systems both in practice and theory has enticed a large number of scholars due to their extensive applications [16], for example, image data processing, circuit analysis, transmission and other areas.

It has been well recognized that time-delay often occurs in practical systems, particularly in 2-D systems due to data transmission and finite speed of information processing among various parts of the system [20]. In addition, the reaction of realistic systems to external signals is seldom instantaneous and always affected by time delays. The time-delay frequently degrades the system performance and even causes the system instability [13]. Therefore, the exploration of time-delay systems stability plays a key role in applied models which has caused quite a stir in recent years.

The $H_\infty$ technique introduced in [9] has been in the spotlight in recent years which attracted researchers, for example [5,7,8,12,17,22,23]. One of it many advantages is that it is insensitive to the exact knowledge of the statistics of the noise signals.


A. Saka et al. (Eds.): CPI 2019, LNME, pp. 65–76, 2021.
https://doi.org/10.1007/978-3-030-62199-5_7

In this paper, 2-D systems are described by Fornasini-Marchesini (FM) second model [6,10,11] and based on the free-weighting matrix approach, proposed in [14,15,21] and by constructing a Lyapunov functional [18], a delay-dependent $H_\infty$ performance analysis is reached by keeping some useful terms from the difference of Lyapunov functions.

This paper is adjusted to five sections: In Sect. 2, the problem under study is formulated. In Sect. 3, new criterion is obtained in terms of LMI, which ensure the asymptotic stability and the $H_\infty$ performance of 2-D discrete systems described by the FM second model. Numerical examples are given to highlight the results in Sect. 4. Finally, some conclusions are provided in Sect. 5.

**Notations:** Throughout the paper, $\mathbb{R}^p$ denotes the p-dimensional real Euclidean space, $\mathbb{R}^{p\times q}$ denotes the set of all $p \times q$ matrices. $\mathbf{0}$ and $I$ represent zero matrix identity matrix respectively. $diag\{...\}$ denotes a block-diagonal matrix in symmetric block matrices or long matrix expressions. $X^T$ stand for the transpose and the matrix $X$. $Q > 0$ ($Q < 0$) means that $Q$ is real symmetric and positive (negative) definite matrix. The notation $||x||$ stands for the Euclidean norm of the vector x.

## 2 Problem Formulation

We consider the 2-D system with time-varying delays described by the following FM second model [10]:

$$\begin{aligned} x(s_1+1,s_2+1) &= A_1x(s_1+1,s_2) + A_2x(s_1,s_2+1) + A_{1d}x(s_1+1,s_2-d_j) \\ &\quad + A_{2d}x(s_1-d_i,s_2+1) + B_1w(s_1+1,s_2) + B_2w(s_1,s_2+1) \\ z(s_1,s_2) &= Cx(s_1,s_2) + Dw(s_1,s_2) \end{aligned} \tag{1}$$

where $x(s_1,s_2) \in \mathbb{R}^n$ is the state vector, $z(s_1,s_2) \in \mathbb{R}^m$ the signal to be estimated, $w(s_1,s_2) \in \mathbb{R}^s$ is the disturbance input. $A_1$, $A_2$, $A_{1d}$, $A_{2d}$, $B_1$, $B_2$, $C$ and $D$ are constant matrices with appropriate dimensions, $d_i$ and $d_j$ are time varying delays along horizontal and vertical directions, respectively, satisfying $\tau_1 \le d_j \le \tau_2$ and $\tau_3 \le d_i \le \tau_4$ where $\tau_1$, $\tau_2$, $\tau_3$ and $\tau_4$ are known positive integers $\tau_v = \tau_2 - \tau_1$ and $\tau_h = \tau_4 - \tau_3$.
The boundary conditions for the system are specified as:

$$\begin{cases} x(s_1,s_2) = \varphi_{s_1,s_2}, & \forall s_1 \geqslant 0, s_2 \in [-\tau_2, 0] \\ x(s_1,s_2) = \psi_{s_1,s_2}, & \forall s_2 \geqslant 0, s_1 \in [-\tau_4, 0] \\ \psi_{0,0} = \varphi_{0,0}. \end{cases} \tag{2}$$

In what follows, the boundary conditions assumed to satisfy:

$$\sum_{s_1=0}^{\infty}\sum_{s_2=-\tau_2}^{0} \varphi^T_{s_1,s_2}\varphi_{s_1,s_2} < \infty, \quad \sum_{s_2=0}^{\infty}\sum_{s_1=-\tau_4}^{0} \psi^T_{s_1,s_2}\psi_{s_1,s_2} < \infty$$

By considering the zero initial conditions, the $H\infty$ norm of the system in (1) is given by:

$$||G(z_1,z_2)||_\infty = sup_{w_1,w_2\in[0,2\pi]}\sigma_{max}[G(e^{jw_1},e^{jw_2})]$$

where $\sigma_{max}$ denotes the maximum singular value of the corresponding matrix, and

$$G(z_1, z_2) = C(z_1z_2I - A_1z_1 - A_2z_2 - A_{1d}z_1z_2^{-d_j} - A_{2d}z_1^{-d_i}z_2)^{-1} \\ (B_2z_2 + B_1z_1) + D$$

is the transfer function from the disturbance input $w(s_1, s_2)$ to the output $z(s_1, s_2)$ for the system in (1).

To get the main results of this paper, the following definition and lemmas are needed.

**lemma1** [4] For given symmetric matrices

$$S = S^T = \begin{bmatrix} S_{11} & S_{12} \\ * & S_{22} \end{bmatrix}$$

where $S_{11}$ and $S_{22}$ are square matrices, the following conditions are equivalent

1. $S < 0$;
2. $S_{11} < 0, \quad S_{22} - S_{12}^T S_{11}^{-1} S_{12} < 0$;
3. $S_{22} < 0, \quad S_{11} - S_{12}^T S_{22}^{-1} S_{12} < 0$;

**Definition 1.** *[24] The 2-D system given in* (1) *with zero boundary conditions in* (2) *is said to have $H_\infty$ disturbance attenuation $\gamma$ if it is asymptotically stable and:*

$$||z(s_1, s_2)||_2 < \gamma||w(s_1, s_2)||_2 \tag{3}$$

For any nonzero $w(s_1, s_2) \in L_2\{[0, \infty), [0, \infty)\}$ where

$$z(s_1, s_2) = \left[ z^T(s_1 + 1, s_2) \; z^T(s_1, s_2 + 1) \right]^T$$
$$w(s_1, s_2) = \left[ w^T(s_1 + 1, s_2) \; w^T(s_1, s_2 + 1) \right]^T$$

## 3 Main Results

In this section, we consider the $H_\infty$ performance analysis problem of system (1). For this case the following theorem holds.

**Theorem 1.** *Given integers $\tau_1 \le d_j \le \tau_2$ and $\tau_3 \le d_i \le \tau_4$ and scalar $\gamma > 0$ the system (1) with time varying $d_i$ and $d_j$ satisfying initial conditions given in (2) is asymptotically stable for all nonzero $w(s_1, s_2) \in L_2\{[0, \infty), [0, \infty)\}$ and (3) is* *satisfied if there exist appropriately dimensioned matrices* $P = P_a + P_b = P^T > 0$, *for* $i = 1,2, j = 1,2$, $Q_{ij} = Q_{ij}^T > 0$, $Z_{1j} = Z_{1j}^T > 0$, $Z_{2j} = Z_{2j}^T > 0$, *for* $j = h, v$ $X_j = \begin{bmatrix} X_{11j} & X_{12j} \\ * & X_{22j} \end{bmatrix} \geq 0$, $Y_j = \begin{bmatrix} Y_{11j} & Y_{12j} \\ * & Y_{22j} \end{bmatrix} \geq 0$, $N_j = \begin{bmatrix} N_{1j} \\ N_{2j} \end{bmatrix}$, $M_j = \begin{bmatrix} M_{1j} \\ M_{2j} \end{bmatrix}$, $S_j = \begin{bmatrix} S_{1j} \\ S_{2j} \end{bmatrix}$, *for* $j = 1,2$ *such that the following matrix inequalities hold:*

$$\beta = \begin{bmatrix} \beta_{11} & \beta_{12} & \beta_{13} & \beta_{14} & \beta_{15} \\ * & \beta_{22} & \beta_{23} & \beta_{24} & \beta_{25} \\ * & * & \beta_{33} & \beta_{34} & \beta_{35} \\ * & * & * & \beta_{44} & \beta_{45} \\ * & * & * & * & \beta_{55} \end{bmatrix} < 0 \tag{4}$$

*and*

$$\Psi_{1j} = \begin{bmatrix} Xi & N_j \\ * & Z_{1j} \end{bmatrix} \geq 0, \quad \Psi_{2j} = \begin{bmatrix} Yi & M_j \\ * & Z_{2j} \end{bmatrix} \geq 0, \quad \Psi_{3j} = \begin{bmatrix} Xi + Yi & S_j \\ * & Z_{1j} + Z_{2j} \end{bmatrix} \geq 0, i = h, v, j = 1,2$$

*where*

$$\beta_{11} = \phi, \quad \beta_{12} = \left[ \sqrt{\tau_4}\phi_1^T Z_{11} \ \sqrt{\tau_h}\phi_1^T Z_{21} \right], \quad \beta_{13} = \left[ \sqrt{\tau_2}\phi_2^T Z_{12} \ \sqrt{\tau_v}\phi_2^T Z_{22} \right]$$

$$\beta_{14} = \left[ \phi_{e1}^T \ \phi_{e2}^T \right], \quad \beta_{15} = \left[ \phi_3^T P_a \ \phi_3^T P_b \right], \quad \beta_{22} = \begin{bmatrix} -Z_{11} & 0 \\ 0 & -Z_{21} \end{bmatrix}$$

$$\beta_{33} = \begin{bmatrix} -Z_{12} & 0 \\ 0 & -Z_{22} \end{bmatrix}, \quad \beta_{44} = \begin{bmatrix} -I & 0 \\ 0 & -I \end{bmatrix}, \quad \beta_{55} = \begin{bmatrix} -P_a & 0 \\ 0 & -P_b \end{bmatrix}$$

$$\beta_{ij} = \begin{bmatrix} 0 & 0 \\ 0 & 0 \end{bmatrix} \ for \quad i = 2,3,4 \quad , j = 3,4,5 \quad and \quad i \neq j$$

*the zero matrix is appropriately dimensioned.*

$$\phi = \begin{bmatrix} \pi_1 & \pi_2 & \pi_3 & \pi_4 & \pi_5 \\ * & \pi_6 & \pi_7 & \pi_8 & \pi_9 \\ * & * & \pi_{10} & \pi_{11} & \pi_{12} \\ * & * & * & \pi_{13} & \pi_{14} \\ * & * & * & * & \pi_{15} \end{bmatrix}$$

$$\pi_1 = \begin{bmatrix} \phi_{v1} & 0 \\ * & \phi_{h1} \end{bmatrix}, \quad \pi_2 = \begin{bmatrix} \phi_{v2} & 0 \\ * & \phi_{h2} \end{bmatrix}, \quad \pi_3 = \begin{bmatrix} S_{12} & -M_{12} \\ 0 & 0 \end{bmatrix}, \quad \pi_4 = \begin{bmatrix} 0 & 0 \\ S_{11} & -M_{11} \end{bmatrix}$$

$$\pi_6 = \begin{bmatrix} \phi_{v3} & 0 \\ * & \phi_{h3} \end{bmatrix}, \quad \pi_7 = \begin{bmatrix} S_{22} & M_{22} \\ 0 & 0 \end{bmatrix}, \quad \pi_8 = \begin{bmatrix} 0 & 0 \\ S_{21} & M_{21} \end{bmatrix}, \quad \pi_{10} = \begin{bmatrix} -Q_{12} & 0 \\ * & -Q_{22} \end{bmatrix}$$

$$\pi_{13} = \begin{bmatrix} -Q_{11} & 0 \\ * & -Q_{21} \end{bmatrix}, \quad \pi_{15} = \begin{bmatrix} -\gamma^2 I & 0 \\ 0 & -\gamma^2 I \end{bmatrix}, \quad \pi_l = \begin{bmatrix} 0 & 0 \\ 0 & 0 \end{bmatrix}, for \quad l = 5, 9, 11, 12, 14$$

*with*

$$\begin{aligned}
\phi_{v1} &= N_{12} + N_{12}^T + (\tau_v + 1)Q_{32} + Q_{12} + Q_{22} + \tau_2 X_{11b} + \tau_v Y_{11b} - P_b \\
\phi_{v2} &= N_{22}^T - N_{12} + M_{12} - S_{12} + \tau_2 X_{12b} + \tau_v Y_{12b} \\
\phi_{v3} &= -N_{22} - N_{22}^T - S_{22} - S_{22}^T + M_{22} + M_{22}^T - Q_{32} + \tau_2 X_{22b} + \tau_v Y_{22b} \\
\phi_{h1} &= N_{11} + N_{11}^T + (\tau_h + 1)Q_{31} + Q_{11} + Q_{21} + \tau_4 X_{11a} + \tau_h Y_{11a} - P_a \\
\phi_{h2} &= N_{21}^T - N_{11} + M_{11} - S_{11} + \tau_4 X_{12a} + \tau_h Y_{12a} \\
\phi_{h3} &= -N_{21} - N_{21}^T - S_{21} - S_{21}^T + M_{21} + M_{21}^T - Q_{31} + \tau_4 X_{22a} + \tau_h Y_{22a}
\end{aligned}$$

*also*

$$\begin{aligned}
\phi_1 &= \begin{bmatrix} A_1 \ (A_2 - I) \ A_{1d} \ A_{2d} \ 0 \ 0 \ 0 \ 0 \ B_1 \ B_2 \end{bmatrix} \\
\phi_2 &= \begin{bmatrix} (A_1 - I) \ A_2 \ A_{1d} \ A_{2d} \ 0 \ 0 \ 0 \ 0 \ B_1 \ B_2 \end{bmatrix} \\
\phi_3 &= \begin{bmatrix} A_1 \ A_2 \ A_{1d} \ A_{2d} \ 0 \ 0 \ 0 \ 0 \ B_1 \ B_2 \end{bmatrix} \\
\phi_{e1} &= \begin{bmatrix} 0 \ C \ 0 \ 0 \ 0 \ 0 \ 0 \ 0 \ 0 \ 0 \ D \end{bmatrix} \\
\phi_{e2} &= \begin{bmatrix} C \ 0 \ 0 \ 0 \ 0 \ 0 \ 0 \ 0 \ D \ 0 \end{bmatrix}
\end{aligned}$$

**Proof.** Let

$$\eta_1(s_1, s_2) = x(s_1 + 1, s_2 + 1) - x(s_1, s_2 + 1) \tag{5}$$

$$\begin{aligned}
\eta_1(s_1, s_2) = {} & A_1 x(s_1 + 1, s_2) + (A_2 - I)x(s_1, s_2 + 1) + A_{1d} x(s_1 + 1, s_2 - d_j) + A_{2d} x(s_1 - d_i, s_2 + 1) \\
& + B_1 w(s_1 + 1, s_2) + B_2 w(s_1, s_2 + 1)
\end{aligned} \tag{6}$$

$$\begin{aligned}
x_{sys}(s_1, s_2) = {} & \big[ x^T(s_1 + 1, s_2) \ x^T(s_1, s_2 + 1) \ x^T(s_1 + 1, s_2 - d_j) \ x^T(s_1 - d_i, s_2 + 1) \\
& .. \ x^T(s_1 + 1, s_2 - \tau_1) \ x^T(s_1 + 1, s_2 - \tau_2) \ x^T(s_1 - \tau_3, s_2 + 1) \ x^T(s_1 - \tau_4, s_2 + 1) \ .. \\
& .. \ w^T(s_1 + 1, s_2) \ w^T(s_1, s_2 + 1) \big]^T
\end{aligned}$$

$\Rightarrow \quad \eta_1(s_1, s_2) = \phi_1 x_{sys}(s_1, s_2)$

The same for the vertical direction:

$\Rightarrow \quad \eta_2(s_1, s_2) = \phi_2 x_{sys}(s_1, s_2)$

Choose a Lyapunov functional candidate to be

$$\begin{aligned}
v(s_1,s_2) &= v_1(s_1,s_2)+v_2(s_1,s_2)+v_3(s_1,s_2)+v_4(s_1,s_2)\\
v(s_1,s_2) &= v_{11}(s_1,s_2)+v_{12}(s_1,s_2)+v_{21}(s_1,s_2)+v_{22}(s_1,s_2)\\
&\quad + v_{31}(s_1,s_2)+v_{32}(s_1,s_2)+v_{41}(s_1,s_2)+v_{42}(s_1,s_2)\\
v_{11}(s_1,s_2) &= x^T(s_1,s_2)P_a x(s_1,s_2), \quad v_{12}(s_1,s_2)=x^T(s_1,s_2)P_b x(s_1,s_2)\\
v_{21}(s_1,s_2) &= \sum_{\theta=-\tau_4+1}^{0}\sum_{l=s_1-1+\theta}^{s_1-1}\eta_1^T(l,s_2)Z_{11}\eta_1(l,s_2)+\sum_{\theta=-\tau_4+1}^{-\tau_3}\sum_{l=s_1-1+\theta}^{s_1-1}\eta_1^T(l,s_2)Z_{21}\eta_1(l,s_2)\\
v_{22}(s_1,s_2) &= \sum_{\theta=-\tau_2+1}^{0}\sum_{l=s_2-1+\theta}^{s_2-1}\eta_2^T(s_1,l)Z_{12}\eta_2(s_1,l)+\sum_{\theta=-\tau_2+1}^{-\tau_1}\sum_{l=s_2-1+\theta}^{s_2-1}\eta_2^T(s_1,l)Z_{22}\eta_2(s_1,l)\\
v_{31}(s_1,s_2) &= \sum_{l=s_1-\tau_3}^{s_1-1}x^T(l,s_2+1)Q_{11}x(l,s_2+1)+\sum_{l=s_1-\tau_4}^{s_1-1}x^T(l,s_2+1)Q_{21}x(l,s_2+1)\\
v_{32}(s_1,s_2) &= \sum_{l=s_2-\tau_2}^{s_2-1}x^T(s_1+1,l)Q_{12}x(s_1+1,l)+\sum_{l=s_2-\tau_1}^{s_2-1}x^T(s_1+1,l)Q_{22}x(s_1+1,l)\\
v_{41}(s_1,s_2) &= \sum_{\theta=1-\tau_4}^{1-\tau_3}\sum_{l=s_1-1+\theta}^{s_1-1}x^T(l,s_2+1)Q_{31}x(l,s_2+1)\\
v_{42}(s_1,s_2) &= \sum_{\theta=1-\tau_2}^{1-\tau_1}\sum_{l=s_2-1+\theta}^{s_2-1}x^T(s_1+1,l)Q_{32}x(s_1+1,l)
\end{aligned}$$

where $P = P_a + P_b = P^T > 0$, $Q_{ij} = Q_{ij}^T > 0$, i=1,2, j=1,2, $Z_{1j} = Z_{1j}^T > 0$ j=1,2, $Z_{2j} = Z_{2j}^T > 0$ j=1,2 are to be calculated. Defining $\Delta V(s_1+1,s_2) = V(s_1+1,s_2) - V(s_1,s_2)$ and $\Delta V(s_1,s_2+1) = V(s_1,s_2+1) - V(s_1,s_2)$ yields

$$\begin{aligned}
\Delta_{v_{11}(s_1,s_2)} &= x_{sys}^T(s_1+1,s_2)\phi_3^T P_a\phi_3 x_{sys}(s_1+1,s_2) - x_{sys}^T(s_1,s_2)P_a x_{sys}(s_1,s_2)\\
\Delta_{v_{12}(s_1,s_2)} &= x_{sys}^T(s_1,s_2+1)\phi_3^T P_b\phi_3 x_{sys}(s_1,s_2+1) - x_{sys}^T(s_1,s_2)P_b x_{sys}(s_1,s_2)
\end{aligned}$$

$$\begin{aligned}
\Delta_{v_{21}(s_1,s_2)} &= x_{sys}^T(s_1,s_2)\phi_1^T(\tau_4 Z_{11}+\tau_h Z_{21})\phi_1 x_{sys}(s_1,s_2) - \sum_{l=s_1-d_i}^{s_1-1}\eta_1^T(l,s_2)Z_{11}\eta_1(l,s_2)\\
&\quad - \sum_{l=s_1-d_i}^{s_1-1-\tau_3}\eta_1^T(l,s_2)Z_{21}\eta_1(l,s_2) - \sum_{l=s_1-\tau_4}^{s_1-1-d_i}\eta_1^T(l,s_2)(Z_{21}+Z_{11})\eta_1(l,s_2)\\
\Delta_{v_{22}(s_1,s_2)} &= x_{sys}^T(s_1,s_2)\phi_2^T(\tau_2 Z_{12}+\tau_v Z_{22})\phi_2 x_{sys}(s_1,s_2) - \sum_{l=s_2-d_j}^{s_2-1}\eta_2^T(s_1,l)Z_{12}\eta_2(s_1,l)\\
&\quad - \sum_{l=s_2-d_j}^{s_2-1-\tau_1}\eta_2^T(s_1,l)Z_{22}\eta_2(s_1,l) - \sum_{l=s_2-\tau_2}^{s_2-1-d_j}\eta_2^T(s_1,l)(Z_{22}+Z_{12})\eta_2(s_1,l)
\end{aligned}$$

$$\begin{aligned}
\Delta_{v_{31}(s_1,s_2)} &= x^T(s_1,s_2+1)(Q_{21}+Q_{11})x(s_1,s_2+1) - \xi^T(s_1-\tau_3,s_2+1)Q_{11}x(s_1-\tau_3,s_2+1)\\
&\quad - x^T(s_1-\tau_4,s_2+1)Q_{21}x(s_1-\tau_4,s_2+1)\\
\Delta_{v_{32}(s_1,s_2)} &= x^T(s_1+1,s_2)(Q_{22}+Q_{12})x(s_1+1,s_2) - x^T(s_1+1,s_2-\tau_1)Q_{12}x(s_1+1,s_2-\tau_1)\\
&\quad - x^T(s_1+1,s_2-\tau_2)Q_{22}x(s_1+1,s_2-\tau_2)
\end{aligned}$$

$$\Delta_{v41(s_1,s_2)} = (\tau_h+1)x^T(s_1,s_2+1)Q_{31}x(s_1,s_2+1) - \sum_{l=s_1-\tau_4}^{s_1-\tau_3} x^T(l,s_2+1)Q_{31}x(l,s_2+1)$$
$$< (\tau_h+1)x^T(s_1,s_2+1)Q_{31}x(s_1,s_2+1) - x^T(s_1-d_i,s_2+1)Q_{31}x(s_1-d_i,s_2+1)$$
$$\Delta_{v42(s_1,s_2)} = (\tau_v+1)x^T(s_1+1,s_2)Q_{32}x(s_1+1,s_2) - \sum_{l=s_2-\tau_2}^{s_2-\tau_1} x^T(s_1+1,l)Q_{32}x(s_1+1,l)$$
$$< (\tau_v+1)x^T(s_1+1,s_2)Q_{32}x(s_1+1,s_2) - x^T(s_1+1,s_2-d_j)Q_{32}x(s_1+1,s_2-d_j)$$

From (5), we have
$\eta_1(s_1,s_2) = x(s_1+1,s_2+1) - x(s_1,s_2+1)$
$\Rightarrow$

$$\sum_{l=s_1-d_i}^{s_1-1} \eta_1(l,s_2) = \sum_{l=s_1-d_i}^{s_1-1} x(l+1,s_2+1) - \sum_{l=s_1-d_i}^{s_1-1} x(l,s_2+1)$$

$\Rightarrow$

$$0 = x(s_1,s_2+1) - x(s_1-d_i,s_2+1) - \sum_{l=s_1-d_i}^{s_1-1} \eta_1(l,s_2)$$
$$0 = x(s_1-d_i,s_2+1) - x(s_1-\tau_4,s_2+1) - \sum_{l=s_1-\tau_4}^{s_1-d_i-1} \eta_1(l,s_2)$$
$$0 = x(s_1-\tau_3,s_2+1) - x(s_1-d_i,s_2+1) - \sum_{l=s_1-d_i}^{s_1-\tau_3-1} \eta_1(l,s_2)$$

the same for vertical direction.
then the following equations are true for any matrices $N_1$, $N_2$, $M_1$, $M_2$, $S_1$ and $S_2$ with appropriate dimensions for $\Delta V(s_1,s_2+1)$ :

$$0 = 2\times[x^T(s_1,s_2+1)N_{11} + x^T(s_1-d_i,s_2+1)N_{21}$$
$$\times [x(s_1,s_2+1) - x(s_1-d_i,s_2+1) - \sum_{l=s_1-d_i}^{s_1-1} \eta_1(l,s_2)] \quad (7)$$

$$0 = 2\times[x^T(s_1,s_2+1)M_{11} + x^T(s_1-d_i,s_2+1)M_{21}$$
$$\times [x(s_1-d_i,s_2+1) - x(s_1-\tau_4,s_2+1) - \sum_{l=s_1-\tau_4}^{s_1-d_i-1} \eta_1(l,s_2)] \quad (8)$$

$$
\begin{aligned}
0 = 2 \times [x^T(s_1, s_2+1)S_{11} + x^T(s_1-d_i, s_2+1)S_{21} \\
\times [x(s_1-\tau_3, s_2+1) - x(s_1-d_i, s_2+1) - \sum_{l=s_1-d_i}^{s_1-\tau_3-1} \eta_1(l, s_2)]
\end{aligned} \tag{9}
$$

The same thing for $\Delta V(s_1+1, s_2)$.

In the other hand, for any appropriately dimensioned matrices $X_i = X_i^T \geq 0$, $Y_i = Y_i^T \geq 0$, for i = h, v, the following equations are true:

$$
\begin{aligned}
0 = \tau_4 \xi_1^T(s_1, s_2)X_h\xi_1(s_1, s_2) - \sum_{l=s_1-d_i}^{s_1-1} \xi_1^T(s_1, s_2)X_h\xi_1(s_1, s_2) \\
- \sum_{l=s_1-\tau_4}^{s_1-d_i-1} \xi_1^T(s_1, s_2)X_h\xi_1(s_1, s_2)
\end{aligned} \tag{10}
$$

$$
\begin{aligned}
0 = \tau_h \xi_1^T(s_1, s_2)Y_h\xi_1(s_1, s_2) - \sum_{l=s_1-d_i}^{s_1-\tau_3-1} \xi_1^T(s_1, s_2)Y_h\xi_1(s_1, s_2) \\
- \sum_{l=s_1-\tau_4}^{s_1-d_i-1} \xi_1^T(s_1, s_2)Y_h\xi_1(s_1, s_2)
\end{aligned} \tag{11}
$$

$$
\begin{aligned}
0 = \tau_2 \xi_2^T(s_1, s_2)X_v\xi_2(s_1, s_2) - \sum_{l=s_2-d_j}^{s_2-1} \xi_2^T(s_1, s_2)X_v\xi_2(s_1, s_2) \\
- \sum_{l=s_2-\tau_2}^{s_2-d_j-1} \xi_2^T(s_1, s_2)X_v\xi_2(s_1, s_2)
\end{aligned} \tag{12}
$$

$$
\begin{aligned}
0 = \tau_v \xi_2^T(s_1, s_2)Y_v\xi_2(s_1, s_2) - \sum_{l=s_2-d_j}^{s_2-\tau_1-1} \xi_2^T(s_1, s_2)Y_v\xi_2(s_1, s_2) \\
- \sum_{l=s_2-\tau_2}^{s_2-d_j-1} \xi_2^T(s_1, s_2)Y_v\xi_2(s_1, s_2)
\end{aligned} \tag{13}
$$

$$
\begin{aligned}
\xi_1(s_1, s_2) = \left[x^T(s_1, s_2+1)\; x^T(s_1-d_i, s_2+1)\right]^T \\
\xi_2(s_1, s_2) = \left[x^T(s_1+1, s_2)\; x^T(s_1+1, s_2-d_j)\right]^T
\end{aligned}
$$

Then, if the terms of the right side of the Eqs. (7)–(13) are added to $\Delta V(s_1, s_2) = \Delta V(s_1+1, s_2) + V(s_1, s_2+1)$, we have:

$$\Delta V(s_1,s_2) + z^T(s_1,s_2)z(s_1,s_2) - \gamma^2 w^T(s_1,s_2)w(s_1,s_2) \le x_{sys}^T(s_1,s_2)(\phi + \phi_1^T(\tau_4 Z_{11} + \tau_h Z_{21})\phi_1$$
$$+ \phi_2^T(\tau_2 Z_{12} + \tau_v Z_{22})\phi_2 + \phi_{e1}^T\phi_{e1} + \phi_{e2}^T\phi_{e2} + \phi_3^T(P_a + P_b)\phi_3)x_{sys}(s_1,s_2)$$
$$- \sum_{l=s_1-d_i}^{s_1-1} \xi_3^T(l,s_2)\Psi_{11}\xi_3(l,s_2) - \sum_{l=s_1-d_i}^{s_1-\tau_3-1} \xi_3^T(l,s_2)\Psi_{21}\xi_3(l,s_2) - \sum_{l=s_1-\tau_4}^{s_1-d_i-1} \xi_3^T(l,s_2)\Psi_{31}\xi_3(l,s_2)$$
$$- \sum_{l=s_2-d_j}^{s_2-1} \xi_3^T(s_1,l)\Psi_{12}\xi_3(s_1,l) - \sum_{l=s_2-d_j}^{s_2-\tau_1-1} \xi_3^T(s_1,l)\Psi_{21}\xi_3(s_1,l) - \sum_{l=s_2-\tau_2}^{s_2-d_j-1} \xi_3^T(s_1,l)\Psi_{32}\xi_3(s_1,l)$$

$$\xi_3(l,s_2) = \left[\xi_1^T(s_1,s_2)\ \eta_1^T(l,s_2)\right]^T$$
$$\xi_3(s_1,l) = \left[\xi_2^T(s_1,s_2)\ \eta_2^T(s_1,l)\right]^T$$

thus if $\Psi_{1j}, \Psi_{2j} \ge 0, \Psi_{3j} \ge 0\ for\ j = 1,2$ and

$$(\phi + \phi_1^T(\tau_4 Z_{11} + \tau_h Z_{21})\phi_1 + \phi_2^T(\tau_2 Z_{12} + \tau_v Z_{22})\phi_2 + \phi_{e1}^T\phi_{e1} + \phi_{e2}^T\phi_{e2} + \phi_3^T(P_a + P_b)\phi_3) < 0$$

which is equivalent to (4) by schur compliments, then

$$\Delta V(s_1,s_2) + z^T(s_1,s_2)z(s_1,s_2) - \gamma^2 w^T(s_1,s_2)w(s_1,s_2) < 0$$

This ensures that (3) holds under zero-initial conditions for all nonzero $w(s_1,s_2) \in L_2\{[0,\infty),[0,\infty)\}$ and a specified $\gamma > 0$ following the similar line in [25]. On the other hand, (4) entails that the following matrix inequality (14) holds, which guarantees $\Delta V(s_1,s_2) < 0$, such that the system (1) with $w(s_1,s_2) = 0$ is asymptotically stable.

$$\beta = \begin{bmatrix} \beta_{11} & \beta_{12} & \beta_{13} & \beta_{15} \\ * & \beta_{22} & \beta_{23} & \beta_{25} \\ * & * & \beta_{33} & \beta_{35} \\ * & * & * & \beta_{55} \end{bmatrix} < 0 \tag{14}$$

where

$$\phi = \begin{bmatrix} \pi_1 & \pi_2 & \pi_3 & \pi_4 \\ * & \pi_6 & \pi_7 & \pi_8 \\ * & * & \pi_{10} & \pi_{11} \\ * & * & * & \pi_{13} \end{bmatrix}$$

$$\phi_1 = \left[A_1\ (A_2 - I)\ A_{1d}\ A_{2d}\ 0\ 0\ 0\ 0\right]$$
$$\phi_2 = \left[(A_1 - I)\ A_2\ A_{1d}\ A_{2d}\ 0\ 0\ 0\ 0\right]$$
$$\phi_3 = \left[A_1\ A_2\ A_{1d}\ A_{2d}\ 0\ 0\ 0\ 0\right]$$

This completes the proof. □

**Remarque:** The terms $\sum_{l=s_1-\bar{h_1}}^{s_1-d_i-1} \eta_1^T(l, s_2)Z_{s_1}\eta_1(l, s_2)$ and $\sum_{l=s_2-\bar{h_2}}^{s_2-d_j-1} \eta_2^T(s_1, l)Z_{s_2}\eta_2(s_1, l)$ are kept in Theorem 1 to overcome the conservativeness. On the other hand, $\tau_2$ and $\tau_4$ are split into two parts like $d_j$ and $\tau_2 - d_j$, $d_i$ and $\tau_4 - d_i$, for vertical and horizontal direction, respectively, in order to prove Theorem 1 thus showing the benefits of the suggested method.

## 4 Numerical Example

In this section, we will give a numerical example to illustrate the applicability of the proposed result.

*Example 1.* [19] Denote the design of 2-D delay-dependent $H_\infty$ performance and filter for a stationary random field in image processing where the disturbances are a random process (noise), using LMI approach proposed in (4), the 2-D system can be converted to the 2-D FM model system (1) with the ensuing parameters:

$$A_1 = \begin{bmatrix} 0.3 & 0 \\ 0 & 0 \end{bmatrix}, \quad A_2 = \begin{bmatrix} 0 & 0 \\ 1 & 0.2 \end{bmatrix}, \quad A_{1d} = \begin{bmatrix} 0 & -0.03 \\ -0.08 & 0 \end{bmatrix}, \quad A_{2d} = \begin{bmatrix} -0.03 & 0 \\ 0 & 0 \end{bmatrix}$$

$$B_1 = \begin{bmatrix} 1 & 0 \\ 0 & 0 \end{bmatrix}, \quad B_2 = \begin{bmatrix} 0 & 0 \\ 0 & 0 \end{bmatrix}, \quad C = \begin{bmatrix} 3 & 1 \end{bmatrix}, \quad D = \begin{bmatrix} 0 & 1 \end{bmatrix}$$

Given $\tau_3 = 3$, $\tau_4 = 3$, $\tau_1 = 2$, $\tau_2 = 2$ by solving the LMIs, the minimum $H_\infty$ norm bound for this example is $\gamma_{opt} = 6.9671$.

Fig. 1 shows the maximum singular values plot of the transfer function matrix of the system (1). In the figure, the grids denote the obtained $H_\infty$ disturbance attenuations and its maximum value is 6.3725, which is below 6.9671.

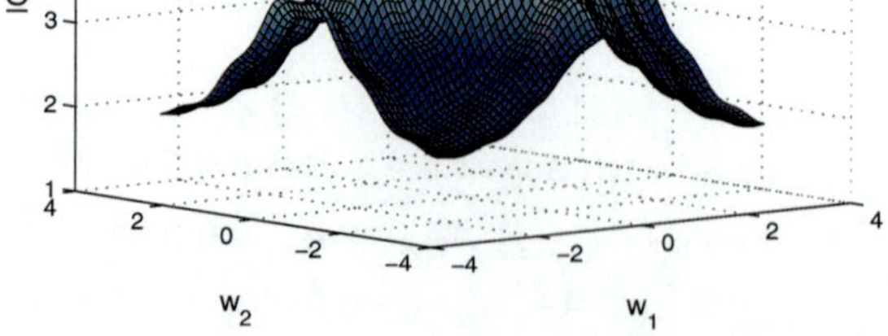


**Fig. 1.** Transfer function (Example 1).

## 5 Conclusion

This paper has explored the problems of stability and delay-dependent $H_\infty$ performance analysis for 2-D discrete systems with time varying delay described by FM second model without ignoring any terms in the derivative of lyapunov functional by considering the relationship between the delay and it upper bound. The new criteria may be extended to systems with uncertainties.

## References

1. Badie, K., Alfidi, M., Chalh, Z.: Improved delay-dependent stability criteria for 2-D discrete state delayed systems. In: 2018 International Conference on Intelligent Systems and Computer Vision (ISCV), pp. 1–6. IEEE (2018)
2. Badie, K., Alfidi, M., Chalh, Z.: New relaxed stability conditions for uncertain two-dimensional discrete systems. J. Control Autom. Electr. Syst. **29**(6), 661–669 (2018)
3. Badie, K., Alfidi, M., Tadeo, F., Chalh, Z.: Correction to: delay-dependent stability and $H_\infty$ performance of 2-D continuous systems with delays. Circ. Syst. Sig. Process. **37**(12), 5688–5689 (2018)
4. Boyd, S., El Ghaoui, L., Feron, E., Balakrishnan, V.: Linear Matrix Inequalities in System and Control Theory, vol. 15. SIAM, Philadelphia (1994)
5. Chen, S.F., Fong, I.K.: Delay-dependent robust $H_\infty$ filtering for uncertain 2-D state-delayed systems. Sig. Process. **87**, 2659–2672 (2007)
6. Dumistrescu, B.: LMI stability tests for the Fornasini-Marchesini model. IEEE Trans. Sig. Process. **56**, 4091–4095 (2008)
7. deSouza, C.E., Xie, L., Wang, Y.: $H_\infty$ filtering for a class of uncertain non linear systems. Syst. Control Lett. **20**, 419–426 (1993)
8. de Oliveira, M.C., Geromel, J.C.: H2 and $H_\infty$ filtering design subject to implementation uncertainty. SIAM J. Control Optim. **44**, 515–530 (2005)
9. Elsayed, A., Grimble, M.J.: A new approach to the $H_\infty$ design of optimal digital linear filters. IMA J. Math. Control Inform. **6**(2), 233–251 (1989)
10. Fornasini, E., Marchesini, G.: State-space realization theory of two-dimensional filters. IEEE Trans. Autom. Control **21**(4), 484–492 (1976)
11. Fornasini, E., Marchesini, G.: Doubly-indexed dynamical systems: State-space models and structural properties. Math. Syst. Theory **12**, 59–72 (1978)
12. Geromel, J.C., de Oliveira, M.C., Bernussou, J.: Robust filtering of discrete-time linear systems with parameter dependent Lyapunov functions. SIAM J. Control Optim. **41**, 700–711 (2002)
13. Gu, K., Kharitonov, V.L., Chen, J.: Stability of Time-delay Systems. Birkhauser, Boston (2003)
14. He, Y., Wu, M., She, J.H., Liu, G.P.: Delay-dependent robust stability criteria for uncertain neutral systems with mixed delays. Syst. Control Lett. **51**, 57–65 (2004)
15. He, Y., Wu, M., She, J.H., Liu, G.P.: Parameter-dependent Lyapunov functional for stability of time-delay systems with polytopic-type uncertainties. IEEE Trans. Autom. Control **49**, 828–832 (2004)
16. Kaczorek, T.: Two-dimensional Linear Systems. Springer, Berlin (1985)
17. Nagpal, K.M., Khargonekar, P.P.: Filtering and smoothing in an H/sup infinity/setting. IEEE Trans. Autom. Control **36**(2), 152–166 (1991)

18. Ooba, T.: On stability analysis of 2-D systems based on 2-D Lyapunov matrix inequalities. IEEE Trans. Circuits Syst. **I**(47), 1263–1265 (2000)
19. Peng, D., Guan, X.: $H_\infty$ filtering of 2-D discrete state-delayed systems. Multidimension. Syst. Signal Process. **20**(3), 265–284 (2009)
20. Richard, J.: Time-delay systems: An over view of some recent advances and open problem. Automatica **39**, 1667–1694 (2003)
21. Wu, M., He, Y., She, J.H., Liu, G.P.: Delay-dependent criteria for robust stability of time-varying delay systems. Automatica **40**, 1435–1439 (2004)
22. Wang, F., Zhang, Q., Yao, B.: LMI-based reliable $H_\infty$ filtering with sensor failure. Int. J. Innovative Comput. Inform. Control **2**, 737–748 (2006)
23. Xie, L., deSouza, C.E., Fu, M.: $H_\infty$ estimation for discrete-time linear uncertain systems. Int. J. Robust Non linear Control **1**, 111–123 (1991)
24. Xu, H., Zou, Y.: Robust $H_\infty$ filtering for uncertain two-dimensional discrete systems with state-varying delays. Int. J. Control Autom. Syst. **8**(4), 720–726 (2010)
25. Zhang, X.M., Han, Q.L.: Delay-dependent robust $H_\infty$ filtering for uncertain discrete-time systems with time-varying delay based on a finite sum inequality. IEEE Trans. Circuits Syst. II: Express Briefs **53**(12), 1466–1470 (2006)

# Observer Design for 2-D Continuous Systems in the Roesser Model

Mohammed Alfidi, Zakaria Chalh(✉), and Mohamed Ouahi

Engineering, Systems and Applications Laboratory,
University Sidi Mohamed Ben Abdellah, Fez, Morocco
alfidi_mohammed@yahoo.fr

**Abstract.** The problem of observer design for two dimensional (2-D) continuous systems represented by Roesser model is investigated in this paper. A new sufficient condition is given to design the suitable 2-D state observer for this system. Based on the use of a Lyapunov function and a simple matrix inequality method, the design of the optimal 2-D state observer is derived in the form of feasibility of a linear matrix inequality (LMI). A simulation example is considered to illustrate the efficiency of the proposed technique.

**Keywords:** 2-D continuous system · Roesser model · Stability · Observer · Linear Matrix Inequality (LMI)

## 1 Introduction

Two dimensional (2-D) systems have attracted a lot of interest in the last years, because they are both theoretically important and are frequently encountered in several practical fields such as data processing and transformation, thermal processes, water steam heating, gas absorption and other fields [1–3]. A great number of studies on analysis and design for 2-D systems in both the continuous and discrete contexts have been published and available in the literature, for instance, by the use of 2-D Lyapunov equations, useful results on the stability analysis have already been obtained in [4–10].

The estimation problem is one of the fundamental problems in various engineering applications and has been extensively investigated in the last three decades especially for one-dimensional systems we can cite for example [11–13]. For 2-D systems, there are little research concerning the observer design problem, for example: The 2-D state observers design problem for 2-D discrete systems represented by the Fornasini-Marchesini (FM) local state-space model was studied in [14]. In [15], the problem of $H_{-}/H_{\infty}$ fault detection observer design for 2-D systems in the Roesser model was addressed. The existence and design of regular unknown input observers for 2-D acceptable singular systems represented by the FM local state-space second model were considered in [16].

Thus, in the present paper, we address the problem of design of 2-D state observers for 2-D continuous systems represented by Roesser model: A new sufficient conditions is developed to the design of state observers using a Lyapunov


A. Saka et al. (Eds.): CPI 2019, LNME, pp. 77–84, 2021.
https://doi.org/10.1007/978-3-030-62199-5_8

function and the linear matrix inequality (LMI) technique, guarantees that the states of the 2-D observer converge to the real values, which means that the error of observation vanishes asymptotically to zero. A simulation example is applied to illustrate the merits of the developed method.

**Notation:** $\mathbb{R}^n$ signifies the n-dimensional real Euclidean space. $\mathbf{0}$ and $I$ represent zero matrix and identity matrix, respectively. The superscript 'T' and '−1' stands for the matrix transpose. $X > 0$ means that $X$ is real symmetric positive definite. In symmetric block matrices, we utilize an asterisk (∗) to symbolize a term that is induced by symmetry.

## 2 Problem Formulation

Consider the following 2-D continuous system represented by the Roesser model

$$\begin{bmatrix} \frac{\partial x^h(t_1,t_2)}{\partial t_1} \\ \frac{\partial x^v(t_1,t_2)}{\partial t_2} \end{bmatrix} = \begin{bmatrix} A_{11} & A_{12} \\ A_{21} & A_{22} \end{bmatrix} \begin{bmatrix} x^h(t_1,t_2) \\ x^v(t_1,t_2) \end{bmatrix} + \begin{bmatrix} B_1 \\ B_2 \end{bmatrix} u(t_1,t_2)$$

$$y(t_1,t_2) = \begin{bmatrix} C_1 & C_2 \end{bmatrix} \begin{bmatrix} x^h(t_1,t_2) \\ x^v(t_1,t_2) \end{bmatrix}, \tag{1}$$

where $x^h \in \mathbb{R}^{n_h}$ and $x^v \in \mathbb{R}^{n_v}$ denote the horizontal and vertical states respectively, $u \in \mathbb{R}^r$ is the control input, and $y \in \mathbb{R}^p$ is the measured output. The matrices $A_{11}$, $A_{12}$, $A_{21}$, $A_{22}$, $B_1$, $B_2$, $C_1$ and $C_2$, are constant with appropriate dimensions.

The initial conditions of system (1) are specified as follows:

$$x^h(0,t_2) = f(t_2)$$
$$x^v(t_1,0) = g(t_1) \tag{2}$$

satisfying

$$\lim_{t_2\to\infty} ||x^h(0,t_2)|| = 0, \qquad \lim_{t_1\to\infty} ||x^v(t_1,0)|| = 0, \tag{3}$$

When there is no control input in (1), we can get the following system

$$\begin{bmatrix} \frac{\partial x^h(t_1,t_2)}{\partial t_1} \\ \frac{\partial x^v(t_1,t_2)}{\partial t_2} \end{bmatrix} = \begin{bmatrix} A_{11} & A_{12} \\ A_{21} & A_{22} \end{bmatrix} \begin{bmatrix} x^h(t_1,t_2) \\ x^v(t_1,t_2) \end{bmatrix} \tag{4}$$

**Definition 1.** *The 2-D system* (4) *with initial conditions* (2)–(3) *is said to be asymptotically stable if*

$$\lim_{(t_1+t_2)\to\infty} ||x^h(t_1,t_2)|| = 0, \qquad \lim_{(t_1+t_2)\to\infty} ||x^v(t_1,t_2)|| = 0, \tag{5}$$

**Lemma 1.** *[17,18] The 2-D continuous system* (4) *is asymptotically stable if there exist symmetric positive definite matrices* $P_h$ *and* $P_v$ *such that the following inequality condition holds*

$$\begin{bmatrix} A_{11} & A_{12} \\ A_{21} & A_{22} \end{bmatrix}^T \begin{bmatrix} P_h & 0 \\ 0 & P_v \end{bmatrix} + \begin{bmatrix} P_h & 0 \\ 0 & P_v \end{bmatrix} \begin{bmatrix} A_{11} & A_{12} \\ A_{21} & A_{22} \end{bmatrix} < 0 \tag{6}$$

*Remark 1.* In many applications engineering systems, the state vector is not always available to measurement, or the output signal is not measurable for lack of technology, which makes the observer design problem very interesting

In this paper, we are interested in designing the following 2-D state observer for 2-D continuous system (1):

$$\begin{aligned} \begin{bmatrix} \frac{\partial z^h(t_1,t_2)}{\partial t_1} \\ \frac{\partial z^v(t_1,t_2)}{\partial t_2} \end{bmatrix} &= \begin{bmatrix} N_{11} & N_{12} \\ N_{21} & N_{22} \end{bmatrix} \begin{bmatrix} z^h(t_1,t_2) \\ z^v(t_1,t_2) \end{bmatrix} + \begin{bmatrix} M_1 \\ M_2 \end{bmatrix} u(t_1,t_2) + \begin{bmatrix} L_1 \\ L_2 \end{bmatrix} y(t_1,t_2) \\ \begin{bmatrix} z^h(t_1,t_2) \\ z^v(t_1,t_2) \end{bmatrix} &= \begin{bmatrix} \hat{x}^h(t_1,t_2) \\ \hat{x}^v(t_1,t_2) \end{bmatrix} \end{aligned} \tag{7}$$

where $\hat{x}^h \in \mathbb{R}^{n_h}$ and $\hat{x}^v \in \mathbb{R}^{n_v}$ denote the estimates of the states vectors $x^h$ and $x^v$, respectively. The matrices $N_{11}$, $N_{12}$, $N_{21}$, $N_{22}$, $M_1$, $M_2$, $L_1$ and $L_2$, are observer gains to be determined.

The initial conditions for 2-D state observer system (7) are given by

$$\begin{aligned} \hat{x}^h(0,t_2) &= \hat{f}(t_2) \\ \hat{x}^v(t_1,0) &= \hat{g}(t_1) \end{aligned} \tag{8}$$

Defining the estimation errors as

$$\begin{bmatrix} e^h(t_1,t_2) \\ e^v(t_1,t_2) \end{bmatrix} = \begin{bmatrix} x^h(t_1,t_2) \\ x^v(t_1,t_2) \end{bmatrix} - \begin{bmatrix} \hat{x}^h(t_1,t_2) \\ \hat{x}^v(t_1,t_2) \end{bmatrix} \tag{9}$$

in addition, combining the 2-D continuous system in (1) and 2-D state observer in (7), we can obtain the dynamics of the observation error as follows:

$$\begin{aligned} \begin{bmatrix} \frac{\partial e^h(t_1,t_2)}{\partial t_1} \\ \frac{\partial e^v(t_1,t_2)}{\partial t_2} \end{bmatrix} &= \begin{bmatrix} N_{11} & N_{12} \\ N_{21} & N_{22} \end{bmatrix} \begin{bmatrix} e^h(t_1,t_2) \\ e^v(t_1,t_2) \end{bmatrix} \\ &+ \begin{bmatrix} A_{11} - N_{11} - L_1C_1 & A_{12} - N_{12} - L_1C_2 \\ A_{21} - N_{21} - L_2C_1 & A_{22} - N_{22} - L_2C_2 \end{bmatrix} \begin{bmatrix} x^h(t_1,t_2) \\ x^v(t_1,t_2) \end{bmatrix} \\ &+ \begin{bmatrix} B_1 - M_1 \\ B_2 - M_2 \end{bmatrix} u(t_1,t_2) \end{aligned} \tag{10}$$

## 3 Observer Design

In the present section, we develop the main result of this paper in the form of following theorem:

**Theorem 1.** *The 2-D state observer in the form of (7) will estimate asymptotically the states vectors of system (1) if there exist symmetric positive definite matrices $P_h$ and $P_v$ and matrices $W_1$ and $W_2$ such that the LMI (11) is feasible*

$$\begin{bmatrix} \Gamma_{11} & \Gamma_{12} \\ * & \Gamma_{12} \end{bmatrix} < 0 \tag{11}$$

*where*

$$\begin{aligned} \Gamma_{11} &= P_h A_{11} - W_1 C_1 + A_{11}^T P_h - C_1^T W_1^T \\ \Gamma_{12} &= P_h A_{12} - W_1 C_2 + A_{21}^T P_v - C_1^T W_2^T \\ \Gamma_{22} &= P_v A_{22} - W_2 C_2 + A_{22}^T P_v - C_2^T W_2^T \end{aligned}$$

*Moreover a suitable observer of from (7) is given by*

$$\begin{aligned} \begin{bmatrix} N_{11} & N_{12} \\ N_{21} & N_{22} \end{bmatrix} &= \begin{bmatrix} A_{11} & A_{12} \\ A_{21} & A_{22} \end{bmatrix} - \begin{bmatrix} P_h^{-1} & 0 \\ 0 & P_v^{-1} \end{bmatrix} \begin{bmatrix} W_1 \\ W_2 \end{bmatrix} \begin{bmatrix} C_1 & C_2 \end{bmatrix}, \\ \begin{bmatrix} M_1 \\ M_2 \end{bmatrix} &= \begin{bmatrix} B_1 \\ B_2 \end{bmatrix}, \\ \begin{bmatrix} L_1 \\ L_2 \end{bmatrix} &= \begin{bmatrix} P_h^{-1} & 0 \\ 0 & P_v^{-1} \end{bmatrix} \begin{bmatrix} W_1 \\ W_2 \end{bmatrix}. \end{aligned} \tag{12}$$

**Proof:** From the dynamics of the observation error (10) The estimation error signals converge to zero only if $N$, $M$ and $L$ are choose to satisfy the following constraints:

- **(i)** $\begin{bmatrix} N_{11} & N_{12} \\ N_{21} & N_{22} \end{bmatrix}$ stable matrix
- **(ii)** $\begin{bmatrix} A_{11} - N_{11} - L_1 C_1 & A_{12} - N_{12} - L_1 C_2 \\ A_{21} - N_{21} - L_2 C_1 & A_{22} - N_{22} - L_2 C_2 \end{bmatrix} = 0$
- **(iii)** $\begin{bmatrix} B_1 - M_1 \\ B_2 - M_2 \end{bmatrix} = 0$

With these constraints, the system (10) becomes:

$$\begin{bmatrix} \frac{\partial e^h(t_1,t_2)}{\partial t_1} \\ \frac{\partial e^v(t_1,t_2)}{\partial t_2} \end{bmatrix} = \begin{bmatrix} A_{11} - L_1 C_1 & A_{12} - L_1 C_2 \\ A_{21} - L_2 C_1 & A_{22} - L_2 C_2 \end{bmatrix} \begin{bmatrix} e^h(t_1,t_2) \\ e^v(t_1,t_2) \end{bmatrix} \tag{13}$$

By applying the stability condition Lemma 1 to (13), the stability condition ensuring the stability asymptotic of the dynamics of the observation error (13) can be derived as follows.

$$\begin{aligned} &\begin{bmatrix} A_{11} - L_1 C_1 & A_{12} - L_1 C_2 \\ A_{21} - L_2 C_1 & A_{22} - L_2 C_2 \end{bmatrix}^T \begin{bmatrix} P_h & 0 \\ 0 & P_v \end{bmatrix} + \\ &\begin{bmatrix} P_h & 0 \\ 0 & P_v \end{bmatrix} \begin{bmatrix} A_{11} - L_1 C_1 & A_{12} - L_1 C_2 \\ A_{21} - L_2 C_1 & A_{22} - L_2 C_2 \end{bmatrix} < 0. \end{aligned} \tag{14}$$

Choosing

$$\begin{bmatrix} W_1 \\ W_2 \end{bmatrix} = \begin{bmatrix} P_h & 0 \\ 0 & P_v \end{bmatrix} \begin{bmatrix} L_1 \\ L_2 \end{bmatrix} \tag{15}$$

The linear matrix inequality (14) can be written as:

$$\begin{bmatrix} P_hA_{11} - W_1C_1 + A_{11}^TP_h - C_1^TW_1^T & P_hA_{12} - W_1C_2 + A_{21}^TP_v - C_1^TW_2^T \\ * & P_vA_{22} - W_2C_2 + A_{22}^TP_v - C_2^TW_2^T \end{bmatrix} < 0 \tag{16}$$

By solving the LMI condition (16) with respect to $P_h$, $P_v$, $W_1$ and $W_2$, the matrices $L_1$ and $L_2$ can be determined by:

$$L_1 = P_h^{-1}W_1$$
$$L_2 = P_v^{-1}W_1$$

Using the conditions (ii) and (iii), $N_{11}$, $N_{12}$, $N_{21}$, $N_{22}$ $B_1$ and $B_2$ are determined by

$$\begin{bmatrix} N_{11} & N_{12} \\ N_{21} & N_{22} \end{bmatrix} = \begin{bmatrix} A_{11} - L_1C_1 & A_{12} - L_1C_2 \\ A_{21} - L_2C_1 & A_{22} - L_2C_2 \end{bmatrix}$$
$$\begin{bmatrix} M_1 \\ M_2 \end{bmatrix} = \begin{bmatrix} B_1 \\ B_2 \end{bmatrix}$$

The proof is thus completed.

*Remark 2.* The problem of 2-D state observers design reduces to obtaining the feasibility of the LMI condition (11), which presents a sufficient LMI criterion for the design of the 2-D observers for systems in the form of (1), leading the observation error signals to converge at zero. It worth noting that, if system (1) reduces to a one-dimensional system, Theorem 1 can be considered as an extension of the existing conditions on the observer design for one-dimensional systems to the 2-D systems.

## 4 Simulation Example

In the present section, we consider a simulation example to show the applicability of the proposed method.

Consider a 2-D continuous system in the form of (1) with matrices as follows:

$$\begin{bmatrix} A_{11} & A_{12} \\ A_{21} & A_{22} \end{bmatrix} = \begin{bmatrix} 0.1 & -0.05 \\ 1 & -1 \end{bmatrix}, \quad \begin{bmatrix} B_1 \\ B_2 \end{bmatrix} = \begin{bmatrix} 1 \\ 1 \end{bmatrix}, \quad \begin{bmatrix} C_1 & C_2 \end{bmatrix} = \begin{bmatrix} 1 & 0 \end{bmatrix},$$

The objective is to design an observer guarantee that the estimated states $\hat{x}^h$ and $\hat{x}^v$ converge asymptotically to the reals values, which means that the errors $e^h$ and $e^h$ vanishes asymptotically at zero. By applying Theorem 1, we obtain the feasible solution and the corresponding observer gains can be obtained based on (12) with:

$$\begin{bmatrix} N_{11} & N_{12} \\ N_{21} & N_{22} \end{bmatrix} = \begin{bmatrix} -0.5000 & -0.0500 \\ 0.0833 & -1.0000 \end{bmatrix}, \quad \begin{bmatrix} M_1 \\ M_2 \end{bmatrix} = \begin{bmatrix} 1 \\ 1 \end{bmatrix}, \quad \begin{bmatrix} L_1 \\ L_2 \end{bmatrix} = \begin{bmatrix} 0.6000 \\ 0.9167 \end{bmatrix},$$

the trajectories corresponding of the resulting observation error system (by choosing the random initial conditions) are shown in Figs. 1 and 2, it can be seen clearly that all of which obviously converge to zero when $(t_1, t_2) \to \infty$.

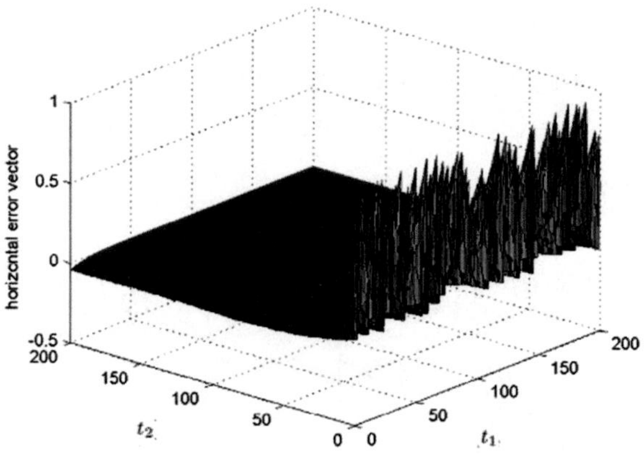


**Fig. 1.** Error $e^h(t_1, t_2)$

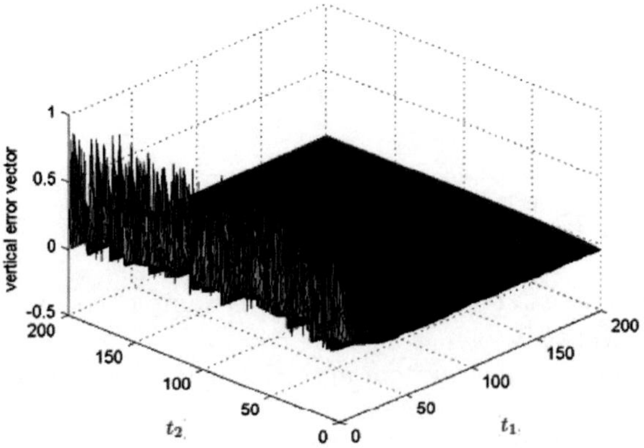


**Fig. 2.** Error $e^v(t_1, t_2)$.

## 5 Conclusion

In this paper, we have addressed the problem of observation of state vector for 2-D continuous systems represented by the Roesser model. A new sufficient LMI criterion for the design of 2-D state observer is presented in terms of an LMIs, witch guarantees that the states of the designed observer converge to the real value is presented. A simulation example is exploited example to show the applicability of the proposed method.

## References

1. Kaczorek, T.: Two-Dimensional Linear Systems. Springer, Berlin (1985)
2. Du, C., Xie, L.: $H_\infty$ Control and Filtering of Two-Dimensional Systems, vol. 278. Springer, Berlin (2002)
3. Roesser, R.: A discrete state-space model for linear image processing. IEEE Trans. Autom. Control **20**(1), 1–10 (1975)
4. Fornasini, E., Marchesini, G.: Stability analysis of 2-D systems. IEEE Trans. Circuits Syst. **27**(12), 1210–1217 (1980)
5. Hinamoto, T.: 2-D Lyapunov equation and filter design based on the Fornasini-Marchesini second model. IEEE Trans. Circuits Syst. I Fundam. Theory Appl. **40**(2), 102–110 (1993)
6. Ooba, T.: On stability analysis of 2-D systems based on 2-D Lyapunov matrix inequalities. IEEE Trans. Circuits Syst. I Fundam. Theory Appl. **47**(8), 1263–1265 (2000)
7. Badie, K., Alfidi, M., Tadeo, F., Chalh, Z.: Delay-dependent stability and $H_\infty$ performance of 2-D continuous systems with delays. Circuits Syst. Sig. Process. **37**(12), 5333–5350 (2018a)
8. Badie, K., Alfidi, M., Chalh, Z.: New relaxed stability conditions for uncertain two-dimensional discrete systems. J. Control Autom. Electr. Syst. **29**(6), 661–669 (2018b)
9. Badie, K., Alfidi, M., Chalh, Z.: Exponential stability analysis for 2D discrete switched systems with state delays. Optim. Control Appl. Meth. **4**, 1–16 (2019). https://doi.org/10.1002/oca.2537
10. Yao, J., Wang, W., Zou, Y.: The delay-range-dependent robust stability analysis for 2-D state-delayed systems with uncertainty. Multidimension. Syst. Sig. Process. **24**(1), 87–103 (2013)
11. Zhang, K., Jiang, B., Shi, P.: Fault estimation observer design for discrete-time Takagi-Sugeno fuzzy systems based on piecewise Lyapunov functions. IEEE Trans. Fuzzy syst. **20**(1), 192–200 (2011)
12. Chadli, M., Karimi, H.R.: Robust observer design for unknown inputs Takagi-Sugeno models. IEEE Trans. Fuzzy Syst. **21**(1), 158–164 (2012)
13. Guerra, T.M., Estrada-Manzo, V., Lendek, Z.: Observer design for Takagi-Sugeno descriptor models: An LMI approach. Automatica **52**, 154–159 (2015)
14. Alfidi, M., Chalh, Z., Ouahi, M.: A constructive design of state observer for two-dimensional systems. WSEAS Trans. Syst. Control **10**, 430–435 (2015)
15. Ding, D.W., Wang, H., Li, X.: $H_-/H_\infty$ fault detection observer design for two-dimensional Roesser systems. Syst. Control Lett. **82**, 115–120 (2015)
16. Wang, L., Xu, H., Zou, Y.: Regular unknown input functional observers for 2-D singular systems. Int. J. Control Autom. Syst. **11**(5), 911–918 (2013)

17. Galkowski, K.: LMI based stability analysis for 2D continuous systems. In: 9th International Conference on Electronics, Circuits and Systems, 2002, vol. 3, pp. 923–926. IEE (2002)
18. Alfidi, M., Hmamed, A.: Robust stability analysis for 2-D continuous-time systems via parameter-dependent Lyapunov functions. WSEAS Trans. Syst. Control **2**(11), 497–503 (2007)

# Miniature 2.45 GHz Rectenna for Low Levels of Power

Abdellah Taybi[1](✉), Abdelali Tajmouati[1], Jamal Zbitou[1], Ahmed Lakhssassi[2], Ahmed Errkik[1], and El Abdellaoui Larbi[1]

[1] LMIET, FST of Settat, Hassan 1st University, Settat, Morocco
taybi.abdellah@gmail.com
[2] University of Quebec in Outaouais, Gatineau, QC, Canada

**Abstract.** Wireless microwave power transmission (WPT) has become a new alternative technology to solve global energy problems, including the depletion of fossil fuel sources. This paper is part of the use of renewable energies such as WPT to power the various sensor networks that surround our environment. In this manuscript, an efficient 2.45 GHz Rectenna system have been designed, optimized and simulated for applications involving remote power supply. As a first step, a 2.45 GHz miniature microstrip antenna have been designed using Defected Ground Structure technology in order to switch the frequency from 5.8 GHz to 2.45 GHz with a gained size reduction reaching 70%. In the other side, a high output voltage 2.45 GHz Rectifier structure using voltage doubler topology have been developed for low power levels. The Rectenna have been simulated using Advanced Design System and CST Microwave Software and we have employed FR4 substrate with dielectric permittivity constant 4.4, thickness of 1.6 mm and loss tangent of 0.025. We have obtained satisfying results in terms of conversion efficiency and output voltage reaching 15 V.

**Keywords:** Wireless power transmission · Rectenna system · Defected Ground Structure · Microstrip antenna

## 1 Introduction

Communicating objects are spreading everywhere in our daily lives. More and more gadgets and portable elements are appearing, with ever more constraints on their dimensions, shapes and uses. The power supply function is therefore increasingly difficult to satisfy and it is becoming crucial to improve the energy autonomy of these objects.

In addition, energy production and transport is a topical subject with multiple issues (economic, environmental and political). Indeed, 70% of electricity is produced by combustion from fossil resources (Petrol, coal…) without forgetting the impact of these combustions on the environment (pollution, global warming, etc.) and the limited stock of these non-renewable resources. One solution seems to be the future for electricity production: the exploitation of renewable energy sources, i.e. the exploitation of energy already existing in our environment. The most widespread and abundant forms are


A. Saka et al. (Eds.): CPI 2019, LNME, pp. 85–94, 2021.
https://doi.org/10.1007/978-3-030-62199-5_9

energies originating from the sun, wind, water masses (dams and oceans) but also - for less important quantities of energy - the recovery of vibratory energy in mechanical, acoustic or electromagnetic form.

Wireless radiofrequency energy transmission can also provide a contribution to solve this problem and can offer new applications and perspectives. Indeed, this radiofrequency energy has invaded our environment in recent years with the increase in wireless applications (GSM, Wi-Fi, WIMAX …). The power level varies from one application to another but this energy is available and permanent. This energy can be used to power microsystems or replace batteries. To do so, this energy must be recovered and transformed into continuous power. These systems require a wave receiving antenna connected to an energy conversion system in continuous power.

The concept of wireless energy transmission made its beginnings with NIKOLA TESLA [1, 2], which was able to demonstrate that energy can be transported without any physical support. These results strongly contributed to the appearance of the Rectenna systems, invented in 1964 by the U.S electrical engineer WILLIAM BROWN [3], who was able to power a helicopter by microwaves transmitted from the ground and received by an attached Rectenna. Figure 1 shows the overall synoptic diagram of a Wireless Power Transmission (WPT) system.

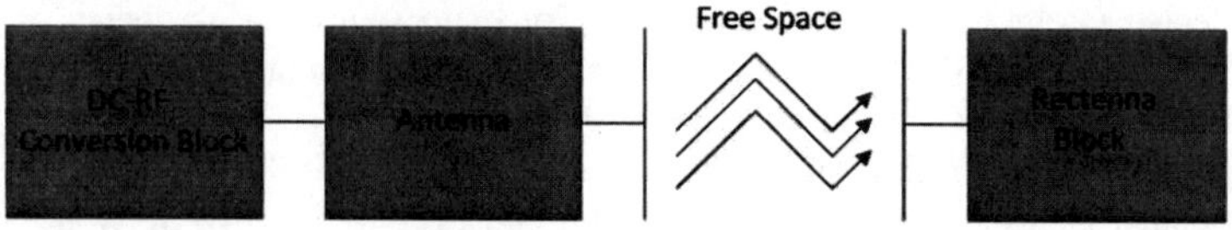


**Fig. 1.** Wireless power transmission system

The key element of a WPT system is called a rectenna (for rectifying antenna) [4–6], A typical rectenna block diagram is shown in Fig. 2. it contains a receiving antenna and an RF-DC conversion circuit. The RF-DC conversion circuit often consists of one or more Schottky diodes, an HF filter at the input and a DC filter at the output. A rectenna is generally loaded by a resistive element, which represents the input impedance of the device to be powered.

First, on the transmitter side, the DC electrical energy is converted into microwave energy using an RF source. Then, this energy is radiated into the free space by a suitable transmitting antenna. Finally, on the receiver side, part of the radiated energy is captured by the Rectenna (Rectifying Antenna) which converts the RF energy into DC power that will flow over a resistive load, which models the circuit to be supplied.

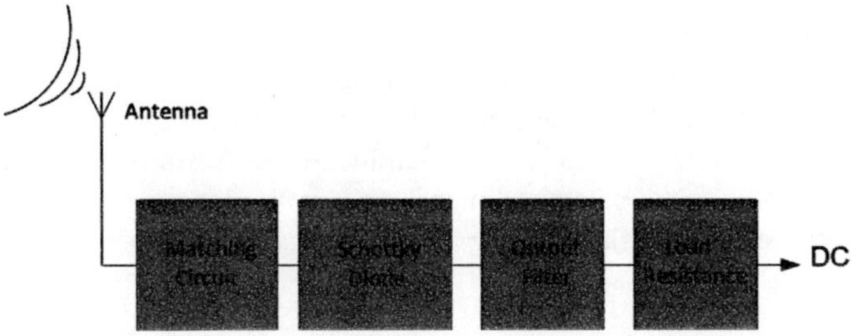


**Fig. 2.** Block diagram of rectenna circuit

The project in which this paper is integrated addresses performance challenges of rectenna systems. It focuses on the feasibility of increasing the recovered power as well as the output voltage of the conversion circuit. The objective of this work is therefore to produce energy recovery circuit or Rectenna printed on FR4 substrate at 2.45 GHz to power wireless devices such as sensors.

The first section of this article concerns the antenna part, we opted for planar technology because of the multitude of advantages it offers. In addition, the Defected Ground Structure technique was used to switch the resonance frequency of our antenna from 5.8 GHz to 2.45 GHz, which allowed us to gain in size. The antenna has been designed, optimized and simulated using the transient solver from CST Microwave Studio, a good matching input impedance of −34 dB have been observed.

The second section deals with the conversion circuit part (rectifier). The developed structure is based on 3 Schottky diodes mounted in voltage doubling topology. Furthermore, it's printed on a FR4 substrate with dielectric permittivity constant 4.4, thickness of 1.6 mm and loss tangent of 0.025 using Microstrip lines. The circuit has a good level of input impedance matching with an output voltage that reaches 15 V and an efficiency of about 80% for low power levels.

## 2 Miniature 2.45 GHz Patch Antenna Using Defected Ground Structure

An antenna is a passive element that transforms electrical power into an electromagnetic wave and vice versa. The role of the antenna in energy recovery is to collect energy from a source, for example from ambient RF energy. In this application, planar antennas are preferred [7, 8]. Indeed, they are characterized by a low mass, a reduced space requirement and ease of construction. Printed antennas are difficult to study because of the inhomogeneity of the medium that supports them (substrate/air).

The boundary conditions become very complex, especially at the air-substrate interface, which makes it difficult to solve numerical equations. Different numerical methods are used to study the electrical characteristics of planar antennas. Some of the methods include the "Finite Difference in Time Domain" method, which is a time domain method and requires a cubic 3D mesh, the "Moments Method" (MoM), which is a frequency domain method requiring a rectangular or triangular mesh and the finite element method ("Finite Element Method") which is a frequency method and requires a 3D mesh of small finite elements, often tetrahedrons. A simple microstrip antenna consists of a very thin metal patch of dimensions L × W placed on a dielectric substrate fixed on a metallic ground plane as shown in Fig. 3.

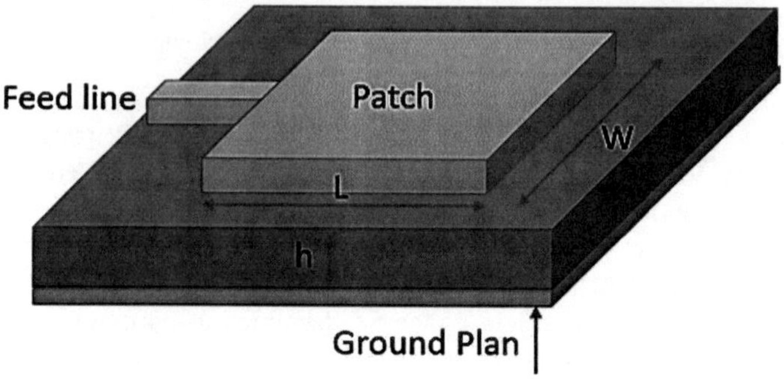


**Fig. 3.** Rectangular micro strip antenna configuration.

The antenna dimensions are calculated theoretically as follow:

**Calculation of Width (W)**

For an effective radiator, practical width of the patch antenna that leads to good radiation efficiencies is given by [9]:

$$W = \frac{C_0}{2F_0}\sqrt{\frac{2}{1_{+}\varepsilon_r}} \tag{1}$$

Where $C_0$ is the free-space velocity of light i.e. $3 \times 10^{-8}$ m/s and $\varepsilon_r$ is the dielectric constant of material.

**Calculation of Effective Dielectric Constant $\varepsilon_{reff}$**

The value of effective dielectric constant is less than dielectric constant of the substrate, because of the fringing fields are not confined in dielectric substrate around the periphery of the patch only, but is also spread in the air. The value of this effective dielectric constant is given by [9]:

$$\varepsilon_{reff} = \frac{\varepsilon_r + 1}{2} + \frac{\varepsilon_r - 1}{2}\left[1 + 12\frac{h}{W}\right]^{-1/2} \tag{2}$$

**Calculation of Length (L)**
The length of the patch determines the resonance frequency thus it is a critical factor for narrowband patch. Since it is not possible to accurately account the fringing field the results are not definite. Below is the equation to calculate the length of the patch [9]:

$$L = \frac{\lambda_{eff}}{2} - 2\Delta L \tag{3}$$

Where ΔL is the length extension because of fringing field, which can be calculated as follow:

$$\Delta L = 0.412\, h \frac{\left(\varepsilon_{reff} + 0.3\right)\left(\frac{W}{h} + 0.264\right)}{\left(\varepsilon_{reff} - 0.258\right)\left(\frac{W}{h} + 0.8\right)} \tag{4}$$

The printed antenna can be supplied in several ways. They can be classified into two categories with contact (Microstrip line feeding and Coaxial probe feeding) and without contact (Slot (or aperture) feeding and feeding using proximity coupling). In contact methods, the radio frequency (RF) current directly supplies the antenna using a connection element. In non-contact techniques, electromagnetic field coupling ensures the transfer of power between the microstrip line and the radiating element.

## 2.1 Initial Antenna Design

In order to optimize the performance of our patch antenna, and to visualize its gain, directivity and reflection coefficient, we will then use the CST Microwave Studio software [10]. First, we determine the dimensions of the feed line so that it is well matched to the antenna input. Then, we simulate the patch antenna in the frequency 5.8 Ghz under the CST software. The proposed patch antenna is depicted in Fig. 4.

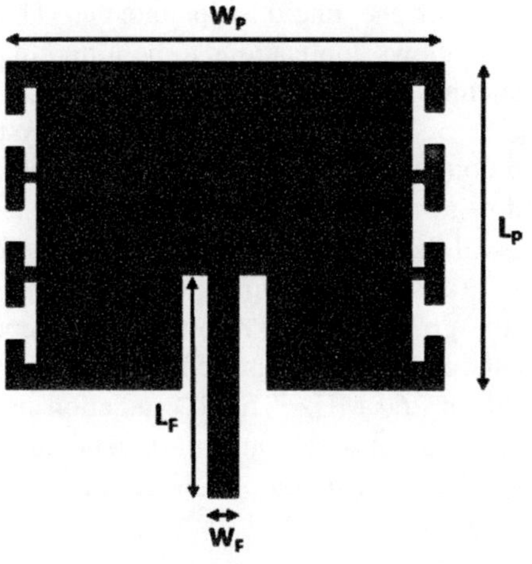


**Fig. 4.** The proposed patch antenna.

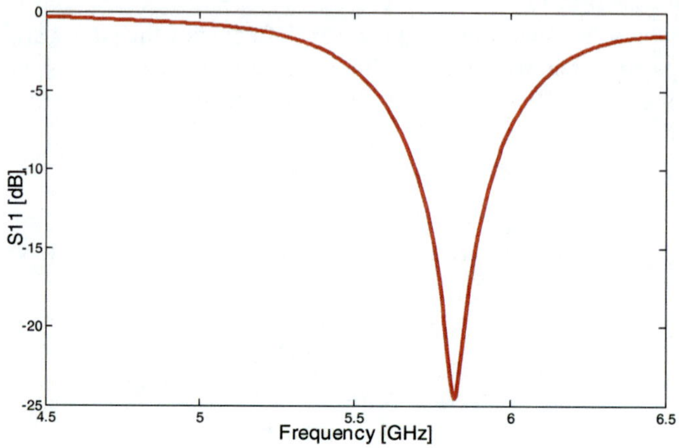


**Fig. 5.** Simulated S11 versus Frequency of the proposed patch antenna.

From Fig. 5, it can be seen that the input impedance matching is very satisfactory with a reflection coefficient of −25 dB at 5.8 GHz.

## 2.2 Defected Ground Structure (DGS)

The purpose being to design an antenna that radiates at 2.45 GHz with improved performance, especially in terms of size. Therefore, the DGS [11, 12] technic was used to switch the frequency of the initial antenna from 5.8 GHz to 2.45 GHz.

DGS is a periodic motif engraved on the ground plan. The name of this technique simply means that a defect is placed on the ground plane which is generally considered to be an anomaly of a perfectly conductive current hole, but this does not make the ground plane defective. Depending on the shape and dimensions of the defect, the distribution of the shielded current in the ground plane is disturbed, this perturbation modifies the characteristics of the micro strip antenna. This shows the resonance property of the DGS which allows applications to be found in a number of microwave circuits such as filters, antennas and amplifiers. In addition, DGS allows a considerable reduction in size required for the miniaturization of the system. The shape can be changed from simple to complicated shape for better performance. After a series of optimizations performed using CST Microwave Studio, we have designed a Defected Ground Structure which will then be engraved in the ground plan of the initial antenna structure (Fig. 6). Figure 7 (a) shows the simulated reflection coefficient of the system, we can conclude from the graph that the resonating frequency is well shifted from 5.8 GHz to 2.45 GHz of the Industrial Scientific Medical band (ISM) with a return loss of −34 dB and bandwidth of 260 MHz. The 2D radiation pattern is given by Fig. 7 (b) in the E-plane, which show a stable and bi-directional radiation pattern for the resonant frequency band. The miniaturization has been achieved with a size reduction reaching 70%.

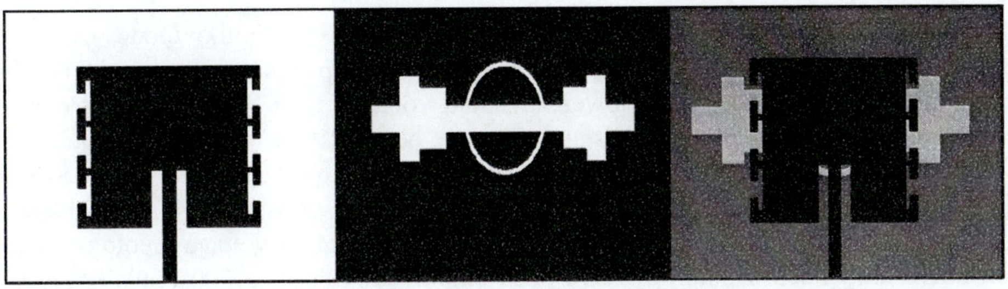

**Fig. 6.** Final antenna design

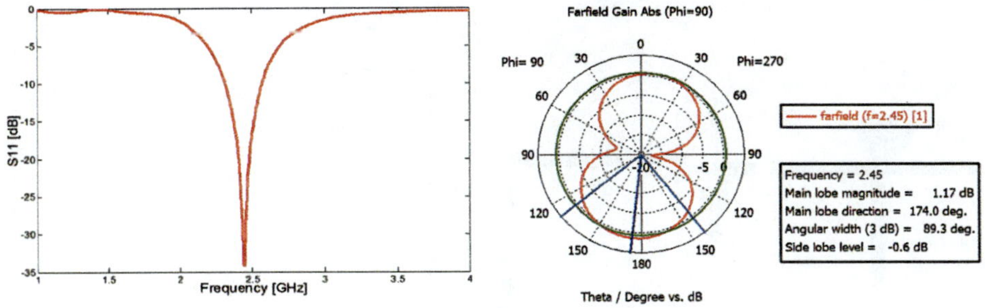


(a) (b)

**Fig. 7.** Simulation results (a) Reflection coefficient versus frequency (b) Antenna 2D Radiation pattern in E-Plane at 2.45 GHz

## 3 Microstrip 2.45 GHz Voltage Doubler Rectifier

The term rectenna is a word composed of "rectifier + Antenna". The rectenna is a device for the recovery of electromagnetic energy. Its principle consists in collecting this energy from one or more sources through a receiving antenna, the recovered power being converted with the greatest possible efficiency into continuous power (DC) through a rectifier circuit consisting of one or more non-linear active semiconductor elements, in order to supply low consumption electronic devices. It should be noted at this stage that, depending on the device to be supplied, it may be necessary to maximize either the direct current, the direct voltage or the DC power itself.

The key element of the rectifier circuit is the non-linear active semiconductor element, which is usually a diode. The choice of diode is very critical for the overall efficiency of the rectenna. The type of diode, frequency range and power level involved are criteria to be considered. For the relatively low powers that are detected in electromagnetic energy recovery applications, we are in the field of "small signals" and therefore we have a strong interest in using Schottky diodes. A Schottky diode differs from a conventional diode in the materials used. Schottky diodes consist of a metal contact and a semiconductor contact, which gives them a considerable advantage over

PN junction diodes in terms of switching speed. This makes Schottky diodes essential for RF detectors where they can operate at frequencies up to several tens of GHz. In addition, Schottky diodes have lower threshold voltages than their PN counterparts, which offers lower switching losses and therefore better RF-DC conversion efficiency.

In designing the proposed Rectifier structure, we have used 3 Schottky HSMS2820 [13] diodes etched on an FR4 substrate having dielectric constant εr = 4.4, substrate thickness h = 1.6 mm and the loss tangent is 0.025. A doubler voltage topology based on micro strip lines technology have been employed. Hence, a 500 ohms' resistive charge modeling the device to be powered have been placed in the output of the circuit. However, simulations and optimizations have been carried out by means of Advanced Design System. The following figure illustrate the final rectifier structure:

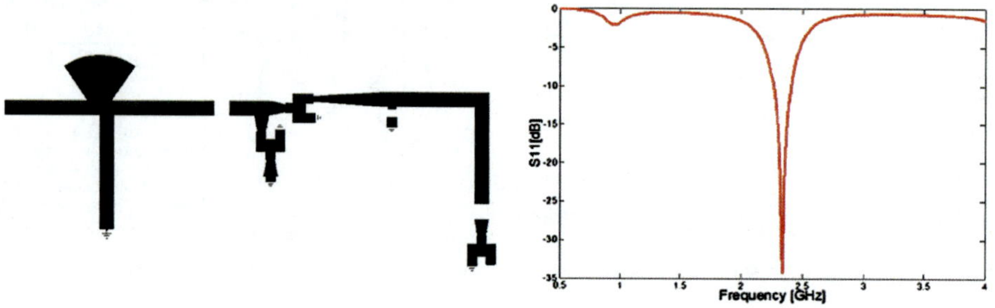


(a) (b)

**Fig. 8.** (a) Final rectifier structure (b) Simulated reflection coefficient versus frequency at 2.45 GHz

The simulation results of the return loss (S11) of the rectifier structure are shown in Fig. 8(b). We can conclude, from the graph, that the micro strip voltage doubler rectifier is well matched at 2.45 GHz.

Generally, a rectenna is characterized by two criteria, namely output voltage and/or conversion efficiency. Conversion efficiency is the ability to recover radiofrequency energy and transform it into continuous signals. There are two types of efficiency, the RF-DC conversion efficiency and the overall efficiency. The first (RF-DC conversion efficiency) corresponds to the ratio of the DC power consumed by the load to the power at the input of the RF-DC conversion circuit. The second (overall efficiency) corresponds to the ratio of the DC power consumed by the load to the power received at the input of the rectenna receiving antenna. To define the performance indicators of the designed rectifier, we have used the Harmonic Balance simulator from ADS [14], Fig. 9 shows the simulated output voltage and conversion efficiency. The circuit have reached an output voltage of 15 V and a conversion efficiency of 80% for low power levels.

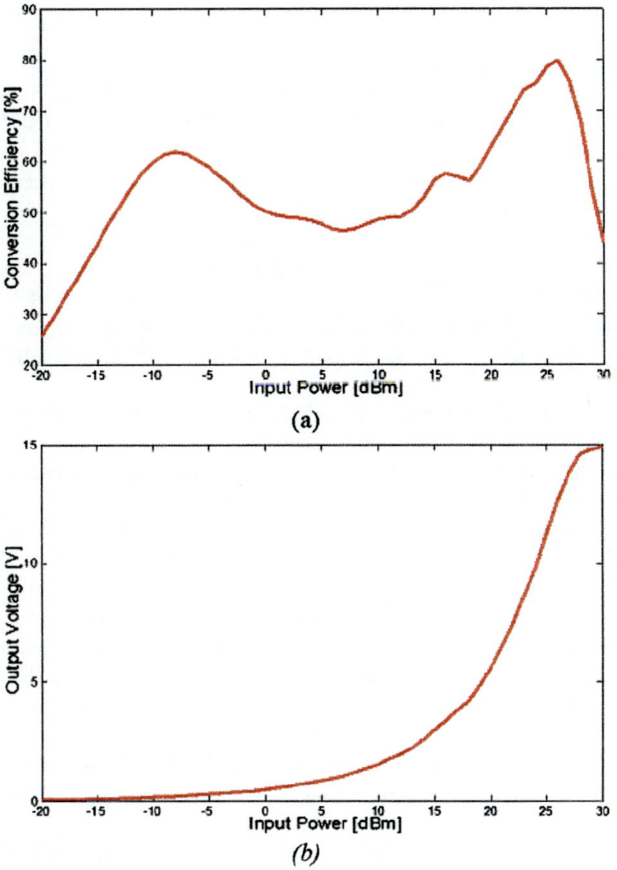


**Fig. 9.** Simulated results of the rectifier (a) Conversion Efficiency versus Input power (b) Output voltage versus input power.

## 4 Conclusion

This paper deals with the problematic of power supply for low consumption devices. The use of renewable energies as solution, especially Wireless Power Transmission, opens up several potential areas of research. In this article, a new efficient 2.45 GHz Rectenna have been presented. Hence, two focal axes have been covered: The first part handle with the design and optimization of a miniature 2.45 GHz patch antenna using Defected Ground Structure as a technic to reach a good size reduction, the Transient solver of CST Microwave Studio have been employed. The second section was dedicated to the Rectifier circuit; it's based on 3 Schottky HSMS 2820 diodes mounted in a voltage doubler topology. In addition, micro strip lines have been etched on an FR4 substrate in order to interconnect the different components of the rectifier structure. Satisfying results in terms of output voltage (15 V) and conversion efficiency (80%) have been reached.

## References


1. Tesla, N.: The transmission of electrical energy without wires. Electr. World Eng. (1904)
2. Tesla, N.: Experiments with Alternate Currents of High Potential and High Frequency. Book Tree (2007)
3. Brown, W.C.: The History of power transmission by radio waves. IEEE Trans. Microw. Theory Tech. MTT **32**(9), 1230–1242 (1984)
4. Zbitou, J., Latrach, M.: Hybrid rectenna and monolithic integrated zero-bias microwave rectifier. IEEE Trans. Microw. Theory Tech. **54**(1), 147–152 (2006)
5. Taybi, A., Tajmouati, A., Zbitou, J., Errkik, A., Latrach, M., El Abdellaoui, L.: A new configuration of a high output voltage 2.45 GHz rectifier for wireless power transmission applications. Telecommun. Comput. Electron. Control. J. (TELKOMNIKA) **16**(5), 1939–1946 (2018)
6. Marian, V., Vollaire, C., Verdier, J., Allard, B.: Rectenna circuit topologies for contactless energy transfer, Eindhoven, Netherlands, June 2011
7. Douyere, A., Lan Sun Luk, J.D., Alicalapa, F.: High efficiency microwave rectenna circuit: modelling and design. Electron. Lett. **44**(24), 1409–1410 (2008)
8. Tudose, D.S, Voinescu, A.: Rectifier antenna design for wireless sensor networks. In: 19th International Conference on Control Systems and Computer Science (CSCS), 29–31 May 2013, pp. 184–188 (2013)
9. Balanis, C.A.: Antenna Theory, Analysis and Design, 3rd edn. Wiley, New York (2005)
10. Computer Simulation Technology (CST) Microwave Studio. www.cst.com
11. Debatosh, G., Chandrakanta, K., Surendra, P.: Improved cross-polarization characteristics of circular microstrip antenna employing arc-shaped defected ground structure (DGS). IEEE Antennas Wirel. Propag. Lett., **8**, 1367–1369 (2009)
12. Chandrakanta, K., Intiyas Pasha, M., Debatosh, G.: Defected ground structure integrated microstrip array antenna for improved radiation properties. IEEE Antennas Wirel. Propag. Lett. **16**, 310–312 (2016)
13. Surface mount RF Schottky barrier diode HSMS-282X series. Agilent Technol., Palo Alto, CA, Tech. Data (2001)
14. Advanced Design System (ADS). http://www.home.agilent.com/agilent/home.jsp

# Stable Computation of Hahn Moments for Large Size 1D Signal Analysis

Achraf Daoui[1(✉)], Omar El Ogri[2], Mohamed Yamni[2], Hicham Karmouni[2], Mhamed Sayyouri[1], and H. Qjidaa[2]

[1] Engineering, Systems and Applications Laboratory, National School of Applied Sciences, Sidi Mohamed Ben Abdellah University, BP 72, My Abdallah Avenue Km. 5 Imouzzer Road, Fez, Morocco
{achraf.daoui,mhamed.sayyouri}@usmba.ac.ma

[2] CED-ST, STIC, Laboratory of Electronic Signals and Systems of Information LESSI, Dhar El Mahrez Faculty of Science, Sidi Mohamed Ben Abdellah-Fez University, Fez, Morocco
{omar.elogri,mohamed.yamni, hicham.karmouni}@usmba.ac.ma, qjidah@yahoo.fr

**Abstract.** Hahn's discrete orthogonal moments are powerful tools for image and signal analysis. Several methods have been proposed to implement Hahn's moments, but their implementation remains limited for high orders by the problem of the propagation of numerical errors. This problem is due to rounding errors during recursive computations of Hahn's polynomials under machine (using Matlab, C++, Java…). In this paper, we propose a stable computation of Hahn polynomials at high orders based on the modified Gram-Schmidt orthonormalization process. the proposed method significantly reduces the propagation of numerical errors and therefore preserves the orthonormality property of Hahn polynomials. The results obtained show the stability of Hahn moments obtained by the proposed method for the reconstruction of large 1D signals.

**Keywords:** 1D reconstruction · Discrete orthogonal moments · Hahn polynomials · Numerical stability of moments

## 1 Introduction

The field of 1D signal processing is a very important field to study and makes possible various other fields such as communications. MP3 music files contain processed, transformed, and compressed music signal data. Speech recognition systems such as dictation software must analyze and process signal data to identify the individual words of a pronounced sentence. Neuronal Interface devices, such as individual medical prostheses (Bio-Signaux), must read complicated signals from the neural interface, process these signals to determine important characteristics. Signal processing allows high-speed data communication, even in noisy environments. 1D signals are generally large in size, so to analyze them effectively it is necessary to use fast and stable methods nu-metrically. Discrete orthogonal moments such as Tchebichef [1, 2],


A. Saka et al. (Eds.): CPI 2019, LNME, pp. 95–102, 2021.
https://doi.org/10.1007/978-3-030-62199-5_10

Krawtchouk [3, 4], Hahn [5], Meixner [6] and Charlier [7] are widely and successfully applied in signal analysis. Among the applications of the latter are classification [8], compression [9], reconstruction [10]. The above-mentioned moments are quickly calculated thanks to the use of recursive formulas and matrix products using special software such as Matlab. But the problem of the numerical instability of the polynomials that define discrete orthogonal moments is still present, especially for high orders. This problem depends on double precision during real computations under machine (using MATLAB, C++, Python…) which destroys the orthogonality property of the polynomials. To solve this problem for Hahn polynomials, we propose in this paper the use of modified Gram-Schmidt ortho-normalization processes (GSOP) to calculate the coefficients of Hahn polynomials. The use of GSOP numerically stabilizes the computation of Hahn coefficients at high orders. The simulation results show the numerical stability of Hahn moments computed via GSOP during the reconstruction of large 1D signals.

## 2 Hahn's Discrete Orthogonal Polynomials and Moments

### 2.1 Recursive Computation of Hahn Polynomials with Respect to the Order *n*

The normalized Hahn polynomials with respect to the order $n$ are defined by the following three-term recursive relation [5]:

$$\tilde{h}_n^{(\alpha,\beta)}(x;N) = \frac{B \times D}{A}\tilde{h}_{n-1}^{(\alpha,\beta)}(x;N) - \frac{C \times E}{A}\tilde{h}_{n-2}^{(\alpha,\beta)}(x;N) \tag{1}$$

$$\text{with } A = \frac{n(\alpha+\beta+n)}{(\alpha+\beta+2n-1)(\alpha+\beta+2)};$$

$$B = x - \frac{\alpha-\beta+2N-2}{4} - \frac{(\beta^2-\alpha^2)(\beta+\alpha+2N)}{4(\alpha+\beta+2n-2)(\alpha+\beta+2n)}$$

$$C = -\frac{(\alpha+n-1)(\beta+n-1)}{(\alpha+\beta+2n-2)} \times \frac{(\alpha+\beta+N+n-1)(N-n+1)}{(\alpha+\beta+2n-1)};$$

$$D = \sqrt{\frac{n(\alpha+\beta+n)(\alpha+\beta+2n+1)}{(N-n)(\alpha+n)(\beta+n)(\alpha+\beta+2n-1)(\alpha+\beta+n+N)}}$$

$$E = \sqrt{\frac{n(n-1)(\alpha+\beta+n)(\alpha+\beta+n-1)(\alpha+\beta+2n+1)}{(\alpha+n)(\alpha+n-1)(\beta+n)(\beta+n-1)(N-n+1)(N-n)(\alpha+\beta+2n-3)(\alpha+\beta+n+N)(\alpha+\beta+n+N-1)}}$$

The zero-order and first-order of the normalized Hahn's discrete orthogonal polynomials are calculated as follows:

$$\widetilde{h}_0^{(\alpha,\beta)}(x;N) = \sqrt{\frac{\omega(x)}{\rho(0)}} \tag{2}$$

$$\widetilde{h}_1^{(\alpha,\beta)}(x;N) = [-(\beta+1)(N-1)+(\alpha+\beta+2)x]\sqrt{\frac{\omega(x)}{\rho(1)}} \tag{3}$$

The following curves present the first six orders and the last six orders of the Hahn polynomials for $\alpha = \beta = 80$ and $N = 1000$.

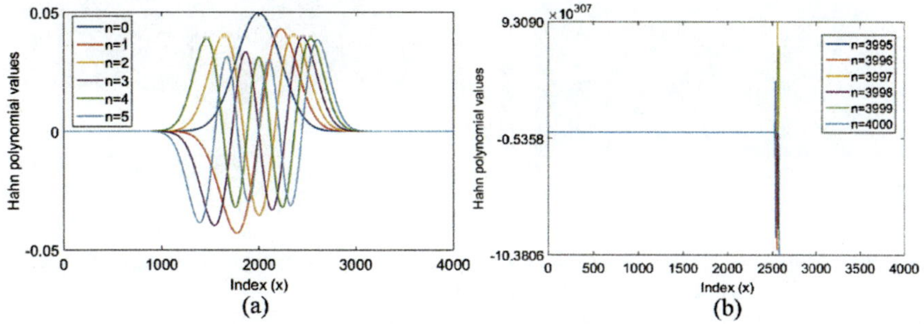


**Fig. 1.** (a): the first five orders and (b): the last five orders of the Hahn polynomials with $\alpha = \beta = 80$.

The figure (Fig. 1) shows that the values of the Hahn polynomials diverge for the last six orders ($\widetilde{h}_n^{(\alpha,\beta)}(x;N) \notin [-1,1]$). The divergence of Hahn polynomial values at higher orders due to the propagation of numerical errors that depend on double precision during real computations under machine (using Matlab, C++, Python…) which destroys the orthogonality property of Hahn polynomials. To solve this problem, we propose to calculate the polynomial coefficients via the modified Gram-Schmidt orthonormalization process (GSOP) which is numerically stable. Therefore, the use of GSOP guarantees the orthogonality property of Hahn polynomials until the last order.

### 2.1.1 Proposed Recursive Computation of Hahn Polynomials via GSOP

In this sub-section, we present the fundamental steps of the proposed algorithm (GSOP). This algorithm consists of three essential steps:

1. The initialization of polynomials to order $n = 0$ and $n = 1$ using the equations (Eq. (3) and Eq. (3)) successively.
2. The calculation of polynomial values for orders $n = 2 : N$ using the Eq. (1).
3. For each polynomial order $n$ the calculated coefficients are stored in a matrix $H_n(1:N) = \widetilde{h}_n(1:N)$ then the coefficients of the $H$ matrix are orthonormalized by the modified Gram-Schmidt orthonormalization procedure (GSOP) by recalculating the coefficients of the matrix by the relationship [11]:

$$\tilde{h}_n(x;N) = \tilde{h}_n(x;N) - \left[\sum_{k=1}^{n-1} r_{n,k}\right] \times \tilde{h}_k(x;N)$$
$$\text{with } r_{n,k} = \frac{H_n(1:N)^T \tilde{h}_n(x;N)}{\left\|\tilde{h}_n(x;N)\right\|}; \ k = 0:n \tag{4}$$

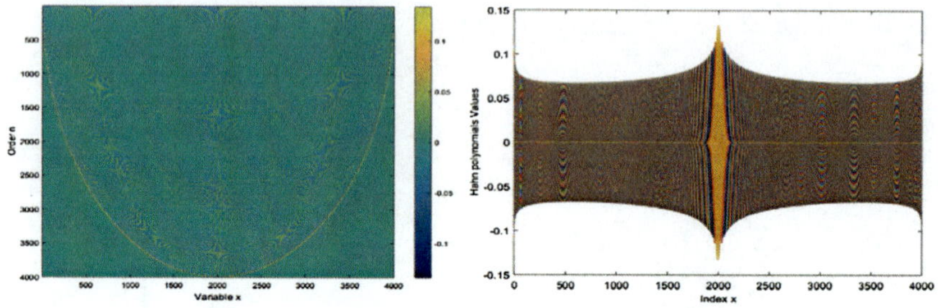


(a) (b)

**Fig. 2.** (a): Matrix of Hahn Polynomials values ($\alpha = \beta = 20$) using GSOP algorithm, (b): the curves of Hahn's polynomials up to the order 4000.

The figure (Fig. 2) shows that the use of the GSOP has thus made it possible to maintain the numerical stability of Hahn's polynomial coefficients until the last order.

After establishing an orthogonalization algorithm for Hahn polynomials using GSOP, we compute Hahn moments for large 1D signals.

### 2.2 1D Hahn Moments

Hahn moments for a one-dimensional signal is defined by the following matrix formulation [9]:

$$HM_n = H_n^T f \text{ with } K_n = \left[\tilde{h}_0^{(\alpha,\beta)}(x), \tilde{h}_1^{(\alpha,\beta)}(x), \ldots, \tilde{h}_{N-1}^{(\alpha,\beta)}(x)\right]^T \tag{5}$$

Where $f$ denotes signal vector length $N$.

The reconstructed signal from the moments of Hahn is calculated by the following relation:

$$\hat{f} = HM_n H_n \tag{6}$$

$\hat{f}(x)$ is the reconstructed signal.

The difference between the reconstructed signal and the original signal is measured by MSE (Mean Squared Error) and PSNR (Peak Signal to Noise Ratio).

In the following section, we present the results of the simulations which show the advantages of the proposed method in terms of quality of reconstruction of a 1D signals.

## 3 Simulation Results and Discussions

In this section, we present the results of simulations that show the stability of the Hahn moments calculated by GSOP. Hahn's moments are used to reconstruct large-size 1D bio-signals. These are selected from the MIT-BIH arrhythmia database [12]. We compare the quality of the reconstructed bio-signals with the proposed moments of Hahn avec les moments de Tchebichef [2], Krawtchouk [4], Meixner [6] and Charlier [7].

### 3.1 Noisy Signal Reconstruction

In the following test, the Cyclic Alternating Pattern (CAP) of EEG activity during sleep type test signal of size $N = 4000$ is reconstructed using Hahn moments calculated by the proposed method (GSOP). Reconstruction is carried out for Hahn's proposed moments with orders ranging from 0 to 4000. Indeed, the figure (Fig. 3) shows the test signal and their reconstructed signals as well as the PSNR curve. The reconstructed signals and the PSNR curve show, on the one hand, that the similarity between the original signal and the reconstructed signal improves gradually with the increase in the order of Hahn's moments up to the last order ($n = 4000$). And on the other hand, the proposed Hahn moments retain their numerical stability at high orders of moments.

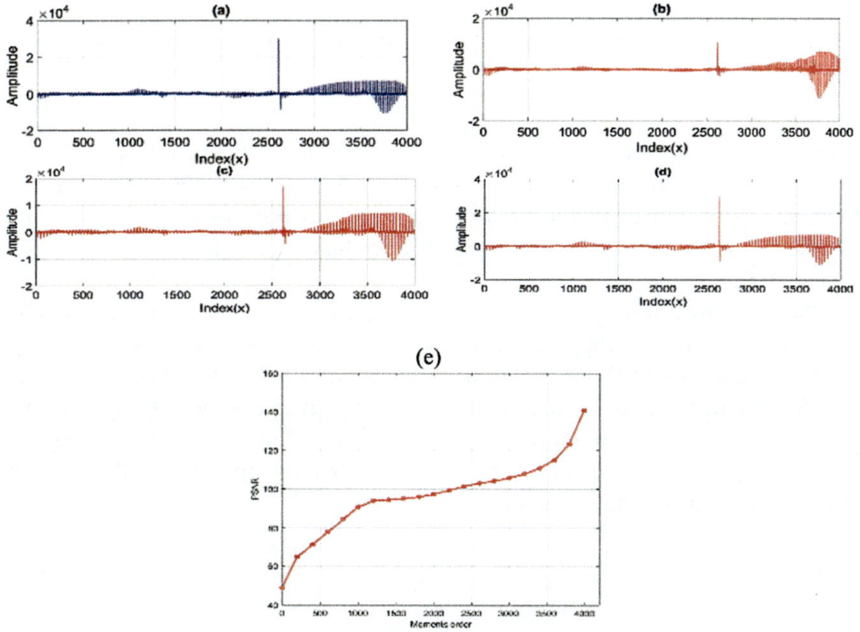


**Fig. 3.** (a): the test signal, (b), (c) and (d): the signals reconstructed at orders 500, 2000 and 4000 successively, (d): the corresponding PSNR curve.

The following test is carried out on a'Blood Pressure' type signal of size $N = 1024$ (Fig. 3). In this test we compare the quality of test signal reconstruction using Thebichef [2], Krawtchouk [4] and Hahn's proposed moments (Fig. 4).

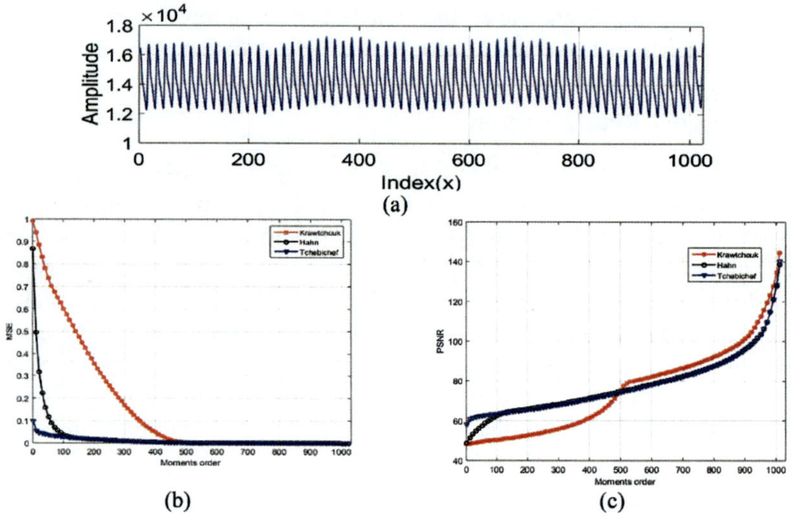


**Fig. 4.** (a): Test signal of type 'Blood Pressure', (b): MSE (a) and (c): PSNR of the reconstructed signal.

The MSE and PSNR curves show that the three methods, including the proposed method, are stable and offer a better quality of reconstruction especially at high orders of moments. It should be noted that the comparison is not made at moments of very high orders (for example $n = 4000$) since Krawchuk moments have limits from the order $n = 1075$.

In the following test, a comparative study in terms of reconstruction quality is carried out between the proposed Chebichef [2], Krawtchouk [4], Hahn [5], Meixner [6] and Charlier [7] and Hahn. To do this, we use a signal of the type "ECG" of size $N = 256$. The test signal is reconfigured by the predicted moments by varying the orders of the moments between 0 and 256. It should be noted that high orders are not considered in this test since Meixner and Charlier's moments present the problem of fluctuating numerical values at high orders.

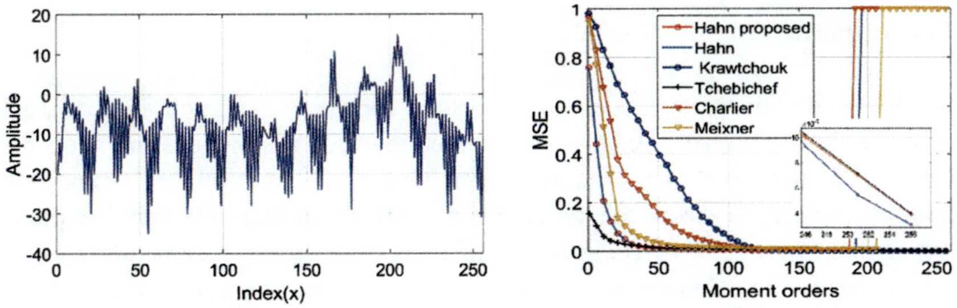


(a) (b)

**Fig. 5.** (a): the test signal 'ECG' and (b): Reconstructed signal MSE.

The MSE curves (Fig. 5(b)) show on the one hand the numerical instability of the Hahn [5], Meixner [6] and Charlier [7] moments when the order of the moments tends towards n = 256. On the other hand, we can see the numerical stability of the moments of Tchebichef [2], Krawchouk [4] and Hahn proposed moments. The latter are stable thanks to the use of GSOP.

### 3.2 Noisy Signal Reconstruction

The next test objective is to test the robustness of the proposed Hahn moments against the noise that affects the signal from different sources. To do this, a phonocardiogram test signal (PCG) of size is assigned by random noise. The original test signal and the noisy signal (Fig. 6 (a)) are reconstructed by the proposed Hahn moments up to the order 1024. The PSNR curves (Fig. 6 (b)) show a slight difference between the PSNR of noiselessly reconstructed signal and the noisy one. This minimal difference justifies the robustness of the Hahn moments calculated by GSOP against noise.

(a) (b)

**Fig. 6.** The original 'ECG' signal without noise and noisy with random noise.

## 4 Conclusion

In this article, we have proposed a stable method for the reconstruction of a large-size 1D signals by using discrete Hahn orthogonal moments. La stalité des moments de The stability of Hahn moments is obtained through the use of modified Gram-Schmidt orthonormatization processes, this process is numerically stable and therefore preserves the orthogonality property of Hahn polynomials at high orders. The results of the tests carried out show the effectiveness of the proposed method when analysing large 1D signals.

## References

1. Abdulhussain, S.H., Ramli, A.R., Al-Haddad, S.A.R., Mahmmod, B.M., Jassim, W.A: On computational aspects of Tchebichef polynomials for higher polynomial order. IEEE Access **5**, 2470–2478 (2017)
2. Camacho-Bello, C., Rivera-Lopez, J.S.: Some computational aspects of Tchebichef moments for higher orders. Pattern Recognit. Lett. **112**, 332–339 (2018)
3. Honarvar, B., Flusser, J.: Fast computation of Krawtchouk moments. Inf. Sci. **288**, 73–86 (2014)
4. Abdulhussain, S.H., Ramli, A.R., Al-Haddad, S.A.R., Mahmmod, B.M., Jassim, W.A.: Fast recursive computation of Krawtchouk polynomials. J. Math. Imaging Vis. **60**(3), 285–303 (2018)
5. Yap, P.T., Paramesran, R., Ong, S.H.: Image analysis using Hahn moments. IEEE Trans. Pattern Anal. Mach. Intell. **29**(11), 2057–2062 (2007)
6. Sayyouri, M., Hmimid, A., Qjidaa, H.: A fast computation of novel set of Meixner invariant moments for image analysis. Circuits Syst. Signal Process. **34**(3), 875–900 (2015)
7. Karmouni, H., Hmimid, A., Jahid, T., Sayyouri, M., Qjidaa, H., Rezzouk, A.: Fast and stable computation of the Charlier moments and their inverses using digital filters and image block representation. Circuits Syst. Signal Process., 1–19 (2018)
8. Zitová, B., Flusser, J.: Invariants to convolution and rotation. In: Rodrigues, M.A. (ed.) Invariants for Pattern Recognition and Classification, pp. 23–46. World Scientific (2000)
9. Zhu, H., Liu, M., Shu, H., Zhang, H., Luo, L.: General form for obtaining discrete orthogonal moments. IET Image Process. **4**, 335–352 (2010)
10. Teague, M.R.: Image analysis via the general theory of moments. J. Opt. Soc. Am. **70**(8), 920–930 (1980)
11. Strang, G.: Linear Algebra and Its Applications. Academic Press, New York (1980)
12. LNCS Homepage. https://www.physionet.org/physiobank/database/mitdb/

# $H_\infty$ Performance and Filtering for 2-D Discrete Systems with Time-Varying Delays

Oubaidi Mohamed(✉), Zakaria Chalh, and Mohammed Alfidi

Laboratory of Engineering, Systems and Applications, National School of Applied Sciences, Sidi Mohamed Ben Abdellah University, Fez, Morocco
{mohamed.oubaidi,zakaria.chalh}@usmba.ac.ma, alfidi_mohammed@yahoo.fr

**Abstract.** This paper is concerned with delay-dependant $H_\infty$ performance and $H_\infty$ filtering for 2-D discrete systems with time-varying delays, described by the Fornasini-Marchesini (FM) second model. By using the free weighting matrices and Lyapunov functional approaches, sufficient conditions are provided for the stability of the system via linear matrix inequality (LMI) feasibility. The delay-dependant $H_\infty$ performance is later tested to assure the stability of the said system. Based on the derived $H_\infty$ performance analysis results, the $H_\infty$ filter is designed. This work is an extended approach from 1-D to 2-D type systems. A numerical example is presented to demonstrate the effectiveness and benefits of the result obtained in this study.

**Keywords:** 2-D discrete time-delayed systems · $H_\infty$ performance · $H_\infty$ filtering · FM second model · LMI

## 1 Introduction

The two-dimensional (2-D) systems have widely attracted interest with their theoretical importance such as linear repetitive control [22] and iterative learning control [17,21] and with a practical significance in process control and image processing [19].

It has been well known that time-delays are inevitable in practical systems, especially in the 2-D class due to the finite speed of information processing and data transmission among various parts of the system. The time-delay often degrades the system performance and even causes the system instability. Therefore, the highlighting of stability of time-delay systems plays an important role in applied models and has been greatly studied in control theory and signal processing fields [23,24].

The $H_\infty$ technique introduced in [8] has attracted the attention of many researchers, for example [6,13,14,27,30]. It's a technique well known in literature to minimize the impact of perturbations on systems.

Over the past decades, considerable attention has been devoted to the problem of state estimation. When a priori information on the external noises is not


A. Saka et al. (Eds.): CPI 2019, LNME, pp. 103–119, 2021.
https://doi.org/10.1007/978-3-030-62199-5_11

precisely known, although the famed Kalman filtering provides an optimal state estimation approach in the sense of error variance, it has been acknowledge that the traditional Kalman filter is considerably sensitive to system parameters [23] but not robust enough against large uncertainties. Therefore, many attempts and efforts have been made to other more robust filtering arrangements, among which, $H_\infty$ filtering is the most concerned one [1–4,9,10,15,26,29]. In which the input signal is assumed to be energy bounded and the main objective is to minimize the $H_\infty$ norm of the filtering error system.

In this paper, by using the free-weighting matrix approach, proposed in [25] and by constructing a Lyapunov functional [20] on a 2-D system described by Fornasini-Marchesini (FM) second model [7,11,12] , a delay-dependant $H_\infty$ performance analysis is established for error systems by retaining some useful terms from the difference of Lyapunov functions. As a result, the $H_\infty$ filter is designed in terms of linear matrix inequalities(LMIs).

This paper is adjusted to five sections: In Sect. 2, the problem under study is formulated. In Sect. 3, new criterion is obtained in terms of LMI, which ensure the $H_\infty$ performance of the 2-D discrete system described by the FM second model and the filtering design is thus established. Numerical example is given to highlight the results in Sect. 4. Finally, some conclusions are also provided in Sect. 5.

**Notations.** Throughout the paper, $\mathbb{R}^p$ denotes the p-dimensional real Euclidean space, $\mathbb{R}^{p\times q}$ denotes the set of all $p \times q$ matrices. **0** and $I$ represent zero matrix identity matrix respectively. $diag\{...\}$ denotes a block-diagonal matrix in symmetric block matrices or long matrix expressions. $X^T$ stand for the transpose and the matrix $X$. $Q > 0$ ($Q < 0$) means that $Q$ is real symmetric and positive (negative) definite matrix. The notation $||x||$ stands for the Euclidean norm of the vector x.

## 2 Problem Formulation

Consider the 2-D system with time-delays described by the following FM second model [11]:

$$\begin{aligned} x(s_1+1,s_2+1) &= A_1x(s_1+1,s_2) + A_2x(s_1,s_2+1) + A_{1d}x(s_1+1,s_2-d_j) \\ &\quad + A_{2d}x(s_1-d_i,s_2+1) + B_1w(s_1+1,s_2) + B_2w(s_1,s_2+1) \\ y(s_1,s_2) &= Cx(s_1,s_2) + Dw(s_1,s_2) \\ z(s_1,s_2) &= Lx(s_1,s_2) \end{aligned} \tag{1}$$

where $x(s_1,s_2) \in \mathbb{R}^n$ is the state vector, $z(s_1,s_2) \in \mathbb{R}^m$ the signal to be estimated, $y(s_1,s_2) \in \mathbb{R}^l$ the measurement output vector, $w(s_1,s_2) \in \mathbb{R}^s$ is the disturbance input. The matrices $A_1$, $A_2$, $A_{1d}$, $A_{2d}$, $B_1$, $B_2$, $C$, $D$ and $L$ are constant matrices with appropriate dimensions, $d_i$ and $d_j$ are positive integers representing varying delays along horizontal and vertical directions, respectively. Where, $\tau_1 \le d_j \le \tau_2$ and $\tau_3 \le d_i \le \tau_4$ with $\tau_1$, $\tau_2$, $\tau_3$ and $\tau_4$ are known positive integers. Denote $\tau_v = \tau_2 - \tau_1$ and $\tau_h = \tau_4 - \tau_3$.

The boundary conditions for the system are described as

$$\begin{cases} x(s_1,s_2)=\varphi_{s_1,s_2}, & \forall s_1 \geqslant 0, s_2 \in [-\tau_2,0] \\ x(s_1,s_2)=\psi_{s_1,s_2}, & \forall s_2 \geqslant 0, s_1 \in [-\tau_4,0] \\ \psi_{0,0}=\varphi_{0,0}. \end{cases} \tag{2}$$

In what follows, the boundary conditions assumed to satisfy:

$$\sum_{s_1=0}^{\infty}\sum_{s_2=-\tau_2}^{0}\varphi_{s_1,s_2}^T\varphi_{s_1,s_2}<\infty, \quad \sum_{s_2=0}^{\infty}\sum_{s_1=-\tau_4}^{0}\psi_{s_1,s_2}^T\psi_{s_1,s_2}<\infty$$

The aim of this paper is to design a linear asymptotically stable filter for system (1) with state-space realization of the form:

$$\begin{aligned} \hat{x}(s_1+1,s_2+1) &= A_{1f}\hat{x}(s_1+1,s_2)+A_{2f}\hat{x}(s_1,s_2+1) \\ &\quad + B_{1f}y(s_1+1,s_2)+B_{2f}y(s_1,s_2+1) \\ \hat{z}(s_1,s_2) &= C_f\hat{x}(s_1,s_2) \end{aligned} \tag{3}$$

$\hat{x}(s_1,s_2)\in\mathbb{R}^p$ is the state vector of the filter and $\hat{z}(s_1,s_2)\in\mathbb{R}^q$ is the estimation of $z(s_1,s_2)$, $A_{1f}$, $A_{2f}$, $B_{1f}$, $B_{2f}$ and $C_f$ filter matrices to be determined. Denote:

$$\xi(s_1,s_2)=\begin{bmatrix} x(s_1,s_2) \\ \hat{x}(s_1,s_2) \end{bmatrix}$$

$$z_e(s_1,s_2)=z(s_1,s_2)-\hat{z}(s_1,s_2)$$

The filtering error dynamics for (1) is obtained and is as follows:

$$\begin{aligned} \xi(s_1+1,s_2+1) &= \bar{A}_1\xi(s_1+1,s_2)+\bar{A}_2\xi(s_1,s_2+1)+\bar{A}_{1d}E\xi(s_1+1,s_2-d_j) \\ &\quad +\bar{A}_{2d}E\xi(s_1-d_i,s_2+1)+\bar{B}_1w(s_1+1,s_2)+\bar{B}_2w(s_1,s_2+1) \\ z_e(s_1,s_2) &= \bar{C}\xi(s_1,s_2) \end{aligned} \tag{4}$$

where

$$\bar{A}_i=\begin{bmatrix} A_i & A_{if} \\ B_{if}C & 0 \end{bmatrix}, \quad \bar{A}_{id}=\begin{bmatrix} A_{id} \\ 0 \end{bmatrix}, \quad \bar{B}_i=\begin{bmatrix} B_i \\ B_{if}D \end{bmatrix}, \quad for \quad i=1,2$$

also

$$\bar{C}=\begin{bmatrix} L & -C_f \end{bmatrix} \quad and \quad E=\begin{bmatrix} I \, 0 \end{bmatrix}$$

By considering the zero initial conditions, the $H_\infty$ norm of the system in (1) is given by:

$$||G(z_1, z_2)||_\infty = sup_{w_1,w_2 \in [0,2\pi]} \sigma_{max}[G(e^{jw_1}, e^{jw_2})]$$

where $\sigma_{max}$ denotes the maximum singular value of the corresponding matrix, and

$$G(z_1, z_2) = \bar{C}(z_1 z_2 I - \bar{A}_1 z_1 - \bar{A}_2 z_2 - \bar{A}_{1d} z_1 z_2^{-d_j} - \bar{A}_{2d} z_1^{-d_i} z_2)^{-1} (\bar{B}_2 z_2 + \bar{B}_1 z_1)$$

is the transfer function from the disturbance input $\bar{w}(s_1, s_2)$ to the filtering error $\bar{z}_e(s_1, s_2)$ for the system in (1).

To get the main results of this paper, the following definition and lemmas are needed.

**lemma 1.** [5] For given symmetric matrices

$$S = S^T = \begin{bmatrix} S_{11} & S_{12} \\ * & S_{22} \end{bmatrix}$$

where $S_{11}$ and $S_{22}$ are square matrices, the following conditions are equivalent

1. $S < 0$;
2. $S_{11} < 0, \quad S_{22} - S_{12}^T S_{11}^{-1} S_{12} < 0$;
3. $S_{22} < 0, \quad S_{11} - S_{12}^T S_{22}^{-1} S_{12} < 0$;

**Definition 1.** *[28] The 2-D system given in* (1) *with zero boundary conditions in* (2) *is said to have $H_\infty$ disturbance attenuation $\gamma$ if it is asymptotically stable and:*

$$||\bar{z}_e(s_1, s_2)||_2 < \gamma ||\bar{w}(s_1, s_2)||_2 \tag{5}$$

for any non zero $w(s_1, s_2) \in L_2\{[0, \infty), [0, \infty)\}$ where

$$\bar{z}_e(s_1, s_2) = \left[ z_e^T(s_1 + 1, s_2) \; z_e^T(s_1, s_2 + 1) \right]^T$$
$$\bar{w}(s_1, s_2) = \left[ w^T(s_1 + 1, s_2) \; w^T(s_1, s_2 + 1) \right]^T$$

## 3 Main Results

In this section, the main results are established with investigating the stability and the $H_\infty$ performance analysis for 2-D systems with time-varying delays, which will play an important role in solving the $H_\infty$ filtering problem next.

### 3.1 $H_\infty$ Performance Analysis

In this subsection, the $H_\infty$ performance analysis problem of the 2-D system (1) is given, then it will be employed next to design the $H_\infty$ filter. For this case the theorem is given as below.

**Theorem 1.** *Given integers $\tau_1 \le d_j \le \tau_2$ and $\tau_3 \le d_i \le \tau_4$ and scalar $\gamma > 0$ the system (1) with time varying $d_i$ and $d_j$ satisfying initial conditions given in (2) is asymptotically stable for all nonzero $w(s_1, s_2) \in L_2\{[0,\infty), [0,\infty)\}$ and (5) is satisfied if there exist appropriately dimensioned matrices $P = P_a + P_b = P^T > 0$, for $i = 1,2, j = 1,2$, $Q_{ij} = Q_{ij}^T > 0$, $Z_{1j} = Z_{1j}^T > 0$, $Z_{2j} = Z_{2j}^T > 0$, for $j = h, v$ $X_j = \begin{bmatrix} X_{11j} & X_{12j} \\ * & X_{22j} \end{bmatrix} \ge 0$, $Y_j = \begin{bmatrix} Y_{11j} & Y_{12j} \\ * & Y_{22j} \end{bmatrix} \ge 0$, for $j = 1,2$ $N_j = \begin{bmatrix} N_{1j} \\ N_{2j} \end{bmatrix}$, $M_j = \begin{bmatrix} M_{1j} \\ M_{2j} \end{bmatrix}$, $S_j = \begin{bmatrix} S_{1j} \\ S_{2j} \end{bmatrix}$, such that the following LMIs hold:*

$$\beta = \begin{bmatrix} \beta_{11} & \beta_{12} & \beta_{13} & \beta_{14} & \beta_{15} \\ * & \beta_{22} & \beta_{23} & \beta_{24} & \beta_{25} \\ * & * & \beta_{33} & \beta_{34} & \beta_{35} \\ * & * & * & \beta_{44} & \beta_{45} \\ * & * & * & * & \beta_{55} \end{bmatrix} < 0 \tag{6}$$

*and*

$$\Psi_{1j} = \begin{bmatrix} Xi & N_j \\ * & Z_{1j} \end{bmatrix} \ge 0, \quad \Psi_{2j} = \begin{bmatrix} Yi & M_j \\ * & Z_{2j} \end{bmatrix} \ge 0, \quad \Psi_{3j} = \begin{bmatrix} Xi + Yi & S_j \\ * & Z_{1j} + Z_{2j} \end{bmatrix} \ge 0, i = h, v, j = 1,2$$

*where*

$$\begin{aligned} \beta_{11} &= \phi, \quad \beta_{12} = \left[ \sqrt{\tau_4}\phi_1^T Z_{11} \ \sqrt{\tau_h}\phi_1^T Z_{21} \right], \quad \beta_{13} = \left[ \sqrt{\tau_2}\phi_2^T Z_{12} \ \sqrt{\tau_v}\phi_2^T Z_{22} \right] \\ \beta_{14} &= \left[ \phi_{e1}^T \ \phi_{e2}^T \right], \quad \beta_{15} = \left[ \phi_3^T P_a \ \phi_3^T P_b \right], \quad \beta_{22} = \begin{bmatrix} -Z_{11} & 0 \\ 0 & -Z_{21} \end{bmatrix} \\ \beta_{23} &= \beta_{24} = \beta_{25} = \beta_{34} = \beta_{35} = \beta_{45} = \begin{bmatrix} 0 & 0 \\ 0 & 0 \end{bmatrix} \\ \beta_{33} &= \begin{bmatrix} -Z_{12} & 0 \\ 0 & -Z_{22} \end{bmatrix}, \quad \beta_{44} = \begin{bmatrix} -I & 0 \\ 0 & -I \end{bmatrix}, \quad \beta_{55} = \begin{bmatrix} -P_a & 0 \\ 0 & -P_b \end{bmatrix} \end{aligned}$$

*The zero matrix is appropriately dimensioned.*

$$\phi = \begin{bmatrix} \pi_1 & \pi_2 & \pi_3 & \pi_4 & \pi_5 \\ * & \pi_6 & \pi_7 & \pi_8 & \pi_9 \\ * & * & \pi_{10} & \pi_{11} & \pi_{12} \\ * & * & * & \pi_{13} & \pi_{14} \\ * & * & * & * & \pi_{15} \end{bmatrix}$$

$$\pi_1 = \begin{bmatrix} \phi_{v1} & 0 \\ * & \phi_{h1} \end{bmatrix}, \quad \pi_2 = \begin{bmatrix} \phi_{v2} & 0 \\ * & \phi_{h2} \end{bmatrix}, \quad \pi_3 = \begin{bmatrix} E^T S_{12} & -E^T M_{12} \\ 0 & 0 \end{bmatrix}$$
$$\pi_4 = \begin{bmatrix} 0 & 0 \\ E^T S_{11} & -E^T M_{11} \end{bmatrix}, \quad \pi_6 = \begin{bmatrix} \phi_{v3} & 0 \\ * & \phi_{h3} \end{bmatrix}, \quad \pi_7 = \begin{bmatrix} S_{22} & M_{22} \\ 0 & 0 \end{bmatrix}$$
$$\pi_8 = \begin{bmatrix} 0 & 0 \\ S_{21} & M_{21} \end{bmatrix}, \quad \pi_{10} = \begin{bmatrix} -Q_{12} & 0 \\ * & -Q_{22} \end{bmatrix}, \quad \pi_{13} = \begin{bmatrix} -Q_{11} & 0 \\ * & -Q_{21} \end{bmatrix}$$
$$\pi_{15} = \begin{bmatrix} -\gamma^2 I & 0 \\ 0 & -\gamma^2 I \end{bmatrix}, \quad \pi_r = \begin{bmatrix} 0 & 0 \\ 0 & 0 \end{bmatrix}, for \quad r = 5, 9, 11, 12, 14$$

*with*

$$\begin{aligned}
\phi_{v1} &= E^T(N_{12} + N_{12}^T + (\tau_v + 1)Q_{32} + Q_{12} + Q_{22} + \tau_2 X_{11b} + \tau_v Y_{11b})E - P_b \\
\phi_{v2} &= E^T(N_{22}^T - N_{12} + M_{12} - S_{12} + \tau_2 X_{12b} + \tau_v Y_{12b}) \\
\phi_{v3} &= -N_{22} - N_{22}^T - S_{22} - S_{22}^T + M_{22} + M_{22}^T - Q_{32} + \tau_2 X_{22b} + \tau_v Y_{22b} \\
\phi_{h1} &= E^T(N_{11} + N_{11}^T + (\tau_h + 1)Q_{31} + Q_{11} + Q_{21} + \tau_4 X_{11a} + \tau_h Y_{11a})E - P_a \\
\phi_{h2} &= E^T(N_{21}^T - N_{11} + M_{11} - S_{11} + \tau_4 X_{12a} + \tau_h Y_{12a}) \\
\phi_{h3} &= -N_{21} - N_{21}^T - S_{21} - S_{21}^T + M_{21} + M_{21}^T - Q_{31} + \tau_4 X_{22a} + \tau_h Y_{22a}
\end{aligned}$$

*also*

$$\begin{aligned}
\phi_1 &= \begin{bmatrix} (A_1 - I)E \; A_2 E \; A_{1d} \; A_{2d} \; 0\,0\,0\,0 \; B_1 \; B_2 \end{bmatrix} \\
\phi_2 &= \begin{bmatrix} A_1 E \; (A_2 - I)E \; A_{1d} \; A_{2d} \; 0\,0\,0\,0 \; B_1 \; B_2 \end{bmatrix} \\
\phi_3 &= \begin{bmatrix} \bar{A}_1 \; \bar{A}_2 \; \bar{A}_{1d} \; \bar{A}_{2d} \; 0\,0\,0\,0 \; \bar{B}_1 \; \bar{B}_2 \end{bmatrix} \\
\phi_{e1} &= \begin{bmatrix} \bar{C} \, 0\,0\,0\,0\,0\,0\,0\,0\,0 \end{bmatrix} \\
\phi_{e2} &= \begin{bmatrix} 0 \, \bar{C} \, 0\,0\,0\,0\,0\,0\,0\,0 \end{bmatrix}
\end{aligned}$$

**Proof**. let

$$\begin{aligned}
\eta_1(s_1, s_2) &= x(s_1 + 1, s_2 + 1) - x(s_1, s_2 + 1) \qquad (7) \\
\eta_1(s_1, s_2) &= A_1 E\xi(s_1 + 1, s_2) + (A_2 - I)E\xi(s_1, s_2 + 1) \\
&+ A_{1d}x(s_1 + 1, s_2 - d_j) + A_{2d}x(s_1 - d_i, s_2 + 1) \\
&+ B_1 w(s_1 + 1, s_2) + B_2 w(s_1, s_2 + 1) \\
\xi_{sys}(s_1, s_2) &= \begin{bmatrix} \epsilon_1 \; \epsilon_2 \; \epsilon_3 \end{bmatrix}^T \\
\epsilon_1 &= \begin{bmatrix} \xi^T(s_1 + 1, s_2) \; \xi^T(s_1, s_2 + 1) \; x^T(s_1 + 1, s_2 - d_j) \; x^T(s_1 - d_i, s_2 + 1) \end{bmatrix} \\
\epsilon_2 &= \big[ x^T(s_1 + 1, s_2 - \tau_1) \, x^T(s_1 + 1, s_2 - \tau_2) \, x^T(s_1 - \tau_3, s_2 + 1) \\
&\quad x^T(s_1 - \tau_4, s_2 + 1) \big] \\
\epsilon_3 &= \begin{bmatrix} w^T(s_1 + 1, s_2) \; w^T(s_1, s_2 + 1) \end{bmatrix} \\
\Rightarrow \quad \eta_1(s_1, s_2) &= \phi_1 \xi_{sys}(s_1, s_2) \\
\Rightarrow \quad \eta_2(s_1, s_2) &= \phi_2 \xi_{sys}(s_1, s_2)
\end{aligned}$$

Introducing Lyapunov functions:

$$\begin{aligned} v(s_1,s_2) &= v_1(s_1,s_2)+v_2(s_1,s_2)+v_3(s_1,s_2)+v_4(s_1,s_2)\\ v(s_1,s_2) &= v_{11}(s_1,s_2)+v_{12}(s_1,s_2)+v_{21}(s_1,s_2)+v_{22}(s_1,s_2)+v_{31}(s_1,s_2)\\ &\quad + v_{32}(s_1,s_2)+v_{41}(s_1,s_2)+v_{42}(s_1,s_2) \end{aligned}$$

with

$$\begin{aligned} v_1(s_1,s_2) &= \xi^T(s_1,s_2)P_a\xi(s_1,s_2)+\xi^T(s_1,s_2)P_b\xi(s_1,s_2)\\ v_{21}(s_1,s_2) &= \sum_{\theta=-\tau_4+1}^{0}\sum_{l=s_1-1+\theta}^{s_1-1}\eta_1^T(l,s_2)Z_{11}\eta_1(l,s_2)+\sum_{\theta=-\tau_4+1}^{-\tau_3}\sum_{l=s_1-1+\theta}^{s_1-1}\eta_1^T(l,s_2)Z_{21}\eta_1(l,s_2)\\ v_{22}(s_1,s_2) &= \sum_{\theta=-\tau_2+1}^{0}\sum_{l=s_2-1+\theta}^{s_2-1}\eta_2^T(s_1,l)Z_{12}\eta_2(s_1,l)+\sum_{\theta=-\tau_2+1}^{-\tau_1}\sum_{l=s_2-1+\theta}^{s_2-1}\eta_2^T(s_1,l)Z_{22}\eta_2(s_1,l)\\ v_{31}(s_1,s_2) &= \sum_{l=s_1-\tau_3}^{s_1-1}x^T(l,s_2+1)Q_{11}x(l,s_2+1)+\sum_{l=s_1-\tau_4}^{s_1-1}x^T(l,s_2+1)Q_{21}x(l,s_2+1)\\ v_{32}(s_1,s_2) &= \sum_{l=s_2-\tau_2}^{s_2-1}x^T(s_1+1,l)Q_{12}x(s_1+1,l)+\sum_{l=s_2-\tau_1}^{s_2-1}x^T(s_1+1,l)Q_{22}x(s_1+1,l)\\ v_{41}(s_1,s_2) &= \sum_{\theta=1-\tau_4}^{1-\tau_3}\sum_{l=s_1-1+\theta}^{s_1-1}x^T(l,s_2+1)Q_{31}x(l,s_2+1)\\ v_{42}(s_1,s_2) &= \sum_{\theta=1-\tau_2}^{1-\tau_1}\sum_{l=s_2-1+\theta}^{s_2-1}x^T(s_1+1,l)Q_{32}x(s_1+1,l) \end{aligned}$$

where $P=P_a+P_b=P^T>0$, $Q_{ij}=Q_{ij}^T>0$, i$=1,2$, j$=1,2$, $Z_{1j}=Z_{1j}^T>0$, $Z_{2j}=Z_{2j}^T>0$ for j$=1,2$ are to be determined. Defining $\Delta V(s_1+1,s_2)=V(s_1+1,s_2)-V(s_1,s_2)$ and $\Delta V(s_1,s_2+1)=V(s_1,s_2+1)-V(s_1,s_2)$ yields

$$\begin{aligned} \Delta_{v_{11}(s_1,s_2)} &= \xi_{sys}^T(s_1+1,s_2)P_a\xi_{sys}(s_1+1,s_2)-\xi_{sys}^T(s_1,s_2)\phi_3^TP_a\phi_3\xi_{sys}(s_1,s_2)\\ \Delta_{v_{12}(s_1,s_2)} &= \xi_{sys}^T(s_1,s_2+1)P_b\xi_{sys}(s_1,s_2+1)-\xi_{sys}^T(s_1,s_2)\phi_3^TP_b\phi_3\xi_{sys}(s_1,s_2) \end{aligned}$$

$$\begin{aligned} \Delta_{v_{21}(s_1,s_2)} &= \xi_{sys}^T(s_1,s_2)\phi_1^T(\tau_4Z_{11}+\tau_hZ_{21})\phi_1\xi_{sys}(s_1,s_2)-\sum_{l=s_1-d_i}^{s_1-1}\eta_1^T(l,s_2)Z_{11}\eta_1(l,s_2)\\ &\quad -\sum_{l=s_1-d_i}^{s_1-1-\tau_3}\eta_1^T(l,s_2)Z_{21}\eta_1(l,s_2)-\sum_{l=s_1-\tau_4}^{s_1-1-d_i}\eta_1^T(l,s_2)(Z_{21}+Z_{11})\eta_1(l,s_2)\\ \Delta_{v_{22}(s_1,s_2)} &= \xi_{sys}^T(s_1,s_2)\phi_2^T(\tau_2Z_{12}+\tau_vZ_{22})\phi_2\xi_{sys}(s_1,s_2)-\sum_{l=s_2-d_j}^{s_2-1}\eta_2^T(s_1,l)Z_{12}\eta_2(s_1,l)\\ &\quad -\sum_{l=s_2-d_j}^{s_2-1-\tau_1}\eta_2^T(s_1,l)Z_{22}\eta_2(s_1,l)-\sum_{l=s_2-\tau_2}^{s_2-1-d_j}\eta_2^T(s_1,l)(Z_{22}+Z_{12})\eta_2(s_1,l) \end{aligned}$$

$$\begin{aligned} \Delta_{v_{31}(s_1,s_2)} &= \xi^T(s_1,s_2+1)(Q_{21}+Q_{11})\xi(s_1,s_2+1)-x^T(s_1-\tau_3,s_2+1)Q_{11}x(s_1-\tau_3,s_2+1)\\ &\quad -x^T(s_1-\tau_4,s_2+1)Q_{21}x(s_1-\tau_4,s_2+1)\\ \Delta_{v_{32}(s_1,s_2)} &= \xi^T(s_1+1,s_2)(Q_{22}+Q_{12})\xi(s_1+1,s_2)-x^T(s_1+1,s_2-\tau_1)Q_{12}x(s_1+1,s_2-\tau_1)\\ &\quad -x^T(s_1+1,s_2-\tau_2)Q_{22}x(s_1+1,s_2-\tau_2) \end{aligned}$$

$$
\begin{aligned}
\Delta_{v41(s_1,s_2)} &= (\tau_h+1)x^T(s_1,s_2+1)Q_{31}x(s_1,s_2+1) - \sum_{l=s_1-\tau_4}^{s_1-\tau_3} x^T(l,s_2+1)Q_{31}x(l,s_2+1)\\
&< (\tau_h+1)\xi^T(s_1,s_2+1)E^TQ_{31}E\xi(s_1,s_2+1) - x^T(s_1-d_i,s_2+1)Q_{31}x(s_1-d_i,s_2+1)\\
\Delta_{v42(s_1,s_2)} &= (\tau_v+1)x^T(s_1+1,s_2)Q_{32}x(s_1+1,s_2) - \sum_{l=s_2-\tau_2}^{s_2-\tau_1} x^T(s_1+1,l)Q_{32}x(s_1+1,l)\\
&< (\tau_v+1)\xi^T(s_1+1,s_2)E^TQ_{32}E\xi(s_1+1,s_2) - x^T(s_1+1,s_2-d_j)Q_{31}x(s_1+1,s_2-d_j)
\end{aligned}
$$

From (7), we have
$\eta_1(s_1,s_2) = x(s_1+1,s_2+1) - x(s_1,s_2+1)$
$\Rightarrow$

$$
\sum_{l=s_1-d_i}^{s_1-1} \eta_1(l,s_2) = \sum_{l=s_1-d_i}^{s_1-1} x(l+1,s_2+1) - \sum_{l=s_1-d_i}^{s_1-1} x(l,s_2+1)
$$

$\Rightarrow$

$$
\begin{aligned}
0 &= E\xi(s_1,s_2+1) - x(s_1-d_i,s_2+1) - \sum_{l=s_1-d_i}^{s_1-1} \eta_1(l,s_2)\\
0 &= x(s_1-d_i,s_2+1) - x(s_1-\tau_4,s_2+1) - \sum_{l=s_1-\tau_4}^{s_1-d_i-1} \eta_1(l,s_2)\\
0 &= x(s_1-\tau_3,s_2+1) - x(s_1-d_i,s_2+1) - \sum_{l=s_1-d_i}^{s_1-\tau_3-1} \eta_1(l,s_2)
\end{aligned}
$$

The same for vertical direction

$$
\begin{aligned}
0 &= E\xi(s_1+1,s_2) - x(s_1+1,s_2-d_j) - \sum_{l=s_2-d_j}^{s_2-1} \eta_2(s_1,l)\\
0 &= x(s_1+1,s_2-d_j) - x(s_1+1,s_2-\tau_2) - \sum_{l=s_2-\tau_2}^{s_2-d_j-1} \eta_2(s_1,l)\\
0 &= x(s_1+1,s_2-\tau_1) - x(s_1+1,s_2-d_j) - \sum_{l=s_2-d_j}^{s_2-\tau_1-1} \eta_2(s_1,l)
\end{aligned}
$$

Then the following equations hold for any matrices $N_1$, $N_2$, $M_1$, $M_2$, $S_1$ and $S_2$ with appropriate dimensions $for \Delta V(s_1,s_2+1):$

$$
\begin{aligned}
0 = 2 \times [&\xi^T(s_1,s_2+1)E^TN_{11} + x^T(s_1-d_i,s_2+1)N_{21}\\
&\times [E\xi(s_1,s_2+1) - x(s_1-d_i,s_2+1) - \sum_{l=s_1-d_i}^{s_1-1} \eta_1(l,s_2)]
\end{aligned}
\qquad (8)
$$

$$0 = 2 \times [\xi^T(s_1, s_2+1)E^T M_{11} + x^T(s_1 - d_i, s_2+1)M_{21}] \times [x(s_1 - d_i, s_2+1) - x(s_1 - \tau_4, s_2+1) - \sum_{l=s_1-\tau_4}^{s_1-d_i-1} \eta_1(l, s_2)] \tag{9}$$

$$0 = 2 \times [\xi^T(s_1, s_2+1)E^T S_{11} + x(s_1 - d_i, s_2+1)^T S_{21} \times [x(s_1 - \tau_3, s_2+1) - x(s_1 - d_i, s_2+1) - \sum_{l=s_1-d_i}^{s_1-\tau_3-1} \eta_1(l, s_2)] \tag{10}$$

The same thing for $\Delta V(s_1+1, s_2)$:

$$0 = 2 \times [\xi^T(s_1+1, s_2)E^T N_{12} + x^T(s_1+1, s_2 - d_j)N_{22} \times [E\xi(s_1+1, s_2) - x(s_1+1, s_2 - d_j) - \sum_{l=s_2-d_j}^{s_2-1} \eta_2(s_1, l)] \tag{11}$$

$$0 = 2 \times [\xi^T(s_1+1, s_2)E^T M_{12} + x^T(s_1+1, s_2 - d_j)M_{22}] \times [x(s_1+1, s_2 - d_j) - x(s_1+1, s_2 - \tau_2) - \sum_{l=s_2-\tau_2}^{s_2-d_j-1} \eta_2(s_1, l)] \tag{12}$$

$$0 = 2 \times [\xi^T(s_1+1, s_2)E^T S_{12} + x(s_1+1, s_2 - d_j)^T S_{22} \times [x(s_1+1, s_2 - \tau_1) - x(s_1+1, s_2 - d_j) - \sum_{l=s_2-d_j}^{s_2-\tau_1-1} \eta_2(s_1, l)] \tag{13}$$

In the other hand for all the matrices with appropriate dimensions: $X_i = X_i^T \geq 0$, $Y_i = Y_i^T \geq 0$, for $i = h, v$ The following equations are true:
for $\Delta V(s_1, s_2+1)$

$$0 = \tau_4 \xi_1^T(s_1, s_2) X_h \xi_1(s_1, s_2) - \sum_{l=s_1-d_i}^{s_1-1} \xi_1^T(s_1, s_2) X_h \xi_1(s_1, s_2) - \sum_{l=s_1-\tau_4}^{s_1-d_i-1} \xi_1^T(s_1, s_2) X_h \xi_1(s_1, s_2) \tag{14}$$

$$0 = \tau_h \xi_1^T(s_1, s_2) Y_h \xi_1(s_1, s_2) - \sum_{l=s_1-d_i}^{s_1-\tau_3-1} \xi_1^T(s_1, s_2) Y_h \xi_1(s_1, s_2) - \sum_{l=s_1-\tau_4}^{s_1-d_i-1} \xi_1^T(s_1, s_2) Y_h \xi_1(s_1, s_2) \tag{15}$$

for $\Delta V(s_1+1, s_2)$

$$0 = \tau_2 \xi_2^T(s_1, s_2) X_v \xi_2(s_1, s_2) - \sum_{l=s_2-d_j}^{s_2-1} \xi_2^T(s_1, s_2) X_v \xi_2(s_1, s_2) - \sum_{l=s_2-\tau_2}^{s_2-d_j-1} \xi_2^T(s_1, s_2) X_v \xi_2(s_1, s_2) \tag{16}$$

$$0 = \tau_v \xi_2^T(s_1, s_2) Y_v \xi_2(s_1, s_2) - \sum_{l=s_2-d_j}^{s_2-\tau_1-1} \xi_2^T(s_1, s_2) Y_v \xi_2(s_1, s_2) - \sum_{l=s_2-\tau_2}^{s_2-d_j-1} \xi_2^T(s_1, s_2) Y_v \xi_2(s_1, s_2) \tag{17}$$

with

$$\xi_1(s_1, s_2) = \left[\xi^T(s_1, s_2+1)\ \xi^T(s_1-d_i, s_2+1)\right]^T$$
$$\xi_2(s_1, s_2) = \left[\xi^T(s_1+1, s_2)\ \xi^T(s_1+1, s_2-d_j)\right]^T$$

Then, if the terms of the right side of the equations (8)–(17) are added into $\Delta V(s_1, s_2) = \Delta V(s_1+1, s_2) + \Delta V(s_1, s_2+1)$, we have:

$$\begin{aligned} &\Delta V(s_1, s_2) + \bar{z}_e^T(s_1, s_2)\bar{z}_e(s_1, s_2) - \gamma^2 \bar{w}^T(s_1, s_2)\bar{w}(s_1, s_2) \le \xi_{sys}^T(s_1, s_2)(\phi + \phi_1^T(\tau_4 Z_{11} + \tau_h Z_{21})\phi_1 \\ &+ \phi_2^T(\tau_2 Z_{12} + \tau_v Z_{22})\phi_2 + \phi_{e1}^T\phi_{e1} + \phi_{e2}^T\phi_{e2} + \phi_3^T(P_a + P_b)\phi_3)\xi_{sys}(s_1, s_2) \\ &- \sum_{l=s_1-d_i}^{s_1-1} \xi_3^T(l, s_2)\Psi_{11}\xi_3(l, s_2) - \sum_{l=s_1-d_i}^{s_1-\tau_3-1} \xi_3^T(l, s_2)\Psi_{21}\xi_3(l, s_2) - \sum_{l=s_1-\tau_4}^{s_1-d_i-1} \xi_3^T(l, s_2)\Psi_{31}\xi_3(l, s_2) \\ &- \sum_{l=s_2-d_j}^{s_2-1} \xi_3^T(s_1, l)\Psi_{12}\xi_3(s_1, l) - \sum_{l=s_2-d_j}^{s_2-\tau_1-1} \xi_3^T(s_1, l)\Psi_{21}\xi_3(s_1, l) - \sum_{l=s_2-\tau_2}^{s_2-d_j-1} \xi_3^T(s_1, l)\Psi_{32}\xi_3(s_1, l) \end{aligned}$$

where

$$\xi_3(l, s_2) = \left[\xi_1^T(s_1, s_2)\ \eta_1^T(l, s_2)\right]^T$$
$$\xi_3(s_1, l) = \left[\xi_2^T(s_1, s_2)\ \eta_2^T(s_1, l)\right]^T$$

Thus if $\Psi_{1j}, \Psi_{2j} \ge 0$ *and* $\Psi_{3j} \ge 0$ *for* $j = 1, 2$ and

$$\begin{aligned} &(\phi + \phi_1^T(\tau_4 Z_{11} + \tau_h Z_{21})\phi_1 + \phi_2^T(\tau_2 Z_{12} + \tau_v Z_{22})\phi_2 \\ &+ \phi_{e1}^T\phi_{e1} + \phi_{e2}^T\phi_{e2} + \phi_3^T(P_a + P_b)\phi_3) < 0 \end{aligned}$$

which is equivalent to 6 by schur compliments, then

$$\Delta V(s_1,s_2) + \bar{z}_e^T(s_1,s_2)\bar{z}_e(s_1,s_2) - \gamma^2\bar{w}^T(s_1,s_2)\bar{w}(s_1,s_2) < 0$$

This ensures that (5) holds under zero-initial conditions for all nonzero $w(s_1,s_2) \in L_2\{[0,\infty),[0,\infty)\}$ and a prescribed $\gamma > 0$ following the similar line in [30]. On the other hand, (6) implies that the following matrix inequality (18) holds, which guarantees $\Delta V(s_1,s_2) < 0$, such that the system (1) with $w(s_1,s_2) = 0$ is asymptotically stable.

$$\beta = \begin{bmatrix} \beta_{11} & \beta_{12} & \beta_{13} & \beta_{15} \\ * & \beta_{22} & \beta_{23} & \beta_{25} \\ * & * & \beta_{33} & \beta_{35} \\ * & * & * & \beta_{55} \end{bmatrix} < 0 \tag{18}$$

where

$$\phi = \begin{bmatrix} \pi_1 & \pi_2 & \pi_3 & \pi_4 \\ * & \pi_6 & \pi_7 & \pi_8 \\ * & * & \pi_{10} & \pi_{11} \\ * & * & * & \pi_{13} \end{bmatrix}$$

$$\begin{aligned} \phi_1 &= \begin{bmatrix} (A_1 - I)E \; A_2E \; A_{1d} \; A_{2d} \; 0\,0\,0\,0 \end{bmatrix} \\ \phi_2 &= \begin{bmatrix} A_1E \; (A_2 - I)E \; A_{1d} \; A_{2d} \; 0\,0\,0\,0 \end{bmatrix} \\ \phi_3 &= \begin{bmatrix} \bar{A}_1 \; \bar{A}_2 \; \bar{A}_{1d} \; \bar{A}_{2d} \; 0\,0\,0\,0 \end{bmatrix} \end{aligned}$$

The proof is thus completed. □

*Remark 1.* The terms $\sum_{l=s_1-\bar{h}_1}^{s_1-d_i-1} \eta_1^T(l,s_2)Z_{s_1}\eta_1(l,s_2)$ and $\sum_{l=s_2-\bar{h}_2}^{s_2-d_j-1} \eta_2^T(s_1,l)Z_{s_2}\eta_2(s_1,l)$ are reserved in *Theorem*1 to overcome conservativeness. On the other hand, $\tau_2$ and $\tau_4$ are separated into two parts like $d_j$ and $\tau_2 - d_j$, $d_i$ and $\tau_4 - d_i$, for vertical and horizontal direction, respectively, in the procedure of proving *Theorem*1 thus showing the benefits of the proposed method.

### 3.2 Filter Design

In this subsection, based on the previous approaches, the $H_\infty$ filter design is formulated. The theorem is as follows.

**Theorem 2.** *Considering the 2-D system with time-varying delays (1) and given the positive scalar $\gamma > 0$, an admissible full-order filter of the forme (4) assuming a prescribed $H_\infty$ performance and the robust stability of the filtering error system exists if there exist matrices* $P_i^k = \begin{bmatrix} P_{i1}^k & P_{i2}^k \\ * & P_{i3}^k \end{bmatrix} > 0$, *for* $i = a, b,$

$Qij^k > 0$, *for* $i = 1,2,3, j = 1,2$, $Zij^k > 0$, *for* $i = 1,2, j = 1,2$ *and* $N_j^k = \begin{bmatrix} N_{1j}^k \\ N_{2j}^k \end{bmatrix}$, $M_j^k = \begin{bmatrix} M_{1j}^k \\ M_{2j}^k \end{bmatrix}$, $S_j^k = \begin{bmatrix} S_{1j}^k \\ S_{2j}^k \end{bmatrix}$, *for* $j = 1,2$, $X_i = \begin{bmatrix} X_{11i} \; X_{12i} \\ * \;\; X_{22i} \end{bmatrix} \geq 0$, $Y_i = \begin{bmatrix} Y_{11i} \; Y_{12i} \\ * \;\; Y_{22i} \end{bmatrix} \geq 0$, *for* $i = h, v$, $H_{ij}$, *for* $i = 1,2, j = 1,2$, $T_1$, $V_1$, $V_2$, $A_{1f}$, $A_{2f}$, $B_{1f}$, $B_{2f}$ *and* $C_f$ *are matrices with appropriate dimensions such that the following LMis are feasible for* $k \in 1, ...n$

$$\Psi_{1j} = \begin{bmatrix} X_i \; N_j^k \\ * \;\; Z_{1j}^k \end{bmatrix} \geq 0, \quad \Psi_{2j} = \begin{bmatrix} Y_i \; M_j^k \\ * \;\; Z_{2j}^k \end{bmatrix} \geq 0, \quad \Psi_{3j} = \begin{bmatrix} X_i + Y_i & S_j^k \\ * & Z_{1j}^k + Z_{2j}^k \end{bmatrix} \geq 0,$$
$$with \quad i = h, v, \quad j = 1,2 \tag{19}$$

also

$$\begin{bmatrix} \theta_{11}^k & \theta_{12}^k & \theta_{13}^k & \theta_{14}^k & \theta_{15}^k \\ * & \theta_{22}^k & \theta_{23}^k & \theta_{24}^k & \theta_{25}^k \\ * & * & \theta_{33}^k & \theta_{34}^k & \theta_{35}^k \\ * & * & * & \theta_{44}^k & \theta_{45}^k \\ * & * & * & * & \theta_{55}^k \end{bmatrix} < 0 \tag{20}$$

$$\theta_{11}^k = \phi^k, \quad \theta_{12}^k = \left[ \sqrt{\tau_4}\phi_1^T H_{11} \; \sqrt{\tau_h}\phi_1^T H_{21} \right], \quad \theta_{13}^k = \left[ \sqrt{\tau_2}\phi_2^T H_{12} \; \sqrt{\tau_v}\phi_2^T H_{22} \right]$$

$$\theta_{14}^k = \left[ \phi_{e1}^T \; \phi_{e2}^T \right], \quad \theta_{15}^k = \left[ \phi_3^T T \; \phi_3^T T \right], \quad \theta_{22}^k = \begin{bmatrix} -\upsilon_{11}^k & 0 \\ 0 & -\upsilon_{21}^k \end{bmatrix}$$

$$\theta_{23}^k = \theta_{24}^k = \theta_{25}^k = \theta_{34}^k = \theta_{35}^k = \theta_{45}^k = \begin{bmatrix} 0 \; 0 \\ 0 \; 0 \end{bmatrix}$$

$$\theta_{33}^k = \begin{bmatrix} -\upsilon_{12}^k & 0 \\ 0 & -\upsilon_{22}^k \end{bmatrix}, \quad \theta_{44}^k = \begin{bmatrix} -I & 0 \\ 0 & -I \end{bmatrix}, \quad \theta_{55}^k = \begin{bmatrix} -\upsilon_a^k & 0 \\ 0 & -\upsilon_b^k \end{bmatrix}$$

The zero matrix is appropriately dimensioned.

$$\phi^k = \begin{bmatrix} \pi_1^k & \pi_2^k & \pi_3^k & \pi_4^k & \pi_5^k \\ * & \pi_6^k & \pi_7 & \pi_8 & \pi_9 \\ * & * & \pi_{10}^k & \pi_{11}^k & \pi_{12}^k \\ * & * & * & \pi_{13}^k & \pi_{14}^k \\ * & * & * & * & \pi_{15}^k \end{bmatrix}$$

$$\pi_1^k = \begin{bmatrix} \phi_{v1}^k & 0 \\ * & \phi_{h1}^k \end{bmatrix}, \quad \pi_2^k = \begin{bmatrix} \phi_{v2}^k & 0 \\ * & \phi_{h2}^k \end{bmatrix}, \quad \pi_3^k = \begin{bmatrix} E^T S_{12}^k & -E^T M_{12}^k \\ 0 & 0 \end{bmatrix}$$

$$\pi_4^k = \begin{bmatrix} 0 & 0 \\ E^T S_{11}^k & -E^T M_{11}^k \end{bmatrix}, \quad \pi_6^k = \begin{bmatrix} \phi_{v3}^k & 0 \\ * & \phi_{h3}^k \end{bmatrix}, \quad \pi_7^k = \begin{bmatrix} S_{22}^k & M_{22}^k \\ 0 & 0 \end{bmatrix}$$

$$\pi_8^k = \begin{bmatrix} 0 & 0 \\ S_{21}^k & M_{21}^k \end{bmatrix}, \quad \pi_{10}^k = \begin{bmatrix} -Q_{12}^k & 0 \\ * & -Q_{22}^k \end{bmatrix}, \quad \pi_{13}^k = \begin{bmatrix} -Q_{11}^k & 0 \\ * & -Q_{21}^k \end{bmatrix}$$

$$\pi_{15}^k = \begin{bmatrix} -\gamma^2 I & 0 \\ 0 & -\gamma^2 I \end{bmatrix}, \quad \pi_l^k = \begin{bmatrix} 0\,0 \\ 0\,0 \end{bmatrix}, for \quad l = 5, 9, 11, 12, 14$$

with

$$\begin{aligned}
\phi_{v1}^k &= E^T(N_{12}^k + N_{12}^{kT} + (\tau_v + 1)Q_{32}^k + Q_{12}^k + Q_{22}^k + \tau_2 X_{11b} + \tau_v Y_{11b})E - P_b^k \\
\phi_{v2}^k &= E^T(N_{22}^{kT} - N_{12}^k + M_{12}^k - S_{12}^k + \tau_2 X_{12b} + \tau_v Y_{12b}) \\
\phi_{v3}^k &= -N_{22}^k - N_{22}^{kT} - S_{22}^k - S_{22}^{kT} + M_{22}^k + M_{22}^{kT} - Q_{32}^k + \tau_2 X_{22b} + \tau_v Y_{22b} \\
\phi_{h1}^k &= E^T(N_{11}^k + N_{11}^{kT} + (\tau_h + 1)Q_{31}^k + Q_{11}^k + Q_{21}^k + \tau_4 X_{11a} + \tau_h Y_{11a})E - P_a^k \\
\phi_{h2}^k &= E^T(N_{21}^{kT} - N_{11}^k + M_{11}^k - S_{11}^k + \tau_4 X_{12a} + \tau_h Y_{12a}) \\
\phi_{h3}^k &= -N_{21}^k - N_{21}^{kT} - S_{21}^k - S_{21}^{kT} + M_{21}^k + M_{21}^{kT} - Q_{31}^k + \tau_4 X_{22a} + \tau_h Y_{22a}
\end{aligned}$$

also

$$\begin{aligned}
\phi_1^k &= \left[ (A_1^k - I)E \; A_2^k E \; A_{1d}^k \; A_{2d}^k \; 0\,0\,0\,0 \; B_1 \; B_2 \right] \\
\phi_2^k &= \left[ A_1^k E \; (A_2^k - I)E \; A_{1d}^k \; A_{2d}^k \; 0\,0\,0\,0 \; B_1 \; B_2 \right] \\
\phi_3^k &= \left[ \bar{A}_1^k \; \bar{A}_2^k \; \bar{A}_{1d}^k \; \bar{A}_{2d}^k \; 0\,0\,0\,0 \; \bar{B}_1 \; \bar{B}_2 \right] \\
\phi_{e1} &= \left[ \bar{C} \; 0\,0\,0\,0\,0\,0\,0\,0\,0 \right] \\
\phi_{e2} &= \left[ 0 \; \bar{C} \; 0\,0\,0\,0\,0\,0\,0\,0 \right]
\end{aligned}$$

with $v_{ij}^k = Z_{ij}^k - H_{ij} - H_{ij}^T$ for i = 1,2, j = 1,2 and $v_i^k = P_i^k - T - T^T, \quad for \quad i = a, b$. $\bar{A}_1^k$, $\bar{A}_2^k$, $\bar{A}_{1d}^k$, $\bar{A}_{2d}^k$, $\bar{B}_1^k$, $\bar{B}_2^k$ and $\bar{C}^k$ for $k = 1, 2, 3....n$ are defined as (4).
A suitable filter of the form (4) is given by
$A_{jf} = V_2^{-1}\bar{A}_{jf}$, $B_{jf} = V_2^{-1}\bar{B}_{jf}$, $C_f = \bar{C}_f, \quad for \quad j = 1, 2$

**Proof**, let:
$T = \begin{bmatrix} T_1 \; T_2 \\ T_4 \; T_3 \end{bmatrix}$
Define

$$J_1 = \begin{bmatrix} I & 0 \\ 0 & T_2 T_4^{-1} \end{bmatrix}$$

$J = diag(J_1, J_1, I, I, I, I, I, I, I, I, I, I, I, I, I, I, J_1, J_1)$
By following the similar line in [16], and by introducing new variables:
$V_1 = T_2 T_4^{-1} T_3$
$V_2 = T_2 T_4^{-1} T_2^T$

$\bar{A}_{jf} = T_2 A_{jf} T_4^{-1} T_2^T, \quad for \quad j = 1, 2$
$\bar{B}_{jf} = T_2 B_{jf}, \quad for \quad j = 1, 2,$
$\bar{C}_f = C_f T_4^{-1} T_2^T$
The matrices (19)–(20) are feasible.
This completes the proof. □

*Remark 2.* Theorem 2 shows that $H_\infty$ filtering problem can be solved via the $H_\infty$ performance analysis results developed in Subsect. 1 using the LMI approach. The effectiveness of the proposed method is reflected in the results of the following example.

## 4 Numerical Example

In this section, a numerical example is employed to show the benefits of the proposed method.

[18] Denote the design of 2-D delay-dependent $H_\infty$ performance and filter for a stationary random field in image processing where the disturbances are a random process (noise), the 2-D system can be converted to the 2-D FM model system (1) with the following parameters:

$$A_1 = \begin{bmatrix} 0.3 & 0 \\ 0 & 0 \end{bmatrix}, \quad A_2 = \begin{bmatrix} 0 & 0 \\ 1 & 0.2 \end{bmatrix}, \quad A_{1d} = \begin{bmatrix} 0 & -0.03 \\ -0.08 & 0 \end{bmatrix}, \quad A_{2d} = \begin{bmatrix} -0.03 & 0 \\ 0 & 0 \end{bmatrix}$$

$$B_1 = \begin{bmatrix} 1 & 0 \\ 0 & 0 \end{bmatrix}, \quad B_2 = \begin{bmatrix} 0 & 0 \\ 0 & 0 \end{bmatrix}, \quad C = \begin{bmatrix} 3 & 1 \end{bmatrix}, \quad D = \begin{bmatrix} 0 & 1 \end{bmatrix} and, \quad L = \begin{bmatrix} 0 & 0.05 \end{bmatrix}$$

Given $\tau_3 = 6$, $\tau_4 = 6$, $\tau_1 = 5$, $\tau_2 = 5$ by solving the LMI (17), the minimum $H_\infty$ norm bound for this example is $\gamma_{opt} = 1.3068$. The filter matrices are as follows:

$$A_{f1} = \begin{bmatrix} -0.0050 & -0.0386 \\ -0.0580 & -0.0708 \end{bmatrix}, \quad A_{f2} = \begin{bmatrix} -0.0050 & -0.0188 \\ 0.0602 & -0.0128 \end{bmatrix}$$
$$B_{f1} = \begin{bmatrix} -0.0460 \\ -0.0784 \end{bmatrix}, \quad B_{f2} = \begin{bmatrix} 0 \\ 0 \end{bmatrix}, \quad C_f = \begin{bmatrix} -0.0730 & -0.2718 \end{bmatrix}$$

The achieved minimum $\gamma_{min}$for various $d_2, d_4$ with given $d_1 = 5, d_3 = 6$

| $d_2 = d_4$ | 7 | 8 | 20 | 37 | 40 |
|---|---|---|---|---|---|
| $\gamma_{min}$ | 1.4554 | 1.534 | 2.602 | 7.19 | Infeasible |

Figure 1 shows the maximum singular value plots of the filtering error system by connecting the filter to the original system in this example. In the figure, the griddings denote the obtained $H_\infty$ disturbance attenuations and its maximum value is 1.1427, which is below 1.3068.

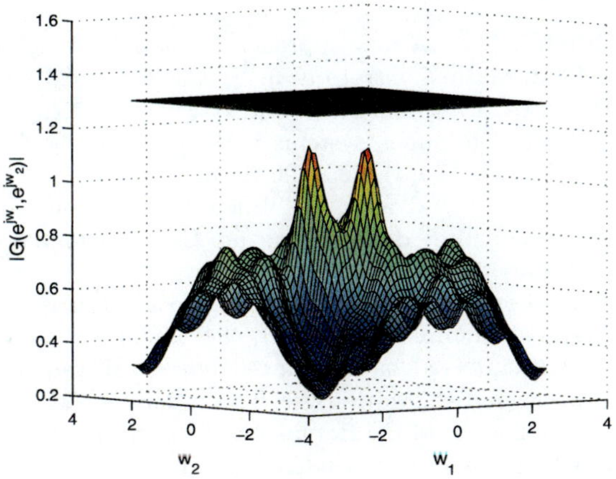


**Fig. 1.** Transfer function

## 5 Conclusion

In this paper, a delay-dependant $H_\infty$ performance analysis result has been established for 2-D discrete systems with time-varying delays using FM second model without ignoring any terms in the derivative of lyapunov functional by considering the relationship between the delay and it upper bound. Based on the derived $H_\infty$ performance analysis result, the $H_\infty$ filter has been designed in terms of LMIs. The new criteria may be extended to systems with uncertainties.

## References

1. Basin, M., Elvira-Ceja, S., Sanchez, E.N.: Mean-square $H_\infty$ filtering for stochastic systems: application to a 2DOF helicopter. Signal Process. **92**(3), 801–806 (2012)
2. Badie, K., Alfidi, M., Chalh, Z.: Improved delay-dependent stability criteria for 2-D discrete state delayed systems. In: International Conference on Intelligent Systems and Computer Vision (ISCV), pp. 1–6. IEEE, April 2018
3. Badie, K., Alfidi, M., Chalh, Z.: New relaxed stability conditions for uncertain two-dimensional discrete systems. J. Control Autom. Electr. Syst. **29**(6), 661–669 (2018)
4. Badie, K., Alfidi, M., Tadeo, F., Chalh, Z.: Delay-dependent stability and $H_\infty$ performance of 2-D continuous systems with delays. Circuits Syst. Signal Process. **37**(12), 5333–5350 (2018)
5. Boyd, S., El Ghaoui, L., Feron, E., Balakrishnan, V.: Linear Matrix Inequalities in System and Control Theory, vol. 15. SIAM, Philadelphia (1994)
6. Chen, S.F., Fong, I.K.: Delay-dependent robust $H_\infty$ filtering for uncertain 2-D state-delayed systems. Signal Process. **87**, 2659–2672 (2007)
7. Dumistrescu, B.: LMI stability tests for the Fornasini-Marchesini model. IEEE Trans. Signal Process. **56**, 4091–4095 (2008)

8. Elsayed, A., Grimble, M.J.: A new approach to the $H_\infty$ design of optimal digital linear filters. IMA J. Math. Control Inf. **6**(2), 233–251 (1989)
9. El-Kasri, C., Hmamed, A., Tissir, E.H., Tadeo, F.: Robust $H_\infty$ filtering for uncertain two-dimensional continuous systems with time-varying delays. Multidimension. Syst. Signal Process. **24**(4), 685–706 (2013)
10. El-Kasri, C., Hmamed, A., Alvarez, T., Tadeo, F.: Uncertain 2D continuous systems with state delay: filter design using an $H_\infty$ polynomial approach. Int. J. Comput. Appl. **44**(18), 13–21 (2012)
11. Fornasini, E., Marchesini, G.: State-space realization theory of two-dimentional filters. IEEE Trans. Autom. Control **21**(4), 484–492 (1976)
12. Fornasini, E., Marchesini, G.: Doubly-indexed dynamical systems: state-space models and structural properties. Math. Syst. Theor. **12**, 59–72 (1978)
13. Geromel, J.C., de Oliveira, M.C., Bernussou, J.: Robust filtering of discrete-time linear systems with parameter dependent Lyapunov functions. SIAM J. Control Optim. **41**, 700–711 (2002)
14. Gu, K., Kharitonov, V.L., Chen, J.: Stability of Time-Delay Systems. Birkhauser, Boston (2003)
15. Gonçalves, E.N., Palhares, R.M., Takahashi, R.H.: H/sub 2//H/sub/spl infin//filter design for systems with polytope-bounded uncertainty. IEEE Trans. Signal Process. **54**(9), 3620–3626 (2006)
16. He, Y., Liu, G.P., Rees, D., Wu, M.: $H_\infty$ filtering for discrete-time systems with time-varying delay. Signal Process. **89**(3), 275–282 (2009)
17. Hladowski, L., Galkowski, K., Cai, Z., Rogers, E., Freeman, C.T., Lewin, P.L.: A 2D systems approach to iterative learning control for discrete linear processes with zero Markov parameters. Int. J. Control **84**(7), 1246–1262 (2011)
18. Peng, D., Guan, X.: $H_\infty$ filtering of 2-D discrete state-delayed systems. Multidimension. Syst. Signal Process. **20**(3), 265–284 (2009)
19. Lu, W.S., Antoniou, A.: Two-Dimensional Digital Filters, vol. 80. CRC Press, Bosa Roca (1992)
20. Ooba, T.: On stability analysis of 2-D systems based on 2-D Lyapunov matrix inequalities. IEEE Trans. Circuits Syst. **I**(47), 1263–1265 (2000)
21. Paszke, W., Merry, R., van de Molengraft, R.: Iterative learning control by two-dimensional system theory applied to a motion system. In: 2007 American Control Conference, pp. 5484–5489. IEEE, July 2007
22. Paszke, W., Gałkowski, K., Rogers, E., Owens, D.H.: $H_\infty$ and guaranteed cost control of discrete linear repetitive processes. Linear Algebra Appl. **412**(2–3), 93–131 (2006)
23. Petersen, I.R., Savkin, A.V.: Robust Kalman filtering for signals and systems with large uncertainties (1999)
24. Richard, J.: Time-delay systems: an over view of some recent advances and open problem. Automatica **39**, 1667–1694 (2003)
25. Wu, M., He, Y., She, J.H., Liu, G.P.: Delay-dependent criteria for robust stability of time-varying delay systems. Automatica **40**, 1435–1439 (2004)
26. Wang, Z., Liu, Y., Liu, X.: $H_\infty$ filtering for uncertain stochastic time-delay systems with sector-bounded nonlineaities. Automatica **44**, 1268–1277 (2008)
27. Xie, L., de Souza, C.E., Fu, M.: $H_\infty$ estimation for discrete-time linear uncertain systems. Int. J. Robust Non linear Control **1**, 111–123 (1991)
28. Xu, H., Zou, Y.: Robust $H_\infty$ filtering for uncertain two-dimensional discrete systems with state-varying delays. Int. J. Control Autom. Syst. **8**(4), 720–726 (2010)

29. Zabari, A., Tissir, E.H., Tadeo, F.: Delay-dependent robust $H_\infty$ filtering with Lossy measurements for discrete-time systems. Arab. J. Sci. Eng. **42**(12), 5263–5273 (2017)
30. Zhang, X.M., Han, Q.L.: Delay-dependent robust $H_\infty$ filtering for uncertain discrete-time systems with time-varying delay based on a finite sum inequality. IEEE Trans. Circuits Syst. II Express Briefs **53**(12), 1466–1470 (2006)

# $H_\infty$ Model Reduction for Discrete 2-D Switched Systems in the Roesser Model

Khalid Badie(✉), Mohammed Alfidi, and Zakaria Chalh

Engineering, Systems and Applications Laboratory,
University Sidi Mohamed Ben Abdellah Fez, Fes, Morocco
khalid.badie@usmba.ac.ma

**Abstract.** The present paper is devoted to solving the problem of $H_\infty$ model reduction for two-dimensional (2-D) discrete switched systems represented by the Roesser model. Given an asymptotically stable system, our purpose is to design a stable reduced-order system such that the corresponding error system is asymptotically stable with a prescribed $H_\infty$ performance level. By employing a switched Lyapunov function technique, a sufficient condition is proposed to ensure the asymptotic stability with an $H_\infty$ noise-attenuation level bound for the error system. Then, the corresponding reduced order model is designed in the form of linear matrix inequalities (LMIs). Numerical example is presented to show the effectiveness of the developed method.

**Keywords:** 2-D switched systems · Reduced-order model · $H_\infty$ performance

## 1 Introduction

Two-dimensional (2-D) systems have drawn significant research interest in control systems theory and applications during the last twenty years due to their potential applications in a large number of fields such as thermal power engineering, image processing, multidimensional signal processing, and so on [1–3]. Some significant results have been available in the literature: to cite a few, the stability and stabilization of 2-D systems have been solved in [4–10] and the $H_\infty$ filtering and control for 2-D linear systems have been considered in [11–16].

On the other hand, switched systems are one of the important kinds of hybrid systems in engineering applications, which consist of a switching signal and a finite number of continuous-time or discrete-time subsystems, where the role of the switching signal is to coordinates the switching between the subsystems. In the last few years, many researchers have focused their attention on the investigation of 2-D switched systems, and many important results have been published in the literature, for example, in [17], the stability and stabilization problems for discrete 2-D switched systems described by the Roesser model have been solved.


A. Saka et al. (Eds.): CPI 2019, LNME, pp. 120–131, 2021.
https://doi.org/10.1007/978-3-030-62199-5_12

The $H_\infty$ state feedback controller design problem for 2-D discrete switched systems with delays described by the second Fornasini and Marchesini (FM) state-space model has been studied in [18]. In [19], the stabilization problem of 2-D discrete switched systems with all modes unstable has been solved.

In addition, it is very known that engineering systems are mostly characterized by their high-order complex mathematical models, which pose major problems in the analysis and synthesis of these systems. Thus, model reduction problem has been an active topic in the past years, and many important results concerning model reduction for various kinds of systems have been reported, for example, fuzzy systems [20,21] and switched systems [22,23] .But, to the best of the authors' knowledge, the relevant problem on model reduction for high-order 2-D switched systems has not been investigated, which is the main motivation for this research.

Thus, in this research, our attention is directed on the study of reduced-order model design for a class of 2-D discrete switched systems, the switched signal is assumed to be arbitrary. Based on the switched Lyapunov function method, a sufficient criterion is developed to ensure asymptotic stability and a prescribed $H_\infty$ disturbance index $\gamma$ for the corresponding 2-D switched error system. Then, the desired reduced order model is designed to achieve the prescribed $H_\infty$ disturbance attenuation index $\gamma$. Numerical example is presented to demonstrate the feasibility and efficacy of the designed reduced order model.

## 2 Problem Formulation

In this paper, we consider a class of discrete 2-D switched system described by the following Roesser model:

$$\begin{bmatrix} x^h(k+1,l) \\ x^v(k,l+1) \end{bmatrix} = A^{\sigma(k,l)} \begin{bmatrix} x^h(k,l) \\ x^v(k,l) \end{bmatrix} + B^{\sigma(k,l)}\omega(k,l), \tag{1}$$

$$y(k,l) = C^{\sigma(k,l)} \begin{bmatrix} x^h(k,l) \\ x^v(k,l) \end{bmatrix} + D^{\sigma(k,l)}\omega(k,l), \tag{2}$$

where $x^h(k,l) \in \mathbb{R}^{n_h}$, $x^v(k,l) \in \mathbb{R}^{n_v}$ represent the horizontal and the vertical states respectively. $\omega(k,l) \in \mathbb{R}^{n_\omega}$ is the input vector which belongs to $\mathcal{L}_2\{[0,\infty),[0,\infty)\}$; $y(k,l) \in \mathbb{R}^{n_y}$ is the measurement output vector; the function $\sigma(k,l)$ is called the switching signal, which takes its values in the finite set $\mathcal{N} := \{1,2,...N\}$ with $N > 1$ is the number of subsystems. $\sigma(k,l) = i$, $(i \in \mathcal{N})$ denote that the $i$-th subsystem is activated. The matrices:

$$A^i = \begin{bmatrix} A^i_{11} \; A^i_{12} \\ A^i_{21} \; A^i_{22} \end{bmatrix}, \quad B^i = \begin{bmatrix} B^i_1 \\ B^i_2 \end{bmatrix}, \quad C^i = \begin{bmatrix} C^i_1 \; C^i_2 \end{bmatrix}, \quad D^i, \qquad i \in \mathcal{N}$$

are known constant matrices with appropriately dimensions denote the $i$-th subsystem. The initial conditions are assumed as:

$$\lim_{r\to\infty} \sum_{q=1}^{r} (|x^h(0,q)|^2 + |x^v(q,0)|^2) < \infty. \tag{3}$$

*Remark 1.* In the present research paper, the switch occurrence between various subsystems can be assumed at each of the sampling points of $k$ or $l$. It is worthwhile mentioning that the value of $\sigma(k,l)$ only relies upon $i+j$, we refer readers to [17] for more details.

In this paper, we are concerned in designing a 2-D reduced order switched system in the form as follow:

$$\begin{bmatrix} \tilde{x}^h(k+1,l) \\ \tilde{x}^v(k,l+1) \end{bmatrix} = \tilde{A}^{\sigma(k,l)} \begin{bmatrix} \tilde{x}^h(k,l) \\ \tilde{x}^v(k,l) \end{bmatrix} + \tilde{B}^{\sigma(k,l)} \omega(k,l), \tag{4}$$

$$\tilde{y}(k,l) = \tilde{C}^{\sigma(k,l)} \begin{bmatrix} \tilde{x}^h(k,l) \\ \tilde{x}^v(k,l) \end{bmatrix} + \tilde{D}^{\sigma(k,l)} \omega(k,l), \tag{5}$$

$\tilde{x}^h(k,l) \in \mathbb{R}^{\tilde{n}_h}$ and $\tilde{x}^h(k,l) \in \mathbb{R}^{\tilde{n}_v}$ are, respectively, the horizontal and vertical states of the reduced-order system with $\tilde{n}_h + \tilde{n}_v < n_h + n_v$. The matrices:

$$\tilde{A}^i = \begin{bmatrix} \tilde{A}^i_{11} & \tilde{A}^i_{12} \\ \tilde{A}^i_{21} & \tilde{A}^i_{22} \end{bmatrix}, \quad \tilde{B}^i = \begin{bmatrix} \tilde{B}^i_1 \\ \tilde{B}^i_2 \end{bmatrix}, \quad \tilde{C}^i = \begin{bmatrix} \tilde{C}^i_1 & \tilde{C}^i_2 \end{bmatrix}, \quad \mathcal{D}^i, \qquad i \in \mathcal{N}$$

are the reduced order system matrices to be determined.

*Remark 2.* It worth noting that, the same switching signals are utilized in system (1)–(2) and reduced-order model (4)–(5). Consequently, when the switching signal $\sigma(k,l)$ changes, the switchings between the subsystems in system (1)–(2) and reduced-order model (4)–(5) occur simultaneously. In addition and as in [24] the switching signal $\sigma(k,l)$ is supposed to be not known a priori, but, in real-time implementation. their instantaneous values is available.

Defining:

$$\bar{x}^h(k,l) = \begin{bmatrix} x^h(k,l) \\ \tilde{x}^h(k,l) \end{bmatrix}, \quad \bar{x}^v(k,l) = \begin{bmatrix} x^v(k,l) \\ \tilde{x}^v(k,l) \end{bmatrix},$$

as the horizontal and vertical state vectors of the resulting system from the augmentation of the 2-D system (1)–(2) to include the states of the model (4)–(5), the error system is given as follows:

$$\begin{bmatrix} \bar{x}^h(k+1,l) \\ \bar{x}^v(k,l+1) \end{bmatrix} = \bar{A}^{\sigma(k,l)} \begin{bmatrix} \bar{x}^h(k,l) \\ \bar{x}^v(k,l) \end{bmatrix} + \bar{B}^{\sigma(k,l)} \omega(k,l), \tag{6}$$

$$e(k,l) = \bar{C}^{\sigma(k,l)} \begin{bmatrix} \bar{x}^h(k,l) \\ \bar{x}^v(k,l) \end{bmatrix} + \bar{D}^{\sigma(k,l)} \omega(k,l), \tag{7}$$

with $e(k,l) = y(k,l) - \tilde{y}(k,l)$ and

$$\bar{A}^i = \begin{bmatrix} \bar{A}^i_{11} & \bar{A}^i_{12} \\ \bar{A}^i_{21} & \bar{A}^i_{22} \end{bmatrix} = \left[\begin{array}{cc|cc} A^i_{11} & 0 & A^i_{12} & 0 \\ 0 & \tilde{A}^i_{11} & 0 & \tilde{A}^i_{12} \\ \hline A^i_{21} & 0 & A^i_{22} & 0 \\ 0 & \tilde{A}^i_{21} & 0 & \tilde{A}^i_{22} \end{array}\right], \quad \bar{B}^i = \begin{bmatrix} \bar{B}^i_1 \\ \bar{B}^i_2 \end{bmatrix} = \left[\begin{array}{c} B^i_1 \\ \tilde{B}^i_2 \\ \hline B^i_2 \\ \tilde{B}^i_2 \end{array}\right],$$

$$\bar{C}^i = \begin{bmatrix} \bar{C}^i_1 & \bar{C}^i_2 \end{bmatrix} = \left[\begin{array}{cc|cc} C^i_1 & -\tilde{C}^i_2 & C^i_2 & -\tilde{C}^i_2 \end{array}\right], \qquad \bar{D}^i = D^i - \tilde{D}^i.$$

The main goal of the present research is to determine the matrices $\tilde{A}^i$, $\tilde{B}^i$, $\tilde{C}^i$, and $\tilde{D}^i$, such that the model in (4)–(5) presents a perfect approximation of the original 2-D system in (1)–(2) in a $H_\infty$ norm sense, that is, given a 2-D switched system (1)–(2) and a prescribed level of noise attenuation $\gamma > 0$; find a reduced order system in the form of (4)–(5) so that the asymptotic stability of the resulting error system is guaranteed with:

$$||e||_2 < \gamma||\omega||_2, \tag{8}$$

for any non-zero $\omega \in \mathcal{L}_2\{[0,\infty),[0,\infty)\}$, and under zero-initial conditions.

## 3 Main Results

### 3.1 $H_\infty$ Performance Analysis

**Theorem 1.** *Let $\gamma > 0$ be a given scalar; the error system (6)–(7) is asymptotically stable with an $H_\infty$ prescribed attenuation level $\gamma$, if there exist symmetric positive definite matrices $P_h^i$, $P_v^i$ and matrices $X^i$ and $Y^i$, with $i \in \mathcal{N}$ such that:*

$$\begin{bmatrix} -sym(X^i) & 0 & \Pi_1 & 0 & 0 & (\bar{A}_{11}^i)^T(X^i)^T & (\bar{A}_{21}^i)^T(Y^i)^T & (\bar{C}_1^i)^T \\ * & -sym(Y^i) & 0 & \Pi_2 & 0 & (\bar{A}_{12}^i)^T(X^i)^T & (\bar{A}_{22}^i)^T(Y^i)^T & (\bar{C}_2^i)^T \\ * & * & -2P_h^i & 0 & 0 & 0 & 0 & 0 \\ * & * & * & -2P_h^i & 0 & 0 & 0 & 0 \\ * & * & * & * & -\gamma^2 I & (\bar{B}_1^i)^T(X^i)^T & (\bar{B}_2^i)^T(Y^i)^T & 0 \\ * & * & * & * & * & -P_h^j & 0 & 0 \\ * & * & * & * & * & * & -P_v^j & 0 \\ * & * & * & * & * & * & * & -I \end{bmatrix} < 0, \quad i,j \in \mathcal{N} \tag{9}$$

*where*

$$\Pi_1 = 0.5(X^i)^T + P_h^i, \qquad \Pi_2 = 0.5(Y^i)^T + P_v^i,$$

*Proof.* Construct a switching Lyapunov function in the following form::

$$V(k,l) = V_h(k,l) + V_v(k,l), \tag{10}$$

where

$$V_h(k,l) = \bar{x}^{hT}(k,l)(X^{\sigma(k,l)})^T(P_h^{\sigma(k,l)})^{-1}X^{\sigma(k,l)}\bar{x}^h(k,l),$$
$$V_v(k,l) = \bar{x}^{vT}(k,l)(Y^{\sigma(k,l)})^T(P_v^{\sigma(k,l)})^{-1}Y^{\sigma(k,l)}\bar{x}^v(k,l),$$

The forward differences of $V_h(k,l)$ and $V_v(k,l)$ can be calculated as:

$$\begin{aligned}\Delta V_h(k,l) &= V_h(k+1,l) - V_h(k,l)\\ &= \bar{x}^{hT}(k+1,l)(X^j)^T(P_h^j)^{-1}X^j\bar{x}^h(k+1,l)\\ &\quad -\bar{x}^{hT}(k,l)(X^i)^T(P_h^i)^{-1}X^i\bar{x}^h(k,l),\\ \Delta V_v(k,l) &= V_v(k,l+1) - V_v(k,l)\\ &= \bar{x}^{vT}(k,l+1)(Y^j)^T(P_v^j)^{-1}X^j\bar{x}^v(k,l+1)\\ &\quad -\bar{x}^{vT}(k,l)(X^i)^T(P_v^i)^{-1}X^i\bar{x}^v(k,l),\end{aligned}$$

For any appropriate dimension matrices $X^i$ and $Y^i$, the following equations are true:

$$\begin{aligned}\Lambda^h &= 2\{\bar{x}^{hT}(k,l)(X^i)^T(P_h^i)^{-1}X^i\bar{x}^h(k,l) + \bar{x}^{hT}(k,l)P_h^i(P_h^i)^{-1}X^i\bar{x}^h(k,l)\\ &\quad -\bar{x}^{hT}(k,l)(X^i)^T(P_h^i)^{-1}P_h^i(P_h^i)^{-1}X^i\bar{x}^h(k,l) - \bar{x}^{hT}(k,l)X^i\bar{x}^h(k,l)\} = 0;\\ \Lambda^v &= 2\{\bar{x}^{vT}(k,l)(Y^i)^T(P_v^i)^{-1}Y^i\bar{x}^v(k,l) + \bar{x}^{vT}(k,l)P_v^i(P_v^i)^{-1}Y^i\bar{x}^v(k,l)\\ &\quad -\bar{x}^{vT}(k,l)(Y^i)^T(P_v^i)^{-1}P_v^i(P_v^i)^{-1}Y^v\bar{x}^v(k,l) - \bar{x}^{vT}(k,l)Y^i\bar{x}^v(k,l)\} = 0;\end{aligned}$$

By combining the above equalities, we obtain:

$$\begin{aligned}\Delta V(k,l) &= \Delta V_h(k,l) + \Delta V_v(k,l),\\ &= \Delta V_h(k,l) + \Delta V_v(k,l) + \Lambda^h + \Lambda^v,\\ &= \xi^T(k,l)\left\{\Pi + \Upsilon_h^T(P_h^j)^{-1}\Upsilon_h + \Upsilon_v^T(P_v^j)^{-1}\Upsilon_v\right\}\xi^T(k,l)\end{aligned}$$

where

$$\xi(k,l) = \begin{bmatrix}\bar{x}^h(k,l)\\ \bar{x}^v(k,l)\\ (P_h^k)^{-1}X^i\bar{x}^h(k,l)\\ (P_v^i)^{-1}Y^i\bar{x}^v(k,l)\end{bmatrix},\quad \Pi = \begin{bmatrix}-sym(X^i) & 0 & \Pi_1 & 0\\ * & -sym(Y^i) & 0 & \Pi_2\\ * & * & -2P_h^i & 0\\ * & * & * & -2P_h^i\end{bmatrix},$$

$$\Upsilon_h = \begin{bmatrix}(\bar{A}_{11}^i)^T(X^i)^T\\ (\bar{A}_{12}^i)^T(X^i)^T\\ 0\\ 0\end{bmatrix}^T,\quad \Upsilon_v = \begin{bmatrix}(\bar{A}_{21}^i)^T(Y^i)^T\\ (\bar{A}_{22}^i)^T(Y^i)^T\\ 0\\ 0\end{bmatrix}^T,$$

Now, by the schur complement Lemma and some simple matrix manipulations, the inequality (9) implies $\Pi + \Upsilon_h^T(P_h^j)^{-1}\Upsilon_h + \Upsilon_v^T(P_v^j)^{-1}\Upsilon_v < 0$, i.e. $\Delta V(k,l) < 0$; thus the asymptotic stability of the system (6)–(7) with $\omega(k,l) = 0$ is guaranteed.

Now, for any non-zero $\omega(k,l) \in \mathcal{L}_2\{[0,\infty),[0,\infty)\}$, we define the performance index $\mathcal{J}$ as follows:

$$\mathcal{J} = \sum_{i=0}^{\infty}\sum_{j=0}^{\infty}\{e^T(k,l)e(k,l) - \gamma^2 w^T(k,l)w(k,l)\},$$

Under zero-initial condition, we obtain:

$$\begin{aligned}\mathcal{J} &\leq \sum_{i=0}^{\infty}\sum_{j=0}^{\infty}\{e^T(k,l)e(k,l) - \gamma^2 w^T(k,l)w(k,l) + \Delta V(k,l) + \Lambda_h + \Lambda_v\} \\ &= \sum_{i=0}^{\infty}\sum_{j=0}^{\infty}\xi_\omega^T(k,l)\{\Pi_\omega + \Upsilon_{\omega h}^T(P_h^j)^{-1}\Upsilon_{\omega h} + \Upsilon_{\omega v}^T(P_v^j)^{-1}\Upsilon_{\omega v} + \mathcal{C}^T\mathcal{C}\}\xi_\omega(k,l)\end{aligned}$$

where

$$\xi_\omega(k,l) = \begin{bmatrix}\xi(k,l)\\ \omega(k,l)\end{bmatrix}, \quad \Pi_\omega = \begin{bmatrix}\Pi & 0\\ * & -\gamma^2 I\end{bmatrix}, \quad \Upsilon_{\omega h} = \begin{bmatrix}\Upsilon_h\\ \bar{B}_1^T(X^i)^T\end{bmatrix},$$
$$\Upsilon_{\omega v} = \begin{bmatrix}\Upsilon_v\\ \bar{B}_2^T(Y^i)^T\end{bmatrix}, \quad \mathcal{C}_\omega = \begin{bmatrix}\mathcal{C}\\ 0\end{bmatrix}, \quad \mathcal{C} = \begin{bmatrix}\bar{C}^T\\ 0\end{bmatrix},$$

Thus, if

$$\Pi_\omega + \Upsilon_{\omega h}^T(P_h^j)^{-1}\Upsilon_{\omega h} + \Upsilon_{\omega v}^T(P_v^j)^{-1}\Upsilon_{\omega v} + \mathcal{C}^T\mathcal{C} < 0,$$

which is equivalent to (9) by the schur complement Lemma, then $\mathcal{J} < 0$. This ensures that (8) holds for all non-zero $\omega(k,l) \in \mathcal{L}_2\{[0,\infty),[0,\infty)\}$ and under zero-initial conditions.

### 3.2 Model Reduction Design

**Theorem 2.** *Given scalar $\gamma > 0$, if there exist symmetric positive definite matrices $P_{1h}^i$, $P_{3h}^i$, $P_{1v}^i$, $P_{3v}^i$ and matrices $X_1^i$, $X_2^i$, $X_3^i$, $Y_1^i$, $Y_2^i$, $Y_3^i$, $\mathcal{A}_{11}^i$, $\mathcal{A}_{12}^i$, $\mathcal{A}_{21}^i$, $\mathcal{A}_{22}^i$, $\mathcal{B}_1^i$, $\mathcal{B}_2^i$, $\mathcal{C}_1^i$, $\mathcal{C}_2^i$ and $\mathcal{D}^i$ with $i \in \mathcal{N}$, such that:*

$$\begin{bmatrix}\Psi_1^i & 0 & \Psi_2^i & 0 & 0 & \Psi_3^i & \Psi_4^i & \Psi_5^i\\ * & \Psi_6^i & 0 & \Psi_7^i & 0 & \Psi_8^k & \Psi_9^i & \Psi_{10}^k\\ * & * & \Psi_{11}^k & 0 & 0 & 0 & 0 & 0\\ * & * & * & \Psi_{12}^i & 0 & 0 & 0 & 0\\ * & * & * & * & \Psi_{13} & \Psi_{14}^k & \Psi_{15}^i & \Psi_{16}^i\\ * & * & * & * & * & \Psi_{17}^j & 0 & 0\\ * & * & * & * & * & * & \Psi_{18}^j & 0\\ * & * & * & * & * & * & * & \Psi_{19}\end{bmatrix} < 0, \qquad i,j \in \mathcal{N} \tag{11}$$

*where*

$$\Psi_1^k = \begin{bmatrix} -sym(X_1^i) & -E_h X_2^i - (X_3^i)^T \\ * & -sym(X_2^k) \end{bmatrix},$$

$$\Psi_2^k = \begin{bmatrix} P_{1h}^i + 0.5(X_1^i)^T & P_{2h}^i + 0.5(X_3^i)^T \\ (P_{2h}^i)^T + 0.5E_h^T(X_2^i)^T & P_{3h}^i + 0.5(X_2^i)^T \end{bmatrix},$$

$$\Psi_3^i = \begin{bmatrix} A_{11}^T(X_1^i)^T & A_{11}^T(X_3^i)^T \\ \mathcal{A}_{11}^T E_h^T & \mathcal{A}_{11}^T \end{bmatrix}, \quad \Psi_4^i = \begin{bmatrix} A_{21}^T(Y_1^k)^T & A_{21}^T(Y_3^i)^T \\ \mathcal{A}_{21}^T E_v^T & \mathcal{A}_{21}^T \end{bmatrix},$$

$$\Psi_5^k = \begin{bmatrix} C_1^T \\ -\mathcal{C}_1^T \end{bmatrix}, \Psi_6^i = \begin{bmatrix} -sym(Y_1^i) & -E_v Y_2^i - (Y_3^i)^T \\ * & -sym(Y_2^i) \end{bmatrix},$$

$$\Psi_7^i = \begin{bmatrix} P_{1v}^i + 0.5(Y_1^i)^T & P_{2v}^i + 0.5(Y_3^i)^T \\ (P_{2v}^i)^T + 0.5E_v^T(Y_2^i)^T & P_{3v}^i + 0.5(Y_2^i)^T \end{bmatrix},$$

$$\Psi_8^i = \begin{bmatrix} A_{12}^T(X_1^i)^T & A_{12}^T(X_3^i)^T \\ \mathcal{A}_{12}^T E_h^T & \mathcal{A}_{12}^T \end{bmatrix}, \quad \Psi_9^i = \begin{bmatrix} A_{22}^T(Y_1^k)^T & A_{22}^T(Y_3^i)^T \\ \mathcal{A}_{22}^T E_v^T & \mathcal{A}_{22}^T \end{bmatrix},$$

$$\Psi_{10}^i = \begin{bmatrix} C_2^T \\ -\mathcal{C}_2^T \end{bmatrix}, \quad \Psi_{11}^i = \begin{bmatrix} -2P_{1h}^i & -2P_{2h}^i \\ * & -2P_{3h}^i \end{bmatrix}, \quad \Psi_{12}^k = \begin{bmatrix} -2P_{1v}^i & -2P_{2v}^i \\ * & -2P_{3v}^i \end{bmatrix},$$

$$\Psi_{13} = -\gamma^2 I, \ \Psi_{14}^i = \begin{bmatrix} B_1^T(X_1^i)^T + \mathcal{B}_1^T E_h^T & B_1^T(X_3^i)^T + \mathcal{B}_1^T \end{bmatrix},$$

$$\Psi_{15}^k = \begin{bmatrix} B_2^T(Y_1^i)^T + \mathcal{B}_2^T E_v^T & B_2^T(Y_3^i)^T + \mathcal{B}_2^T \end{bmatrix}, \quad \Psi_{16}^i = (D^i)^T - (\mathcal{D}^k)^T,$$

$$\Psi_{17}^j = \begin{bmatrix} -P_{1h}^j & -P_{2h}^j \\ * & -P_{3h}^j \end{bmatrix}, \quad \Psi_{18}^j = \begin{bmatrix} -P_{1v}^j & -P_{2v}^j \\ * & -P_{3v}^j \end{bmatrix}, \quad \Psi_{19} = -I,$$

$$E_h = \begin{bmatrix} I_{\tilde{n}_h} & 0_{\tilde{n}_h, n_h - \tilde{n}_h} \end{bmatrix}^T, \quad E_v = \begin{bmatrix} I_{\tilde{n}_v} & 0_{\tilde{n}_v, n_v - \tilde{n}_v} \end{bmatrix}^T.$$

*Then, there exist a reduced-order model system* (4)–(5) *guarantees the asymptotic stability and satisfies the* $H_\infty$ *performance specification* (8) *for the error system* (6)–(7)*. Moreover, the parameters of reduced-order system* (4)–(5) *can be designed as:*

$$\left[\begin{array}{cc|c} \bar{A}_{11}^i & \bar{A}_{12}^i & \bar{B}_1^i \\ \bar{A}_{21}^i & \bar{A}_{22}^i & \bar{B}_2^k \\ \hline \bar{C}_1^i & \bar{C}_2^i & \bar{D}^i \end{array}\right] = \left[\begin{array}{cc|c} (X_2^k)^{-1} & 0 & 0 \\ 0 & (Y_2^i)^{-1} & 0 \\ \hline 0 & 0 & I \end{array}\right] \left[\begin{array}{cc|c} \mathcal{A}_{11}^i & \mathcal{A}_{12}^i & \mathcal{B}_1^k \\ \mathcal{A}_{21}^i & \mathcal{A}_{22}^i & \mathcal{B}_2^i \\ \hline \mathcal{C}_1^i & \mathcal{C}_2^i & \mathcal{D}^i \end{array}\right], \tag{12}$$

*Proof.* In accordance with Theorem 1, the error system (6)–(7) is asymptotically stable and satisfies the $H_\infty$ performance specification (8), if there exist symmetric positive definite matrices $P_h^k$, $P_v^i$ and matrices $X^i$, $Y^i$, for $i \in \mathcal{N}$ satisfying (9). Then, in order to resolve the model reduction design problem, we partition these matrices as:

$$P_h^i = \begin{bmatrix} P_{1h}^i & P_{2h}^i \\ * & P_{3h}^i \end{bmatrix}, P_v^k = \begin{bmatrix} P_{1v}^i & P_{2v}^i \\ * & P_{3v}^i \end{bmatrix}, X^i = \begin{bmatrix} X_1^i & E_h X_2^i \\ X_3^k & X_2^i \end{bmatrix}, Y^k = \begin{bmatrix} Y_1^i & E_v Y_2^i \\ Y_3^i & Y_2^i \end{bmatrix}.$$

By substituting corresponding partitioned parts into (9) and defining the following matrces variables for $i \in \mathcal{N}$,

$$\begin{aligned} \mathcal{A}_{11}^k &= X_2^i \tilde{A}_{11}^i, & \mathcal{A}_{12}^i &= X_2^i \tilde{A}_{12}^k, & \mathcal{A}_{21}^i &= Y_2^i \tilde{A}_{21}^i, \\ \mathcal{A}_{22}^k &= Y_2^i \tilde{A}_{22}^i, & \mathcal{B}_1^i &= X_2^i \tilde{B}_1^i, & \mathcal{B}_2^i &= Y_2^i \tilde{B}_2^i, \\ \mathcal{C}_1^i &= \tilde{C}_1^k, & \mathcal{C}_2^i &= \tilde{C}_2^i, & \mathcal{D}^i &= \tilde{D}^i. \end{aligned}$$

the LMIs in (11), can be obtained readily. The proof is thus completed.

## 4 Numerical Example

In this section, we provide an example to apply our result, and to demonstrate the effectiveness of the proposed technique.

Consider a discrete 2-D switched system in the form in (1)–(2) composing of two subsystems ($N = 2$) with ($n_h = 1, n_v = 4$). For subsystem 1, the matrices are given by:

$$A^1 = \left[\begin{array}{c|cccc} 0.2 & -1.1 & 0 & 0 & 0 \\ \hline 0.15 & 0.13 & 0.8 & 0.1 & 0 \\ -0.1 & 0 & 0 & 1 & 0 \\ 0 & 0.14 & 0 & 0.1 & 0.1 \\ 0.10 & 0 & 0.2 & 0 & 0.1 \end{array}\right], \quad B^1 = \left[\begin{array}{c} 1 \\ \hline 0 \\ 1 \\ 0.5 \\ 0 \end{array}\right],$$
$$C^1 = \left[0.2 | -0.8 \; 0 \; 1.9 \; 0.2\right], \quad D^1 = 1.$$

For subsystem 2, the matrices are given by:

$$A^2 = \left[\begin{array}{c|cccc} 0.2 & -0.9 & 0.6 & 0.1 & 0 \\ \hline 0.25 & 0.1 & 1.1 & 0 & 0 \\ -0.1 & 0 & 0.3 & 1 & 0 \\ 0 & 0.1 & 0.04 & 0.1 & 0.05 \\ 0.1 & 0 & 0.2 & 0 & 0.1 \end{array}\right], \quad B^2 = \left[\begin{array}{c} 1.7 \\ \hline 0.4 \\ 0.2 \\ 0 \\ 0.1 \end{array}\right],$$
$$C^2 = \left[0.2 | -1 \; 0 \; 1 \; 0.1\right], \quad D^2 = 1.$$

Our purpose is to design a reduced-order model in the form of (4)–(5) assuring the asymptotic stability with a prescribed $H_\infty$ performance $\gamma$ of the resulted error system. By using Theorem 2, we obtain the different minimum noise attenuation level bounds $\gamma$ for different $\tilde{n}_v$ as shown in Table 1.

**Table 1.** Minimum values of $H_\infty$ performance $\gamma_{min}$ for different $\tilde{n}_v$ .

| Dimensions | $\gamma_{min}$ | Ndv | Time(s) |
|---|---|---|---|
| $(\tilde{n}_h = 1, \tilde{n}_v = 3)$ | 0.0465 | 193 | 6.561 |
| $(\tilde{n}_h = 1, \tilde{n}_v = 2)$ | 0.6504 | 143 | 4.437 |
| $(\tilde{n}_h = 1, \tilde{n}_v = 1)$ | 1.7060 | 103 | 3.324 |

And we obtain the reduced-order models for different $\tilde{n}_v$.
For $(\tilde{n}_h = 1, \tilde{n}_v = 3)$:

$$\tilde{A}^1 = \left[\begin{array}{c|ccc} 0.2004 & -1.0985 & -0.0061 & 0.0145 \\ \hline 0.1505 & 0.1347 & 0.7837 & 0.1380 \\ -0.1063 & -0.0064 & 0.0191 & 0.9578 \\ 0.0059 & 0.1692 & -0.0113 & 0.1244 \end{array}\right], \quad \tilde{B}^1 = \left[\begin{array}{c} -0.9999 \\ \hline -0.0002 \\ -0.9995 \\ -0.5006 \end{array}\right],$$

$$\tilde{C}^1 = \left[\begin{array}{c|ccc} -0.2005 & 0.7509 & -0.0034 & -1.8925 \end{array}\right], \quad \tilde{D}^1 = 1.0003,$$

$$\tilde{A}^2 = \left[\begin{array}{c|ccc} 0.2000 & -0.8997 & 0.5995 & 0.0999 \\ \hline 0.2503 & 0.1000 & 1.0976 & 0.0021 \\ -0.1000 & 0.0002 & 0.2998 & 0.9989 \\ -0.0001 & 0.1147 & 0.0349 & 0.1005 \end{array}\right], \quad \tilde{B}^2 = \left[\begin{array}{c} -1.7001 \\ \hline -0.4004 \\ -0.2001 \\ -0.0001 \end{array}\right],$$

$$\tilde{C}^2 = \left[\begin{array}{c|ccc} -0.2003 & 0.9703 & 0.0136 & -1.0047 \end{array}\right], \quad \tilde{D}^2 = 1.0007.$$

For $(\tilde{n}_h = 1, \tilde{n}_v = 2)$:

$$\tilde{A}^1 = \left[\begin{array}{c|cc} 0.1724 & -1.0609 & -0.0209 \\ \hline 0.1455 & -0.0234 & 0.8590 \\ -0.0567 & 0.0668 & 0.4011 \end{array}\right], \quad \tilde{B}^1 = \left[\begin{array}{c} -0.9303 \\ \hline 0.0509 \\ -1.0999 \end{array}\right],$$

$$\tilde{C}^1 = \left[\begin{array}{c|cc} -0.1445 & 0.8791 & -0.8364 \end{array}\right], \quad \tilde{D}^1 = 1.0380,$$

$$\tilde{A}^2 = \left[\begin{array}{c|cc} 0.1994 & -0.8656 & 0.5296 \\ \hline 0.2499 & 0.1029 & 0.9313 \\ -0.1067 & 0.0473 & 0.3862 \end{array}\right], \quad \tilde{B}^2 = \left[\begin{array}{c} -1.7407 \\ \hline -0.4660 \\ -0.2346 \end{array}\right],$$

$$\tilde{C}^2 = \left[\begin{array}{c|cc} -0.1755 & 0.9235 & -0.1668 \end{array}\right], \quad \tilde{D}^2 = 1.0727.$$

For $(\tilde{n}_h = 1, \tilde{n}_v = 1)$:

$$\tilde{A}^1 = \left[\begin{array}{c|c} 0.0998 & -0.9795 \\ \hline 0.0594 & 0.2584 \end{array}\right], \quad \tilde{B}^1 = \left[\begin{array}{c} -0.8157 \\ \hline 1.2536 \end{array}\right],$$

$$\tilde{C}^1 = \left[\begin{array}{c|c} -0.0281 & 0.4759 \end{array}\right], \quad \tilde{D}^1 = 1.0348,$$

$$\tilde{A}^2 = \left[\begin{array}{c|c} 0.0911 & -0.3834 \\ \hline 0.1734 & 0.2648 \end{array}\right], \quad \tilde{B}^2 = \left[\begin{array}{c} -1.1174 \\ \hline -0.6063 \end{array}\right],$$

$$\tilde{C}^2 = \left[\begin{array}{c|c} -0.1253 & 0.7511 \end{array}\right], \quad \tilde{D}^2 = 1.4325.$$

*Remark 3.* According to Table 1, it can be observed clearly that the number of decision variables Ndv is related to the order of the reduced-reduced-order model and it will increase if $\tilde{n}_h$ and $\tilde{n}_v$ increase, which leads the increased computational cost and the execution time. In the other hand, the increase of ($\tilde{n}_h$, $\tilde{n}_v$) makes possible to further reduce the conservatism. Therefore, $\tilde{n}_h$ and $\tilde{n}_v$ should be selected suitably by taking into account the compromise among the computational cost and the conservatism of the results.

Given the initial conditions:

$$x^h(0,j) = \begin{cases} 0.1, 0 \leq j \leq 30, \\ 0, \quad otherwise. \end{cases}$$

$$x^v(i,0) = \begin{cases} \left[0.3\ 0\ 0.2\ 0.1\right]^T, 0 \leq i \leq 30, \\ 0, \qquad\qquad\qquad otherwise. \end{cases}$$

and the disturbance input is assumed that:

$$\omega(k,l) = \begin{cases} -0.2, 0 \leq i \leq 10, \quad 0 \leq j \leq 10, \\ 0, \qquad otherwise. \end{cases}$$

In the case $\tilde{n}_v = 1$, Fig. 1 (A). presents the difference between the outputs of the original system and the reduced-order system under the arbitrary switching signal shown in Fig. 1 (B). It can be observed from the simulation result that the error $e(k,l)$ tends to zero, i,e., the signal $\tilde{y}(k,l)$ follows $y(k,l)$ well, which confirm that the designed reduced-order model is efficient.

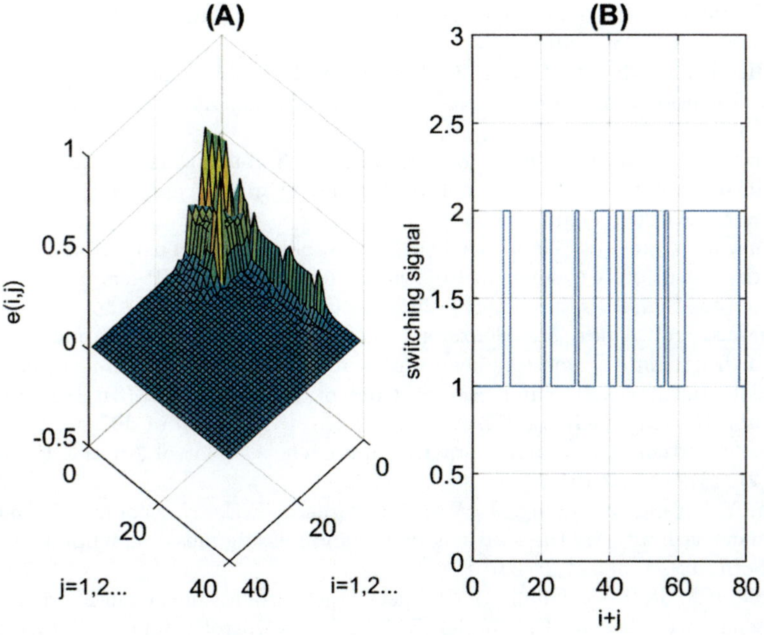


**Fig. 1.** Simulation result.

## 5 Conclusions

The problem of $H_\infty$ model reduction for a class of 2-D discrete switched systems under arbitrary switching signal is investigated. A sufficient condition is developed to ensures not only the asymptotic stability of the error system, but the $H_\infty$-norm of the system is less than a prescript level $\gamma$. The reduced-order model can be designed by solving a set of LMIs. Numerical example is used to prove the feasibility and effectiveness of the developed technique.

## References

1. Roesser, R.: A discrete state-space model for linear image processing. IEEE Trans. Autom. Control **20**(1), 1–10 (1975)
2. Kaczorek, T.: Two-Dimensional Linear Systems. Lecture Notes in Control and Information Sciences, Berlin (1985)
3. Du, C., Xie, L.: $H_\infty$ Control and Filtering of Two-Dimensional Systems, vol. 278. Springer Science & Business Media (2002)
4. Gao, H., Lam, J., Xu, S., Wang, C.: Stability and stabilization of uncertain 2-D discrete systems with stochastic perturbation. Multidimension. Syst. Signal Process. **16**(1), 85–106 (2005)
5. Badie, K., Alfidi, M., Tadeo, F., Chalh, Z.: Delay-dependent stability and $H_\infty$ performance of 2-D continuous systems with delays. Circuits Syst. Signal Process. **37**(12), 5333–5350 (2018)
6. Badie, K., Alfidi, M., Chalh, Z.: New relaxed stability conditions for uncertain two-dimensional discrete systems. J. Control Autom. Electr. Syst. **29**(6), 661–669 (2018)
7. Badie, K, Alfidi, M, Chalh, Z.: Exponential stability analysis for 2D discrete switched systems with state delays. Optim. Control Appl. Meth. 1–16 (2019). https://doi.org/10.1002/oca.2537
8. Wang, L., Wang, W., Gao, J., Chen, W.: Stability and robust stabilization of 2-D continuous-discrete systems in Roesser model based on KYP lemma. Multidimension. Syst. Signal Process. **28**(1), 251–264 (2017)
9. Hmamed, A., Alfidi, M., Benzaouia, A., Tadeo, F.: Robust stabilization under linear fractional parametric uncertainties of two-dimensional systems with Roesser models. Int. J. Sci. Techniq. Autom. Control Comput. Eng. **1**(3), 336–348 (2007)
10. Chaibi, R., Hmamed, A., Tissir, E.H., Tadeo, F.: Control of discrete 2-D Takagi-Sugeno systems via a sum-of-squares approach. J. Control Autom. Electr. Syst. **30**(2), 137–147 (2019)
11. Luo, Y., Wang, Z., Liang, J., Wei, G., Alsaadi, F.E.: $H_\infty$ control for 2-D fuzzy systems with interval time-varying delays and missing measurements. IEEE Trans. Cybern. **47**(2), 365–377 (2016)
12. Badie, K., Alfidi, M., Chalh, Z.: Robust $H_\infty$ control for uncertain 2D state-delayed systems in the second FM model. Int. J. Syst. Control Commun. **10**(3), 218–234 (2019)
13. Gao, C.Y., Duan, G.R., Meng, X.Y.: Robust $H_\infty$ filter design for 2D discrete systems in Roesser model. Int. J. Autom. Comput. **5**(4), 413–418 (2008)
14. de Souza, C.E., Xie, L., Coutinho, D.F.: Robust filtering for 2-D discrete-time linear systems with convex-bounded parameter uncertainty. Automatica **46**(4), 673–681 (2010)
15. El-Kasri, C., Hmamed, A., Alvarez, T., Tadeo, F.: Uncertain 2D continuous systems with state delay: filter design using an $H_\infty$ polynomial approach. Int. J. Comput. Appl. **44**(18), 13–21 (2012)

16. El-Kasri, C., Hmamed, A., Tissir, E.H., Tadeo, F.: Robust $H_\infty$ filtering for uncertain two-dimensional continuous systems with time-varying delays. Multidimension. Syst. Signal Process. **24**(4), 685–706 (2013)
17. Benzaouia, A., Hmamed, A., Tadeo, F., Hajjaji, A.E.: Stabilisation of discrete 2D time switching systems by state feedback control. Int. J. Syst. Sci. **42**(3), 479–487 (2011)
18. Duan, Z., Xiang, Z., Karimi, H.R.: Delay-dependent $H_\infty$ control for 2-D switched delay systems in the second FM model. J. Franklin Inst. **350**(7), 1697–1718 (2013)
19. Shi, S., Fei, Z., Sun, W., Yang, X.: Stabilization of 2-D switched systems with all modes unstable via switching signal regulation. IEEE Trans. Autom. Control **63**(3), 857–863 (2017)
20. Li, H., Yin, S., Pan, Y., Lam, H.K.: Model reduction for interval type-2 Takagi-Sugeno fuzzy systems. Automatica **61**, 308–314 (2015)
21. Su, X., Wu, L., Shi, P., Song, Y.D.: $H_\infty$ model reduction of takagi-sugeno fuzzy stochastic systems. IEEE Trans. Syst. Man Cybern. Part B (Cybern.) **42**(6), 1574–1585 (2012)
22. Zhang, L., Shi, P., Boukas, E.K., Wang, C.: $H_\infty$ model reduction for uncertain switched linear discrete-time systems. Automatica **44**(11), 2944–2949 (2008)
23. Zhang, L., Boukas, E.K., Shi, P.: $\mu$-dependent model reduction for uncertain discrete-time switched linear systems with average dwell time. Int. J. Control **82**(2), 378–388 (2009)
24. Wang, D., Wang, W., Shi, P.: Design on $H_\infty$-filtering for discrete-time switched delay systems. Int. J. Syst. Sci. **42**(12), 1965–1973 (2011)

# Modeling and Developing Control Strategies for the Spherical Inverted Pendulum

Soukaina Krafes[(✉)], Zakaria Chalh, and Abdelmjid Saka

Laboratoire d'Ingénierie, Systèmes et Applications,
Université Sidi Mohammed Ben Abdellah, Fez, Morocco
{soukaina.krafes,zakaria.chalh,abdelmjid.saka}@usmba.ac.ma

**Abstract.** In this article, we are interested in the modeling and the development of nonlinear control strategies of the underactuated mechanical systems. In this respect, we used the Inverted Spherical Pendulum system as a case study to conduct our analysis based on the simulation results performed under Matlab/Simulink.

**Keywords:** Spherical inverted pendulum · Nonlinear control · Underactuted mechanical system

## 1 Introduction

Underactuated mechanical systems refer to those with more number of degrees of freedom than the control inputs. In other words, these systems have generalized coordinates that are not operated. These can be controlled indirectly by dynamically coupled coordinates.

Underactuation is can be found in many applications and fields [22], as an example, robotics that includes, among other things, Underactuated robot manipulators (humanoid robots), the acrobot [10], the pendubot [30], the aerospace systems; vertical take-off and landing aircraft [25], underwater vehicles [38], robotic prosthetic knees for leg amputees and other robots [43]. The underactuation condition in mechanics, robotics, mechatronics or dynamic systems implies that many system configuration variables cannot be controlled directly, the thing that make the construction of control algorithms more difficult [10]. Nevertheless, it is possible to obtain an approximation of the linearized model around a point where the signals are sufficiently small. Due to the powerful methods that have been developed over the last years for linear systems, the first thing in the analysis of nonlinear systems was generally to linearize around an point and to analyze the resulting linear model. Several designs of controllers and linear system analysis techniques have been proposed. However, there are several fundamental limitations of linearization. Where one can cite, the fact that linearized systems are characterized by an unstable equilibrium and nonlinear dynamics that is unmodelled, where there are external disturbances that


A. Saka et al. (Eds.): CPI 2019, LNME, pp. 132–150, 2021.
https://doi.org/10.1007/978-3-030-62199-5_13

vary over time and that cannot be accurately modeled. Adding to this, The dynamics of linear system is poor in front of nonlinear one the thing that can exist only in the presence of nonlinearity. Therefore, this dynamic cannot be predicted and/or described by linear motion equations [15] and therefore modeling errors could weaken the control method based on the linearized system [24,27]. Because of this, control based on a linear model cannot be considered as the best choice because dynamics is usually not available. The presence of disturbances can destabilize the operation of the controller based on a linear model and lead to unstable equilibrium. Unmodelled dynamics and environmental disturbances are important problems in the command of underactuated systems. Indeed, these are fundamentally nonlinear control problems that require ideas and new methods to construct new control strategies. The thing that has attract the attention of the community of researchers in the robotics and nonlinear control fields.

In recent decades, several tools and analysis of control methods have been designed for underactuated mechanical systems. For example, the Lyapunov method for the stability analysis of the nonlinear system [9], the feedback linearization [56], the sliding mode control [19], the Lyapunov redesign [44], backstepping [1] and passivity-based control [48]. In addition, the control of these nonlinear underactuated systems represents a very broad and interesting field in research work today. Where the construction of controllers for underactuated systems is not limited to the interest of controlling them, but it has an advantage for systems that are fully actuated and that also need to have controllers of this kind. These controllers are implemented due to security in case one of the actuators of these systems fails. The inverted pendulum as an example is an underactuated mechanical system [33] that has been considered for more than seven decades to be the most knowen in robotics and control theory. And because of its very simple structure that allows testing of control algorithms, and performing experimental studies is one of the main goal for its use in education and research [18]. Table 1 summarizes the different control strategies used to control a class of UMS. However, despite the importance of the results achieved to date in the field of underactuated systems, these remain applicable only on simple models and having a particular structure. More complex systems such as the inverted spherical pendulum, which is the subject of our study, is a system that has been considered very little and for which we find few results concerning its control in the literature.

One of the reasons for this is that it is theoretically known that inverted and forced spherical pendulums can develop orderly and complex and sometimes chaotic motion patterns. It is also known by the impossibility of measuring its state at its spherical connection. It should be noted that there are no sensors capable of performing this task. It will therefore be interesting to develop nonlinear control strategies capable of controlling the spherical pendulum in the unstable vertical position while ensuring a fairly wide stability domain.

The rest of this manuscript is organized as follows. In Sect. 2, the dynamics of the SIP is developed by the Bond Graph approach. Section 3 discusses control strategies. The results of the simulation and a conclusion are given in Sects. 4 and 5.

**Table 1.** Classification and control stategies of the underactuated mechanical systems

| Classification | Systems | Controllers | | |
|---|---|---|---|---|
| | | Linear | Nonlinear | Intelligent |
| Class I | 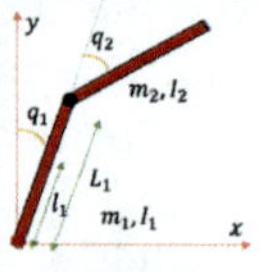 | | [31, 36, 49] | |
| | 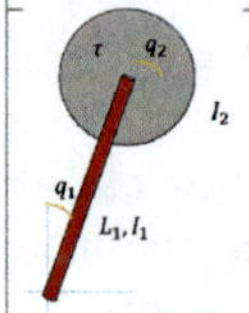 | | [2, 4, 28, 41] | |
| | 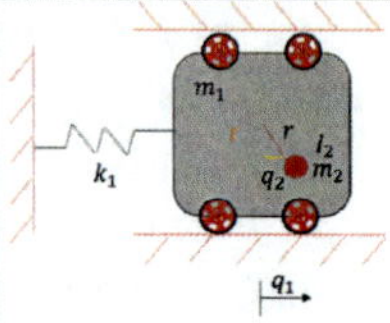 | | [7, 14, 17, 50] | [7, 8, 39] |
| Class II | 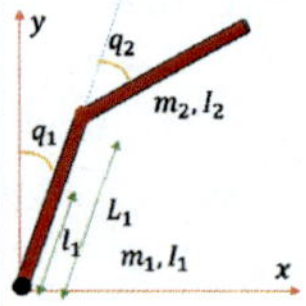 | [52, 54] | [46, 57] | [51, 52] |
| | 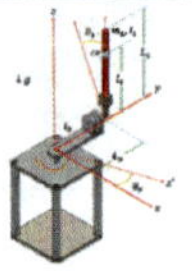 | [3, 42, 53] | [3, 23, 47] | [42] |
| | 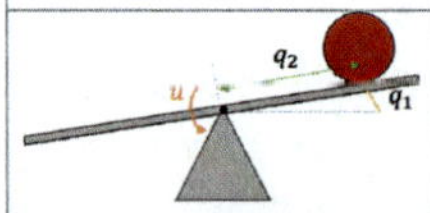 | | [6, 13, 45] | [11–13, 16, 37] |
| Class III | 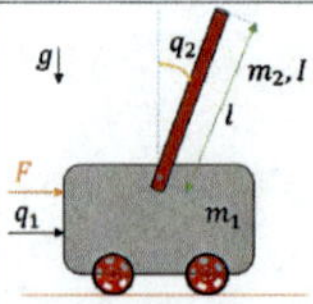 | | [5, 7, 14, 40, 41] | [7, 32] |

## 2 Graphical Modeling of the Spherical Inverted Pendulum

This article deals with the control of the underactuated mechanical system which is the spherical inverted pendulum. The command synthesis for this system is fraught with difficulties from multiple sources. Namely, its nonlinearity, since this nonlinearity makes it impossible to use linear control methods. Although it is always possible to partially linearize the dynamics of an under-actuated mechanical system by state feedback, it is not guaranteed in the general case that a state feedback stabilizing the linearized part also stabilizes the internal dynamics of the system. Its instability, underactuation multi-input multi-output, and adding to this the impossibility of stabilizing it from an initial position $\frac{\pi}{2}$ of the angle of the pendulum. For all these reasons, control synthesis for the spherical inverted pendulum [34,35] system is a vast area of unresolved research.

The spherical inverted pendulum considered in this article consists of an inverted pendulum which is connected via a spherical joint to the center of mass of a base that can move both in the horizontal and vertical planes 1.

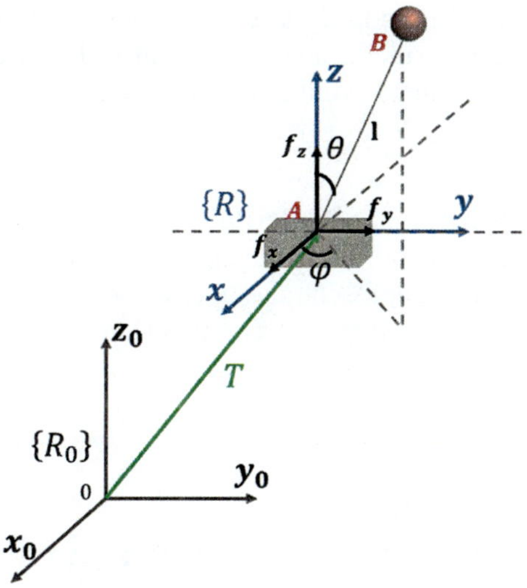


**Fig. 1.** The spherical inverted pendulum

Figure 1 provides the geometric relationships in (1), and the speed constraints (2) can be modeled by the Bond Graph in Fig. 2.

Let $R_0$ a fixed frame and $R$ a moving frame.

Let an spherical inverted pendulum which is connected to $R$ in $A$ and subject to forces $f_x$, $f_y$ and $f_z$ along the axes $x$ axis, $y$ axis and $z$ axis respectively.

The expression of the position vector A and B with respect to frame $R_0$ are given as follow:

$$^{R_0}A = \begin{bmatrix} x_0 \\ y_0 \\ z_0 \end{bmatrix}, {}^{R_0}B = \begin{bmatrix} x_0 + l\cos\varphi\sin\theta \\ y_0 + l\sin\varphi\sin\theta \\ z_0 + l\cos\theta \end{bmatrix} \tag{1}$$

Their velocity vectors:

$$
\begin{aligned}
{}^{R_0}\dot{A} &= \begin{bmatrix} \dot{x}_0 \\ \dot{y}_0 \\ \dot{z}_0 \end{bmatrix}, \\
{}^{R_0}\dot{B} &= \begin{bmatrix} \dot{x}_0 + l\dot{\theta}\cos\varphi\cos\theta - l\dot{\varphi}\sin\varphi\sin\theta \\ \dot{y}_0 + l\dot{\theta}\cos\theta\sin\varphi + l\dot{\varphi}\cos\varphi\sin\theta \\ \dot{z}_0 - l\dot{\theta}\sin\theta \end{bmatrix}
\end{aligned} \tag{2}
$$

The masses of the ball attached to the pendulum as well as the base are represented by the element I. The velocities in the system are represented by 1-junction and kinematic relations between velocities are modeled using the element MTF and the element 0-junction. The forces $f_x$, $f_y$ and $f_z$ applied on the basis are modeled by modulated effort sources (element MSe).

Both flow sources are artificial flow sources introduced by [29]. Where these flow sources resolve the causal conflicts at the junctions 1 to which they are connected.

**Bond Graph Approch for the Calculation of the Lagrange Equations.** The lagrange equations can be determined from a BG by summing the forces of all the junctions 1 which represent the derivative of a generalized coordinate. By calculating the sum of the forces at the junctions 1:$\dot{\theta}$ and 1 : $\dot{\varphi}$, we can find:

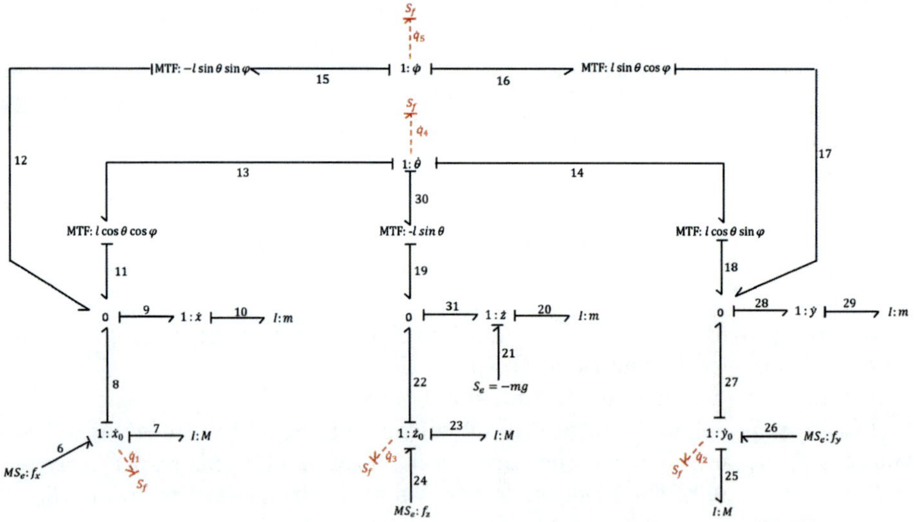

$$
\begin{cases} 0 = l\cos\varphi\cos\theta\,(m\ddot{x}) + l\sin\varphi\cos\theta\,(m\ddot{y}) - l\sin\theta\,(m\ddot{z} + mg) \\ 0 = -l\dot{\varphi}\sin\varphi\sin\theta\,(m\ddot{x}) + l\cos\varphi\sin\theta\,(m\ddot{y}) \end{cases} \tag{3}
$$

With,

$$\begin{cases} \ddot{x} = \begin{matrix} \ddot{x}_0 + l\ddot{\theta}\cos\varphi\cos\theta - l\dot{\theta}^2\cos\varphi\sin\theta - l\dot{\theta}\dot{\varphi}\sin\varphi\cos\theta - l\ddot{\varphi}\sin\varphi\sin\theta \\ -l\dot{\varphi}\sin\theta\dot{\theta}\sin\varphi - l\dot{\varphi}^2\cos\varphi\sin\theta \end{matrix} \\ \ddot{y} = \begin{matrix} \ddot{y}_0 + l\ddot{\theta}\sin\varphi\cos\theta - l\dot{\theta}^2\sin\varphi\sin\theta + l\dot{\theta}\dot{\varphi}\cos\varphi\cos\theta + l\ddot{\varphi}\cos\varphi\sin\theta \\ +l\dot{\varphi}\dot{\theta}\cos\varphi\cos\theta - l\dot{\varphi}^2\sin\varphi\sin\theta \end{matrix} \\ \ddot{z} = \ddot{z}_0 - l\ddot{\theta}\sin\theta - l\dot{\theta}^2\cos\theta \end{cases} \tag{4}$$

Replacing the expressions of the Eqs. (4) in (3), we find:

$$\begin{cases} 0 = \begin{matrix} ml^2\ddot{\theta} - ml^2\dot{\varphi}^2\sin\varphi\cos\theta - mgl\sin\theta + ml^2\sin\varphi\cos\theta\ddot{y}_0 \\ +ml^2\cos\varphi\cos\theta\ddot{x}_0 - ml^2\sin\theta\ddot{z}_0 \end{matrix} \\ 0 = m\left(l\sin\theta\right)^2\ddot{\varphi} + ml^2\sin\varphi\sin\theta\ddot{x}_0 + ml^2\cos\varphi\sin\theta\ddot{y}_0 + 2ml^2\dot{\theta}\dot{\varphi}\sin\theta\cos\theta \end{cases} \tag{5}$$

By calculating the sum of the forces at the junctions $1 : x_0$,$1 : y_0$ et $1 : z_0$, we find:

$$\begin{cases} 0 = f_x - M\ddot{x}_0 \\ 0 = f_y - M\ddot{y}_0 \\ 0 = f_z - M\ddot{z}_0 \end{cases} \tag{6}$$

From (5) and (6), we can deduce the representation of the spherical inverted pendulum system as shown below:

$$\begin{cases} \ddot{x}_0 = f_x \\ \ddot{y}_0 = f_y \\ \ddot{z}_0 = f_z \\ \ddot{\theta} = \dot{\varphi}^2\sin\varphi\cos\theta - \frac{1}{l}\cos\varphi\cos\theta f_x - \frac{1}{l}\sin\varphi\cos\theta f_y + \frac{1}{l}\sin\theta f_z + \sin\theta \\ \ddot{\varphi} = \frac{1}{\sin\theta^2}\left(\frac{1}{l}\sin\varphi\sin\theta f_x - \frac{1}{l}\cos\varphi\sin\theta f_y - 2\dot{\theta}\dot{\varphi}\sin\theta\cos\theta\right) \end{cases} \tag{7}$$

The mathematical model (7) can be rewritten as $\dot{x} = f(x) + \sum_i^3 g_i(x)u_i$ with:

$$f(x) = \begin{pmatrix} x_2 \\ 0 \\ x_4 \\ 0 \\ x_6 \\ 0 \\ x_8 \\ x_{10}^2\cos x_7\sin x_7 - \frac{g}{l}\sin x_7 \\ x_{10} \\ \left(\frac{1}{\sin x_7})^2\right)(-2\dot{x}_7\dot{x}_9\cos x_7\sin x_7) \end{pmatrix} \quad g_1(x) = \begin{pmatrix} 0 \\ 1 \\ 0 \\ 0 \\ 0 \\ 0 \\ 0 \\ -\frac{1}{l}\cos x_7\cos x_9 \\ 0 \\ \left(\frac{1}{\sin x_7})^2\right)\left(\frac{1}{l}\sin x_7\sin x_9\right) \end{pmatrix} \tag{8}$$

$$g_2(x) = \begin{pmatrix} 0 \\ 0 \\ 0 \\ 0 \\ 1 \\ 0 \\ 0 \\ -\frac{1}{l}\sin\varphi\cos\theta \\ 0 \\ \left(\frac{1}{\sin x_7})^2\right)\left(-\frac{1}{l}\sin x_7 \cos x_9\right) \end{pmatrix} \quad g_3(x) = \begin{pmatrix} 0 \\ 0 \\ 0 \\ 0 \\ 0 \\ 0 \\ 0 \\ \frac{1}{l}\sin x_7 \\ 0 \\ 0 \end{pmatrix} \tag{9}$$

## 3 Design of Nonlinear Control Laws for the Spherical Inverted Pendulum

After determining the system's motion equations, this section presents two control laws to control it namely one of the most important developments in the theory of stability which is the Lyapunov stability concept and which is used to develop a controller based on the Lyapunov function to stabilize the pendulum angle from an initial position of $\frac{\pi}{2}$. The simulation results will then be compared to the same system with two planar inputs $f_x$ and $f_y$. As well as a controller is developed using the Backstepping method in such a way as to bring the pendulum from any initial state to the vertically position and the base to the origin coordinates. In addition, a Linear Quadratic Regulator based on the linearized model is also made to compare controller performance.

### 3.1 Lyapunov Approch

Based on the Lyapunov function that have been proposed in [26] for the Spherical inverted Pendulum, we propose to follow the same steps to the determinate the Lyapunov function in the case of SIP with 5 dof:

From [26], the Lyapunov function is given as:

$$\begin{aligned} V(q) &= \tfrac{1}{2}\dot{\xi}^2 + \tfrac{1}{2}\dot{\mu}^2 + \tfrac{1}{2}\dot{\eta}^2 + \tfrac{k_1}{2}\xi^2 + \tfrac{k_2}{2}\mu^2 + \tfrac{k_3}{2}\eta^2 \\ &+ l\left(\tfrac{h_1}{2}\dot{\theta}^2 + \tfrac{h_2}{2}\dot{\varphi}^2\sin\theta^2 + \tfrac{\dot{x}^2}{2} + \tfrac{\dot{y}^2}{2} + \tfrac{\dot{z}^2}{2} + k_p(1-\cos\theta)\right) \end{aligned} \tag{10}$$

### 3.2 Checking the Positiveness of the Proposed Lyapunov Function

Equation (10) can be written as:

$$V = K_p(q) + k_c(q) \tag{11}$$

where:

$$\begin{aligned} K_p(q) &= \tfrac{k_1}{2}\left(x + k_p\cos\varphi\sin\theta\right)^2 + \tfrac{k_2}{2}\left(k_p\sin\varphi\sin\theta\right)^2 \\ &+ \tfrac{k_3}{2}\left(z + k_p\cos\theta\right)^2 + lk_p(1-\cos\theta) \end{aligned} \tag{12}$$

$$\begin{aligned} K_c(q) = \tfrac{1}{2}\left(\dot{x} + k_p\dot{\theta}\cos\varphi\cos\theta - k_p\sin\varphi\sin\theta\dot{\varphi}\right)^2 \\ +\tfrac{1}{2}\left(\dot{y} + k_p\cos\theta\dot{\theta}\sin\varphi + k_p\dot{\varphi}\cos\varphi\sin\theta\right)^2 \\ +\tfrac{1}{2}\left(\dot{z} - k_p\dot{\theta}\sin\theta\right)^2 + l\tfrac{\dot{x}^2}{2} + l\tfrac{\dot{y}^2}{2} + l\tfrac{\dot{z}^2}{2} - \tfrac{lh_1}{2}\dot{\theta}^2 - \tfrac{lh_2}{2}\dot{\varphi}^2\sin\theta^2 \end{aligned} \tag{13}$$

For all $q \neq 0$, we have $K_p(q) > 0$
$K_c(q) = q^T Qq$, $Q > 0$ if $detQ_i > 0$; for $i = 1, \ldots, 5$.

$$Q = \begin{bmatrix} 1+l & 0 & 0 & k_p\cos\varphi\cos\theta & -k_p\sin\varphi\sin\theta \\ 0 & 1+l & 0 & k_p\sin\varphi\cos\theta & +k_p\sin\theta\cos\varphi \\ 0 & 0 & 1+l & -k_p\sin\theta & 0 \\ k_p\cos\varphi\cos\theta & k_p\sin\varphi\cos\theta & -k_p\sin\theta & k_p(k_p-l) & 0 \\ -k_p\sin\varphi\sin\theta & -k_p\cos\varphi\sin\theta & 0 & 0 & k_p\sin\theta^2(k_p-l) \end{bmatrix} \tag{14}$$

Clearly, $det(Q_1) = (1+l)$, $det(Q_2) = (1+l)^2$ and $det(Q_3) = (1+l)^3$ are $> 0$. And by computing the determinant of $Q_4$ and $Q_5$, we get:

$$det(Q_4) = -k_p l(l+1)^2(l-kp+1),\ Q_4 > 0,\ for\ l - kp + 1 < 0$$

$$det(Q_5) = k_p^2 l^2 \sin\theta^2(l+1)(l-k_p+1)^2, Q_5 > 0$$

### 3.3 Derivation of the Control Law

From the proposed Lyapunov function, we determine the derivative of V (10) along the trajectories of system (7). After some calculations:

$$\begin{aligned} \dot{V}(q) = & \nabla V\left(f(q) + g(q)u\right) \\ \dot{V}(q) = & [\xi\tfrac{\partial\dot{\xi}}{\partial q}\ \mu\tfrac{\partial\dot{\mu}}{\partial q}\ \eta\tfrac{\partial\dot{\eta}}{\partial q}]\,[f(q) + g(q)u] \\ \dot{V}(q) = & \dot{\xi}\left(\tfrac{\partial\dot{\xi}}{\partial q}f(q) + \left(\tfrac{\partial\dot{\xi}}{\partial q}g(q) + (l,0,0)\right) * u + k_1\xi\right) \\ & +\dot{\mu}\left(\tfrac{\partial\dot{\mu}}{\partial q}f(q) + \left(\tfrac{\partial\dot{\mu}}{\partial q}g(q) + (0,l,0)\right) * u + k_2\mu\right) \\ & +\dot{\eta}\left(\tfrac{\partial\dot{\eta}}{\partial q}f(q) + \left(\tfrac{\partial\dot{\eta}}{\partial q}g(q) + (0,0,l)\right) * u + k_3\eta\right) \end{aligned} \tag{15}$$

Based on 15, we propose a control law such as:

$$\left(\frac{\partial\dot{\xi}}{\partial q}f(q) + \left(\frac{\partial\dot{\xi}}{\partial q}g(q) + (l,0,0)\right) * u + k_1\xi\right) = -\dot{\xi}, \tag{16}$$

$$\left(\frac{\partial\dot{\mu}}{\partial q}f(q) + \left(\frac{\partial\dot{\mu}}{\partial q}g(q) + (0,l,0)\right) * u + k_2\mu\right) = -\dot{\mu} \tag{17}$$

and

$$\left(\frac{\partial\dot{\eta}}{\partial q}f(q) + \left(\frac{\partial\dot{\eta}}{\partial q}g(q) + (0,0,l)\right) * u + k_3\eta\right) = -\dot{\eta} \tag{18}$$

which leads to

$$\dot{V}(q) = -\dot{\xi}^2 - \dot{\mu}^2 - \dot{\eta}^2 \tag{19}$$

### 3.4 Backstepping Controller

To proceed with the controller design, backstepping technique will be first employed to find the virtual controls.

$1^{st}$ **step:** This step is to control the $x$ position. We define the error by:

$$e_1 = x_1 - x_{1d} \tag{20}$$

Our goal is to construct a virtual command that gives $e_1 = 0$. To do that, we define the Lyapunov function as:

$$V_1 = \frac{1}{2}e_1^2 \tag{21}$$

So,

$$\dot{V}_1 = e_1\dot{e}_1 = e_1(x_2 - \dot{x}_{1d}) \tag{22}$$

This allows as to have the first virtual command:

$$x_{2d} = -c_1 e_1 + \dot{x}_{1d} \tag{23}$$

With: $c_1 > 0$.
(13) becomes:

$$\dot{V}_1 = e_1(x_2 - \dot{x}_{1d}) = -c_1 e_1^2 + e_1 e_2 \tag{24}$$

$2^{nd}$ **step:** By using the second Lyapunov function and expression in (9), the first control law is deduced

We define: $e_2 = x_2 - x_{2d}$ The second Lyapunov function is given as:

$$V_2 = V_1 + \frac{1}{2}e_2^2 \tag{25}$$

So,

$$\begin{aligned} \dot{V}_2 &= \dot{V}_1 + e_2\dot{e}_2 \\ &= -\sum_{i=1}^{2} c_i e_i^2 + e_2(e_1 + c_2 e_2 + u_1 - \ddot{x}_{2d}) \end{aligned} \tag{26}$$

The controller is given by:

$$u_1 = -e_1 - c_2 e_2 + \ddot{x}_{2d} \tag{27}$$

To control the $y$ and $z$ positions, the same steps will be considered as in $1^{st}$ and $2^{nd}$ steps demonstrated below. Which gives the control law for $y$ and $z$ positions respectivelly:

$$u_2 = -e_3 - c_4 e_4 + \ddot{x}_{4d} \tag{28}$$

$$u_3 = -e_5 - c_6 e_6 + \ddot{x}_{6d} \tag{29}$$

**Backstepping Controller for the Angle $\theta$.** In order to control the angle $\theta$ and its angular velocity $\dot{\theta}$, the fourth subsystem in (9) will be considered as a group of 3 subsubsystems.

**Subsubsystem 1:** It will be assumed that only $u_1$ is available control effort:

$$\begin{cases} \dot{x}_7 = x_8 \\ \dot{x}_8 = x_{10}^2 \cos x_7 \sin x_7 + \frac{g}{l} \sin x_7 - \frac{1}{l} u_1 \cos x_7 \sin x_9 \end{cases} \tag{30}$$

$1^{st}$ **step:** We define the error by:

$$e_7 = x_7 - x_{7d} \tag{31}$$

Our goal is to construct a virtual command that gives $e_7 = 0$. To do that, we define the Lyapunov function as:

$$V_7 = \frac{1}{2} e_7^2 \tag{32}$$

So,

$$\dot{V}_7 = e_7 \dot{e}_7 = e_7(x_8 - \dot{x}_{7d}) \tag{33}$$

This allows as to have the first virtual command:

$$x_{8d} = -c_7 e_7 + \dot{x}_{7d} \tag{34}$$

With: $c_7 > 0$
(24) becomes:

$$\dot{V}_7 = e_7(e_8 - c_7 e_7) \tag{35}$$

$2^{nd}$ **step:** By using the second Lyapunov function and expression in (9), the first control law is deduced

We define: $e_8 = x_8 - x_{8d}$ The second Lyapunov function is defined as:

$$V_8 = V_7 + \frac{1}{2} e_8^2 \tag{36}$$

So,

$$\begin{aligned} \dot{V}_8 &= \dot{V}_7 + e_8 \dot{e}_8 \\ &= \begin{aligned} &-\sum_{i=1}^{8} c_i e_i^2 + e_8(e_7 + c_8 e_8 + x_{10}^2 \cos x_7 \sin x_7 \\ &+ \tfrac{g}{l} \sin x_7 - \tfrac{1}{l} u_1 \cos x_7 \sin x_9 - \ddot{x}_{8d}) \end{aligned} \end{aligned} \tag{37}$$

From the equation below, the controller is given by:

$$u_1 = \frac{l}{\sin x_9 \cos x_7} [e_7 + c_8 e_8 + x_{10}^2 \cos x_7 \sin x_7 + \frac{g}{l} \sin x_7 - \ddot{x}_{8d}] \tag{38}$$

**Subsubsystem 2:** It will be assumed that only $u_1$ and $u_2$ are available control efforts:

$$\begin{cases} \dot{x}_7 = x_8 \\ \dot{x}_8 = \begin{array}{l} x_{10}^2 \cos x_7 \sin x_7 + \frac{g}{l} \sin x_7 \\ -\frac{1}{l} u_1 \cos x_7 \sin x_9 - \frac{1}{l} u_2 \cos x_7 \cos x_9 \end{array} \end{cases} \tag{39}$$

By adopting the same steps as $1^{st}$ and $2^{nd}$ steps in **subsubsystem 1**, the controller is given by:

$$u_2 = \frac{l}{\cos x_7 \cos x_9}[e_7 + c_8 e_8 + x_{10}^2 \cos x_7 \sin x_7 + \frac{g}{l} \sin x_7 - \frac{1}{l} u_1 \cos x_7 \sin x_9 - \ddot{x}_{8d}] \tag{40}$$

**Subsubsystem 3:** It will be assumed that $u_1$, $u_2$ and $u_3$ are available control efforts:

$$\begin{cases} \dot{x}_7 = x_8 \\ \dot{x}_8 = \begin{array}{l} x_{10}^2 \cos x_7 \sin x_7 + \frac{1}{l} \sin x_7 \\ -\frac{1}{l} u_1 \cos x_7 \sin x_9 - \frac{1}{l} u_2 \cos x_7 \cos x_9 \\ +\frac{1}{l} u_3 \sin x_7 \end{array} \end{cases} \tag{41}$$

By adopting the same steps as $1^{st}$ and $2^{nd}$ steps in **subsubsystem 1**, the controller is given by:

$$u_3 = \frac{l}{\sin x_7}[-e_7 - c_8 e_8 - \frac{g}{l} \sin x_7 - x_{10}^2 \cos x_7 \sin x_7 + \frac{1}{l} u_1 \cos x_7 \sin x_9 + \frac{1}{l} u_2 \cos x_7 \cos x_9 + \ddot{x}_{8d}] \tag{42}$$

**Backstepping Controller for the Angle $\varphi$.** In order to control the angle $\varphi$ and its angular velocity $\dot{\varphi}$, the fifth subsystem in (9) will be considered as a group of 2 subsubsystems.

**Subsubsystem 1:** It will be assumed that only $u_1$ is available control effort:

$$\begin{cases} \dot{x}_9 = x_{10} \\ \dot{x}_{10} = (\frac{1}{\sin x_7})^2[-\frac{1}{l} u_1 \sin x_7 \cos x_9 - 2 x_8 x 10 \cos x_7 \sin x_7] \end{cases} \tag{43}$$

$1^{st}$ **step:** We define the error by:

$$e_9 = x_9 - x_{9d} \tag{44}$$

Our goal is to construct a virtual command that gives $e_9 = 0$. To do that, we define the Lyapunov function as:

$$V_9 = \frac{1}{2} e_9^2 \tag{45}$$

So,

$$\dot{V}_9 = e_9 \dot{x}_9 = e_9(x_{10} - \dot{x}_{9d}) \tag{46}$$

This allows as to have the first virtual command:

$$x_{10d} = -c_9 e_9 + \dot{x}_{9d} \tag{47}$$

With: $c_9 > 0$
(12) becomes:

$$\dot{V}_9 = e_9(e_{10} - c_9 e_9) \tag{48}$$

$2^{nd}$ **step:** By using the second Lyapunov function and expression in (9), the first control law is deduced

We define: $e_{10} = x_{10} - x_{10d}$ The second Lyapunov function is defined as:

$$V_{10} = V_9 + \frac{1}{2}e_{10}^2 \tag{49}$$

So,

$$\begin{aligned} \dot{V}_{10} &= \dot{V}_9 + e_{10}\dot{e}_{10} \\ &= -\sum_{i=1}^{10} c_i e_i^2 + e_{10}(e_9 + c_{10}e_{10} - 2x_8x_{10}\frac{\cos x_7}{\sin x_7} - \ddot{x}_{10d}) \end{aligned} \tag{50}$$

From the equation below, the controller is given by:

$$u_1 = \frac{\sin x_7}{\cos x_9}\left[e_9 + c_{10}e_{10} - 2x_8x_{10}\frac{\cos x_7}{\sin x_7} - \ddot{x}_{10d}\right] \tag{51}$$

**Subsubsystem 2:** It will be assumed that $u_1$ and $u_2$ are available control efforts:

$$\begin{cases} \dot{x}_9 = x_{10} \\ \dot{x}_{10} = \begin{matrix} (\frac{1}{\sin x_7})^2[-\frac{1}{l}u_1 \cos x_9 \sin x_7 - 2x_8x10 \cos x_7 \sin x_7 \\ +\frac{1}{l}u_2 \sin x_7 \sin x_9] \end{matrix} \end{cases} \tag{52}$$

By adopting the same steps as $1^{st}$ and $2^{nd}$ steps in **subsubsystem 1**, the controller is given by:

$$u_2 = \frac{\sin x_7}{\sin x_9}\left[-e_9 - c_{10}e_{10} + 2x_8x_{10}\frac{\cos x_7}{\sin x_7} + u_1\frac{\cos x_9}{\sin x_7} + \ddot{x}_{10d}\right] \tag{53}$$

### 3.5 LQR

In this section, we will explain the second approach called Linear Quadratic Regulation (LQR).

The Quadratic Linear Regulation called LQR consists in the search for a gain matrix $K$ such as the command by return state:

$$u = -Kx \tag{54}$$

controls the system and minimizes the criterion:

$$J = \int_0^\infty (x^T Qx + u^T Ru)dt \tag{55}$$

where the weighting matrices Q and R satisfy:

$$Q = Q^T \geq 0 \; R = R^T \geq 0 \tag{56}$$

The LQR method insist a linear dynamic. Thus, the nonlinear dynamic model of the inverted spherical pendulum given in 7 is linearized according to [20,21,55].

$$\begin{cases} Q = diag[5,1,5,1,5,1,5,1,5,1] \\ R = diag[3,3,3] \end{cases} \tag{57}$$

In the Eq. (57), the matrices $Q$ and $R$ are respectively state weight and control weight matrices. The diagonals of the matrices $Q$ and $R$ are chosen by tests, in order to obtain good tracking and control performance. The function "*lqr*" attempts to determinate the value of the control gain $K$ in such a way that the state return law $u = -Kx$:

$$K = lqr(A, B, Q, R)$$

$$K = \begin{bmatrix} 0.9110 & 1.2630 & -0.0585 & -0.4886 & -0.0585 & -0.4886 & -0.7328 & 0.1102 & 0.9110 & 1.2591 \\ 0.0413 & -0.2026 & 1.2897 & 2.1759 & -0.0013 & 0.4685 & 0.8518 & -0.0283 & 0.0413 & -0.2040 \\ 0.0413 & -0.2038 & -0.0013 & 0.4685 & 1.2897 & 2.1759 & 0.8518 & -0.0283 & 0.0413 & -0.2028 \end{bmatrix} \tag{58}$$

Therefore, the $u$ control law is given by the Eq. (54).

# 4 Simulations and Results

In the simulations, we assume that there is no friction.

## 4.1 In the Case of Lyapunov Function

To examine the performance of the developed method and controller, we apply the controllers (16), (17) and (18) to the system (7). The positives constants that have been used in the paper were set to:

$$k_1 = 10 \; k_2 = 10 \; k_3 = 1 \; k_p = 20 \; l = 10$$

The initial conditions were set to 0, except the angle of the pendulum that was set to $\theta = \frac{\pi}{2}$. The Fig. 2 show the responses of the angle of the pendulum and its angular velocity with 3 inputs and 2 inputs respectively. As it can be seen, the proposed controller stabilizes the pendulum in its unstable vertical position in the case of 3 inputs, while the pendulum oscillates in the case of 2 inputs. In both situations, there is a decrease of the amplitudes oscillation, but in the case of 3 inputs, the system gives much better results than with 2 inputs Fig. 2.

## 4.2 In the Case of the LQR and the Backstepping Controllers

To examine the performance of the proposed development and backstepping controller, we apply the controllers from (27) to (53) to the system (7).

The simulation results are shown in Fig. 4 and 3. As it can be seen, the Backstepping is done to control the system ensures the stability of the system to achieve a steady state in a short period of time when the pendulum oscillates slightly in the vertical position of $\pm 0.03 rad$ in the angle of $\varphi$. While, for the LQR control, is unable to stabilize the angle $\varphi$ and the position of the base along $x$ axis.

Compared to the LQR method, the Backstepping controller not only performs a fast response but also controls the inverted spherical pendulum in a vertical position and the base in the origin of the coordinates.

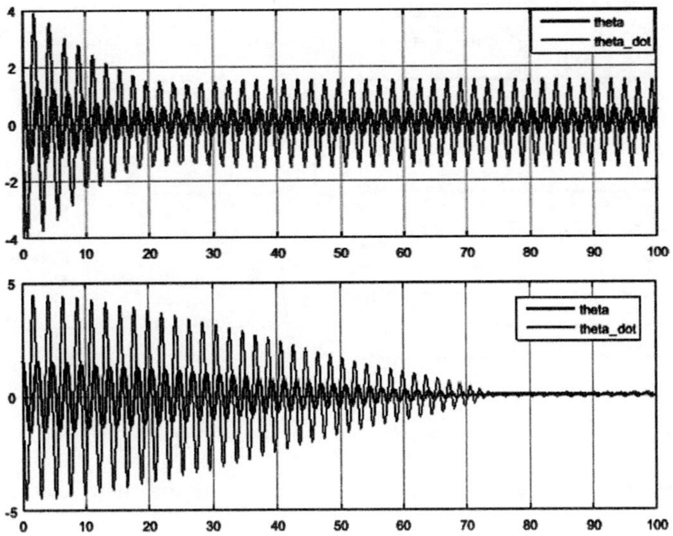


**Fig. 2.** The pendulum angle and its angular velocity with 3 inputs and 2 inputs respectively

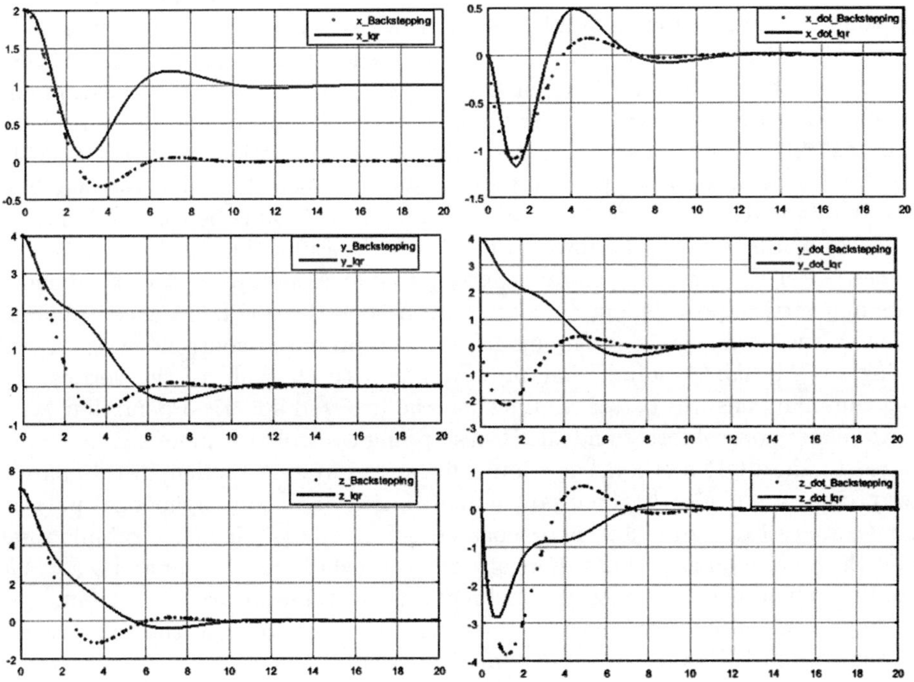


**Fig. 3.** The position of the base and its velocity

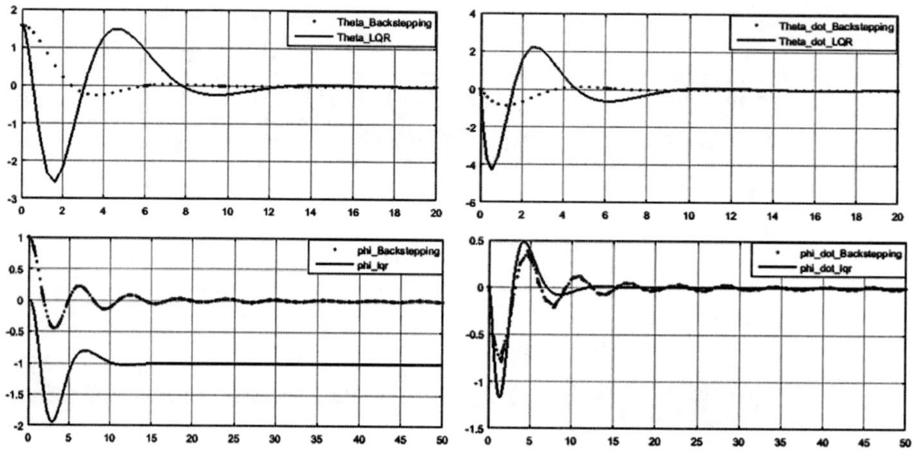


**Fig. 4.** The angular position of the pendulum and their velocities

# 5 Conclusion

In this paper, we have defined some underactuted mechanical systems with their different control strategies that have been proposed. We after that have focused on the inverted spherical pendulum system, where we have presented its graphical modeling for the determination of motion equations of the system. In order to control the system, we have proposed a controller based on the Lyapunov function in such a way as to bring the pendulum of $\frac{\pi}{2}$ as an initial condition to the unstable vertical position. The results of the simulation were compared to the same system with two plane inputs $f_x$ and $f_y$ which show that the energy is slowly dissipated in the case of 2 inputs while it is zero in the case of 3 entries. In other words, adding the force $f_z$ gives better results than those with planar forces. After that, a backstepping controller is developed to control the pendulum in a vertical position and to control the base in the origin of the coordinates. Finally, the performance of the model and the proposed controller were tested using the Matlab/Simulink that proved that with the help of the vertical $f_z$, the command ensures better results than the inverted spherical pendulum with both planar control forces and the Backstepping controller is more efficient compared to the LQR method based on a linearized system model. In this paper, we present a set of orthogonal polynomials based on Jacobi orthogonal polynomials, are called Shifted Jacobi orthogonal polynomials, this helps to build a set of orthogonal moments that are invariant to translation, rotation and scale. We apply a new object classification technique using these invariant moments and an improvement of fuzzy k-means clustering algorithm. To show the effectiveness of our approach we present some numerical results.

## References


1. Adhikary, N., Mahanta, C.: Integral backstepping sliding mode control for underactuated systems: swing-up and stabilization of the cart-pendulum system. ISA Trans. **52**, 870–880 (2013)
2. Aguilar, L.T., Iriarte, R., Orlov, Y.: Variable structure tracking control-observer for a perturbed inertia wheel pendulum via position measurements. IFAC-Papers OnLine **50**(1), 7151–7156 (2017). https://doi.org/10.1016/j.ifacol.2017.08.575
3. Aguilar-Avelar, C., Moreno-Valenzuela, J.: A composite controller for trajectory tracking applied to the furuta pendulum. ISA Trans. **57**, 286–294 (2015). https://doi.org/10.1016/j.isatra.2015.02.009
4. Aguilar-Avelar, C., Rodriguez-Calderon, R., Puga-Guzman, S., Moreno-Valenzuela, J.: Effects of nonlinear friction compensation in the inertia wheel pendulum. J. Mech. Sci. Technol. **31**(9), 4425–4433 (2017). https://doi.org/10.1007/s12206-017-0843-4
5. Aguilar-Ibanez, C., Mendoza-Mendoza, J., Davila, J.: Stabilization of the cart pole system: by sliding mode control. Nonlinear Dyn. **78**(4), 2769–2777 (2014). https://doi.org/10.1007/s11071-014-1624-6
6. Almutairi, N.B., Zribi, M.: On the sliding mode control of a ball on a beam system. Nonlinear Dyn. **59**(1–2), 222–239 (2010). https://doi.org/10.1007/s11071-009-9534-8
7. Azimi, M.M., Koofigar, H.R., Edrisi, M.: Stabilization of underactuated mechanical systems with time-varying uncertainty using adaptive fuzzy sliding mode. Mediterr. J. Meas. Control, p. 8 (2017)
8. Bachir Bouiadjra, R., Khelfi, M.F., Salem, M., Sedraoui, M.: Nonlinear ($h_\infty$) control via measurement feedback using neural network. J. Braz. Soc. Mech. Sci. Eng. **39**(4), 1109–1118 (2017). https://doi.org/10.1007/s40430-016-0597-4
9. Behal, A., Dixon, W., Dawson, D.M., Xian, B.: Lyapunov-Based Control of Robotic Systems (Automation and Control Engineering). CRC Press, Boca Raton (2009)
10. Blajer, W., Dziewiecki, K., Kolodziejczyk, K., Mazur, Z.: Dynamics and control of underactuated mechanical systems: analysis and simple experimental verification. Proc. Appl. Math. Mech. **9**, 107–108 (2009). https://doi.org/10.1002/pamm.200910028
11. Castillo, O., Liz'Arraga, E., Soria, J., Melin, P., Valdez, F.: New approach using ant colony optimization with ant set partition for fuzzy control design applied to the ball and beam system. Inf. Sci. **294**, 203–215 (2015). https://doi.org/10.1016/j.ins.2014.09.040
12. Chang, Y., Chan, W., Chang, C.: T-s fuzzy model-based adaptive dynamic surface control for ball and beam system. IEEE Trans. Ind. Electron. **60**(6), 2251–2263 (2013). https://doi.org/10.1109/TIE.2012.2192891
13. Chang, Y., Chang, C., Tao, C., Lin, H., Taur, J.: Fuzzy sliding-mode control for ball and beam system with fuzzy ant colony optimization. Expert Syst. Appl. **39**(3), 3624–3633 (2012). https://doi.org/10.1016/j.eswa.2011.09.052
14. Chen, Y., Huang, A.: Controller design for a class of underactuated mechanical systems. IET Control Theor. Appl. **6**(1), 103–110 (2012). https://doi.org/10.1049/iet-cta.2010.0667
15. Chernous'ko, F.L., Ananievski, I.M., Reshmin, S.A.: Control of Nonlinear Dynamical Systems: Methods and Applications. Springer, Berlin Heidelberg (2008). https://doi.org/10.1007/978-3-540-70784-4

16. Chien, T., Chen, C., Tsai, M., Chen, Y.: Control of Amira's ball and beam system via improved fuzzy feedback linearization approach. Appl. Math. Model. **34**(12), 3791–3804 (2010). https://doi.org/10.1016/j.apm.2010.03.020
17. Choukchou-Braham, A., Cherki, B., Djemai, M.: A backstepping procedure for a class of underactuated system with tree structure. In: Proceedings of the 2011 International Conference on Communications, Computing and Control Applications, CCCA. pp. 1–6 (2011). https://doi.org/10.1109/CCCA.2011.6031527
18. Dauphin-Tanguy, G.: The Inverted Pendulum in Control Theory and Robotics. IET control, robotics and sensors series, vol. 111 (Olfa Boubaker and Rafael Iriarte)
19. Davila, J., Fridman, L., Levant, A.: Second-order sliding-mode observer for mechanical systems. IEEE Trans. Autom. Control **50**, 1785–1789 (2005)
20. Duan, C., Wu, F.: Output-feedback control for switched linear systems subject to actuator saturation. Int. J. Control **85**(10), 1532–1545 (2012). https://doi.org/10.1080/00207179.2012.691611
21. Duan, C., Wu, F.: Analysis and control of switched linear systems via dwell-time min-switching. Syst. Control Lett. **60**, 8–16 (2014). https://doi.org/10.1016/j.sysconle.2014.05.004
22. Fantoni, I., Lozano, R.: Non-linear Control for Underactuated Mechanical Systems. Springer, London (2002). https://doi.org/10.1007/978-1-4471-0177-2
23. Garcia-Chavez, G., Munoz-Panduro, E.: Global control for the furuta pendulum based on partial feedback linearization and stabilization of the zero dynamics. In: Proceedings of the 2017 13th IEEE Conference on Automation Science and Engineering, pp. 334–339 (2017). https://doi.org/10.1109/COASE.2017.8256125
24. Grasser, F., Arrigo, A., Colombi, S., Rufer, A.: Joe: a mobile, inverted pendulum. IEEE Trans. Ind. Electron. **49**, 107–114 (2002)
25. Gruszka, A., Malisoff, M., Mazenc, F.: Tracking control and robustness analysis for planar vertical takeoff and landing aircraft under bounded feedbacks. Int. J. Robust Nonlinear Control **22**, 1899–1920 (2011). https://doi.org/10.1002/rnc.1794
26. Gutiérrez F., O.O., Aguilar Ibéñez, C., Sossa A.H.: Stabilization of the inverted spherical pendulum via Lyapunov approach. Asian J. Control **11**(6), 587–594 (2009). https://doi.org/10.1002/asjc.140
27. Ha, Y., Yuta, S.: Trajectory tracking control for navigation of the inverse pendulum type self-contained mobile robot. Robot. Auton. Syst. **17**, 65–80 (1996)
28. Iriarte, R., Aguilar, L.T., Fridman, L.: Second order sliding mode tracking controller for inertia wheel pendulum. J. Franklin Inst. **350**(1), 92–106 (2013)
29. Karnopp, D.: Lagrange's equations for complex bond graph systems. J. Dyn. Syst. Meas. Control **99**, 300–306 (1977). https://doi.org/10.1115/1.3427123
30. Kien, C., Son, N., Huy Anh, H.: Swing up and balancing implementation for the pendubot using advanced sliding mode control. In: Proceedings of the 2015 International Conference on Electrical, Automation and Mechanical Engineering, pp. 389–392 (2015). https://doi.org/10.2991/eame-15.2015.108
31. Kozlowski, K., Michalski, M., Parulski, P.: Stabilization of acrobot after landing. In: Proceedings of the 17th International Conference on Climbing and Walking Robots and the Support Technologies for Mobile Machines, pp. 617–624 (2014). https://doi.org/10.1142/9789814623353_0072
32. Krafes, S., Chalh, Z., Saka, A.: Review: linear, nonlinear and intelligent controllers for the inverted pendulum problem. In: 2016 International Conference on Electrical and Information Technologies (ICEIT), pp. 136–141 (2016). https://doi.org/10.1109/EITech.2016.7519577

33. Krafes, S., Chalh, Z., Saka, A.: A review on the control of second order underactuated mechanical systems. Complexity **2018**, 17 (2018). https://doi.org/10.1155/2018/9573514
34. Krafes, S., Chalh, Z., Saka, A.: Vision-based control of a flying spherical inverted pendulum. In: 2018 4th International Conference on Optimization and Applications (ICOA), pp. 1–6 (2018). https://doi.org/10.1109/ICOA.2018.8370509
35. Krafes, S., Chalh, Z., Saka, A.: Visual servoing of a spherical inverted pendulum on a quadrotor using backstepping controller. Int. Rev. Aerospace Eng. (IREASE) **11**(1), 6–14 (2018). https://doi.org/10.15866/irease.v11i1.13460
36. Lai, X., She, J., Yang, S.X., Wu, M.: Comprehensive unified control strategy for underactuated two-link manipulators. IEEE Trans. Syst. Man Cybern. Part B Cybern. **39**(2), 389–398 (2009). https://doi.org/10.1109/TSMCB.2008.2005910
37. Li, X., Yu, W.: Synchronization of ball and beam systems with neural compensation. Int. J. Control Autom. Syst. **8**(3), 491–496 (2010). https://doi.org/10.1007/s12555-010-0301-x
38. Li, Z., Yang, C., Ding, N.: Robust adaptive motion control for underwater remotely operated vehicles with velocity constraints. Int. J. Control Autom. Syst. **10**, 421–429 (2011)
39. Liu, C., Gao, B., Zheng, G., Sun, G.: Fuzzy control design of oscillating trajectory tracking for underactuated tor. Electr. Mach. Control **22**(5), 117–122 (2018). https://doi.org/10.15938/j.emc.2018.05.015
40. Liu, P., Yu, H., Cang, S.: Modelling and control of an elastically joint-actuated cart-pole underactuated system. In: Proceedings of the 20th International Conference on Automation and Computing, ICAC, pp. 26–31 (2014). https://doi.org/10.1109/IConAC.2014.6935455
41. Lopez-Martinez, M., Acosta, J.A., Cano, J.M.: Non-linear sliding mode surfaces for a class of underactuated mechanical systems. IET Control Theor. Appl. **4**(10), 2195–2204 (2010). https://doi.org/10.1049/iet-cta.2008.0583
42. Oltean, S.: Swing-up and stabilization of the rotational inverted pendulum using PD and fuzzy-PD controllers. Procedia Technol. **12**, 57–64 (2014). https://doi.org/10.1016/j.protcy.2013.12.456
43. Oryschuk, P., Salerno, A., Al-Husseini, A., Angeles, J.: Experimental validation of an underactuated two-wheeled mobile robot. IEEE/ASME Trans. Mechatron. **14**, 252–257 (2009)
44. Ravichandran, M.T., Mahindrakar, A.D.: Robust stabilization of a class of underactuated mechanical systems using time scaling and Lyapunov redesign. IEEE Trans. Ind. Electron. **58**, 4299–4313 (2011)
45. Ravichandran, M.T., Mahindrakar, A.D.: Robust stabilization of a class of underactuated mechanical systems using time scaling and Lyapunov redesign. IEEE Trans. Ind. Electron. **58**(9), 4299–4313 (2011). https://doi.org/10.1109/tie.2010.2102318
46. Rudra, S., Barai, R.K.: Design of block backstepping based nonlinear state feedback controller for pendubot. In: Proceedings of the 1st IEEE International Conference on Control, Measurement and Instrumentation, pp. 479–483 (2016). https://doi.org/10.1109/CMI.2016.7413794
47. Rudra, S., Barai, R.K., Maitra, M.: Stabilization of furuta pendulum: a backstepping based hierarchical sliding mode approach with disturbance estimation. In: Proceedings of the 7th International Conference on Intelligent Systems and Control, pp. 99–105 (2013). https://doi.org/10.1109/ISCO.2013.6481130
48. Shiriaev, A., Ludvigsen, H., Egeland, O., Pogromsky, A.: On global properties of passivity based control of the inverted pendulum. In: IEEE Conference on Decision and Control, pp. 2513–2518 (1999)

49. Wang, W., Yi, J., Zhao, D., Liu, X.: Adaptive sliding mode controller for an underactuated manipulator. In: Proceedings of 2004 International Conference on Machine Learning and Cybernetics, pp. 882–887 (2004). https://doi.org/10.1109/ICMLC.2004.1382310
50. Wang, Y., Li, S., Chen, Q.: Stabilization of the translational oscillator with a rotational actuator. J. Inf. Comput. Sci. **8**(8), 1439–1448 (2011)
51. Xia, D., Chai, T., Wang, L.: Fuzzy neural-network friction compensation-based singularity avoidance energy swing-up to non-equilibrium unstable position control of pendubot. IEEE Trans. Control Syst. Technol. **22**(2), 690–705 (2014). https://doi.org/10.1109/tcst.2013.2255290
52. Xia, D., Wang, L., Chai, T.: Neural-network-friction compensation-based energy swing-up control of pendubot. IEEE Trans. Ind. Electron. **61**(3), 1411–1423 (2014). https://doi.org/10.1109/TIE.2013.2262747
53. Xin, X., Liu, Y.: Rotational pendulum. Control Design and Analysis for Underactuated Robotic Systems pp. 109–125 (2014). https://doi.org/10.1007/978-1-4471-6251-3_6
54. Xin, X., Tanaka, S., She, J., Yamasaki, T.: New analytical results of energy-based swing-up control for the pendubot. Int. J. Non-Linear Mech. **52**, 110–118 (2013). https://doi.org/10.1016/j.ijnonlinmec.2013.02.003
55. Zhai, D., Lu, A., Li, J., Zhang, Q.L.: State and dynamic output feedback control of switched linear systems via a mixed time and state-dependent switching law. Nonlinear Anal. Hybrid Syst. **22**, 228–248 (2016). https://doi.org/10.1016/j.nahs.2016.04.007
56. Zhang, A., She, J., Lai, X., Wu, M.: Motion planning and tracking control for an acrobot based on a rewinding approach. Automatica **49**, 278–284 (2013)
57. Zhang, M., Tarn, T.: Hybrid control of the pendubot. IEEE/ASME Trans. Mechatron. **7**(1), 79–86 (2002). https://doi.org/10.1109/3516.990890

# Modeling and Simulation of DFIG Based Wind Turbine System Using Sliding Mode Control

Yahya Dbaghi(✉), Sadik Farhat, and Mohamed Mediouni

Laboratoire des Sciences de l'Ingénieur et Management de l'Energie, ESTA Ibn Zohr University, BP 33/S, 80000 Agadir, Morocco
yahya.dbaghi@gmail.com, sadikfarhat@gmail.com

**Abstract.** This paper describes a study, modeling and control of a doubly fed induction generator (DFIG) based on wind turbine system. The main aim is to regulate the stator active and reactive power of the DIFG to track their references generated by the wind turbine system, in this sense a sliding mode control method is proposed. This technique has the advantage of making any system either linear or not resistant to external disturbances and internal parameters changes. The results of simulations obtained show the good performances and robustness of the proposed SMC controller In comparison with the linear PI controller.

**Keywords:** Inverter · DFIG · Sliding Mode Control (SMC) · PI controller · Wind turbine

## 1 Introduction

The doubly fed induction generator is known as adequate solution to implement in a wind energy system [7]. It offers a lot of advantages, such a high performances with strong stability in variable speed, and reduction of inverter cost. The association of DFIG to static converters gives many ways and possibilities to drive it. Researchers have proposed and tested a various control strategies in literature, like torque control, speed control, and power control. All these strategies have the same aim is to track to maximum power extracted from the wind turbine system [1].

This paper presents a sliding mode control of DFIG. This technique is one of the variable structure control methods; it's a robust control strategy, it consists in converging the state trajectory of a system towards the sliding surface where the system can find stability [6].

This paper begins by presenting the modeling of the DFIG. The third section presents the vector control technology used to control the stator active and reactive power. The Sects. 4 and 5 describe the synthesis of the proportional integrator and the sliding mode controllers. Finally, a comparative study between the proposed controls under different conditions is illustrated by numerical simulations using Matlab/Simulink software in the 6 section.


A. Saka et al. (Eds.): CPI 2019, LNME, pp. 151–160, 2021.
https://doi.org/10.1007/978-3-030-62199-5_14

## 2 Modeling of DFIG

DFIG electrical equations in the park frame are expressed as [4, 5].

$$\begin{cases} V_{ds} = R_s I_{ds} + \frac{d}{dt}\varphi_{ds} - \omega_s \varphi_{qs} \\ V_{qs} = R_s I_{qs} + \frac{d}{dt}\varphi_{qs} + \omega_s \varphi_{ds} \\ V_{dr} = R_r I_{dr} + \frac{d}{dt}\varphi_{dr} - \omega_r \varphi_{qr} \\ V_{qr} = R_r I_{qr} + \frac{d}{dt}\varphi_{qr} + \omega_r \varphi_{dr} \end{cases} \quad (1)$$

Where: Rr and Rs are respectively the rotor and stator resistances per phase.
The rotor and stator flux can be defined as:

$$\begin{cases} \varphi_{ds} = L_s I_{ds} + M I_{dr} \\ \varphi_{qs} = L_s I_{qs} + M I_{qr} \\ \varphi_{dr} = L_r I_{dr} + M I_{ds} \\ \varphi_{qr} = L_r I_{qr} + M I_{qs} \end{cases} \quad (2)$$

Where: Idr, Iqr, Ids, Iqs, are respectively the (d, q) rotor and stator currents. Lr, Ls are respectively the rotor and stator inductances. M is the machine mutual inductance.
The stator active and reactive powers are defined as [1]:

$$\begin{cases} P_s = V_{ds} I_{ds} + V_{qs} I_{qs} \\ Q_s = V_{qs} I_{ds} - V_{ds} I_{qs} \end{cases} \quad (3)$$

The equation of electromagnetic torque is given as:

$$\mathrm{T_{em}} = \frac{\mathrm{pM}}{\mathrm{L_s}}\left(\mathrm{I_{dr}}\varphi_{\mathrm{qs}} - \mathrm{I_{qr}}\varphi_{\mathrm{ds}}\right) \quad (4)$$

P is the pair pole number

## 3 Vector Control of the DFIG

By adjusting the d-axis in the direction of the stator flux [1, 7] we get:

$$\varphi_{ds} = \varphi_s, \ \varphi_{qs} = 0 \quad (5)$$

Therefore by using Eqs. (5) and (6) we obtain:

$$\mathrm{T_{em}} = \frac{\mathrm{pM}}{\mathrm{L_s}} \mathrm{I_{qr}}\varphi_{\mathrm{ds}} \quad (6)$$

We can express the stator flux by:

$$\begin{cases} \varphi_s = L_s I_{ds} + M I_{dr} \\ 0 = L_s I_{qs} + M I_{qr} \end{cases} \tag{7}$$

The stator resistance is neglected due to its small value, in this case the stator voltages can be written as:

$$\begin{cases} V_{ds} = 0 \\ V_{qs} = V_s = \omega_s \varphi_s \end{cases} \tag{8}$$

Stator currents are expressed in terms of rotor currents as:

$$\begin{cases} I_{ds} = \frac{\varphi_s}{L_s} - \frac{M}{L_s} I_{dr} \\ I_{qs} = -\frac{M}{L_s} I_{qr} \end{cases} \tag{9}$$

From (3), (8) and (9) the stator active and reactive powers become:

$$\begin{cases} \mathrm{P_s} = -\mathrm{V_s} \frac{\mathrm{M}}{\mathrm{L_s}} \mathrm{I_{qr}} \\ \mathrm{Q_s} = \frac{\mathrm{V_s}}{\omega_s \mathrm{L_s}} - \mathrm{V_s} \frac{\mathrm{M}}{\mathrm{L_s}} \mathrm{I_{dr}} \end{cases} \tag{10}$$

From the previous equations we obtain the rotor voltages (d, q) in terms of the rotor currents (d, q):

$$\begin{cases} V_{dr} = \left[R_r + S\left(L_r - \frac{M^2}{L_s}\right)\right] I_{dr} - \omega_s g \left(L_r - \frac{M^2}{L_s}\right) I_{qr} \\ V_{qr} = \left[R_r + S\left(L_r - \frac{M^2}{L_s}\right)\right] I_{qr} + \omega_s g \left(L_r - \frac{M^2}{L_s}\right) I_{dr} + \frac{gMV_s}{L_s} \end{cases} \tag{11}$$

Where: g is the slip.

By using these equations we can establish a simplified block diagram of the DFIG that can be regulated (Fig. 1):

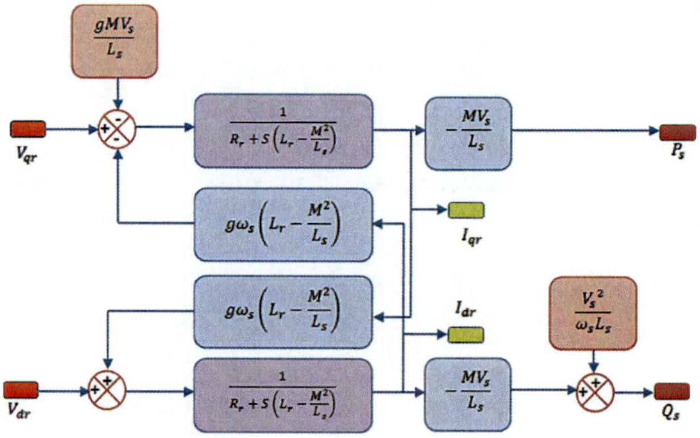


**Fig. 1.** Block diagram DFIG

This block diagram shows that these two controls variables Vdr and Vqr equations are coupled. The decoupling is obtained by compensation in order to control separately Ps and Qs.

## 4 PI Control of the DFIG

The process is first order transfer function whose input is the rotor current signal and its output is the rotor voltage. We make two cascade control loops on each power with an independent regulator (PI) while decoupling rotor voltages and compensating the disturbance terms [7] (Fig. 2).

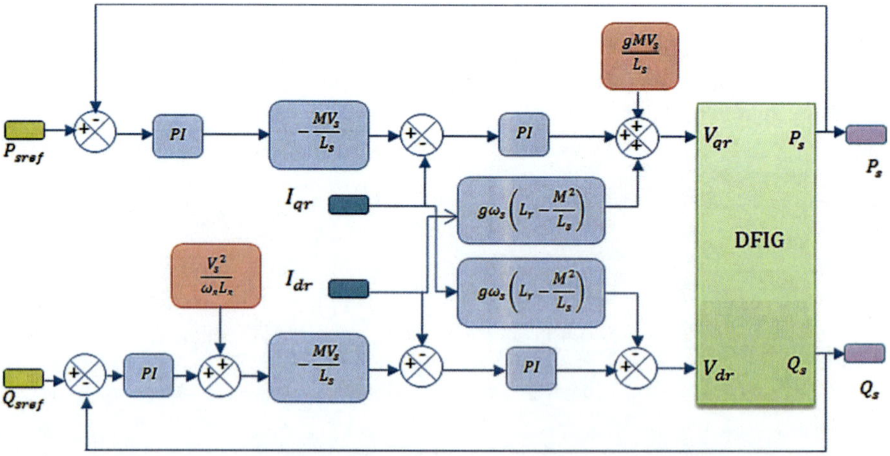


**Fig. 2.** Block diagram of a Proportional-Integral control of DFIG

The Proportional Integral controller (PI) remains the most popularly controller used to control the DFIG, moreover in other industrial control systems. It is simple and easy to implement whereas offering acceptable performances [7]. The synthesis of PI controllers implemented is the pole-zero elimination method is projected in [1] and refined with Simulink tuner at last.

# 5 Sliding Mode Control of the DFIG

The sliding mode control is a nonlinear control method which known with its strong robustness. In this control the system stability is obtained by keeping the system's states on the sliding surface [1, 2].

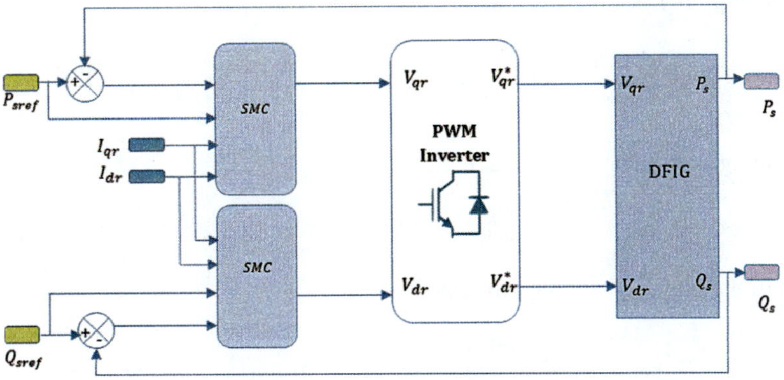


**Fig. 3.** Block diagram of a sliding mode control of DFIG

Figure 3 shows the block diagram where SMC controllers are integrated into the vector control block of DFIG. The active and reactive powers are controlled independently, each with its own Sliding mode controller.

## 5.1 Active Power Control by SMC

For a first order sliding the expressions of the active power control surface become [1, 6]:

$$S(P_s) = \varepsilon = P_s^{ref} - P_s \tag{12}$$

The Lyapunov equation is defined by:

$$V(S_{P_s}) = \frac{1}{2} S_{P_s}^2 \tag{13}$$

By deriving the active power control surface we obtain:

$$\dot{\varepsilon} = \dot{P}_s^{ref} - \dot{P}_s = \dot{P}_s^{ref} + \frac{V_s M}{L_s}\dot{I}_{qr} \tag{14}$$

$$\dot{\varepsilon} = \dot{P}_s^{ref} + \frac{V_s M}{\sigma L_s L_r}\left(V_{qr} - R_r I_{qr} - g\sigma L_r \omega_s I_{dr} - g\frac{V_s M}{L_s}\right) \tag{15}$$

Where: $\sigma = 1 - \frac{M^2}{L_s L_r}$.

We take:

$$V_{qr} = V_{qr_{eq}} + V_{qr_n} \tag{16}$$

Replacing (16) in (15) we obtain again:

$$\dot{\varepsilon} = \dot{P}_s^{ref} + \frac{V_s M}{\sigma L_s L_r}\left(V_{qr_{eq}} + V_{qr_n} - R_r I_{qr} - g\sigma L_r \omega_s I_{dr} - g\frac{V_s M}{L_s}\right) \tag{17}$$

In sliding mode control we have:

$$\dot{\varepsilon} = \varepsilon = 0, V_{qr_n} = 0 \tag{18}$$

Where the control variable equivalent is:

$$V_{qr_{eq}} = -\dot{P}_s^{ref}\frac{\sigma L_s L_r}{V_s M} + R_r I_{qr} + g\sigma L_r \omega_s I_{dr} + g\frac{V_s M}{L_s} \tag{19}$$

To ensure the condition of convergence mode $\dot{\varepsilon} \times \varepsilon < 0$ we should have:

$$V_{qr_n} = -K_{vq}.sign(S(P_s)) \tag{20}$$

With $k_{vq}$ is a positive constant.

## 5.2 Reactive Power Control by SMC

The surface of active stator power is defined by:

$$S(Q_s) = \varepsilon = Q_s^{ref} - Q_s \tag{21}$$

Lyapunov equation:

$$V(S_{Q_s}) = \frac{1}{2}S_{Q_s}^2 \tag{22}$$

By deriving the active power control surface we obtain:

$$\dot{\varepsilon} = \dot{Q}_s^{ref} - Q_s = \dot{Q}_s^{ref} + \frac{V_s M}{L_s} \dot{I}_{dr} \tag{23}$$

$$\dot{\varepsilon} = \dot{Q}_s^{ref} + \frac{V_s M}{\sigma L_s L_r} \left( V_{dr} - R_r I_{dr} + g \sigma L_r \omega_s I_{qr} \right) \tag{24}$$

We take:

$$V_{dr} = V_{dr_{eq}} + V_{dr_n} \tag{25}$$

Replacing (25) in (24) we obtain again:

$$\dot{\varepsilon} = \dot{Q}_s^{ref} + \frac{V_s M}{\sigma L_s L_r} \left( V_{dr_{eq}} + V_{dr_n} - R_r I_{dr} + g \sigma L_r \omega_s I_{qr} \right) \tag{26}$$

In sliding mode control we have:

$$\dot{\varepsilon} = \varepsilon = 0, V_{dr_n} = 0 \tag{27}$$

Where the control variable equivalent is:

$$V_{dr_{eq}} = -\dot{Q}_s^{ref} . \frac{\sigma L_s L_r}{V_s M} + R_r I_{dr} - g \sigma L_r \omega_s I_{qr} \tag{28}$$

To ensure the condition of convergence mode $\dot{\varepsilon} \times \varepsilon < 0$ we should have:

$$V_{dr_n} = -K_{vd}.sign(S(Q_s)) \tag{29}$$

With $k_{vd}$ is a positive constant.

# 6 Results and Discussion

Simulation results are shown by MATLAB/Simulink software. The DFIG used in this work is a 7.5 kW

## 6.1 Normal Tracking Test

We apply a stator active power reference step (−7000 W to 7500 W) at t = 1.5 also for the stator reactive power we apply a reference step (7500 W to 0 W) at t = 1. The simulation results are presented in Figs. 4, 5. The active and reactive powers track perfectly their references. The results show that the sliding mode control yields a strong decoupling between those two powers.

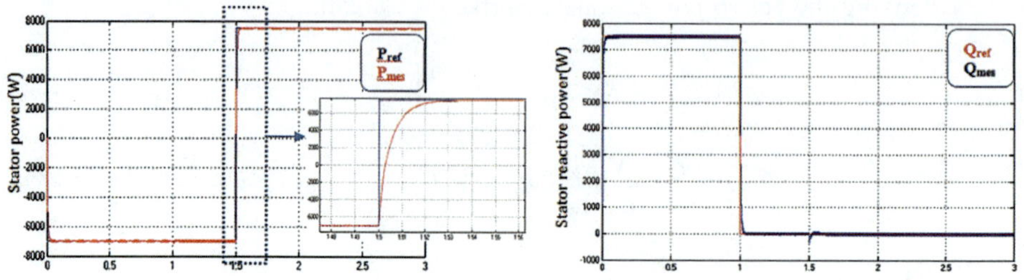


**Fig. 4.** PI controller-Stator active and reactive powers responses

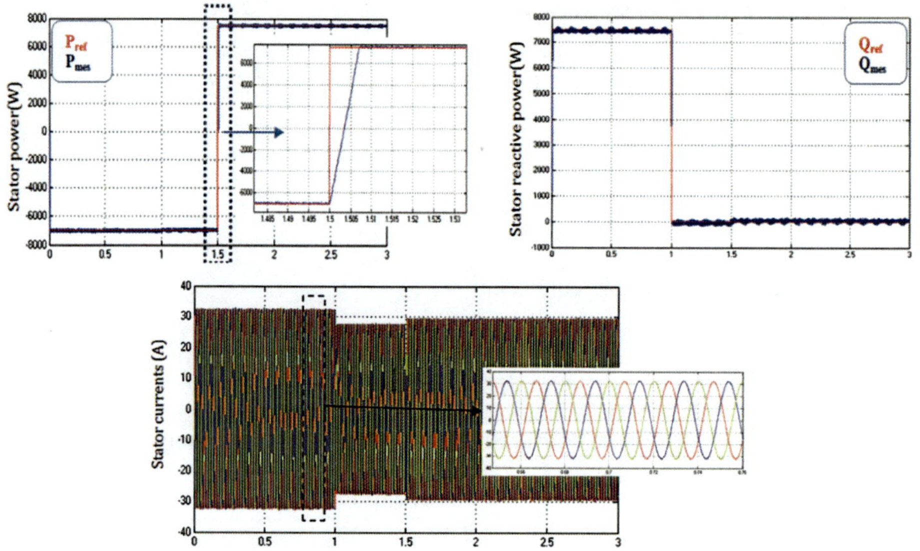


**Fig. 5.** Sliding mode controller-Stator active and reactive powers responses

The active and reactive powers follow with brilliant success their references for the two controllers (PI and SMC) but we can observe that the results of sliding mode controller have a good response time and a perfect decoupling between those two stator powers.

### 6.2 Robustness Test

In this test we have changed the value of the stator inductance as below (Figs. 6 and 7):

$$L_s^{'} = 1.5 \times Ls$$

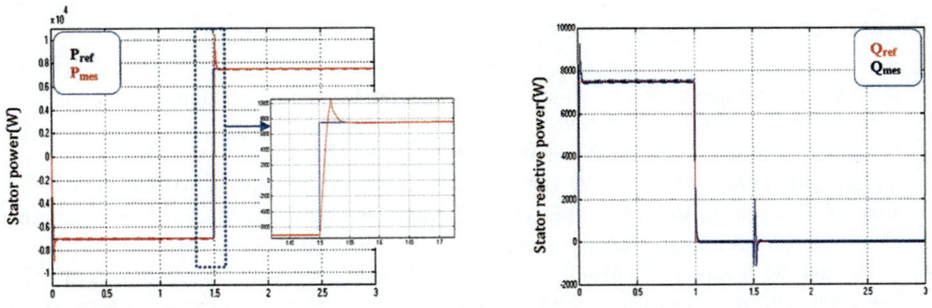


**Fig. 6.** PI controller-Stator active and reactive powers responses ($Ls' = 1.5 \times Ls$)

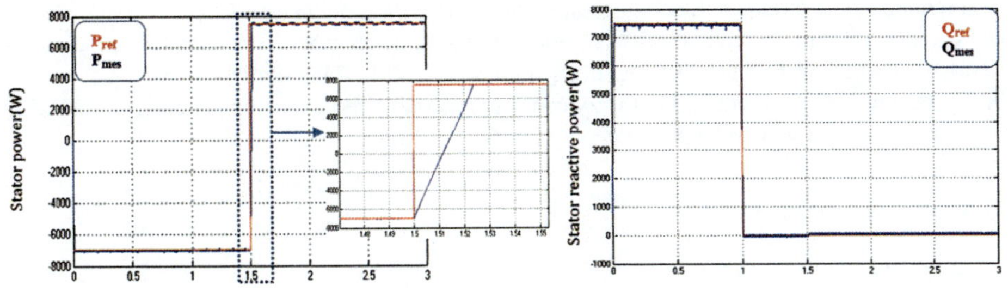


**Fig. 7.** Sliding mode controller-Stator active and reactive powers responses ($Ls' = 1.5 \times Ls$)

Figures show the robustness test response of stator active and reactive power for each controller (PI and SMC). The results responses of Sliding mode controller show that the variation of stator inductance $L_S$ has a slight effect on the response time, a limited variation on the amplitude reference tracking, and at t = 1.5 s the coupling between the stator active and reactive power stayed neglected. But for the PI controller their responses show a fluctuation of amplitude and the coupling between $P_s$ and $Q_s$ is shown again.

## 7 Conclusion

This paper describes a modeling of a doubly fed induction generator whose purpose is to control their stator active and reactive powers. We applied the vector control technique with two types of controller: a linear proportional-integral and nonlinear sliding mode control in order to demonstrate its performance. The results showed the power of the SMC controllers, perfect robustness resistant to machine parameter variations.

## References


1. Kelkoul, B., Boumediene, A.: Nonlinear sliding mode control of DFIG based on wind turbines. In: ICAIRES 2018, LNNS, vol. 62, pp. 206–215 (2019)
2. Bekakra, Y., Ben Attous, D., Bennadji, H.: Sliding mode control of DFIG driven by wind turbine with SVM inverter. In: ICAIRES 2018, LNNS, vol. 62, pp. 192–198 (2019)
3. Abad, G., Lopez, J., Rodriguez, M., et al.: Doubly Fed Induction Machine: Modeling and Control for Wind Energy Generation Applications. Wiley-IEEE Press (2011)
4. El Aimani, S., Francois, B., Minne, F., Robyns, B.: Modeling and simulation of doubly fed induction generators for variable speed wind turbines integrated in a distribution network. In: 10th European Conference on Power Electronics and Applications, Toulouse, France (2003)
5. Bekakra, Y., Attous, D.B.: Active and reactive power control of a DFIG with MPPT for variable speed wind energy conversion using sliding mode control. World Acad. Sci. Eng. Technol. Int. J. Comput. Electr. Autom. Control Inf. Eng. **5**(12) (2011)
6. Kiruthiga, B.: Implementation of first order sliding mode control of active and reactive power for DFIG based wind turbine. Int. J. Inf. Future Res. **2**(8), 2487–2497 (2015)
7. El Azzaoui, M., Mahmoudi, H.: Fuzzy control of a doubly fed induction generator based wind power system. Int. J. WSEAS Trans. Power Syst. **12** (2017)

# Towards a Comparative Assessment Between Physical and Characteristic of Tire of Two-Wheeler Vehicle

Mouad Garziad(✉) and Abdelmjid Saka

Laboratoire d'Ingénierie, Systèmes et Applications, Fez, Morocco
{mouad.garziad, abdelmjid.saka}@usmba.ac.ma

**Abstract.** The tire is one of the most important and critical components in the two-wheeled vehicle as in automotive vehicle. It has direct impact on the performance of the vehicle and the rider, his safety and comfort which are massively influenced by the physical and geometric characteristic of tires. This paper deals with a quantitative and qualitative approach and a compare between four well-known models of tire model such Pacejka tire, modified Dugoff Model, TMeasy tire model, and LuGre tire model. Another objective is to determine the reliable model between these four models to integrate this model in a realistic model of two-wheeler vehicles.

**Keywords:** Tire models · Modeling · Pacejka · Modified Dugoff · TMeasy · LuGre · Comparaison

## 1 Introduction

Tire serves as an important component of the two and four-wheeled vehicles. Their analytical modeling has attracted the attention of numerous scholars in for decades. This component represents the relation between the vehicle and the road, as the only element that physically connects the vehicle to the road. So, there are three essential basic functions to be guaranteed by the tires: the first one is to reinforce the vertical load and reduce the irregularities of the ground, the development the longitudinal efforts that accelerate and brake the vehicle, and develop cross-sectoral efforts to the vehicle to turn. There are plenty of models that exist in the literature. Some of these are simplified while others are more complicated. Further, we can distinguish between three kinds of the tire which are the physical, semi-empirical and the empirical models. The first model which concerns physical models such as Dugoff [1], Brush [2, 3] and TreadSim [4] are based on the theory of the tire physical structure behavior, while the analytical model such LuGre model [3, 6] used to capture crucial phenomena such Stribeck effect. The semi-empirical models such Magic Formula [7] developed by Pacejka, TMeasy [8, 9] developed by professors Rill and Hirschberg, and MF-Swift tire model [10] which is the extension of MF-Tire developed by TNO Automotive. These models utilized broadly a small number of physical parameters to estimate steady-state tire force and moment properties. The empirical tire model is Pacejka's Magic Formula introduced by Pacejka [11] was presented, at first, in 1987 and then extended in 1989


A. Saka et al. (Eds.): CPI 2019, LNME, pp. 161–177, 2021.
https://doi.org/10.1007/978-3-030-62199-5_15

based on a physical background [17]. This approach is able to generate the characteristics that approximately match measured curves for the side force and longitudinal force as a function of their respective slip. Moreover, it described the characteristics of side force, brake force and self-aligning torque with accuracy.

The first part of this article is a selective review of the literature which is going to study tire models and their categories. The major models such as Pacejka's Magic Formula, TMeasy, and Dugoff and LuGre tire model will be analyzed by plotting the forces and moments using Matlab. Section two illustrates a quantitative and qualitative comparison between four models of the tire models and some results and outcomes after some demonstrated simulations. Last but not least, we are going to provide a conclusion that discusses the results and the findings of our development.

## 2 Review of Tire Models

### 2.1 System Context

This section deals with the main tire models in the literature describing the forces and moments produced at the tire-ground interface.

The modeling of the forces of the tire-ground contact is extremely complex. As stated earlier, numerous models exist in the literature as cited below:

- The physical model such as Brush model and Dugoff;
- The analytical model such LuGre model;
- A semi-empirical model such as Pacejka and TMeasy model;
- Empirical model such Magic Formula model.

The most crucial model for each category such as Pacejka Magic Formula, TMeasy, the LuGre tire model and Dugoff model have been analyzed.

The primary requirement of tire models is to estimate the tire forces and moments in the three directions between the tire and the road. Figure 1 shows the forces and torques transmitted.

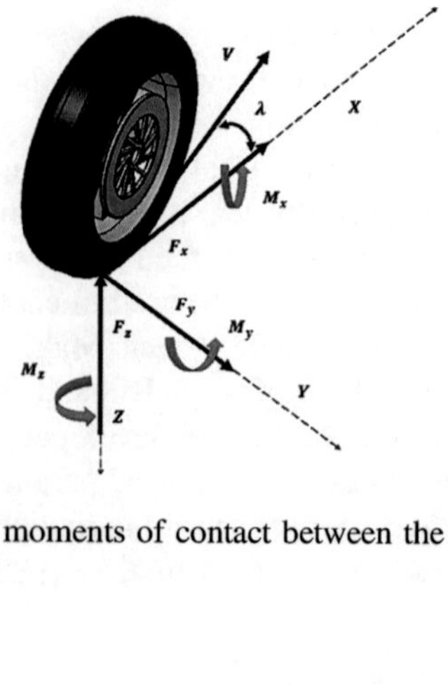


**Fig. 1.** Forces and moments of contact between the tire and the ground.

The forces acting on the tire are the longitudinal force $F_x$, the lateral force $F_y$ and the vertical force $F_z$. The moments acting on the tire are the overturning torque $M_x$, the wheel torque $M_y$ and the Self-aligning torque $M_z$. The longitudinal force $F_x$ and the rolling resistance moment $M_y$ is generated when braking or accelerating. Cornering force $F_y$ is the lateral force produced by a vehicle tire during cornering. The vertical force $F_z$ is the load of the vehicle.

In the upcoming part, we are going to describe the forces and moments produced and developed between the tire and the road using tire models Pacejka, modified Dugoff model, TMeasy Model and LuGre Model, so, For all the models studied, the formulation for longitudinal and lateral forces has been presented and illustrated by physical way.

## 2.2 Pacejka Model

The Pacejka model [7, 11–17] became a prominent model in the vehicle field. Pacejka has produced a range of tire design models: Pacejka Magic Formula 96 Model, Pacejka Magic Formula 2002 Model, Pacejka Magic Formula 2002 Model and finally Pacejka Magic Formula 2006 Model, all in over the past 2 decades. The model offers explanations both theoriticlly and pratically on the modeling of the interaction between the road tire and its influence on the dynamic performance of the vehicle and on the overall behavior and safety of the vehicle. The Pacejka model is based on the mathematical presentation of the dynamic behavior of the tire using analytical functions. This empiric model is able to identify the parameters that correspond to physical characteristics of the pneumatic-ground couple such as longitudinal and transverse stiffness, the coefficient of friction. In fact, it made it possible to calculate the longitudinal and lateral forces as a function of the longitudinal sliding, the slip angle, the camber angle and the normal force as shown in Fig. 2 that illustrates the inputs and the outputs of MF Tire [22].

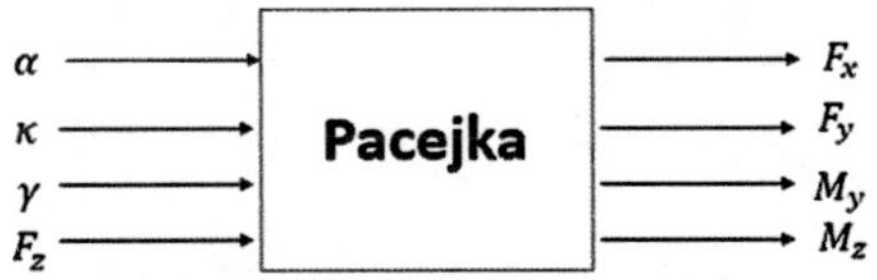


**Fig. 2.** Input and output of Pacejka Model

The model expresses the longitudinal and lateral forces, as the aligning torque in the following:

$$y(x) = D\sin[C arctan[(1 - E)x + (E/B)\arctan(BX)]] \quad (1)$$

With

$$\mathbf{Y(X) = y(x) + S_v}$$

$$\mathbf{x = X + S_h}$$

$X$ illustrates the input variable: the longitudinal slip $\kappa$ as input for longitudinal forces or the lateral slip $\alpha$ as input for lateral forces.

The parameters of equation are:

- **B**: Stiffness factor;
- **C**: Shape factor;
- **D**: Peak value;
- **E**: Curvature factor;
- $\mathbf{S_h}$: Horizontal shift;
- $\mathbf{S_v}$: Vertical shift;

Must be identified from the experimental data using the nonlinear curve-fitting algorithms:

$\boldsymbol{y(x)}$ is the output variable defined, the longitudinal force $F_x$, the lateral force $F_y$, or the self-aligning torque $\boldsymbol{M_z}$ [12] (Fig. 3).

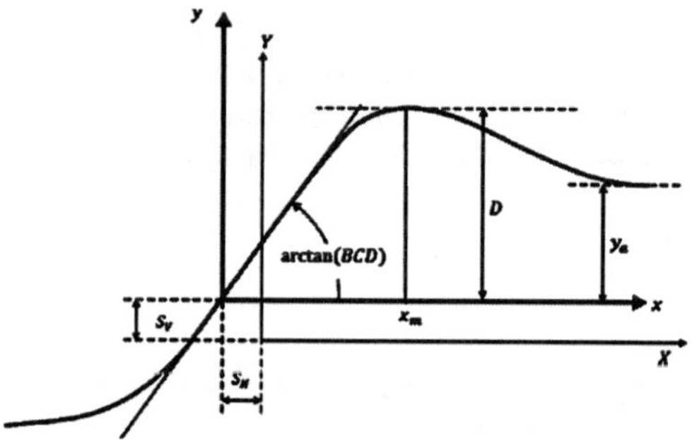


**Fig. 3.** Typical characteristics of Magic Formula

The mathematical formula of Pacejka's model has been described using Matlab to calculate the steady-state force response to side slip and longitudinal slip. The results are obtained by applying a vertical force from $2\,KN$ to $10\,KN$. The curves slope is lower for the lateral force than for the longitudinal force [22] (Figs. 4 and 5).

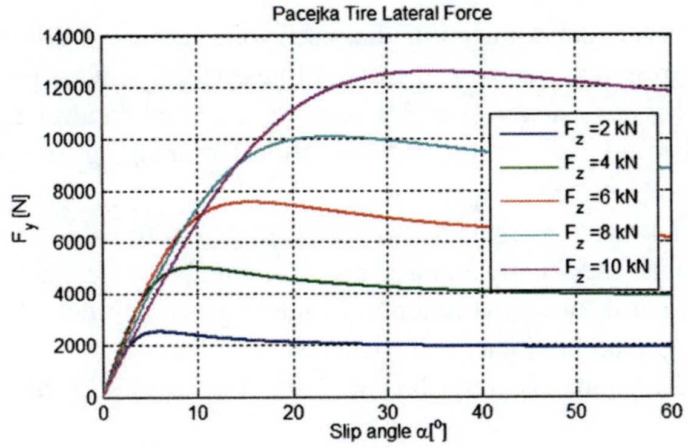


**Fig. 4.** Lateral force versus slip angle for different vertical loads

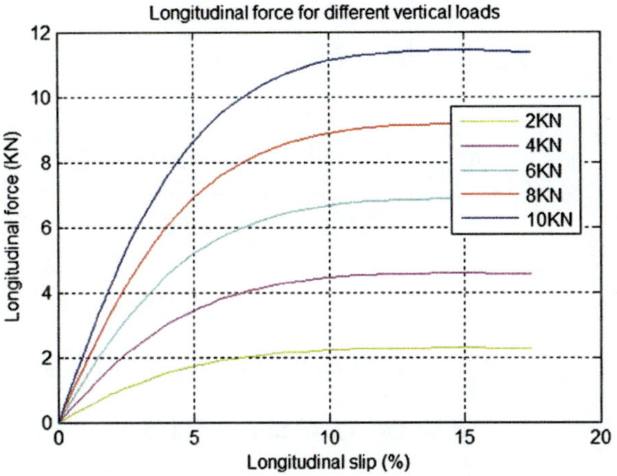


**Fig. 5.** Longitudinal force vs slip angle for different vertical loads

In this model, it is observed that the peak value for the longitudinal force curves is higher than the other two cases. However, In all cases, as the vertical force increases, the forces studied also increase. It is also observable that the shape of the two forces after the peak is different due to the different values of the curve factors.

### 2.3 Modified Dugoff Model

The physical model, which is a part of Dugoff's tire model, was presented by Dugoff et al. in [1]. In this model, a regular vertical pressure distribution is presumed on the tire contact patch. The Dugoff model [1] makes it possible to achieve a simple formulation, for stresses longitudinal or lateral, or coupled longitudinal and lateral simultaneously. The Modified model of Dugoff was proposed by adjusting the experimental results, it used to explicit the dynamic characteristics of tire-road friction forces under pure longitudinal, pure lateral, and combined longitudinal, lateral wheel slip situation.

This model reveals the connections between longitudinal and lateral force and the slip as a function of two parameters; the tire stiffness ($C_x$ $C_y$), which describes the slope of the force-slip curve in low slip region and the tire-road friction coefficient ($\mu$) that draws its curvature and peak value. The correction factors $g_x$, $g_y$ are introduced for the longitudinal and lateral forces equations.

Dugoff's model needs a small number of parameters to estimate the lateral forces [23]. This nonlinear model, by nature, gives an analytical connection between the longitudinal and lateral force as a function of the angle of camber $\gamma$, the slip rate $\lambda$ and the load $F_z$ with a fixed coefficient of friction.

The modified formula as described in [18] can be expressed by the following equations:

$$\begin{cases} F_x = C_x \frac{\lambda}{1-\lambda} f(\theta) g_x \\ F_y = C_y \frac{\tan\alpha}{1-\lambda} f(\theta) g_y \end{cases} \tag{2}$$

$$f(\theta) = \begin{cases} \theta(2-\theta) & \theta \leq 1 \\ 1 & \theta > 1 \end{cases}$$

$$\theta = \frac{\mu F_z \left(1 - \varepsilon v \sqrt{\lambda^2 + tan^2(\alpha)}\right)(1-\lambda)}{\sqrt{2C_x^2\lambda^2 + C_y^2 tan^2(\alpha)}} \tag{3}$$

With

$$g_x = (1,15 - 0,75\mu)\lambda^2 - (1,63 - 0,75\lambda)1,5$$

$$g_y = (\mu - 1,6)\tan(\alpha) + 1,5$$

$$V_\lambda = u\sqrt{\lambda^2 + tan^2(\alpha)}$$

$$\mu = \mu_0(1 - A_\lambda V_\lambda)$$

Where:

- $C_x$ and $C_y$ represent the stiffness of tire;
- $\alpha$ represent side slip angle of tires;
- $\lambda$ is longitudinal slip ratio of tires;
- $\mu$ represent the maximum friction coefficient;
- $u$ is velocity component in the wheel plane;
- $A_S$ is friction reduction factor;
- $g$ presents correction factor.

The mathematical formula of this model has been described using Matlab to calculate the steady-state force response to side slip and longitudinal slip [23].

The results are obtained by applying a vertical force from 2000 N to 8000 N.

The longitudinal and lateral forces under pure slip conditions are shown in Fig. 6 and 7.

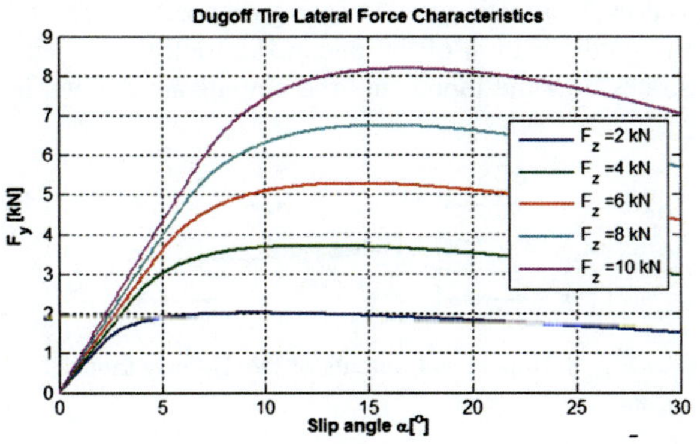


**Fig. 6.** Lateral force versus slip angle for different vertical loads

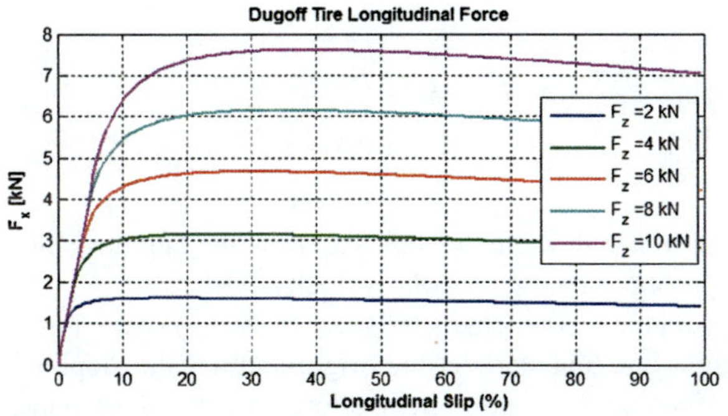


**Fig. 7.** Longitudinal force versus Longitudinal Slip for different vertical loads

We can notice above that the variation of the tyre forces in the tyre load $F_z$ is different from results. So, Different vertical loads have been applied to the previous plots of force characteristics. As the figures suggest, the shape of the lateral and longitudinal forces can be quite different the same vertical force. The reason for this is that the longitudinal force is significantly larger than the lateral force.

## 2.4 TMeasy Tire Model

TMeasy was established based on a simple aspects represented in a practical applicability in vehicle dynamics. The first version was developed in 1994 by Georg Rill. It is a semi-physical tire model used in low-frequency applications for vehicle-dynamics and handling analyses. It divines with reasonable efficiency the steady-state characteristics of the forces and the self-aligning torque amongst the tire and the road. The longitudinal and lateral contact forces are determined by many physical parameters that take into consideration the affection of decreasing tire load [9]. The self-aligning torque is expressed by the multiplication of the pneumatic trail and the lateral force.

Figure 8 illustrates how the inputs and the outputs are defined in this model.

**Fig. 8.** Inputs and outputs of the TMeasy model.

TMeasy is featured by comprising between the model complexity and efficiency in computation time, user-friendliness, and precision in representation [19].

In this model, the slip is described as follows during acceleration or deceleration situations.

$$\begin{aligned} S_x &- \frac{(V_x - r\omega)}{r|\omega|} \\ S_y &\frac{-V_y}{r|\omega|} \end{aligned} \tag{4}$$

Where $r$ is the dynamic radius of the tyre, $\omega$ is the angular velocity of the wheel, and $V_x$ and $V_y$ describe the contact point velocity [19].

TMeasy approximates the characteristics of steady-state tyre forces, The first interval is a straight line that starts at the origin and has the slope of $dF_0$. The second and third intervals where the maximum tyre force is located are parabolas. Finally, the sliding area is a straight line.

Within handling models the steady state tire forces in longitudinal as lateral direction are almost by appropriate functions.

The function $F = F(s)$ is described in intervals by a broken rational function, a cubic polynomial and a constant $F^G$

$$\begin{aligned} F_x^S &= F_x^S(S_x, S_y, \ldots) \\ F_y^S &= F_y^S(S_x, S_y, \ldots) \end{aligned} \tag{5}$$

$$F(S) = \begin{cases} dF^0 & 0 \leq S \leq s^0 \\ as^2 + bs + c & s^0 \leq S \leq s^M \\ ds^2 + es + f & s^M \leq S \leq s^S \\ F^S & s \geq S^S \end{cases} \tag{6}$$

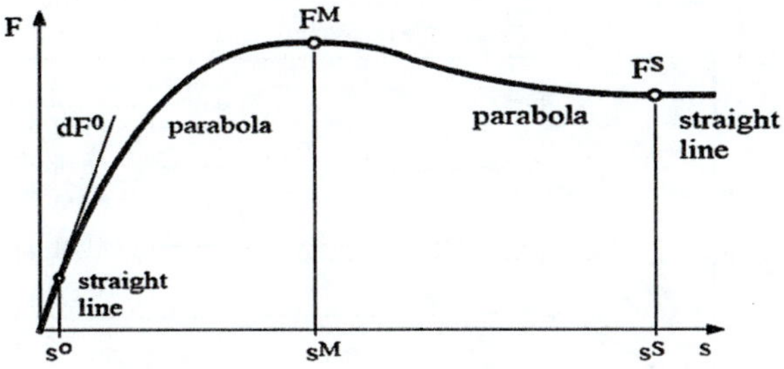


**Fig. 9.** Generalized Tire Characteristics [19]

The maximal forces $F^M$ for different wheel loads $F_z$ are calculated by quadratic functions. For the maximum force it reads as (Fig. 9):

$$F^M(F_z) = \frac{F_z}{F_z^N}\left[2F^M(F_z^N) - \frac{1}{2}F^M(2F_z^N) - \left(F^M(F_z^N) - \frac{1}{2}F^M(2F_z^N)\right)\frac{F_z}{F_z^N}\right]$$
$$S^M(F_z) = S^M(F_z^N) + S^M(2F_z^N) - S^M(F_z^N))\left(\frac{F_z}{F_z^N} - 1\right) \tag{7}$$

Different vertical loads have been applied to the previous plots of force characteristics.

As the Figs. 10 and 11, the shape of the lateral and longitudinal forces can be seen different from the same vertical force. The reason behind that is showed in the longitudinal force which is significantly larger than the lateral force.

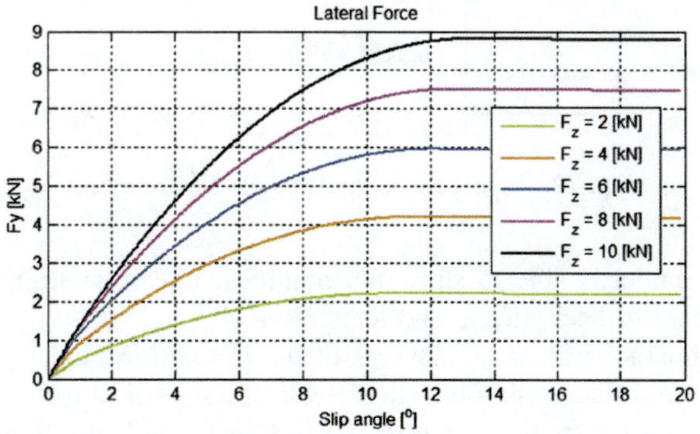


**Fig. 10.** TMeasy Lateral steady-state tire characteristics

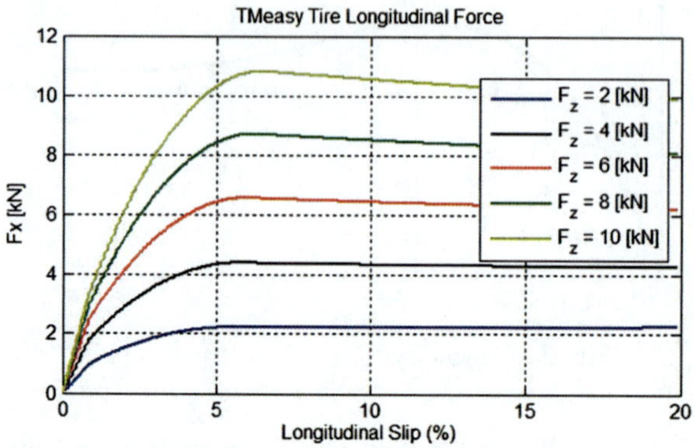


**Fig. 11.** Longitudinal force vs Longitudinal Slip for different Vertical loads

## 2.5 LuGre Tire Model

Wit et al. [3] were the first to represent LuGre tire model, the physics-based dynamic tire model in 1995. The aim was to show the dynamic process when applying a brake torque on the tire.

The surfaces are presumed to be in an interaction through elastic bristles. It represents friction effects of interactions between two contacting surfaces. This dynamic friction model estimates the transient action of the tire-road forces under varying velocity circumstances [20]. Figure 12 illustrates how the inputs and the outputs are defined in this model.

**Fig. 12.** Inputs and outputs of the LuGre model

The **LuGre** model is able to show the nonlinear effects to friction, as diverging break-away forces, Stribeck effects and hysteresis.

One of the most significant advantages of the LuGre tire model is centered on its ability to reflect the surface conditions, the effects of tire vertical force and the effects of the vehicle speed on the friction force [8]. This Model has multiple input that include Tire normal force, Translational speed of wheel center, Angular velocity of wheel and Tire slip angle where outputs of the model estimate Tire longitudinal force, Tire lateral force and Tire aligning moment; There is an interaction between the lateral force and the longitudinal force acting on the tire. To understand the LuGre model, the longitudinal and lateral motion is examined at the first.

$$S = \begin{cases} \frac{r\omega}{v} - 1\, if\, v > r\omega & v \neq 0\, for\, braking \\ 1 - \frac{v}{r\omega} if\, v < r\omega & \omega \neq 0\, for\, driving \end{cases} \tag{8}$$

In the LuGre tire force model, the friction state amongst a tire tread and road which are defined by an interior friction parameter

$z(\xi, t)$ described in time and space, and $\xi$ is a spatial coordinate defined along the contact patch $(0 \leq \xi \leq L)$.

The tire force is described using the equations, showed below, for the longitudinal and lateral friction forces Expression for the tire aligning moment:

$$\frac{d(\xi, t)}{dt} = V_r - \frac{\sigma_0 |V_r|}{g(V_r)} z(\xi, t) \tag{9}$$

$$F = \int_0^L (\sigma_0 z(\xi, t) + \sigma_1 \frac{\partial z}{\partial t}(\xi, t) + \sigma_2 V_r) f_n(\xi) d\xi \tag{10}$$

$$M_z(t) = \int_0^L \mu_y(t, \xi) * f_n(\xi) * \left(\frac{L}{2} - \xi\right) * d\xi \tag{11}$$

The instantaneous friction coefficient $\mu_{i,j}$.

Where $v_r$ is the slip velocity; $\sigma_0$ and $\sigma_1$ are coefficients related to $z$ and $\dot{z}$, respectively; $\sigma_2$ is a viscous damping coefficient, which can be used for defining viscous friction impact between the tread and road. $f_n(\xi)$ is the regular interaction force per unit length.

The function $g(V_r)$ embodies the slip-dependent friction coefficient and is given by:

$$g(V_r) = \mu_c + (\mu_s - \mu_c) \exp\left(-\left|\frac{V_r}{V_s}\right|^{\alpha}\right) \tag{12}$$

where $\mu_c$ and $s$ $\mu_s$ are, respectively, coefficients of Coulomb friction and static friction; $v_s$ is Stribeck velocity; and $\alpha$ is Stribeck exponent (Fig. 14).

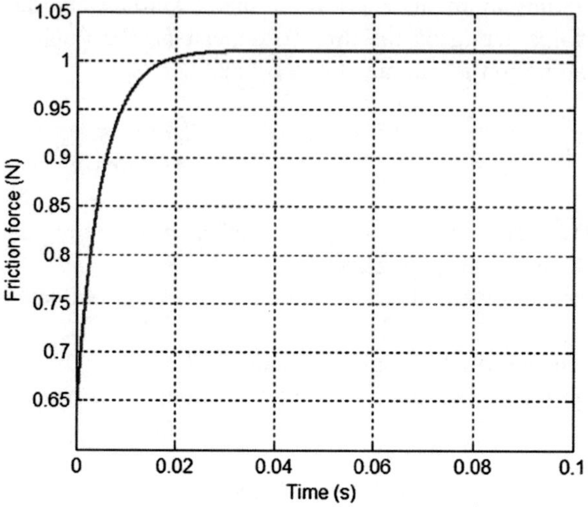


**Fig. 13.** Steady-state distributed LuGre model with uniform vertical force distribution under Velocity

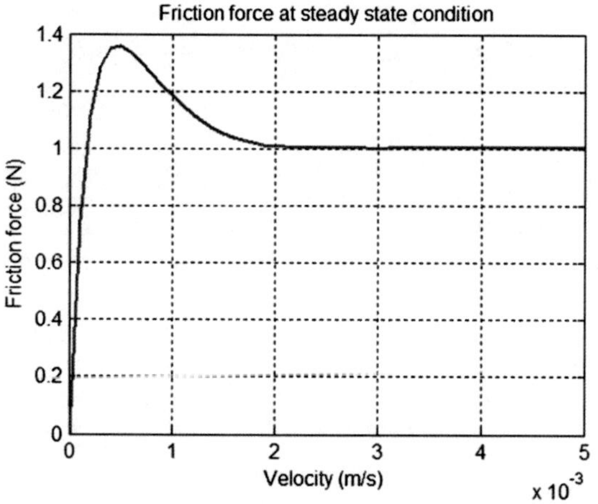


**Fig. 14.** Friction force versus Vs Velocity

## 3 Qualitative Comparison Among Various Tire Models

The following results illustrate the behavior of the tire. Figures 15 and 16 illustrate a comparison between the two models; Modified Dugoff Model and Pacejka Model, on calculating the longitudinal and lateral tire force with the tire vertical load. We can notice that level of the growing linear part, the two models are progressing linearly in the same way, for unstable zone, models acts differently.

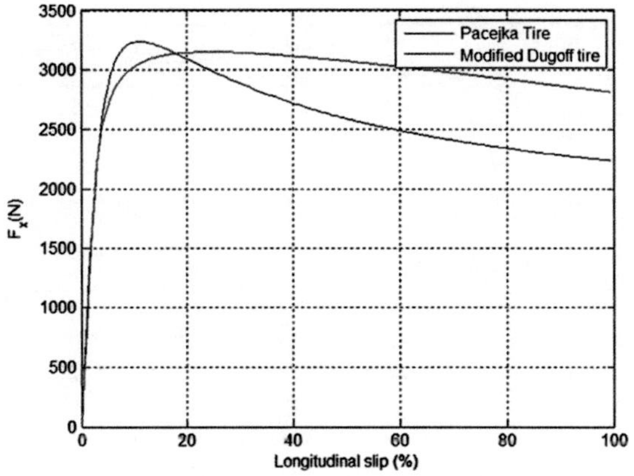


**Fig. 15.** Comparison of modified Dugoff Model with Magic Formula for Longitudinal Force (Fx) vs. slip ratio (S).

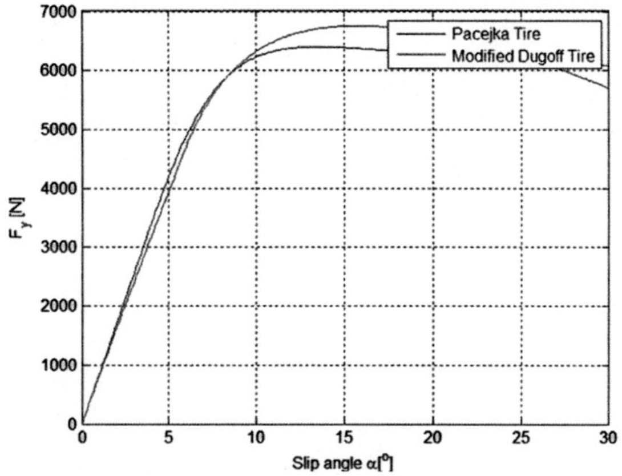


**Fig. 16.** Comparison of Modified Dugoff Model with Magic Formula for Lateral Force ($F_y$) vs. slip angle ($\alpha$).

Figure 17 and 18 represent the comparison of TMeasy model with the Magic Formula on calculating the longitudinal tire force with the tire vertical load of 2 $KN$. So, the curves for the longitudinal have tiny shapes as illustrated in Figs. 13.

However, in the case of the lateral force, they diverge a little bit, as the TMeasy model is not able to approximate the lateral force in the same way as Pacejka.

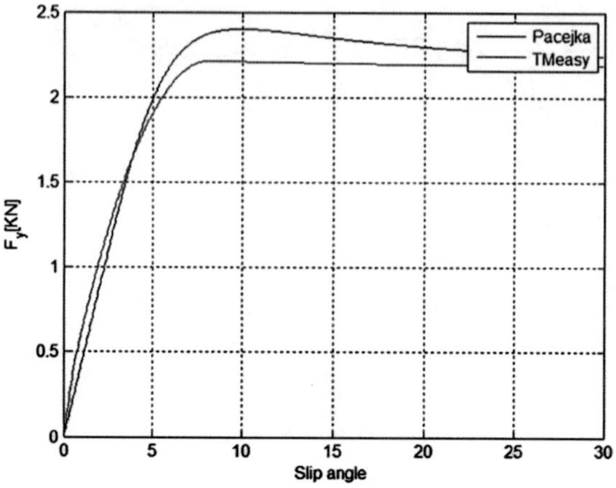


**Fig. 17.** Lateral force comparison between the TMeasy and Pacejka models

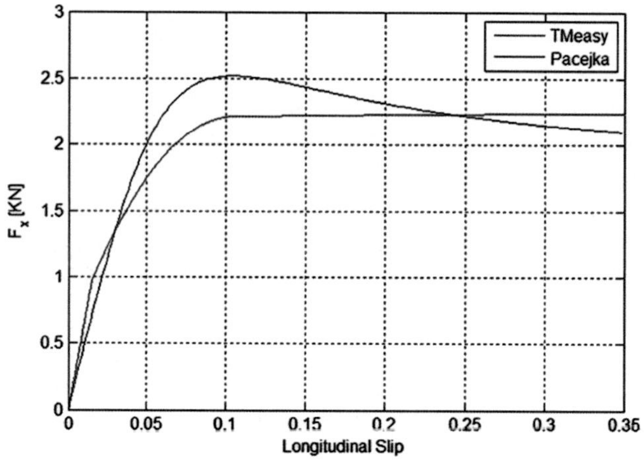


**Fig. 18.** Longitudinal force comparison between the TMeasy and Pacejka models

# 4 Qualitative Comparison Among Various Tire Models

The main outcomes drawn from this study are summarized and categorized in the below table using the scale that shows the strengths and weaknesses of each models. The strengths of Pacejka's Magic Formula are that:

- The model best describes steady-state tire behaviour.
- There is a lot of information about this model, and many extensions have been done.

Regarding the TMeasy tire model, its strengths are that:

- The equations describing it are easy to implement.
- The model parameters have a physical meaning.

The LuGre tire model strength is that:

- It describes both steady-state and dynamic tire behaviour.
- Mechanics of the Tire

The Modified Dugoff tire model strength is that:

- It is able to show the dynamic characteristics of tire-road friction forces beneath pure longitudinal, pure lateral, and combined longitudinal/lateral wheel slip situation.

| Tire models | Advantage | Limitation | Accuracy/Capability | Vehicle systems analysis | Parameter identifications |
|---|---|---|---|---|---|
| MF Tire Model | – Easy to construct and apply<br>– Capture the nonlinearity in tire behavior at very low to very high slip conditions | | High accuracy and efficiency up to 80 Hz Can be used in both steady state and transient state conditions; | – Steering system, Comfort, stability<br>– Handling validation<br>– Control design<br>– Test Rig | – Parameters are identified from experimental tests |
| TMeasy Tire Model | – Easy to Construct and apply efficient;<br>– High accuracy at applicable region | – Restrict to pure slip case;<br>– only apply to certain types of tire; difficult to predict tire behavior beyond the experimental range<br>– cannot handle the large slip or high tire deformation cases | – High accuracy within applicable range | – driving stability<br>– Test Rig | – 10 parameters are identified through experiments, and each of them has actual physical interpretations |
| Modified Dugoff Tire Model | Presents the nonlinear dynamics and combined-slip forces<br>Lateral and longitudinal forces are straightly related to the tire road friction coefficient in clear equations | Describing the dynamic characters of tire forces with the change of wheel slip in many conditions<br>Not accurate for tiny slip angles | More accuracy when is achieved accounting the longitudinal tire dynamics as well as the lateral dynamics<br>Expressed the behavior of the brake force for large slip | Performance Control design | – Model uses less parameters, and is less reliant on accurate tire parametrization the cornering stiffness and friction coefficient |
| LuGre tire | Dynamics of longitudinal and lateral tire friction forces describing special cases of friction situations utilized in vehicle state estimation problems and in tire slip control design | – Cannot reproduce friction characteristics that depend to time | – Capture the complex friction phenomena occurring between the tires and the ground | – Performance<br>– Control design | – The LuGre model contains only a few parameters |

## 5 Conclusion

The tire reprsent one of the most significant components in the vehicle. This article went through dealing with the essential parameters of the tire to exhibit this importance as to show the forces, moments and sideslip angle which are considred as the essential parameters for improving vehicle safety, handling, steering and comfort. This paper used a quantitative and qualitative approaches and provided the differences and similarities between the four models of the tire such Pacejka tire, Dugoff Model, TMeasy tire model, and LuGre tire model.

## References

1. Dugoff, P.F.H., Segel, L.: An analysis of tire traction properties and their influence on vehicle dynamic performance. SAE Trans. **3**, 1219–1243 (1970)
2. Deur, J.: A brush type dynamic tire friction model for non-uniform normal pressure distribution. IFAC Proc. Vol. **35**, 409–414 (2002)
3. Svendenius, J., Wittenmark, B.: Brush tire model with increase flexibility. In: European Control Conference. ECC03, Cambridge, UK (2003)
4. Uil, R.T.: Tyre Models for Steady-State Vehicle Handling Analysis. Eindhoven Univerity of Technology, Eindhoven (2007)
5. Deur, J.: Modeling and analysis of longitudinal tire dynamics based on the LuGre friction model. IFAC Proc. Vol. **34**(1), 91–96 (2001)
6. Yamashita, H., Matsutani, Y., Sugiyama, H.: Longitudinal tire dynamics model for transient braking analysis: ANCF-LuGre tire model. J. Comput. Nonlinear Dyn. **10**(3), 031003 (2015)
7. TNO Automotive: MF-Tool 6.1 User Manual. TNO Automotive, The Netherlands (2008)
8. Rill, G.: First order tire dynamics. In: Proceedings of the 3rd European Conference on Computational Mechanics Solids, Structures and Coupled Problems in Engineering, Lisbon, Portugal (2006)
9. Rill, G.: TMeasy–a handling tire model based on a three-dimensional slip approach. In: Proceedings of the XXIII International Symposium on Dynamic of Vehicles on Roads and on Tracks (IAVSD 2013), Quingdao, China, 2013 August
10. Schmeitz, A.J.C., Besselink, I.J.M., de Hoogh, J., Nijmeijer, H.: Extending the magic formula and SWIFT tyre models for inflation pressure changes. In: Reifen, Fahrwerk, Fahrbahn – VDI Conference, Hannover, Germany, pp. 201–225 (2005)
11. Bakker, E., Nyborg, L., Pacejka, H.B.: Tyre modelling for use in vehicle dynamics studies. SAE Technical Paper 870421 (1987)
12. Pacejka, H.B.: Tyre and Vehicle Dynamics, 2nd edn. Butterworth-Heinemann, Oxford (2006)
13. Pacejka, H.B., Besselink, I.J.M.: Magic formula tyre model with transient properties. Suppl. Veh. Syst. Dyn. **27**, 234–249 (1997)
14. de Vries, E.J.H., Pacejka, H.B.: Motorcycle Tyre Measurements and Models. Delft University of Technology, The Netherlands (1998)
15. Wong, J.Y.: Theory of Ground Vehicles, 3 edizione. Wiley, Hoboken (2001)
16. Pacejka, H.: Tyre and Vehicle Dynamics, 2nd edn. (2006)
17. Pacejka, H.B., Sharp, R.S.: Shear force development by pneumatic tyres in steady state conditions: a review of modelling aspects. Veh. Syst. Dyn. **20**, 121176 (1991)
18. Seddiki, L., Rabhi, A., M'Sirdi, N.K., Delanne, Y.: Analyse comparative des modeles de contact pneu chaussée. e. STA, 3(1) (2006)
19. Gim, G., Nikravesh, P.: An analytical model of pneumatic tires for vehicle dynamic simulations. Part2: comprehensive slips. Int. J. Veh
20. Riedel, A., van Oosten, J.J.M.: Standard tyre interface, release 1.4. Presented at 2nd International Colloquium on Tyre Models for Vehicle Dynamics Analysis, 20–21 February 1997. Issued by the TYDEX - Working Group

21. Canudas-de-Wit, C., Tsiotras, P., Velenis, E., et al.: Dynamic friction models for road/tire longitudinal interaction. Veh. Syst. Dyn. **39**, 189–226 (2003)
22. Garziad, M., Saka, A.: Development and modeling of a PTW vehicle: co-simulation approach. J. Européen des Systèmes Automatisés **52**(2), 115–121 (2019)
23. Seddiki, L., et al.: Analyse comparative des modeles de contact pneu chaussée. e. STA 3.1 (2006)

# Parameters Extraction of Single Diode PV Model and Application in Solar Pumping

Mustapha Errouha[1(✉)], Saad Motahhir[2], Quentin Combe[3], and Aziz Derouich[1]

[1] Laboratory of Production Engineering, Energy and Sustainable Development, Higher School of Technology, USMBA, km5 Rte Imoouzzer, BP 2427, Fez, Morocco
errouha.mustapha@gmail.com
[2] ENSA, SMBA University, Fez, Morocco
[3] LEMTA, University of Lorraine, Vandœuvre-lès-Nancy, France

**Abstract.** This work aims to propose a technique giving a good compromise between accuracy and simplicity to identify the parameters of a single diode photovoltaic (PV) panel. the proposed extraction of the DC parameters of solar panel is based on experimental measurement and the manufacturer data. This method requires the measuring of the shunt resistance, the resolution of the equations related to this technique allowed to determine the other parameters such as (saturation current, light-generated current, series resistance and ideality factor). The suggested model permitted to predict the behavior of the PV panel. To validate this method, the I-V and P-V curves are compared with experimental and manufacturer data, under different meteorological conditions. The results show the simplicity, efficiency and the precision of the developed model. A dimensioning is done thereafter to determine the necessary field to feed a centrifugal pump.

**Keywords:** PV cell · Solar radiation · Modeling · Experimental validation · PSIM · COMPASS

## 1 Introduction

The big increase of fuel and gas prices make the production of electricity more expansive, for this reason, the use of other sources of energy production becomes an obligation. Renewable energy sources such as solar, hydro, wind and biomass energies have become more important contributor to the total energy consumed in the world and considered as potential solutions to protect the environment [1]. In fact, the use of solar energy grew to 25% over the past 20 years [2]. The superiority of photovoltaic systems to other types of renewable energy is due to its clean production, silent behavior and free cost because it depends only on solar irradiation and temperature.

Photovoltaic cell is a sensor made from semiconductor material which allows to convert the absorbed light energy into an electric current [3]. The principle of this


A. Saka et al. (Eds.): CPI 2019, LNME, pp. 178–191, 2021.
https://doi.org/10.1007/978-3-030-62199-5_16

operation is based on the absorption properties of light radiation by semiconductor materials [4]. The power delivered by the PV cell is low, which makes it insufficient for most domestic or industrial applications, for this, the cells can be connected either in parallel or in series according to the needs of the intended application [5]. When the cells are assembled in series, the voltages add up, if they are connected in parallel, currents add up, this operation form PV generator which named as a PV panel [6]. The photovoltaic cell can be realized by technologies: monocrystalline, polycrystalline and thin layers [7].

PV panels are used in several applications, including standalone pump stations, hybrid cars ... But to fully exploit the PV energy, it is necessary first to master the principle of operation of the PV panel, for this, modeling is an essential step to achieve. The researchers proposed different models for PV panels [6–8]. In particular the model with a single diode and the model with two diodes, and with consequently the choice of one model or another is a little difficult. Two diode model is very precise and gives satisfactory results when the irradiation is low [9], but it's complex considering the number of parameters intervening in this model [10]. The four parameters model (4-p), in which the effect of the shunt resistance is negligible [11], this model is used to simulate photovoltaic panels and improved the performance of pumping system installations [10]. The five parameters model (5-p) introduces a parallel resistance in the equivalent circuit of the photovoltaic cell [4], the leakage current of the P-N junction is described by this resistance [12]. This model is the most used because it presents a compromise between the precision and the simplicity [11].Comparative studies have been made between the 4-p and 5-p models, the results have shown that the 4-p model is less accurate than the 5-p model [5]. So we can be satisfied with this model to do our research.

In order to use the chosen model, the elements of the equivalent circuit model of the cell must be determined. In the literature, analytical, numerical methods evolutionary computational methods are used to calculate model parameters [12, 13]. In this work, a new method to determine the missing parameters of the used model is proposed. To validate the proposed extraction strategy, P-V and I-V curves are compared with I-V and P-V experimental and manufacturer curves under different weather conditions. Moreover, a theoretical dimensioning is presented in this paper to define the photovoltaic field necessary to feed a centrifugal pump. The dimensioning result obtained is validated by Compass Lorentz software.

## 2 PV Cell

Figure 1 illustrates the adopted model of PV cell which contains two resistors, a current source and a diode

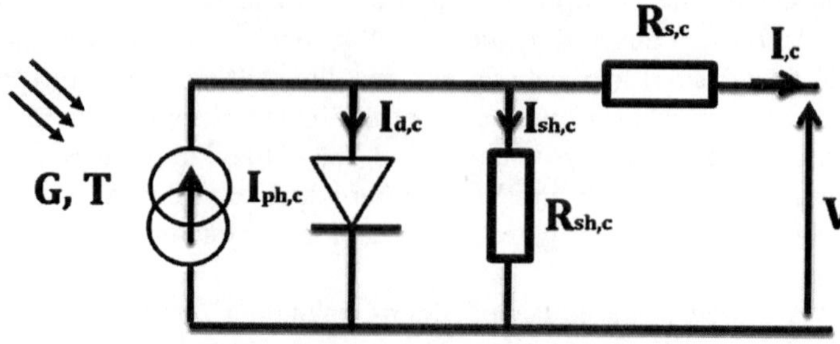


**Fig. 1.** Equivalent model of a PV cell

The mathematical model for the current-voltage characteristic of a PV cell is:

$$I_{,c} = I_{ph,c} - \mathrm{I}_{o,c}\left(\exp\frac{q(V_{,c} + R_{s,c}I_{,c})}{aKT} - 1\right) - \frac{(V_{,c} + I_{,c}R_{s,c})}{R_{sh,c}} \tag{1}$$

Where

$I_{ph,c}$: the photocurrent of a cell,
$I_{0,c}$: the reverse saturation current of the diode,
q: the electron charge
a: the ideality factor of the diode
T: the junction temperature
K: the Boltzmann's constant respectively
$V_{,c}$: the output voltage of PV cell,
$I_c$: the output current of PV cell
$R_{s,c}$: the series resistance of the cell
$R_{sh,c}$: the shunt resistance of the cell.

## 3 Model of the PV Panel

The PV panel used in this study consists of Ns cells coupled in series in order to increase the voltage, and therefore the functioning of our photovoltaic panel can be modeled by the following approximate analytical expression:

$$I = I_{ph} - \mathrm{I}_O\left(\exp\frac{q(V + R_s I)}{aKTN_s} - 1\right) - \frac{(V + IR_s)}{R_{sh}} \tag{2}$$

$$I_{ph} = I_{sc} + K_i\,(T - 298.15)\frac{G}{1000} \tag{3}$$

$$I_o = \frac{I_{sc} + K_i(T - 298.15)}{\exp\left(\frac{q(V_{oc} + K_v(T - 298.15))}{akTN_s}\right) - 1} \tag{4}$$

Where:

$I_{ph}$: the photocurrent
$I_0$: the reverse saturation current
$N_s$: the number of cells coupled in series
$I$: the current generated by the panel
G: the solar irradiation
$R_{sh}$: the shunt resistance
$R_s$: the series resistance

## 4 Proposed Extraction Method

To adjust the PV panel, electrical parameters such as (ideality factor a, saturation current, the series resistance *Rs*, the light-generated current *Iph*, and shunt resistance *Rsh*) should be determined because there are not given by the manufacturer. To do that, extraction parameters method is used. The proposed technique is based on two steps. Firstly, *Rsh* is measured on obscurity and no wind as shown in Fig. 3.

The second step consists of using the manufacturer data to establish the system for parameter determination [14]:

$$\begin{cases} 0 = I_{ph} - I_0\{\exp[AV_{oc}] - 1\} - \frac{V_{oc}}{R_{sh}} \\ I_{sc} = I_{ph} - I_0\{\exp[AI_{sc}R_s] - 1\} - \frac{R_s I_{sc}}{R_{sh}} \\ I_m = I_{ph} - I_0\{\exp[A(V_m + R_s I_m)] - 1\} - \frac{V_m + R_s I_m}{R_{sh}} \end{cases}$$

The saturation current I0 value at temperature of 298.18 K is given by:

$$I_o = \frac{I_{SC} - \frac{V_{oc} - R_s I_{SC}}{R_{sh}}}{\exp(AV_{oc}) - \exp(AI_{SC}R_s)} \tag{5}$$

Based on the equations cited in previous system, the Eqs. 6 and 7 are obtained:

$$\frac{I_{SC} - \frac{V_{oc} - R_s I_{SC}}{R_{sh}}}{\exp(AV_{oc}) - \exp(AI_{SC}R_s)} = \frac{I_m - \frac{V_{oc} - V_m - R_s I_m}{R_{sh}}}{\exp(AV_{oc}) - \exp[A(V_m + R_s I_m)]} \tag{6}$$

$$\frac{I_m - \frac{V_{oc} - V_m - R_s I_m}{R_{sh}}}{\exp(AV_{oc}) - \exp[A(V_m + R_s I_m)]} = \frac{\frac{I_m}{V_m - R_s I_m} - \frac{1}{R_{sh}}}{A\exp[A(V_m + R_s I_m)]} \tag{7}$$

Where

$$\mathrm{A} = \frac{q}{qKTNs}$$

By revolving the eq the missing parameters are determined and represented in Table 2. Figure 2 shows the steps taken to determine the missing parameters.

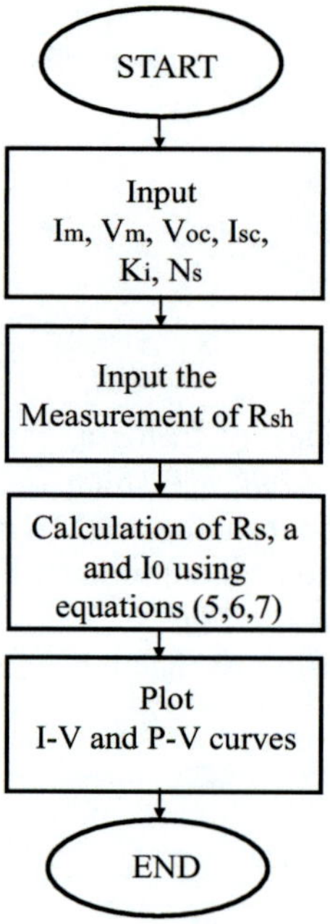


**Fig. 2.** Diagram of proposed method

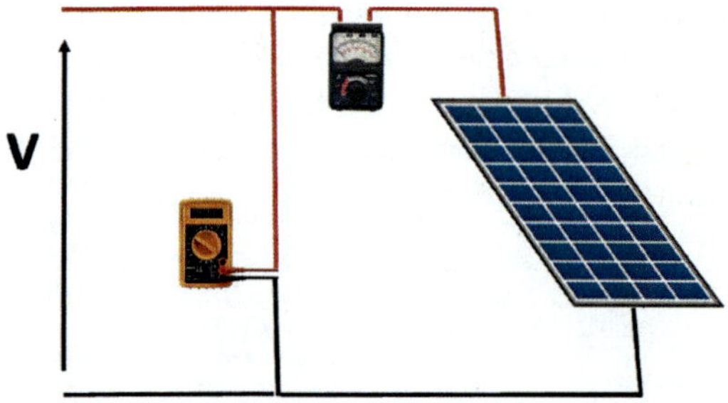


**Fig. 3.** Measurement of the shunt resistance

The characteristics of our photovoltaic panel adopted in this contribution are presented in Table 1:

**Table 1.** Electrical characteristics of the CSUN235-60P Photovoltaic module

| | |
|---|---|
| Maximum power | 235 W |
| Open circuit voltage | 36.8 V |
| Short circuit current | 8.59 A |
| Maximum power voltage | 29.5 V |
| Maximum power current | 7.97 A |
| Number of cell | 60 |
| Temperature coefficient of open-circuit Photovoltaic Panel CSUN235-60P voltage $V_{oc}$, $K_V$ | −0.37%/°C |
| Temperature coefficient of short-circuit current $I_{sc}$, $K_i$ | 0.07%/°C |

**Table 2.** Parameters of the panel CSUN235-60P

| | |
|---|---|
| Photocurrent $I_{ph}$ | 8.607 A |
| Saturation current $I_0$ | 4.1381 $e^{-10}$ A |
| series resistance $R_s$ | 0.34814 Ω |
| Shunt resistance $R_{sh}$ | 180 Ω |
| Ideality factor a | 1.0058 |

On the basis of Eq. (2), (3) and (4), the model developed in Fig. 4 is obtained using the software PSIM:

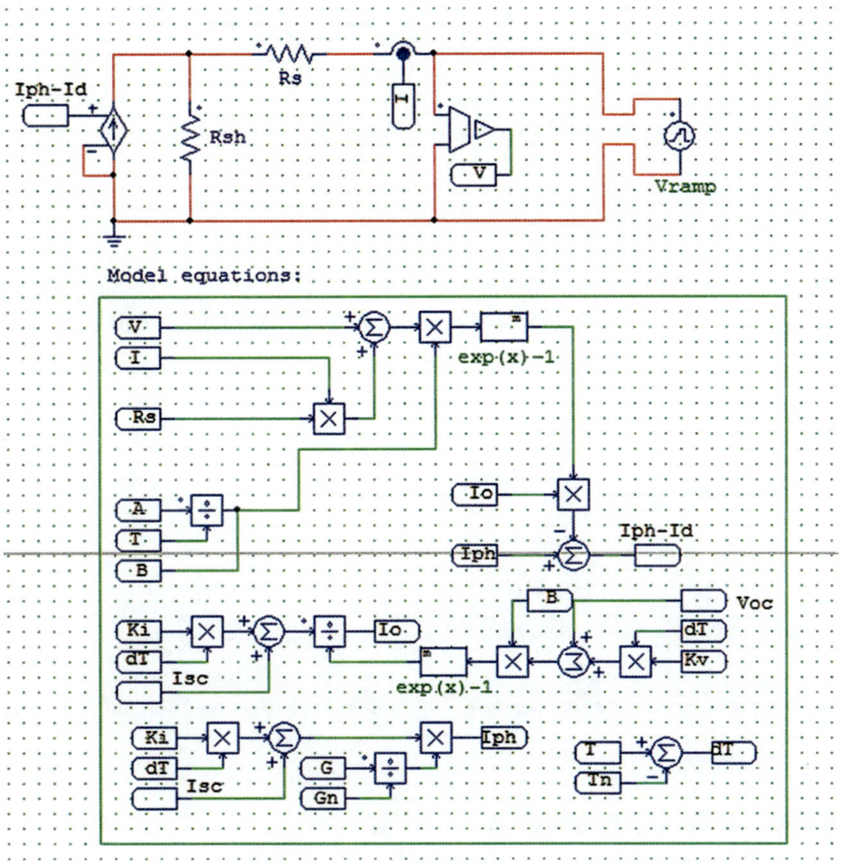


**Fig. 4.** Model of the photovoltaic panel

## 5 Effect of Solar Radiation Variation

Figure 5 clearly shows that the current produced by the PV depends strongly on the solar irradiation. Figure 6 presents P-V curve, the voltage raises by 1.5 V when solar irradiation has raised from 400 W/m$^2$ to 1000 W/m$^2$.

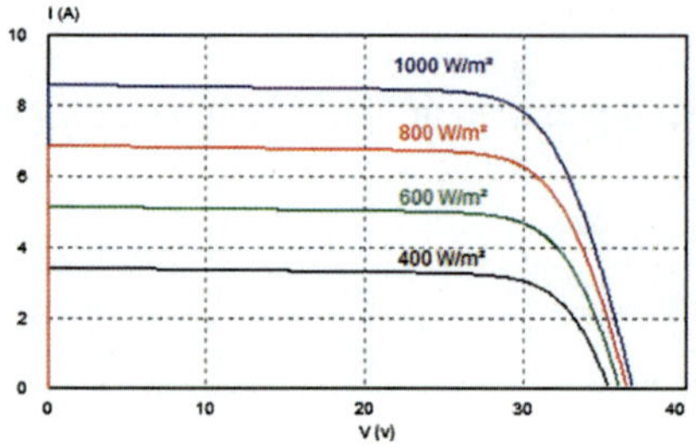


Fig. 5. I-V curves for different solar radiation

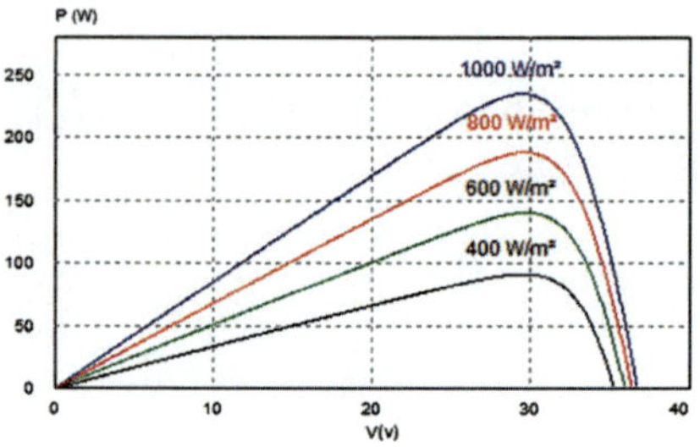


Fig. 6. P-V curves for different solar radiation

## 6 Effect of Temperature Variation

Mostly, for a particular solar radiation, when the temperature increases, the open-circuit voltage $V_{oc}$ decreases slightly, while the short-circuit current raises. This comportment is validated and illustrated in Figs. 7 and 8.

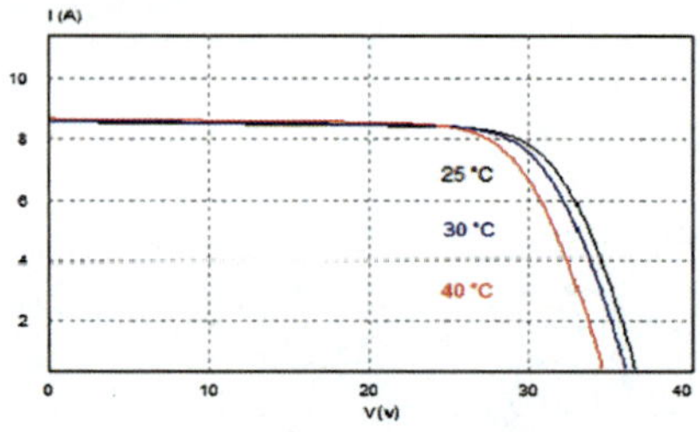


Fig. 7. I-V curve for different temperature values

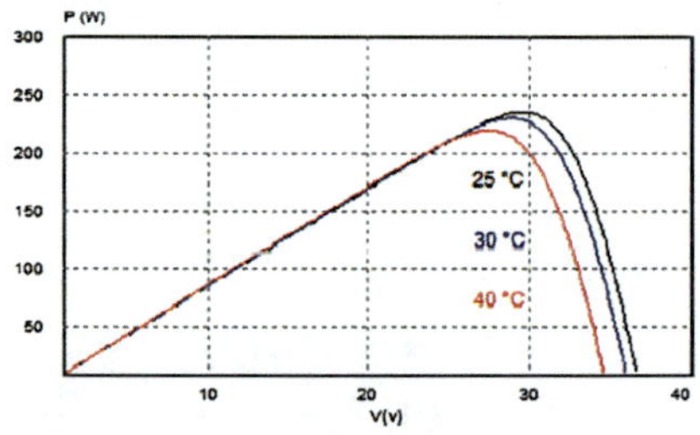


Fig. 8. P-V curves for different temperature values

## 7 Effect of Series Resistor Variation

The value of the series resistance is very low, and in some cases it can be neglected. Nevertheless, to make the model suitable for any given photovoltaic panel. We can vary this resistance to predict its influence on the outputs of the PV panel. According to Figs. 9 and 10, the variation of $R_s$ affects the angle of curve I-V by causing a deviation of the slope of the maximum power point.

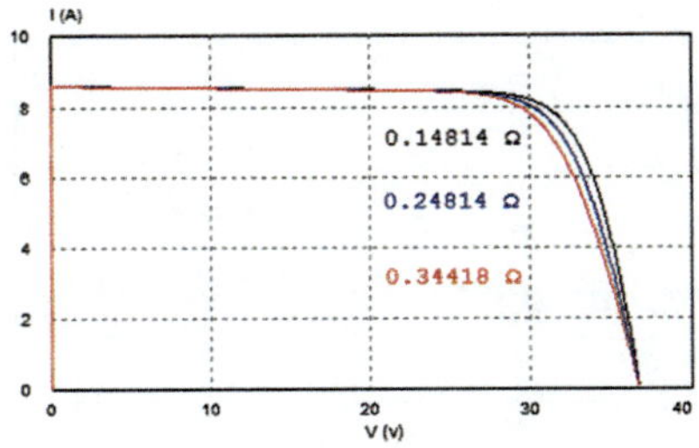


**Fig. 9.** I-V curves for different values of $R_s$

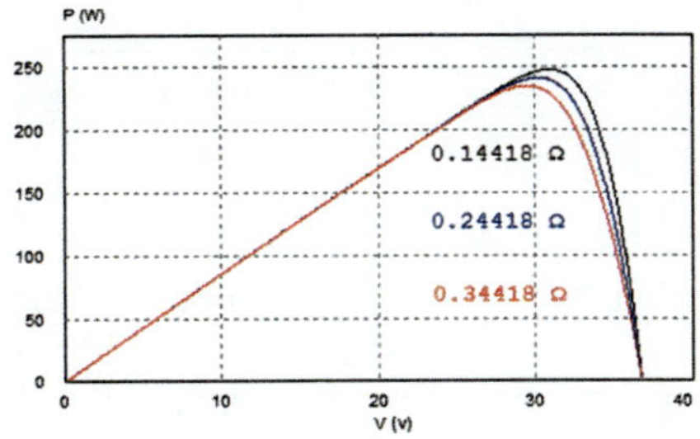


**Fig. 10.** P-V curves for different values of $R_s$

The simulation was done for three different Rs values, 0.14418 Ω, 0.24418 Ω and 0.34418 Ω. We have shown that the higher $R_s$ values reduce the output power. The fill factor, given by Eq. (8), decreases as $R_s$ increases.

$$FF = \frac{P_{max}}{V_{oc} * I_{sc}} \quad (8)$$

## 8 Effect of Shunt Resistance Variation

The shunt resistance must be large enough to have better output power and a good fill factor. For a low shunt resistance, the current collapses more strongly so that there will be a high-power loss and a low fill factor as shown in Figs. 11 and 12

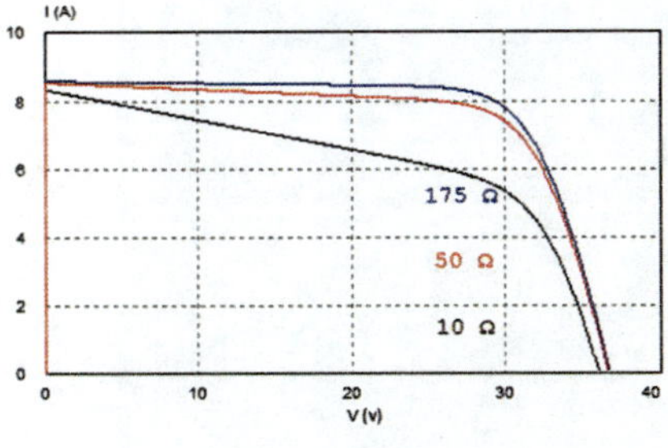


**Fig. 11.** I-V curves for different values of $R_{sh}$

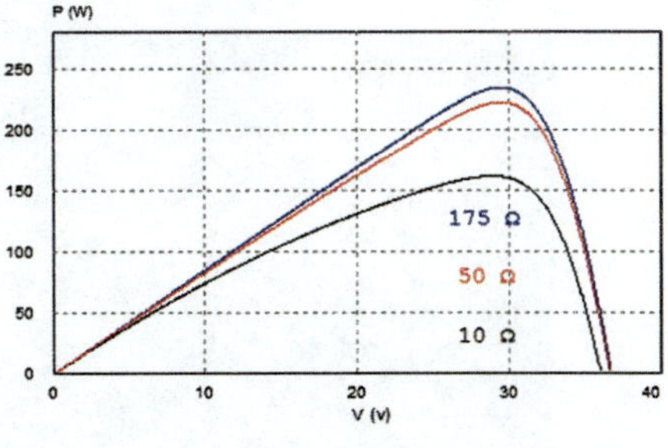


**Fig. 12.** P-V curves for different values of $R_{sh}$

## 9 Effect of Reverse Saturation Current Variation

The curves of Figs. 13 and 14 show the behavior of the PV panel for different saturation current values.

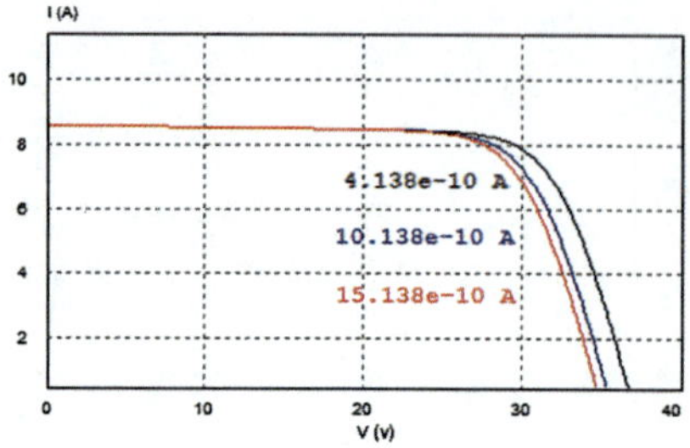


**Fig. 13.** I-V curves for different saturation current values

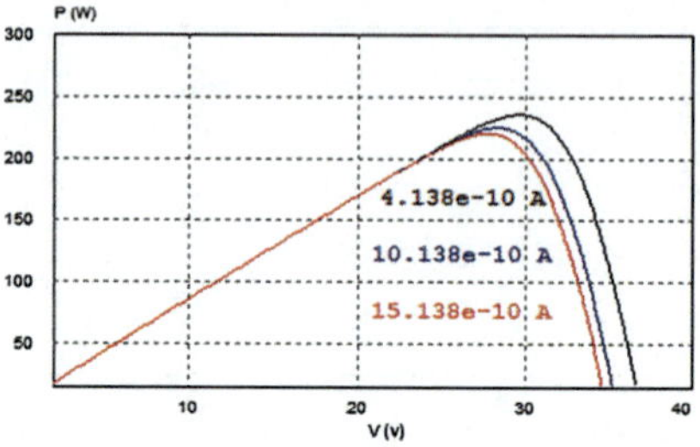


**Fig. 14.** P-V curves for different saturation current values

# 10 Experimental Results

**Fig. 15.** Experimental hardware setup

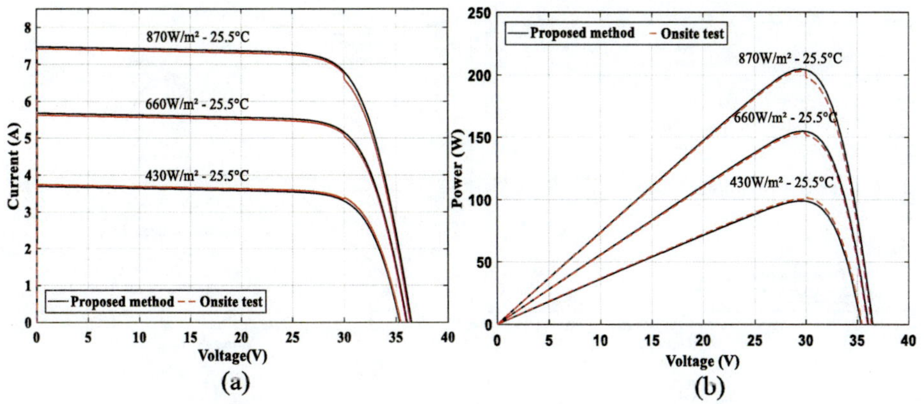


**Fig. 16.** Comparison between the proposed strategy and experimental data I-V (a) and P-V (b) characteristics of CSUN235_60P with different irradiations at 25,5 °C.

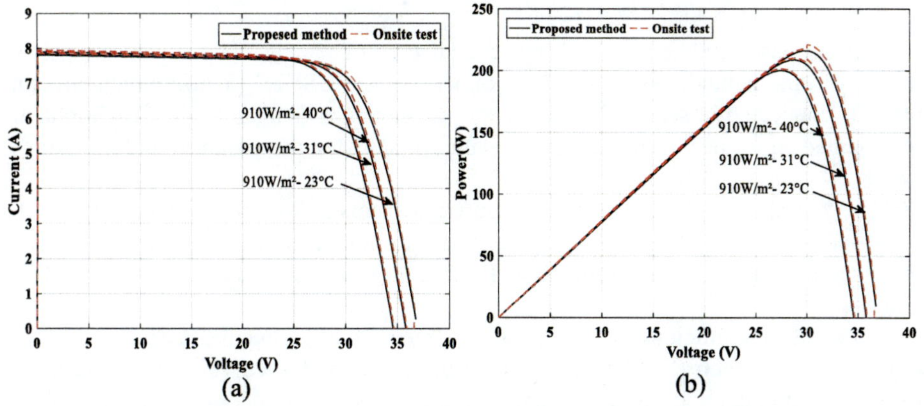


**Fig. 17.** Comparison between the proposed strategy and experimental data I-V (a) and P-V (b) characteristics of CSUN235_60P with different values of temperatures at 910 W/m$^2$.

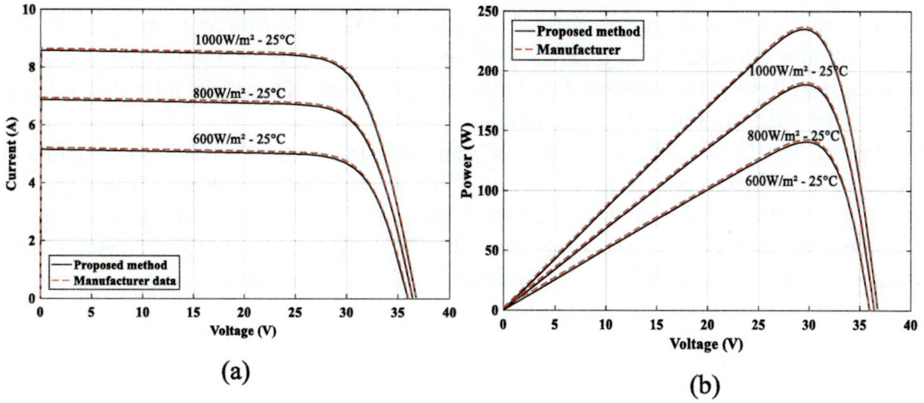


**Fig. 18.** Comparison between the proposed strategy and Manufacturer data I-V (a) and P-V (b) curves of CSUN235_60P with different values of irradiances at 25 °C.

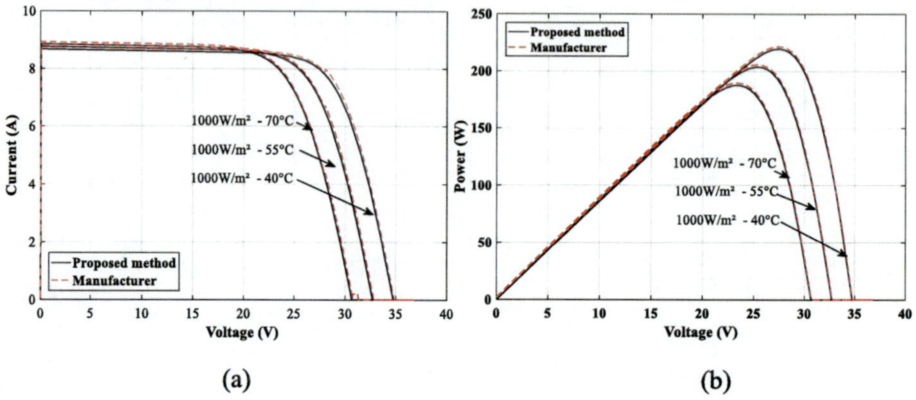


**Fig. 19.** Comparison between the proposed strategy and Manufacturer data I-V (a) and P-V (b) characteristics of CSUN235_60P with different values of temperatures at 1000 W/m$^2$

The proposed method can be used for all type of panels. To evaluate the performance of this method. We chose the monocrystalline CSUN235_60P which contains 60 cells connected in series. The datasheet parameters at Standard Test Conditions are represented in Table 1. The numerical value of Rsh for the monocrystalline Csun is 180 Ω.

The datasheet values and the measured resistance $R_{sh}$ are implemented in the proposed method at MATLAB script in order to obtain the missing parameters. The extracted parameters are represented in Table 2 and there are used to plot P-V and I-V characteristics.

To assess the efficiency of the proposed method, experimental measurements are done on a test bench (Fig. 15) to examine the I-V and P-V curves with the obtained experimental data under different values of solar irradiation and temperature. The P-V and I-V curves are also compared to characteristics provided by the manufacturer [15].

Figure 16 shows the I-V and the P-V curves of the CSUN235-60 for different values of temperature and at a solar irradiance of G = 910 W/m$^2$. Figure 17 presents the P-V and the I-V characteristics of the CSUN235-60 for different values of irradiance and at a temperature of G = 25.5 °C. As indicated in these figures, the proposed technique and the experimental data have similar curves. There is a little discordance around the MPP due to sudden variation of irradiation.

Figure 18 illustrates the I-V and P-V characteristics of the CSUN235-60P panel at three different irradiations (1000 W/m$^2$, 800 W/m$^2$ and 600 W/m$^2$) and constant temperature (25 °C). Figure 19 shows the P–V and I-V curves of the CSUN 250 panel at three different temperatures conditions (40 °C, 55 °C, and 70 °C) and constant solar irradiation G = 1000 W/m$^2$. It can be seen that, the curves of the proposed strategy agree with the one provided by the manufacturer.

## 11 Design of a PV Array to Supply a Pumping Station

The aim of this section is to determine the number of PV panels needed and how to connect them in order to supply our school's pumping station. A sizing of the components of the studied system was carried out in order to provide a flow rate of 20 $m^3/s$ and a height of 10 m. Table 3 presents the sizing steps of the PV system.

**Table 3.** Steps of sizing PV water pumping system

| Symbols | Expressions | Results |
|---|---|---|
| Electric power $E_{ele}$ | $E_{ele}$ = ($C_H$ *Q*HMT)/$R_{mp}$ | 3096.59 Wh/day |
| The peak power of the PV generator | $P_{pvtot}$ = $E_{ele}$/[$E_{ns}$ *(1− pertes)] | 1046,14 W |
| Number of panels series | $N_s$ = ($P_{pvtot}$/$P_{pv}$) | 5 |
| Number of panels parallel | $N_p$ = $I_{pv}$/$I_{pvt}$ | 1 |

Where: $C_H$, g, HMT, Q, $R_{mp}$, $E_{ns}$, are respectively the hydraulic constant, the acceleration of gravity, the total head, the water flow, the performance Motor pump group and the hours of radiation per day.

By the reason of characteristics of the induction motor ($P_n$ = 600 W, $I_n$ = 3.3 A), the panels must be coupled in series.

Another dimensioning was done by COMPASS LORENTZ software in order to validate the calculation made previously. The first step is to seize the data from the pumping station Fig. 20. Figure 21 shows the characteristics of the PV studied. Figure 22 shows that the pumping station requires 5 PV connected in series and this coincides with the results found previously (Table 3).

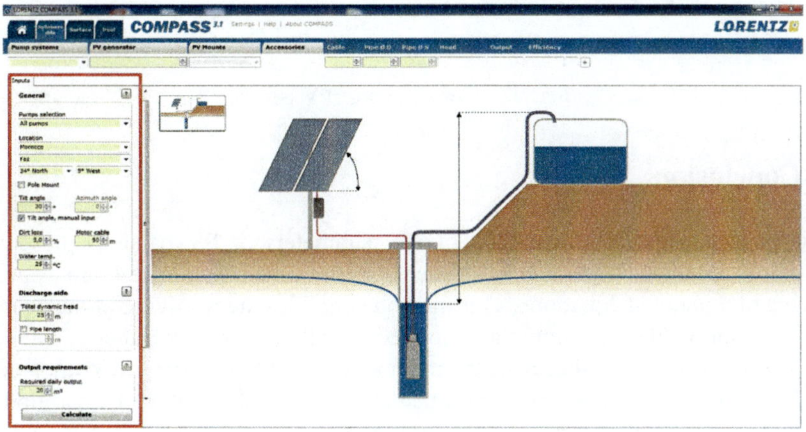


**Fig. 20.** Characteristics of the pumping station

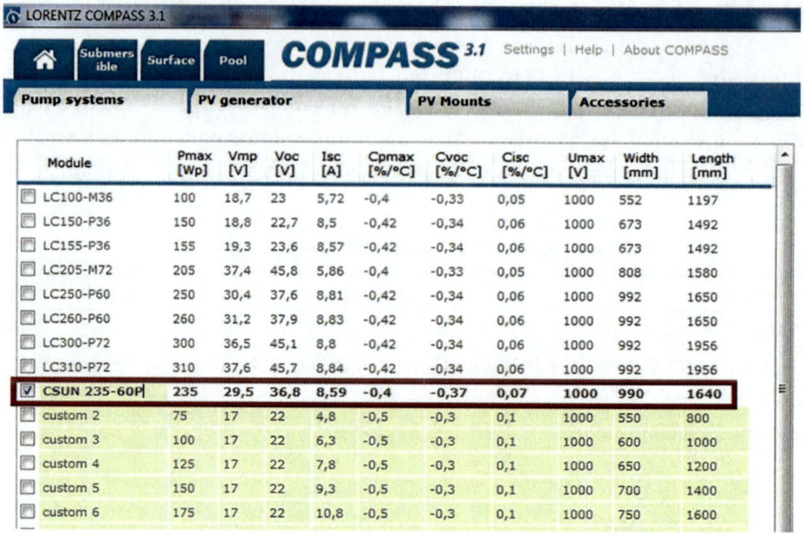


| Module | Pmax [Wp] | Vmp [V] | Voc [V] | Isc [A] | Cpmax [%/°C] | Cvoc [%/°C] | Cisc [%/°C] | Umax [V] | Width [mm] | Length [mm] |
|---|---|---|---|---|---|---|---|---|---|---|
| LC100-M36 | 100 | 18,7 | 23 | 5,72 | -0,4 | -0,33 | 0,05 | 1000 | 552 | 1197 |
| LC150-P36 | 150 | 18,8 | 22,7 | 8,5 | -0,42 | -0,34 | 0,06 | 1000 | 673 | 1492 |
| LC155-P36 | 155 | 19,3 | 23,6 | 8,57 | -0,42 | -0,34 | 0,06 | 1000 | 673 | 1492 |
| LC205-M72 | 205 | 37,4 | 45,8 | 5,86 | -0,4 | -0,33 | 0,05 | 1000 | 808 | 1580 |
| LC250-P60 | 250 | 30,4 | 37,6 | 8,81 | -0,42 | -0,34 | 0,06 | 1000 | 992 | 1650 |
| LC260-P60 | 260 | 31,2 | 37,9 | 8,83 | -0,42 | -0,34 | 0,06 | 1000 | 992 | 1650 |
| LC300-P72 | 300 | 36,5 | 45,1 | 8,8 | -0,42 | -0,34 | 0,06 | 1000 | 992 | 1956 |
| LC310-P72 | 310 | 37,6 | 45,7 | 8,84 | -0,42 | -0,34 | 0,06 | 1000 | 992 | 1956 |
| **CSUN 235-60P** | **235** | **29,5** | **36,8** | **8,59** | **-0,4** | **-0,37** | **0,07** | **1000** | **990** | **1640** |
| custom 2 | 75 | 17 | 22 | 4,8 | -0,5 | -0,3 | 0,1 | 1000 | 550 | 800 |
| custom 3 | 100 | 17 | 22 | 6,3 | -0,5 | -0,3 | 0,1 | 1000 | 600 | 1000 |
| custom 4 | 125 | 17 | 22 | 7,8 | -0,5 | -0,3 | 0,1 | 1000 | 650 | 1200 |
| custom 5 | 150 | 17 | 22 | 9,3 | -0,5 | -0,3 | 0,1 | 1000 | 700 | 1400 |
| custom 6 | 175 | 17 | 22 | 10,8 | -0,5 | -0,3 | 0,1 | 1000 | 750 | 1600 |

**Fig. 21.** Characteristic of the PV panel

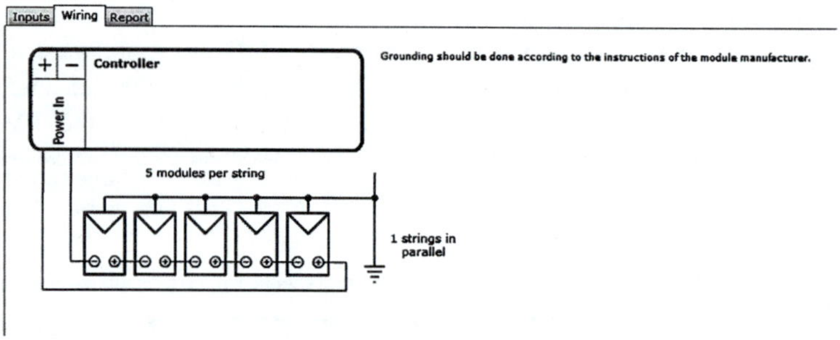


**Fig. 22.** Connection of PV panels

## 12 Conclusion

In this paper, an extraction method of the parameters of PV panel is presented, it consists to measure the parallel resistance $R_{sh}$, the others parameters are defined by resolving of numerical equations. This method simplifies the resolving of the equations. This extraction method is simple and applied for all types of PV panels.

In order to prove the efficacy of the method, I-V and P-V curves are compared with the Manufacturer data and the experimental curves under different conditions, the results showed a good agreement. The detailed sizing is also done using the Compass Lorentz Software. As perspectives, we consider to install the pump system which will be alimented by PV panel studied in this work.

## References

1. Zeng, J., Li, M., Liu, J.F., Wu, J., Ngan, H.W.: Operational optimization of a stand-alone hybrid renewable energy generation system based on an improved genetic algorithm. In: Power Energy Society General Meeting EEE, pp. 1–6 (2010)
2. Salmi, T., Bouzguenda, M., Gastli, A., Masmoudi, A.: MATLAB/Simulink based modelling of solar photovoltaic cell. Int. J. Renew. Energy Res. **2**(2), 6 (2012)
3. Campana, P.E., Li, H., Zhang, J., Zhang, R., Liu, J., Yan, J.: Economic optimization of photovoltaic water pumping systems for irrigation. Energy Convers. Manag. **95**, 32–41 (2015)
4. Chaibi, Y., Salhi, M., Eljouni, A., Essadki, A.: A new method to determine the Parameters of a photovoltaic Panel equivalent circuit. Sol. Energy **163**, 376–386 (2018)
5. Nezihe, Y., Tacer, E.: Identification of photovoltaic cell single diode discrete model parameters based on datasheet values. Solar Energy **127**, 175–183 (2016)
6. Villalva, M., Gazoli, J., Filho, E.: Comprehensive approach to modeling and simulation of photovoltaic arrays. IEEE Trans. Power Electron. **24**(5), 1198–1208 (2009)
7. Chandel, S.S., Nagaraju Naik, M., Chandel, R.: Review of solar photovoltaic water pumping system technology for irrigation and community drinking water supplies. Renew. Sustain. Energy Rev. **49**, 1084–1099 (2015)
8. Barth, N., Jovanovic, R., Ahzi, S., Khaleel, M.A.: PV panel single and double diode models: optimization of the parameters and temperature dependence. Sol. Energy Mater. Sol. Cells **148**, 87–98 (2016)
9. Ishaque, K., Salam, Z., Syafaruddin: A comprehensive MATLAB Simulink PV system simulator with partial shading capability based on two-diode model. Sol. Energy **85**(9), 2217–2227 (2011)
10. Ma, T., Yang, H., Lu, L.: Solar photovoltaic system modeling and performance prediction. Renew. Sustain. Energy Rev. **36**, 304–315 (2014)
11. Khezzar, R., Zereg, M., Khezzar, A.: Modeling improvement of the four parameter model for photovoltaic modules. Sol. Energy **110**, 452–462 (2014)
12. Chin, V.J., Salam, Z., Ishaque, K.: Cell modelling and model parameters estimation techniques for photovoltaic simulator application: a review. Appl. Energy **154**, 500–519 (2015)
13. Hansen, C.W.: Parameter estimation for single diode models of photovoltaic modules. Sandia report, no. SAND2015–2065, pp. 1–68 (2015)
14. Park, J.Y., Choi, S.J.: A novel datasheet-based parameter extraction method for a single-diode photovoltaic array model. Sol. Energy **122**, 1235–1244 (2015)
15. https://www.pekat.com.my/pdf/CSUN250-60P.pdf

# 5G Network Conception in Fez City Center

Fatima Zahra Hassani-Alaoui(✉) and Jamal El Abbadi

Smart Communications Research Team (ERSC), E3S Research Center, EMI, Mohammed V University Rabat, Rabat, Morocco
fatimazahra.alaoui.h@gmail.com, elabbadi@emi.ac.ma

**Abstract.** The latest vision of the 5G architecture has been defined, few months before launching the future generation of cellular networks. Even though this revolutionary milestone, the researches and the works are still in progress.

The main goal of this perceivable work is to construct a 5G network in the center of Fez City. Our construction study is based on the visualization of the signal-to-interference-and-noise ratio (SINR) on this environment, in order to test the power and the efficiency of the signal and the transmissions of the proposed schemes.

This paper highlights two scenarios options for 5G conception: Standalone (SA) mode and Non-Standalone (NSA) mode. This work compares both of them, in order to choose the convenable one for our study environment. We begin by defining the substantial elements that will characterize our 5G network. Using this framework, we evaluate the winner solution of the 5G conception in Fez City center.

**Keywords:** 5G · 5G New Radio (5G NR) · Standalone (SA) mode · Non-Standalone (NSA) mode · MIMO · Beamforming · mmWave · SINR · Fez city

## 1 Introduction

Nowadays, the technology is getting more and more developed; as a consequence, consumers and businesses anticipate seeing more opportunities in the technologies that are coming to be deployed, this one has to be faster and have the ability to accomplish many services.

The fifth generation of mobile networks will be part of the major important fields of industries, namely: education, healthcare, smart homes and entertainment, and also smart transportation. Researchers and Industrials have started to clarify the 5G architecture after leading many studies and projects. The 5G conception includes a set of new and successful technologies that will be used in its networks.

The first set of 5G standards, Release 15, was delivered in December 2017, and it marked a revolutionary milestone on the way to the deployment of the future generation [1]. The 5G planning has been developed rapidly than expected. Release 17 will be defined at the end of 2019. According to the 3rd Generation Partnership Project (3GPP), the commercial deployment of the 5G mobile network has already been launched in 2019, and it will go until 2020.

In the first stage of deployment, there will be ameliorations in LTE-Advanced and LTE Pro technologies. Those improvements will make the existence of 5G.


A. Saka et al. (Eds.): CPI 2019, LNME, pp. 192–205, 2021.
https://doi.org/10.1007/978-3-030-62199-5_17

Hereafter, a significant step-up with the introduction of a new air interface: 5G New Radio (5G NR) (see Fig. 1), will take place:

- In Phase 1 of 5G conception (Release 15), the use of the Non-Stand Alone (NSA) option takes place. For this case, devices will use the existing LTE radio and core network.
- In Phase 2 of 5G conception (Release 16), the use of the Stand Alone (SA) option is occurred. This conception type implies full user and control plane capability using the new 5G core network architecture.

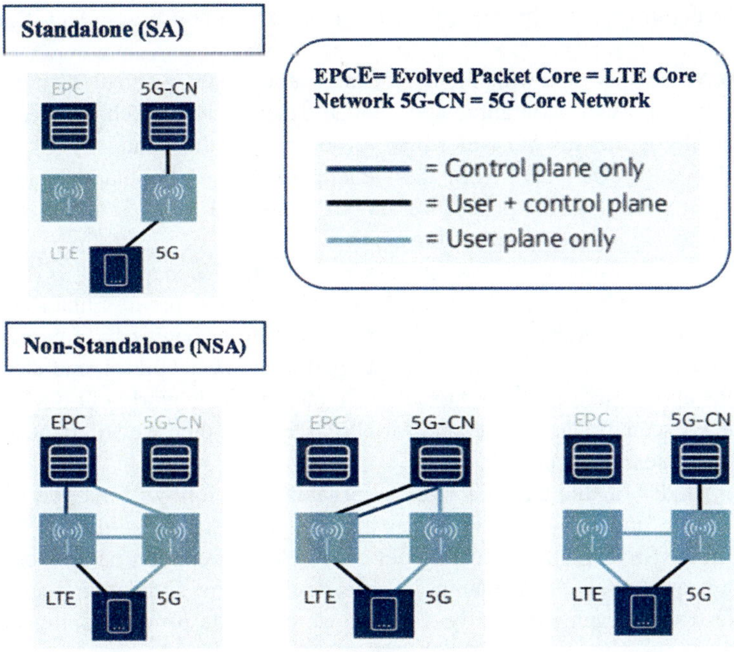


**Fig. 1.** SA and NSA 5G NR architecture

In Morocco, as the other countries in the world, 4G of the mobile network demonstrates how greatly wireless technology can support mobile broadband and other advanced technologies as Internet of Things (IoT). This generation provides the intended capacity, enhance throughput rate, reduce latency, and making new applications that never exist before with wireless networks. However, this situation will not last forever. As the number of connected devices grows, the frequency bands used for mobile communications are becoming overcrowded, as a result, the actual networks will not respond to the consumer demands in the near future. During rush hours periods, consumers will experience lack of services, delays in connections, poor throughput rate, and other issues.

To solve those issues, 5G has been emerged as a key to enhance the future wireless network. This generation will ameliorate the network capacity and throughput with peak data speeds up to 20 Gbps downlinks and up to 10 Gbps uplinks. The latency will be reduced to One millisecond. The energy efficiency will be over 100 times over the 4G. The 5G network will manage over 1 million devices, while leading to new services for customers. On the other hand, security has always been a first priority with all previous mobile generations. The 5G mobile network takes that security priority to higher level, with a large diversity of advanced safeguards, to protect both the networks and customers.

Morocco still has the second best 4G coverage in Africa, according to a study done by OpenSignal. As the world of technology is developing in many countries, Morocco is also concerned to develop its mobile network, as it experiences evolution opportunities in the coming years, this country will be one of the best investment destinations in Africa.

The Moroccan city Fez, which is one of the best touristic destinations in the world, is also a very important geographic spot, and it deserves to be technologically developed; especially is that the 5G is destined to be developed for industry.

According to the last data from the General Census of Population and Housing done in September 2014, the population of Fez is about 1 150 131 in an area of 332 $km^2$, with a density of 3 464 HAB./$km^2$.

The main goal of this paper is to invent a construction scenario of a 5G network in the center of Fez City, which is considered as a dense urban Enhanced Mobile Broadband (eMBB) environment. Our construction study begins by creating a positioning scheme of antennas in the whole studied area. Then, the second step will be choosing the appropriate configuration of each antenna. This work will be done twice, the first for SA scenario and the second for NSA scenario; then we will choose the best solution for the test environment.

The remainder of this detailed work is organized as follow. First, we present an overview of the fundamental technologies and elements that make the 5G NR. Then, we highlight the SINR mathematical model used in this simulation-based work, as well as the antennas model. From that point, we describe the simulation scenarios and results. We discuss each case study. Finally, we conclude by mentioning our future work.

## 2 Background

### 2.1 mmWave Communication

Nowadays, as the number of technological devices is highly growing, the current frequency bands [3 kHz–6 GHz] is starting to get overcrowded. Therefore, the communications suffer from dropped connections and overdue services. To solve this complication, it is required to transmit signals on new frequency bands which called centimeter waves (cmWave) [3 GHz–30 GHz] and millimeter waves (mmWave) [30 GHz–300 GHz]. The most important benefits for using this new range of spectrum are to furnish high capacity, high data rates and ultra-high reliability [2].

Contrariwise, using the mmWave spectrums, present many challenges to overcome. The successful and the most important projects demonstrated in their accomplished measurements (that were carried out in various scenarios, in different frequencies and across many materials) [3–8], that those new frequency bands have some shortcomings namely: higher sensitivity to obstacles, high pathloss and the decreased diffraction.

### 2.2 Massive MIMO Beamforming

Massive multiple input multiple output (MIMO), which use a multiple controllable antenna, has been employed in the current 4G network and it has demonstrated its effectiveness. In other word, MIMO makes a big advancement in wireless systems. In 2017, Full-Dimension MIMO has been deployed. This type is a $4 \times 4$ MIMO which supports 32 antennas at the base station. While in the mmWave bands, a base station supports hundreds of antennas [9, 10].

The principal benefits of Massive MIMO are improving the coverage and limiting pathloss by using high antenna gain. In addition to that, MIMO takes an important part in limiting interference.

To overcome all the mmWave communication issues, that we have cited before, namely: penetration losses, sensitivity to blockage and diminished diffraction; a cluster network concept is envisioned to solve those issues. This concept is composed of a set of coordinated access points (APs) work together to furnish omnipresent coverage. In the scenario of blockage, one AP will quickly handoff the User Equipment (UE) to another AP in the cluster. Those Handoffs may be done while hand movement, moving obstacles, or changing orientation. The APs will fit in small boxes that are easy to install [11].

The MIMO implementation can be configured in various options, the implication of beamforming is a particular technique used to join multiple antenna elements to condense the power in a determined direction. The beamforming technology is considered as an essential tech in the 5G, because it removes the interference, accordingly, upgrade the SNIR value [12].

### 2.3 Deployment Cells Size

The criterion of the new spectrum characteristics preconditions the choice of the deployment environment of the 5G. The shortcoming of the mmWave conduct us to think about the cell characteristics of the 5G network. The necessary solution taken for that, is to deploy small cells and macro cells, and ensuring the cooperation between those two kinds of cells [13]. In this proposed scenario, small base stations provide high data rates, while macro-cells provide wide area coverage.

## 3 Methods

### 3.1 SINR Model

In the current sub-section, we will pinpoint the SINR mathematical model, that will be used in the planification of the 5G network, with all its essential elements.

The SINR is defined as the quantity used to give the theoretical superior limits on channel capacity in a wireless communication system. In other term, the SINR presents the power of a signal divided by the sum of the interference power and the power of some background noise. SINR is used to evaluate the quality of the network connections. Table 1. presents the signal power in each SINR range of values.

In our case study, we have particularly used the SINR model in the simulation of the 5G network construction, in order to detect the strongest radio links, and the weak ones.

**Table 1.** SINR values descriptions.

| SINR values | Signal description |
|---|---|
| >=20 dB | Excellent |
| 13 dB to 20 dB | Good |
| 0 dB to 13 dB | Fair to poor |
| <=0 dB | No signal |

The general definition of the SINR is given by the following expression:

$$SINR = \frac{Signal\,power}{Interference\,power + Noise} \tag{1}$$

According to this Eq. (1), the effective transmission depends on the potency of the received signal, the interference made by the simultaneous transmissions, and the noise level [14].

Being more specific, the SINR is detailed as the expression below:

$$SINR(l_x) = \frac{P_{r_x}(t_x)}{N + I_{r_x}(L)} = \frac{P(t_x)G(t_x, r_x)}{N + \sum_{x \neq y} P(t_y)G(t_y, r_x)} \tag{2}$$

$$SINR(l_x) = \frac{\frac{P(t_x)}{D(t_x, r_x)^{\sigma}}}{N + \sum_{x \neq y} \frac{P(t_y)}{D(t_y, r_x)^{\sigma}}} \tag{3}$$

Where:

$P_{r_x}(t_x)$ is the received power of a transmitted signal sent by $t_x$ for an intended receiver $r_x$;
$G(t_x, r_x)$ is the propagation attenuation;
$\sigma$ is the path-loss exponent, which depends on the environment conditions;
$l_x = (t_x, r_x)$ is a pair of transmitter and receiver;
$I_{r_x}(L)$ is the interference sum gotten by the set L of nodes emitting simultaneously, excluding the chosen transmitter $t_x$.

### 3.2 Antennas

In this sub-section, we will point out the antenna characteristics for either Base Station (BS) antennas or UE antennas, which will be applied in this work.

**BS Antenna.** BS antennas have one or multiple antenna panels (depending on the environments and situation) placed vertically, horizontally or in a two-dimensional array within each panel. Mathematically, each antenna panel possesses M × N antenna elements, whither N is the number of columns and M is the number of antenna elements with the same polarization in every column. The M × N elements could be single polarized or dual polarized.

In the case where the BS has multiple antenna panels, a uniform rectangular panel array is patterned, including $M_xN_x$ antenna panels whither $M_x$ is number of panels in a column and $N_x$ is number of panels in a row. Antenna panels are evenly spaced with a center-to-center spacing of $d_{x,H}$ in the horizontal direction and $d_{x,V}$ in the vertical direction [15]. The Fig. 2 presents a BS antenna model illustration.

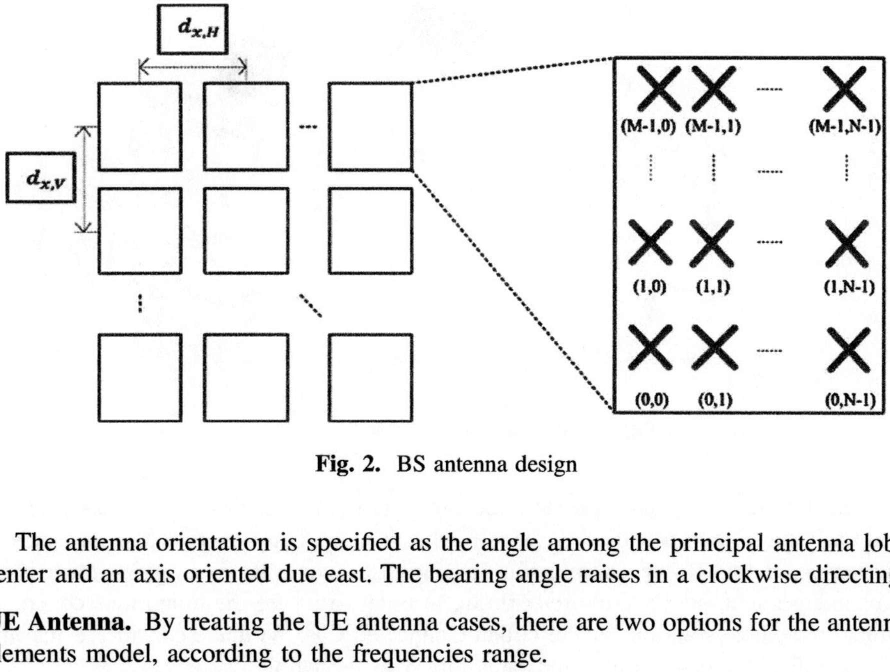


**Fig. 2.** BS antenna design

The antenna orientation is specified as the angle among the principal antenna lobe center and an axis oriented due east. The bearing angle raises in a clockwise directing.

**UE Antenna.** By treating the UE antenna cases, there are two options for the antenna elements model, according to the frequencies range.

- The Omnidirectional antenna element is supposed to be used in 700 MHz and 4 GHz cases.
- The directional antenna panel is supposed to be used in 30 GHz and 70 GHz cases, for that the $M_xN_x$ antenna panels can have various orientations.

## 4 Simulation Scenarios

The main goal of this work is to construct a 5G network in Fez City. In this case study, we have limited our work at the center of the city and the neighborhood region (More details on Fig. 4), as it is the best geographic environment to apply the Dense Urban Enhanced Mobile Broadband (eMBB) scenario, which is one of the three principal 5G NR use cases, this one is characterized with the high user density and traffic loads.

Our construction study is based on the visualization of the SINR on this environment, using MathWorks software, in order to test the power and the efficiency of the signal and the transmissions.

The 5G topographical construction network in this case, is based on a regular scheme, following the hexagonal form [15].

The chosen study environment consists of two layers:

- A macro layer: where the base stations are placed in regular position (Fig. 3).
- A micro layer: where they are three sites installed in each cell area (Fig. 3).

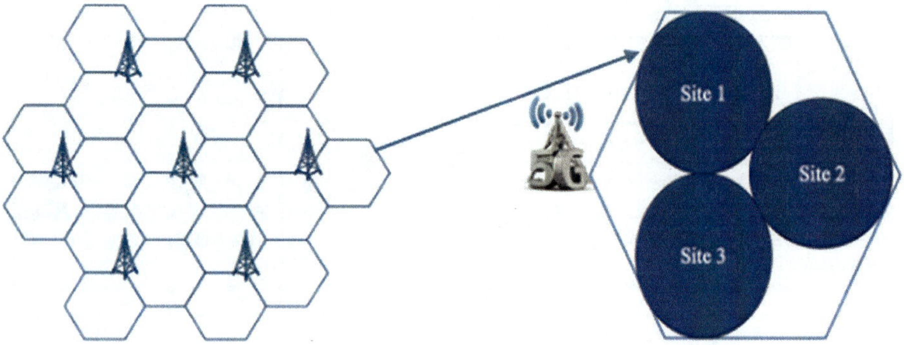


**Fig. 3.** Structure of macro and micro layer

The simulation of our case study consists on a configuration of 25 sites (Fig. 4), and each site is composed of three cells, the distance between two adjacent sites is chosen depending to the environment, in our case, where the test environment is an urban dense area, we have chosen 500 m, in order to insure the ubiquitous coverage, which is the best solution for the Urban Connected Car. We have considered that the UEs are distributed uniformly and randomly over the whole test area.

The first task of the simulation was the creation of the locations corresponding to cell sites in the network plan, using the “Place Ahmed El Mansour “ as the center location (Fig. 4).

**Fig. 4.** Cell site localization

The second task was the determination of the configuration parameters of each cell site and each transmitter site.

The third task was the determination of the antenna characteristics for either base station antennas, or UE antennas. We have created a uniform rectangular phased array System MIMO 8 × 8 (Fig. 5). We defined the characteristics in a way to increase the directional gain and the peak SINR values.

Each selected test environment is characterized by its configuration parameters. In our case, and for an Urban Dense eMBB environment, there are 2 elementary configuration that warrant the spectral efficiency and the mobility. To evaluate the optimum solution, we will test and compare both of the configurations. Configuration I is presented in Table 2. while Configuration II is presented in Table 3.

We have adapted the major configuration part, proposed by The International Mobile Telecommunications-2020 (IMT-2020), to our case study [15].

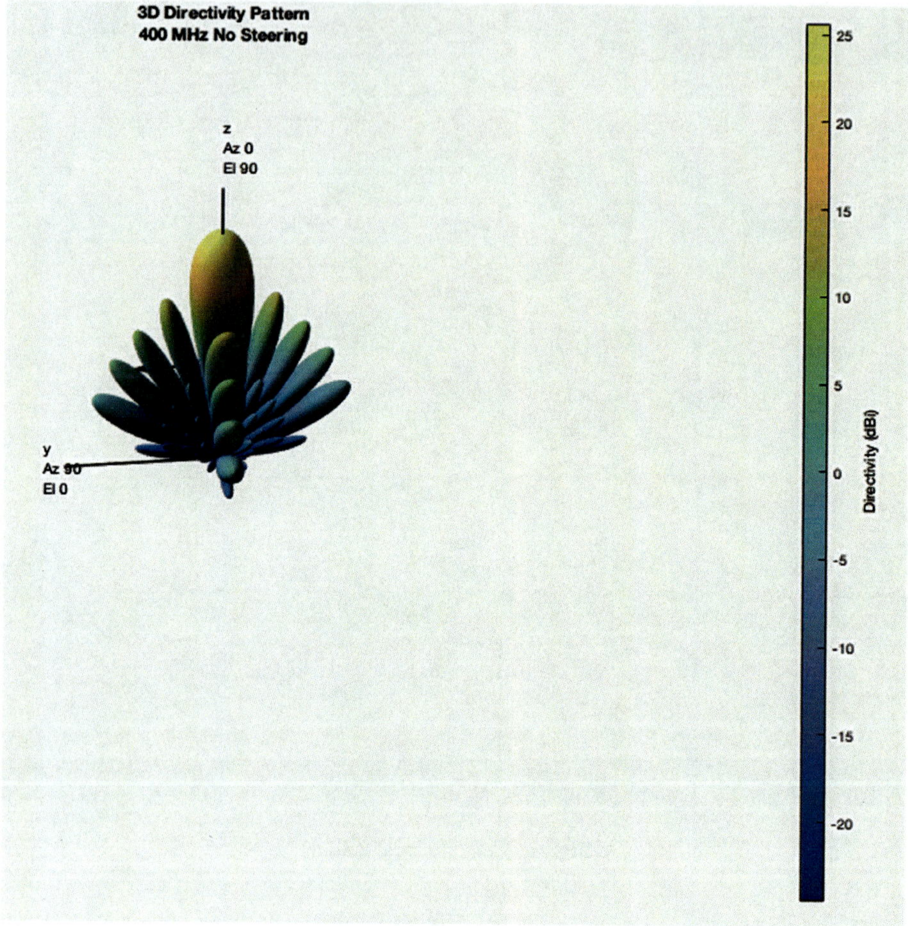


**Fig. 5.** Antennas model

**Table 2.** Simulation parameters of configuration I.

| Parameters | Value |
|---|---|
| Carrier frequency | 4 GHz |
| Bandwidth | 20 MHz |
| Total transmit power | 44 dBm |
| TX antenna array | 8 × 8 |
| TX antenna height | 25 m |
| TX antenna element gain | 8 dBi |
| TX azimuth beam width | 65° |
| TX elevation beam width | 65° |
| Tilt angle | 0° |

*(continued)*

**Table 2.** (*continued*)

| Parameters | Value |
|---|---|
| Down tilt | 15° |
| RX antenna height | 1.5 m |
| RX antenna element gain | 0 dBi |
| RX noise figure | 7 dB |
| TX noise figure | 5 dB |
| Maximum backward attenuation | 30 dB |
| Thermal noise level | −174 dBm/Hz |
| Simulation bandwidth for TDD | 20 MHz |

**Table 3.** Simulation parameters of configuration II.

| Parameters | Value |
|---|---|
| Carrier frequency | 30 GHz |
| Bandwidth | 40 MHz |
| Total transmit power | 37 dBm |
| TX antenna array | 8 × 8 |
| TX antenna height | 25 m |
| TX antenna element gain | 8 dBi |
| TX azimuth beam width | 65° |
| TX elevation beam width | 65° |
| Tilt angle | 0° |
| Down tilt | 15° |
| RX antenna height | 1.5 m |
| RX antenna element gain | 5 dBi |
| RX noise figure | 10 dB |
| TX noise figure | 7 dB |
| Maximum backward attenuation | 30 dB |
| Thermal noise level | −174 dBm/Hz |
| Simulation bandwidth for TDD | 80 MHz |

## 5 Simulation Results

The rectangular antenna array used in this work, for both configurations, have demonstrated its effectiveness comparing to a single antenna element; the SINR values have been peaked.

The obtained results of the both configurations, show that the Configuration I (Fig. 6) is more efficient for ensuring an efficient transmission and a powerful signal. In other word, in the configuration I, the SINR is strong, and the network coverage is insured comparing to the configuration II (Fig. 7).

For this reason, it is necessary to reduce the ISD value for the second configuration, and to increase the number of antennas, as the frequency used is higher.

**Fig. 6.** Configuration I result

**Fig. 7.** Configuration II result

As the Table 2 and Table 3 show, the both configurations depend especially to the used frequencies. According to that, the 5G New Radio NR conception can be done in two different options:

- For low frequencies: The Non-standalone deployment is required (Configuration I).
- For mm-wave frequencies: The Stand-alone deployment is required (Configuration II).

In the standalone deployments, and to decrease the problems of coverage and mobility related to the use of the mm-wave frequencies, it is indispensable to optimize the coordination of the Access Point. In contrast, for the non-standalone scenarios, there will be a co-deployment between nodes operated in mm-wave (high) frequencies and low frequencies. This conception affords more optimization opportunities for the system transmission, particularly when low frequency warrant initial access. This big advantage makes the non-standalone deployment the best option in this work.

## 6 Conclusion and Future Trends

This paper aims to propose a scenario of construction of 5G network in Fez City Center, one of the best touristic destinations in the world. Due to the importance of this region, we have studied two planning options, in order to choose the best conception. The idea gotten from this work describes the importance of optimizing the non-standalone deployment option, which demonstrated its effectiveness.

While academics and industrials are presently working on what define 5G wireless networks, it is obligatory to think about Green Communications, which warrant energy efficient, low cost and especially safe for the human body.

Recently, operators and mobile industries have started to leading projects about 6G. Even if it is too early to thing about this generation, it presents the objective of a future work.

## References

1. RAN Rel-16 progress and Rel-17 potential work areas. (n.d.). https://www.3gpp.org/news-events/2058-ran-rel-16-progress-and-rel-17-potential-work-areas
2. Nokia: 5G new radio network Uses cases, spectrum, technologies and architecture (n.d.). https://onestore.nokia.com/asset/205407
3. Document Number: H2020-ICT-671650-mmMAGIC/D2.2, Project Name: Millimetre-Wave based Mobile Radio Access Network for Fifth Generation Integrated Communications: Measurement Results and Final mmMAGIC Channel Models Measurement Results and Final mmMAGIC Channel Models (2017)
4. mmMAGIC – mm-Wave based Mobile Radio Access Network for 5G Integrated Communications (n.d.). https://5g-mmmagic.eu/
5. Rappaport, T.S., Sun, S., Mayzus, R., Zhao, H., Azar, Y., Wang, K., Gutierrez, F.: Millimeter wave mobile communications for 5G cellular: it will work! IEEE Access **1**, 335–349 (2013)
6. Maccartney, G.R., Rappaport, T.S., Samimi, M.K., Sun, S.: Millimeter-wave omnidirectional path loss data for small cell 5G channel modeling. IEEE Access **3**, 1573–1580 (2015)
7. Al-Dabbagh, R.K., Al-Raweshidy, H.S., Al-Aboody, N.A.: Performance comparison of exploiting different millimetre-wave bands in 5G cellular networks. In: 2017 International Conference on Performance Evaluation and Modeling in Wired and Wireless Networks (PEMWN), pp. 1–6 (2017)
8. MacCartney, G.R., Rappaport, T.S., Ghosh, A.: Base station diversity propagation measurements at 73 GHz millimeter-wave for 5G coordinated multipoint (CoMP) analysis. In: IEEE Globecom Workshops (GC Wkshps), pp. 1–7. IEEE (2017)
9. Abidin, I.S.Z., Brown, T.W.C.: Improving MIMO multiplexing for mmWave static links. In: Australian Microwave Symposium (AMS), pp. 41–42 (2018)
10. Kong, H., Wen, Z., Jing, Y., Yau, M.: A compact millimeter wave (mmWave) mid-field over the air (OTA) RF performance test system for 5G massive MIMO devices. In: 2018 IEEE MTT-S International Wireless Symposium (IWS), pp. 1–4 (2018)
11. The 5G mmWave revolution Using mmWave spectrum in practical 5G and ultra-dense networks (n.d.). www.nokia.com

12. Berraki, D.E., Armour, S.M.D., Nix, A.R.: Benefits of the sparsity of mmWave outdoor spatial channels for beamforming and interference cancellation. Phys. Commun. **27**, 170–180 (2018)
13. Hassani-Alaoui, F.Z., El Abbadi, J.: What will millimeter wave communication (mmWave) be?, pp. 119–131. Springer, Cham (2019)
14. Dolev, S. (ed.): ALGOSENSORS 2009. LNCS, vol. 5804. Springer, Heidelberg (2009). https://doi.org/10.1007/978-3-642-05434-1
15. Draft new Report ITU-R M.[IMT-2020.EVAL] – "Guidelines for evaluation of radio interface technologies for IMT-2020" (n.d.)

# Microstrip Antenna Array with Dumbbell Defected Ground Structure for Ka-Band Radar Application

Salaheddine Aourik[1](✉), Ahmed Errkik[1], Jamal Zbitou[1], Ahmed Lakhssassi[2], Abdelali Tajmouati[1], and Larbi Elabdellaoui[1]

[1] LMEET, FST of Settat, Hassan 1st University, Settat, Morocco
aouriksalaheddine@gmail.com

[2] Laboratory of Advanced Microsystems Engineering, Department of Computer Science and Engineering, University of Quebec at Outaouais, Gatineau, Canada

**Abstract.** In this paper, a broadband microstrip antenna array matched with quarter wave microstrip line matching and Defected Ground Structure technical is proposed. The purpose of this paper is to use Defected Ground Structure (DGS) in microstrip antenna array to achieve an antenna array for ka-band application. The DGS position which is under the quarter wave line matching has an effect that has been presented in this paper. This work started with the single element antenna matched to a band [28–31.5] Ghz. After That, we have constructed and designed the antenna array in order to increase the gain and to obtain a radar antenna with high performances. The antenna array structure, fed by microstrip line and matched by quarter wave transformer, is printed on a Duroid 5880 substrate with dielectric constant of $\epsilon_r = 2.22$ and, tangent value 0.004 and 0.4 mm thickness. The simulations of this study are performed on CST Microwave Studio.

**Keywords:** Defected Ground Structure · Ka-band · Dumbbells shaped · Antenna array

## 1 Introduction

Recently, the communication system development is gradually going up having the wider bandwidth in a single gadget with higher data rates. Due to the technology development, the need of constructing capable antennas to cover both the short and long range using light weight, we find that compact wireless devices are quickly increasing. Uncomplicated design, low cost and low complexity are some of the main attractive characteristics of the microstrip patch antenna which is used in croissant technology communication systems by radar and satellite. There are a lot of organisms and fields (eg. military, aeronautic, aerospace as well as others) that need low profile, low weight and conformal radar antenna. The Patch antenna technology which was suggested is the best solution to solve these


A. Saka et al. (Eds.): CPI 2019, LNME, pp. 206–219, 2021.
https://doi.org/10.1007/978-3-030-62199-5_18

issues. There are several researches [1,2] in this topic to obtain a high performance and precise radar antenna. Recently, efforts have been done towards the Ka-band which a lot of recommendations [3] advise about because it is precise at the level of detection. This work is about design and experimental analysis of single element antenna for Ka-band which will be in an antenna array. Likewise, we will obtain the broadband and good matching for Ka-band in order to get a high distance resolution $\Delta R$: it is the ability of a radar system to distinguish among two or more targets in the same direction but at different distances. The quality of the resolution depends on the bandwidth $\Delta f$ of the transmitted pulse [4].

$$\Delta R = \frac{c}{2\Delta f} \quad With(\Delta f = \frac{\int_{-\infty}^{+\infty} |s(f)|^2}{|s(f)^2|} dx) \tag{1}$$

In the case of an unmodulated pulse, $c/2\Delta f$ the distance resolution is increased by a radar antenna with a broadband based on the transmission of very short pulses, often less than the nanosecond, we speak here about margin of gigahertz. The radar equation not only reflects [4] the influence of physical phenomena on the radiated power from the propagation of the wave until the reception of the reflected signal (echo), but also allows to estimate the performance estimated of the radar system. The total power captured by the radar which depends on the equivalent surface $A_e$ of the radar antenna is:

$$P_r = \frac{P_t * G^2 * \sigma^2 * \lambda^2}{(4\pi)^2 * R^2 * L_p} \tag{2}$$

This expression shows that the echo signal intensity is proportional to the radiation power, the square of the gain, the effective cross section of the object and to the square of the wavelength. So, the maximum range of the radar depends on the square root of the Gain. This results in the performance of the antenna which has a very great influence on the radar performance. To reach a long range, we have to increase the gain by putting the single element antenna in the array with the obtain of good matching in operating frequency. We choose to work in the frequency [30 Ghz] that belongs to ka-band due to its importance in the radar future system. In this work, we used the technique of quarter wave transformer matching reinforced to filter and eliminate the other frequencies in the Ka-band.

## 2 The Single Element Antenna Structure

The geometry of the suggested antenna Fig. 1 is a rectangular structure matched with the notched technique to the central frequency 30 Ghz, with a bandwidth [28 to 31.5] Ghz. This antenna is burned on substrate Duroid 5880 with dielectric constant of $\epsilon_r = 2.22$ and, tangent value of 0.004 with a partial ground plan situated under the feed-line. The patch dimensions are calculated using equations available in the literature [5] for rectangular patch antenna.

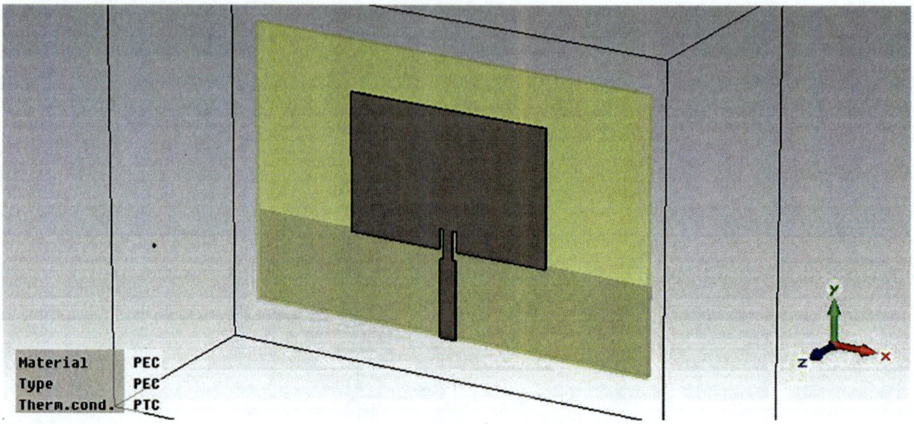


**Fig. 1.** Structure of single antenna.

A metal patch ($L_p = 6.12\,\text{mm}$) and ($W_p = 9.14\,\text{mm}$) was connected to $50\,\Omega$ feed line with an inset in the top of the substrate. The dimensions of inset feed were ($L_i = 3.5\,\text{mm}$) and ($W_i = 0.76\,\text{mm}$). Figure 2 shows the simulated result of reference antenna, S11 versus frequency indicating fundamental resonance frequency 30 Ghz and large bandwidth [28–31.5] Ghz.

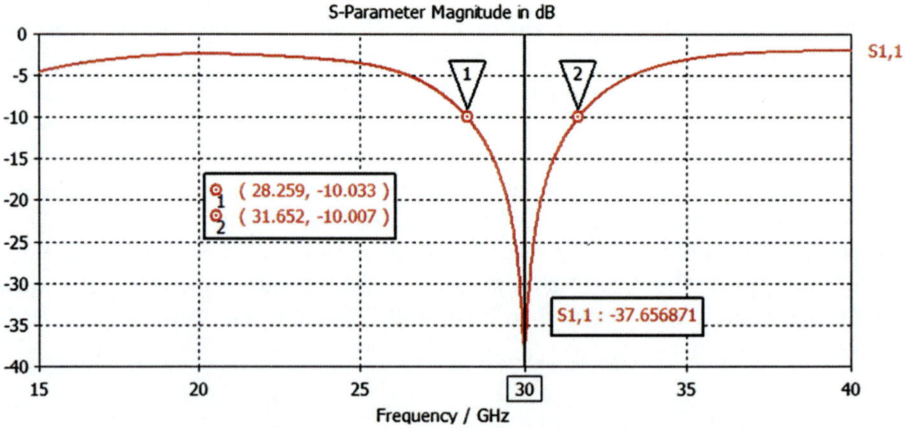


**Fig. 2.** Simulated return loss of single antenna.

The suggested antenna presented a very convincing result at the level of matching with a wide bandwidth. To enhance the gain (Fig. 3) and to get a radar antenna that replies to the requirements of matching for the Ka-band frequency which we need, we should put the single element antenna in the array.

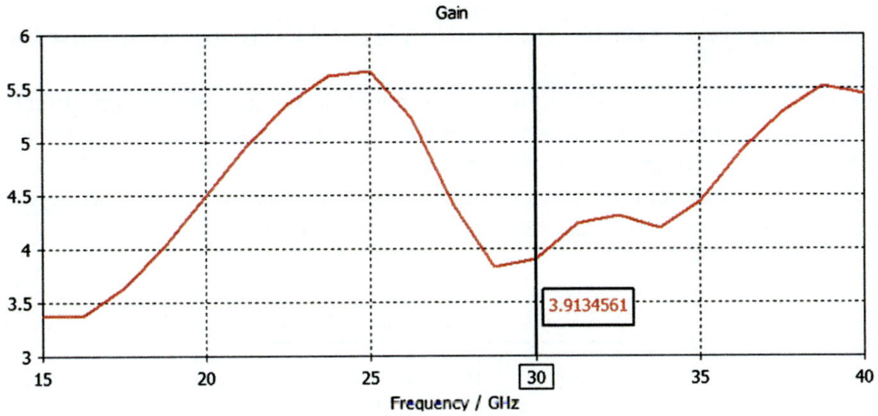


**Fig. 3.** Gain of single element antenna in 30 Ghz.

Figure 4 shows the measured far-field E-plane and H-plane radiation patterns at the resonance frequency, which shows us the radius of the single antenna characterized by the multi direction for both E and H plans.

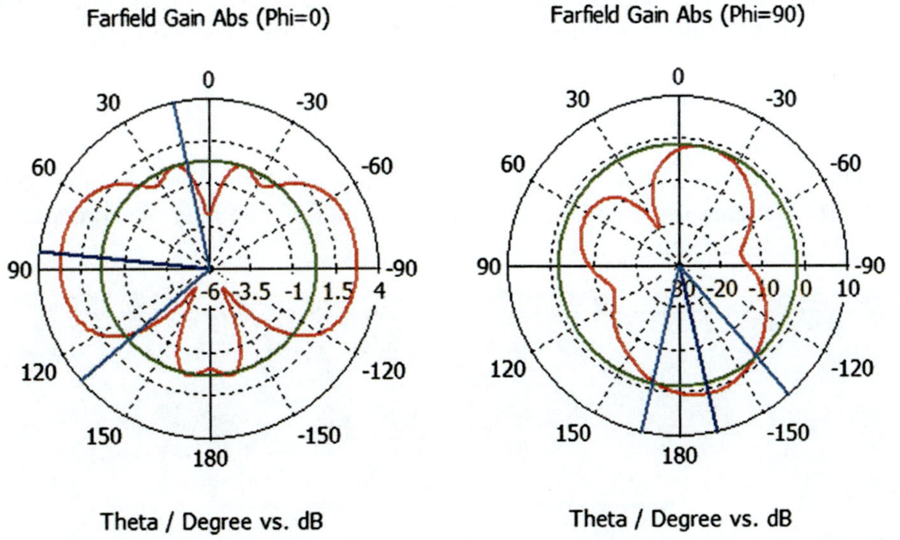


**Fig. 4.** 2D radiation pattern E-plane and H-plane at the resonance frequency 30 Ghz.

# 3 Microstrip Patch Antenna Array Without Dumbbells

In order to enhance the performance of the antenna like increasing gain, directivity, and other functions which are difficult to do with the single element; we have created the antenna array with quarter-wave transformers that are used to make the impedance matching (Fig. 5) calculated by Linecalc tool. The quarter-wave transformers are used to make the impedance matching 30 Ghz.

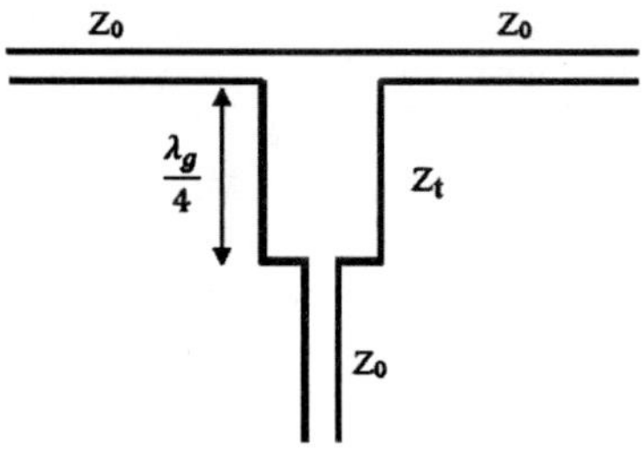


**Fig. 5.** Power divider by two with adapter at the entrance.

We see that $Z_0$ in parallel with $Z_0$ brings $Z_0/2$, we use the relation of matching between two lines, the impedance of the quarter-wave section is written:

$$Z_t = \sqrt{\frac{1}{2} Z_0 Z_0} = \frac{Z_0}{\sqrt{2}} \tag{3}$$

The Fig. 6 presents the geometry of antenna array with partial ground plan, the patch antenna separated by $0.8\lambda_g$ to avoid mutual coupling phenomenon.

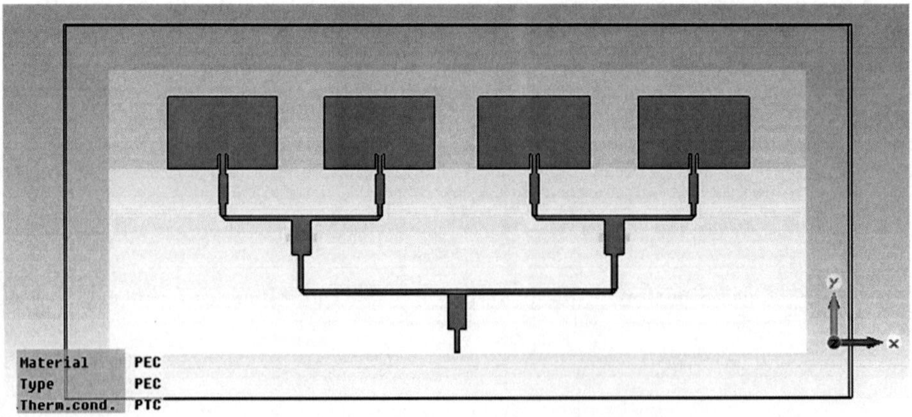


**Fig. 6.** Structure of antenna array proposed.

The Fig. 7 shows matching the frequency 30 Ghz that we focused on. However, it gives us replies [36 Ghz] inside the Ka-band which we study. Practically speaking, this is not efficient when we exploit it at radar systems that select the frequency in band, we will have echoes in the same band not in the frequency we want. That is why we choose to delete this frequency by the help of DGS technique.

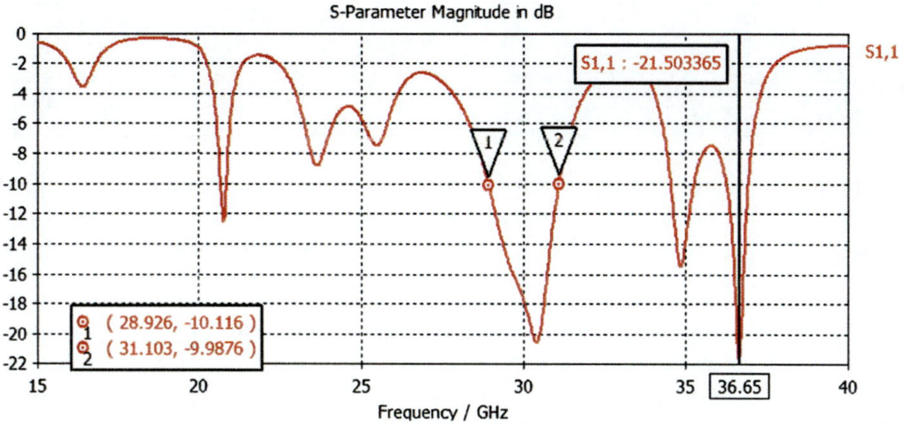


**Fig. 7.** Simulated S11 versus frequency indicating fundamental 30 Ghz resonant frequency and higher harmonics.

The suggested power divider that we used to design the antenna array is not enough to keep the matching frequency in the band since it created other unwanted frequencies. The challenge is to use a method to find a matching just in a certain field for the Ka-band.

## 4 Microstrip Patch Antenna Array Using Two Dumbbells

So as to diminish higher order harmonics, DGS is introduced in microstrip patch antenna array. From the outset, dumbbell shaped DGS structures were introduced below the Power divider and its impact on the antenna properties that were studied in [8] (Fig. 8).

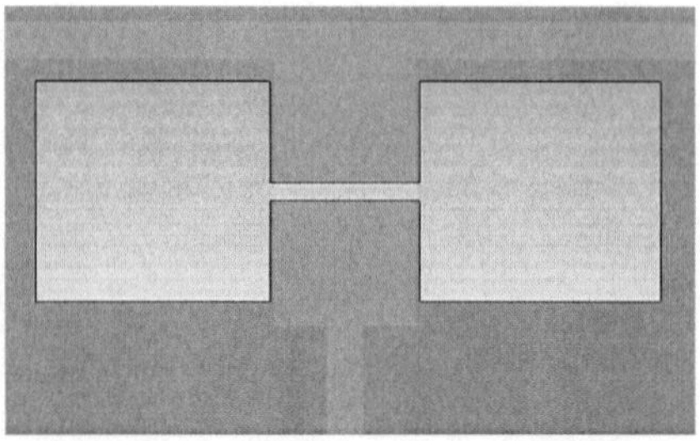

**Fig. 8.** Proposed antenna array with dumbbell shape DGS below Power divider

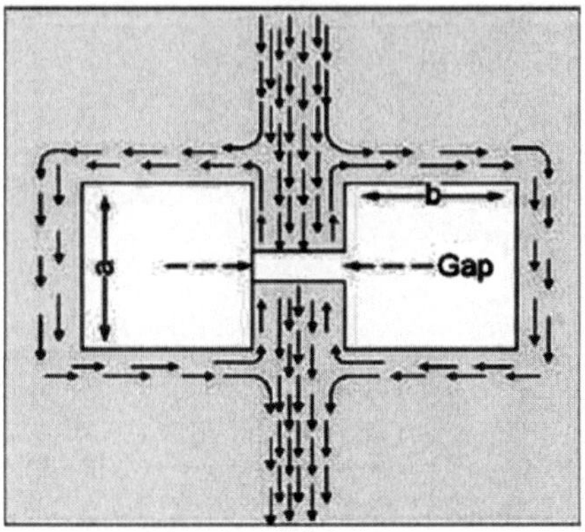


**Fig. 9.** Current distribution in the Ground Plane of DGS microstrip line conventional design and analysis method of DGS [6].

The current path is improved due to the rectangular parts of dumbbell DGS; therefore, effective inductance and effective capacitance of microstrip line are altered. The two rectangular slots of dumbbell DGS are responsible to give a plus to a capacitive effect and a thin rectangular defected slot that links both the rectangular shaped defects calculated for adding the inductance to the total impedance [7]. The return path of the current is fully disturbed using the DGS and this current is confined to the periphery of the perturbation and returns to the underneath of the microstrip line once the perturbation is over as shown in Fig. 9.

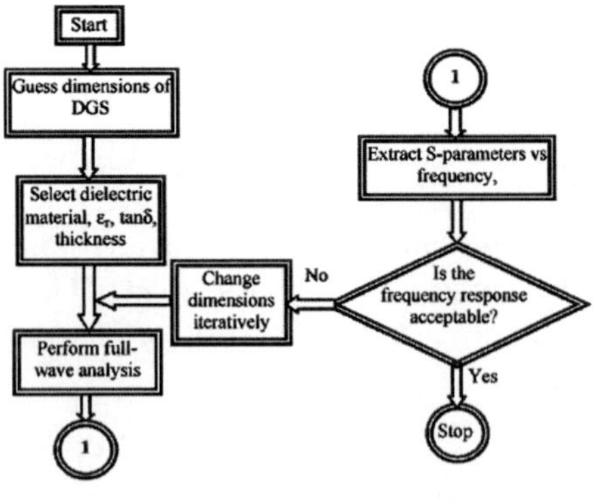


**Fig. 10.** Conventional design and analysis method of DGS.

The flowchart in Fig. 10 shows the conventional design and analysis methods of DGSs [8]. It can be seen in Fig. 10 that the design phase of a DGS starts with the design specifications of stop band frequencies.

The DGS below the quarter-wave transformer shows the stop band characteristics at a certain frequency and that has been adjusted to diminish the higher order harmonics [8]. Owing to this LC circuit, a resonance is shown up at a certain frequency. The slotted area of the DGS is proportional to the effective inductance and inversely proportional to the effective capacitance. A diminution in slotted DGS area gives a reduction to the effective inductance; so, results in a higher cut-off frequency. A decrement in the DGS area reduces the effective capacitance, hence decreasing the resonant frequency.

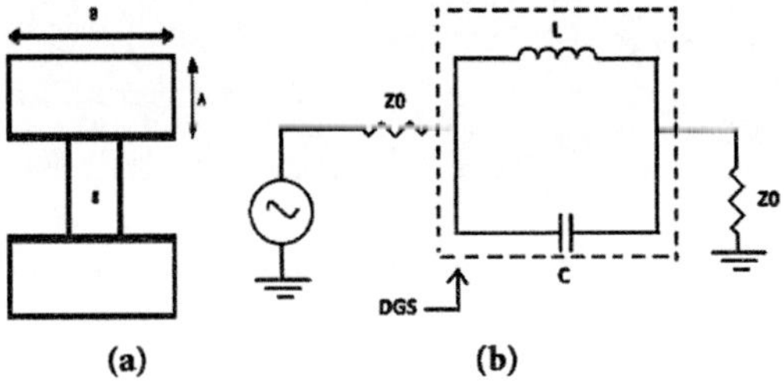


**Fig. 11.** (a) DGS unit cell. (b) L-C equivalent.

Figure 11(a) and (b) introduces the simple and frequently used dumbbell shape DGS and its equivalent circuit composed by an inductance (L) and capacitance (C) in parallel [8]. The head area (A * B) is utilized in variation of inductance (L) and the slot (g) for capacitance (C). The parametric study has been done for optimum size and location of DGSs as well as other parameter of antenna [8]. The optimized parameter of DGS are A = 0.8 mm, B = 0.42 mm and g = 0.2 mm. Figure 12 and Fig. 13 presents the front and back view respectively of microstrip patch antenna array with DGS.

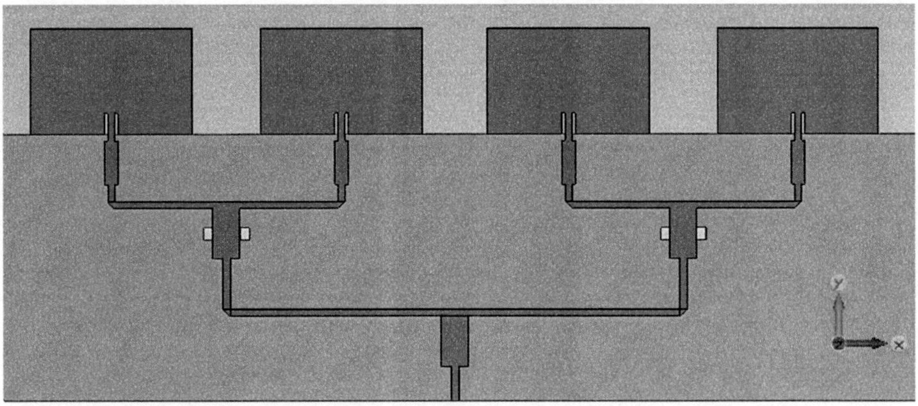

**Fig. 12.** Microstrip patch antenna array with two dumbbell shapes DGS.

**Fig. 13.** Back Microstrip patch antenna array with two dumbbell shapes DGS.

Applying with different size dumbbell shape DGS, the simulation effects we got for the return loss and Gain is as shown in Fig. 14 and Fig. 15 respectively. It is noticed that the return loss has been affected by the existence of size DGS (Fig. 16).

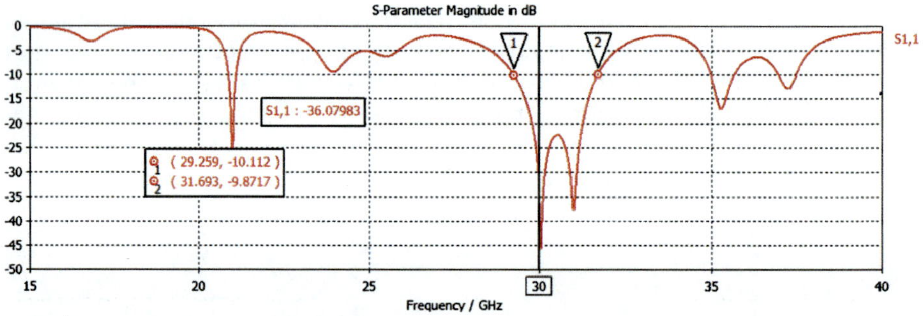


**Fig. 14.** Simulated S11 of antenna array with two dumbbell shapes DGS.

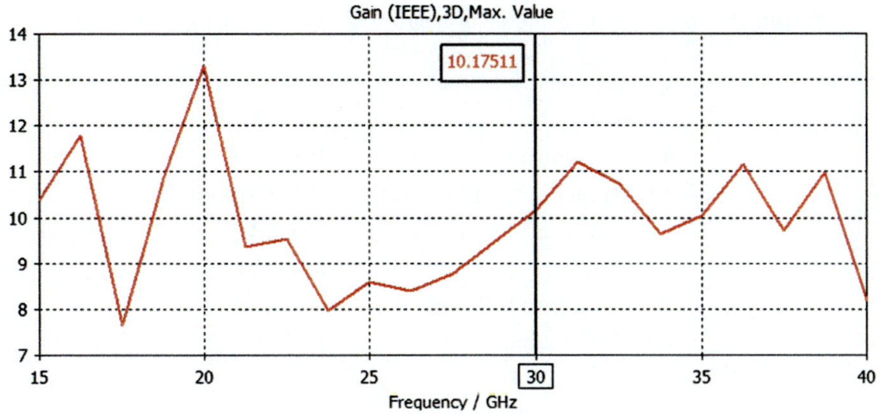


**Fig. 15.** Simulated Gain of antenna array in 30 Ghz.

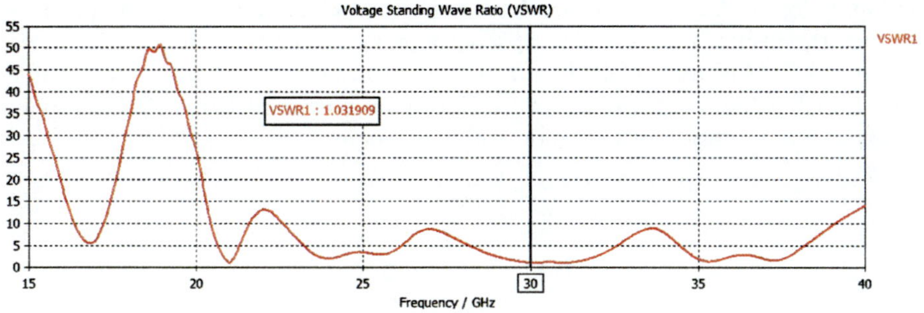


**Fig. 16.** Simulated voltage standing wave ratio (VSWR) of antenna.

The head area (A * B) is useful in variation of inductance (L) and the slot (g) for capacitance (C). The parametric study has been done for optimum size and location of DGSs and other parameter of antenna. The change of the return loss in comparison with the head area of DGS are given in Table 1 and 2.

**Table 1.** Optimized parameter of head area (A * B) with g = 0.4 mm.

| g = 0.4 mm | | |
|---|---|---|
| A * B | S11 (36 Ghz) | S11(30 Ghz) |
| 1.5 * 1.5 (mm) | −25.33 dB | −7 dB |
| 1 * 1 (mm) | −19.05 dB | −10.13 dB |
| 0.5 * 0.5 (mm) | −15.01 dB | −10 dB |
| 1 * 0.5 (mm) | −13.7 dB | −12.89 dB |
| 0.8 * 0.5 (mm) | −13.01 dB | −18.8 dB |
| 0.8 * 0.4 (mm) | −11.09 dB | −21.01 dB |

**Table 2.** Optimized parameter of head area (A * B) with g = 0.2 mm.

| g = 0.2 mm | | |
|---|---|---|
| A * B | S11 (36 Ghz) | S11(30 Ghz) |
| 1.5 * 1.5 (mm) | −21.07 dB | −18 dB |
| 1 * 1 (mm) | −19.05 dB | −19.6 dB |
| 0.5 * 0.5 (mm) | −14.31 dB | −19.6 dB |
| 1 * 0.5 (mm) | −11.07 dB | −21.39 dB |
| 0.8 * 0.5 (mm) | −8.6 dB | −30.04 dB |
| 0.8 * 0.4 (mm) | −7.01 dB | −36.07 dB |

To validate the result we chose to use another electromagnetic simulator which is Advanced Design System (ADS). Figure 17 presents the structure in ADS and Fig. 18 illustrates return losses obtained by simulation of the proposed antenna, the result presents an acceptance agreement between the two simulation software CST-MW and ADS. It is a matching in the central frequency studied 30 Ghz with a bandwidth of 29 to 31 Ghz, and it is remarkable that the result presents an elimination of the undesirable frequency.

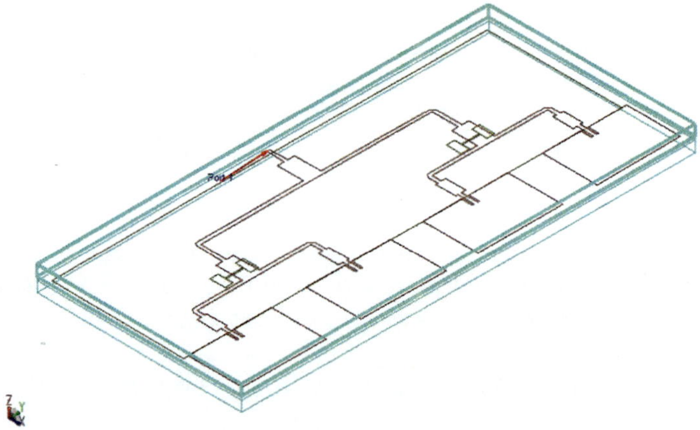

**Fig. 17.** 3D view of patch antenna array with two dumbbells shape DGS in ADS.

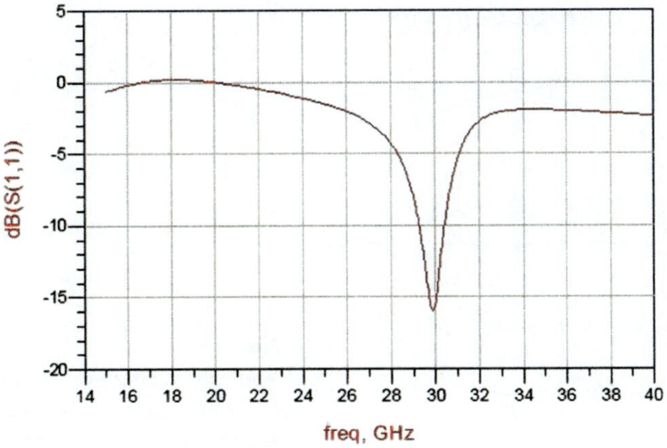


**Fig. 18.** Return losses of patch antenna array with two dumbbells obtained by ADS.

To illustrate the performance of the proposed antenna array in free space, a radiation patterns 2D have been measured. The measured far-field E-plane radiation pattern at the resonance frequency [30 Ghz] is presented in Fig. 19, whereas Fig. 20 displays the measured H-plane radiation pattern. It is observed that E-plane radiation pattern is concentrated in a specific direction, with an overall radiation pattern compatible with radar application.

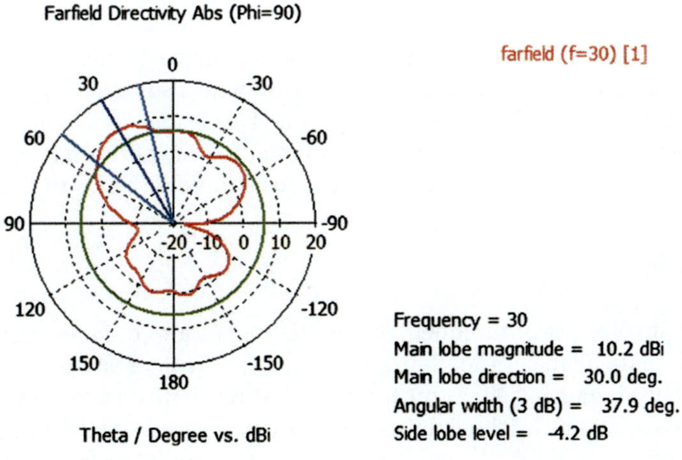


**Fig. 19.** 2D radiation patterns at H plane of the proposed antenna array.

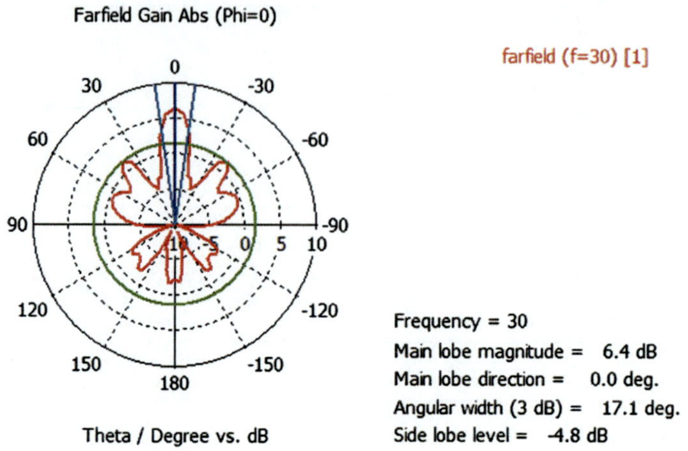


**Fig. 20.** 2D radiation patterns at E plane of the proposed antenna array.

The main objective of this study is to eliminate other unwanted frequencies in a band by using different size & position of DGS structure. The dumbbell shape DGS of different size and position as detailed in Table 1 and 2 was added to the antenna under the quarter wave transformer. A number of optimization has been implemented for final DGS size and position. The result we have found is that the technique used by Dumbbell was able to delete the unwanted frequency and let us the central frequency that we want with a highlight of the bandwidth [29–31.5 Ghz].

Many new researches [10,11] about dumbbell shape technique were used as Bandstop, Highpass, and Bandpass Filters, it was also used to get a specific radiation patterns [12]. The originality about this paper is the use of a power divider controlled by the dumbbell shape to monitor the performance of Antenna array in different levels (the frequency response, matching and radiation).

## 5 Conclusion

This paper involves two dumbbell shape DGS structure of same size and position to eliminate higher order harmonics present in microstrip patch antenna array without significant effect on fundamental resonance frequency. The reference antenna with and without DGS structure was simulated by two simulation software programs CST-MW and ADS. We start our design by implementing a single element antenna matched to the center frequency 30 Ghz then we inserted the single element in the antenna array with the help of quarter wave microstrip line matching power divider which resulted other undesirable frequencies. After that, we used the dumbbell shapes in order to filter frequencies and keep the useful ones in ka-band. This technique served to obtain a good gain, bandwidth, radiation results. This work went in the path to give us a result of a millimeter antenna in a certain frequency for the Ka-band that responds to the following criteria: distance resolution by a wide band, ameliorating the range by the gain augmentation and eliminating wrong echoes that come from unwanted frequencies.

## References


1. du Preez, J., Sinha, S.: Millimeter-Wave Antennas: Configurations and Applications. Signals and Communication Technology. Springer International Publishing, Switzerland (2016)
2. Singh, I., Tripathi, V.S.: Micro strip patch antenna and its applications: a survey. Int. J. Comput. Tech. Appl. **2**(5), 1595–1599 (2011)
3. Melezhik, P.N., et al.: High-efficiency millimeter-wave coherent radar for airport surface movement monitoring and control. In: Proceedings of 5th European Radar Conference, pp. 361–364 (2011). https://doi.org/10.3846/16487788.2011.596673
4. Ottersten, B., Viberg, M., Stoica, P., Nehorai, A., Huang, T.S., Kohonen, T., Schroeder, M.R., Lotsch, H.K.V., Haykin, S., Litva, J., Shepherd, T.J.: Exact and large sample maximum likelihood techniques for parameter estimation and detection in array processing. In: Radar Array Processing, vol. 25, pp. 99–151. Springer, Berlin, Germany (1993)
5. Balanis, C.A.: Antenna Theory: Analysis and Design. Wiley (1997)
6. Khandelwal, M.K., Kanaujia, B.K., Kumar, S.: Defected ground structure: fundamentals, analysis, and applications in modern wireless trends. J. Int. J. Antennas Propag, vol. 2017, 22 pp. (2017)
7. Karmakar, N.C., Roy, S.M., Balbin, I.: Quasi-static modeling of defected ground structure. IEEE Trans. Microw. Theory Tech. **54**(5), 21602168 (2006)
8. Ahn, D., Park, J.-S., Kim, C.-S., Kim, J., Qian, Y., Itoh, T.: A design of the low-pass filter using the novel microstrip defected ground structure. IEEE Trans. Microw. Theory Tech. **49**(1), 8693 (2001)
9. Dal Ahn, M., Park, J.-S., Kim, C.-S., Kim, J., Yongxi Qian, M., Itoh, T.: Design of the low-pass filter using the novel Microstrip defected ground structure. IEEE Trans. Microw. Theory Tech. **49**(1) (2001)
10. Ebrahimi, A., Baum, T., Ghorbani, K.: Differential bandpass filters based on Dumbbell-shaped defected ground resonators. IEEE Microw. Wirel. Compon. Lett. **28**(2), 129–131 (2018)
11. Yuan, W., Liu, X., Lu, H., Wu, W., Yuan, N.: Flexible design method for microstrip bandstop, highpass, and bandpass filters using similar defected ground structures. IEEE Access, vol. 7, pp. 98453–98461 (2019)
12. Arul Selvam, V., Arun, M., Ashok, N., Giridharan, L., Sheriba, T.S.: Beam-repositioning system using microstrip patch antenna array for wireless applications. Int. J. Sci. Res. Dev. **6**(01), 2321–0613 (2018)

# Ceramic Paste Extruder of 3D Printing: Status, Types, and Prospects

Jihad El Mesbahi[1(✉)], Irene Buj-Corral[2], and Abdelilah El Mesbahi[1]

[1] EIMIS-Faculty of Sciences and Technologies, Mechanical Engineering, Tangier, Morocco
jihad.elmesbahi@gmail.com, elmesbahi_abdelilah@gmail.com

[2] ETSEIB-Universitat Politècnica de Catalunya, Mechanical Engineering, Barcelona, Spain
Irene.buj@upc.edu

**Abstract.** Additive manufacturing (AM) is an important, rapidly emerging, manufacturing technology that takes the information from a computer-aided design (CAD) and builds parts in a layer-by-layer style. Since about three decades ago, AM technologies have been used to fabricate prototypes or parts mostly from polymeric or metallic materials. However, 3D printing of ceramics is getting growing attention. It is an appropriate method to answer the increased demand for ceramic structures with complicated morphology by fabricating ceramic parts with high resolution and good surface quality. There are several 3D printing technologies available to 3D print ceramic material as paste extrusion, powder sintering, binder jetting, and photopolymerization. This paper presents the classification of 3D printing technologies used for the manufacturing of ceramic components, the historical origin of these technologies and their technical principle. This work focuses on ceramic pasta extruder systems. It presents the different types of material extrusion systems for ceramic, some examples of their current machines, and their advantages and limitations. Then, after an extensive categorization of the different extrusion systems with existing technical solutions, some prospects to improving these systems are presented and that will be the work of the next paper.

**Keywords:** Additive Manufacturing (AM) · 3D printing · Ceramic components · Polymer material · Extrusion system · Direct Ink Writing (DIW)

## 1 Introduction

Owing to their various excellent properties, ceramics are used in a wide range of applications, including the chemical industry, machinery, electronics, aerospace and biomedical engineering [1]. However, geometrically complex ceramic parts are difficult to manufacture using traditional processes. Therefore several Additive manufacturing (AM) technologies have been developed in recent years that can fabricate ceramic components having complex geometries [2]. The introduction of 3D printing (another term to talk about AM) into the manufacturing of ceramic components offers entirely new possibilities for addressing the above-mentioned problems and challenges. Several


A. Saka et al. (Eds.): CPI 2019, LNME, pp. 220–229, 2021.
https://doi.org/10.1007/978-3-030-62199-5_19

processes, such as Ink-jet Printing, Selective Laser Sintering, Stereolithography, Laminated Object Manufacturing, etc., have evolved to include the ability to fabricate ceramic parts. However, extrusion-based methods are among the most popular approaches for freeform fabrication of ceramic components due to the simplicity and low cost of their fabrication system, high density of their fabricated parts, their capability of producing parts with multiple materials and/or as functionally graded materials, and the low amount of material wasted during processing [3]. This paper will focus on the material extrusion system for ceramic or Direct Ink Writing (DIW). It is organized in three subsequent sections following this introduction. Section 2 discusses the classification of 3D printing technologies used for the manufacturing of ceramic components. Definitions of each technology are given, and the historical origins of each type of technology are specifically emphasized. Section 3 explains the different types of material extrusion systems for ceramic currently available. Advantages and disadvantages of these methods are also discussed. Finally, in Sect. 4, the perspectives for further improving the extruder subsystem are presented.

## 2 Classification of Ceramic 3D Printing Technologies

The ceramic 3D printing technologies can be classified according to the form of the pre-processed feedstock prior to manufacturing [1]. Hence, there are liquid-, powder-, and bulk solid-based methods, as summarized in Fig. 1. The slurry-based ceramic 3D printing technologies lead to the solidification of a liquid or semi-liquid. Process acting on powder-based use a thermal source provided by a laser beam containing loose ceramics particles as feedstock. For bulk solid-based process, they use solid plates which can be bonded with a laser or with an adhesive.

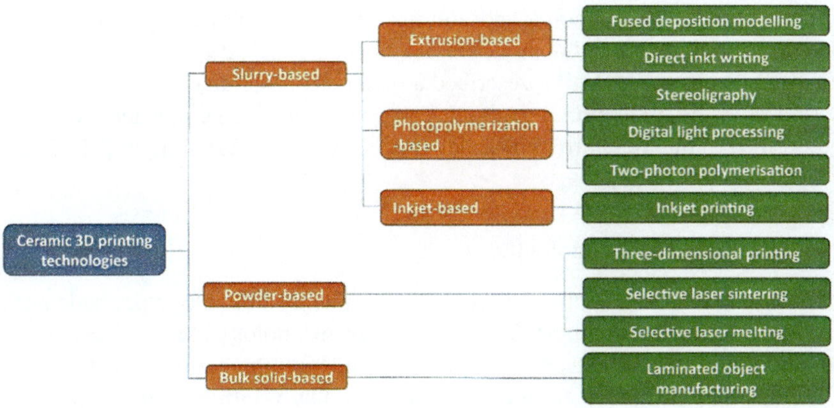


**Fig. 1.** Classification of ceramic 3D printing technologies (Adapted from [1])

## 2.1 Slurry-Based Technologies

In this section, the liquid-based technologies dispersed with fine ceramic particles as feedstock are presented. They refer to a number of AM (additive manufacturing) methods patterning either by ceramic extrusion with Direct Ink Writing (DIW) or with Fused Deposition Modeling (FDM). Other processes refer to photopolymerization of ceramic powders suspended in a polymerizable medium such as stereolithography (SL) and its derivatives, i.e. digital light processing (DLP), along with inkjet-based inkjet printing (IJP).

As for extrusion processes, in the DIW process, a ceramic is extruded through a syringe and deposited on a printing bed. It was first filed as a patent by Cesarano and co-workers at Sandia National Laboratories in 1997. The technique was originally developed for processing concentrated materials such as ceramic slurries with little organic content [1]. In FDC, ceramic powder is mixed with a plastic filament. The material is printed in a similar way to the fused deposition modeling process (FDM), also known as fused filament fabrication (FFF). FDM was first developed by Crump et al., with a patent filed in 1989 [4], and later commercialized by Stratasys Inc. in 1990. In a FDM process, the pasta is continuously fed into a syringe so that it can be easily extruded via the nozzle to form layers, and upon extrusion, the material solidifies immediately over the previously printed layer [1].

For photopolymerization of ceramic powders technologies, the SL was first proposed and developed by Hull in 1986 and was commercialized in 1987 by 3D Systems Inc [3]. It uses a build platform submerged into a translucent tank filled with liquid photopolymer resin. Various monolithic materials such as ZrO2, Al2O3, SiC, SiOC, SiO2, and TiC have been fabricated with ceramic suspension [5]. For DLP concept, it was first proposed by Nakamoto and Yamaguchi in 1996 using physical masks, and it was further developed and improved on by Bertsch et al. in 1997 [1]. The use of the DLP technique as a 3D printing tool in ceramics fabrication has been widely explored. High-density (97–99%) zirconia and alumina structural parts comparable with those prepared using conventional approaches have been produced [6].

In 1992, Sachs et al. at MIT described a method for manufacturing ceramic casting cores and shells by inkjet printing a binder phase onto a ceramic powder bed [6]. It was first described in the literature by Blazdell and co-workers beginning in [7] using ZrO2 and TiO2 ceramic inks.

## 2.2 Powder Based Technologies

Powder-based ceramic 3D printing technologies mainly utilize powder beds normally containing loose ceramic particles as feedstock. The technology used to inscribe the layer information depends on the specific process considered: three of the most well-known and world-spread processes are the "three-dimensional printing (3DP)", the "selective laser sintering (SLS) and the "Selective laser melting (SLM). The 3DP method was first devised by Sachs et al. at the Massachusetts Institute of Technology (MIT), who filed a patent for it in 1989 [7]. Its application to ceramics fabrication was originally reported in the 1990s with the use of alumina and silicon carbide particles as powdered materials and colloidal silica as a binder [8]. For the SLS process, it was invented by Deckard and

Beaman at the University of Texas at Austin, with the very first patent filed in 1986 [9], and was further extended by the DTM company, which was acquired by 3D Systems in 2001. The SLM process was developed at the Fraunhofer Institute for Laser Technology (ILT) in Germany in 1996, resulting in German patent DE 19649865 [9]. Its application to a ceramic powder involves full melting of the powder to form a solid part by high-energy-density laser scanning layer by layer, without binders or post-sintering owing to the full melting and fusion of the powder.

### 2.3 Bulk Solid-Based Technologies

LOM was first reported by Kunieda in 1984, then further developed by Helysis Corporation in 1986 and commercialized in 1991. The application of this technique to ceramics manufacturing was first demonstrated by Griffin and co-workers in 1994 based on tape-cast alumina and zirconia green sheets [10].

## 3 Types of Material Extrusion System for Ceramic

The extrusion subsystem is the core of the entire system and greatly affects the properties of the final products [3]. The basic principle of material extrusion additive technology involves the loading of the material, moving the material through a nozzle by applying force or pressure, plotting liquefied material according to a pre-defined path in a controlled manner, and layer-by-layer bonding of the material to itself or a secondary build material to form a coherent solid structure. After a layer is completed, the build platform moves down or the extrusion head moves up, and a new layer of material is deposited and adhered onto the previous layer [11]. After reviewing the existing of 3D printing extrusion machines in the ceramic industries, the authors have identified three main constructive designs for paste and clay extrusion devices, which will be described in the following section and schematically shown in Fig. 2.

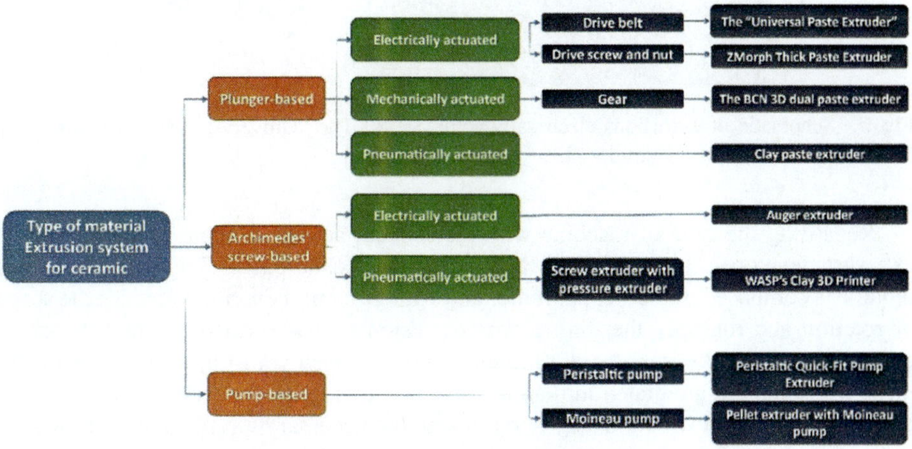


**Fig. 2.** Different types of extrusion-based AM for ceramic.

### 3.1 Plunger-Based Extrusion

Based on plunger extrusion, several open-source designs have become available online for retrofitting consumer 3D ceramic printers. It is similar to a syringe, electrically, mechanically or pneumatically actuated. For the type of this device that uses an electric motor (Fig. 3), the rotation (the axis of the motor rotates) is transformed into linear translation using a drive belt mechanism (a), or a drive-screw and nut mechanism (b). The "Universal Paste Extruder" by RichRap is shown in Fig. 3a. It is a sort of syringe pushed by a drive belt mechanism to drive the plunger that presses the feedstock through a nozzle. It simply uses the existing extruder motor output from printer electronics, and it can print with ceramic Clays [12]. The limitations of this design related to several aspects. Due to the reduced amount of feedstock, the extruder tube can hold, and because of the inability to feed more material during fabrication, the productivity of this extruder is limited to fabricating small-volume objects. However, the worm drive control machine such as ZMorph Thick Paste Extruder (b) 30 to 100 ml syringes as their material reservoirs, but what they gain in volume, they lose in performance and precision [13].

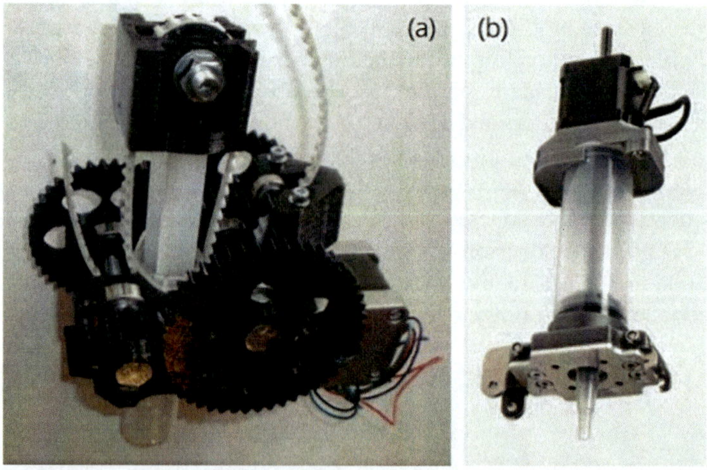


**Fig. 3.** Schematic of extrusions electrically actuated: (a) The "Universal Paste Extruder" [12], (b) ZMorph Thick Paste Extruder [13]

Regarding the type of machine with mechanical actuation, there is the dual paste extruder developed by CIM-UPC and BCN3D Technologies (Fig. 4). The power module is composed of the stepper motor and with gears of reduction to reduce the speed of rotation and multiply the torque. The extrusion unit is formed by the ml syringe currently used and the nozzles of different diameters used with Luer-Lock system. All these two units are integrated into an outer casing that holds the whole assembly. The current transmission which manages to transmit the necessary torque to the gear wheels to multiply it consists of a gear train with an output rack that provides linear movement to the plunger of the syringe and transmits the necessary force to operate it.

**Fig. 4.** The BCN 3D dual paste extruder [14]

The third type is a clay paste extruder (Fig. 5), which was inspired by an air-pressure driven solder paste depositor. The body of the extruder is a syringe filled with clay paste. When pressurized, the paste is pushed out of the extruder at a near-constant rate. Two valves are required to equalize the pressure in the extruder to atmospheric when you want to stop extrusion. Without a relief valve, clay will continue to ooze out of the extruder even when the pressure valve is closed [15].

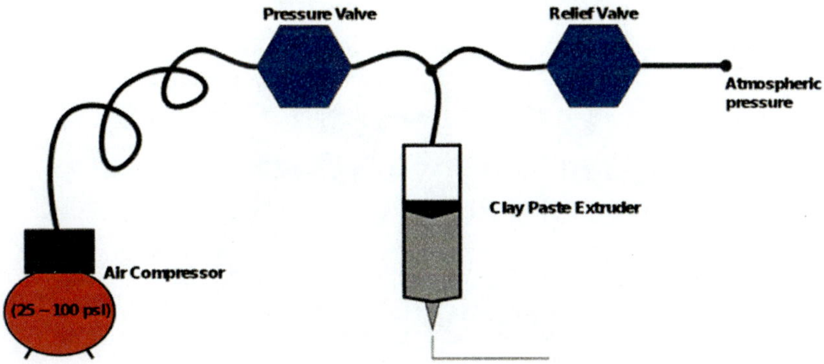


**Fig. 5.** Clay paste extruder [15]

### 3.2 Screw-Based Extrusion

The second constructive design uses an auger screw to push material through the nozzle. The auger extruder mechanism (Fig. 6) shows pressurized feedstock is stored in a separate container and fed to the extruder. Extrusion is achieved by rotating the auger using a servo motor [16]. While this design allows for the feeding of material during the fabrication process, and thus enables the fabrication of large objects, this feature also requires additional drive and control elements. Consequently, it increases the complexity of the AM machine.

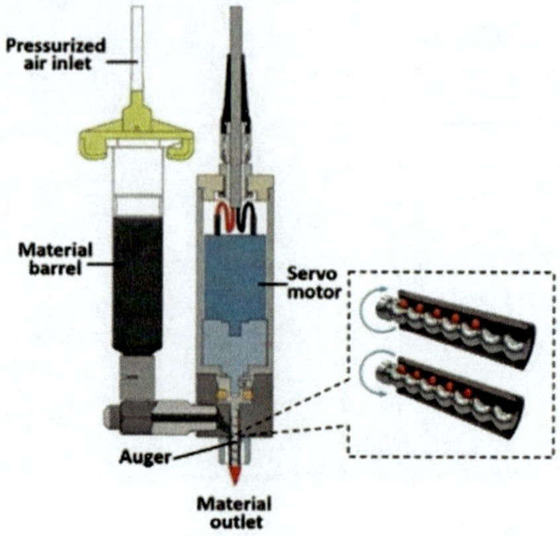


**Fig. 6.** Auger extruder [16]

The second example is the Italian 3D printer manufacturer WASP (Fig. 7). The new LDM Wasp Extruder can reaches a level of precision which is very close to that of plastic polymers extruders thanks to the combination of a screw extruder and a pressure extruder. Among the most notable innovations presented by the new ceramic extruder is the extrusion system itself. It eliminates air bubbles out of the mixture, has an extrusion controlling system with a retraction option to interrupt deposition [17].

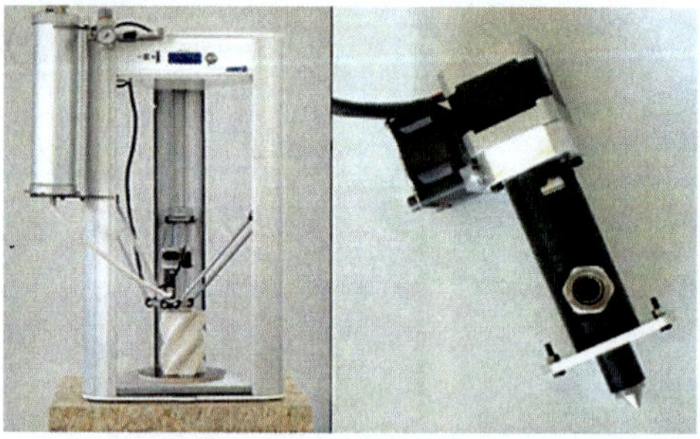

**Fig. 7.** WASP's clay 3D printer [17]

Other extrusion mechanisms that work with pump, for example [18] shows the design for a geared peristaltic pump designed for 3D printers to enable them to extrude various materials for building objects. The main advantage is that the pumped medium is completely enclosed with no possibility of leakage. Another advantage is the continuous pumping of the material and also a low cost of the equipment [19]. Another more interesting way of feeding thick liquids onto a build platform is a progressive cavity pump (Moineau pump). The first idea to implement a plastic extruder using a Moineau pump was inspired by an article appeared in Issue 3 of "RepRap Magazine" at page 26, entitled "The road to better paste extrusion" [20]. It is a solution for professional businesses and research institutes developing advanced food applications, high tech ceramics or medical/bioprinting.

In addition, there are other limitations that relate to several aspects. The productivity of the machines that use the syringe as an extruder is limited to the manufacturing of small-volume objects because the syringes may contain a reduced amount of feedstock. Syringe-based paste extruders typically use large nozzle diameters and layer heights to fabricate large objects; this can have a negative impact on the surface finish. In this context, we designed a paste extrusion device that will be presented in a future paper. It is able to increase productivity and print large parts.

## 4 New Development and Prospects

As future work, it is expected to improve the extrusion subsystem to fabricate ceramic parts, to adapt in REP RAP machine. The new 3D print-head is made for the BCN machine but it is suitable for any 3D printer. With the idea of improving print quality and increasing productivity, two studies are underway:

- It introduces a technical solution based on the use of a larger syringe (more than 20 ml) as housing for the extruder. Currently, the use of a diameter less than 15 mm is limited. In this way, the size of the printed parts could be enlarged. For syringes, the addition of an additional restraint point could be assessed. Currently, they are only submitted from the top. By adding a second attachment, the syringe receives greater stability. This can have a significant influence on both the dimensional error and the surface condition.
- Upgrade the force reduction system or replace it with a new one that offers greater pressure. When the ceramic is dense, it takes more force to flush it through the nozzle. Unable to choose the material of the syringe, on some occasions, the zipper skipped a step because it cannot move forward. These situations can cause the breakage of a gear system tooth.

The solution must be open-source and compatible with open-source hardware and software, making it inexpensive and widely accessible to the 3D printing community. This solution will enable us to achieve the goal to increase productivity, print continuity and the flexibility of material change during printing.

## 5 Conclusion

In this paper, we have broadly classified the 3D printing technologies for ceramic, the slurry-based, powder-based, and bulk solid-based processes. We have given the definitions, the historical origins of each AM technologies, and some applications. After that, we have investigated the different types of a material extrusion system for ceramic currently available, i.e., plunger-based extrusion, screw-based extrusion, and extrusion with pump. We have presented the advantages and disadvantages of these methods. Finally, we have dedicated the last section of our paper to present the perspectives for further improving machine for extrusion ceramic. Currently, we are working on 3D drawing on Solidworks to realize the prototype of the proposed design.

**Acknowledgements.** The authors thank the Spanish Ministry of Industry, Economy, and Competitiveness for financial help of project DPI2016-80345R.

## References

1. Chen, Z., Li, Z., Li, J., Liu, C., Lao, C., Fu, Y., Liu, C., Li, Y., Wang, P., He, Y.: 3D printing of ceramics: a review. J. Eur. Ceram. Soc. **39**(4), 661–687 (2019)
2. Deuser, B., Landers, M., Leu, C., Hilmas, E.: Hybrid extrusion force-velocity control using freeze-form extrusion fabrication for functionally graded material parts. J. Manuf. Sci. Eng. **135**, 041015 (2013)
3. Ghazanfari, A., Li, M., Leu, C., Hilmas, G: A novel extrusion-based additive manufacturing process for ceramic parts. In: Proceedings of the 26th Annual International Solid Freeform Fabrication Symposium, pp. 1509–1529 (2016)
4. Kai, C., Kah Fai, L., Chu, L.: Rapid Prototyping: Principles and Applications. Wiley, New York (2010)
5. Yang, L., Miyanaji, H: Ceramic additive manufacturing: a review of current status and challenges. In: Proceedings of the 26th Annual International Solid Freeform Fabrication Symposium, pp. 652–679 (2017)
6. He, R., et al.: Fabrication of complex-shaped zirconia ceramic parts via a DLP-stereolithography-based 3D printing method. Ceram. Int. **44**, 3412–3416 (2018)
7. Xiang, Q.F., Evans, J.R.G., Edirisinghe, M.J., Blazdell, P.F.: Solid free-forming of ceramics using a drop-on-demand jet printer. Proc. Inst. Mech. Eng. Part B J. Eng. Manufact. **211**, 211–214 (1997)
8. Sachs, E., Cima, M., Cornie, J.: Three-dimensional printing: rapid tooling and prototypes directly from a CAD model. CIRP Ann. **39**, 201–204 (1990)
9. Deckard, C.: Method and apparatus for producing parts by selective sintering. US4863538A (1989)
10. Griffin, C., Daufenbach, J., McMillin, S.: Solid Freefonn Fabrication of Functional Ceramic Components Using a Laminated Object Manufacturing Technique (1994)
11. Gonzalez, J., Cano, S., Schuschnigg, S., Kukla, C., Sapkota, J., Holzer, C.: Additive manufacturing of metallic and ceramic components by the material extrusion of highly-filled polymers: a review and future perspectives. Materials **11**(5), 840 (2018)
12. LNCS Homepage. https://www.thingiverse.com/thing:20733. Accessed 19 May 2019
13. Pusch, K., Hinton, T., Feinberg, A.: Large volume syringe pump extruder for desktop 3D printers. HardwareX **3**, 49–61 (2018)
14. LNCS Homepage. https://3dfabprint.com/bcn3d-technologies-develops-dual-paste-extruder/lncs. Accessed 19 May 2019
15. Arnone, L.: Designing a ceramic printhead for a RepRap 3D printer. Report, Swarthmore College Department of Engineering, (2014)
16. Li, W., Ghazanfari, A., Leu, M., Landers, R.: Extrusion-on-demand methods for high solids loading ceramic paste in freeform extrusion fabrication. Virt. Phys. Prototyping **12**(3), 193–205 (2017)
17. LNCS Homepage. https://www.3dwasp.com/en/wasp-launches-the-new-professional-clay-extruder/lncs. Accessed 21 May 2019
18. Benchoff, B.: 3D Printering: Pastestruders. Hackaday (2013)

19. Vojtech, L.: Conception, design and materialisation of a pumping-based extrusion system for food 3D-printing. Dissertation, University Polytechnic of Barcelona (2015)
20. Canessa, E., Baruzzo, M., Fonda, C.: Study of Moineau-based pumps for the volumetric extrusion of pellets. Addit. Manufact. **17**, 143–150 (2017)

# Reviews of Mechanical Design and Electronic Control of Multi-material/Color FDM 3D Printing

Mohammed Boulaala[1(✉)], Driss Elmessaoudi[2], Irene Buj-Corral[3], Jihad El Mesbahi[2], Mohamed Mazighe[1], Abdelali Astito[1], Mhamed El Mrabet[1], and Abdelilah Elmesbahi[2]

[1] LIST, Faculty of Sciences and Technologies, Computer Science, Systems and Telecommunications Laboratory, Tangier, Morocco
m.boulaala@gmail.com, mohamed_mazighe@outlook.com, abdelali_astito@yahoo.com

[2] EIMIS, Faculty of Sciences and Technologies, Mechanical Engineering, Tangier, Morocco
driss_elmessaoudi@yahoo.fr, jihad.elmesbahi@gmail.com, elmesbahi_abdelilah@gmail.com

[3] ETSEIB, Universitat Politècnica de Catalunya, Mechanical Engineering, Barcelona, Spain
Irene.buj@upc.edu

**Abstract.** An extruder is used to receive and dispense the material in extrusion additive manufacturing process. Multiple extruders cover a large variety of application of interest to 3D printing: enhancing productivity, multiple colors of the object, the variation of printing materials, interchangeability In case of a failure…etc. However, to support multiple extruders, the open sources electronic boards available in the market are very limited in this contest and still far from meeting the needs of some specific3D printers. Then this paper gives an overview of existing electronic boards able to support more than one extruder and their limitations.

**Keywords:** Additive Manufacturing (AM) · 3D printing · Fused Deposition Modeling (FDM) · Multi-extruder · Multi-material · Multi-color

## 1 Introduction

Additive manufacturing (AM) is the industrial production name for 3D printing, a computer-controlled process that creates three-dimensional objects by depositing materials layer by layer. According to the ASTM and ISO standards, AM processes are divided into seven categories [1], including stereo-lithography, fused deposition modeling, powder bed fusion, laminated object manufacturing, and direct energy deposition and Fused Deposition Modeling (FDM). Despite the astonishing progress of additive manufacturing and its original aspect to obtain complex geometries never manufactured even before, the greatest dream of users and industrials to create a productive-competitive machine able to print parts and assemblies with defined characteristics and


A. Saka et al. (Eds.): CPI 2019, LNME, pp. 230–238, 2021.
https://doi.org/10.1007/978-3-030-62199-5_20

properties, for any material and color and without additional post-processing, is still, up to now, far from being reached.

Then, based in open-source Prusa I3M 3D printer project with single extruder [2], our goal is to develop a piece of new integrated technical solutions (mechanical design, electronic board, and personalized software) to overcome the weakness of this process. Now; we are mainly focused on integrated-design of a new multi-extruder system and its electronic board control unit able to extend the number of deposed materials and mix different colors while improving the productivity rate. So, this paper gives a preliminary study that aims to summarize the different possibilities of mechanical design and simultaneous control of multi-tasking extruders (materials, colors, nozzles …).

## 2 Classification of Multi-material/Color FDM 3D Printers

### 2.1 Internal Mechanism

a. **Dual Extruder**

The extrusion principle is the same as for a single-nozzle extruder, but there are two nozzles instead of on. Dual extruders can take full advantage of use dissolvable or removable support to hold some region of the part are in overhangs but mainly to get rid of the supports intelligently without affecting the fineness of the final surface of par. This category can be of parallel or independent type as illustrated in the following figure [3] (Fig. 1).

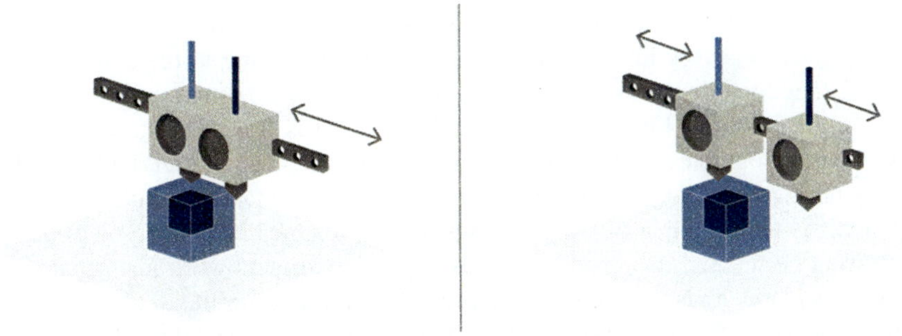

**Fig. 1.** Types of Dual extruder system: a) Parallel: Two or more print heads placed together in the same carriage and b) Independent Dual Extruder: Two or more print heads moving independently in the X-axis same [3]

In the parallel dual extruder, two nozzles are embedded on the same body (print-head). Then are they are constrained to describe parallel paths, while in independent Dual Extruder, there are two print heads moving independently in the X-axis. The advantages and disadvantages of each solution are summarized in the following Table 1.

**Table 1.** The advantages and disadvantages of parallel and independent dual extruder.

| Performances | Parallel dual extruder | Independent dual extruder |
|---|---|---|
| Productivity | Unchanged because only one extruder is active in print | Can clone the part and also print distinct objects. Then the production capability is doubled |
| Accuracy | The high mass of two print heads implies high inertias when moving at high speed from one point to another, which leads to less accuracy | Design and the operating regime are identical to that of a single extruder |
| Calibration | A special calibration is not necessary | Difficult calibration if we plan to work simultaneously on the same slicing of the same piece |
| Quality of print, when compared to single extruder | Inactive nozzle can drop molten plastic while waiting | The same as one extruder |
| Workspace of 3D printer | It is not affected | Is divided between independent extruder |

b. **Multi-extruder**

By the implementation of multi-extruders with several nozzles, it is possible to feed materials of different types or with multi-color, simultaneously or alternatively, in an automatic way and without interrupt the printing process.

Design of available FDM printers provides different technical solutions for multi-material/color extruder. Although there are several classifications of extruders in the literature [4–6], it is very interesting to classify them according to the relevance of the additional functions that they offer for the initial extrusion system, mainly the multi-material/color function (Fig. 2). The most classical [5] is based on the duplication of the extruder and operates according to the same principle that parallel dual extruder, so the head carriage becomes fitted with many identical extruders. The number of added components is multiple that of extruder uses to be implemented. This type of design is disappearing because of its very limited advantages (extruders work just separately), it is very bulky and adds additional weight and increase as well as the price of the machine. Because these extruders work alternately, some designers [4, 7–9] have optimized the weight and space occupied by this extrusion system through the use of a single wire drive system (feeding mechanism with it stepper motor). This engagement/disengagement process, similar to that of the clutch-gearbox, allows the application or not of the pressing force against the filament to be selected. The advantage to have multiple independent nozzles is the possibility to print various materials with different melting temperature and colors.

Independent extruder [3, 10] is another class of multiples nozzle with a different vision for dealing with distinct material and enhancing productivity. Design of this class generally limits the number of extruders to two (dual extruder) because the complexity of their implementation and difficulty of calibration. Duplication of the extruders, which simultaneously and independently share the same space, will considerably reduce the volume-printed of the pieces and increase the probability of collision if no measure is taken. This kind of solution is suitable for cloning parts but can enable the collaborative working of the deposition process of the same object.

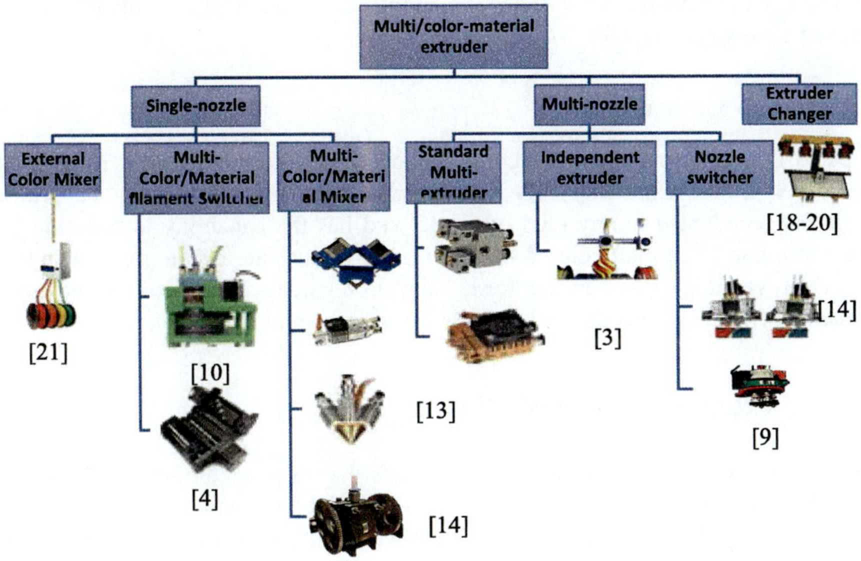


**Fig. 2.** Classification of multi-color/material extruder of Fused Deposition Modeling (FDM) 3D printer

By using the system of multi-extrusion with a single nozzle, several filaments are supposed to pass through the same orifice, which solves the majority of the problems mentioned for classical or independent extruders (compact design, no need to calibration and reduction of the number of extruder system).

The first design with one nozzle (Multi-Color/Material filament Switcher) [11–14] uses a switch box with the ultimate function is to allows one filament to switch between a printing active state (passing through the unit nozzle orifice) to a standby mode automatically. The second [15–17] allows several filaments to borrow the same nozzle indefinitely or at the same time offering the opportunity of color mixing, which is not possible in the first category. However, for both solutions, productivity is still the same as in a normal printer. Moreover, the filament must be of the same material, or have, at least, the same melting temperature, which considerably limits their abilities to obtain interesting graded-physical properties.

Other creators have proposed a completely different solution; called tool-changer [18–20], for the deposition of multi-materials, by designing an extruder changer that gives the printer the ability to extract the hotend. The system is very similar to the tool changers of CNC milling machines. The hotend are stored on a rack, and the code allows the deposit of the old hot-end and the loading of the new hotend. The advantage of this new design is that the mass and volume of the cart are kept to a minimum while allowing you to use many hot ends. Through this modular design, the parameters of each hotend can be configured individually and there is even the possibility of using the hotend of diversified models.

### 2.2 External Mechanism

The pallet [12, 21] is an external feeding- splicer system to the 3D printer that uses a cutter called Splice Core that collects and splices the filament during the printing step. It can be adapted to any kind of printer and has the capability to combine four filaments in one for a multicolored mosaic print without altering the machine's work space. However, this system, except for its outsourcing quality, can be classified among the filament Switcher extruder and possess the same limited advantages.

## 3 Multi-extruder Electronic Boards

Although the mechanical design remains the task that requires the most reflection for the creation of a new extrusion system multi-material/color with additional features over the existing, the realization of the system will be dependent on the electronic part and software that will guarantee its digital control. Then, these two components must be redesigned in advance, in parallel with mechanical design. Because, standard electronics of 3D printers is configured to drive just a single-extruder although some board, as RAMPS 1.4, has the accommodation for 2 extruders. So, according to the complexity of the mechanism designed, the electronics and code need to ensure it. Control may be of standard or custom type requiring itself to be created. Actually, there are several open-source electronic boards [22, 23] designed to control multiple extruders. Among the famous electronic boards are mentioned in Table 2.

The number of stepper drivers is one of the most decisive characteristics for the choice of the electronic board, to ensure the control of a multi-extrusion system. This number is two for the RAMPS board and three for both megatronics and rumba. However, if the board allows, we can expand this number by simply adding a driver expander board, as is the case of Duet3D board that reached nine drivers per coupling with DueX5.

If we focus on use of multi-stepper drivers, the standard and independent extruders need more stepper drivers to ensure feeding of the different filaments. While for Multi-Color/Material Mixer, the number of drivers is equal to three to four, following as color system is of RGB or CMK type. Finally, the classes that require the least number of stepper drivers seem to be switcher filament extruder.

**Table 2.** Popular electronic boards for monitoring multi-extruder.

| Board type | Image | CPU | Stepper driver | **Number of extruders** | **Expansion possibility Of extruders** |
|---|---|---|---|---|---|
| RAMPS |  | Atmega 2560 | A4988, panneaux Pololu | *2* | ***4** with RAMPSXB* *2 with ExtrudrBoard* *4 with CNC Shield* |
| Duet WiFi or Duet Ethernet | 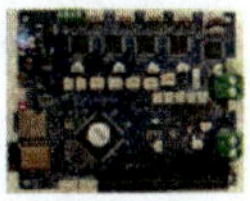 | ARM 32 bits SAM4E8 E | Trinamic TMC2660, A5984 Digital current control | *2* | *7 with DueX5 expansion board* |
| Rumba | 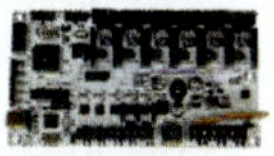 | Atmega 2560 | A4988, DRV8825 | *3* | no expansion found |
| Megatronics 3.0 |  | Atmega 2560 | A4988 , Pololu pannel ,DRV8825 | 3 | no expansion found |
| Rambo | 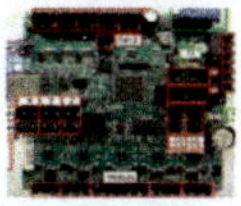 | Atmega 2560, Atmega 32u2 | A4982, A5984 Digital current control | 2 | no expansion found |

## 4 Need for Development of a New Extrusion System for FDM 3D Printing

In recent years, Multi-material and Multi-color prints have been extensively developed because of their promising properties for a new applications [24], ranging from integrated sensors and electronic circuits [25], robotic joints [26] and prosthetics [12]. Actually, multi-color and multi-material models can be easily printed by using commercial 3D printers with distinct technologies. As an example, polyjet and voxel are able to produce wonderful color prints and graded material but these two high technologies necessitate digital processing materials and the price of machine and printing material still very expensive that restrict their accessibility. As described above, multiple material/color capability can also be seen with material extrusion technologies and FDM is a promising cheap technology in this contest. A large variety of materials (Acrylonitrile Butadiene Stirene-ABS, Polylactic Acid-PLA, Polyethilene Terephtalate-PET, High Impact Polystirene-HIPS, Polycarbonate-PC, Thermoplastic Polyurethane-TPU, Polyvinil Alcohol-PVA, nylon, metal PLA, and carbon-fiber PLA) is successfully tested and this variety is expanding rapidly. However, the solutions proposed so far for the deposition of multiple materials have only partial advantages and therefore a great deal of work remains to be provided for the design of an optimal technical solution. Figure 3 summarizes some of the news functions that can be added by reviewing the extrusion system of filament.

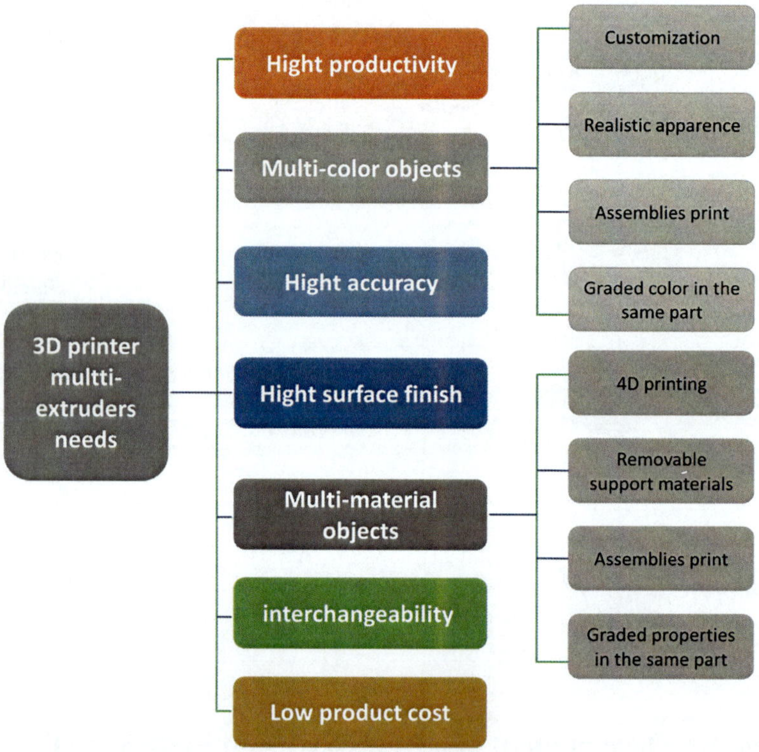


**Fig. 3.** New functionalities that can be obtained by adding Multi-extruders to FDM 3D printer

So, in the immediate future, we plan to merge the functionalities for regrouping most of advantages cited by redesigning the extruder system. Because, as already demonstrated, this component has the key role in the material depositing process and it is responsible for feeding filament through a nozzle and melting it as it's deposited into the bed where the part is made. Also, most of the physical 3D printing parameters depend on this smart system.

## 5 Conclusion

This study presents an overview of the current state of diverse extruders designed for FDM multi-material/color, and provide some interesting conclusions on their performances and weakness. Extrusion systems with several nozzles offer the opportunity of depositing multi-materials regardless of their melting temperatures, but they are less suitable for the mixing operation, are very cumbersome and their balancing is very difficult. On the other hand, a system with a single-nozzle is very compact and makes it possible to obtain colored objects with a surprising degradation, but since it is endowed with a single source of heating, it forces the use of a same grade of materials. However the two classes of extrusion system far from meeting the needs of the industry in terms of productivity, which stimulates us to seek a new design with technical solutions adapted for the emergence of the different desirable features for the 3D printing of materials of different shade/color.

**Acknowledgements.** The authors thank the Spanish Ministry of Industry, Economy and Competitiveness for financial help of project DPI2016-80345R.

## References

1. ASTM International, International Organization for Standardization (ISO). Additive manufacturing-General principles-Terminology. ISO/ASTM 52900 (2015)
2. https://www.thingiverse.com/thing:1939507. Accessed 19 Aug 2019
3. Guide on how to choose a professional desktop 3D printer. https://www.bcn3dtechnologies.com/documents. Accessed 19 Aug 2019
4. Ali, M.H., Mir-Nasiri, N., Ko, W.L.: Multi-nozzle extrusion system for 3D printer and its control mechanism. Int. J. Adv. Manufact. Technol. **86**, 999–1010 (2015). https://doi.org/10.1007/s00170-015-8205-9
5. Laureto, J.J., Pearce, J.M.: Open source multi-head 3D printer for polymer-metal composite component manufacturing. MDPI, Technol. **5**(36), 1–22 (2017)
6. Mhatre, P.S.: Process Planning for Concurrent Multi-nozzle 3D Printing, thesis. https://scholarworks.rit.edu/theses/10075. Accessed 20 Aug 2019
7. [Stratasys 2007] Stratasys, Inc.: Single-motor extrusion head having multiple extrusion lines. USA. International publication number WO 2007/130229 A2. Accessed 15 Nov 2007
8. Keller, P.: Designing a compact dual head for FLM 3D printing technology. MM Sci. J. 1560–1564 (2016)
9. Andersen, K.E.: Optimization of a Low-melting Alloy for Fused Filament Fabrication, thesis. SIMON FRASER University Summer (2015)
10. Miller, R.: Autodesk Looks to Future of 3D Printing with Project Escher. https://techcrunch.com/2016/04/10/autodesk-looks-to-future-of-3d-printing-with-project-escher. Accessed 20 Aug 2019
11. Prusa, J.: Prusa I3 MK3S MULTI-MATERIAL UPGRADE 2.0 ADDON (2019). https://www.prusa3d.com/original-prusa-i3-mk3. Accessed 20 Aug 2019
12. https://www.kickstarter.com/projects/dglass3d/next-generation-3d-printer-extruders-the-rugged-hp. Accessed 20 Aug 2019
13. Pascale, D., Simion, I.: Multi-material 3D printer extruder concept. J. Ind. Des. Eng. Graph. **13**(28), 25 (2018)

14. Miroslav Kajzr, Bc.: Remote control of 3D printer, thesis. Technical University Brno (2018)
15. Guan, Y., Shen, B., Zhang, Y., Fu, Z.: Design of color mixing 3D printing system based on LabVIEW. J. Comput. **28**(6), 277–287 (2017)
16. ORD Solutions: RoVa3D 5 Extruder 3D Printer Package (2019). https://www.ordsolutions.com/rova3d-5-extruder-3d-printer-package. Accessed 20 Aug 2019
17. The crane quad 3D printer, within color-calibrated CMYK filaments. https://store.printm3d.com/products/the-crane-quad-3d-printer?variant=21155064053838. Accessed 21 Aug 2019
18. Lina, W., Shena, H., Xua, G., Zhanga, L., Fua, J., Dengc, X.: Single-layer temperature-adjusting transition method to improve the bond strength of 3D-printed PCL/PLA parts. Compos. A **115**, 22–30 (2018)
19. E3D. 2019. E3D tool-changer and motion system: beta 30, https://e3d-online.com/blog/2019/04/18/toolchanger-the-update-youve-all-been-waiting-for. Accessed 21 Aug 2019
20. Mark2. 2017. The smart way to multi-extrusion. https://magnetic-tool-changer.com. Accessed 21 Aug 2019
21. Simple, multi-material 3D printing, Palette 2. https://www.mosaicmfg.com. Accessed 21 Aug 2019
22. Search keyword: Ramps, RAMBO, RUMBA…etc., https://reprap.org/wiki/. Accessed 21 Aug 2019
23. https://reprap.org/wiki/Comparison_of_Electronics. Accessed 21 Aug 2019
24. Pei, E., Monzón, M., Bernard, A.: Additive Manufacturing Developments in Training and Education. Springer, Cham (2019). ISBN 978-3-319-76084-1
25. Saari, M., Cox, B., Richer, E., Krueger, P.S., Cohen, A.L.: Fiber encapsulation additive manufacturing: an enabling technology for 3D printing of electromechanical devices and robotic components. 3D Printing Addit. Manufact. **2**(1), 32–39 (2015)
26. Zapciu, A., Constantin, G., Popescu, D.: Elastomer overmolding over rigid 3D-printed parts for rapid prototypes. Proc. Manuf. Syst. **13**(2), 75–80 (2018)

# Electronic States in $GaAs/Ga_{0.6}Al_{0.4}As$ Multi-quantum Wells with Two Defect Layers

Fatima-Zahra Elamri(✉), Farid Falyouni, and Driss Bria

Laboratoire des Matériaux, Ondes, Energie et Environnement, Faculté des sciences, Université Mohamed Premier, Oujda, Morocco
f.elamri@ump.ac.ma

**Abstract.** We have studied a $GaAs/Al_{0.4}Ga_{0.6}As$ multi-quantum wells (MQWs) placed between two semi-finite medium (substrate made of GaAs) containing two defects layer: the first one is a well defect layer GaAs with a thickness $d_{01}$ and the second one is presented as a barrier defect layer $Al_{0.4}Ga_{0.6}As$ with a thickness $d_{02}$. It is shown that slight fluctuations in the two defect layer thicknesses are likely to cause strong localization of electronic states. Therefore, we have found two localized modes inside the gaps for defect thicknesses $d_{01}$ and $d_{02}$ belong to the thickness range [54 A°–70 A°] and one mode for defect thickness $d_{01} = d_{02} < 20$ A° and $d_{01} = d_{02} > 70$ A°. These modes are localized in the $3^d$ gaps and they are considered as defect modes. All these electronic states are strongly depending on defect layer parameters. Moreover, these modes induced by the two defect layers give rise to two closer modes for all thickness in the interval [60 A°–70 A°]. The transmission and the quality factor of MQWs are depending also on the presence of the two defect layer in the structure.

**Keywords:** Transmission · Band structure · Well defect · Barrier defect · MQWs · Electronic states

## 1 Introduction

Semiconductors are widely used in a variety of applications such as laser, solar cells, light-emitting diodes and photodetectors [1]. They are fabricated in very thin layers, where the quantum confinement leads to the discrete energy levels and the formation of subbands. These energy levels can be obtained by changing the geometry of the barrier (widths and heights), and the well (widths), and by modifying the shape of a well with different composition profiles. Studying the band structures of semiconductor materials not only reveals the physical and electronic properties of semiconductors but their vital role in high-speed optoelectronic devices and nanostructure. As well, bandgap energy is one of the most important parameters that characterize a semiconductor and determines many gross electronic and optical properties [2, 3]. Previous studies about defects have usually investigated the ones caused during the growth of the Multi Quantum Well (MQW) layers. For the structural defects, M. Shafi et al. who studied the electrical properties of defect states present within the bandgap of Si-doped GaAs/AlGaAs Multi Quantum Wells [4]. Also, J. J. Liu et al. investigated the modification of optical


A. Saka et al. (Eds.): CPI 2019, LNME, pp. 239–248, 2021.
https://doi.org/10.1007/978-3-030-62199-5_21

transition probability in the superlattices (SLs) due to the presence of a structural defect and the variation of layer thickness [5]. In our previous work, we investigated localized modes associated with a Multi-Quantum Wells (MQW) sandwiched between two substrates with a single defect layer in the middle; these modes show significant variations inside the gap bands as a function of the Aluminium concentration [6]. In this paper, a theoretical study of the transmission and the band structure for a Multi Quantum Wells structure is presented. The structure in question consists of two periodic semiconductor materials, in the presence of two defective layers.

Defects in bulk semiconductors and heterostructures can have considerable influence on different applications and corresponding operating temperatures. Whether intentionally incorporated or not, defects can affect the structure, carrier lifetime, radiative efficiency, and other properties [7, 8]. One of these studies has investigated the Mg doping effect in the barriers, in which they found that it enhances photoluminescence intensity, thermal stability, and internal quantum efficiency of LEDs [9]. Recently, non defective GaAs/GaAlAs structure has been fabricated and used in thermal imaging [10, 11]. This kind of composite multilayer system has shown an efficiently performance in quantum cascade lasers, THz devices, and high-speed electronics [12]. In other applications, those systems are used in photodetectors [13], solar cells [14], high-Q electron energy filters [15], as well as in the detection of radiation in the region of the wide band from microwave to THZ [16, 17].

In what follow, we present in the Sect. 2, the model and formalism based on the Green Function approach to obtain the different physical properties of the system such the transmission coefficient and the band structure. Move on to the results and discussions section, where we give the variation of the transmission and the band structure in the case of MQWs containing two geometrical defects (Barrier and Well). We investigate also the quality factor for the created localized states. To end the paper a conclusion of the founded results will be given.

## 2 Model and Formalism

The structure used in this study is GaAs/$Ga_{0.6}Al_{0.4}As$ multi-quantum wells, with a GaAlAs barrier defect with the same concentration (x = 0.4) and GaAs defect well. The number of layers is equal to N = 9. As shown in Fig. 1, the GaAs wells and the GaAlAs barrier are presented with thicknesses equal to 40 A°.

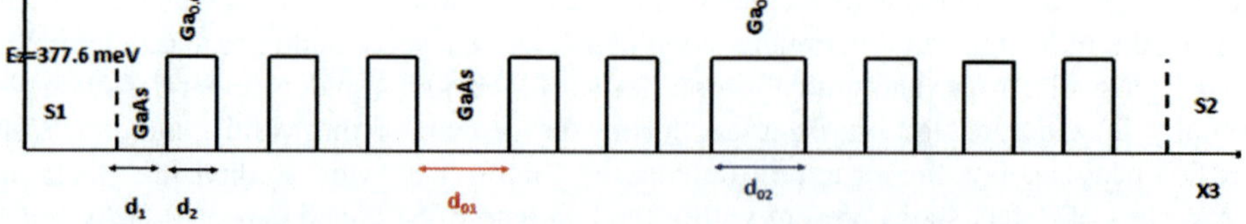


**Fig. 1.** GaAs/$Ga_{0.6}Al_{0.4}As$ MQWs consisting of N = 9 layers with two defects.

The barrier and the well defects have a variable thickness and their height $E_{01}$, $E_{02}$ and the effective mass depend on the Al concentration [18, 19]. The theoretical calculations of the transmission, the band structure, and the quality factor were done using the Green's function method.

These two defect layers are made of GaAs well and GaAlAs barrier (geometrical defects) with a thickness $d_{01} = d_{02}$ placed in two sites: $J_{def1} = 3$ and $J_{def2} = 5$.

Our theoretical model based on the interface response theory. This theory allows the calculation of the Green's function of a MQWs structure in terms of its elementary constituents. The object of this theory is to calculate the Green's Function of a composite system containing a large number of interfaces that separate different media. The knowledge of this Green's function enables us to obtain different physical properties of the composite system such as the transmission coefficients and the band structure.

We replace the layer (i = 1) located at the site ($n = j_{def1}$) by a layer (i = 01) of a thickness $d_{01}$ and the layer (i = 2) located at the site ($n = j_{def2}$) by a layer (i = 02) of a thickness $d_{02}$. Therefore, there are 8 disturbing states $M_s$:

$$M_s = \{\left(n=-1, i=1, \tfrac{d_1}{2}\right), \left(n=0, i=1, \tfrac{d_1}{2}\right), \left(n=j_{def1}, i=01, -\tfrac{d_{01}}{2}\right),$$
$$\left(n=j_{def1}, i=01, \tfrac{d_{01}}{2}\right), \left(n=j_{def2}, i=02, -\tfrac{d_{02}}{2}\right), n=j_{def2}, i=02, -\tfrac{d_{02}}{2},$$
$$\left(n=N, i=1, \tfrac{d_1}{2}\right), \left(n=N+1, i=1, \tfrac{d_1}{2}\right).$$

The calculation of the response function in the space of the interfaces $M_m$ (where $M_m$ is the real space $\{\left(n, i=1, -\frac{d_1}{2}\right), \left(n', i\ = 1, \frac{d_1}{2}\right)\}$) of the MQWs, thus obtained requires establishing the disturbance operator that leads to its construction. The cleavage operator $V(M_sM_s)$ is the following $(8 \times 8)$ square matrix defined in the interface $M_s$ by:

$$V(M_sM_s) = d^{-1}(M_mM_m) - g^{-1}(M_mM_m) \tag{1}$$

Where $d(M_m; M_m)$ is the Green function of final MQWs containing the two defects, its inverse is $d^{-1}(M_m; M_m)$, and $g^{-1}(M_mM_m)$ is the inverse Green function of the infinite system.

The cleavage operator $V(M_sM_s)$, allows us to deduce the elements of the response function of the finite structure necessary for the calculation of the transmission coefficient.

The interface response operator $A(M_sM_s)$ is written as follows:

$$A(M_sM_s) = \sum_{M_s} V(M_sM_s)g(M_sM_s) \tag{2}$$

The function $g(M_sM_s)$ is computed from the four equations below:

$$g\left(n,1,-\frac{d_1}{2};n',1,-\frac{d_1}{2}\right)=\left(\frac{C_1S_2}{F_2}+\frac{C_2S_1}{F_1}\right)\frac{t^{|n-n'|+1}}{t^2-1} \tag{2.a}$$

$$g\left(n,1,-\frac{d_1}{2};n',1,+\frac{d_1}{2}\right)=\frac{S_2}{F_2}\frac{t^{|n-n'|+1}}{t^2-1}+\frac{S_1}{F_1}\frac{t^{|n-n'-1|+1}}{t^2-1} \tag{2.b}$$

$$g\left(n,1,\frac{d_1}{2};n',1,-\frac{d_1}{2}\right)=\frac{S_2}{F_2}\frac{t^{|n-n'|+1}}{t^2-1}+\frac{S_1}{F_1}\frac{t^{|n-n'+1|+1}}{t^2-1} \tag{2.c}$$

$$g\left(n,1,+\frac{d_1}{2};n',1,+\frac{d_1}{2}\right)=\left(\frac{C_1S_2}{F_2}+\frac{C_2S_1}{F_1}\right)\frac{t^{|n-n'|+1}}{t^2-1} \tag{2.d}$$

The operator $\Delta$ $(M_sM_s)$ is given by the following relation:

$$\Delta(M_sM_s)=I(M_sM_s)+A(M_sM_s) \tag{3}$$

With I $(M_sM_s)$ it's the diagonal matrix in the interface space $M_s$.

After calculating the operator $\Delta$ $(M_sM_s)$, we write this operator in the space $M_0=\left\{\left(\mathrm{n}=0,\mathrm{i}\ =1,-\frac{d_1}{2}\right),\left(\mathrm{n}=\mathrm{j}_{\mathrm{def1}},\mathrm{i}=01,-\frac{d_{01}}{2}\right),\left(\mathrm{n}\ =\mathrm{j}_{\mathrm{def1}},\mathrm{i}=01,\frac{d_{01}}{2}\right),\left(\mathrm{n}=\mathrm{j}_{\mathrm{def2}},\mathrm{i}=02,-\frac{d_{02}}{2}\right),\left(\mathrm{n}\ =\mathrm{j}_{\mathrm{def2}},\mathrm{i}=02,\frac{d_{02}}{2}\right),\left(\mathrm{n}\ =\mathrm{N},\mathrm{i}\ =1,\frac{d_1}{2}\right)\right\}$.

The Green function of the interface $\overleftrightarrow{d}(M_0M_0)$ for a MQWs structure is defined in space $M_0$ by the following equation:

$$d(M_0M_0)=g(M_0M_0)\Delta^{-1}(M_0M_0) \tag{4}$$

With $\Delta^{-1}(M_0M_0)$ is the inverse of the operator $\overleftrightarrow{\Delta}(M_0M_0)$. The elements ($d_{ij}$, i, j = 1, 6 ) are calculated numerically.

$$d(M_0,M_0)=\begin{bmatrix} d_{11} & d_{12} & d_{13} & d_{14} & d_{15} & d_{16}\\ d_{21} & d_{22} & d_{23} & d_{24} & d_{25} & d_{26}\\ d_{31} & d_{32} & d_{33} & d_{34} & d_{35} & d_{36}\\ d_{41} & d_{42} & d_{43} & d_{44} & d_{45} & d_{46}\\ d_{51} & d_{52} & d_{53} & d_{54} & d_{55} & d_{56}\\ d_{61} & d_{62} & d_{63} & d_{64} & d_{65} & d_{66} \end{bmatrix} \tag{5}$$

We deduce the truncated matrix $d_b(M_0M_0)$ in the space $M_0'=\{0,N\}$. The inverse of that matrix is written as follows:

$$d_b^{-1}\left(M_0',M_0'\right)=\begin{bmatrix} A_{11} & A_{12}\\ A_{21} & A_{22} \end{bmatrix} \tag{6}$$

Where $A_{11}, A_{12}, A_{21}$ and $A_{22}$ are the elements (1, 1), (1, 6), (6,1) and (6,6) of the inverse matrix. And we find the Green function of finite MQWs contains two defects $\overleftrightarrow{d_f}^{-1}(M_0' M_0')$ located between two semi-infinite media (GaAs).

We invert the matrix $d_f^{-1}(M_0' M_0')$, and we deduced the transmission coefficient through MQWs containing two defects.

$$T = -2F_1 \overleftrightarrow{d_f}(s, e) \tag{7}$$

With $e$: The interface between the first substrate and MQWs structure.
$s$: The interface between the second substrate and the MQWs structure.

## 3 Results and Discussions

We know that the introduction of defect layers inside a perfect structure leads to the presence of localized states inside the bandgaps. Consequently, in order to investigate the effect of the barrier and the well defects, we plot the transmission spectrum as a function of the energy for different defects thickness $d_{01} = d_{02}$ starting from 20 A° to 65 A° with fixed defect positions $J_{def1} = 3$ and $J_{def2} = 5$. Figure 2a shows that for a thickness $d_{01} = d_{02} = 20$ A° one can see the apparition of only one mode inside the 3$^{rd}$ gap, with a low transmission rate. Figure 2b represents the transmission spectrum in the case of perfect structure. Moreover, for $d_{01} = d_{02} = 50$ A° (Fig. 2c), we found one mode inside the bandgaps which is closer to the permissible band. Furthermore, this mode has a higher transmission rate. If we stepwise the defect thickness to 60 A° and 65 A° (Fig. 2e–d), we have then two modes with higher transmission rates (T = 0.96 for both modes in Fig. 2d and T = 0.95, T = 0.6 for the modes in Fig. 2e). The resonant modes in the permissible bands decrease when we increase the defect thicknesses due to the loss of the energy, i.e. attenuation of the structure. So the thickness of the perfect defect which leads to well-defined defect modes is $d_{01} = d_{02} = 60$ A°.

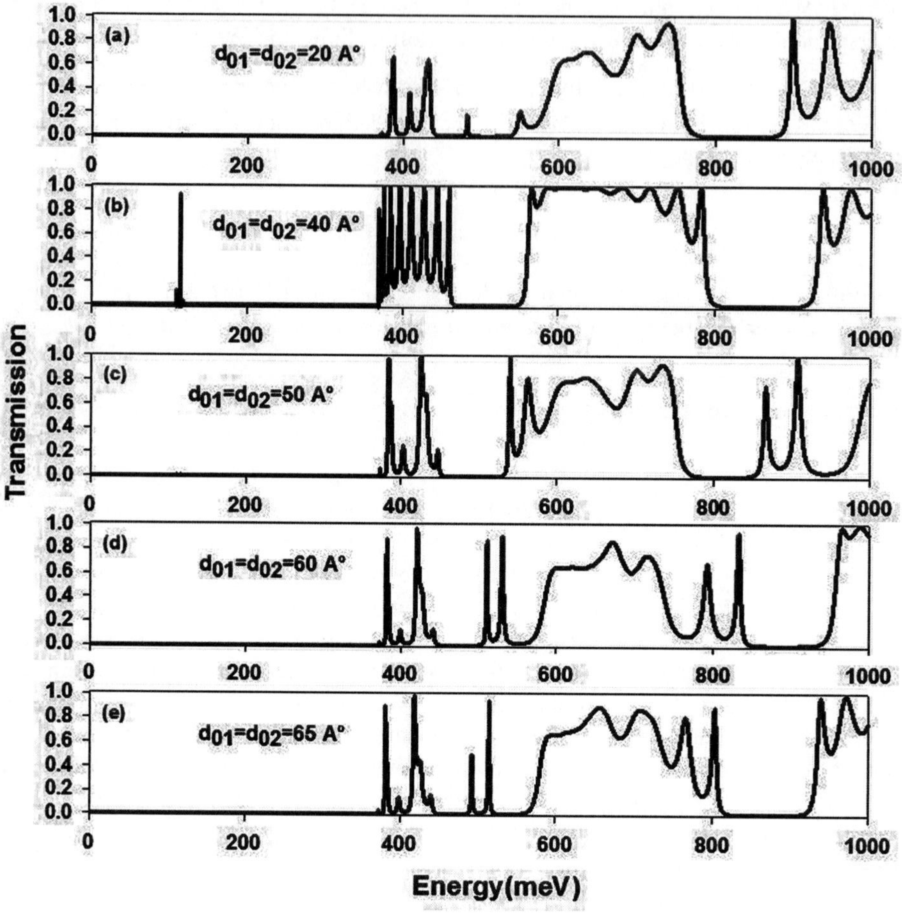


**Fig. 2.** Transmission spectrum as a function of the energy for different defects thickness values: (a) $d_{01} = d_{02} = 20$ A°, (b) $d_{01} = d_{02} = 40$ A°, (c) $d_{01} = d_{02} = 50$ A°, (d) $d_{01} = d_{02} = 60$ A°, (e) $d_{01} = d_{02} = 65$A°.

In the 4$^{th}$ gap, one can see one mode for the defect thicknesses equal to 20A° and two modes for $d_{01} = d_{02}$ belong to the thickness range 50 A°–60 A°. These two modes shift to the permissible bands when the thickness is $d_{01} = d_{02} = 60$A°.

We have investigated also, the effect of the barrier and the well defect positions on the creation of localized states. As shown in the Fig. 3, we have studied the transmission for different positions with a well defect fixed in the position $j_{def1} = 3$ and, a varying positions for the barrier defect from $j_{def2} = 4$ to $j_{def2} = 7$. Using the same defect thicknesses $d_{02} = d_{01} = 60$ A° (as shown in Fig. 2d).

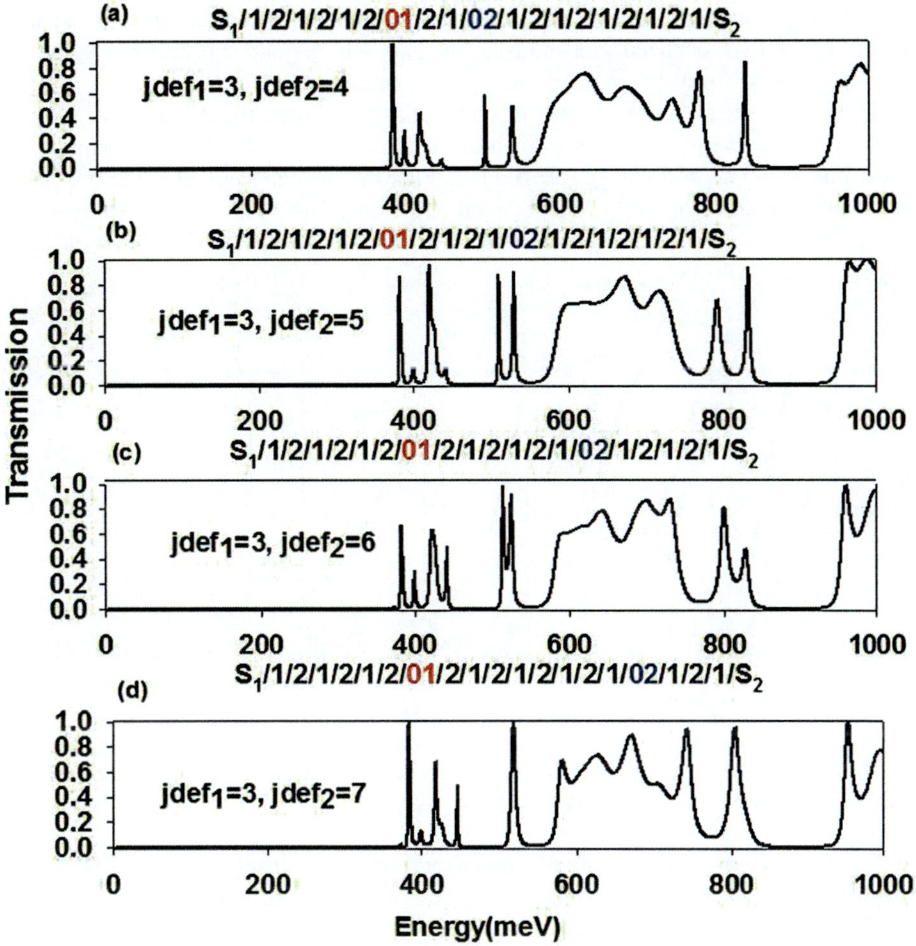


**Fig. 3.** Transmission spectrum as a function of the energy for different defect positions: (a) $j_{def1}$, $j_{def2}$ = 3,4, (b) $j_{def1}$, $j_{def2}$ = 3,5, (c) $j_{def1}$, $j_{def2}$ = 3,6, (d) $j_{def1}$, $j_{def2}$ = 3,7.

Figure 3a and 3b, designed the cases where the second defect position is equal to $j_{def2} = 4$ and $j_{def2} = 5$ respectively. In those cases, we have two modes appear inside the 3rd gap, these two modes have a lower transmission with an energetically distance equal to 35 meV (Fig. 3a) and a higher transmission rate with also an energetically distance equal to 19 meV (Fig. 3b). However, if we move the second defect to the position $j_{def2} = 6$ (Fig. 3c), two modes appear inside the gaps with good transmission rate, in this case, the two modes become too closer to each other. If we stepwise the position to $j_{def2} = 7$ (Fig. 3d), the two defect modes become closer and tend to be one defect mode with a higher transmission rate. So more, we distanced the two defects, the number of modes increase.

Then from the Fig. 2 and Fig. 3, the best parameters of the barrier and the well defect to have two localized modes inside the gap, with a good transmission value are $d_{01} = d_{02} = 60A°$, and defect positions $j_{def1} = 3$ and $j_{def2} = 5$.

To have an overview of the two modes that appear inside the 3[d] gaps, we plot the variation of the energy levels as a function of the two defect thicknesses, namely for defect positions $j_{def1} = 3$ and $j_{def2} = 5$. From Fig. 4a, one can see that when the defect thicknesses are less than 20A°, we obtained one mode. This mode is induced by a well defect (Fig. 4c). The same results are obtained for thicknesses varying between 20 A ° $\leq d_{01} = d_{02} < 40$ A°.

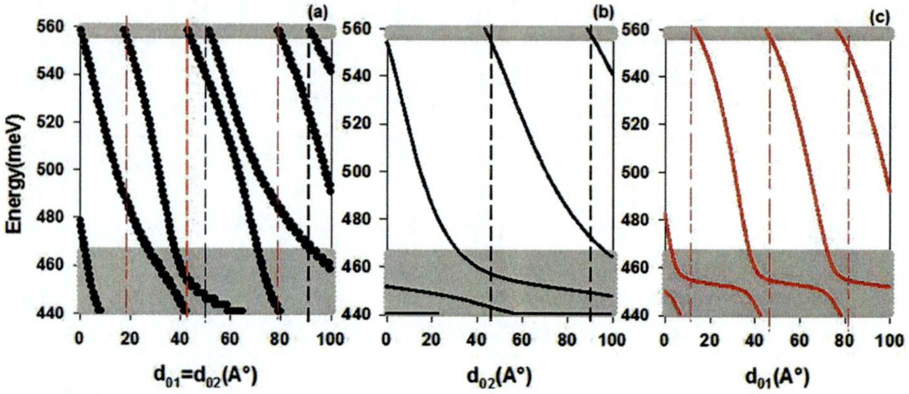


**Fig. 4.** Variation of the energy levels as a function of the thickness $d_{01} = d_{02}$ (a) of the two defects, (b) the barrier defect, (c) the well defect.

From 40 A° to less than 50 A°, no branch appeared inside the gaps. However, starting from $d_{01} = d_{02} = 54$ A° to 70 A°, we notice the appearance of two branches: the first one is the mode induced by the barrier defect (the first black dotted line), and the second one is induced by the well defect (the second red dotted line).

For higher thicknesses varying from 80 A° to 90 A°, we have one branch inside the gap which is induced by the well defect (Fig. 4c). Moreover, two branches appear for $d_{01} = d_{02}$ higher than 90 A°, in which the upper one induced by the barrier defect (Fig. 4b) and the lower one by the Well defect (Fig. 4c).

From Fig. (4a – 3a), one can see the existence of two modes in the interval [54 A° −70 A°] with an almost constant energetically distance equal to 19 meV.

When $d_{01} = d_{02} =$ [54 A°−70 A°], the transmission of the upper branch decreased slightly from T = 0.99 to T = 0.9 for the interval varying from 54 A° to almost 60 A° (Fig. 5a–b), while the transmission of the lower branch increases from T = 0.79 to T = 0.9. Moreover, for $d_{01} = d_{02} = 60$ A°, we note that the two modes have the same transmission values, which are equal to T = 0.85. For $d_{01} = d_{02}$ between 60 A° and 70 A°, the transmission of the upper branch increases for thickness 64A° and decreases after reaching T = 0.6. In this investigation, we have also determined the quality factor for both modes (Fig. 5c). Let us recall that the quality factor is defined as the ratio between the central energy and the full width at half-maximum of the transmittance modes. The upper mode has $Q_2 = 161.94$ and the lower one $Q_1 = 264.26$.

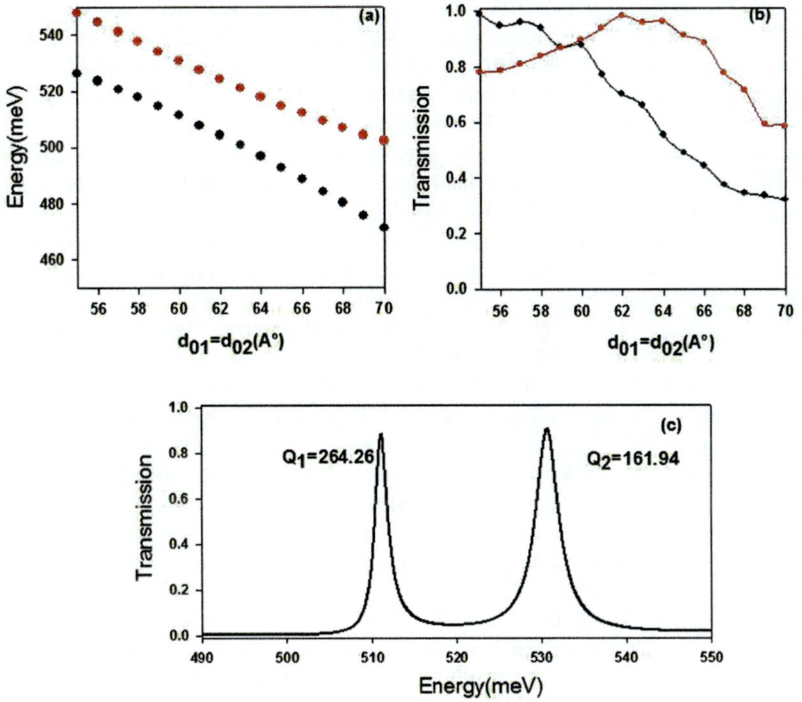


**Fig. 5.** (a) Variation of the energy levels as a function of the thickness of the defects [54 A°–70 A°], (b) Variation of the transmission as a function of the thickness $d_{01} = d_{02}$, (c) the transmission spectrum for the thickness of the defects $d_{01} = d_{02} = 60$ A°.

## 4 Conclusion

The effect of the presence of two defective layers on the creation of localized states for a multi-quantum wells GaAs/Al$_{0.4}$Ga$_{0.6}$As placed between two semi-finite medium (substrate made of GaAs) is investigated theoretically. By taking the defect positions in $j_{def1} = 3$, $j_{def2} = 5$, and varying the thickness of the two defects, we have found one mode for lower thicknesses (less than/or equal to 20 A° and also for $d_{01} = d_{02} > 70$ A °) and two modes inside the gaps for thicknesses between 54 A° and 70 A°. These modes are localized in the 3$^{d}$ gaps. These modes induced by the two defect layers give rise to two closer modes for all thickness in the interval [60 A°–70 A°]. Their transmission and quality factor depend strongly on the two geometrical defects parameters.

## References

1. Yamaguchi, M., Takamoto, T., Raki, K., Ekins-Daukes, N.: Multi-junction III-V solar cells: current status and future potential. Solar Cells **79**, 78 (2005)
2. Charles, K.: Introduction to Solid State Physics, 8th edn. Wiley, New York (2005)

3. Rozale, H., Lazreg, A., Chahed, A., Ruterana, P.: Structural, electronic and optical properties of the wide-gap $Zn_{1-x}Cd_xTe$ ternary alloys. Superlattices Microstruct. **46**(4), 554–562 (2009)
4. Shafi, M., Mari, R.H., Khatab, A., Taylor, D., Henini, M.: Deep-level Transient Spectroscopy of GaAs/AlGaAs multi-quantum Wells Grown on (100) and (311)B GaAs Substrates. Nanoscale Res. Lett. **5**, 1948–1951 (2010)
5. Liu, J.Q., Wang, L.-L., Fang, J.-N., Huang, W.-Q.: Modification of optical transition probability in semiconductor superlattices: effects of structural defect and layer thickness. Int. J. Modern Phys. B **25**(32), 4533–4541 (2011)
6. Elamri, F.Z., Falyouni, F., Tahri, Z., Bria, D.: Localized states in $GaAs/Ga_{1-X}Al_xAs$ multi-quantum-wells. In: International Conference on Electronic Engineering and Renewable Energy, pp. 137–145. Springer, Singapore (2018)
7. Weber, E.R.: Understanding defects in semiconductors as key to advancing device technology. Physica B **1**, 340–342 (2003)
8. Alkauskas, A., McCluskey, M.D., Van de Walle, C.G.: Tutorial: defects in semiconductors —combining experiment and theory. J. Appl. Phys. **119**, 181101 (2016)
9. Han, S.H., Cho, C.Y., Lee, S.J., Park, T.Y., Kim, T.H., Park, S.H., Kang, S.W., Kim, J.W., Kim, Y.C., Park, S.J.: Effect of Mg doping in the barrier of InGaN/GaN multiple quantum well on optical power of light-emitting diodes. Appl. Phys. Lett **96**, 051113 (2010)
10. Li, N., Zhen, H.l., Wang, W.-p., Wang, J., Chen, X.-s., Li, Z.-f., Wang, W.-x., Chen, H., Liu, F.-qi, Lu, W.: Quantum structure optimization for infrared detection. Sci. China Ser. G Phys. Mech. Astron **39**(3), 336–343 (2009)
11. Guo, F.M., Li, N., Xiong, D.Y., Zhen, H.L., Xu, X.Y., Hou, Y., Ding, R .J., Lu, Q., Huang, W., Zhou, J.M.: The theory and experiment of very-longwavelength 256 $\times$ 1 $GaAs/Al_xGa_{1-x}As$ quantum well infrared detector linear arrays. Sci. China Ser. G-Phys. Mech. Astron **51** (7), 805–812 (2008)
12. Agnew, G., Grier, A., Taimre, T., Bertling, K., Lim, Y.L., Ikoni´c, Z., Dean, P., Valavanis, A., Indjin, D., Raki´c, A.D.: Frequency tuning range control in pulsed terahertz quantum-cascade lasers: applications in interferometry. IEEE J. Quantum Electron. **54** (2), 1-8 (2018)
13. Wang, H.X., Fu, Z.L., Shao, D.X., Zhang, Z.Z., Wang, C., Tan, Z.Y., Guo, X.G., Cao, J.C.: Broadband bias-tunable terahertz photodetector using asymmetric GaAs/AlGaAs step multi-quantum well. Appl. Phys. Lett. **113**, 171107 (2018)
14. Feng, C., Zhang, Y., Qian, Y., Wang, Z., Liu, J., Chang, B., Shi, F., Jiao, G.: High-efficiency $Al_xGa_{1-x}As/GaAs$ cathode for photon-enhanced thermionic emission solar energy converters. Opt. Commun. **413**, 1–7 (2018)
15. Shen, M., Cao, W.: Electronic band-structure engineering of GaAs/AlxGa1_xAs quantum well superlattices with substructures. Mater. Eng. **B103**, 122–127 (2003)
16. Gradauskas, J., Sužiedėlis, A., Ašmontas, S., Širmulis, E., Kazlauskaitė, V., Lučun, A., Vingelis, M.: Sensitive planar semiconductor detector from microwave to infrared applications. IEEE Sens. J. **10**(3), 662–667 (2010)
17. Sužiedėlis, A., Ašmontas, S., Gzradauskas, J., Šilėnas, A., Čerškus, A., Lučun, A., Paškevič, Č., Anbinderis, M., Žalys, O.: Planar asymmetric dual diode for millimetre wave detection and power measurement. Lithuanian J. Phys. **57**(4), 225–231 (2017)
18. Ohno, H., Mendez, E.E., Alexandrou, A., Hong, J.M.: Tamm states in superlattices. Surf Sci. **267**, 161 (1992)
19. Ohno, H., Mendez, E.E., Brum, J.A., Hong, J.M., Agullo-Rueda, C.L.L., Esaki, L.: Observation of "Tamm states" in superlattices. Phys. Rev. Lett. **64**(21), 2555 (1990)

# Using Ontologies to Improve New Product Development Process - Case Study

Imane Zahri[1(✉)], Souhail Sekkat[2], Ibtissam El-Hassani[2], El-Moukhtar Zemmouri[1], and Mohammed Douimi[1]

[1] Research Team in Modeling and Knowledge Extraction (ModEC), ENSAM – UMI, Meknes, Morocco
imane.zahri@edu.umi.ac.ma, e.zemmouri@yahoo.fr, dutinfo_ensam@yahoo.fr

[2] Artificial Intelligence Team for Engineering Sciences (IASI), ENSAM – UMI, Meknes, Morocco
s.sekkat@ensam-umi.ac.ma, i.elhassani@ensam.umi.ac.ma

**Abstract.** Companies are under great pressure to innovate and industrialize new products. In addition, the customization and complexity of good and services as well as the competitive environment require the optimization of new product development process (NPD), and choice and manufacturing process configuration.

PLM or Product Lifecycle Management is the tool that allows creating, managing, sharing and capitalizing all the information related to the product throughout its lifecycle to gain collective efficiency. Nevertheless, the choice of the manufacturing process configuration always requires a lot of effort, time and expertise. The Plug & Produce concept means that the production process is quickly configured, and put into service without the need for excessive manual effort, and that is what we want to have when we industrialize a new product. Moreover, the reconfiguration of a production system following the NPD process is a difficult function to ensure, it is considered among the challenges to be taken into account in the design of a production system.

In this work, we propose an approach bottom-up, based on ontology and the Semantic Web, to automate the process of industrialization of a new product. To illustrate this approach, a case study of a PLM module of the flexible manufacturing and assembly cell is presented.

**Keywords:** New Product Development (NPD) · Industrialization · Product Lifecycle Management (PLM) · Ontology · Semantic web

## 1 Introduction

Nowadays, consumer needs have evolved and products have become more complex and highly personalized, and they must be renewed as quickly as possible. In fact, currently, the product's life cycle has become ever shorter. The rapid sharing of information through the integration of PLM and control software enables cost and time reduction of the New Product Development process [1]. To be competitive, companies must optimize production management and product development.


A. Saka et al. (Eds.): CPI 2019, LNME, pp. 249–260, 2021.
https://doi.org/10.1007/978-3-030-62199-5_22

This paper proposes a new approach, to automate the new products industrialization process. This approach is used to develop a PLM module within the flexible machining cell in our laboratory. It is based on the ontology concept for the modeling of product and production resource data and proposes semantic Web technologies for knowledge sharing within the computer control system.

In Sect. 2, we explain the general context of our study. Then, Sect. 3 describes the flexible machining cell and its computer control system. In Sect. 4, we present the proposed approach based on products ontology. In Sect. 5, we develop an Ontology describing the products within the flexible cell to automate the industrialization process. Finally, in Sect. 6, we present conclusions and future work of this research.

## 2 General Context

### 2.1 Evolvable Production Systems

In 2002, the concept of Evolvable Production Systems emerged as a paradigm of production systems [2]. Evolvable production systems (EPS/EAS (Evolvable Assembly Systems)) are built based on flexible and adaptive production systems and contribute to the fourth industrial revolution [3]. The evolvability of systems resides in the ability of system components to deal with changing operating conditions, as well as the evolution of these components over time [1]. Indeed, EPS uses the Plug-and-produce concept, which allows Manufacturing Systems components to be added and removed according to demand changes. It is an extension of Reconfigurable Manufacturing Systems (RMS) proposed by Yoram Koren team at the University of Michigan in 1995. RMS modularity achievement requires connection interfaces of modules standardization [4]. Therefore, the overall EAS/EPS system approach aims to improve the ability of a production environment to respond quickly to changes in product, process or market.

The reconfiguration of a production system following the development of a new product is a difficult function to ensure. It is considered among the challenges in designing an evolutionary production system [5]. It is in this context where the problematic of our study takes place.

### 2.2 PLM in the New Product Development Process

The product goes through several stages from conception to recycling; this cycle can be divided into three main phases: the BOL (Beginning of Life), where there is the conceptualization, the design, the development, the prototyping, the launch and then production. After that, comes the MOL (Middle of Life) and the EOL (End of Life) (see Fig. 1) [6].

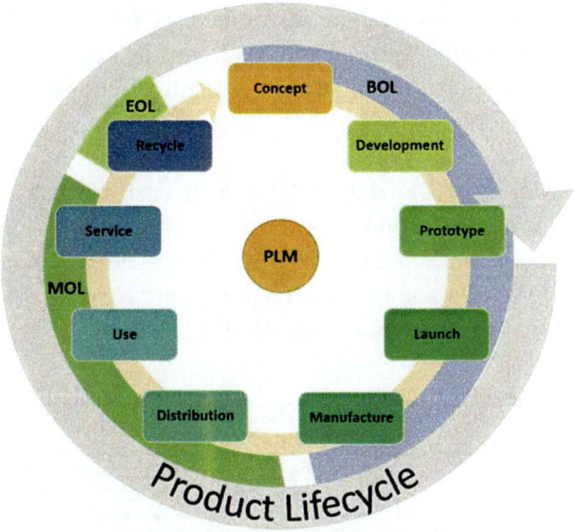


**Fig. 1.** The life cycle of a product [7]

PLM is a software tool that allows companies to manage the entire life cycle of a product from its concept to its end of life by integrating multiple information systems (IS) such as CAD (Computer-Aided Design), Enterprise Resource Planning (ERP), Manufacturing Execution System (MES), Product Data Management (PDM), etc. [8, 9]. This serves to efficiently optimize and manage the NPD process through all of its phases namely: idea generation and evaluation, product development and launch (see Fig. 2). In fact, these information systems benefit from the feedback of each phase of the product life cycle in order to improve and optimize product design and industrialization, especially to minimize the time to market, reduce development costs, have a quality product, reduce risks, etc. [10].

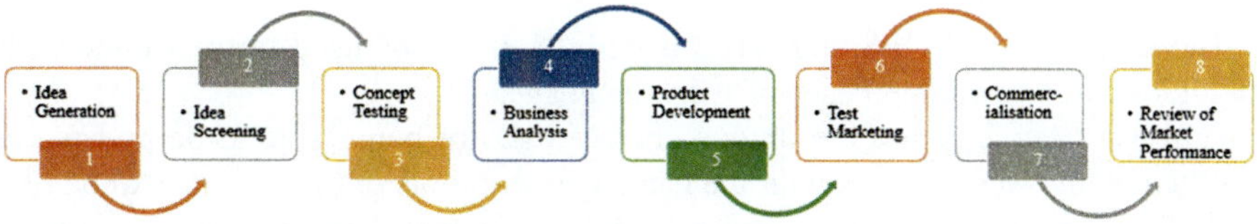


**Fig. 2.** NPD process phases

### 2.3 Semantic Web

Nowadays the techniques of web, semantic web, and web services are employed for the development of production computer systems, whether it is the software of control (MES, supervision…) or the product engineering software (PLM, CAD/CAM, etc.) [8]. The Semantic Web aims to improve the current web for the processing and connection of web data by computers in order to allow users to find the required knowledge.

Parallel to the huge hypertext system distributed WWW (World Wide Web); the Semantic Web is about sharing and reuse of data, not just documents; it forms a huge system based on distributed knowledge. It is a result of cooperation led by the W3C (World Wide Web Consortium).

The Semantic Web stack is illustrated in Fig. 3.

The Extensible Markup Language (XML) layer ensures that a common syntax is used in the Semantic Web. Resource Description Framework (RDF) is a basic data representation format for the Semantic Web. It illustrates resource information in the form of graphs. The OWL language allow creating ontologies that are more detailed, it is integrated at the level of the syntax with RDF [11].

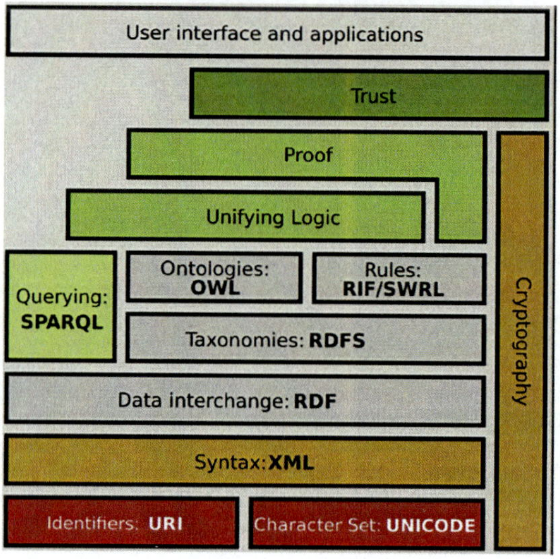


**Fig. 3.** Semantic Web Stack [12]

The semantics of RDFS and OWL is defined; it can be used for reasoning within ontologies and knowledge bases of these languages.

However, ontology has the unique characteristic of being able to be updated and consulted by the web services with the languages and tools of the semantic web [13]. It is therefore more adapted than a simple database to represent and manage web services. Ontology can interact with web services more easily, whereas conventional databases require a rather difficult configuration of communication [14].

### 2.4 Related Works

We conducted a literature study of articles dealing with the same concepts studied in this paper. We used the following keywords for our research: PLM, NPD, Configuration, Ontology, Case study. We have been interested in articles published between 2006 and 2019. The following Table 1 summarizes the articles we considered most relevant:

**Table 1.** A set of articles dealing with the same concepts studied in this paper

| Paper | PLM | NPD | Configuration | Ontology | Case study |
|---|---|---|---|---|---|
| Product configuration knowledge modeling using ontology web language [15] | | x | x | x | x |
| Generic modeling and configuration management in Product Lifecycle Management [16] | x | | x | | |
| Development of a product configuration system with an ontology-based approach [17] | | | x | x | x |
| Improving the Industrialization of a New Product in an International Production Network: A Case Study from the Machinery Industry [18] | | x | | | x |
| Development of Modular Integration Framework between PLM and ERP Systems [19] | x | | x | | |
| A knowledge sharing framework for black, grey and white box supplier configurations in new product development [20] | | x | x | | x |
| Supply chain dynamic configuration as a result of new product development [21] | | x | x | | |
| Ontology-based service product configuration system modeling and development [22] | | | x | x | x |
| Research on industrial product–service configuration driven by value demands based on ontology modeling [23] | | | x | x | x |
| Configuration lifecycle management maturity model [24] | x | | x | | x |

We note that there are no articles, to our knowledge, that discuss the configuration of a production system for a new product by integrating PLM, using ontologies in a real case study. This justifies the orientation of our research.

## 3 Supervision of the Flexible Machining Cell

Since we are going to approach the process of developing a new product in a case study, we describe, in this section, the flexible machining cell, the computer system used for its management and the paradigm deployed for its development.

The studied cell (see Fig. 4) is formed by:

- Conveyor transport system with 4 mating points with workstations.
- StockSingle with a 3-axis robot that ensures the storage of raw materials, components, and finished products with a dynamic addressing system.
- MILL 55 CNC 3 axis milling machine,

- RV-2AJ 5 axis industrial robot, which supports the loading/unloading of the CNC machine.
- RV-2FB 6 axis that assembles the final products from components purchased or manufactured in the same flexible cell.

**Fig. 4.** ENSAM flexible machining cell

The computer control system of the cell (see Fig. 5) is composed by:

- CIROS Supervision software for supervising and driving flexible production cells where we can design production processes and visualize them. It also provides communication functions with multiple systems and equipment through open communication interfaces.
- MES ICIM Production Manager software.

To ensure the integration of the control computer system, an integration platform is developed on CIROS [25], it communicates with ICIM Manager Server through the ODBC driver and with the machines through the TCP/IP and RS232 drivers. The SQL language is used for communication with ICIM Manager and server client programs. CIROS Software, allows developing programs and storing them in a library. The libraries are used for a central management of shared project components. They may be linked into several projects. The changes, which are made also, have an effect on the projects, which include components of the library. In the integration framework, there is a hierarchy of programs. The high-level programs allow launching a production order at the cell level of CIM Pyramid [26]. The macro process task level programs are involving more than one machine. The Prog operation or task level consists in a task carried out by a single machine. The device command level is the lowest level; it enables to implement the communication protocol between the supervisor and the machines.

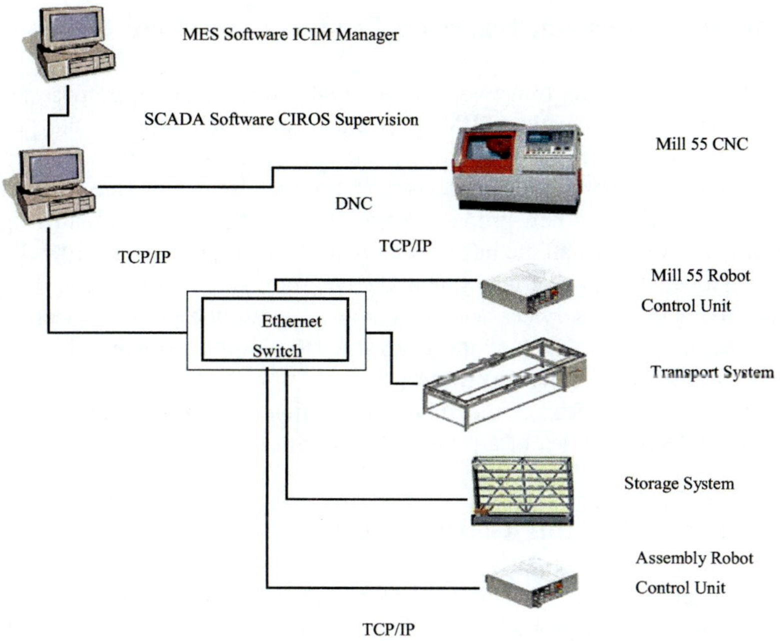


**Fig. 5.** Control system of the ENSAM flexible machining cell.

The architecture of the flexible cell facilitates the re-configurability of the system. In fact, the use of the XML language enables to structure the data describing the production cell. When adding new production equipment, a new production component is added to the XML file and updates the control program of the cell.

However, this platform does not ensure the reconfiguration of a new product.

Indeed, during the industrialization of a product, the most difficult step is the programming and the configuration of product Process Plan within the supervision software. It consists in creating the programs of the different levels of this product and connecting them with each other and with the programs of the machines, the databases of ICIM Manager and the screen of supervision of the cell. This implies an integration of the different levels of the production system and an expertise for their configuration for the new product.

Thus, we develop an approach based on knowledge base technologies for automatic design of a production system. It represents knowledge about the new products and an inference engine that can reason about that knowledge and use rules and other forms of logic to deduce the design and the configuration of the production system.

## 4 Proposed Approach Based on Products Ontology

Having clearly defined the framework of our study, we, therefore, propose to use the semantic web tools to develop PLM software in order to automate the process of developing a new product [4]. Based on an ontology of the products to realize, we can develop a knowledge base of the products [27]. For that, we can use the XML language and each time we define a new product to realize, the semantic web tools will generate an XML file that contains all the information related to this product. This XML file will update the knowledge base of the system; the control layer, therefore, can recognize it automatically. The web services will be able to generate the production programs of this new product at the level of the control system and to connect them with the machines and the management software.

To concretize this idea, in the following section, we propose an ontology that presents the different entities of a production system.

## 5 Ontology of Flexible Cell Products

In order to automate the new product development process, more specifically the product industrialization process, we propose a new approach that modelizes information about products and production resources and develops software agents using ontology.

We present, initially, an ontology of production systems, specific to the ENSAM flexible machining cell, using the UML language. Then, using OWL and Protégé, we propose an ontology of this UML model for the configuration of the system following the development of a new product.

We use, by now, the UML model of a production system, proposed in this study, and based on knowledge management, we aim to develop shared knowledge management services enabling the automation of the industrialization process of a new product.

Our UML model, shown in Fig. 6, presents the main concepts proposed within a production system. The proposed UML diagram illustrates the different relationships between all the entities of a production system namely the factory, the workshop, the flexible cell, the product, the software, etc.

This model, thus, integrates the different partners of the ENSAM flexible cell. This modelization will thus allow automating the configuration of a new product within its production system.

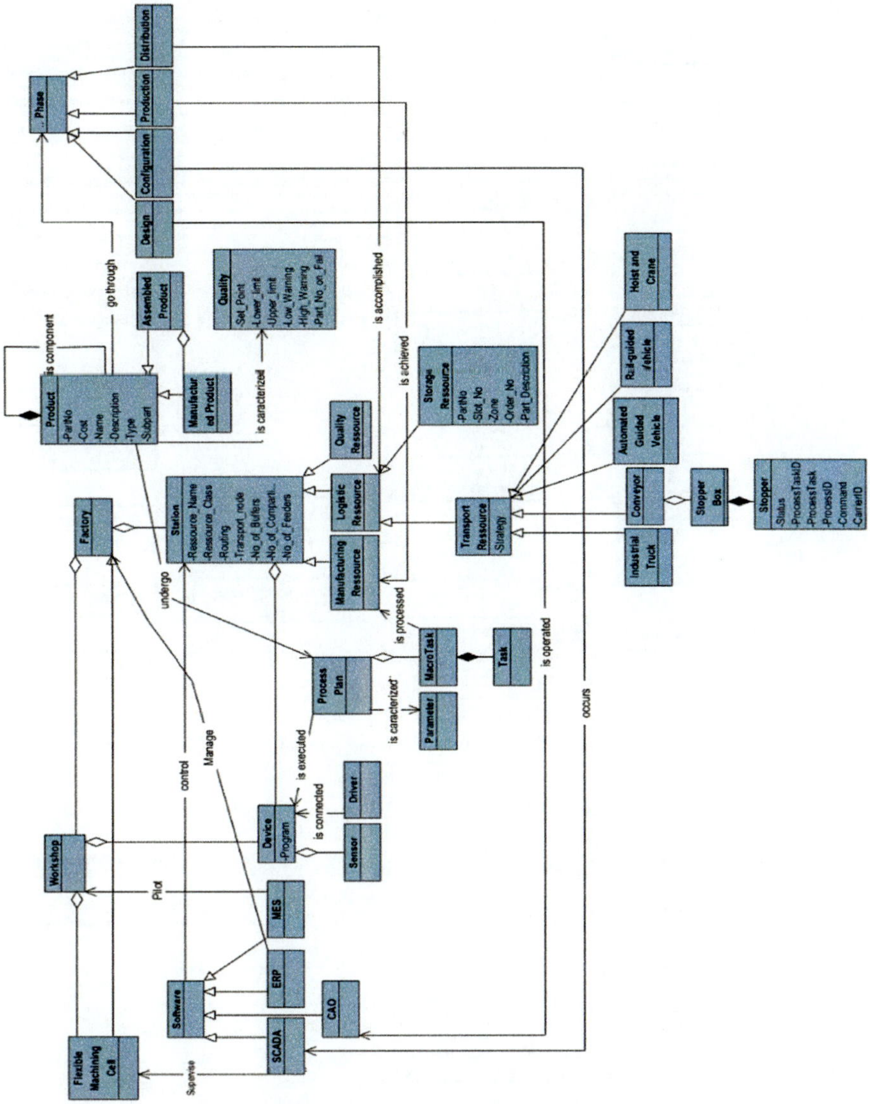


**Fig. 6.** Proposed UML model

We have converted this UML model into an ontology as illustrated in Fig. 7. Our knowledge model is expressed using OWL and Protégé, an open source platform for creating the knowledge model of a production computer system.

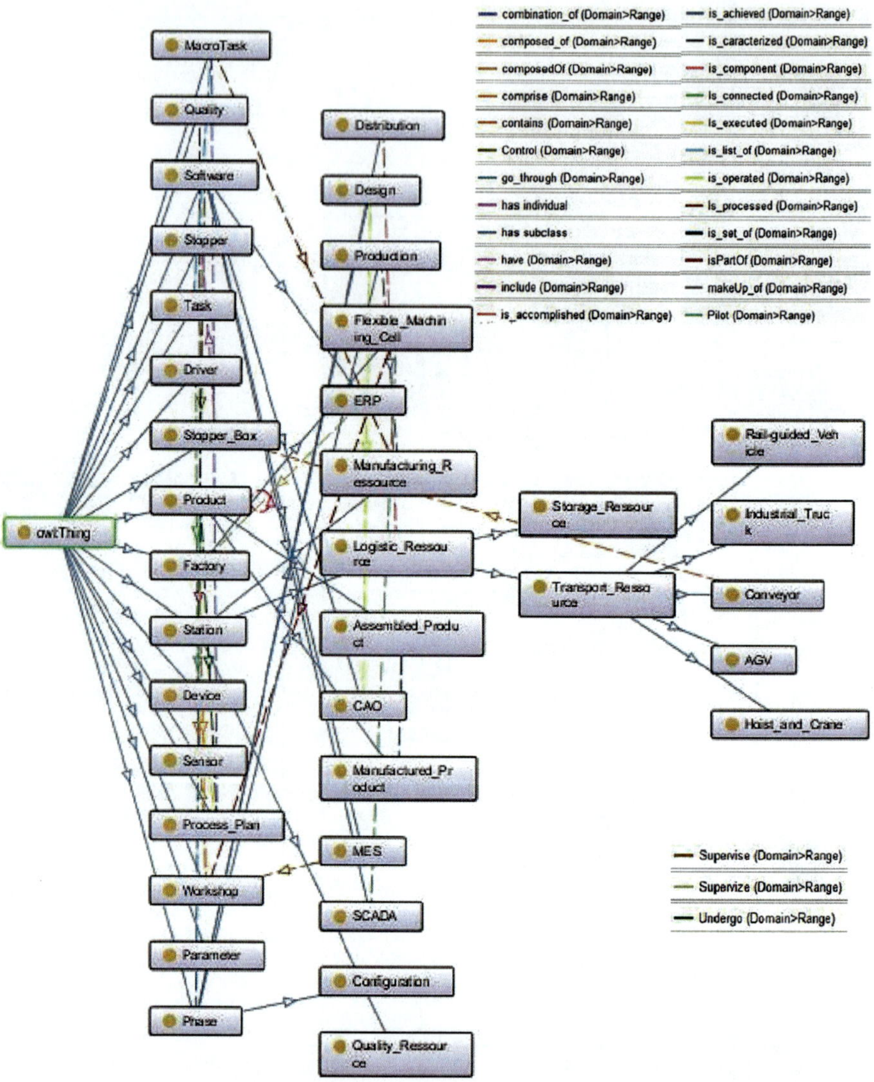


**Fig. 7.** Visualization of the ontology developed under Protégé Editor 5.5

The UML model and ontology proposed in this study are under development and improvement, for the purpose of, subsequently generating an XML file that will include all the data related to the product. Based on Semantic Web tools and ontologies, we aim to automate the configuration of this product within a production system and more specifically within the ENSAM flexible machining cell.

## 6 Conclusion and Future Works

The integration of PLM and computer control systems can optimize costs, control of production and the time of the new product development process. The development of a software agent for a production computer system within PLM software is based on ontology.

In this paper, we proposed an ontology for automating the configuration of a new product within the ENSAM flexible machining cell, using the UML and OWL languages.

As a future work, we suggest using inductive reasoning to generalize our problem and extend it to a class of problems, and thus address the problem of product re-configurability, posed by the research community in the field of production systems.

## References

1. Oh, J., Lee, S., Yang, J.: A collaboration model for new product development through the integration of PLM and SCM in the electronics industry. Comput. Ind. **73**, 82–92 (2015)
2. Onori, M., Semere, D., Lindberg, B.: Evolvable systems: an approach to self-x production. In: Proceedings of the DET/CIRP Conference on Digital Enterprise Technology, Hong Kong (2009)
3. Kagermann, H., Wahlster, W., Helbig, J.: Securing the future of German manufacturing industry: recommendations for implementing the strategic initiative Industrie 4.0. Final Report of the Industrie 4.0 Working Group. Acatech – National Academy of Science and Engineering, Berlin (2013)
4. Lamrani, S., Hasan, R., Martin, P.: Processus de conception des Systèmes Manufacturiers Reconfigurables (SMR): Approche à l'aide des Réseaux de Pétri (RdP). In: 5éme Conférence Internationale Conception et Production Intégrées. CPI 2003, N° 62, ENSAM Meknes (2003)
5. Koren, Y., Heisel, U., Jovane, F., Moriwaki, T., Pritschow, G., Ulsoy, G., Van Brussel, H.: Reconfigurable Manuf. Syst. CIRP Ann. **48**(2), 527–540 (1999)
6. PLM Technology Guide. What is PLM? http://plmtechnologyguide.com/site/?page_id = 435. Accessed 20 June 2019
7. Paavel, M., Karjust, K., Majak, J.: PLM maturity model development and implementation in SME. Procedia CIRP **63**, 651–657 (2017)
8. Garetti, M., Fumagalli, L., Lobov, A., Martinez, J. L.: Open automation of manufacturing systems through integration of ontology and web services. In: 7th IFAC Conference on Manufacturing Modelling, Management, and Control, International Federation of Automatic Control, Saint Petersburg, Russia, vol. 46, pp. 198–203 (2013)
9. Siemens: Defining PLM. http://www.plm.automation.siemens.com/en_us/plm/definition. Accessed 20 June 2019
10. Lentes, J., Eckstein, H., Zimmermann, N.: A platform to integrate manufacturing engineering and product lifecycle management. In: Proceedings of the 14th IFAC Symposium on Information Control Problems in Manufacturing, Bucharest, Romania, vol. 45, pp. 1071–1076 (2012)
11. Luczak, E.: A guide to the semantic web. In: Leading Edge Forum Technology Grant (2004)
12. Horrocks, I., Parsia, B., Patel-Schneider, P., Hendler, J.: Semantic web architecture: stack or two towers?. In: International Workshop on Principles and Practice of Semantic Web Reasoning, pp. 37–41. Springer, Heidelberg (2005)

13. Gruber, T.: Towards principles for the design of ontologies. In: Formal Ontology in Conceptual Analysis and Knowledge Representation (1993)
14. Ribeiro, L., Barata, J., Onori, J., Amado, A.: OWL ontology to support evolvable assembly systems. In: 9th IFAC Workshop on Intelligent Manufacturing Systems (2008)
15. Yang, D., Miao, R., Wu, H., Zhou, Y.: Product configuration knowledge modeling using ontology web language. Expert Syst. Appl. **36**(3), 4399–4411 (2009)
16. Zina, S., Lombard, M., Lussent, L., Henriot, C.: Generic modeling and configuration management in product lifecycle management. Int. J. Comput. Commun. Control **1**(4), 126–138 (2006)
17. Yang, D., Dong, M., Miao, R.: Development of a product configuration system with an ontology-based approach. Comput. Aided Des. **40**(8), 863–878 (2008)
18. Corti, D., Choudhury, S.: Improving the industrialization of a new product in an international production network: a case study from the machinery industry. IFIP Adv. Inf. Commun. Technol. **398**, 495–502 (2013)
19. Prashanth, B.N., Venkataram, R.: Development of modular integration framework between PLM and ERP systems. Mater. Today Proc. **4**(2), 2269–2278 (2017)
20. Le Dain, M.A., Merminod, V.: A knowledge sharing framework for black, grey and white box supplier configurations in new product development. Technovation **11**, 688–701 (2014)
21. Jafarian, M., Bashiri, M.: Supply chain dynamic configuration as a result of new product development. Appl. Math. Model. **38**(3), 1133–1146 (2014)
22. Dong, M., Yang, D., Su, L.: Ontology-based service product configuration system modeling and development. J. Exp. Syst. Appl. **38**(9), 11770–11786 (2011)
23. Wang, P.P., Ming, X.G., Wu, Z.Y., Zheng, M.K., Xu, Z.T.: Research on industrial product–service configuration driven by value demands based on ontology modelling. Comput. Ind. **65**, 247–257 (2014)
24. Myrodi, A., Randrup, T., Hvam, L.: Configuration lifecycle management maturity model. Comput. Ind. **106**, 30–47 (2019)
25. Festo Didactic: CIROS Supervisions - User Manual (2008)
26. AMICE, CIMOSA: CIM Open System Architecture. 2nd edn. Springer, Heidelberg (1993)
27. Lemaignan, S., Siadat, A., Dantan, J.Y., Semenenko, A.: MASON: a proposal for an ontology of manufacturing domain. In: Proceedings of the IEEE Workshop on Distributed Intelligent Systems: Collective Intelligence and its Applications (DIS 2006) (2006)

# Production Planning and Its Impact on Quality in the Automotive Industry

Samiha Mansouri(✉), Latifa Ouzizi, Youssef Aoura, and Mohammed Douimi

CED-Recherche et Innovation pour les sciences de l'ingénieur, Université Moulay Ismail _ Ecole Nationale Supérieure d'arts et Métiers, Meknes, Morocco
samihamansouri@gmail.com, louzizi@yahoo.fr, y_aoura@yahoo.fr, mdouimi@yahoo.fr

**Abstract.** Recently, the automotive industry tends toward the economy of scope, as it gives cost reduction's advantage, but its manufacturing's management is difficult. Planning several products in the same line has an impact on the products quality and output if it's not performed correctly. However, the objective of this paper is to develop a mathematical model for production planning in the operational horizon. This model includes several automotive industry constraints in terms of capacity, dynamic customer orders, and similarities between references, cycle time and the balance between shifts.

**Keywords:** Economy of scope · Production planning · Product quality · Mathematical model

## 1 Introduction

Today, to give customers a multiple choice of products, the market tends towards diversity and increasing variation of products, as well as the decrease of the product life cycle, since the customers think quickly that it's no more fashionable [1].

While trying to seek customer satisfaction, companies moves from the economy of scale to the economy of scope by saving cost due to producing several products by the same operations, which provides product customization with lower cost [2]. Therefore, to follow the trend, companies must have a dynamic and flexible system that meets customer's changes at any time.

In this regard, we introduce dynamic production planning as a method of tracking the customer's order in a reactive and reliable way. A method or more precisely a complex system that must take into account several aspects and factors involved in the realization of multiple products ordered in a fixed time interval through ensuring the quality of products with the defined cost.


A. Saka et al. (Eds.): CPI 2019, LNME, pp. 261–268, 2021.
https://doi.org/10.1007/978-3-030-62199-5_23

## 2 Economy of Scope

The automotive industrial competitiveness level is increasing significantly, so companies face challenges to satisfy their customers. In such manner, the manufacturers tend towards improvement and creativity of their products and creating more and more diversity, by reducing the average cost to produce multiple products in the same production line with common resources [3].

This solution is profitable from a marketing view, as the total cost to produce multiple products individually is much higher than producing them using the same processes and resources. But to ensure its realization and management, the task becomes more complex. The difficulty in managing diversity reveals in product design and structure, process planning, and production planning and control [4].

## 3 Production Planning

The above passage describes the difficulties faced under the strategy of the economy of scope, however, in this paper, we focus on the process related to the production planning and control, as it's the one that connect the product realization with the customer order.

Planning represents the advanced organization and assurance of all necessary means and resources to manufacture the desired product in terms of time and cost. It can be carried out across multiple horizons, depending on the available information and the strategy or company direction [5].

a. Strategic planning:

The strategic planning is reflected in the long term horizon (2 or 3 years), it includes the commercial vision of the company, as the subcontractors of production, supplier selection and development, extension or creation of new plants…

b. Tactical planning:

Also known by MPS (Master Production Schedule), It represents the middle term horizon of the production planning, where we plan for each period the quantities to be produced by product family, while considering the lead time, minimum and maximum stock and the work centre capacity… the figure below shows the different factors to be considered during the preparation of the MPS (Fig. 1).

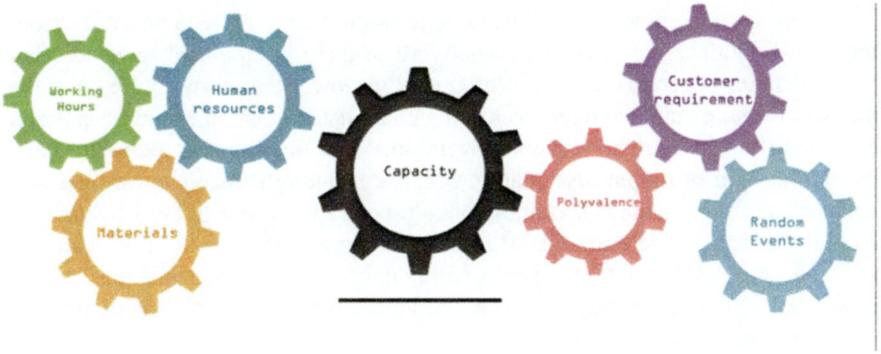


**Fig. 1.** Different factors to be considered in the MPS

a. Operational planning:

It refers to the short term planning horizon, where the volume to be produced is specified for each reference in a short period, based on the output of the MPS. The update frequency of the operational planning is higher because it depends on the details of the production process's operations.

Our study is performed on the operational planning which refers to the daily production planning, where the planning responsible do not have an automatic system that generates the optimal quantities to be produced by reference for each shift, it's a manual process that depends on the expertise of the planner, which decreases the efficiency of this operation.

By referring to the MPS, the planner's first task is to continuously update the stock level, the quantities in transit as well as the quantities in progress, in order to have visibility on the volume to be produced in a daily basis. Second, he starts to affect the planned batches for each shift based on the references priority list. As a result, we have a sequence of product with the related quantity to be manufactured by shift.

## 4 Production Management

The organization is working with a pull system, as we only produce what is needed when demanded by the customer, using KANBAN system. *"Kanban system is one of the tools under lean manufacturing system that can achieve minimum inventory at any one time... It helps to improve the company's productivity and at the same time minimize waste in production. The Kanban system requires production only when the demand of products is available"* [6].

The production orders affected by the planner are considered as one of the production KPI (Key Performance Indicator) that evaluate its effectiveness. Each shift has its capacity level (Maximum that it can produce), and in a KPI point of view, we need to balance the volumes planned for each shift.

In a flow shop environment illustrated in figure below [7], composed by more than forty workstations, one product passes by all of them in a defined sequence. Though, when planning several references with quantities lower than forty in the same line, we might have more than two references being constructed in the line, and it‘s possible that each one have an extremely distinct cycle time, and that impacts the performances of the line in term of output and quality, because if one reference's cycle time is 4 h, it means that all workstations is running with a speed appropriate to the 4 h, and when the second reference's cycle time is 10 h, it needs a speed relative to 10 h, which is very hard to ensure, as the first reference is still being produced in the line, there for we will have a delay in most second reference's products, then we will not be aligned with the production planning (Fig. 2).

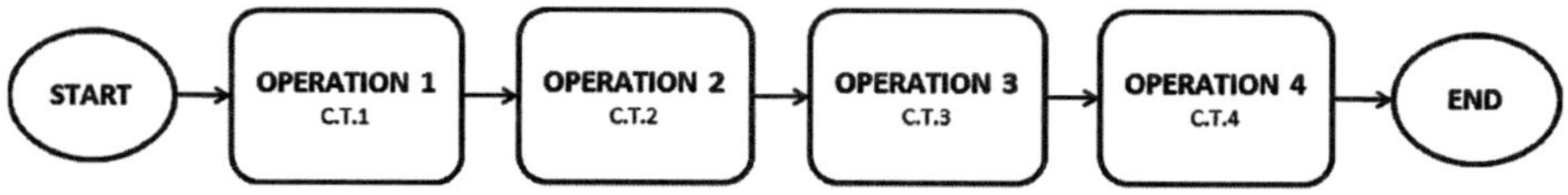


**Fig. 2.** Flow shop illustration.

## 5 Quality Aspect/Issues

In one hand, the discussion above emphasizes the link between production management and planning. Which indirectly impact the product quality, through planning two references with different cycle time (the previous example), this will push some operators to work faster, and as a consequence, it increases the possibility to make quality defect. On the other hand, we need to consider the similarity between references in terms of components and design, as one workstation produce N different references, we have N different operation relative to each reference, in case of a high similarity level between two successive references, the risk of incorrect operation performed by the operator is high. So to rework the defectives parts, there are some defects that require a long time to be corrected. This might delay the delivery, and also increase the scrap value relatively the production cost.

As explained before, the diversity of product in one production line is beneficial from marketing and economic vision, but we need to master our process to deliver the desired products with required quality and minimum costs [2].

## 6 Research Methodology

The problematic introduced earlier is linked to mainly three axes: Production Planning, Production Management and Product quality. Which lead us to analyze in detail the process of each one of them in operational level, and conclude the link between them. As results, we had developed a mathematical model that includes all the factors mentioned earlier: cycle time, capacity, similarity, customer demand, and balancing the qualities for all shifts.

a. Hypothesis:

To build the mathematical model, we set several hypotheses to be converted into constraints in order to concretize all aspects considered in this paper.

- To ensure the balance between shifts, we need to plan approximately the same total quantity per shift.
- Customer demand satisfaction
- ordering of references based on
- Take into consideration the similarity of references: if two references are similar, they must be separated.
- Scheduling of references from min to max for odd shifts and max to min for even shifts.

b. Model parameters:

**n**: Total number of references.
**i, j**: Reference indices.
**K**: Total number of shifts.
**k**: Shift indices.
$\mathbf{S_{ij}}$: the cycle time subtraction of two references i and j.
$\mathbf{EDI_i}$: customer demand of the reference i.
**CAP**: Production line capacity.
$\mathbf{d_{ij}}$: Similarity level between references i and j expressed in percentage.
s: similarity threshold.

c. Decision variables

$\mathbf{X_{ijk}}$:

$$\begin{cases} 1 \; \textit{if the reference j follows the reference i in the shift k} \\ 0 \; \textit{other} \end{cases}$$

**q**ik: quantity of the reference i to be produced in the shift k

d. Objective function and model constraints

$$\boldsymbol{min} \sum_{i=1}^{n} \sum_{j=1}^{n} \sum_{k=1}^{K} \boldsymbol{Xijk * sijaveci \neq j}$$

to improve the sequence of planned products by minimizing the cycle time subtraction of two consecutive references.

- Customer demand Satisfaction:

$$\sum_{k=1}^{K} qik = EDIi \quad \forall i \in \{1\ldots n\}$$

- Same total planned quantity approximately per shift.

$$\sum_{i=1}^{n} qik = \sum_{i=1}^{n} qik' \quad \forall k, k' \in K \,\&\, k \neq k'$$

- Line capacity:

$$\sum_{i=1}^{n} qik \leq CAPchain \quad \forall k \in K$$

- Scheduling of references from min to max cycle time for odd shifts and max to min cycle time for even shifts.

$$Sij * Xijk \leq 0 \quad \forall i,j \in \{1..n\} \quad si\, k\%2 \neq 0$$

$$Sji * Xijk \leq 0 \quad \forall i,j \in \{1..n\} \quad si\, k\%2 = 0$$

- If the degree of similarity of i and j: dij is greater than the threshold s, the reference j will be shifted, otherwise i and j may succeed one another.

$$\text{dij} * \text{Xijk} \leq \text{s} * \text{Xijk}$$

# 7 Results

To simulate our model, we used CPLEX software to solve a mathematical model considering 3 references and 2 shifts

As results we have:

| Reference | CT (Hours) |
|---|---|
| Ref 1 | 5.29 |
| Ref 2 | 4.72 |
| Ref 3 | 4.81 |

Similarity level by combination of references is:

| Reference | Ref 1 | Ref 2 | Ref 3 |
|---|---|---|---|
| Ref 1 | 1 | 0.2 | 0.4 |
| Ref 2 | 0.2 | 1 | 0.6 |
| Ref 3 | 0.4 | 0.6 | 1 |

Customer demand:

| Reference | Customer demand |
|---|---|
| Ref 1 | 17 |
| Ref 2 | 40 |
| Ref 3 | 60 |

- Results:

The table below shows the affected quantities by reference for both shifts. The total planned quantity is low than the line capacity, also the customer demand is satisfied.

| References | Shift 1 | Shift 2 |
|---|---|---|
| Ref 1 | 15 | 2 |
| Ref 2 | 38 | 2 |
| Ref 3 | 7 | 53 |

For the order of planned references, the next table represent the sequence of reference per shift, by respecting the cycle time flow.

| | | | |
|---|---|---|---|
| Shift 1 | Ref 2 | Ref 3 | Ref 1 |
| Shift 2 | Ref 2 | Ref 3 | Ref 1 |

## 8 Conclusion

The operational planning issue is due to missing efficient system that takes into consideration the entire variables, as the automotive industry is a highly flexible and dynamic field. The solution was to develop a model that includes all constraints and limits of the planning process. Also we introduced a new constraints related to product quality which the similarity level between references, when they are highly similar and we have a risk to have quality defect, the model separate them to not be planned successively. The model was tested for tree references, our perspective is to improve it by considering a high number of references.

## References


1. Iwaarden, J.D.: Changing quality controls: Veranderingen in kwaliteitsbeheersing: de effecten van toenemende product variëteit en korter wordende product levenscycli: the effects of increasing product variety and shortening product life cycles. Erasmus University Rotterdam, Rotterdam (2006)
2. ElMaraghy, H., et al.: Product variety management. CIRP Ann. **62**(2), 629–652 (2013)
3. Surbhi, S.: Difference Between Economies of Scale and Economies of Scope (with Comparison Chart). Key Differences, 19 Jan 2016
4. ElMaraghy, H., Azab, A., Schuh, G., Pulz, C.: Managing variations in products, processes and manufacturing systems. CIRP Ann. **58**(1), 441–446 (2009)
5. Aide à la planification dans les chaînes logistiques en présence de demande flexible, p. 157
6. Rahman, N.A.A., Sharif, S.M., Esa, M.M.: Lean manufacturing case study with Kanban system implementation. Procedia Econ. Finance **7**, 174–180 (2013)
7. Khatami, M., Salehipour, A., Hwang, F.J.: Makespan minimization for the m-machine ordered flow shop scheduling problem. Comput. Oper. Res. S0305054819301728 (2019)

# Evaluation of the Form Error of Partial Spherical Part on Coordinate Measuring Machine

Abdelilah Jalid(✉), Mohammed Oubrek, and Abdelouahab Salih

Normal High School for Technical Teaching of Rabat (ENSAM of Rabat), Applied Mechanics and Technologies Research Laboratory, Mechanical Engineering Department, University Mohammed V, Street Royal Army, Madinat Al Irfane, 10100 Rabat, Morocco
a.jalid@um5s.net.ma

**Abstract.** In dimensional metrology, the coordinate measuring machine (CMM) is the privileged tool using to control the mechanical part. To evaluate a form error of a spherical feature, discrete points are preleved on a feature to evaluates, the software of the machine fit a perfect geometric form to estimates a form error, least-squares fitting and minimum-zone fitting are two widely adopted algorithms to evaluate the form. Several methods for the evaluation of sphericity error have been proposed by different researchers. But when the feature is measured partially, the problem will be difficult, and an initial value must be introduced to the algorithm to ensure convergence to the solution. The proposed paper presents an efficient algorithm based on orthogonal distance regression (ODR) to estimates the parameters of the substitute sphere and the sphericity error. The model proposed tack into account the uncertainty of coordinates points. The obtained results are compared after to NIST reference values.

**Keywords:** Coordinate measuring machine · Sphericity error · Orthogonal distance regression · NIST comparison

## 1 Introduction

On Coordinate Measuring Machine (CMM), to control a manufactured part, a discrete points are preleved on the features to be evaluated, the software of the machine fit a perfect geometric form to estimates a form error. Least-squares fitting and minimum-zone fitting are two widely adopted algorithms to evaluate the form, several methods for the evaluation of sphericity error have been proposed by different researchers. Xiong proposed as in [1] a general mathematical model for computer-aided measurement of profile error of any surface and curve. His method is very powerful and can be applied to the verification of sphericity tolerance. Kanada as in [2] evaluated sphericity error using two methods: (i) an iterative least-square method; and (ii) a simplex method assuming that the sphericity function is unimodal. Shunmugam proposed in [3, 4] a different criterion, called the Minimum Average Deviation to evaluate the sphericity error, Dhanish and Shunmugam in [5] used the linear Chebyshev approximation


A. Saka et al. (Eds.): CPI 2019, LNME, pp. 269–275, 2021.
https://doi.org/10.1007/978-3-030-62199-5_24

method to calculate the sphericity error based on the same criterion and model. Elmaraghy et al. introduced in [6] a fine-tuning technique during the final step of their algorithms in order to improve performance. But when the feature is measured partially, the problem will be difficult, and an initial value must be introduced to the algorithm to ensure convergence to the solution. The proposed paper presents an efficient algorithm to estimates the parameters of the substitute sphere and the sphericity error. The obtained results are compared to NIST reference values.

## 2 Mathematical Modeling

It sometimes happens that we only have one part of a sphere, either the other part has been damaged or for functional reasons the piece has just a spherical part. The parameterization adopted for the substitute geometric elements is the same as described in the standard ISO 10360-6 [7], one recalls the characteristic parameters below.

C $(x_C, y_C, z_C)^T$ as a center point of sphere and R its radius (see Fig. 1).

All points M with coordinates $(x_M, y_M, z_M)^T$ belonging to this sphere confirms: $\|\overrightarrow{AM}\| = R$;

$$\text{This gives : } (x_M - x_C)^2 + (y_M - y_C)^2 + (z_M - z_C)^2 = R^2 \tag{1}$$

The vector of the parameters to be determined is $\beta = (x_C, y_C, z_C, R)$.

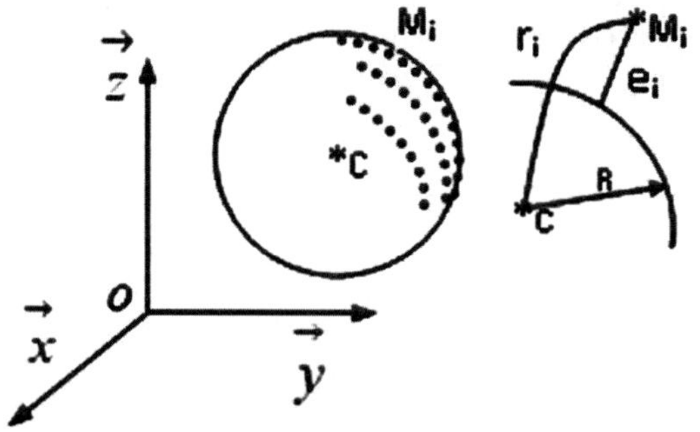


**Fig. 1.** Cloud of points on a partial sphere

$e_i$ represents the difference between the measured point $M_i$ and the desired model given by $e_i = r_i - R$

The least square criterion is to minimize the quantity

$$W = \sum (ei)^2 \tag{2}$$

### 2.1 Model Taking Account of Uncertainty on the Coordinates Points

In dimensional metrology and particularly on a CMM, the uncertainty of acquisition of a point is not constant and depends on the position of the measured point, the evaluation of this uncertainty requires tests of repeatability and reproducibility which are complicated procedures and demand a good knowledge of the machine, in this study it is assumed that all variables xi and yi are respectively assigned with error $\delta \in R^1$ and $\varepsilon \in R^1$. The representative model is written as follows:

$$y_i = f(x_i + \delta_i; \beta) - \varepsilon_i \tag{3}$$

The problem is to find the parameters $\beta$ that minimize the sum of squared distances orthogonal to the measured points and the theoretical model as shown in Fig. 2.

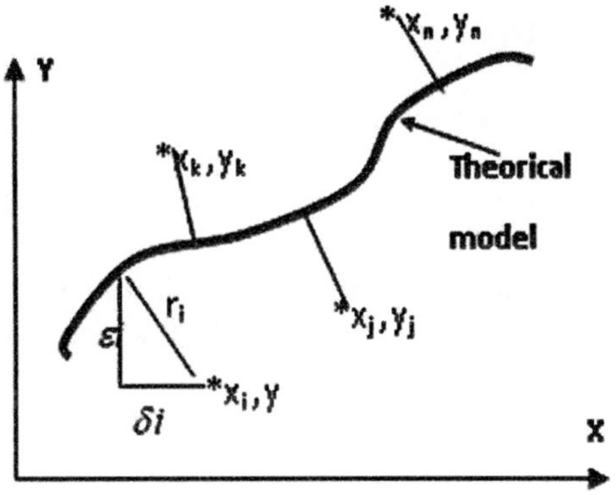


**Fig. 2.** Orthogonal distance regression

With $r_i$ distance is from the measured point $P_i$ ($x_i$, $y_i$) to the desired model. The coordinate's points are random, assumed and distributed according to a normal distribution with standard deviation $\sigma_i$. Moreover, $\omega_i = \frac{1}{\sigma_i^2}$ is a weighting introduced to compensate the case where the measurement accuracy is not the same for all points.

Generalized to the case of several explanatory variables, $x_i \in R^m$, with $m = 3$, has several functions and modeled output $f_i: \ R^{p+m} \rightarrow \ R^q$ , with p as the number of parameters $\beta$ and $q$ as the number of functions $f$. The problem to solve is then expressed by the following equation:

$$\begin{cases} \min\limits_{\beta,\delta} \frac{1}{2}\sum_{i=1}^{n} [\delta_i^T][\omega_{\delta_i}][\delta_i] \\ \text{With constraints:} \\ \left.\begin{array}{l} f1_i(x_i+\delta_i;\beta)=0 \\ f2_i(x_i+\delta_i;\beta)=0 \\ fq_i(x_i+\delta_i;\beta)=0 \end{array}\right\} \quad i=1 \ldots n \end{cases} \tag{4}$$

Where

$[\delta_i]$: The vector where the elements represent the variations according to $x_i$ between the data points and the corresponding points on the fitted model.
$[\omega_{\delta_i}]$: The matrix of weights of order $(m, m)$.
$[fq_i]$: The value given by the equation $q$ to the ith observation.

The solution of the problem is iteratively found using a trust region Levenberg-Marquardt method, see Levenberg [9], Marquard [10] and Nocedal [11]. The Jacobian matrices, i.e., the matrices of first partial derivatives of $f_{li}(x_i + \delta_i; \beta)$, $l = 1\ldots\ldots q$, *and* $i = 1\ldots\ldots\ldots n$, with respect to each component of $\beta$ and $\delta$ are computed at every iteration by either finite differences.

The algorithm of this method is implemented in ODRPACK, Boggs [12, 13].

## 3 Determination of the Sphericity Error

To solve this problem we relied on the model presented by Jalid as in [8], from the collected points this model estimated the parameters of the substitute feature. The method used is based on the orthogonal distance regression (ODR) as explained in the previous paragraph.

To determine initial estimates for the center and radius, Forbes [14] presents a method by simplifying the objective function. The strategy is to approximate E as shown:

$$E = \sum_{i=1}^{n} (e_i)^2 = \sum_{i=1}^{n} (r_i - R)^2 \approx \sum_{i=1}^{n} (r_i^2 - R^2) \tag{5}$$

The parameters $R$ and $r_i$ are given in Fig. 2, This formulation of $E$ can now be re-written in matrix form AP = B, where

$$A = \begin{bmatrix} 2x(1) & 2y(1) & 2z(1) & -1 \\ 2x(2) & 2y(2) & 2z(2) & -1 \\ \ldots\ldots\ldots & \ldots\ldots\ldots & \ldots\ldots & \\ 2x(n) & 2y(n) & 2z(n) & -1 \end{bmatrix}, \quad P = \begin{bmatrix} x_c \\ y_c \\ z_c \\ \rho \end{bmatrix}, \quad B = \begin{bmatrix} x(1)^2 + y(1)^2 + z(1)^2 \\ x(2)^2 + y(2)^2 + z(2)^2 \\ \ldots\ldots \quad \ldots\ldots \\ x(n)^2 + y(n)^2 + z(n)^2 \end{bmatrix} \tag{6}$$

$$\textit{Where} \quad \rho = x_c^2 + y_c^2 + z_c^2 - R^2 \tag{7}$$

after convergence of the algorithm the parameters of the substitute feature will be known, we can then determine the sphericity error by the following formula:

$$\mathrm{SE} = \max(\mathrm{e_i}) - \min(\mathrm{e_i}) \tag{8}$$

## 4 Validation of the Model

To validate our algorithm we have treated a cloud of 10 points taken on a partial sphere (sph 26), the data sets is provided by the National Institute of Standards and Technology (NIST) as shown in Table 1.

**Table 1.** Cloud of points

| Points | x | y | z |
|---|---|---|---|
| 1 | 8.17975 | 3.03306 | 15.84002 |
| 2 | −2.45794 | 13.79265 | −11.32302 |
| 3 | −5.19478 | −16.40815 | −4.42448 |
| 4 | 1.68745 | 3.99905 | 17.49733 |
| 5 | −8.19088 | 13.99244 | −7.73239 |
| 6 | −10.73838 | −14.05683 | −1.32163 |
| 7 | −6.2899 | 4.42986 | 16.21618 |
| 8 | −13.43246 | 11.65546 | −2.02934 |
| 9 | −15.28145 | −8.62731 | 2.60925 |
| 10 | −15.81164 | 3.30997 | 7.546 |

In order to see the distribution of points in space, a representation is given below (see Fig. 3).

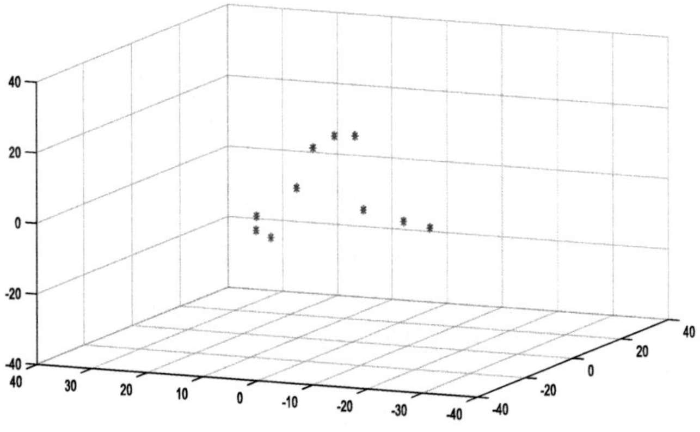


**Fig. 3.** Distribution of data sets treated

The choice of these data highlights the problems most often encountered in optimization (convergence, local optimum, computing time etc.…).

By applying the proposed model, the parameters of the substitute sphere obtained are compared to the reference values provided by NIST as shown in Table 2.

**Table 2.** Comparison of the obtained results and the NIST reference values in mm.

| Parameters | NIST Results | Obtained Results | Comparison |
|---|---|---|---|
| xc | 0,18068578771571805286 | 0,18068578900103400000 | −1,285315937e-9 |
| yc | 0,13271395840993977558 | 0,13271395808149900000 | 3,2844077457e − 10 |
| zc | 0,02846130069278626305 | 0,02846130074751700000 | −0,5473073730e − 10 |
| D = 2 * R | 35,910212698726290632 | 35,9102126997648000000 | −1,03854347344e − 9 |
| SE | 4.222493e − 03 | 4.222256e − 03 | −0,23718 e − 6 |

The sphericity error is calculated by the Eq. (8), SE = 0.00422 mm, and that of NIST is calculated from the provided parameters. Given the differences obtained, it can be concluded that the proposed approach estimates well the parameters and the sphericity of a partial feature.

We also made the comparison with the work of F.W.Meng [15]. In this study the criterion used is the minimum circumscribed sphere (MCS), the authors estimates the parameters of the substitute sphere and the sphericity error. After treatment of the same points named dataset1 and dataset2, the results obtained are compared. The results are presented in the Table 3.

**Table 3.** Comparison with the minimum circumscribed sphere criterion (MCS)

| Parameters | Comparison of MCS and obtained results | |
|---|---|---|
| | MCS | Obtained results |
| DataSet1 xc | 40.0003592 | 40.00068121 |
| yc | 30.0006796 | 30.00080621 |
| zc | 60.000867 | 60.00041213 |
| R | 20.00413 | 20.0028846 |
| Sphericity error | 0.00349195 | 0.00340 |
| DataSet2 xc | 0.00410626 | 0.00405968 |
| yc | 0.00248367 | 0.00327214 |
| zc | 0.00378684 | 0.00346281 |
| R | 1.00352 | 0.99941521 |
| Sphericity error | 0.00834573 | 0.009080 |

From these results, it can be seen that the results obtained by the two criteria are very close together, and that the model developed is well suited to the estimation of the parameters of a sphere whether it is partially measured or totally.

## 5 Conclusion

The present article presents a new method to estimates parameters of substitute sphere, and evaluates the form error of partial spherical feature; the same approach can also be applied to the total sphere. We relied on the orthogonal distance regression algorithm (ODR) to estimate the parameters of the substitute feature. To determine initial estimates for the center and radius to ensure convergence towards the global minimum, we use the approximation as in [14]. The solution of the problem is resolved by an iterative calculation according to the Levenberg Marquard optimization method. To validate our method we compared the results obtained to those provided by NIST considered as reference values.

## References


1. Xiong, Y.L.: Computer aided measurement of profile error of complex surfaces and curves: theory and algorithm. Int. J. Mach. Tools Manuf. **30**(3), 339–357 (1990)
2. Kanada, T.: Evaluation of spherical form errors – computation of sphericity by means of minimum zone method and some examinations with using simulated data. Precis. Eng. **17** (4), 282–289 (1995)
3. Shunmugam, M.S.: On assessment of geometric errors. Int. J. Prod. Res. **24**(2), 413–425 (1986)
4. Shunmugam, M.S.: New approach for evaluating form errors of engineering surfaces. Comput. Aided Des. **19**(7), 368–374 (1987)
5. Dhanish, P.B., Shunmugam, M.S.: An algorithm for form error evaluation - using the theory of discrete and linear Chebyshev approximation. Comput. Methods Appl. Mech. Eng. **92**, 309–324 (1991)
6. Elmaraghy, W.H., Elmaraghy, H.A., Wu, Z.: Determination of actual geometric deviations using coordinate measuring machine data. Manuf. Rev. **3**(1), 32–39 (1990)
7. ISO 10360-6, Geometrical Product Specification (GPS) - Acceptance and reverification tests for coordinate measuring machines (CMM) - Part 6: Estimation of errors in computing Gaussian associated features (2002)
8. Jalid, A., Hariri, S., Senelaer, J.P.: Estimation of form deviation and the associated uncertainty in coordinate metrology. Int. J. Qual. Reliab. Manage. **32**(5), 456–471 (2015)
9. Levenberg, K.: A method for the solution of certain problems in least squares. Quart. Appl. Math. **2**, 164–168 (1944)
10. Marquardt, D.: An algorithm for least-squares estimation of nonlinear parameters. SIAM J. Appl. Math **11**, 431–441 (1963)
11. Nocedal, J., Wright, S.J.: Numerical Optimization. Livre. Springer, New York (1999)
12. Boggs Paul, T., Byrd Richard, H., Janet E. Rogers and Schnabel Robert B., Users Reference Guide for ODRPACK version 2.01, Software for Weightes Orthogonal Distance Regression (1992)
13. Boggs Paul, T., Byrd Richard, H., Schnabel Robert, B.: A stable and efficient algorithm for nonlinear orthogonal distance regression. SIAM J. Sci. Statis. Comput. **8**, 1052–1078 (1987)
14. Forbes, A.B.: Least-Squares Best-Fit Geometric Elements, NPL Report DITC 140/89. National Physical Laboratory, Teddington (1989)
15. Meng, F.W., et al.: Sphericity evaluation based on minimum circumscribed sphere method. Adv. Mater. Res. **433–440**, 3146–3151 (2012)

# The Development of the Decision-Making Aspect of the Manufacturing Executing System

Sarah Adnane[1], Fatima Bennouna[2](✉), Aicha Sekhari[3], Driss Amegouz[1], and Aurélie Charles[3]

[1] PE2D, Sidi Mohamed Ben Abdellah University, Fez, Morocco
sarahadnanel@gmail.com, amegouz@yahoo.fr
[2] LISA, Sidi Mohamed Ben Abdellah University, Fez, Morocco
bennouna.ensa@gmail.com
[3] DISP, Lumière Lyon 2 University, Bron, France
{aicha.sekhari,A.Charles}@univ-lyon2.fr

**Abstract.** The value potential of the IoT announced by analysts and suppliers seems gigantic: several thousand billion euros per year. The IoT actually affects all sectors of activity mainly industries (Industry 4.0). Industry 4.0 is designed to lead the digitization of manufacturing and production processes. The Manufacturing Execution System (MES) is a system for performing, monitoring, tracking and reporting operations on the factory floor in real time. It is a very critical element of industry 4.0 that will act as an action factor to achieve the goal of end-to-end scanning by acting as a bridge between the different factory floor systems. This article presents a literature review of the Internet of Things and its application areas. It also intends to describe MES as an application of IoT in industries, its definition and to identify major issues related to MES. We also discuss limitations and research challenges that need to be explored in order to improve the decision-making process and the growth of production performance.

**Keywords:** CPS · Quality control · Manufacturing · MES

## 1 Introduction

The internet of Things (IoT) is a new paradigm that is becoming one of popular modern wireless telecommunications trends. It is visible by its variety of applications in different domains of everyday life. These applications are gaining such an important place because of the strength of the IoT concept to transform classic objects that surround us to interconnected objects able to communicate and interact with each other and to share information and results with their environment.

The need for creating an intelligent network of interconnected objects is the result of the big quantity of data generated by domestic and working processes, this information needs to be stored, analyzed and shared easily.

According to Gubbi, the Internet of Things demands three important elements:

A comprehension of the situation by the devices and their users.

A software architecture and communication networks.


A. Saka et al. (Eds.): CPI 2019, LNME, pp. 276–285, 2021.
https://doi.org/10.1007/978-3-030-62199-5_25

The analytics tools to analyze the information in an autonomous way.

We can say that IoT is transforming the current internet into a fully integrated Future Internet.

## 2 Literature Review

**Background**

"The Internet of Things has the potential to change the world, just as the Internet did. Maybe even more so." (Ashton, n.d.).

Ashton (2009).

As Kevin said, the Internet of Things is a technology that represent a pillar for the future development of all domains. It's a new revolution in communication which means that every physical object, being, real or virtual environment will be connected, interact with each other and share information with the external environment.

Because of the importance of the Internet of Things, each researcher and manufacturer defined this concept according to their use of this technology and according to the sector of application.

The very first definition of IoT derives from a "Things oriented" perspective; the considered things were very simple items: Radio-Frequency IDentification (RFID) tags. The terms "Internet of Things" is, in fact, attributed to The Auto-ID Labs [4]. According to this definition, IoT involves "things with virtual identities and personalities operating in intelligent spaces using intelligent interfaces for communication in social, environmental and user contexts" (Luigi et al. n.d.). In other words, "at any time, any connectivity for anyone, we will now have connectivity for anything."

From another vision "Things oriented", CASAGRAS consortium proposes a vision of IoT as a global infrastructure which connects both virtual and physical generic objects and highlights the importance of including existing and evolving Internet and network developments in this vision (Luigi et al. n.d.).

Finally, the third definition of IoT is generated from the "Semantic oriented" vision. The idea behind it is that the number of items involved in the Future Internet is destined to become extremely high. Therefore, issues related to how to represent, store, interconnect, search, and organize information generated by the IoT will become very challenging. In this context, semantic technologies could play a key role. In fact, these can exploit appropriate modeling solutions for things description, reasoning over data generated by IoT, semantic execution environments and architectures that accommodate IoT requirements and scalable storing and communication infrastructure. ("Internet of Things (IoT) – A Technological Analysis and Survey on Vision, Concepts, Challenges, Innovation Directions, Technologies, and Applications (*An Upcoming or Future Generation Computer Communication System Technology*)," n.d.).

So, we can adopt the definition of IoT that converge from the three visions which is "in a Smart environment is the presence of intelligent objects that are able to detect in real time the information, analyze it and represent the result of the analysis to the user by a representation interface".

Another concept that is closely related to IoT is Cyber-Physical System (CPS), defined as a system that integrates the 3C: Computation, Communication and Control, and realizes the interaction between the physical world and the cyber world. CPS can provide real-time sensing, dynamic control, information feedback, and other services.

Although both IoT and CPS are aimed at increasing the connection between the cyber space and the physical world by using the information sensing and interactive technology, they have obvious differences: the IoT emphasizes the networking, and is aimed at interconnecting all the things in the physical world, thus it is an open network platform and infrastructure; the CPS emphasizes the information exchange and feedback, where the system should give feedback and control the physical world in addition to sensing the physical world, forming a closed-loop system (Ma 2011).

The progress of IoT is driving the adoption of more and more applications of this innovative technology. IoT applications are becoming increasingly common in public and private sector industries and organizations that save our time, resources and efforts. The applications of IoT were classified in the literature on the basis of different classification criteria and factors placing them in several areas.

**IoT's Areas of Application**

Three major areas of IoT applications are identified: industry, the environment and society. These fields are consistently and interdependently linked with others and cannot be isolated. In each wide field, more and more applications can be identified. The basic requirements of these applications in these areas are often the same with a marginal difference depending on the main functionality of the application (Abdul-Qawy et al. 2015).

Approaching this new concept in professional fields, we notice that IoT is profitable in areas where development is also faster than product quality. One of these areas is manufacturing, which will be our field of study, and the Internet of Industrial Objects (IoT) has transformed it now through the implementation of new approaches such as big data, artificial intelligence (AI) and machine learning.

Most industrial controls and monitoring processes have human interventions, for decision-making in solving problems or failures, which can have human error. As installation in industries are too expensive, human error and one of the damages is not affordable. This is one of the open-ended questions asked in the context of IoT in industries and is the issue chosen for our thesis study.

Manufacturing process is both the core stage of achieving product quality and the main period of quality issues exposing extensively. At present in quality management of manufacturing process, enterprises may have the following aspects of the problems in varying degrees: due to lack of pre-planning, various quality information are recorded either incompletely or irregularly so that cannot provide sufficient data support for the follow-up query and statistical analysis; products in the process with defects are in lack of standardized handling process, of which the corresponding tracking, monitoring and recording work is not in place, which means that the quality of process control is relatively weak; the quality data is still manually handled, rarely using statistical techniques and tools, causing the formed statistical results neither timely nor effective in the face of large data, production organizations and decision-making more impressions, lacking of scientific.

In recent years, with the rapid development of information technology, the information construction is widely promoted in manufacturing enterprises, which introduce information systems such as ERP, PDS in production management. Among all the management information services, the manufacturing execution system (MES) is a shop floor oriented management information system, which integrates all the static and dynamic data of the manufacturing process between the upper planning management system and the underlying production control system, providing an effective realization of ideas and support platform for the quality of management (Zhang et al. 2011).

## 3 Manufacturing Executing System (MES) Background

For more than 25 years, companies have invested in information systems to achieve productivity gains that have caused the enterprise information system market to grow steadily. But for most of this period, the information system specialists have not paid attention to the shop floor (Ben Khedher et al. 2011).

And very quickly, manufacturers were able to realize that neither the systems at their disposal for production (control-command and supervision), nor the management packages (especially the Enterprise Resource Planning ERPs) allowed it. So, there was an uncovered area.

The difficulty of integrating multiple point systems has brought software providers to package multiple execution management components into single and integrated solutions. These systems, commonly referred to as Manufacturing Execution System (MES), provide a common user interface and data management system.

The concept of MES began to evolve from the late seventies. It has been started to solve the individual problem of a single function of MES system like quality management system, equipment monitoring system and production process tracking system. In the absence of overall integrated information system, the enterprise works in an island of automation. The Enterprise Resource Planning (ERP), known before as Material Resource Planning (MRP) and Distributed control system (DCS) were operated in an isolated mode and there exists a missing link between these two systems.

Since the mid-1990s, manufacturers began to recognize the need for a control-execution system integration to fill the gap between the plant level and the control level.

We can schematize the gap between the plant level and the control level as below (Fig. 1):

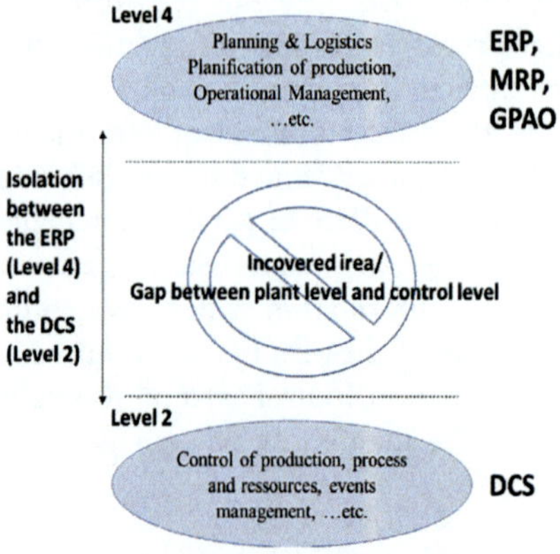


**Fig. 1.** The Computer Integrated Manufacturing (CIM) levels

MES being as one of the software solutions used to bridge between production planning and equipment control system. We can represent the role of MES as a link between the two systems ERP and DCS as below in Fig. 2:

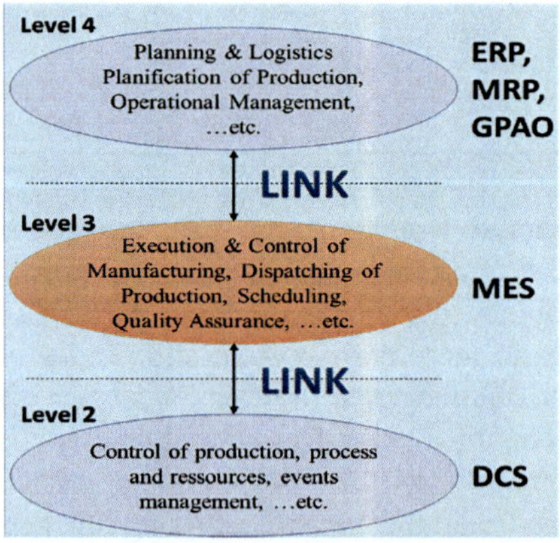


**Fig. 2.** Manufacturing Execution Systems positioning in the Computer Integrated Manufacturing (CIM) context

MES is one of the software solutions used to bridge between production planning and equipment control system.

In the early 1990s, an American association created the term Manufacturing Execution System to refer to this new domain, which gave its name to the association itself (MESA).

The MES is a production scheduling and tracking system used to analyze and report resource availability and status, schedule orders, collect real time data such as material usage, WIP parameters, work order and equipment status, and other critical information. It is used for real-time shop floor reporting and monitoring that feeds activity data back to the ERP (Younus et al. 2010). Since the appearance of the concept of MES, each manufacturer and software editor had defined MES according to his use of this new technology or to the sector of application.

The MESA organization made the first step toward the MES standardization by gathering the major actors of the market and by proposing a formal definition of MES (MESA 1997c):

"MES deliver information that enables the optimization of production activities from order launch to finished goods. Using current and accurate data, an MES guides, initiates, responds to and reports on plant activities as they occur. The resulting rapid response to changing conditions, coupled with a focus on reducing non-value-added activities, drives effective plant operations and processes. The MES improves the return on operational assets as well as on time delivery, inventory turns, gross margin and cash-flow performance. An MES provides mission-critical information about production activities across the enterprise and supply chain via bidirectional communications." (Saenz de Ugarte et al. 2009).

The first steps of the MESA resulted in the recognition of eleven functions that he was led to cover. And for a long time, these eleven functions were practically the definition of the MES itself.

The 11 functions of the MES:

Resource management
Ordering
Product and batch tracking
Document management
Data collection and acquisition
Personnel management
Quality Management
Process management
Maintenance management
Product traceability and genealogy
Performance analysis

From these functions, we can remark that some are directly related to the process and its performance, such as scheduling and quality control, while others, such as resource management and traceability, are considered as support functions.

Each enterprise organized these functions according to their needs and to their internal architecture but with a common agreement on putting the manufacturing control function in a central position. All functions contribute to improve the performance of the process, the on-time realization and the total satisfaction to customer order.

According to MESA's survey (MESA 1997c), MES systems have provided manufacturing enterprises with some of the most impressive benefits of any manufacturing software, such as an average 45% reduction in manufacturing cycle time, a significant improvement of the flexibility to respond to customer demands, the realization of certain degrees of agile manufacturing and customer satisfaction (Saenz de Ugarte et al. 2009).

The recent advances in computing power and memory have opened up the possibility of on-line simulation-based optimization. A typical area of application for online simulation, also called real-time simulation, is proactive decision support for scheduling problems in manufacturing systems (Siemiatkowski and Przybylski 2006).

According to Saenz De Ugarte, the main objective of development of MES is to implement real time decision-making support components as well as feedback loop mechanisms integrating optimization and simulation techniques in ERP and MES applications allowing connecting the shop floor to the rest of the enterprise. The proposed platform responds in real time to various events occurring on the shop floor and may be extended beyond the scheduling issue (Ugarte n.d.).

## 4 Use Case

The objective of the MES concept is to optimize manufacturing processes and resources. The first step is, of course, to measure the performance of the current system. MES is naturally strongly linked to manufacturing processes. It is the best tool to measure in real time performance indicators such as material utilization, batch productivity, and machine failure. Then, it is possible to create a steering dashboard by evaluating indicators such as overall performance, time between failures or average repair time.

MES is also a decision-making support by its alarms and its results from indicators of performance of the process.

For example, in this use case, we can see that MES system contributes to different process as raw materials inspection, production control, quality control, packing, warehouse storage and expedition.

But we can remark that this MES is providing just in the Executing part of these process, it helps in the control and the support of production and quality. But how can we develop MES to be a decision-making tool and overshoot its standard limits of the operative execution and the control of process to its autonomy in making choices and decisions (Fig. 3).

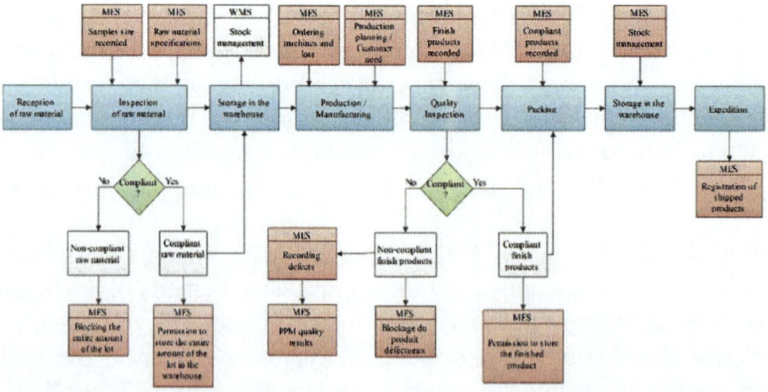


**Fig. 3.** The initial positioning of MES in a manufacturing company

So, to contribute more to the decision-making process, MES must have access to use operational research and analytical methods and not only data analysis.

We can propose the development of the MES system to do others steps beside the pure execution and control, these steps can be modeled in the figures as below (Fig. 4):

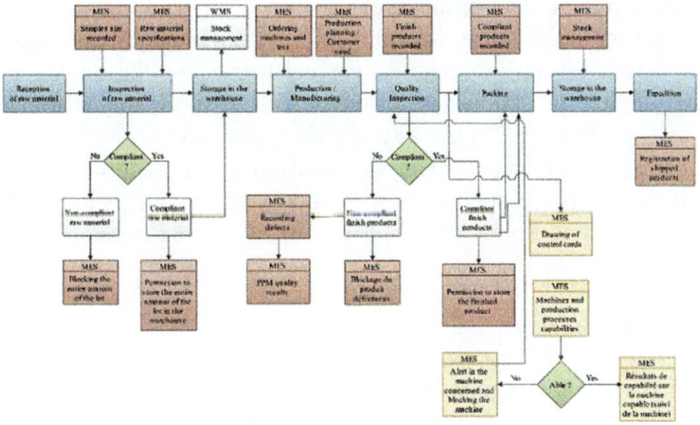


**Fig. 4.** The proposed modeling of development of MES (Added steps are in the orange color)

MES might therefore be able to draw process control cards of machines and production process based on the recorded data, it could decide the reaction it should make based on the capability of machine/process (Fig. 5).

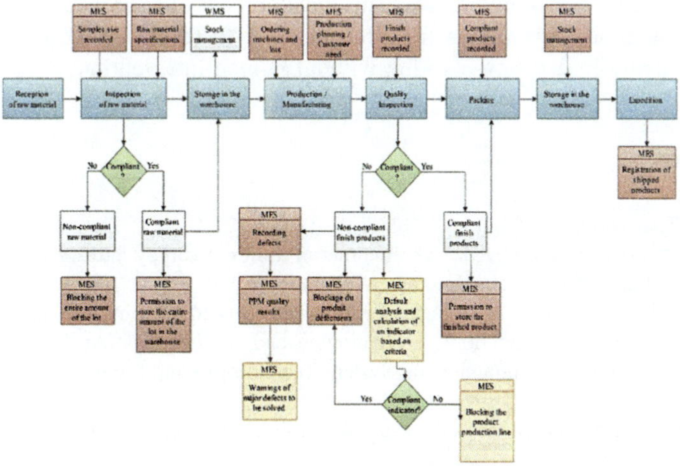


**Fig. 5.** The second proposed modeling of development of MES (Added steps are in the orange color)

MES is developed to use quality PPM results and be able to display alerts about major quality defects to be resolved.

The reconfiguration of the MES gives it the opportunity to react to each defect, to decide whether it is a serious defect that requires the shutdown of the entire production chain, or it is a defect that does not require this shutdown and that will be treated as a normal case of non-compliant product.

This decision on the seriousness of the defect will depend on an indicator that will be calculated based on several criteria: the occurrence of the defect, the customer claims on that defect, the seriousness of the default vis-à-vis the customer…

So, we can say that MES is the appropriate application not only to access critical data in real time, but also to propagate the decision taken. To do this, MES solutions must combine different resolution tools with multiple data sources to develop solutions within a reasonable time frame consistent with the planning horizon.

## 5 Conclusion

MES has been evolving with more and more intelligence being built into it. Internet of Things has been envisioned as the integral part of manufacturing industry that will impact all the areas from planning till delivery.

This article provided a brief introduction to the main concept of IoT. It briefly introduced the Manufacturing Executing System as an important manifestation of IoT in manufacturing industries. This paper presented that an MES offers functional and technological benefits that facilitate an environment for continuous improvement for manufacturers.

MES will require more than a production control tool but also a technology that will support decision-making process. This development of MES need an integration if analysis methods and operational research in order to result an MES which is independent relatively in decision-marking without human intervention.

## References

Ashton, K.: That “Internet of Things” Thing. RFID J. **22**(7), 97–114 (2009)

Luigi, A., Antonio, I., Giacomo, M..: The Internet of things: a survey. https://www.researchgate.net/publication/22257757_The_Internet_of_Things_A_Survey

Internet of Things (IoT) – A Technological Analysis and Survey on Vision, Concepts, Challenges, Innovation Directions, Technologies, and Applications (An Upcoming or Future Generation Computer Communication System Technology). http://pubs.sciepub.com/ajeee/4/1/4/index.html

Ma, H.-D.: Internet of Things: objectives and scientific challenges. J. Comput. Sci. Technol. **26**(6), 919–924 (2011)

Abdul-Qawy, A.S., Pramod, P.M., Magesh, E., Srinivasulu, T.: The Internet of Things (IoT): an overview. Int. J. Eng. Res. Appl. **1**(5), 71–82 (2015)

Zhang, J., Jiang, Y., Jiang, W.: Quality management of manufacturing process based on manufacturing execution system. In: Proceedings of the AIP Conference, vol. 1829, no 1, p. 020025, avr. 2017

Ben Khedher, A., Henry, S., Bouras, A., et al.: Integration between MES and product lifecycle management. In: ETFA2011, Toulouse, France, pp. 1–8 (2011)

Younus, M., Peiyong, C., Hu, L., Yuqing, F.: MES development and significant applications in manufacturing -a review. In: 2nd International Conference on Education Technology and Computer, Shanghai, China, p. V5-97–V5-101 (2010)

Saenz de Ugarte, B., Artiba, A., Pellerin, R.: Manufacturing execution system – a literature review. Prod. Plan. Control **20**(6), 525–539 (2009)

Siemiatkowski, M., Przybylski, W.: Simulation studies of process flow with in-line part inspection in machining cells. J. Mater. Process. Technol. **171**, 27–34 (2006)

Ugarte, B.S.D.: AIDE À LA PRISE DE DÉCISION EN TEMPS RÉELDANS UN CONTEXTE DE PRODUCTION ADAPTATIVE, p. 201

# New Method for Estimating Form Defect of a Feature by Coupling the Genetic Algorithm and the Interior Points Method: Case of Flatness

Mohamed Zeriab Es Sadek[1] and Abdelilah Jalid[2](✉)

[1] Research Center STIS, M2CS, ENSET, Mohammed V University in Rabat, Rabat, Morocco
essadekzeriab@gmail.com
[2] Research Center STIS, PCMT, ENSET, Mohammed V University in Rabat, Rabat, Morocco
a.jalid@um5s.net.ma

**Abstract.** The function of a part in a mechanism depends on its dimensional and geometric quality. On a coordinate measuring machine, a cloud of points is taken from the surface to be inspected and the shape defect is estimated after an optimization calculation. In this paper, we present a new method based on the coupling between a genetic algorithm and the method of internal points, after a comparison with NIST the proposed approach gives better results.

**Keywords:** Flatness · Genetic algorithm · Hybrid algorithm · Form defect

## 1 Introduction

On a three-dimensional measuring machine, a cloud of points is taken from the surface to be controlled and the form defect is estimated after an optimization calculation. The methods used, frequently, are the least squares method (LSM) and the minimum zone method (MZM). Boggs et al. [2] have presented a method based on orthogonal distance regression, Jalid et al. [5] have worked with the ODR and estimate the parameters of the associated model and their uncertainty. Generally, the modelling of a form defect problem is a non linear and non convex optimization problem, which the standard algorithms of optimization cannot give the global optimum but only a local one. In this paper, we present a new hybrid algorithm based on the coupling between the genetic algorithm which is an evolutionary algorithm and the interior points method which is a deterministic algorithm.


A. Saka et al. (Eds.): CPI 2019, LNME, pp. 286–291, 2021.
https://doi.org/10.1007/978-3-030-62199-5_26

## 2 Modelling

The norm ISO 1101 [4] define the flatness as the minimal gap between two parallel planes $P_1$ and $P_2$ containing the points taken (Fig. 1).

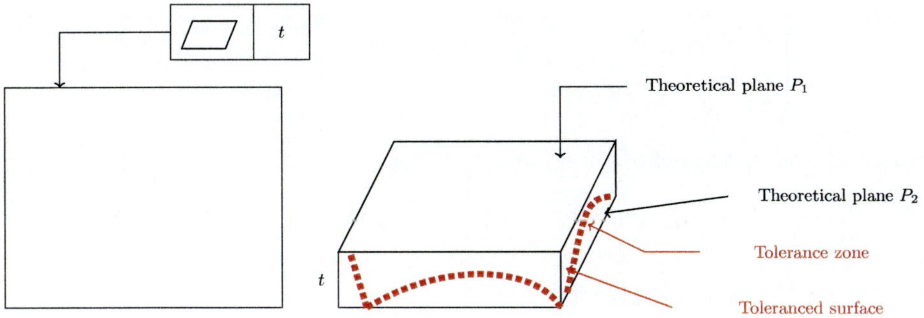


**Fig. 1.** Flatness defect of a feature

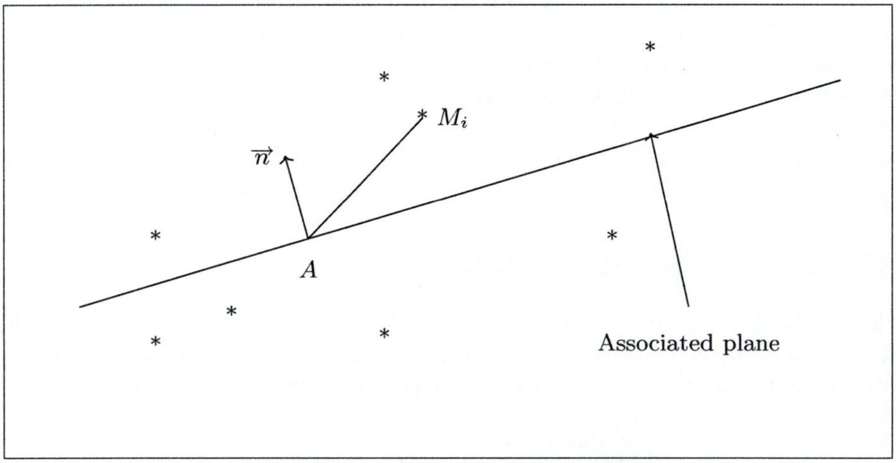


**Fig. 2.** Parameters of an associated plane

The parameters of a plane are a point $A(x_A, y_A, z_A)$ and a vector normal to the associated plane $\overrightarrow{n}(n_x, n_y, n_z)$ (Fig. 2).

The form defect can be approximated by

$$dp = \left|\overrightarrow{Mm} \cdot \overrightarrow{n}\right| = |(x_m - x_M)n_x + (y_m - y_M)n_y + (z_m - z_M)n_z|$$

The plane is defined by a point $M$ and a normal $\overrightarrow{n}$, then a vector $X = (x_1, x_2, x_3, x_4, x_5, x_6)'$ is sought, where $(x_1, x_2, x_3)$ are the coordinates of $M$ and $(x_4, x_5, x_6)$ the coordinates of $\overrightarrow{n}$.

Let $d_{i,j}$, $i = 1, \dots, n$ be the data of the coordinate $j = 1, 2, 3$ collected by the MMT, the problem is modelled in the following form:

$$\begin{cases} \min \left[ \sum_{i=1}^{n} (d_{i,1} - x_1)x_4 + (d_{i,2} - x_2)x_5 + (d_{i,3} - x_3)x_6 \right]^2 \\ a_1 \le x_1 \le b_1 \\ a_2 \le x_2 \le b_2 \\ a_3 \le x_3 \le b_3 \\ x_4^2 + x_5^2 + x_6^2 = 1 \end{cases} \tag{1}$$

where $a_j$ and $b_j$ are defined for $j = 1, 2, 3$, by:

$$a_j = \{\min d_{i,j} \ ; \ i = 1, \dots, n\} \quad b_j = \{\max d_{i,j} \ ; \ i = 1, \dots, n\}$$

# 3 Resolution Method

## 3.1 Introduction

The optimization problem (1) is not convex and not linear, then the deterministic algorithms of optimization cannot give the global optimum but only a local one. Also, the evolutionary algorithms like the genetic algorithm [3] don't give a good result if the field of research is wide.

To remedy the problems described above, some authors have thought to couple the genetic algorithm with a deterministic algorithm [1]. The algorithm that we used is a coupling of the genetic algorithm and the method of the interior points, applying as a mutation operator the method of the interior points.

## 3.2 Method of the Interior Points

The interior points method is a so-called penalty optimization method, which means that instead of solving a constrained optimization problem, a problem of optimization without constraints that is easier is solved. Let the problem

$$\begin{cases} \min f(x) \\ g_i(x) \le 0 \ , \quad i = 1, \dots, m \\ l_i(x) = 0 \ , \quad i = m+1, \dots, p \end{cases} \tag{2}$$

The problem (2) is transformed into the optimization problem without constraints:

$$\min \phi(x) \tag{3}$$

Where the function $\phi$ is called the penalty function, it is defined for $X \in \mathbb{R}^n$ and $r > 0$ by [6]:

$$\phi(X, r) = f(X) - r \sum_{i=1}^{m} \frac{1}{g_i(X)} + \frac{1}{\sqrt{r}} \sum_{i=m+1}^{p} l_i^2(X)$$

The problem (3) is resolved by using an optimization algorithm without constraints as the conjugate gradient algorithm.

This method ensures convergence towards a global optimum only in the case where the problem is convex, otherwise it provides, generally, a local optimum.

### 3.3 Hybrid Algorithm

The genetic algorithm is based on four major operators which are:

- Generation
- Crossover
- Mutation
- Selection

The approach adopted for the hybrid algorithm consists of modifying these operators as follows:

**Generation**
$N$ points are generated randomly. To make these points feasible, we apply the correction operator introduced by Rosen [7]. Indeed, every no feasible point can be corrected towards the border, using a projection matrix.

**Crossover.** For every two points $X$ and $Y$ randomly taken from the population, we generate two new points $X_{new}$ and $Y_{new}$:

$$Xnew = \frac{|X - Y|}{2}$$

$$Ynew = \frac{|X - Y|}{3}$$

To be sure that these new elements satisfy the constraints, we apply the correction operator.

**Mutation.** At each point $X$, we apply the method of the interior points by using as initial point $X$.

**Selection.** We take the $N$ best elements, obtained after the crossover and the mutation.

### 3.4 Results

In our approach to validation of the method, we treated clouds of points provided by the National Institute of Standards and Technology (NIST).

The results and the comparison are shown in the Table 1.

**Table 1.** Comparison between results found and those of NIST

| | Sample | NIST results | Hybrid algorithm results | Gap |
|---|---|---|---|---|
| Plan2 8 points | Point $A$ | 520,99994875 | 521.1307709702896300 | 0,13082222 |
| | | −430,5527825 | −430.21104702690957 | 0,341735473 |
| | | 731,412905 | 730.9903530205760900 | 0,422551979 |
| | Normal $\vec{n}$ | 0,205572461998438 | 0.2055724271449133 | $3,48535 \cdot 10^{-08}$ |
| | | 0,728842915703706 | 0.7288429037108722 | $1,19928 \cdot 10^{-08}$ |
| | | 0,653091086370362 | 0.6530911107250301 | $2,43547 \cdot 10^{-08}$ |
| | $f(X)$ | 0,0000006375773718 | 0.0000006375773823 | $1,05 \cdot 10^{-14}$ |
| | Defect | 0,000888153583016 | 0,0008881367925407 | $1,67905 \cdot 10^{-08}$ |
| Plan6 30 points | Point $A$ | 0,0935786666666666 | 0.0928336884387363 | 0,000744978 |
| | | −0,864989666666666 | −0.8632708775257116 | 0,001718789 |
| | | 1,02021766666666 | 1.0214066302971267 | 0,001188964 |
| | Normal $\vec{n}$ | −0,754202107469195 | 0.7542021084781786 | $1,00898 \cdot 10^{-09}$ |
| | | −0,561915157740874 | 0.5619151576508896 | $8,998510^{-11}$ |
| | | 0,339750697703583 | −0.3397506956126001 | $2,09098 \cdot 10^{-09}$ |
| | $f(X)$ | 0,0000000001697214 | 0.0000000001697216 | $2 \cdot 10^{-16}$ |
| | Defect | 0,000009502749043 | 0,0000095015837698 | $1,16527 \cdot 10^{-09}$ |
| Plan17 36 points | Point $A$ | 102,924305 | 102.8816885017847200 | 0,042616498 |
| | | −663,372305833333 | −663.37230161409047 | $4,21924 \cdot 10^{-06}$ |
| | | 650,287965277777 | 652.8124294678800700 | 2,52446419 |
| | Normal $\vec{n}$ | −0,00001622795052309 | 0.0000236818496444 | $7,4539 \cdot 10^{-06}$ |
| | | −0,999999999868302 | 0.9999999997193421 | $1,4896 \cdot 10^{-10}$ |
| | | $2,19 \cdot 10^{-07}$ | 0.0000006970039011 | $4,77662 \cdot 10^{-07}$ |
| | $f(X)$ | 0,0003573901110887 | 0.0003574702754382 | $8,01643 \cdot 10^{-08}$ |
| | Defect | 0,0124729960709469 | 0,0124819003103514 | $8,90424 \cdot 10^{-06}$ |
| Plan19 110 points | Point $A$ | −91,8525067272727 | −91.4317354744796230 | 0,420771253 |
| | | 506,198875454545 | 506.1787626729232600 | 0,020112782 |
| | | −451,56514390909 | −451.13678767636731 | 0,428356233 |
| | Normal $\vec{n}$ | −0,700310125962299 | 0.7003099051812104 | $2,20781 \cdot 10^{-07}$ |
| | | −0,225400689612005 | 0.2254004202459256 | $2,69366 \cdot 10^{-07}$ |
| | | 0,677318430723024 | −0.6773187486391059 | $3,17916 \cdot 10^{-07}$ |
| | $f(X)$ | 1,08986440316221 | 1.0898643984428247 | $4,71939 \cdot 10^{-09}$ |
| | Defect | 0,418292542292985 | 0,418277706985051 | $1,48353 \cdot 10^{-05}$ |

## 4 Conclusion

The studied case is an estimate of flatness of a plane surface from a palpated point cloud. The problem is resolved using a hybrid algorithm based on the coupling between the genetic algorithm and the interior points method.

This method has the advantage of being a robust algorithm, because it does not need an initial value to converge, and it gives the global optimal even the problem is not linear and not convex.

The results found are compared with those of the international organization NIST, and we remark that they are very close together and the difference between the different values is of the order of $10^{-5}$ to $10^{-8}$ which is very encouraging for a start.

## References


1. Belkourchia, Y., Azrar, L., Es-sadek, M.Z.: Hybrid optimization procedure applied to optimal location finding for piezoelectric actuators and sensors for active vibration control. Appl. Math. Model. (2018)
2. Boggs, P.T., Byrd, R.H., Schnabel, R.B.: A stable and efficient algorithm for non-linear orthogonal distance regression. SIAM J. Sci. Stat. Comput. (1987)
3. Cui, C., Li, B., Huang, F., Zhang, R.: Genetic algorithm-based form error evaluation. Measur. Sci. Technol. (2007)
4. ISO 1101: Geometrical product specifications (GPS) - Geometrical tolerancing - Tolerances of form, orientation, location and run-out (2004)
5. Jalid, A., Hariri, S., Senelaer,J.P.: Estimation of form deviation and the associated uncertainty in coordinate metrology. Int. J. Qual. Reliab. Manage. (2015)
6. Rao, S.: Engineering Optimization. Wiley (1996)
7. Rosen, J.B.: The gradient projection method for nonlinear programming. Part II. Nonlinear constraints. J. Soc. Indust. Appl. Math. (1961)

# New Maturity Model of Industrial Performance for SME - Creation and Case Study

Badr Elwardi(✉), Anwar Meddaoui, Ahmed Mouchtachi, and Hakim Nissoul

Laboratory Structural Engineering, Intelligent Systems and Electrical Energy, National School of Arts and Crafts, Casablanca, Morocco
badr.elwardi@gmail.com

**Abstract.** In this science research explosion century, the competitive economic environment and manufacturers are forced to follow the development and changes to protect their existence with respect of delay, quality and cost challenges, therefore lean manufacturing concepts become part of solution, contrariwise some manufacturers, especially small and medium –sized enterprises (SME) suffer to implement these concepts, this paper purpose to define an advanced model of lean manufacturing system (LMS) based on ISO:9001 version 2015 principles and VDA6.3 version 2016 norm, suggest a maturity model to evaluate the maturity level of LMS which has been confirmed through a statistical studies, Nevertheless, few works related to this field make academic research and industry closer, our work falls within this context through the industrial productivity improvement. This paper proposes also some research perspectives aiming to complete this research axis by enhancing the presented model through other industrial experiments and studies.

**Keywords:** ISO9001 · Lean manufacturing system · Key performance indicators · Maturity model · VDA · LSRO model

## 1 Introduction

The world recognizes a considerable jump on industry army performance within second world that create a new world vision about performance which disseminate strictly after war achievement to other Industrial manufacturers sectors, the quality, delay and cost are the most key element to establish competitive position and keep the manufacturing business, the lean manufacturing philosophy come to fix, improve and monitor those key elements.

The aim of Lean Manufacturing is to eliminate wastes in every area of production including customer relations, product design, supplier networks, and factory management. Its goal is to incorporate less human effort, less inventory, less time to develop products, and less space to become highly responsive to customer demand while producing top quality products in the most efficient and economical manner possible. (Kariuki 2013).


A. Saka et al. (Eds.): CPI 2019, LNME, pp. 292–304, 2021.
https://doi.org/10.1007/978-3-030-62199-5_27

Lean has gained great prominence as a world class business philosophy in the turn of the last century. Its excellent performance as a world class strategic operations and business philosophy was brought to the fore by the report on the Massachusetts Institute of Technology International Motor Vehicles Program (IMVP). The study which became a bestselling book titled The Machine that Changed the World (Womack et al. 1990) was cross disciplinary and multinational as it involved visits to 90 assembly plants in 17 countries. The study showed that success of Asian production systems especially that of Japan was not due to the erroneous believe of low wages and government assistance but due to the type of production system used regardless of the country.

Moreover, Moroccan small and medium enterprises characterized by less than 200 peoples employed and a turnover less than 5 Million DHs on creation phase, 20 million DH for the growth phase and 50 million DH for the maturity phase have lack of technical, financial resources and poor of managerial human capital, training and skills (Mouhallab and Jianguo 2016) therefore their management are looking for the quick benefit and agree only the investment with results reached on short term benefit.

In business strategy, the cost leader ship represents a competitive advantage by getting the lowest cost of operation in the industries, some studies highlights that both differentiation and cost leader ship strategies are positively related to performance and refines the understanding of the effect of entrepreneurial orientation on small and medium enterprises (Lechner and Gudmundsson 2013).

Presented paper will propose a new model of lean manufacturing system (LMS) applied to mass producer SME taking in consideration the current achievement on management system based on ISO9001 version2015 and adapts the AFNOR's maturity model from quality management system to Lean manufacturing system (LMS), 10 of Moroccan SME will evaluate their LMS using our solution, the inter-correlation of variables study between the survey's articles and famous industrial performance indicators KPIs will demonstrate the importance and efficiency of the solution and chance for the enterprises to identify their position regarding Lean maturity level and define a strategic solution from one side.

Other side the LSRO implementation will held in Moroccan SME to evaluate the recent performance, define weaknesses, generate an action plan and check effectiveness.

## 2 Literature Review and Research Methodology

**Literature Review**

The Lean manufacturing maturity level treated using Lean LPPO Model focused quality management principles reflected on Leadership-People-Process-Output axes, the correlation study carried out on 7 industrial companies using a questionnaire with 5 ranks for each, and a significant positive link concluded between the Leadership, involvement of People, Process approach, System approach to management, continual management factual approach for decision making and mutual beneficial supplier relationship with Lean manufacturing outcome (Dibia and Dhakal 2012).

The controversial point, interest on Lean short term "Lean practice" orientation exclusive focus on lean tools and lean long-term orientation surrounded on lean mindset "lean thinking and culture" requires a strong effort to transform the organizational culture (Guimarães and de Carvalho 2014), example change from professional bureaucracy to collective behaviours.

Lean business system created founded on 5 main axes strategy deployment, people enabled process, value stream management, application of lean tools and techniques, and extended enterprise, (Setianto and Haddud 2016) identified the synergy between lean practices and other management approaches as an effort to improve manufacturing competence attributes.

Some research conducts lean manufacturing on visual management, 5S and work standardization as the most utilized lean techniques (Akugizibwe and Clegg 2014), the finding 'Poor understanding by the leadership team' and 'passive and active resistance to change' were the greatest barriers to lean implementation. 'Engaging with management and leadership' and 'employee ownership/workforce led change' were the critical success factors but this stays weak from system approach point of view as the management aren't part of analysis.

The JCQEV model represents lean manufacturing on Just in time, Continuous improvement, Quality, Elimination of wastes, Visual management (Lyonnet et al. 2010), this remains poor model as it misses the systematic approach, also LOC Model refer only to Last Order Costing which analyze the process and value added without involve the management axes.

The aim of some research contributes the managerial factors as successful implementation of lean manufacturing within the small and medium scale enterprises (SMEs). For this purpose, shop floor commitment and employee trust, business strategy based on customer demand, analysis of data, linking lean into the business were labelled as four constructs with 19 variables to measure the managerial factor. it adopts SEM Model framework to measure the managerial contributing factors towards the success of lean implementation. After the analyses of the results, shows that these four constructs with 14 variables are significant to the managerial involvement (Ramadas et al. 2016).

Recently the lean manufacturing system presented on following axes, waste reduction and value added, people, process, risk management, continuous improvement, systematical approach (Nordin et al. 2012).

The Lean manufacturing has been represented to process approach, this is highlighted on article (En-Nhaili et al. 2015) by applying the value stream mapping method to each defined sub-process and calculate the time driven activity based on costing TDABC, the main idea of this approach to introduce also the investment project cost using the formula:

$$\text{OEE-TE} = \text{Availability} \times \text{Performance} \times \text{Quality} * \text{ACP}$$

As:

Availability = Operating Time/Planned Production Time,

ACP = CDR Budget/CDR Actual

Performance = Ideal Cycle Time/(Operating Time/Total Pieces),

Quality = Good Pieces/Total Pieces

Our contribution aims to define a new Lean manufacturing systems refers to above articles and contributions.

**Research Methodology**

Methodology followed in this article is structured into five main phases as shown in Fig. 1 below:

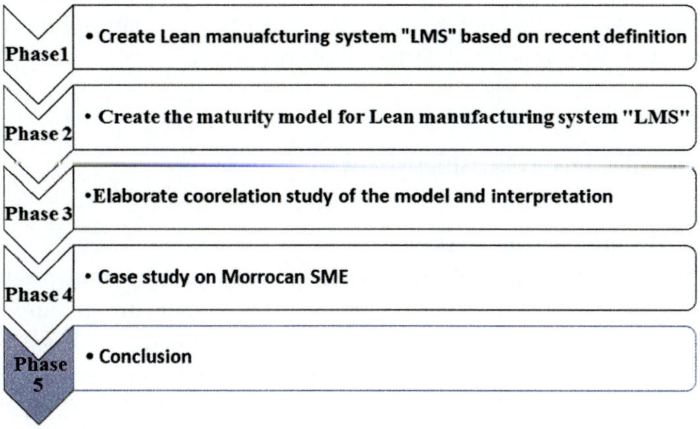


**Fig. 1.** Methodology of research

The methodology phases showed in Fig. 1 illustrates the following points:

- Phase 1: This section suggests a Lean manufacturing system model LSRO based on quality management system ISO9001: version2015, VDA 6.3 "norm applied on automotive" covering the context of enterprise, leadership, support, execution of operational activity and continuous improvement process also describes the requirement for each rubric.
- Phase 2: It adopts the famous maturity model questions proposed by AFNOR from quality management system evaluation to lean manufacturing system based on our proposed model LSRO.
- Phase 3: Data collection survey, examines on 10 industrial Moroccan SMEs referring to LSRO model, identify their maturity level and determines representable industrial performance for lean implementation.

  The purpose of this phase is to evaluate the efficiency of the proposed model LSRO using inter-correlation study of following variables LSRO model survey results among performance industrial KPIs.

  Also evaluate the efficiency of the proposed model LSRO using inter-correlation study of following variables LSRO model survey results among performance industrial KPIs.
- Phase 4: It is about a Moroccan brickyard manufacturing company, indexed on SME category, which strives to improve their industrial performance using lean manufacturing system. The management are aware about a high competitiveness in this field, and we benefit to use LSRO maturity model as they are preparing to be

certified ISO9001:2015, In this section, initial performance measured by OEE of third quarter of 2017.
Further, the advanced LSRO maturity model offer validate the recent LMS position and develop clear road map to reach a better this level.
The global action plan will be implemented and review the KPI and compare if better level of LMS is achieved.

- Phase 5: The presented maturity model will simplify the implementation of lean manufacturing system and define easily and precisely the maturity level and related strategic action plan and facilitate the implementation.

## 3 Creation of New Maturity Model

**Create Lean Manufacturing System**
On the literature the meaning of lean manufacturing turns around production lines performance, some authors consider peoples and management are the important fact of lean manufacturing, others introduce the improvement process as key element, Now we're forced to standardize lean manufacturing definition, in our purpose we propose a system Lean manufacturing based on ISO9001:2015, VDA6.3 and cited articles as shown in below new concept (Fig. 2).

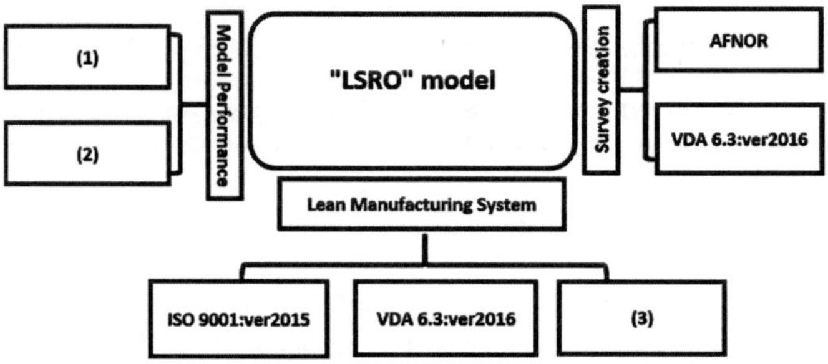


**Fig. 2.** Design of LSRO model

(1) Ahmed En-Nhaili*, Anwar Meddaoui* and Driss Bouami, 2015
(2) Ifechukwude K. Dibia* and Hom Nath Dhakal, 2012
(3) Norani Nordin, Baba Md Deros, Dzuraidah Abdul Wahab and Mohd Nizam Ab. Rahman No. 1, 2012

**Create the Maturity Model**
The lean manufacturing system will represent a new "LSRO" model focus on Leadership, Support, Realization and outcome (Fig. 3).

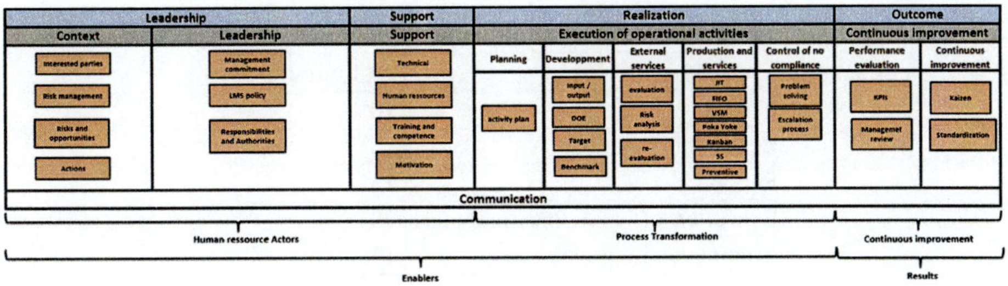


**Fig. 3.** LSRO model axis

Our proposed maturity model developed based on ("ISO9001: version 2015 Systèmes de management de la qualité – Exigences", édition Afnor, www.afnor.org, 25-Juillet-2014) and VDA6.3 Version 2016 guidelines checklist to help the organization to perform self-evaluation of compliance for their own lean manufacturing system, the model can be designed on PDCA schema as shown in below (Fig. 4):

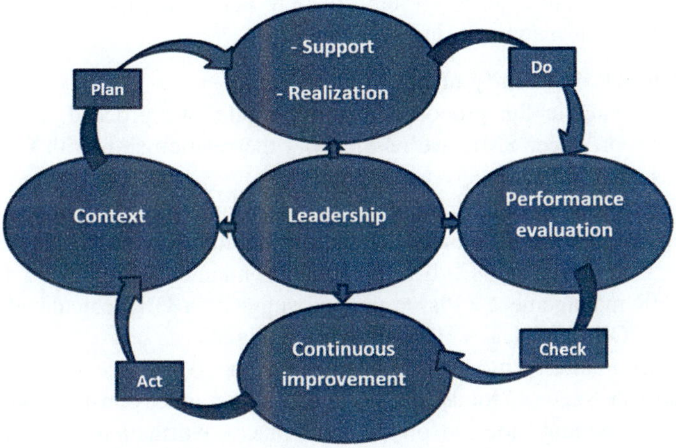


**Fig. 4.** LSRO presented in PDCA model

It also serves as a dashboard with which it can comment and evaluate the progress of the lean manufacturing maturity system.

It's developed based on LSRO Lean manufacturing model contains five levels which contains 69 questions divided as below Table 1:

**Table 1.** Question distribution on LSRO model

| Variable | # of questions |
|---|---|
| Context | 10 |
| Leadership | 5 |
| Support | 13 |
| Execution of operational activities | 28 |
| Continuous improvement | 13 |

The self-evaluation is done by answer to the questions as there are 4 score possible for each question:

False: The action is not performed or very randomly with score of 0%

Rather False: The action is carried out a few times informally with score of 30%.

Rather True: The action is formalized and carried out quite convincingly with score of 70%

True: The formalized action is realized, improved and traced with score of 100%

At the end the model provide the score of each article of the model and the global score of the lean manufacturing system, display also a dashboard to determine the weakness of the lean manufacturing system.

**Elaborate Correlation Study of the Model**

To apply and evaluate the proposed maturity model we proceed to perform it to Moroccan mass producer SME, we're checking the relation-ship with Lean manufacturing performance indicators.

To do this, we define first a new mathematic formulary of the famous indicator of Lean OEE as we can see differently defined on literature review displaying OEE-TE.

Our target is to define an OEE based on quality, maintenance; logistic, productivity, project and HR management KPIs, at the end suggest an OEE calculated as below:

OEELMS = OEEClassical × [1-Turnover] × logistic

The assigned maturity model is implemented on 10 Moroccan SME worked on different industrial sectors "foods, plastic injection, ceramic, construction, confection, polymers and recycling" located on different place "North, Est, West and middle" characterized with less than 200 people employed and a turnover less than 5 Million DHs on creation phase, 20 million DH for the growth phase and 50 million DH for the maturity phase, their above Lean defined KPIs calculated.

Due to confidentiality of the activity and companies we will use Xn as audited company number n (Table 2).

**Table 2.** Organization details and number of given and return questions

| Organisation | Size | No. of staff | Q.given | Q.returned |
|---|---|---|---|---|
| X1 | Medium | 65 | 69 | 69 |
| X2 | Medium | 57 | 69 | 69 |
| X3 | Small | 32 | 69 | 66 |
| X4 | Medium | 62 | 69 | 50 |
| X5 | Small | 47 | 69 | 69 |
| X6 | Small | 39 | 69 | 69 |
| X7 | Medium | 89 | 69 | 46 |
| X8 | Medium | 189 | 69 | 67 |
| X9 | Small | 42 | 69 | 58 |
| X10 | Medium | 135 | 69 | 69 |
| Total | | | 552 | 505 |

The aim of this chapter is to define the efficiency of the maturity model by analyzing the inter-correlations of the variables and relations ship between model articles and different lean manufacturing KPIs, to reach this we apply first Cronbach's coefficient Alpha to examine the reliability of the data per research, we accept data collection with Cronbach's coefficient Alpha > 0.7 (Table 3).

**Table 3.** reliability test

| Cronbach alpha | Number of element |
|---|---|
| ,964 | 11 |

The calculated Cronbach's coefficient Alpha is about 0,964 which reflect a high reliability of the collected data.

Note: Correlation is significant at 0.01 level (2-Tailed), in our study we interpret only the correlations with an index r > 0,55 (Table 4).

**Table 4.** Matrix of inter-correlation between variables

| | Context | Leadership | Support | Realization | Outcome | Maturity rate | OEE LMS | OEE classic | 1-Turn over | Logistic | No of staff |
|---|---|---|---|---|---|---|---|---|---|---|---|
| Context | 1.000 | | | | | | | | | | |
| Leadership | .824 | 1.000 | | | | | | | | | |
| Support | .868 | .849 | 1.000 | | | | | | | | |
| Realization | .892 | .878 | .900 | 1.000 | | | | | | | |
| Outcome | .800 | .860 | .947 | .893 | 1.000 | | | | | | |
| Maturity rate | .930 | .925 | .969 | .960 | .949 | 1.000 | | | | | |
| OEE LMS | .767 | .816 | .880 | .915 | .900 | .899 | 1.000 | | | | |
| OEE classic | .742 | .810 | .806 | .885 | .895 | .866 | .937 | 1.000 | | | |
| 1-Turn over | .507 | .459 | .733 | .551 | .569 | .605 | .633 | .350 | 1.000 | | |
| Logistic | .440 | .437 | .552 | .584 | .445 | .518 | .678 | .395 | 0.821 | 1.000 | |
| No of staff | .364 | .362 | .427 | .246 | .456 | .397 | .324 | .399 | .092 | -0.057 | 1.000 |

The online version of the volume will be available in LNCS Online. Members of institutes subscribing to the Lecture Notes in Computer Science series have access to all the pdfs of all the online publications. Non-subscribers can only read as far as the abstracts. If they try to go beyond this point, they are automatically asked, whether they would like to order the pdf, and are given instructions as to how to do so.

Please note that, if your email address is given in your paper, it will also be included in the meta data of the online version.

**Diagnostic and Result**

All "LSRO" model component is on significant relationship with them-self, the Model maturity rate and Lean manufacturing system performance "OEE LMS", which prove the importance and impact of each component from the "LSRO" model on Lean performance.

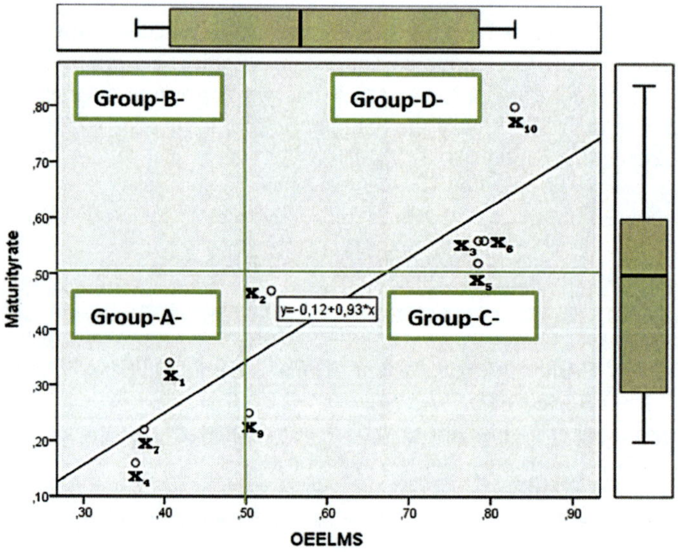


**Fig. 5.** Correlation between the maturity Model rate and lean performance

Based on the Fig. 5 the Maturity rate of the model "LSRO" could be displayed on function of Lean performance as below:

Maturity "LSRO" = 0.93xOEELMS −0.12

Other hand, Fig. 5 displays four different areas (Hohmann 2004):

Group-A-: Includes enterprises {X4; X7; X1}, represents danger area with low maturity and performance, the direction should define an urgent emergency plan and focus on staff competences training and make in place the new competences.

Group-B-: Reflects theatrical enterprises with high level of maturity and low level of performance, the direction should define an urgent emergency plan to make in place the staff knowledge.

Group-C-: Demonstrates enterprises {X9; X2} with effective level of performance "high level of performance and low level of maturity", enterprises reach the performance with high resources, the direction should improve the competence of staff and apply lean manufacturing tools.

Group-D-: Demonstrates enterprises {X3; X6; X10} with efficiency level of performance "high level of maturity and performance" enterprises reach the performance with minimum of resources, the enterprises should sustain its position and focus on continuous improvement process.

**Case Study on Moroccan SME**

Studied company is a leading Moroccan actor in the brickyard industry categorized as medium enterprise, their management are aware about the market compactivity and define a new strategic focus to install the quality system and prepare for certification of ISO9001:2015 (Fig. 6).

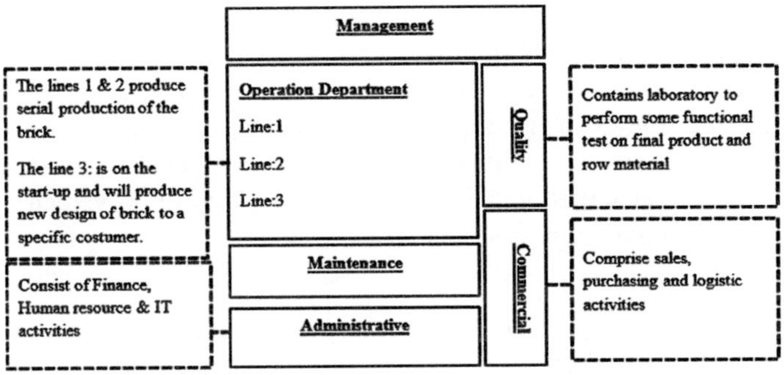


**Fig. 6.** Plant schema and department definition

The activity chosen for this improvement project is the whole plant processes, on the production side for the line 1 & 2 we will evaluate the process development and product realization however the line 3 we will take advantage to check the product and process development and project approach based on LSRO model.

Referring to the global plant performance of the third quarter "Q3" of 2017 as showed in Table 3 provides overview about different departments performance like quality, maintenance, Production, Human resource & logistic that are defined as below:

$$\mathbf{OEE_{LMS} = OEE_{Classical} \times [1\text{-}Turnover] \times logistic}$$

The diagnostics indicates that the lean manufacturing system in the organization correspond to level-2 with 34% of LMS rate. See below Table 5.

**Table 5.** KPIs overview and target

| Organization | Context | Leadership | Support | Execution of operational activities | Continuous Improvement | LSRO Maturity rate |
|---|---|---|---|---|---|---|
| Brickyard Manufacturer | 38.00% | 46.00% | 24.00% | 35.00% | 29.00% | 34.00% |
| Levels | Level-2 | Level-2 | Level-2 | Level-2 | Level-2 | Level-2 |
| Target | 70.00% | 70.00% | 70.00% | 70.00% | 70.00% | 70.00% |
| Level | Level-3 | Level-3 | Level-3 | Level-3 | Level-3 | Level-3 |

This evaluation aims to give a realistic picture of the standards of international industrial excellence assessed against a fixed target of 70 and not 90 which remains very difficult to achieve soon (See Table 5).

The gap between the initial and final aimed situation is described through 26%. The first discrepancy interpret that the company's problems observed on all levers starting from management to continuous improvement.

Moreover, after the current situation analysis, we reacted by holding special meetings to get into the main factors behind the observed weakness.

These initial agitations are followed by a detailed diagnostic report containing, in addition to what has been mentioned above, a proposal for improvement projects and targeted actions to fix the current situation.

**Diagnostic and Result**

After implementation of about 70% from the global action plan due to budget and implementation time constraints we reassess the LMS system using LSRO maturity model and compare with established Q3 LMS KPI to evaluate effectiveness of the global action plan.

The Table 6 display the new maturity level based on LSRO model and recognize a significant improvement presented in all levers and enhanced from level-2 to level-3.

It concluded the new LMS conducts an improvement of 127% and reaches 77.40%.

**Table 6.** Benefit of action plan

| | Context | Leadership | Support | Execution of operational activities | Continuous Improvement | Maturity rate |
|---|---|---|---|---|---|---|
| Initial assessment | 85.00% | 76.00% | 73.00% | 77.00% | 76.00% | 77.40% |
| After action plan implementation | 38.00% | 46.00% | 24.00% | 35.00% | 29.00% | 34.00% |
| Benefit | 47.00% | 30.00% | 49.00% | 42.00% | 47.00% | 43.40% |

The OEE LMS and OEEclassic with benefit for Q3-2018 compared with the same quarter of last year and it's clearly presented a benefit on all KPIs of Quality, maintenance, productivity turn-over and logistic.

Therefore, the final OEELMS increased by 80% and reaches 74.52% (Table 7).

**Table 7.** Benefit of action plan 2018 vs 2017

| | Quality | Availability | Productivity | 1-Turn over | Logistic | OEE LMS | OEE classic |
|---|---|---|---|---|---|---|---|
| Initial assessment | 99.50% | 96.00% | 89.00% | 94.00% | 92.00% | 73.52% | 85.01% |
| After action plan implementation | 95.00% | 75.00% | 72.00% | 90.00% | 88.00% | 40.63% | 51.30% |
| Benefit | 4.50% | 21.00% | 17.00% | 4.00% | 4.00% | 32.89% | 33.71% |

## 4 Conclusion and Perspective

The LSRO conceptual model for lean implementation, from one side, come with new philosophy around enterpriser's Context including interested parties "stakeholders, customers, suppliers…" requirements, Risk management "defining and act on risks and opportunities" and demonstrates their direct impact on lean manufacturing system, other side it re-affirm the importance of each article of Leadership "direction commitment", Support "infrastructure, technical and human resources, staff implication and motivation", Execution of operational " planning, development of product and process, Mastery of production…" and continues improvement "performance evaluation and continuous improvement process".

This paper offers to industrial enterprises especially SMEs, the optimum flow based on PDCA approach to implement an advanced Lean manufacturing system through the LSRO model and generate an efficient action plan without requires a special training for staffs or high investment, other side the value added of this study appears on shirt term that is defined around six months based on this case study.

This contribution re-confirms the importance of the LSRO model to achieve a high maturity on industrial performance and we advise the Moroccan SME to choose it to improve quickly their situation.

The LSRO is developed based on lean manufacturing system to evaluate and define easily the weaknesses therefore establish an optimum action plan to reach manufacturing greatness however it bases on cost leadership strategy for SME, our perspective on next work is to adapt it to other business strategy like differentiation strategy.

## References

Kariuki, B.M.: Role of lean manufacturing on organization competitiveness. Ind. Eng. Lett. Inderscience Enterprises Ltd. (2013)

Lechner, C., Gudmundsson, S.V.: Entrepreneurial orientation, firm strategy and small firm performance. Scientific Library of Moscow State University, pp. 1–25 (2013)

Mouhallab, S., Jianguo, W.: Small and medium enterprises in Morocco: definition's issues and challenges. J. Wuhan Univ. Technol. 12 (2016)

Dibia, I.K., Dhakal, H.N.: Lean: a continuous improvement philosophy in agile systems based on quality management principles. Int. J. Agile Syst. Manag. **5**(4) (2012). Inderscience Enterprises Ltd.

Guimarães, C.M., de Carvalho, J.C.: Cultural change in healthcare organisations through lean practices. Eur. J. Cross-Cult. Comp. Manag. **3**(3/4) (2014). Inderscience Enterprises Ltd.

Setianto, P., Haddud, A.: A maturity assessment of lean development practices in manufacturing industry. Int. J. Adv. Oper. Manag. (IJAOM) **8**(4) (2016). Inderscience Enterprises Ltd.

Akugizibwe, A.M., Clegg, D.R.: Int. J. Lean Enterp. Res. (IJLER) **1**(2) (2014). Inderscience Enterprises Ltd.

Ramadas, T., Satish, K.P., Girish, T.: Int. J. Lean Enterp. Res. (IJLER) **2**(1) (2016). Inderscience Enterprises Ltd.

Nordin, N., Deros, B.Md., Wahab, D.A., Rahman, M.N.Ab.: A framework for organisational change management in lean manufacturing implementation. Int. J. Serv. Oper. Manag. **12**(1) (2012)

En-Nhaili, A., Meddaoui, A., Bouami, D.: Effectiveness improvement approach basing on OEE and lean maintenance tools. Int. J. Process Manag. Benchmark. **5**(2), 171–193 (2015)

Womack, J.P., Jones, D.T., Roos, D.: The Machine That Changed the World. Macmillan Publisher, New York (1990)

Hohmann, C.: Audit combiné qualité/supply chain. Sci. Dir. **151**(1–2), 187–191 (2004)

Lyonnet, B., Pillet, M., Pralus, M.: Lean manufacturing in the screw cutting sector: assessment of maturity level. Int. J. Rapid Manuf. **1**(3), 256–277 (2010)

# RTCM's Role in Green Building and the Green Economy

Basma M'lahfi[1](✉), Driss Amegouz[2], and Mostafa El Qandil[1]

[1] LERDD, FS, Université Sidi Mohamed Ben Abdelah, Fès, Morocco
mlahfi.basma@gmail.com, mostafelquandil@yahoo.fr
[2] LPEDD, EST, Université Sidi Mohamed Ben Abdelah, Fès, Morocco
amegouz@yahoo.fr

**Abstract.** The building sector is ranked second in the world after transportation in terms of energy consumption, mainly distributed in air conditioning and heating, these expenses generate more and more high energy bills that encourage all the countries on a global scale find radical solutions.

To be able to its global counterparts, Morocco has recently launched the Moroccan Construction Thermal Regulation "RTCM". This regulation was launched by the National Agency for the Development of Renewable Energies and Energy Efficiency (ADEREE) and it showed according to simulation studies by TRNSYS software a great efficiency in the final economy of energy related particularly to air conditioning and heating despite the extra cost that will be added to the final price of the investment. This extra cost is shown to be the lowest in the category of luxury housing, which shows the interest of its application in this category of housing.

In this respect, the present article mainly examines the results of an energy simulation of a building according to the requirements of the RTCM and its impact on the gain and the energy bill. This simulation has shown its effectiveness despite a surcharge that remains surmountable for all climatic zones and some categories of buildings, hence the interest of making more known the RTCM among professionals in the field through surveys and surveys, to know the degree of its application in Moroccan constructions, the obstacles to overcome and the proposals to be studied so that the RTCM is a practical means of assistance for the simulation of energy-efficient buildings and so that it will be more applied in the buildings of tomorrow.

In addition, we will propose insulating construction materials based on previous literatures that will be adapted for residential and tertiary buildings in each climatic zone.

**Keywords:** RTCM · Green building · Energy efficiency · Green economy · Insulating materials

## 1 Introduction

The objective announced by the Moroccan Government is to achieve a primary energy saving of around 12% to 15% by 2020 [1] through the implementation of an energy efficiency plan in the different regions. economic sectors. Among the energy efficiency


A. Saka et al. (Eds.): CPI 2019, LNME, pp. 305–321, 2021.
https://doi.org/10.1007/978-3-030-62199-5_28

plans, the elaboration of the Construction Thermal Regulation in Morocco (RTCM) which aims essentially to improve thermal performance by reducing the heating and air conditioning needs of buildings, by improving the comfort of non-residential buildings. air-conditioning and the reduction of the power of the heating and air-conditioning equipment to be installed;

Compliance with RTCM requirements in the housing sector allows for final energy savings of around 22 kWh/m$^2$/year and a significant reduction in CO2 gas emissions [1–3], variable according to climatic zoning. These savings generate substantial gains estimated on average at 18 Dh/m$^2$/year [1], compared to the usual energy bill of the final consumer.

On the other hand, the constraints related to the application of this regulation apply automatically on the extra cost of the construction, the qualification of the labor, the choice of the insulating materials, and the non disposition of a strategic plan and adequate communication tools to mobilize and sensitize stakeholders, including governments, businesses, professionals and the general public, to energy efficiency measures in buildings. Hence the need to work on these last points, while proposing the optimal insulating materials for this type of construction.

## 2 Literature Review of RTCM

### 2.1 Benchmark on the Energy Consumption of the Building Sector

The results of global statistics on energy consumption in 2010 showed that the building sector alone accounts for around 28% of final energy consumption and accounts for around a third of $CO_2$ emissions [1], as shown in Fig. 1:

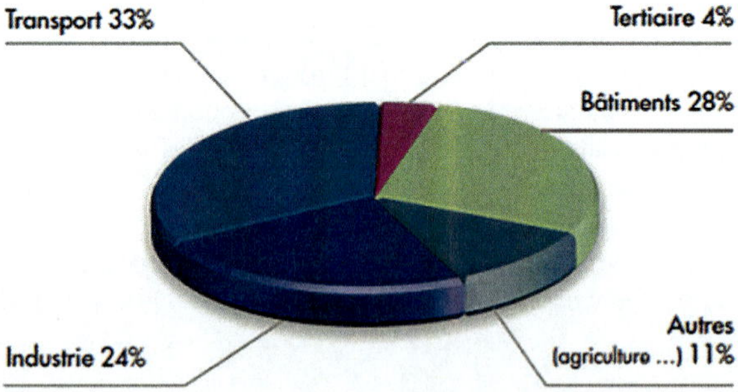


**Fig. 1.** World building sector final energy consumption in 2010 [1]

The southern Mediterranean region is no exception to this observation since, on average, the building sector accounts for around 38% of the energy consumed (this percentage varies between 27% and 65% depending on the country). Compared to other sectors of activity, it represents the largest source of savings and is often around 40% in most countries in the region [1].

In Morocco, the building is the second largest energy consumer after transport with a 25% share of the total energy consumption of the country, of which 18% is reserved for residential and the rest for the tertiary sector as shown in Fig. 2 energy consumption due to various sources leads to an increase in the release of CO2 gas and subsequently an imbalance of our ecosystem [3].

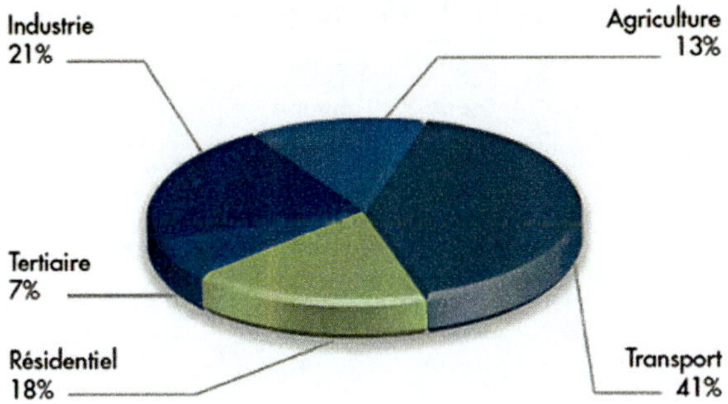


**Fig. 2.** Structure of consumption by sector [1]

## 2.2 RTCM Requirements

The RTCM covers the housing sector and the tertiary buildings [1] and applies initially only to new buildings and gives priority to the urban sector since energy consumption in rural areas is generally low, as well, the application of RTCM in these areas is difficult due to the dispersal of habitat and its informal nature.

In the habitat the regulation targets all socio-economic categories of buildings, namely:

- Economic;
- Standing.

For the tertiary sector, four segments are particularly covered:

- The hotels;
- Administrative buildings (offices);
- Buildings of education and higher education;
- Hospitals.

## 2.3 Moroccan Climate Zoning

The climate zoning works were carried out in close coordination between the National Meteorology Department (DMN) and ADEREE, with the support of international expertise [1].

The Moroccan territory has been subdivided into homogeneous climatic zones based on the analysis of climatic data recorded by 37 weather stations over the period 1999–2008 (10 years). The construction of the zones was done according to the criterion of the number of winter degree days and the number of summer day degrees [1, 4].

Two types of zoning were established by the DMN [1, 4]:
• Zoning based on heating degree days at 18 °C;
• Zoning based on cooling degree days at 21 °C.

The 6 climate zones are listed in Table 1 and the final zonation map includes the six climate zones shown in Fig. 3 [2].

**Table 1.** Climate zones [2]

| | |
|---|---|
| Zone 1 | Agadir |
| Zone 2 | Tanger |
| Zone 3 | Fès |
| Zone 4 | Ifrane |
| Zone 5 | Marrakech |
| Zone 6 | Errachidia |

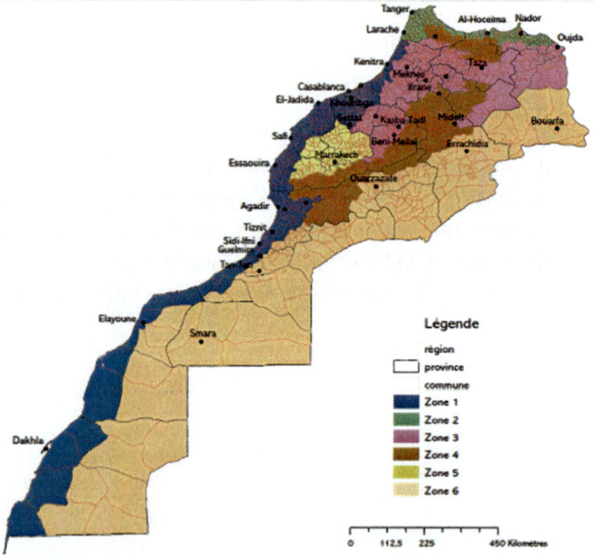


**Fig. 3.** Morocco's climate zoning adapted to Morocco's Construction Thermal Regulation [2]

Morocco's thermal building regulations target both the residential and tertiary sectors. The requirements of the first sector are the same for the different socio-economic categories of the buildings, they differ only from one climatic zone to another by fixing for each zone the minimum technical specifications, namely the rate of the bay windows, the minimum resistance of the floors on the ground, the solar factor of glazing and the thermal transmission coefficient of exposed roofs, external walls and glazings [1, 2].

On the other hand, the thermal regulation targets specifically in the tertiary sector the four main branches of the tertiary sector, namely:

- Administrative buildings;
- Schools;
- Hospital buildings;
- Hotels [1, 2].

In addition, the technical specifications of the by-law and its impacts are determined for each of the branches and also for each of the zones similar to the residential sector, namely, the rate of the bay windows, the minimum resistance of the floors on the ground, the solar factor of the glazings and the thermal transmittance of exposed roofs, exterior walls and glazing [1, 2].

## 2.4 Economic Impacts of the RTCM on the Construction of Green Buildings

In order to determine the economic impact of the RTCM on economical construction, a series of thermal simulations were carried out on reference buildings using the TRNSYS software, whose objective is to establish the optimal technical options to significantly improve the thermal performance of the target buildings compared to the current situation, considered as a reference. For this, the choice of reference buildings is as follows:

* For residential buildings:

- An economic group;
- A semi-standing collective;
- An individual villa of economic type [1].

* In the case of tertiary buildings:

- A hotel;
- A hospital;
- A school;
- An administrative building [1].

### 2.4.1 Impact on Heat Requirements in Heating and Air Conditioning

In the residential sector, the application of the thermal regulation should allow gains of 40% to 65% depending on the zones compared to the reference situation, as shown in Fig. 4 [1].

And in the case of the tertiary sector, the thermal simulations show that the gain in thermal requirements for heating and cooling varies according to the climatic zones compared to the reference situation. Figure 5 shows a gain that varies between 52% to 74% for administrative buildings [1].

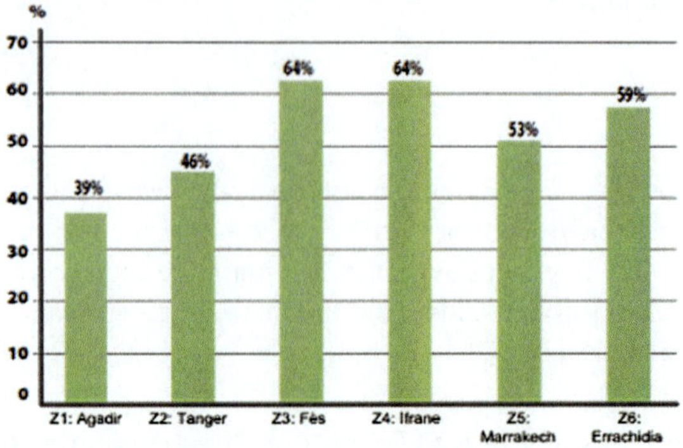


**Fig. 4.** Impact of thermal regulation on the reduction of heating and cooling needs of residential buildings (Ti = 26 °C in summer) in Morocco (% reduction)

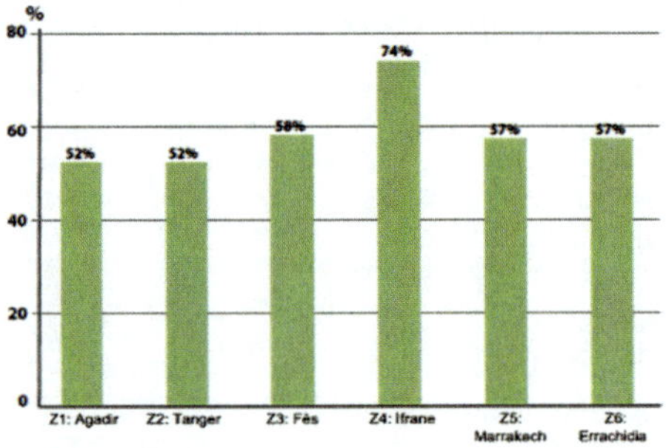


**Fig. 5.** Impact of thermal regulation on the reduction of the heating and cooling needs of administrative buildings in Morocco (% reduction)

Figure 6 shows a gain that varies between zones between 32% and 70% for schools [1]

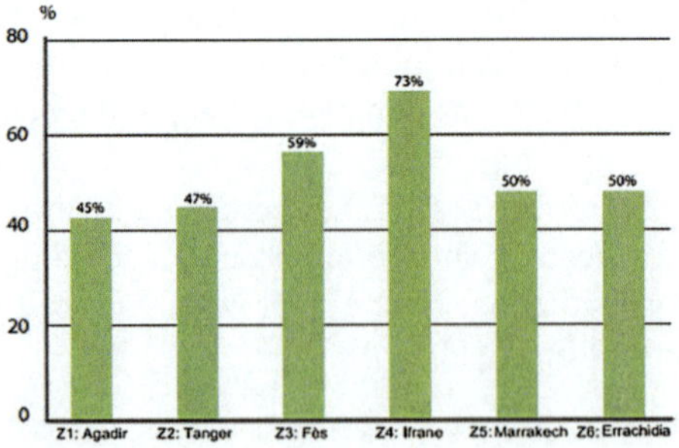


**Fig. 6.** Impact of thermal regulation on the reduction of heating and cooling needs of school buildings in Morocco (% reduction)

Figure 7 shows a gain that varies between 40% to 73% for hospital buildings

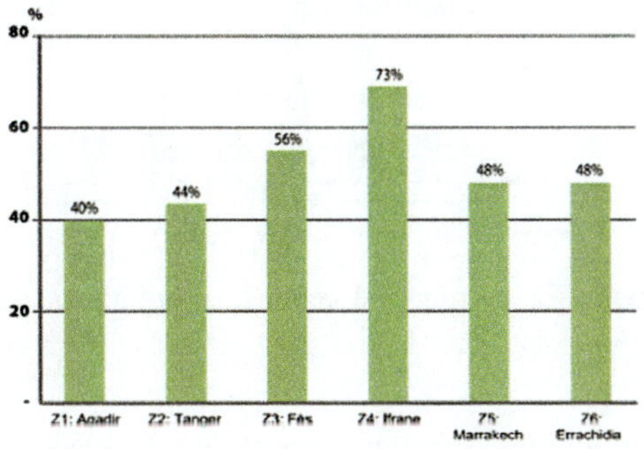


**Fig. 7.** Impact of thermal regulation on the reduction of heating and cooling requirements of hospital buildings in Morocco (% reduction)

Figure 8 shows a gain that varies between zones between 32% and 70% for hotels [1].

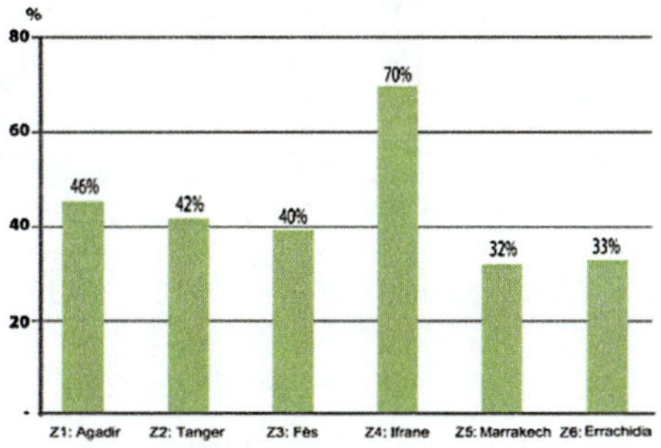


**Fig. 8.** Impact of thermal regulation on the reduction of heating and cooling requirements of hotel establishments in Morocco (% reduction)

Generally, the largest gains are observed in cold areas like Ifrane and Fez for both sectors.

### 2.4.2 Impact on Final Energy Consumption

The thermal regulation requirements allow for final energy savings for the consumer of around 22 kWh per year per $m^2$ of covered building. Figure 9 shows savings for the residential sector ranging from 8 kWh/$m^2$/year (zone Z1) to 75 kWh/$m^2$/year (zone Z4) [1].

In the case of the tertiary sector, and taking into account the heating and cooling modes as well as the efficiency of the equipment used, the final energy gain varies according to the sectors and climatic zones.

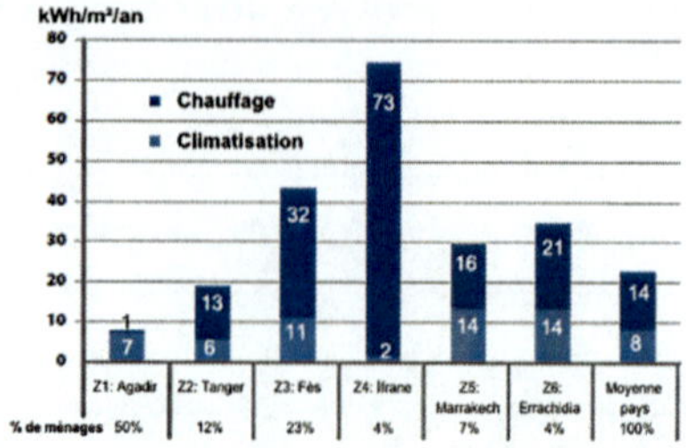


**Fig. 9.** Final energy saving for heating and cooling according to climate zones: Residential sector

– For administrative buildings, the gain varies from 31 kWh/m$^2$/year to 52 kWh/m$^2$/year (Fig. 7) [1];

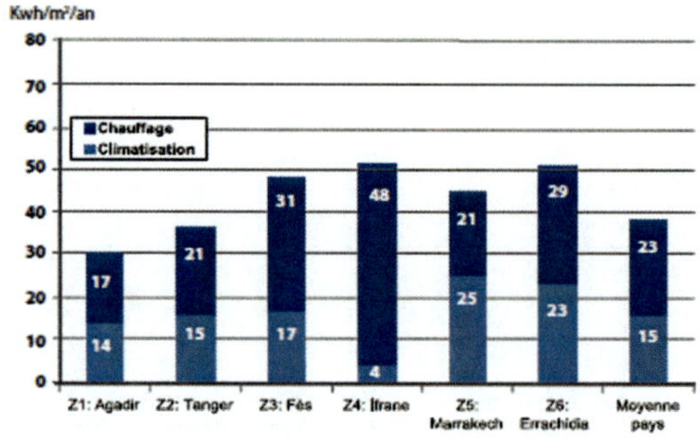


**Fig. 10.** Final energy saving for heating and cooling according to climatic zones: administrative buildings.

– For school buildings, the gain varies from 23 kWh/m$^2$/year to 202 kWh/m$^2$/year and an average of 48 kWh/m$^2$/year (Fig. 8) [1];

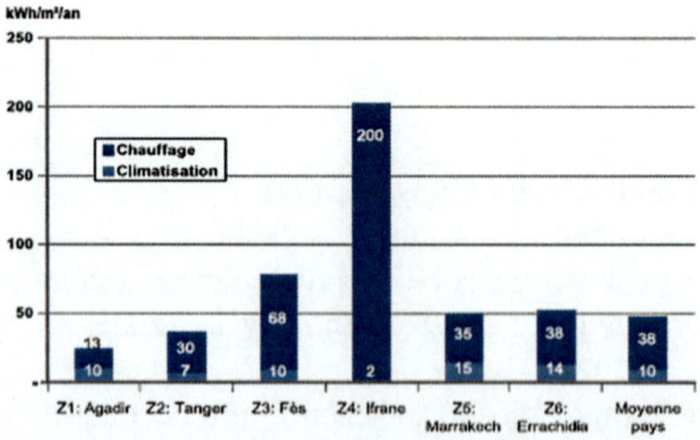


**Fig. 11.** Final energy saving for heating and cooling according to climatic zones: schools

– For hospital buildings, the gain ranges from 23 kWh/m$^2$/year to 115 kWh/m$^2$/year and an average of 39 kWh/m$^2$/year including 22 kWh for air conditioning (Fig. 9) [1];

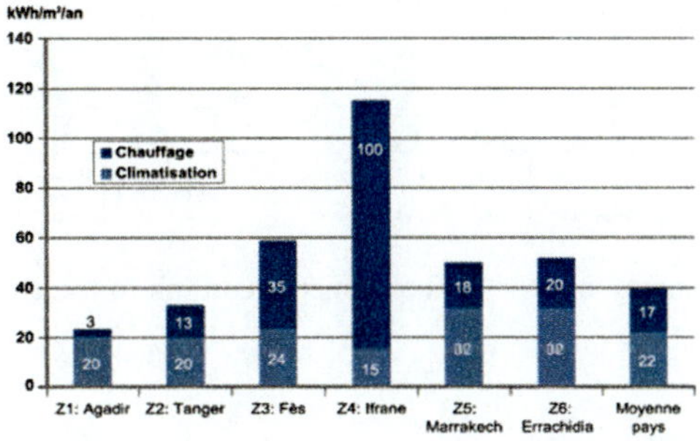


**Fig. 12.** Final energy saving for heating and cooling according to climatic zones: hospital building

– For hotels, the gain varies from 19 kWh/m$^2$/year to 80 kWh/m$^2$/year (Fig. 10) [1].

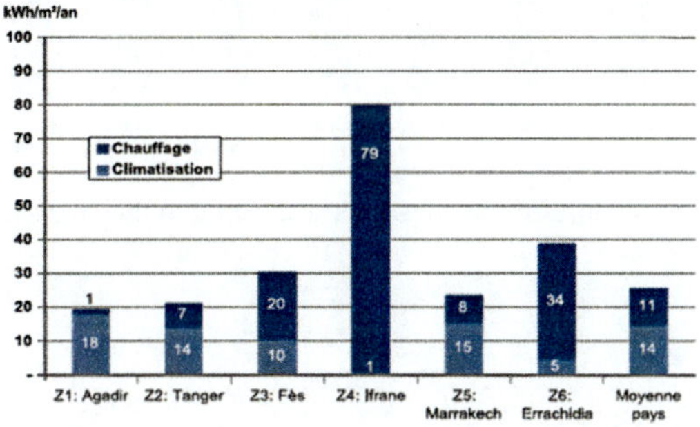


**Fig. 13.** Final energy saving for heating and cooling according to climate zones: hotel establishment

*PS: These savings were evaluated by taking into account the prevailing modes of heating and cooling in the different regions of Morocco and the average efficiency of the equipment used [RTCM].*

### 2.4.3 Impact on the Energy Bill of the Final Consumer

Given the current energy tariffs, these savings imply a gain for the final consumer in the energy bill for heating and cooling.

This gain is estimated at an average of 18 Dh/m$^2$/year and varies from 11 Dh/m$^2$/year in zone Z1, which represents more than 50% of dwellings, to 30 Dh/m$^2$/year in zone Z4 which does not represents only 4%. Figures 11, 12, 13, 14 and 15 show the different gains on energy bills, listed for each zone and sector [1] (Figs. 18 and 19).

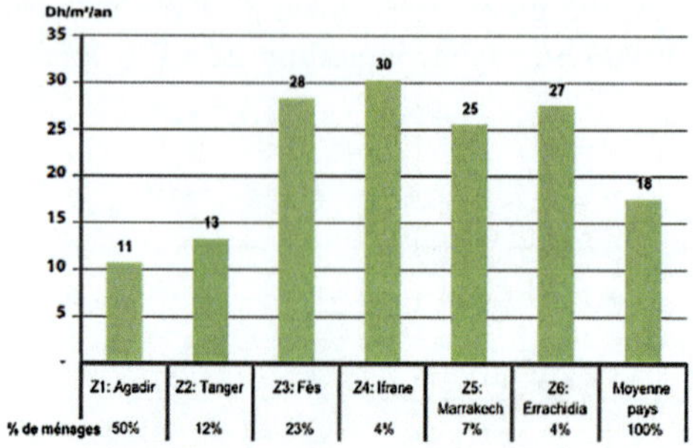


**Fig. 14.** Gains on the energy bill for the consumer according to the climate zones: Residential sector

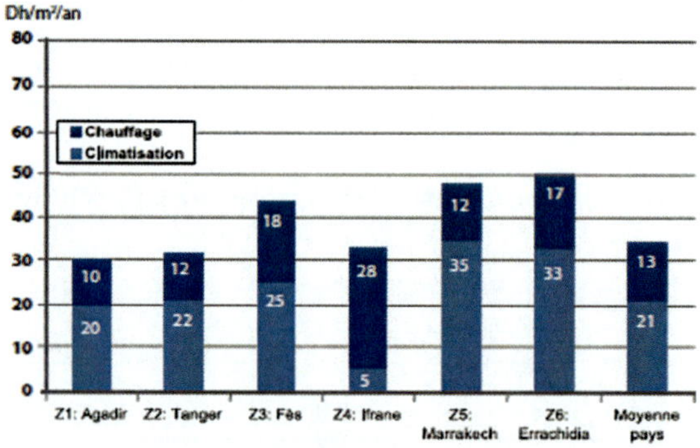


**Fig. 15.** Gain on the energy bill for the consumer according to the climatic zones: administrative buildings.

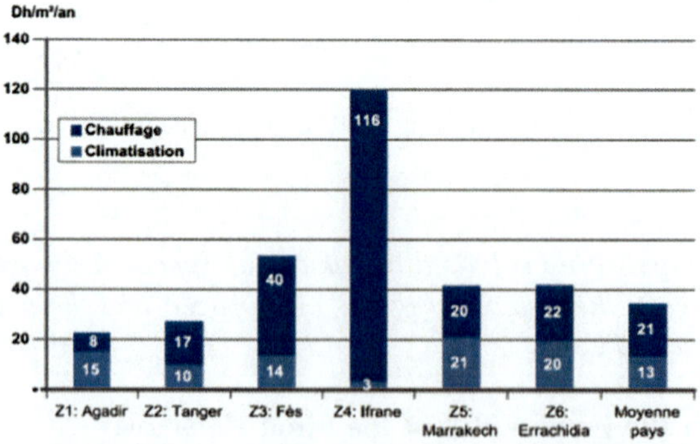


**Fig. 16.** Gain on the energy bill for the consumer according to the climatic zones: schools

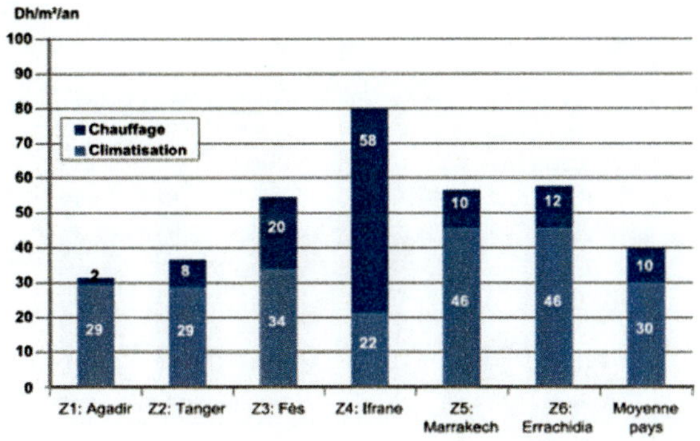


**Fig. 17.** Gain on the energy bill for the consumer according to the climatic zones: hospital building

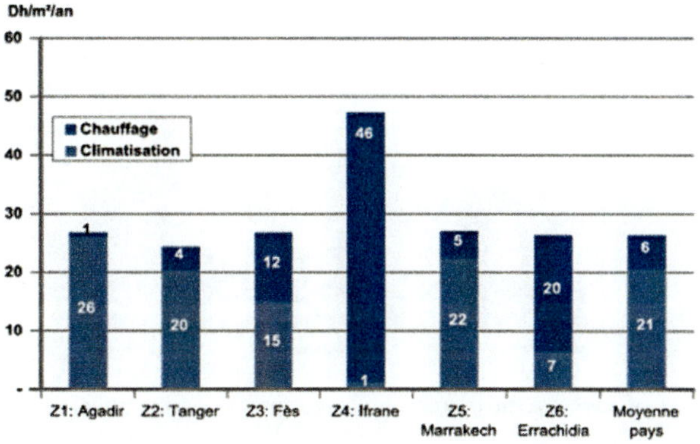


**Fig. 18.** Savings on the energy bill for the consumer according to the climate zones: hotel establishment

### 2.4.4 Incremental Cost of Compliance with the Thermal Regulation

Compliance with the technical specifications of the regulation implies an average investment cost increase of approximately 112 Dh/m$^2$, i.e. an average of 3.2% of the average construction cost.

This additional cost is higher or lower depending on the zone and the category of habitat, given the difference in the measures to be implemented. It thus varies from 43 Dh/m$^2$ in the Agadir area for luxury apartments to 315 Dh/m$^2$ for economic villas in the areas of Ifrane and Fez.

In relative terms, this extra cost represents a particularly high percentage of the construction cost for the category of economic housing, especially off-shore (Z1 and Z2).

Figures 16 and 17 show the simulated extra cost for the 3 categories of "Economic apartment, luxury apartment and economic villa" for the 6 climatic zones generated by the application of the RTCM in Dh/m$^2$ and in % [1, 4].

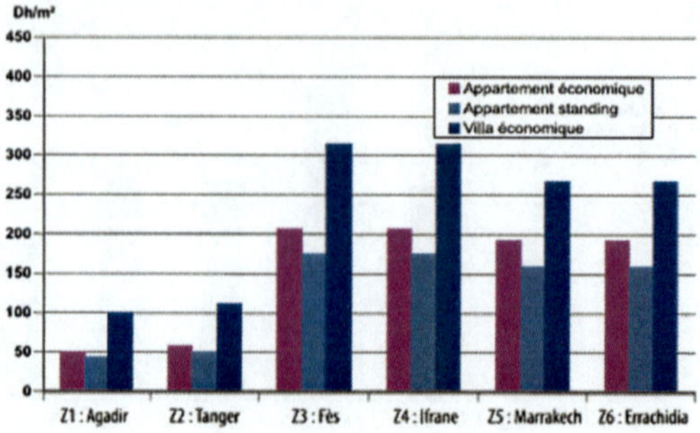


**Fig. 19.** Investment costs generated by the RTCM according to climate zones

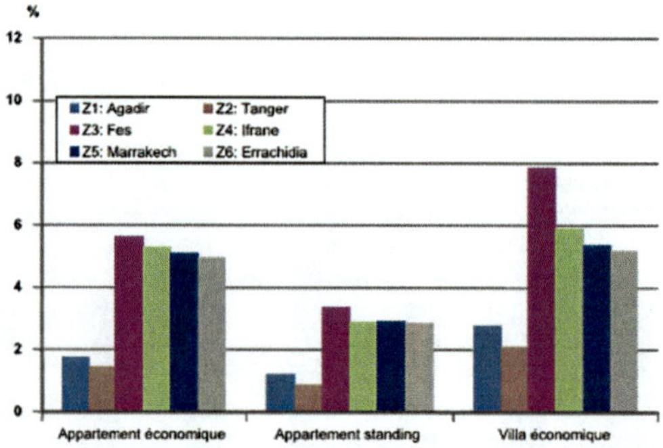


**Fig. 20.** Percentage of investment incremental due to RTCM by housing category and region

We note that zones 3 and 4 (Fes and Ifrane) generate the highest additional cost in the 3 housing categories [1] (Figs. 20 and 21).

**Table 2.** Surcharge for administrative buildings in %

| | | |
|---|---|---|
| Z1 | Agadir | 0,42 |
| Z2 | Tanger | 0,46 |
| Z3 | Fès | 2,72 |
| Z4 | Ifrane | 2,62 |
| Z5 | Marrakech | 2,48 |
| Z6 | Errachidia | 2,48 |

Similarly for the tertiary sector, the zones Z3 (Fes) and Z4 (Ifrane) in the 4 establishments under study of the tertiary sector generate the highest additional cost by the application of the RTCM as indicated in the Tables 2, 3, 4 and 5 [1].

**Table 3.** Incremental cost for schools in %

| | | |
|---|---|---|
| Z1 | Agadir | 1.93 |
| Z2 | Tanger | 2.05 |
| Z3 | Fès | 5.23 |
| Z4 | Ifrane | 5.23 |
| Z5 | Marrakech | 4.13 |
| Z6 | Errachidia | 4.13 |

**Table 4.** Surcharge for hospital establishments in %

| | | |
|---|---|---|
| Z1 | Agadir | 1.05 |
| Z2 | Tanger | 1.15 |
| Z3 | Fès | 2.65 |
| Z4 | Ifrane | 2.65 |
| Z5 | Marrakech | 2.05 |
| Z6 | Errachidia | 2.05 |

**Table 5.** Hotel surcharges in %

| | | |
|---|---|---|
| Z1 | Agadir | 0.45 |
| Z2 | Tanger | 0.52 |
| Z3 | Fès | 1.85 |
| Z4 | Ifrane | 1.85 |
| Z5 | Marrakech | 1.65 |
| Z6 | Errachidia | 1.65 |

### 2.5 Avoided Emissions of CO2 by the Application of the RTCM in the Constructions of Tomorrow

Taking into account the emission factors considered in Morocco for fuels and for the electricity grid, the cumulative avoided emissions over the next 20 years would be around 20 MCO2 for the residential sector.

And for the tertiary sector, averages of avoided CO2 emissions are summarized for each sector:

- 16 kgeCO2/m$^2$/year for administrative buildings;
- 6 kgeCO2/m$^2$/year for schools;
- 20 kgeCO2/m$^2$/year for hospital buildings;
- 13 kgeCO2/m$^2$/year for hotels [1].

## 3 Insulating Materials Adapted to the RTCM

The Construction Thermal Regulation in Morocco (RTCM) provides a pragmatic solution to the problem of excessive energy consumption in the building sector, be it residential or tertiary, and this mainly at the level of the construction plan using known insulating materials by a low thermal conductivity adapted to each climatic zone.

The integration of these insulating building materials will significantly reduce the need for air conditioning and heating. This reduction alone covers 50% of energy savings. Figure 17 is a study of the southern Mediterranean region over the period 2010–2030 which shows that we can gain 35% in heating and 14% in air conditioning [1] if we apply an effective envelope and a thermal insulation appropriate to using insulating materials guaranteeing low thermal inertia.

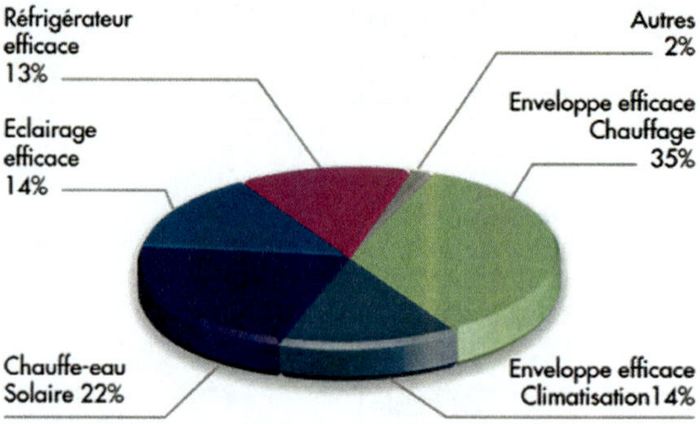


**Fig. 21.** Structure of energy efficiency potential in the southern Mediterranean region over the period 2010–2030 [1].

Good thermal insulation means a good insulating material introduced to the building envelope. Insulation materials are usually listed in four families. This classification is made according to their natures and origins [2]:

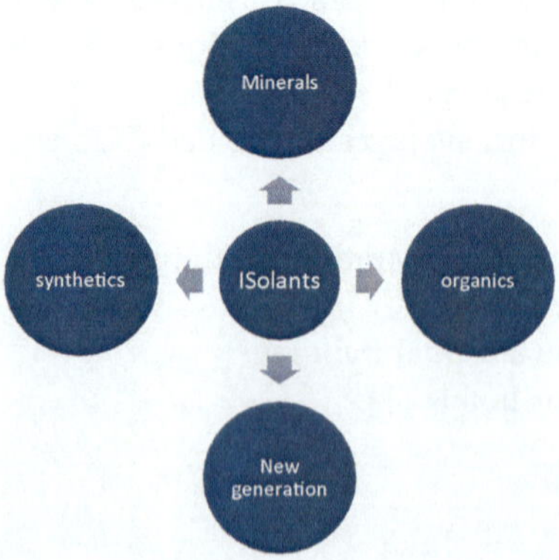

Examples of mineral insulators with their conductivity:

- Glass wool: 0.03 to 0.045
- Rockwool: 0.032 to 0.04
- Cellular glass: 0.035 to 0.048
- Perlite: 0.045 to 0.06
- Vermiculite: 0.045 to 0.08
- Expanded clay: 0.09 to 0.16
- Pozzolan: 0.1 to 0.2

Examples of organic insulators with their conductivity:

- Cork: 0.032 to 0.045
- Wood fiber: 0.037 to 0.049
- Cotton wool: 0.037 to 0.042
- Linen wool: 0.037 to 0.042
- Cellulose wadding: 0.037 to 0.044
- Hemp: 0.04 to 0.046
- The coconut fiber: 0.037 to 0.050
- The chènevotte: 0.048 to 0.06
- Sheep's wool: 0.035 to 0.042
- Duck feather: 0.033 to 0.042

Examples of synthetic insulators with their conductivity:

- Expanded polystyrene: 0.029 to 0.035
- Extruded polystyrene: 0.028 to 0.037
- Polyurethane: 0.022 to 0.030
- Phenolic foam: 0.018 to 0.035

Examples of new generation insulators with their conductivity:

- Airgel foam: 0.011 to 0.013
- Insulating panels under: 0.0042 to 0.0050
- Reflective insulation

Beyond the insulation materials whose main role in a construction is to improve the thermal performance of the walls, it was recently developed materials originally used to ensure the structure of the building, with improved thermal performance. This allows these building materials to also provide an insulation function. For example:

- Hemp brick: 0.048 to 0.06
- Cellular concrete: 0.035
- Monomour brick: 0.12 to 0.18 up to 0.07 if the cells are filled with an insulator;
- Agglo monomur in pumice stone: 0.09 to 0.12
- Foam concrete: 0.07 to 0.20

In connection with this type of materials, several studies have been developed to study the possibility of substitution or incorporation of the results of this research into conventional building materials in order to have ecological and energy-efficient buildings.

In this context, the article by Rahim and all 2017 [5] presented the results of the testing of a new concrete based on hemp and rapeseed, these materials have shown some response to the requirements of the RTCM given their low thermal conductivity of about 0.12 and 0.09 W/mK for hemp and rapeseed respectively.

And in Bourhaneddine Haba's thesis, 2017 [6], the author investigated the effect of adding date palm wood in concrete, the results showed that the dry thermal conductivity of date palm concrete is lower. at 0.3 $Wm^{-1}K^{-1}$. However, despite the fact that this type of concrete has a low thermal conductivity, the hygrometry problem "rapid absorption of water" blocks their expanded use until today, since this phenomenon influences the mechanical strength of the final product.

On the other hand, the article by "Cinzia Buratti, Elisa Moretti, Elisa Belloni, Fabrizio Agosti" [7] studied the use of innovative insulation panels based on natural basalt fibers.

After measuring the thermal conductivity of this type of panels, it is included between 0.030–0.034 W/mK. Compared with other traditional solutions such as rock wool or glass wool, this mineral material will be a good insulation solution that meets both the sanitary and mechanical requirements of the required building products.

As well, the authors "F. Asdrubali, AL Pisello, F. D'Alessandro, F. Bianchi, M. Cornicchia, C. Fabiani" [8] developed in their article tests on the thermal and acoustic performances of the panels of corrugated board commonly used in the packaging industry. The results show that this type of cardboard panels has promising performances in terms of acoustic insulation capacity and thermal insulation performance, slightly lower than the commonly used insulation panels. The results of the tests showed that 50 and 75 mm matching C flute samples had a thermal conductivity of $0.053 \pm 0.001$ W/mK and concordant E flute samples had slightly different $\lambda$ values, both within their limits. maximum permissible error, namely $0.058 \pm 0.001$ and $0.060 \pm 0.001$ W/mK, respectively for 50 and 75 mm.

## 4 Conclusion

The simulations applied previously on buildings applying the requirements of the RTMC have shown that this type of construction will allocate an additional cost that will add to the cost of the average investment, this extra cost will mostly be amortized on energy gains related to air conditioning and heating.

And by comparing this extra cost in the 3 housing categories and in the 6 zones, we find that this extra cost is the lowest in the category of standing buildings. From where it leads to a reflection to apply this regulation in a practical way to this category of housing because it will not increase the cost of the construction in a crucial way and what will lead to have by the savings in energy bills and reimbursement of additional expenses. Because if the household heats up and/or air-conditioning these gains will result in a saving on the final energy consumption, and if the household does not heat up or air-conditioning, the application of the thermal regulation will result in an improvement of the comfort thermal.

Through this reflection, we will participate in the construction of sustainable energy-efficient buildings and support the country in the development of the green economy. However, the level of application of this regulation is still poor since the players in the field of construction are not yet trained in the application of the RTCM, as well, the information to its application remains little disseminated to convince its interest to users.

Hence the need for sensible action in this direction will enable the expected energy savings in the building sector to be achieved quickly. Among the proposed actions for an effective implementation of the RTCM:

- Establishment of a strategic plan and adequate communication tools to mobilize and sensitize stakeholders, including governments, businesses, professionals and the general public, to energy efficiency measures in buildings;
- Support and technical assistance to professionals and administrations responsible for monitoring the application of thermal performance requirements, in order to strengthen their capacities in this area;
- Encourage architects, engineers and contractors to use high-performance thermal design principles of the building envelope;
- Make available to project owners, public decision-makers and funders a work tool integrating the requirements of the RTCM to improve their competence and productivity in this area.

The next perspective considered for the advancement of our research is to launch a survey among architects, engineers, real estate developers, design offices, project managers and all those involved in the construction field in order to have their feedback on the technical applicability of the RTCM in the buildings of tomorrow and also to know the causes of the hardness in its application in sustainable building.

## References

1. Règlement Thermique de Construction au Maroc- Version simplifié par l'ADEREE
2. CETEMCO-Guide technique AZEL, Décembre 2018
3. Bouroubat, K.: La construction durable: étude juridique comparative/Maroc-France. Université Paris-Saclay, Droit (2016)
4. Allouhi, A., El Fouih, Y., Kousksou, T., Jamil, A., Zeraouli, Y., Mourad, Y.: . Energy consumption and efficiency in buildings: current status and future trends. J. Clean. Prod. **109**, 118–130 (2015)
5. Rahim, M., Douzane, O., Tran Le, A.D., Promis, G., Langlet, T.: Experimental investigation of hygrothermal behavior of two bio-based building envelopes. Energy Builds. **139**, 608–615 (2017)
6. Bourhaneddine,, H.A.B.A.: Etude in-situ des performances énergétiques et mécaniques des matériaux biosourcés et locaux. Université El-Hadj Lakhdar - BATNA 1 (2017)
7. Buratti, C., Moretti, E., Belloni, E., Agosti, F.: Thermal and acoustic performance evaluation of new basalt fiber insulation panels for buildings. Energy Procedia **78**, 303–308 (2015)
8. Asdrubali, F., Pisello, A.L., D'Alessandro, F., Bianchi, F., Cornicchia, M., Fabiani, C.: Innovative cardboard based panels with recycled materials from the packaging industry: thermal and acoustic performance analysis. Energy Procedia **78**, 321–326 (2015)

# Supply Chain 4.0 Risk Management: Bibliometric Analysis and a Proposed Framework

Kamar Zekhnini(✉), Anass Cherrafi, Imane Bouhaddou, and Youssef Benghabrit

LM2I Laboratory, ENSAM, Moulay Ismail University, 50500 Meknes, Morocco
kamar.zekhnini@gmail.com, a.cherrafi@ensam.umi.ac.ma, b_imane@yahoo.fr, you_benghabrit@yahoo.fr

**Abstract.** Currently, the concept of Internet of Things, as well as Industry 4.0 and Smart Manufacturing, are attracting considerable attention. Digital technologies and the industry's rapid progress have stimulated enormous advances in the development of manufacturing technology. In the era of digitalization, the role of new technologies in enhancing the competitiveness of industries has become more crucial. So these novel technologies have created a new potential of supply chain risks. It mainly includes information and cybersecurity risks. The present paper aims to present a literature review through a bibliometric analysis of related works about risk management in supply chain 4.0. This study discusses the supply chain 4.0 risks. It also provides a framework about Risk management in supply chain 4.0. The suggested work is helpful for academics as well as professionals as it emphasizes the importance of implementing new methodologies of risk management in the age of intense competition. Furthermore, it presents a roadmap to confront new risks facing new interconnected and dynamic supply chains.

**Keywords:** Supply chain management 4.0 · Supply chain 4.0 risks · Risk management · Supply chain 4.0 risks framework

## 1 Introduction

Supply chain digitization is one of today's major business value and growth possibilities, with the ability to improve the performance and make future prospects. In fact, digital technologies have changed deeply how the information is shared. Consequently, they altered the way supply chains (SCs) exchange information. These novel technologies affected every industry [1].

There are many sources of risks in SCs. It includes supply interruptions, volatile exchange rate, dynamic markets, work incidents, political instability, uncertain demand, natural disasters and terrorism [2]. In addition to these risks sources, many others sources have been appeared with the digitalization of SCs.


A. Saka et al. (Eds.): CPI 2019, LNME, pp. 322–332, 2021.
https://doi.org/10.1007/978-3-030-62199-5_29

Many scientists have concentrated on supply chain risk management (SCRM) over the years by contributing in the fields of risk definition, operationalization and mitigation [3]. Until now, little importance has been given to risk management in SC 4.0 [4, 5]. Hence, this study aims to fill the gap of the literature by a proposed framework to implement risk management in digital SCs. It also aims to present a literature review about SCRM in the digital era by a bibliometric analysis. It discusses the new type of risks born with the digitalization of the supply chains. Moreover, this paper intends to provide readers with another perspective of supply chain 4.0 risk management.

This article is organized as follows: in the following section, we describe the review methodology used for the study. Section 3 illustrates the bibliometric analysis of literature review related to SCM 4.0 risks. And finally, Sect. 4 discusses SC 4.0 risks and presents the proposed framework for risk management in SC4.0.

## 2 Methodology

The objective of this research is to define and review risk management in SC 4.0. For any research study, a literature review seems to be crucial. It enables writers to assess and analyze the important literature, define the field's conceptual content and improve the theory. The study outlined in this article was carried out using various types of writing as key sources of data, namely, scientific papers, articles and business reports from companies. The following databases were used to search for appropriate journals in order to address a wide variety of academic and conference productions:

- Springer (https://www.springer.com/gp);
- IEEE (https://ieeexplore.ieee.org/);
- Emerald (http://www.emeraldinsight.com);
- Elsevier (www.sciencedirect.com);
- Taylor & Francis (http://www.taylorandfrancis.com);
- And the Google Scholar.

The keywords used by the authors dropped into the following: Risk management & SCM 4.0, Risk management & Digital SCM, Risk management & Smart SCM, Cybersecurity & SCM and Information Security & SCM. In this article, the complete amount of papers assessed is 31.

## 3 Bibliometric Analysis

For the bibliometric analysis, we used the free software Vosviewer. It aims to create, visualize and explore maps based on network data.

### 3.1 Co-authorship Analysis

Figure 1 presents the relatedness of determined items based on their number of co-authored documents. The largest set of connected authors consisted of 7 authors from 76. The weight of each author determined the size of its label and its circle. We can observe that circles and labels have the same size. In addition, the thickness of lines and distance between two nodes are the same. This similarity means that those authors have the same weight. It's clear that there is a weak link between authors because of the weak thickness of lines between authors. So, it can be concluded that authors have weak collaboration between each other. This weak relationship between authors working on supply chain 4.0 risk management is due to the subject's novelty. Also, it can explain the fact of the lack of authors' continuity in the research field.

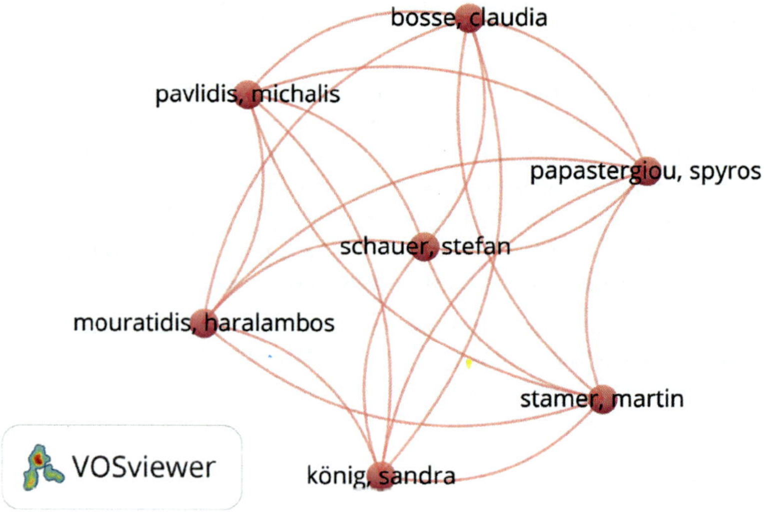


**Fig. 1.** Co-authorship network

### 3.2 Keywords Occurrence

Applying 2 as the minimum number of occurrence of SC 4.0 risk management papers' keywords; it is found that of the 89 keywords 9 meet the threshold. Figure 2 illustrates a network of 3 clusters. The biggest cluster of the author keywords network is the red one. It includes 'Supply chain risk management', 'supply chain management', 'risk management' and 'data analytics'. This cluster is connected to the green cluster gathering the digitalization concept.

It can be concluded that the green cluster has a stronger link with the blue one, in comparison with the orange cluster including "resilience" and "supply chain". This is because the line between the green and the blue cluster is thicker than the line between the green and the orange cluster. Similarly, the distance between the green cluster and the blue one is closer than the distance between the green and orange cluster.

From this frames network, it can be deducted that data or information and the digitalization concepts are related to risk management in the SCs. However, there is a weak redundancy of these words which means that there is a weak presence of works discussing the risk management in SC 4.0. So, it is highly recommended to discuss the risk management in the era of digitalization.

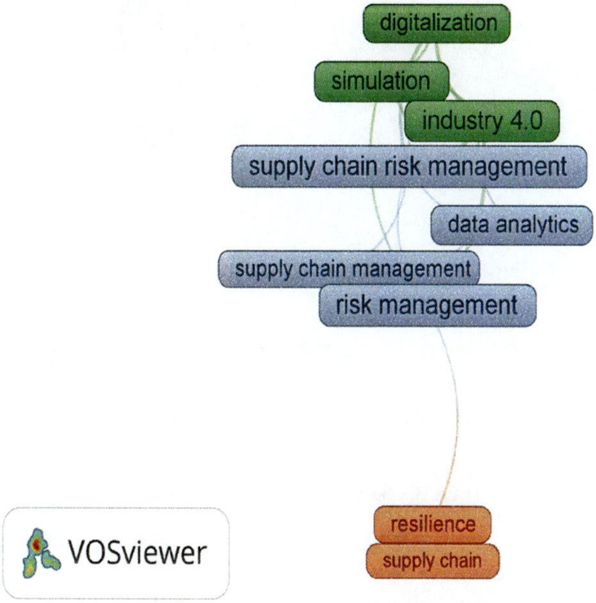


**Fig. 2.** Keywords co-occurrence

### 3.3 Title Occurrence

For the title occurrence analysis, 2 was set as the minimum number of occurrence of the title's common words. Of the 79 expressions, 7 fulfill the criterion. The largest connected title expressions are 6 items. (Figure 3) illustrates the title most common words in 2 clusters.

We can observe the existence of two clusters. The red cluster includes "risk management", "industry", "supply chain" and "framework". While the green cluster includes "supply chain risk management" and "risk matrix". It can be seen that the red cluster has a weak link with the green one. This is because the line between the green and the red cluster is very thin. Also, the distance between the two clusters is important.

It is clear that the words describing the digital transformation are absent. All authors focus on Supply chain risk management far away from highlighting on the novel risks generated from the emergence of novel technologies.

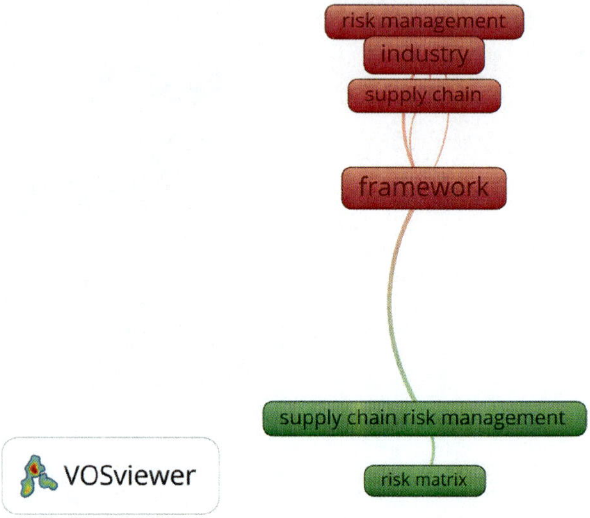


**Fig. 3.** Title co-occurrence network

### 3.4 Abstract Occurrence

The abstract co-occurrence represents the relatedness of abstract words on the basis of the amount of papers they contain. Figure 4 presents the abstract occurrence of analyzed papers. The minimum number of occurrences of a term is 6. Of the 806 terms, 21 meet the criterion. 15 expressions have been chosen. There are three clusters describing the connection between words. So in the abstract we found the presence of 'technology' and 'information risk'. Thus, authors are linking the relation of novel technology with risk management. Even so, this presence is not strong enough for such a complicated subject. To sum up from the four bibliometric analysis, we can conclude that there is a lack of discussing the novel era impact on SCs and the way to manage the new born risks that especially focus on information security.

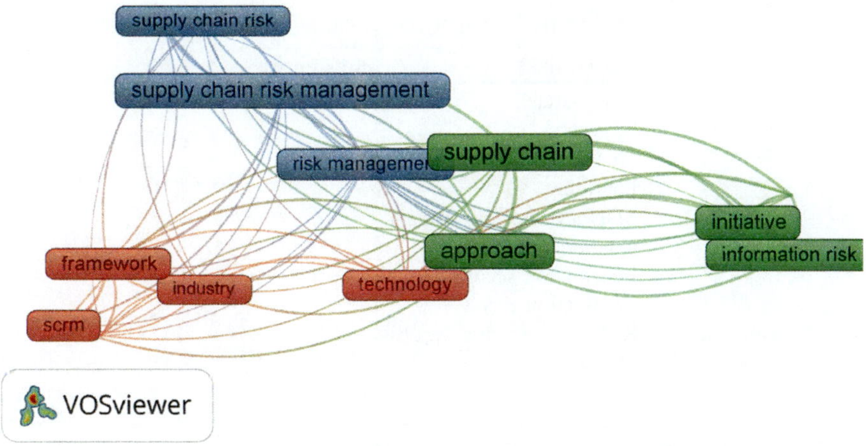


**Fig. 4.** Abstract occurrence network

# 4 Discussion

## 4.1 Risks in SC 4.0

Digital Supply Chain (DSC) is a savvy, productive procedure to create new types of income and business esteem for associations [6]. It is the new interconnected business framework which reaches out from segregated, neighborhood, and single-organization applications to store network wide methodical autonomous executions.

Digitalized SC enables firms to monitor material flows in real time [7]. However, compared to other intra-organizational structures, more external factors influence the implementation of digital supply chain systems [8]. The consequences of these external factors impact have typically caused productivity losses, increased costs, and lost revenue [9].

In every SC process, radical innovations, fresh entrants and external variables are about to interrupt the current system [10]. Due to the rise in vulnerability and threats, the notion of Industry 4.0 produces fresh risk categories in the SCs [11]. Thus, the novel technologies have different security problems and, when incorporated into the SCs, they may pose numerous difficulties in keeping safety requirements such as confidentiality, integrity and availability [12].

There are many risks that could threat firms such as IT failures, piracy, theft, and cyberattacks. These risks are among the leading causes of reputation crises and reputation value losses. They can cause interruptions and can harm the SC performance [13]. Table 1 presents identified risks from the studied papers.

**Table 1.** Supply chain 4.0 risks

| SC 4.0 risks | References |
|---|---|
| Cyber-attack<br>Spyware<br>Phishing<br>Pharming<br>Malware and loss of data<br>Confidentiality risk<br>Information security risks<br>Transaction risk<br>Risks to computer-security | [1, 8, 12–17] |

### 4.2 Framework for Risk Management in SC 4.0

Due to heavy automation using Cyber-Physical System, end-to-end digitization, distributed and well-connected components using IoT, there will be a challenge to securing SC systems [12]. Consequently, firms need risk management system to face threats. Hence, risk management is one of the nine expertise fields that the Project Management Institute (PMI) has transmitted and is likely the most challenging element of project management [11]. Risk management is a systematic process that helps organizations know what the risk is, who is at danger and what the adequate controls for such risks [11].

Supply chain risk management stages are complicated. They involve a high amount of technical skills [18]. The proposed framework is inspired from the ISO 31000 V2018 standard. It is about four phases aiming to ensure an efficient risk management system for SC 4.0 threats. (Figure 5) represents SCRM 4.0 including four stages namely: Identifying SC 4.0 risks, analyzing SC 4.0 risks, implementing SC 4.0 risk management and finally reviewing the SC 4.0 risk management implemented system.

The framework illustrates the SC 4.0 risk management. The main component of the framework is the four processes for the risk management.

The first step is to identify SC 4.0 risks. It is about producing an extensive list of risks based on incidences that could discourage, degrade or delay reaching objectives [11]. This phase is very sensitive because it needs a certain amount of understanding of the SC network structure and the network's physical and data flows [2]. Hence, it is crucial to identify and map risks, including vulnerabilities and cybersecurity risks in the SC 4.0. Moreover, these risks should be prioritized and classified by severity impact on SCs. In this phase, different approaches for classifying risks can be used.

The second step is about the analysis of founded SC 4.0 risks. In this stage, the previously identified supply chain risks will be deeply analyzed through various methods like Cause-effect analysis to determine the direct impact of each risk on digital SC. In addition, the identified risks have to be studied in term of SC costs/performance by using KPIs [19]. By analyzing SC 4.0 risks, it is possible to assess the potentials of using novel technology for managing SC 4.0 risks [19, 20]. This phase will result in providing policies, procedures, processes and practices for SC 4.0 risk management.

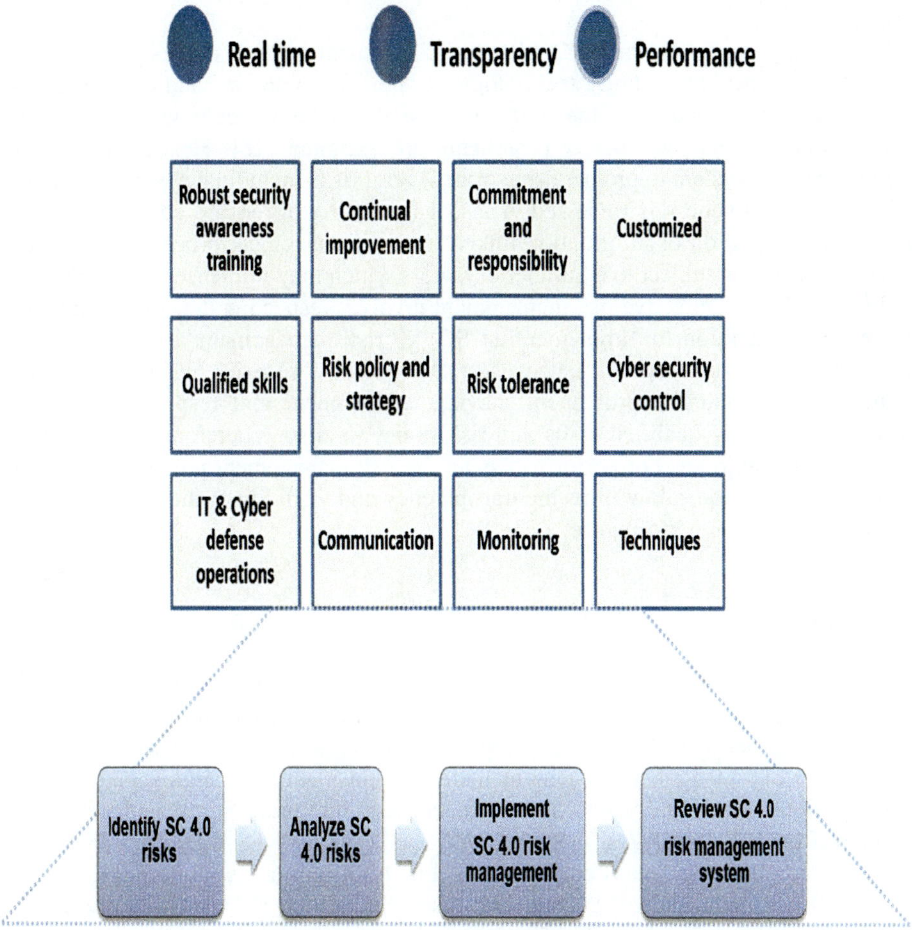


**Fig. 5.** SC 4.0 risk management framework

The third step is to implement the adequate SC 4.0 risk management solutions. In fact, efficient management of SC data and real time risk monitoring is needed for effective SCs [14, 18, 21–23]. A solid IT and cyber risk management program can protect the supply chain's strategic objectives, prevent company disruptions and maintain the reputation of all stakeholders [13]. To improve security of SCs a SC proactive risk management based on risk-related information transparency is needed. In other words, IT and cyber risk management would have the strategic advantages of improving the ability of companies to ensure a tolerable level of general risk for all stakeholders and continuously do so within their true risk appetite [13, 14, 19]. In addition, it is wide important to manage unplanned SC 4.0 incidences. Therefore, managers should monitor whether and how appropriate assets and interactions with upstream and downstream supply chain partners may be threatened by IT and cyber risks [17].

The final step is to review SC 4.0 risk management solutions. This phase aims to evaluate implemented methods, technologies or policies to prevent and manage SC 4.0 risks. The SC performance of the company depends on the degree to which the SCRM plan is implemented and the commitments are assigned [18]. Hence, after implementing SCRM system to protect firms from several risks including cyber security, risk mitigation effectiveness is reviewed. Thus, it is required to ensure continuous monitoring and evaluation of information linked to risk management procedures in SCs.

Each system is subject to a number of factors which may influence its advancement and results. Hence, effective risk management involves respecting the risk management processes. In addition, by implementing SC 4.0 risk management, it is important to address many keys of success such as ensuring robust security awareness training, guarantying continual improvement, having commitment and responsibility, being customized, having qualified skills and risk policy, and so. Therefore, to promote an organization's high level of performance, all the risk management processes should be executed in real time, allow firms the transparency and visibility of the whole activities and ensure the SCs performance.

## 5 Conclusion

Supply chain risk management has shown excellent interest in manufacturing studies and this study is an effort to outline the current risk management for the digital supply chains. Bibliometric and network analysis instruments have been used to evaluate quantitative SC 4.0 Risk Management literature and identify emerging patterns. The identification of key problems, study trajectories and future study paths could be acquired from these analyzes.

Our study discusses many major points. Firstly, it presents a bibliometric analysis of SC 4.0 risk management. Secondly, it discusses SC 4.0 risks such as information risks and cybersecurity risks. Thirdly, it proposed a conceptual framework for identifying and implementing SC 4.0 risk management. This framework can be used in further work as a roadmap.

As a practical contribution, this study proposes a conceptual framework that will help guiding managers and research beginners to identify and implement SC 4.0 risk management. The purpose of the framework is to give a consistent set of standards for managers to satisfy the need of highly protected supply chains. The presented conceptual framework can be adapted to suit firm's requirements. In addition, the SC 4.0 risk management framework provides a starting point and a prevalent vocabulary that can be editable to fit manager's objectives in term of facing supply chain 4.0 risks.

This study also presents some limitations due to the small number of papers used for the analysis of the subject. Our results remain to be discussed and reinforced in future studies. In addition, due to the various new risks that could arise in the supply chain 4.0, it is desirable to review the standard steps in the proposed framework so far and consider introducing additional actions. Also, in the future, additional mitigation measures need to be further developed. Therefore, we will apply the SC 4.0 risk management framework in an industrial real-case implementation such as an automotive digital SC.

## References


1. Ivanov, D., Dolgui, A., Sokolov, B.: The impact of digital technology and Industry 4.0 on the ripple effect and supply chain risk analytics. Int. J. Prod. Res. **57**, 829–846 (2019). https://doi.org/10.1080/00207543.2018.1488086
2. Kara, M.E., Fırat, S.Ü.O., Ghadge, A.: A data mining-based framework for supply chain risk management. Comput. Ind. Eng. 105570 (2018). https://doi.org/10.1016/j.cie.2018.12.017
3. Ho, W., Zheng, T., Yildiz, H., Talluri, S.: Supply chain risk management: a literature review. Int. J. Prod. Res. **53**, 5031–5069 (2015). https://doi.org/10.1080/00207543.2015.1030467
4. Gottlieb, S., Ivanov, D., Das, A.: Case studies of the digital technology impacts on supply chain disruption risk management. In: Schröder, M., Wegner, K. (eds.) Logistik im Wandel der Zeit – Von der Produktionssteuerung zu vernetzten Supply Chains. Springer Fachmedien Wiesbaden, Wiesbaden, pp 23–52 (2019)
5. Radanliev, P., Treall, L., Santos, O., Montalvo, R.M.: Cyber risk from IoT technologies in the supply chain-discussion on supply chains decision support system for the digital economy (2019). https://doi.org/10.13140/rg.2.2.17286.22080
6. Büyüközkan, G., Göçer, F.: Digital supply chain: literature review and a proposed framework for future research. Comput. Ind. **97**, 157–177 (2018). https://doi.org/10.1016/j.compind.2018.02.010
7. Schlüter, F.: Procedure model for supply chain digitalization scenarios for a data-driven supply chain risk management. In: Zsidisin, G.A., Henke, M. (eds.) Revisiting Supply Chain Risk, pp. 137–154. Springer International Publishing, Cham (2019)
8. Xue, L., Zhang, C., Ling, H., Zhao, X.: Risk mitigation in supply chain digitization: system modularity and information technology governance. J. Manag. Inf. Syst. **30**, 325–352 (2013). https://doi.org/10.2753/MIS0742-1222300110
9. Wildgoose, N.: Supply chain risk management. In: Enterprise Risk Management, pp. 75–87. Elsevier, Amsterdam (2016)
10. Bamberger, V., Nansé, F., Schreiber, B., Zintel, M.: Logistics 4.0 – Facing digitalization-driven disruption. 14 (2017)
11. Tupa, J., Simota, J., Steiner, F.: Aspects of risk management implementation for Industry 4.0. Procedia Manuf. **11**, 1223–1230 (2017). https://doi.org/10.1016/j.promfg.2017.07.248
12. Chhetri, S.R., Rashid, N., Faezi, S., Faruque, M.A.A.: Security trends and advances in manufacturing systems in the era of Industry 4.0. In: 2017 IEEE/ACM International Conference on Computer-Aided Design (ICCAD), pp. 1039–1046. IEEE, Irvine (2017)
13. Gaudenzi, B., Siciliano, G.: Managing IT and cyber risks in supply chains. In: Khojasteh, Y. (ed.) Supply Chain Risk Management, pp. 85–96. Springer Singapore, Singapore (2018)
14. Colicchia, C., Creazza, A., Menachof, D.A.: Managing cyber and information risks in supply chains: insights from an exploratory analysis. Supp. Chain Manag. **24**, 215–240 (2019). https://doi.org/10.1108/SCM-09-2017-0289
15. Jaradat, O., Sljivo, I., Habli, I., Hawkins, R.: Challenges of safety assurance for Industry 4.0. In: 2017 13th European Dependable Computing Conference (EDCC), pp. 103–106. IEEE, Geneva (2017)
16. Leonhardt, F., Wiedemann, A.: Realigning risk management in the light of Industry 4.0. SSRN J. (2015) https://doi.org/10.2139/ssrn.2678947

17. Niesen, T., Houy, C., Fettke, P., Loos, P.: Towards an integrative big data analysis framework for data-driven risk management in Industry 4.0. In: 2016 49th Hawaii International Conference on System Sciences (HICSS), pp. 5065–5074. IEEE, Koloa (2016)
18. Hamburg International Conference of Logistics: Digitalization in Supply Chain Management and Logistics: Smart and Digital Solutions for an Industry 4.0 Environment, 1st edn. epubli GmbH, Berlin (2017)
19. Schlüter, F., Sprenger, P.: Migration framework for decentralized and proactive risk identification in a steel supply chain via Industry 4.0 technologies (2016). https://doi.org/10.13140/rg.2.1.2332.8883
20. Schlüter, F., Henke, M.: Smart Supply Chain Risk Management - A Conceptual Framework. 22
21. Baryannis, G., Dani, S., Validi, S., Antoniou, G.: Decision support systems and artificial intelligence in supply chain risk management. In: Zsidisin, G.A., Henke, M. (eds.) Revisiting Supply Chain Risk, pp. 53–71. Springer International Publishing, Cham (2019)
22. Harland, C.M.: Risk in complex supply chains, networks and systems. In: Zsidisin, G.A., Henke, M. (eds.) Revisiting Supply Chain Risk, pp. 439–455. Springer International Publishing, Cham (2019)
23. Kayis, B., Dana Karningsih, P.: SCRIS: a knowledge-based system tool for assisting manufacturing organizations in identifying supply chain risks. J. Manuf. Technol. Manag. **23**, 834–852 (2012). https://doi.org/10.1108/17410381211267682

# Supply Chain Network Design Under Different Paradigms: Literature Review and Future Research Areas

Chaimaa Arfach[1,2(✉)], Said Elfezazi[1,2], and Anass Cherrafi[3,4]

[1] Cadi Ayyad University, Marrakech, Morocco
arfach.chaimaa@gmail.com
[2] Laboratoire LAPSII, EST de Safi, Safi, Morocco
[3] Moulay Ismail University, Meknes, Morocco
[4] Laboratoire LAMMI, ENSAM de Meknes, Meknes, Morocco

**Abstract.** This paper aims to investigate the relationship and links between green, lean and agile supply chain (GLASC) practices for designing a robust supply chain network with different objectives and dimensions (social, environmental, economic and agile). Thus, understanding what is the best way for companies to design a system of practices that meets, on the one hand, The social and envirmental considerations (green) and, on the other hand, the need to be Lean and agile is critical. Our paper develops a literature review addressing the integration of Green, lean and agile supply chain management paradigms. For that, we develop a framework of GLASC which contain as elements practices of each paradigm, integration category and finally the objectives reached by gathering lean, green and agile SC. Finally, we discuss the interaction effects between green, lean and agile supply chain practices.

Our result is based on the analysis of 33 papers which are combining at least two paradigms (green, Lean and agile SC), it will be a roadmap for researchers to identify challenges, and also opportunities for future studies.

Both academicians and practitioners will find our research useful because it outlines the major lines of research in the field of supply chain management.

**Keywords:** Supply chain network design · Lean supply chain · Green supply chain · Agile/responsive supply chain · Supply chain management · Flexibility · Integrated model

## 1 Introduction

Over the last two decades, a large number of relevant publications have highlighted the importance of Supply Chain Network Design (SCND) (Govindan 2017). The optimization of a Supply Chain has caught more attention recently; the objective functions are different in various studies. Traditionally, the objective function in most problems is primarily monetary (for example, minimizing total cost or maximizing overall profit). However, new paradigms have recently emerged in the field of the supply chain (Farrahani 2013). Moreover, Established supply chain management paradigms such as leanness, agility, and sustainability have received increased attention in the literature,


A. Saka et al. (Eds.): CPI 2019, LNME, pp. 333–349, 2021.
https://doi.org/10.1007/978-3-030-62199-5_30

but mainly as separate topics (Ciccullo et al. 2018). Thus, it is necessary to study the effects of "Greenness", "sustainability" and "responsiveness" on the supply chain (Farahani 2013). Most researchers have studied the established paradigms (lean/agile) and the green/sustainable paradigms as separate entities (e.g. Mollenkopf et al. 2010; Dües et al. 2013). Rothenberg et al. (2001) argue that the complex links between supply chain practices and environmental and social performance are still to be uncovered, and this is an interesting area for future research. Thus, understanding what is the best way for companies to design and implement a system of practices that meets, on the one hand, the environmental and social requirements of a wide set of stakeholders and, on the other hand, the need to be lean (efficient and waste free) and/or agile (fast and flexible to the needs of the market place) is critical, (Ciccullo et al. 2018). The integration of lean-green-agile manufacturing strategies would be a complete and comprehensive manufacturing system which is the need of the 21st century. However, this integration was rarely explored in the field of supply chain (e.g., Ciccullo et al. 2018).

With the aim of improving SC performance, previous studies consider different paradigms (lean, agility, green, resilience) and their influence on the SC (Carvalho et al. 2011; Carvalho and Cruz-Machado 2011; Govindan et al. 2015). Other studies also reveal that these practices may affect SC sustainability (Govindan et al. 2014). However, the joint influence of green and lean and agile practices in SC performance at different levels has not yet been examined in depth.In the past these dimensions have been studied individually, so the focus lies on green SC or lean SC or agileSC separately. Now it's time to integrate lean, green and agile SC to build an unanimous and a robust structure. Our study will allow us to answer the following questions; what are the common objectives achieved by companies that implement lean and green and agile practices in their SC? Is there any relationship between lean and green and agile SC practices that may help in the implementation of such practices?

To address the aforementioned questions, this paper is organized as follows. First, we define each paradigm separately, then we present the research methodology and a descriptive analysis is explored to illustrate the trend of research. Next, we discuss the theoretical elements of our framework; the practices of each paradigm, the integration categories then common objectives are developed. Finally, the conclusion and the limitations as well as future extensions.

## 2 Basic Terminology

For the purpose of enriching our literature review, in this paragraph the main paradigms are highlighted and explained separately.

**Green Supply Chain**

A new paradigm entitled "Green Supply Chain" is developed and many studies in the SC literature have focused on designing product recovery networks called reverse or backward networks to ensure reducing the natural disruptions caused by wastes of used products, (Farahani 2013). Some authors, among others (Huo et al. 2019), (Sarkis et al. 2011) and (Tseng et al. 2018) have defined GSCM as the integration of environmental issues into the business supply chain activities. Govindan (2017) argues that green supply chain management (GSCM) aims to merge economic and environmental objectives/factors in the design of SC networks. According to Lam et al. (2015), GSCM exposes the applications of the most important sustainable development issues. It demonstrates how green technologies and practices can be implemented and, in line with this, the motivation of saving money and increasing efficiency. In the context of supply chain, it is important to integrate management practices as they improve organizational, overall supply chain performance and they take into consideration the economic, environmental, and social dimensions (Beske 2014). Additionally, Vinodh et al. (2011) argue that the adoption of green practices in supply chains can have a positive impact on sustainability. Similarly, due to the complexity of SC, Stakeholders often insist upon the adoption of green supply chain management principles (Tachizawa and Wong 2015). It not only improve the ecological efficiency but also it contribute to the company's competitive advantage by reducing its environmental risks and impacts (Zhu et al. 2007; Ortas et al. 2014).

**Lean Supply Chain**

(Farahani et al. 2013) present that the major concern of a lean SC is economic considerations and it focuses on minimizing the overall costs of SC. According to (Huo et al. 2019) LSCM emphasized efficient resource utilization and waste reduction along the supply chain. LSCM requires the involvement of partners to make continuous improvements throughout the supply chain (e.g., Campos and Vazquez-Brust 2016; Garza-Reyes 2015b). More importantly, lean SC represents a strategy based on cost reduction and flexibility and it embraces all the processes starting with the product design to the product sale (Carvalho et al. 2010). (Longoni and Cagliano 2015a; Shah and Ward 2003) argue that LSCM enhance value by eliminating more generalized waste (e.g., waiting, overprocessing, overproduction, inventory, correction, and defects) from purchasing to final delivery. LSCM improve the efficiency with less cost and time by emphasizing the participation of supply chain partners in continuous improvement practices (Rothenberg et al. 2001). Similarly, there are benefits for the supply chain when it adopts lean techniques, it improve manufacturing efficiencies, profitability, and flexibility (Azevedo et al. 2012; Vonderembse et al. 2006). Moreover, leanness in a SC maximizes profits through cost reductions (Singh and Pandey 2015). According to Vonderembse et al. (2006) a lean supply chain employs continuous improvement efforts that focus on eliminating waste or non-value steps along the chain. Anand and Kodali (2008) argue that a LSCM uses resources effectively; lean practices are integrated into upstream and downstream activities and it reduce demand variation by simplifying, optimizing, streamlining, and creating capabilities. In addition, for eliminating inefficiencies along the SC many organizations have implemented lean practices and methods such as time compression (Carvalho et al. 2016). It shorten the distances between supply chain partners (Tachizawa and Wong 2015). Thus, Lean supply chains are generally intended to reduce wastes in all forms as far as possible to construct a level schedule assuming a certain market demand.

**Agile/Responsive Supply Chain**

Gunasekaran et al. (2008) argue that agile/responsive SC is an interaction of a collaborative network of partners, information technology and systems and knowledge management. Serrador and Pinto (2015) define the agile SC as methods focus on people, technology and processes while collaborating with customers and adapting to change. This idea has been extended beyond organisation's boundaries to include the activities of the supply chain, emphasizing the need for strategic alliances, knowledge transfer, information sharing, aligning resource capabilities and effective leadership practices across supply chain (Dyer et al. 2018; Dubey et al. 2018). Thus, Agile supply chains are a class of supply chains responsive and flexible enough to handle market fluctuations. An agile supply chain focuses on the enrichment/satisfaction of customers (Lin et al. 2006). More importantly, agile supply chains is about being responsible and adaptable to the customer requirements while the risk of supply chain disruptions is avoided. Supply chain agility is the ability of firm to sense short term, temporary changes in supply chain and market environment as well as to quickly adjust to those changes (Aslam et al. 2018; Eckstein et al. 2015). Gunasekaran (1998) adds that agile supply chain requires the capability to survive and prosper in a competitive environment of continuous and unpredictable change by reacting quickly and effectively to changing markets, driven by customer-designed products and services. Aslam et al. (2018) identified market sensing capability, supply chain alignment and adaptability as some of the key precursors of supply chain agility. Similarly, SC responsiveness as the ability of a SC to produce innovative products, meet short lead-times, cope with a wide range of products, and meet a high service level (Chopra and Meindl 2013). Lee (2002) states that agile supply chains are aimed at being responsive and flexible to the customers while the risk of supply shortage or related to upstream disruptions is hedge thanks to resource and inventory pooling and/or redundancy.

## 3 Research Methodology

In this section we describe the methodology used in this research. Figure 1 presents the step-by-step process followed in the research. We first start by searching the keywords: green, lean and agile supply chain. Then, we used Boolean operator (AND) in searches of keywords, titles, abstracts and full article text in different databases (Sciencedirect Emerald, Springer, tandfonline, Googlescholar). The period of the researched articles is 2002 up to now, knowing that 2002 was the date of publishing the first article including a combination of two paradigms in SC. To ensure that the selected articles were relevant to our review an examination of the content was carried out. 33 articles were identified. With the aim to analyze the existing literature in green, lean and agile supply chain and the links between those paradigms, and to review the relevant articles in this field. Finally, to identify challenges and areas for future researches.

A summary of the process followed is shown in Table 1.

## 4 Descriptive Analysis

The following section presents a descriptive statistics based on form analysis of the papers found.

### 4.1 Literature Over Time

This study carried out an analysis of the distribution of publications per year during the period studied in order to identify the trend of research over the years.

Figure 1 presents the year wise distribution of all papers from 2002 to 2019.

Although, the first selected article appeared in 2002 (Herer et al. 2002), as part of the research effort.

The aim of his paper is: "To achieve supply chain leagility". Similarly, the first paper combining green and lean SC was appeared in 2006 "A multiple attribute utility theory approach to lean and green supply chain management" and concluded that there was a relationship between lean and agile supply chain practices then, lean and green SCM. The graphical representation of Fig. 1 clearly indicates the increasing number of research articles published over the last decade of the period: 81.8% of the articles were published between 2013 and 2019. The year 2018 has the highest number (11) of published articles and the year 2019 has the second highest by 4 papers up to now; most of those papers study only the combination of two paradigms.

**Table 1.** Summary of the research process.

| | Process | Papers found |
|---|---|---|
| Year | 2002–2019 | **33** |
| Keywords | Green supply chain management, sustainable supply chain management, lean supply chain management, agile/responsive supply chain management<br>Boolean operator «AND» for searching combination and integration | |
| Criteria | We maintained only articles:<br>• Type: Research and review articles<br>• English language | |
| | Including at least a combination of two paradigms | |
| | Articles which were out of the scope of the research were excluded | |

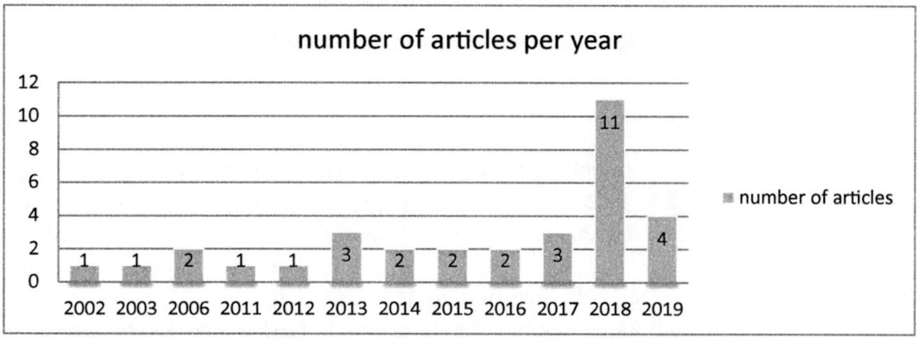


**Fig. 1.** Number of articles per year

In 2017 a special issue entitled "Adoption of Integrated Lean-Green-Agile Strategies for Modern Manufacturing Systems" was the first published paper which studies the integration of lean-green-agile manufacturing strategies without studying the impact of this integration on the whole supply chain. In the same year, F. Ciccullo et al. have discussed the integration of environmental and social sustainability pillars into the lean and agile supply chain management paradigms;

## 4.2 Literature Per Journal

Figure 2 presents the journal wise distribution of lean, Green and Agile SC.

More than 72% of the research articles were published in three journals: International Journal of Production Economics (IJPE), Journal of Cleaner Production (JCLP) and Journal of Resources, Conservation and Recycling. The highest number of relevant articles was published in IJPE (13). The second highest number of articles was published in JCLP (eight papers). The IJPE played a leading role in this field because it published 39.3% of the papers considered for this review. Also, the first paper gathering green, lean and agile was published in "Procedia CIRP" and it only focuses on the integration's impact at the manufacturing level and not on the whole supply chain, and another paper which discuss the integration of sustainable SC with agile and lean SC was published in "JCLP".

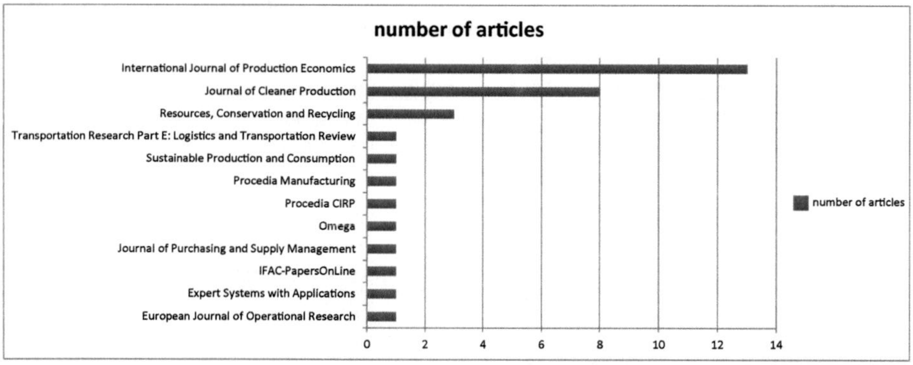


**Fig. 2.** Number of articles per journal

## 4.3 Research Streams

An analysis of the studies reviewed shows that the majority (97%) of the papers has treated only two paradigms (see Fig. 3). 48% of papers combining Green and Lean SC, 21% of papers gathering Green and Agile SC and 27% of papers has treated the relationship between Lean and Agile SC. while only 3% are combining three paradigms, which it shows clearly the gap of studies in this field.

Although, some authors, including Farahani et al. (2014) and Govindan et al. (2017) have recognized the need to investigate the effect of greenness, sustainability,

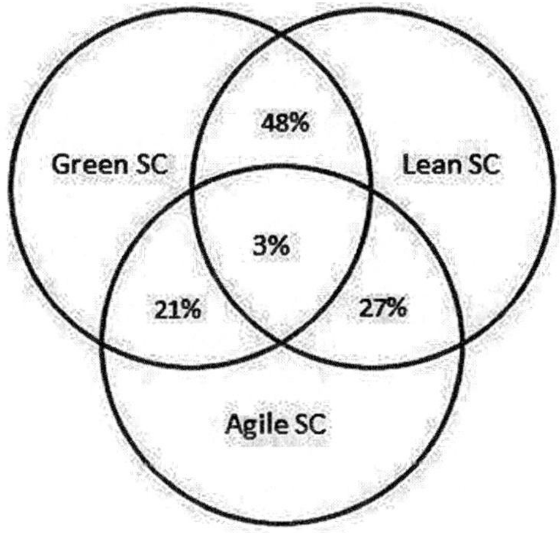


**Fig. 3.** Results of combined paradigms

and responsiveness in the brand reputation of the supply chain. Similarly, The current adverse global economic conditions have highlighted the need for organisations to integrate various supply chain paradigms, in order to meet customer demand efficiently and effectively. Companies's strategy should simultaneously consider and balance economic, environmental and social goals (Carter and Rogers 2008). This recognition will provides the opportunity of generating new studies not only regarding the integration of Green, Lean and Agile supply chain but also with other performance enhancement paradigms. For all these reasons, in the next section, our literature review will primarily focus on Integration of Green, Lean and Agile supply chain.

# 5 Theoretical Elements of Our Framework

In order to answer the research question, we develop an integrated model which combine green SC, Lean SC and Agile SC. The theoretical elements of our framework are; GLASC practices, Integration categories and common objectives of GLASC.

Figure 4 presents the framework of GLASC.

## 5.1 Practices of GLASC

Because there are several trade-offs among lean and green practices in the supply chain context, the selection of the best set of practices is not a trivial choice (Carvalho et al. 2016). Lean practices are usually adopted for a pull type production system and are generally seen in association with Just-in-Time (JIT) flows and certain quality practices. Reduced buffers in terms of inventory and capacity could be appreciated as an enhanced feature of lean supply chains, Rajesh (2018). Zhu and Sarkis (2004) have

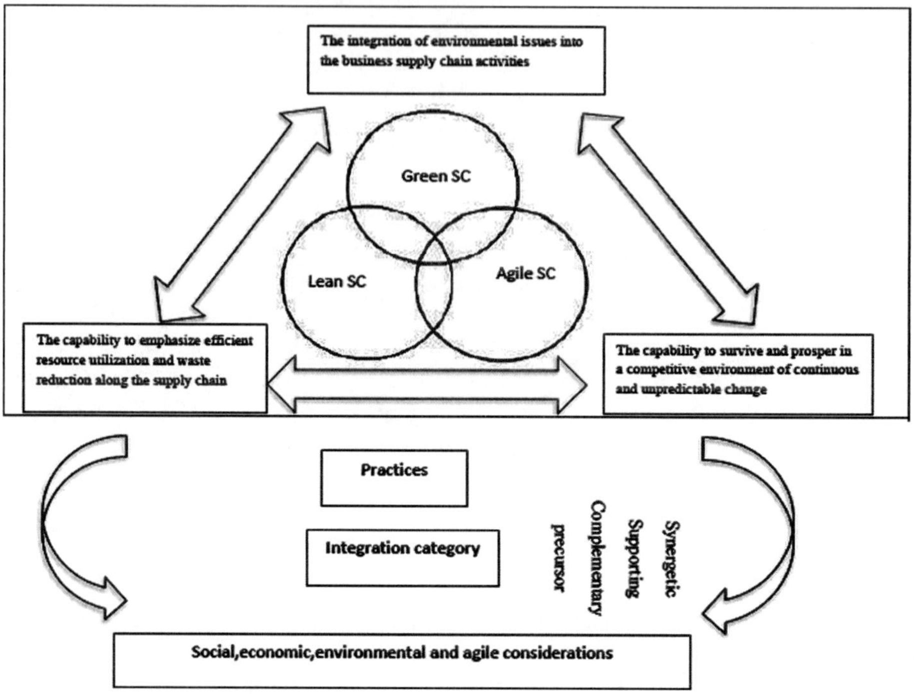


**Fig. 4.** The framework of GLASC

considered JIT and quality management as moderator of the relationship between green supply chain practice and organizational performance. The supply chain management perspective considers the green and lean management practices essential to improving a company's environmental and economic performance (Azevedo et al. 2012; Martínez-Jurado and Moyano-Fuentes 2013; Garza-Reyes 2015). According to Cherrafi et al. (2017a), the organizations that have simultaneously implemented lean and green practices have achieved better results than those organizations that have only focused on either of the initiatives. The integration of lean and green will need a new approach for addressing economic and environmental issues. Azevedo et al. (2013) group the green supply chain practices into four categories: supply practices, advanced practices, product-based practices, and delivery practices.

| Paradigm | Practices | Sources |
|---|---|---|
| Green SC | - inter-organizational practices,<br>- supply practices<br>- advanced practices<br>- product-based practices<br>- delivery practices. | Azevedo et al. (2013); Formentini and Taticchi (2016) |

(*continued*)

(*continued*)

| Paradigm | Practices | Sources |
|---|---|---|
| | - sustainable products design,<br>- sustainable procurement,<br>- environmental management system,<br>- investment recovery<br>- social sustainability practices | |
| Lean SC | Some practices related to suppliers:<br>- procurement consolidation;<br>- supplier certification;<br>- supplier evaluation and rating;<br>some practices related to customers' relationship:<br>- supplier involvement in product development;<br>some practices related to operations:<br>- electronic-enabled supply chains;<br>- JIT delivery practices;<br>- pull production systems;<br>- inter-organizational value stream mapping;<br>- training in lean initiatives;<br>- just-in-time (JIT)<br>- total quality management (TQ)<br>- total preventive maintenance (TPM)<br>- human resource management (HRM)<br>- just-in-sequence(JIS) | Longoni and Cagliano (2015a); Shah and Ward (2003); Carvalho et al. (2010); Wiengarten et al. (2013); Kou and Lee (2015); So and Sun (2010); Agus and Hajinoor (2012); Marodin et al. (2016); Hines and Rich (1997); Lopes de Sousa Jabbour et al. (2014); Rajesh (2018) |
| Agile SC | - strategic alliances<br>- knowledge transfer<br>- information sharing<br>- aligning resource capabilities<br>- effective leadership practices<br>- inter-firm collaboration, | Dyer and Singh (1998); Gunasekaran et al. (2018); Dyer et al. (2018); Dubey et al. (2018); Gunasekaran et al. (2018); Naylor et al. (1999); Lee (2002); Goldman et al. (1995) |

(*continued*)

(*continued*)

| Paradigm | Practices | Sources |
|---|---|---|
| | - Market sensitivity<br>- Employee empowerment<br>- Process alignment<br>- Technology integration<br>- Supply Network<br>- Transparent customisation<br>- Intelligent automation | |

LSCM is a combination of synergetic practices, including pull production systems, just-in-time (JIT), total quality management (TQM), total preventive maintenance (TPM), and human resource management (HRM) (Longoni and Cagliano 2015a; Shah and Ward 2003). Similarly, there are several kind of lean's practice; Some practices are related to suppliers (i.e., procurement consolidation, supplier certification; supplier evaluation and rating) (Wiengarten et al. 2013); some related to customers' relationship (i.e., supplier involvement in product development) (Kou and Lee 2015); and some practices are related to operations, for example, electronic-enabled supply chains (So and Sun, 2010), JIT delivery practices (Agus and Hajinoor 2012), pull production systems (Marodin et al. 2016), inter-organizational value stream mapping (Hines and Rich 1997) and training in lean initiatives (Lopes de Sousa Jabbour et al. 2014).

ASCM is about emphasising the need for strategic alliances, knowledge transfer, information sharing, aligning resource capabilities and effective leadership practices across supply chain (Dyer et al. 2018; Dubey et al. 2018). Moreover, Gunasekaran et al. (2018) defined agile supply chain practices in term of five enabling competencies of transparent customisation, supply network, intelligent automation, total employee empowerment and technology integration.

## 5.2 Integration Categories

Lean-sustainable, agile-sustainable, or even lean agile sustainable supply chain paradigms can be achieved by leveraging different integration categories. Similarly, there is a spectrum of diverse ways to integrate the established lean and agile supply chain paradigms with the sustainable paradigm (Ciccullo et al. 2018). Narasimhan et al. (2006) argue that established paradigms can be *competing*, *precursor* and *complementary* to each other.

| Paradigms | Integration category | Sources |
|---|---|---|
| Lean and green SC | - Supporting<br>- Synergetic<br>- Lean SC Precursor to Green SC | Ciccullo et al. (2018); Narasimhan et al. (2006); Rothenberg et al. (2001); Yang et al. (2011); Mollenkopf et al. (2010); Kurdve et al. (2015) |

(*continued*)

(*continued*)

| Paradigms | Integration category | Sources |
|---|---|---|
| | - Complementary | |
| Lean and agile SC | - Synergetic<br>- Lean SC precursor to Agile SC | Ciccullo et al. (2018); Christopher and Towill (2001); Yang et al. (2011); Mollenkopf et al. (2010) |
| Green and agile SC | - Supporting<br>- Synergetic<br>- Agile SC precursor to green SC | Ciccullo et al. (2018); Yang et al. (2011); Mollenkopf et al. (2010) |

Moreover, the integration between lean and sustainable supply chain paradigms occur under the *supporting, complementary*, *synergistic*, *precursor*, and *competing* integration categories. The integration between agile and sustainable supply chain paradigms instead occurs under a smaller set of integration categories, namely *supporting*, *precursor*, *and synergistic* (Ciccullo et al. 2018).

Lean or Agile paradigms *supporting* sustainable paradigm. Also, lean or agile practices are considered as part of the sustainable paradigm (Ciccullo et al. 2018). In addition, we found that the lean and agile paradigms can be *supportive* of sustainable paradigm as well as the other way around. Interestingly, we found that paradigms can be *synergistic* too. Carvalho and Cruz-Machado (2011) argue that the three paradigms independently (i.e. by means of different practices) *synergistically* affect the same supply chain attribute (e.g. integration level) and ultimately the same performance indicator (service level). According to Godwin et al. (2019) sustainable supply chain practices are drivers of agile capabilities. Thus, Lean paradigm as *complementary* with sustainable paradigm: a lean practice is modified to make it coherent with sustainable principles (Rothenberg et al. 2001). Lean and Agile paradigms as *precursors* of the sustainable paradigm. Similar as the concept of lean as antecedent of agile (Christopher and Towill 2001).

## 5.3 Objectives of GLASC

Dües et al. (2013) piloted literature examinations to link lean processes with green practices and have recognized that lean practices are catalysts for greening the supply chains. The supply chain management perspective considers the green and lean management practices essential to improve a company's environmental and economic performance (Azevedo et al. 2012; Martínez-Jurado and Moyano-Fuentes 2013; Garza-Reyes 2015). Dües et al. (2013); Garza-Reyes (2015b) argue that the combination of green supply chain management (GSCM) and lean supply chain management (LSCM) are suggested as two approaches to achieve economic return with lower environmental harm and greater social welfare. Additionally, the implementation of lean and green practices along the supply chain enables an enhanced economic, environmental, and social performance (Azevedo et al. 2012). On the customer side, lean processes are directly link to social, environmental, and economic performance, while green processes improve only environmental performance. On the supplier side, green processes improve social and economic performance,while lean processes improve only economic performance Huo et al. (2019).

| Paradigm | Objectives | Sources |
|---|---|---|
| Green and Lean SC | • Economic, environmental, and social performance;<br>• Economic return with lower environmental harm and greater social welfare;<br>• Saving money and increasing efficiency;<br>• Sustainable performance;<br>• Organizational performance;<br>• Ecological efficiency and competitiveness;<br>• Efficiency, reduced costs, profitability and flexibility; | Azevedo et al. (2012a); Martínez-Jurado and Moyano-Fuentes (2013); Garza-Reyes (2015); .Dües et al. (2013); Garza-Reyes (2015b); Lam et al. (2015); Huo et al. 2019; Carvalho et al. (2016); Ng et al. (2015); Zhu et al. (2007); Vonderembse et al. (2006); |
| Green and Agile SC | • Organisational performance.<br>• The visibility of supply chain operations;<br>• Responsiveness and delivery speed;<br>• Flexibility and adaptability;<br>• Collaboration and strategic partnerships;<br>• Transparency and reverse logistics; | Aslam et al. (2018); Eckstein et al. (2015); Yusuf et al. (2004); Bruce et al. (2004); Mishra et al. (2017); Green et al. (2012); Tang and Tomlin (2008); Rajesh (2018) |
| Lean and Agile SC | • Faster product to market at a minimum total cost;<br>• Responsiveness and lead time reduction;<br>• Virtual cooperation and information sharing;<br>• Flexibility and efficiency;<br>• Rapid reconfiguration and process streamlining;<br>• Competency and complexity reduction; | Gunasekaran et al. (2007); Mason-Jones et al. (2000); Lin et al. (2006); Rajesh (2018) |

(Lam et al. 2015) demonstrate how green technologies and practices can be implemented on the motivation of saving money and increasing efficiency".) Similarly, LSCM is a process oriented and therefore focuses on enhancing efficiency by eliminating more generalized waste (Qi et al. 2009). GSCM and LSCM play different roles in achieving sustainable performance (Huo et al. 2019). So, Green and lean SC share a common objectives which are increasing 'efficiency' and sustainable performance. According to (Carvalho et al. 2016; Ng et al. 2015) the integration of lean and green practices can have a more important, positive impact on organizational performance when implemented together. Some researchers (e.g. Cherrafi et al. 2016a; Hajmohammad et al. 2013) highlighted the benefits of integrating lean and green principles include reduced lead time and costs, improved process flow, better relationships with suppliers, customers and other stakeholders, improvement of environmental quality, as well as employee morale, and commitment.

Green supply chains are classically focused on reducing environmental impacts. Generally this is achieved through reducing the wastes over stages in supply chain that sequentially emanates as the primary focus of lean supply chains. Green supply chains consider the principles of waste reduction on a wider scope than what the lean paradigm essentially does. Green supply chains can be considered as advanced forms of lean supply chains with improved focus on environmental considerations while upholding the core principles of inventory reduction, resource conservation and efficiency as seen in lean supply chains (Rajesh 2018).

The focus on rapid reconfigurations to maintain market buoyancy as seen in leagile supply chains contradicts the principles of resource conservations for greening the supply chains, since rapid reconfigurations are possible with the availability of resource buffers (Rajesh 2018). The greater the degree of reach and range practices the better the visibility of supply chain operations (Yusuf et al. 2004). Lee (2002) states that agile supply chains are aimed at being responsive and flexible to the customers while the risk of supply shortage or related to upstream disruptions is hedged thanks to resource and inventory pooling and/or redundancy.

Extending lean principles from manufacturing to SCM can leverage the SC's competitiveness further with increased responsiveness to demand changes and reduced operating costs (Oliver et al. 1993). Companies are now seriously exploring the potential of the concept of supply chain management (SCM) to improve their revenue growth. In particular, they are attempting to develop agile supply chains to get their product to market faster at a minimum total cost. The waste in lean production is an essential buffer in agile supply chains. So the concepts of leagile supply chains were introduced; in which the supply chain adopts a lean manufacturing approach upstream and once the leanness is achieved, it follows the principles of an agile supply chain in the downstream that is capable of delivering to an unpredictable market place. Thus leagile supply chains are progressive forms of agile supply chains, with optional buffer levels available (Sheffi and Rice 2005). According to Rajesh (2018), Leagile supply chains uphold the core principles of agile supply chains like building flexibilities and improving proficiencies for rapid reconfigurations, thus reducing the uncertainties associated with the supply chains along with the considerations of waste reduction as far as possible; hence can be regarded as enriched forms of agile supply chains. Similarly Flexibility is built throughout the supply chain by incorporating flexible product, process and pricing strategies. Flexibility enables the supply chain to be competitive enough to capture market attention and allows the supply chain to reconfigure quickly incorporating robust changes (Gunasekaran et al. 2007). Additionally, Lee (2002) highlights visibility as a fundamental dimension of both lean and agile paradigms. To reach visibility, companies rely on information sharing with supply chain partners. In lean supply chains, the information sharing is meant to facilitate the implementation of cost optimisation. In agile supply chains, visibility helps in capturing customer requirements and timely communicating orders along the supply chain.

Rajesh (2018) argue that when the target focus is on the reduction of non-value added processes, the lean supply chains are said to be efficiency focused. Information sharing enhances the capabilities of being lean and this also helps to reduce the complexity of the supply chain to the maximum end possible. More importantly, Leagile supply chains have a dual focus on waste reduction as well as uncertainty reduction and are responsive enough to reduce the lead times of delivery. With high levels of integration, these supply chains are flexible enough to reconfigure quickly to handle sudden market perturbations (Rajesh 2018; Gunasekaran et al. 2007).

**Conclusion, Limitations and Future Areas**

This study seeks to answer the question of integrating different paradigms (lean green and agile) in a supply chain. We start by identifying 33 papers published between 2002 up to now. Then we analysed the content of this papers for the objective to developing an integrated model for GLASC. Thus, we present the theoretical elements of our integrated model starting by a list of practices of each paradigm, then we discuss the integration categories; Supporting, synergetic, precursor and complementary and finally we explored the objectives of GLASC. On one hand, we found that lean and green SC share some common objectives (e.g. Saving money and increasing efficiency, Sustainable performance, Organizational performance). On the other hand, Lean and agile SC also share some common objectives (e.g. Faster product to market at a minimum total cost, Responsiveness and lead time reduction, Flexibility and efficiency).We concluded that lean SC is precursor to agile SC and green SC. So, it is necessary to focus on the implementation of the lean's practices at first. this study trying to cover the existing gap in the literature on the relationship/effect of lean and green and agile paradigms in the whole supply chain.

An additional contribution of this study lies in the higher number of practices considered in each paradigm to better capture the existing links between practices, that may have been previously ignored.

Despite of the contribution of our study, our integrated model is a theoretical model it needs a case study for knowing if it's applicable or not. Also, it explores only the practices and integration categories without shading light on drivers/barriers and synergies/conflicts. Therefore, we recommend the following further research areas to be investigated in the near future:

- Mathematical model of an integrated system f GLASC paradigms;
- Empirical longitudinal studies on the evolution of supply chains of companies that have already integrated leanness, greenness and responsiveness.

## References

Huo, B., Gu, M., Wang, Z.: Green or lean? A supply chain approach to sustainable performance (2019)

Cherrafi, A., Garza-Reyes, J., Kumar, V., Mishra, N., Ghobadian, A., Elfezazi, S.: Lean, green practices and process innovation: a model for green supply chain performance (2018)

Jakhara, S., Rathoreb, H., Mangla, S.: Is lean synergistic with sustainable supply chain? An empirical investigation from emerging economy(2018)

Carvalho, H., Govindan, K., Azevedoc, S.G., Cruz-Machado, V.: Modelling green and lean supply chains: an eco-efficiency perspective (2016)
Ramirez-Peña, M., Sánchez Sotano, A.J., Fernandez, V., Abad, F.J., Batista, M.: Achieving a sustainable shipbuilding supply chain under I4.0 perspective (2019)
Orji, I.J., Liu, S.: A dynamic perspective on the key drivers of innovation-led lean approaches to achieve sustainability in manufacturing supply chain (2018)
Zhu, Q., Shah, P., Sarkis, J.: Addition by Subtraction: Integrating Product Deletion with Lean and Sustainable Supply Chain Management (2018)
Das, K.: Integrating lean systems in the design of a sustainable supply chain model (2018)
López, C., Ruiz-Benítez, R.: Multilayer analysis of supply chain strategies' impact on sustainability (2019)
Dües, C.M., TaKainuma, Y., Tawara, N.: A multiple attribute utility theory approach to lean and green supply chain management (2006)
Tan, K.H., Lim, M.: Green as the new lean: how to use lean practices as a catalyst to greening your supply chain (2013)
Kainuma, Y., Tawara, N.: A multiple attribute utility theory approach to lean and green supply chain management (2006)
Li, A.Q., Found, P.: Lean and green supply chain for the product-services system (PSS): The Literature Review and A Conceptual Framework (2016)
Martínez-Jurado, P.J., Moyano-Fuentes, J.: Lean Management, Supply Chain Management and Sustainability: A Literature Review (2014)
Fahimnia, B., Sarkis, J., Eshragh, A.: A tradeoff model for green supply chain planning: a leanness-versus-greenness analysis (2015)
De, D., Chowdhury, S., Dey, P.K., Ghosh, S.K.: Impact of Lean and Sustainability Oriented Innovation on Sustainability Performance of Small and Medium Sized Enterprises: A Data Envelopment Analysis-Based Framework (2018)
Martínez-Jurado, P.J., Moyano-Fuentes, J.: Lean Management, Supply Chain Management and Sustainability: A Literature Review (2014)
Fahimnia, B., Sarkis, J., Eshragh, A.: A tradeoff model for green supply chain planning: a leanness-versus-greenness analysis (2015)
De, D., Chowdhury, S., Dey, P.K., Ghosh, S.K.: Impact of Lean and Sustainability Oriented Innovation on Sustainability Performance of Small and Medium Sized Enterprises: A Data Envelopment Analysis-Based Framework (2018)
Hajmohammad, S., Vachon, S., Klassen, R.D., Gavronski, I.: Reprint of lean management and supply management: their role in green practices and performance (2013)
Gey, D.G., Yusuf, Y., Menhat, M.S., Abubakar, T., Ogbuke, N.J.: Agile capabilities as necessary conditions for maximising sustainable supply chain performance: an empirical investigation (2019)
Fahimnia, B., Jabbarzadeh, A., Sarkis, J.: Greening versus resilience: a supply chain design perspective. Transportation Research Part E: Logistics and Transportation Review (2018)
Rajesh, R.: On sustainability, resilience, and the sustainable–resilient supply networks (2018)
Azevedo, S.G., Govindan, K., Carvalho, H., Cruz-Machado, V.: Ecosilient Index to assess the greenness and resilience of the upstream automotive supply chain (2013)
Rosário Cabrita, M.do, Duarte, S., Carvalho, H., Cruz-Machado, V.: Integration of Lean, Agile, Resilient and Green Paradigms in a Business Model Perspective: Theoretical Foundations (2016)
Mujkić, Z., Qorri, A., Kraslawski, A.: Consumer Choice and Sustainable Development of Supply Chains (2018)
Abdollahi, M., Arvan, M., Razmi, J.: An integrated approach for supplier portfolio selection: Lean or agile? Expert Systems with Applications (2015)

Ruiz-Benítez, R., López, C., Real, J.C.: The lean and resilient management of the supply chain and its impact on performance (2018)
Ciccullo, F., Pero, M., Caridi, M., Gosling, J., Purvis, L.: Integrating the environmental and social sustainability pillars into the lean and agile supply chain management paradigms: a literature review and future research directions (2018)
Agarwal, A., Shankar, R., Tiwari, M.K.: Modeling the metrics of lean, agile and leagile supply chain: an ANP-based approach (2006)
Purvis, L., Gosling, J., Naim, M.M.: The development of a lean, agile and leagile supply network taxonomy based on differing types of flexibility (2014)
Azevedo, S.G., Govindan, K., Carvalho, H., Cruz-Machado, V.: An integrated model to assess the leanness and agility of the automotive industry (2012)
Stratton, R., Warburton, R.D.: The strategic integration of agile and lean supply (2003)
Kisperska-Moron, D., de Haan, J.: Improving supply chain performance to satisfy final customers: "Leagile" experiences of a polish distributor (2011)
Mohammaddust, F., Rezapour, S., Farahani, R.Z., Mofidfar, M., Hill, A.: Developing lean and responsive supply chains: a robust model for alternative risk mitigation strategies in supply chain designs (2017)
Herer, Y.T., Tzur, M., Yücesan, E.: Transshipments: an emerging inventory recourse to achieve supply chain leagility (2002)
Ruiz-Benitez, R., López, C., Real, J.C.: Environmental benefits of lean, green and resilient supply chain management: the case of the aerospace sector (2017)
Mittal, V.K., Sindhwani, R., Kalsariya, V., Salroo, F., Sangwan, K.S., Singh, P.L.: Adoption of Integrated Lean-Green-Agile Strategies for Modern Manufacturing Systems (2017)
Farahani, R.Z., Miandoabchi, E., Szeto, W.Y., Rashidi, H.: A review of urban transportation network design problems. Eur. J. Oper. Res. **229**, 281–302 (2013)
Sarkis, J., Zhu, Q., Lai, K.: An organizational theoretic review of green supply chain management literature. Int. J. Prod. Econ. **130**, 1–15 (2011)
Mollenkopf, D.F., et al.: Association of dry cow therapy with the antimicrobial susceptibility of fecal coliform bacteria in dairy cows. Prev. Vet. Med. **96**, 30–35 (2010)
Vonderembse, M.A., Uppal, M., Huang, S.H., Dismukes, J.P.: Designing supply chains: towards theory development. Int. J. Prod. Econ. **100**, 223–238 (2006)
de A. Machado, A.E., Dos Santos, H.F., de Almeida, W.B.: Enhanced nonlinearites of functionalized single wall carbon nanotubes with diethynylsilane derivatives. Chem. Phys. Lett. **514**, 134–140 (2011)
Vinodh, S., Jayakrishna, K.: Environmental impact minimisation in an automotive component using alternative materials and manufacturing processes. Mater. Des. **32**, 5082–5090 (2011)
Zhu, Q., Dou, Y.: Evolutionary game model between governments and core enterprises in greening supply chains. Syst. Eng. - Theory Pract. **27**, 85–89 (2007)
Aslam, M.K., Javed, M.S., Hussain, S., Xu, J., Chen, C.: Facile synthesis of cobalt ferrite nanoparticles (CFO-NPs) as anode material with enhanced lithium storage capability. Mater. Sci. Eng. B **236–237**, 162–169 (2018)
Govindan, K., Darbari, J.D., Agarwal, V., Jha, P.C.: Fuzzy multi-objective approach for optimal selection of suppliers and transportation decisions in an eco-efficient closed loop supply chain network. J. Clean. Prod. **165**, 1598–1619 (2017)
Brust, S., Röttger, A., Theisen, W.: High-temperature stability and interfacial reactions of Ti and TiN thin films on $Al_2O_3$ and $ZrO_2$. Surf. Coat. Technol. **307**, 47–55 (2016)
Masteri-Farahani, M., Abednatanzi, S.: Immobilized molybdenum–Schiff base complex on the surface of multi-wall carbon nanotubes as a new heterogeneous epoxidation catalyst. Inorg. Chem. Commun. **37**, 39–42 (2013)

Carvalho, L.V., et al.: Immunological parameters related to the adjuvant effect of the ordered mesoporous silica SBA-15. Vaccine **28**, 7829–7836 (2010)

Govindan, K., Azevedo, S.G., Carvalho, H., Cruz-Machado, V.: Impact of supply chain management practices on sustainability. J. Clean. Prod. **85**, 212–225 (2014)

Reyes-Contreras, C., Vidal, G.: Methanogenic toxicity evaluation of chlortetracycline hydrochloride. Electron. J. Biotechnol. **18**, 445–450 (2015)

Tseng, C.-H., Chen, L.-L., Yeh, P.-C.: Modeling contamination conditions in small-scale industrial areas to estimate health savings benefits associated with remediation. Heliyon **4**, e00995 (2018)

Rothenberg, M.E., Mishra, A., Collins, M.H., Putnam, P.E.: Pathogenesis and clinical features of eosinophilic esophagitis. J. Allergy Clin. Immunol. **108**, 891–894 (2001)

Beske, P., Land, A., Seuring, S.: Sustainable supply chain management practices and dynamic capabilities in the food industry: a critical analysis of the literature. Int. J. Prod. Econ. **152**, 131–143 (2014)

Lam, R.K.K., Han, W., Yu, K.N.: Unirradiated cells rescue cells exposed to ionizing radiation: activation of NF-κB pathway in irradiated cells. Mutat. Res. Mol. Mech. Mutagen. **782**, 23–33 (2015)

Gunasekaran, S., Ko, S., Xiao, L.: Use of whey proteins for encapsulation and controlled delivery applications. J. Food Eng. **83**, 31–40 (2007)

Brito, P., et al.: Y-SNP analysis in an Angola population. Forensic Sci. Int. Genet. Suppl. Ser. **3**, e369–e370 (2011)

# Integrated Procurement, Production and Distribution Under Mass-Customization: Case of Moroccan Automotive Industry

Mouad Benbouja(✉), Achraf Touil, Abdelkabir Charkaoui, and Abdelwahed Echchatbi

Laboratory of Engineering, Industrial Management and Innovation, Faculty of Sciences and Technology, Hassan 1st University, PO Box 577, Settat, Morocco
mouad.benbouja@gmail.com

**Abstract.** The market environment tends toward satisfying customer individual needs. In this scope, product customization is an appropriate approach to foresee a customer focus which has a development ability besides mass production toward mass customization concept in order to outfit customers' product specifications for a mass market. This article develops a tactical integrated procurement, production and distribution deterministic model of a 3-echelon supply chain within mass customization context according to Leader and Followers tradeoffs. A case study from the automotive industry is proposed to emphasize the operations' schemes within a business process driven by the customer that involve: (a) Original Equipment Manufacturer, identified as Leader (Upper level) and (b) First-tier Supplier: Wiring Harnesses Manufacturer (c) Second-tier Supplier: Raw Material supplier, identified as followers (Lower level). Findings from this study highlight the customization background that lead to product variety management strategy and mass customization link and provide an integrated operations management framework within a manufacturing overview arises from partner's interdependency network with dyadic relationships interfaces.

**Keywords:** Mass customization · Product variety · Integrated supply chain · Base stock

## 1 Introduction

In order to compete in a growing business market, supply chain actors have to develop a sustainable customer focus with high agility level. In fact, product customization allows an advanced customer service level which should be perceived with cost effectiveness approach for a global performance. In such a context, the standalone positions will hinder success factors based on collaboration activities throughout all stakeholders with regards to performance measurement that include mainly three expectations as stated by [2] (a) Resources: related to cost efficiency (Inventory, Manufacturing, Distribution…) (b) Output: customer service level (c) Flexibility: the functional ability to deal with a changing environment (i.e. internal operations).


A. Saka et al. (Eds.): CPI 2019, LNME, pp. 350–364, 2021.
https://doi.org/10.1007/978-3-030-62199-5_31

The underpinning of this paper is inspired by a real-word case from the automotive industry since the Original Equipment Manufacturers (OEMs) have to deal with technology development for security, environmental and entertainment standpoints. However, customers' requirements remain a must, thus, multiple configurations came out according to different features that a vehicle depicts. The focus is to identify the impact of Product Variety Management Strategy (PVMS) on collaboration activities of a 3-echelon supply chain by a case study of the automotive industry evolving toward Mass Customization. Hence, an integrated optimization decision model of production planning, inventory and distribution through a deterministic environment is proposed to describe the merging structure of mass customization while considering supply chain stakeholders' tactical constraints illustrated from the hierarchical business control. The considered supply chain in this case study is based on a) Original Equipment Manufacturer (OEM): Automaker (b) First-tier Supplier: Wiring Harnesses Manufacturer (c) Second-tier Supplier: Raw Material suppliers, while the network interaction framework is driven by OEMs. Thereof, this dependency structure mentioned allows the leader position to OEMs, and the First-tier & Second-tier suppliers will be pointed as followers. However, the decision making considered in the model is designed in a centralized endeavor to maintain global performance through the supply chain while driving tradeoffs between actors with explicit constraints.

To summarize, the main contributions of this paper are as follows. First, we describe and model a real-life problem for which procurement, production and distribution decisions have to be considered simultaneously. Second, and integrated optimization model for a deterministic case has been developed considering mass customization constraints and global supply chain performance. Third, we present a set of perspectives to strengthen the model positioning.

The paper is structured as follows. It begins with a summary of a literature review and supply chain positioning. Section 3 introduces the supply chain structure and describes the model assumptions then it presents the mathematical formulation. Section 4 presents the computational experiments results. Section 5 concludes the paper.

## 2 Literature Review

Product customization is a concept oriented towards customers' specifications that should be considered in the product or service design. This market development has been introduced on 1989 by [6], where he indicated also that the mass customization challenge is to keep the customer service level similar to the mass market conditions with an individually treatment. In fact, the result is to elaborate many 'Units of One'. In our actual context, firm's competitiveness is largely improved by this strategy while providing the customer the choice to configure its own product to fit its own needs. However, this increased product variety requires an important agility and flexibility levels from all supply chain partners to ensure cost effectiveness, as the essential value of mass customization is to satisfy customers' individual preference at a low cost brought by mass production [14]. [4] proposed a set of mass customization success factors coming from a literature review, they are classified in two main divisions which involve many important operations aspects such as: (a) Market: variety of customer demand

(b) Organization: support willingness by supply chain actors and the existence of systems to develop mass customization environment. These success factors have been further developed towards modeling and case studies by the multiple academic articles published in that area as stated by the same authors on 2012 [5]. [11] proposed a conceptual model where PVMS has been linked to supply chain responsiveness. Firstly, the capability to reconfigure internal and external competencies which represent respectively supply chain flexibility and supply chain agility, and secondly, according to the level of customization, check the supply chain performance from cost efficiency and customer service fields. This conceptual model verified by a survey tends to confirm that supply chain flexibility (internally) lead to supply chain agility (externally) by ensuring a high customer service level. In fact, [13] explains that the challenge turns around supply chain's configuration in this environment, in order to support that, they developed a conceptual framework with 4 decision areas: (a) Relationship management (b) Postponement (c) Customization level (d) Modularity level. Thereupon, supply chain integration is widely adopted to drive commitments raised from real-word features. [12] developed two mixed integer linear programming formulations for a tactical level to address two-stage production system, the first stage prepares what is called a resource which is the input of the second stage for end products manufacturing. The difference between the models is addressed by the following assumptions (a) a set of pre-defined production sequence is given at the first stage (b) production sequence at the first stage is driven by a product-oriented model. In order to find best solutions, two general heuristic algorithms have been adopted (a) relax-and-fix (b) variable neighborhood search with an adapted fix-and-optimize improvement. [7] proposed a mixed integer linear programming for an integrated production and distribution operations, it considers a set of orders requested by unique customers which the delivery time should be respected to avoid neither an early delivery nor a late one. The joint scheduling system includes machine scheduling and vehicles routing data for an optimal assignment. [9] addresses an integrated production, inventory and distribution model for a multi-plant, multi-period and multi-echelon system. The delivery time window is considered to fulfill demand within a specific timeframe with the assumption that a geographically dispersed distribution centers are available to be rented for a specific time period. The integrated problem is then solved by an exact method and a metaheuristic approach.

Another key point with a specific characteristic, finding the best result between customer service and the aggregate inventory levels is considered as a performance pledge. In the literature, two main approaches are discussed (a) stochastic service (b) guaranteed service. [1] affirms that the stochastic service has been largely studied, nevertheless, most studies of the guaranteed service approach are relatively recent. Hence, the guaranteed characteristic should be considered as an organizational capability upon mass customization environment. [10] included capacity constraint to a multi-echelon supply chain in order to determine the base stock level within guaranteed service approach, it has been found that the model places safety stock at constraint level and upstream. [8] performed a real case study from the automotive industry for safety stock assessment with base stock definition within a guaranteed service approach, the model has been solved with dynamic programing algorithm, it has been highlighted that the safety stock cost increase according to the guaranteed time window. [3] studied inventory optimization for supply chain planning according to (r, Q) and (s, S) policies.

In order to cope with demand uncertainties, the following safety stock formulations have been proposed a) proportional to throughput b) proportional to throughput with risk-pooling effect, c) explicit risk-pooling d) guaranteed service time. According to constraints structure, the model formulations are addressed with mixed integer programming and mixed integer non-linear programming.

## 3 Supply Chain Structure

For model formulation, we consider a 3-echelon supply chain with leader and followers structure globally connected with customer/supplier interactions according to a collaboration mechanism based on contractual clauses. It highlights the key characteristic of mass customization approach trough the make to order business process. The illustration is carried out from the automotive industry grouped into (a) OEMs: Customers (b) $1^{st}$ tier supplier: wire harnesses manufacturer and (c) $2^{nd}$ tier suppliers: subset of upstream raw material's supplier. The supply chain model will consider a set of OEMs in order to reflect as much as possible a general framework and keep the possibility for potential adoption of such structure. In this study, the focus is on the $1^{st}$ tier supplier operations, the choice is motivated by its major processing function within the supply chain according to its role of the first interface with the OEMs. It appears clearly that the OEM has the prevailing position through the supply chain as all downstream activities are driven by OEM's product definition coming also from consumers' requirements which are represented by product modularization design. The decision making system is centralized due to collaboration endeavor for a global supply chain performance. The consumers' preferences are selected by a configurator leverage that is proposed to customize each car. Hence, the corresponding wiring system is impacted directly and supply chain partners should align with this dynamic complexity for an effective collaboration. The supply chain structure is presented in Fig. 1.

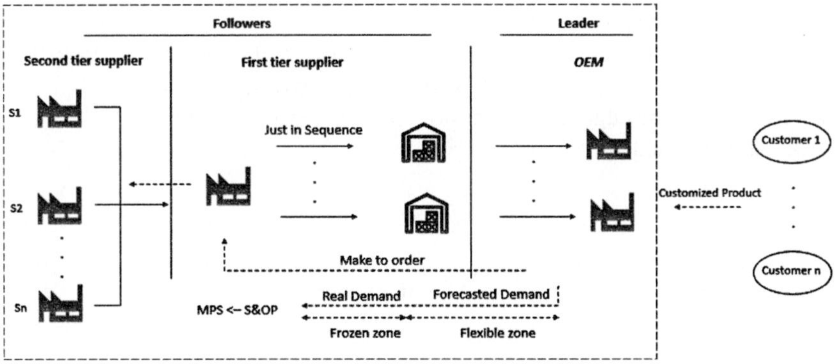


**Fig. 1.** Supply chain structure

Sales and operations planning (S&OP) in addition to the resulted master production schedule (MPS) are deemed for involving a broad planning framework to share demand information raised upon two levels: (a) frozen horizon: confirmed demand (b) flexible horizon: forecasted demand. Therefore, the 1st tier supplier has to ensure production following make-to-order process according to Order Penetration Point and delivery in Just in Sequence concept to avoid any disruptions in the OEMs' production lines.

## 3.1 Mathematical Formulation

The decision making considered in the model is designed in a centralized endeavor to maintain global performance through the supply chain while driving tradeoffs between actors with explicit constraints. In fact, the frozen horizon corresponds to OEM real customized orders which should be produced by the 1st tier supplier with make to order policy, thus, the OEM cannot provide any update, unlike the flexible horizon where the OEM has the ability to update its forecast but with a contractual flexibility hurdle. The interaction between the OEM and the 1st tier supplier allow backlogs for the unsatisfied demand within a period through a defined service level, however, any lost sales should not occur for the confirmed orders since the 1st tier supplier is considered as the exclusive supplier, in addition to the highly cost of any eventual disruption on the automaker's production scheduling due to customized product. To summarize, the following assumptions are considered for the model:

1. Three-echelon supply chain in multi periods of time is considered;
2. The production at the 1st tier supplier is performed according to make to order (start production after receiving a confirmer demand)
3. Product design is under modularization approach, the end product is assembled from multiple defined; basic and optional modules, and the basic module is always selected for product's configuration. Each module has its own bill of material;
4. Procurement is performed exclusively with forecasted demand;
5. The unsatisfied demand are classified as backorders for each period then aggregated to backlogs to be produced;
6. Capacity hurdles are addressed across interactions between supply chain stakeholders, it is addressed by demand levels;
7. The level of the required raw materials is assessed through an order-up to level strategy (base stock) over supplier's lead time horizon;
8. The distribution throughout the supply chain is performed by a 3PL. No capacity hurdle is concerned.

**Indices and Sets**

| | |
|---|---|
| $t \in T$ | Set of time periods |
| $o \in O$ | Set of Original Equipment Manufacturers |
| $p \in P$ | Set of products |
| $m \in M$ | Set of modules |
| $f \in F$ | Set of 2nd tier suppliers |
| $c \in C$ | Set of raw materials |
| $v \in V$ | Set of distribution centers |

**Parameters**

*Sales*

| | |
|---|---|
| $DR_{o,p,t}$ | Real demand of product p from OEM o in period t |
| $DF_{o,p,t}$ | Forecasted demand of product p from OEM o in period t (t > F) |
| $D^{min}_{o,p,t}$ | Minimum demand quantity of product p of OEM o in period t |
| $F$ | Frozen horizon |
| $PS_{o,m,p,t}$ | Parameter; 1 if the module m ϵ M has been selected by the OEM o for product p in period t; 0 otherwise |

*Production*

| | |
|---|---|
| $PCPS_{p,t}$ | Production cost of product p in period t |
| $BO0_{p}$ | Initial backorder level of product p |
| $BL0_{p}$ | Initial backlog level of product p |
| $BLC_{p,t}$ | Backlog cost of product p in period t |
| $BOC_{p,t}$ | Backorder cost of product p in period t |
| $IHPS_{p,t}$ | Production cost of product p of OEM o in period t |
| $IPS0_{p}$ | Initial inventory level of product p |
| $PQ^{\max}$ | Maximum production capacity per period |
| $M$ | Big number |

*Procurement*

| | |
|---|---|
| $IHCS_{c,t}$ | Inventory holding cost of raw material c in period t |
| $TCCF_{f,c,t}$ | Purchase price of component c from supplier f in period t |
| $ICS0_{c}$ | Initial inventory level of raw material c |
| $QFC^{\min}_{f,c,t}$ | Minimum contracted demand for raw material c from supplier f in period t |
| $QF^{\max}_{f,t}$ | Maximum contracted demand for supplier f in period t |
| $LTF_{c,f}$ | Lead time for raw material c from supplier f |
| $H_{f,c,t}$ | Parameter; 1 if replenishment from supplier f of raw material c in period t is allowed and 0 otherwise |
| $\alpha_{c,m}$ | The needed quantity of raw material c in module m |

*Distribution*

| | |
|---|---|
| $IHPV_{v,p,t}$ | Inventory holding cost of product p in period t at the advanced warehouse v |
| $TCPV_{v,p,t}$ | Shipping cost of product p to the advanced warehouse v in period t |
| $TCPO_{o,v,p,t}$ | Shipping cost of product p from the advanced warehouse v to the OEM o in period t |
| $IPV0_{v,p}$ | Initial inventory level of product p at the advanced warehouse v |

**Decision Variables**

*Production*

| | |
|---|---|
| $XQT_{p,t}$ | Total produced quantity of product p in period t |
| $XQD_{p,t}$ | Produced quantity of product p in period t to satisfy demand |
| $XQB_{p,t}$ | Produced quantity of product p in period t to satisfy backlogs |
| $BL_{p,t}$ | Backlog quantity of product p in period t |
| $BO_{p,t}$ | Backorder quantity of product p in period t |
| $IPS_{p,t}$ | Inventory level of product p in period t |
| $Z_{(p,t)}$ | Binary variable, 1 if product p is produced in period t; 0 otherwise |

*Procurement*

| | |
|---|---|
| $ICS_{c,t}$ | Inventory level of raw material c in period t |
| $BS_{c,t}$ | Required quantity to purchase of raw material c in period t |
| $S_{c,t}$ | Required base stock level of raw material c in period t |
| $QTCFS_{f,c,t}$ | Purchased quantity from supplier f of raw material c in period t |

*Distribution*

| | |
|---|---|
| $IPV_{v,p,t}$ | Inventory level at the advanced warehouse v of product p in period t |
| $QTPSV_{v,p,t}$ | Shipping quantity to the advanced warehouse v of product p in period t |
| $QTPVO_{o,v,p,t}$ | Shipping quantity to the customer o from the advanced warehouse v of product p in period t |
| $RR_{(p,t)}$ | Binary variable, 1 if demand is more than the contracted minimum; 0 otherwise |

**Constraints**

*Production*

The total production quantity XQT of a period t is carried out by XQD and XQB in order to perform production of the confirmed demand and the previously generated backlogs respectively. The total production can be described by constraint 1 as follows:

$$XQT_{p,t} = XQD_{o,p,s,t} + XQB_{o,p,s,t}, \forall p \in P, t \in T, \forall o \in O \tag{1}$$

Constraint 2 ensures production trigger for demand fulfillment and convert the unsatisfied ones to period's backorders to be served later, it is expressed by:

$$XQD_{p,t} = \sum_{o \in O} DR_{o,p,t} - BO_{p,t}, \forall p \in P, t \in T \tag{2}$$

But within a period, the realized production of the confirmed demand and the related backorder should not be greater than the received demand and the previous backlog, as stated by constraint 3. In this case, the backorder is used to indicate the unsatisfied demand of each period, while the backlog allows a coupling constraint between the new arising backorders, backlogs of the previous period and the realized backlog's production to decrease its level, this connection is defined by constraint 4 as follows:

$$XQD_{p,t} + BO_{p,t} \leq \sum_{o \in O} DR_{o,p,t} + BL_{p,t-1}, \forall p \in P, t \in T \tag{3}$$

$$BL_{p,t} = BL_{p,t-1} - XQB_{p,t} + BO_{p,t}, \forall p \in P, t \in T \tag{4}$$

Constraint 5 expresses the backorder hurdle according to the addressed demand. In order to give an alignment with the reality, it defines the service level over the planning horizon. Due to customization context, the overall backorder should be kept as much as possible at a higher level, for this reason, it has been defined as 10%. This constraint is represented as follows:

$$\sum_{t \in T} BO_{p,t} \leq 0.1. \sum_{o \in O} \sum_{t \in T} DR_{o,p,t}, \forall p \in P \tag{5}$$

Constraint 6 guarantees the backorder limit which should be kept less than the confirmed demand, i.e.:

$$BO_{p,t} \leq \sum_{o \in O} DR_{o,p,t}, \forall p \in P, t \in T \tag{6}$$

Constraint 7 states the backlog's production level should be lower than or equal to the backlog of the previous period, i.e.:

$$XQB_{p,t} \leq BL_{p,t-1}, \forall p \in P, t \in T \tag{7}$$

Constraint 8 involves the binary decision variable Z, it represents that total production maximum capacity of each period should not exceed a defined level. It is indicated by:

$$XQT_{p,t} \leq PQ^{\max} \cdot Z_{p,t}, \forall p \in P, t \in T \tag{8}$$

Constraint 9 represents the product inventory balance at the 1$^{st}$ tier supplier plant, i.e.:

$$IPS_{p,t} = IPS_{p,t-1} + XQT_{p,t} - \sum_{v \in V} QTPSV_{v,p,t}, \forall p \in P, t \in T \tag{9}$$

*Procurement*

Constraint 10 asses the inventory level of raw materials by joining the remaining inventory from the previous period and the 2nd tier suppliers deliveries minus the consumption of raw materials performed by the completed production, i.e.:

$$ICS_{c,t} = ICS_{c,t-1} + \sum_{f \in F}^{F} QTCFS_{f,c,t} - \sum_{o \in O} \sum_{p \in P} \sum_{m \in M} PS_{o,p,m,t} \cdot XQT_{p,t} \cdot \alpha_{c,m}, \forall c \in C, t \in T \quad (10)$$

Constraint 11 aims to evaluate the required inventory level derived from product demand forecast over suppliers' lead time, and it represents the coming consumption level. This description is indicated as follows:

$$S_{c,t} = \sum_{o \in O} \sum_{p \in P} \sum_{q=t}^{q=t+LT1} DF_{(o,p,q)} \sum_{m \in M} PS_{o,p,m,q} \cdot \alpha_{c,m}, \forall c \in C, t \in T \quad (11)$$

Constraint 12 calculates the net raw material base stock by excluding the available quantity in the manufacturer plant from the previous required inventory evaluation, it is considered as a basis of purchasing orders to ensure an optimal raw material inventory position within model horizon:

$$BS_{c,t} = S_{c,t} - ICS_{c,t-1}, \forall c \in C, t \in T \quad (12)$$

Constraints 13 determines the purchased quantity provided by the base stock calculation, while constraints 14–15 correspond, respectively, to the conditional statements of the minimum order quantity contracted with suppliers and the upper bound restraint. These elements are described as follows:

$$QTCFS_{f,c,t} \geq BS_{c,t} \cdot H_{f,c,t}, \forall f \in F, c \in C, t \in T \quad (13)$$

$$QTCFS_{f,c,t} \geq QFC_{f,c,t}^{min} \cdot H_{f,c,t}, \forall f \in F, c \in C, t \in T \quad (14)$$

$$\sum\nolimits_{c=1}^{C} QTCFS_{f,c,t} \leq QF_{f,t}^{max}, \forall f \in F, t \in T \quad (15)$$

In this model, it is supposed that the ordered quantity will be fully received, so as no uncertainty is considered.

The policy of the replenishment frequency contracted with suppliers is formulated by constraint 16, the associated reviewing selection is defined by a binary matrix while the purchased quantity variable is activated when the parameter has a true value, i.e.:

$$QTCFS_{f,c,t} \leq M \cdot H_{f,c,t}, \forall f \in F, c \in C, t \in T \quad (16)$$

*Distribution*

The appropriate product inventory assessment according to the real case should consider also the level at the 3PL warehouse, it is defined by the following constraint 17 that connect product's flow from the plant and deliveries to the OEM:

$$IPV_{v,p,t} = IPV_{p,t-1} + QTPSV_{v,p,t} - \sum_{o \in O} QTPVO_{o,v,p,t}, \forall v \in V, p \in P, t \in T \quad (17)$$

Constraints 18–19 define the set-up of deliveries to OEMs, they should not exceed the affirmed demand while ensuring a contracted minimum delivery threshold to avoid an overstock at the distribution center, i.e.:

$$\sum_{v \in V} QTPVO_{o,v,p,t} \leq DR_{o,p,t} \cdot RR_{o,p,t}, \forall o \in O, p \in P, t \in T \tag{18}$$

$$\sum_{v \in V} QTPVO_{o,v,p,t} \geq D^{min}_{o,p,t} \cdot RR_{o,p,t}, \forall o \in O, p \in P, t \in T \tag{19}$$

Constraint 20 aims to address minimum contracted demand level, it is represented as follows:

$$D^{min}_{o,p,t} \leq DR_{o,p,t} + M \cdot \left(1 - RR_{o,p,t}\right), \forall o \in O, p \in P, t \in T \tag{20}$$

**Objective Function**

The objective function tends to minimize the costs of the inventory level and the purchased quantity of raw materials. Further than the total produced quantity and the inventory level of end product at the 1st tier supplier and at the distribution center in order to shed light on the adopted vendor managed inventory. The unmet demand costs represented by backlogs and backorders. The distribution costs are illustrated by the transportation quantity between the 1st tier supplier and distribution centers in addition to final deliveries from distribution centers to OEMs.

$$\begin{aligned} TC = & \sum_{t=1}^{T} \sum_{c=1}^{C} IHCS_{c,t} \cdot ICS_{c,t} \\ & + \sum_{t=1}^{T} \sum_{o=1}^{O} \sum_{p=1}^{P} IHPS_{p,t} \cdot IPS_{p,t} + \sum_{t=1}^{T} \sum_{v=1}^{V} \sum_{p=1}^{P} IHPV_{v,p,t} \cdot IPV_{v,p,t} \\ & + \sum_{t=1}^{T} \sum_{p=1}^{P} PCPS_{p,t} \cdot XQT_{p,t} + \sum_{t=1}^{T} \sum_{p=1}^{P} BLC_{p,t} \cdot BL_{p,t} + \sum_{t=1}^{T} \sum_{p=1}^{P} BOC_{p,t} \\ & \cdot BO_{p,t} \\ & + \sum_{t=1}^{T} \sum_{f=1}^{F} \sum_{c=1}^{C} TCCF_{f,c,t} \cdot QTCFS_{f,c,t} \\ & + \sum_{t=1}^{T} \sum_{v=1}^{V} \sum_{p=1}^{P} TCPV_{v,p,t} \cdot QTPSV_{v,p,t} \\ & + \sum_{t=1}^{T} \sum_{o=1}^{O} \sum_{v=1}^{V} \sum_{p=1}^{P} TCPO_{o,v,p,t} \\ & \cdot QTPVO_{o,v,p,t} \end{aligned} \tag{12}$$

## 4 Computational Experiments and Results

This section presents the conducted computational experimentation and results for the mathematical model. The deterministic parameters have been generated randomly within a specified interval in order to respect the aforementioned model formulation. Namely, the OEMs' demands are generated between [2000, 3500] with a minimum demand between [100, 200] while the 1st tier supplier production capacity has been set to 3000, the lower delivery bound of raw materials' suppliers is set to 600, and the upper bound varying from 6000 to 8000. It is worth mentioning that the initial level of backlogs, backorders and product inventory has been set to zero due to customization context triggered by make to order policy. According to product modularization structure, the module affectation is driven by a random binary function considering the permanent presence of the basic module. The corresponding bills of materials are declared to drive component penetration and consumption level. The inventory at the distribution center is managed by a 3PL and its level is under the 1st tier supplier responsibility following the adopted vendor managed strategy, thus, the affectation link between distribution centers and OEMs assuming product assignment also is handled by a defined binary matrix. The computational experiment of the MILP model is coded in GAMS 22.5/CPLEX 12.2 optimization software and all numerical experiments are solved using a Core i5 2.49 GHz computer with 8 GB RAM. The obtained results show that products' inventory at the 1st tier supplier assumes a complete transfer to the distribution centers, which tend to highlight an optimal result for mass customization environment, in addition to a real constraint of the limited storage capacity at the 1st tier supplier plant. This trend is also emphasized for the inventory level at the distribution centers. In fact, it could be perceived that the defined production capacity outlines a prevailing parameter for the model, the contracted level between the 1st tier supplier and OEMs depicts the flows' shape through the supply chain. From procurement standpoint, the replenishment frequency from each 2nd tier supplier is applied to consider the lead time for the net base stock calculation in contrast to the continuous replenishment policy. The base stock definition follows the (s, S) inventory policy which assess the projected consumption level in order to be introduced for the net base stock and build up an accurate procurement data. The numerical results are shown in the Tables 1, 2, 3, 4, 5 and 6. In order to develop a global model cost assessment, three instances have been run. The first instance represents the aforementioned results, however, the modified parameters for the second instance are: (a) Increase the products variety set from 8 to 12 (b) Increase the OEMs set from 4 to (c) Increase the distribution centers from 4 to 8.

We can conclude that production, storage, distribution and backlogs costs have been increased with a normal trend due to product's portfolio growth, which rise up the related operational cost. The production and distribution costs represent respectively +52% and +36%, which explain the optimization leverage of demand fulfilment. This is aligned with the evolution trend of storage cost that is around 21%. For the third instance, the model modification concerns the planning horizon which has been increased to 15 periods instead of 10 periods, all others parameters have not been changed (Fig. 2). We can conclude that the cost evolution follows the same trend, which demonstrate the validity of the proposed model.

**Table 1.** Purchased quantity from supplier f of raw material c in period t

| | | QTCFS(f,c,t) | | | | | | | | | |
|---|---|---|---|---|---|---|---|---|---|---|---|
| | | t1 | t2 | t3 | t4 | t5 | t6 | t7 | t8 | t9 | t10 |
| f1 | c1 | 9793,10 | 10726,89 | 15395,04 | 600,00 | 17600,75 | 600,00 | 8712,67 | 13594,68 | 11314,61 | 7430,08 |
| f2 | c2 | 29905,77 | | 26369,64 | | 36469,26 | | 24232,75 | | 24657,83 | |
| f3 | c3 | 25588,07 | | 20575,60 | | 26382,89 | | 20618,51 | | 22791,73 | |
| f4 | c4 | 28729,32 | | 22947,53 | | 30387,84 | | 22691,33 | | 25464,12 | |

**Table 2.** Inventory level, net base stock and base stock of raw material c in period t

| | ICS(c,t) | | | | | | | | | |
|---|---|---|---|---|---|---|---|---|---|---|
| | t1 | t2 | t3 | t4 | t5 | t6 | t7 | t8 | t9 | t10 |
| c1 | 1862,50 | 45,26 | 4585,66 | 310,18 | 8907,43 | 624,95 | 886,66 | 4476,68 | 995,34 | 4408,14 |
| c2 | 15960,57 | | 10833,09 | | 19464,88 | | 14196,45 | | 8947,17 | |
| c3 | 15491,73 | | 7388,86 | | 13600,13 | | 10166,24 | | 6212,31 | |
| c4 | 17434,15 | | 8421,56 | | 15998,59 | | 10825,27 | | 7198,87 | |
| | BS(c,t) | | | | | | | | | |
| c1 | 9793,10 | 10726,89 | 10095,77 | 600,00 | 9278,81 | 600,00 | 8712,67 | 13594,68 | 11314,61 | 7430,08 |
| c2 | 15603,36 | 360,89 | 14620,72 | | 16727,86 | 1650,00 | 9578,89 | 7536,77 | 16814,04 | 8340,22 |
| c3 | 11962,60 | 49,52 | 12352,56 | 72,80 | 13074,49 | 1053,31 | 10794,32 | 7564,87 | 17248,99 | 5853,82 |
| c4 | 13279,63 | | 13645,98 | | 14597,35 | 1211,40 | 12093,46 | 8910,51 | 18854,96 | 6692,06 |
| | S(c,t) | | | | | | | | | |
| c1 | 9893,10 | 12589,40 | 10141,03 | 5185,66 | 9588,99 | 9507,43 | 9337,62 | 14481,34 | 15791,30 | 8425,42 |
| c2 | 15703,36 | 16321,46 | 14620,72 | 10833,09 | 16727,86 | 21114,88 | 9578,89 | 21733,22 | 16814,04 | 17287,38 |
| c3 | 12062,60 | 15541,24 | 12352,56 | 7461,66 | 13074,49 | 14653,44 | 10794,32 | 17731,11 | 17248,99 | 12066,13 |
| c4 | 13379,63 | 17434,15 | 13645,98 | 8421,56 | 14597,35 | 17209,99 | 12093,46 | 19735,78 | 18854,96 | 13890,93 |

**Table 3.** Total production quantity of product p in period t

| | XQT(p,t) | | | | | | | | | |
|---|---|---|---|---|---|---|---|---|---|---|
| | t1 | t2 | t3 | t4 | t5 | t6 | t7 | t8 | t9 | t10 |
| p1 | 3000,00 | 2233,26 | 2981,14 | 2992,33 | 2291,55 | 2545,28 | 2935,84 | 468,79 | 2620,96 | 2236,24 |
| p2 | 2018,78 | 2015,26 | 3000,00 | 3000,00 | 3000,00 | 3000,00 | 3000,00 | 2074,60 | 2829,55 | 1197,65 |
| p3 | 3000,00 | 2855,48 | 3000,00 | 2069,49 | 3000,00 | 2524,53 | 2176,01 | 1109,25 | 2668,29 | 2618,48 |
| p4 | 3000,00 | 2692,45 | 2336,26 | 2760,09 | 3000,00 | 2491,15 | 2223,18 | 3000,00 | 2376,55 | 276,06 |
| p5 | 2465,22 | 2060,30 | 3000,00 | 2578,19 | 2615,04 | 2453,87 | 2667,38 | 3000,00 | 595,43 | 2196,79 |
| p6 | 2241,87 | 2473,45 | 2858,09 | 2403,08 | 2054,59 | 3000,00 | 3000,00 | 82,05 | 3000,00 | 2265,17 |
| p7 | 3000,00 | 3000,00 | 3000,00 | 2189,04 | 3000,00 | 2833,13 | 2909,27 | 2110,11 | 1297,78 | 2499,07 |
| p8 | 2127,03 | 2858,25 | 2033,08 | 3000,00 | 3000,00 | 2841,23 | 3000,00 | 3000,00 | 2769,52 | 1056,07 |

**Table 4.** Transferred quantity of product p in period t from distribution center v to OEM o

| | | | QTPVO(o,v,p,t) | | | | | | | | |
|---|---|---|---|---|---|---|---|---|---|---|---|
| | | | t1 | t2 | t3 | t4 | t5 | t6 | t7 | t8 | t9 |
| o1 | v1 | p1 | 3000,00 | 2233,26 | 2916,07 | 2992,33 | 2291,55 | 2545,28 | 2935,84 | | |
| o1 | v1 | p2 | 2018,78 | 2015,26 | 3000,00 | 3000,00 | 3000,00 | 3000,00 | 2212,42 | 2074,60 | |
| o2 | v2 | p3 | 3000,00 | 2855,48 | 2459,31 | 2131,10 | 2645,75 | 2524,53 | 2176,01 | 1942,57 | |
| o2 | v2 | p4 | 3000,00 | 2320,68 | 2336,26 | 2813,50 | 2946,58 | 2491,15 | 2223,18 | | |
| o3 | v3 | p5 | 2465,22 | 1828,55 | 3231,75 | 2346,44 | 2615,04 | 2453,87 | 2667,38 | | |
| o3 | v3 | p6 | 2241,87 | 2473,45 | 2858,09 | 2403,08 | 2054,59 | 2988,05 | 3011,95 | | 3082,05 |
| o4 | v4 | p7 | 3000,00 | 3000,00 | 3000,00 | 2189,04 | 2424,55 | 2833,13 | 2620,99 | 2110,11 | |
| o4 | v4 | p8 | 2127,03 | 2858,25 | 2033,08 | 3000,00 | 3000,00 | 2841,23 | 2709,24 | 3076,35 | |

**Table 5.** Production quantity to satisfy demand (XQD) and backlog (XQB), levels of backlog (BL) and backorder (BO) of product p in period t

| XQD(p,t) | t1 | t2 | t3 | t4 | t5 | t6 | t7 | t8 | t9 | t10 |
|---|---|---|---|---|---|---|---|---|---|---|
| p1 | 3000,00 | 2233,26 | 2916,07 | 2992,33 | 2291,55 | 2545,28 | 2935,84 | 468,79 | 2620,96 | 2236,24 |
| p2 | 2018,78 | 2015,26 | 3000,00 | 3000,00 | 3000,00 | 3000,00 | 2212,42 | 2074,60 | 2829,55 | 1197,65 |
| p3 | 3000,00 | 2855,48 | 2459,31 | 2069,49 | 2645,75 | 2524,53 | 2176,01 | 1109,25 | 2668,29 | 2618,48 |
| p4 | 3000,00 | 2320,68 | 2336,26 | 2760,09 | 2946,58 | 2491,15 | 2223,18 | 3000,00 | 2376,55 | 276,06 |
| p5 | 2465,22 | 2060,30 | 3000,00 | 2346,44 | 2615,04 | 2453,87 | 2667,38 | 3000,00 | 595,43 | 2196,79 |
| p6 | 2241,87 | 2473,45 | 2858,09 | 2403,08 | 2054,59 | 3000,00 | 3000,00 | 82,05 | 3000,00 | 2265,17 |
| p7 | 3000,00 | 3000,00 | 3000,00 | 2189,04 | 2424,55 | 2833,13 | 2620,99 | 2110,11 | 1297,78 | 2499,07 |
| p8 | 2127,03 | 2858,25 | 2033,08 | 3000,00 | 3000,00 | 2841,23 | 2709,24 | 3000,00 | 2769,52 | 1056,07 |
| **XQB(p,t)** | | | | | | | | | | |
| p1 | | | 65,07 | | | | | | | |
| p2 | | | | | | | 787,58 | | | |
| p3 | | | 540,69 | | 354,25 | | | | | |
| p4 | | 371,77 | | | 53,42 | | | | | |
| p5 | | | | 231,75 | | | | | | |
| p7 | | | | | 575,45 | | 288,29 | | | |
| p8 | | | | | | | 290,76 | | | |
| **BL(p,t)** | | | | | | | | | | |
| p1 | 65,07 | 65,07 | | | | | | 2628,30 | 2628,30 | 2628,30 |
| p2 | | | 428,05 | 893,07 | 1342,54 | 1626,96 | 839,38 | 839,38 | 839,38 | 1917,78 |
| p3 | 491,25 | 849,40 | 308,71 | 370,32 | 16,07 | 16,07 | 16,07 | 1785,80 | 1785,80 | 1785,80 |
| p4 | 371,77 | | | 53,42 | | | | 393,72 | 393,72 | 2211,54 |
| p5 | | | 231,75 | | | | | 74,00 | 2368,30 | 2368,30 |
| p6 | | | | | | 29,59 | 41,53 | 2457,68 | 2597,59 | 2597,59 |
| p7 | 23,72 | 33,20 | 280,03 | 863,74 | 288,29 | 288,29 | | | 1911,23 | 1911,23 |
| p8 | | | | 113,06 | 470,71 | 470,71 | 179,95 | 256,29 | 256,29 | 2530,84 |
| **BO(p,t)** | | | | | | | | | | |
| p1 | 65,07 | | | | | | | 2628,30 | | |
| p2 | | | 428,05 | 465,02 | 449,48 | 284,42 | | | | 1078,40 |
| p3 | 491,25 | 358,15 | | 61,61 | | | | 1769,73 | | |
| p4 | 371,77 | | | 53,42 | | | | 393,72 | | 1817,82 |
| p5 | | | 231,75 | | | | | 74,00 | 2294,30 | |
| p6 | | | | | | 29,59 | 11,95 | 2416,15 | 139,91 | |
| p7 | 23,72 | 9,48 | 246,82 | 583,71 | | | | | 1911,23 | |
| p8 | | | | 113,06 | 357,65 | | | 76,35 | | 2274,55 |

**Table 6.** Inventory level and the transferred quantity to the distribution center v of product p in period t

| IPV(v,p,t) | t2 | t3 | t4 | t5 | t6 | t7 | t8 | t9 | t10 |
|---|---|---|---|---|---|---|---|---|---|
| v1 p1 | | 65,07 | 65,07 | 65,07 | 65,07 | 65,07 | 533,86 | 3154,82 | 3154,82 |
| v1 p2 | | | | | | 787,58 | 787,58 | 3617,13 | 3617,13 |
| v2 p3 | | 540,69 | 479,08 | 833,33 | 833,33 | 833,33 | | 2668,29 | 2668,29 |
| v2 p4 | 371,77 | 371,77 | 318,36 | 371,77 | 371,77 | 371,77 | 3371,77 | 5748,32 | 5748,32 |
| v3 p5 | 231,75 | | 231,75 | 231,75 | 231,75 | 231,75 | 3231,75 | 3827,18 | 6023,97 |
| v3 p6 | | | | | 11,95 | | 82,05 | | |
| v4 p7 | | | | 575,45 | 575,45 | 863,74 | 863,74 | 2161,52 | 2161,52 |
| v4 p8 | | | | | | 290,76 | 214,42 | 2983,93 | 4040,00 |
| **QTPSV(v,p,t)** | | | | | | | | | |
| v1 p1 | 3000,00 | 2233,26 | 2981,14 | 2992,33 | 2291,55 | 2545,28 | 2935,84 | 468,79 | 2620,96 |
| v1 p2 | 2018,78 | 2015,26 | 3000,00 | 3000,00 | 3000,00 | 3000,00 | 3000,00 | 2074,60 | 2829,55 |
| v2 p3 | 3000,00 | 2855,48 | 3000,00 | 2069,49 | 3000,00 | 2524,53 | 2176,01 | 1109,25 | 2668,29 |
| v2 p4 | 3000,00 | 2692,45 | 2336,26 | 2760,09 | 3000,00 | 2491,15 | 2223,18 | 3000,00 | 2376,55 |
| v3 p5 | 2465,22 | 2060,30 | 3000,00 | 2578,19 | 2615,04 | 2453,87 | 2667,38 | 3000,00 | 595,43 |
| v3 p6 | 2241,87 | 2473,45 | 2858,09 | 2403,08 | 2054,59 | 3000,00 | 3000,00 | 82,05 | 3000,00 |
| v4 p7 | 3000,00 | 3000,00 | 3000,00 | 2189,04 | 3000,00 | 2833,13 | 2909,27 | 2110,11 | 1297,78 |
| v4 p8 | 2127,03 | 2858,25 | 2033,08 | 3000,00 | 3000,00 | 2841,23 | 3000,00 | 3000,00 | 2769,52 |

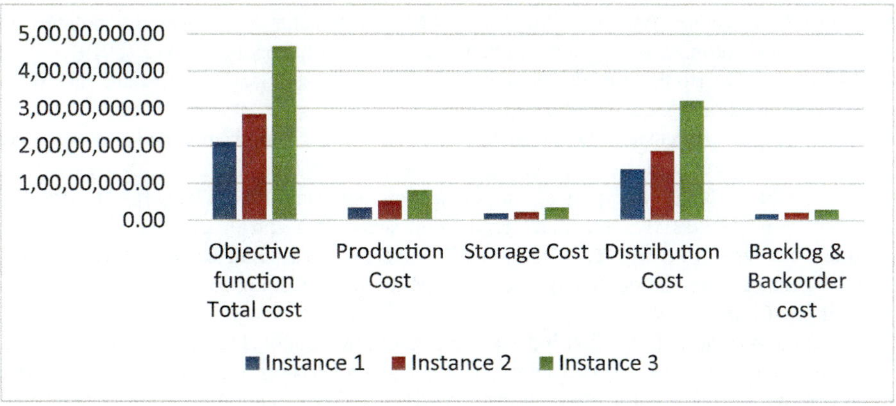


**Fig. 2.** Cost structure obtained from the different instances

# 5 Conclusion

A deterministic model formulation has been developed for a 3-echelon supply chain with mass customization foundation. The proposed model is an integrated procurement, production and distribution tactical planning according to a centralized decision-making process with contractual features across partners which are presented as model's assumptions. The studied real case arises from the automotive industry and due to customization abilities, the OEMs have been pointed out as supply chain leaders while their downstream stakeholders as followers. The model formulation focused on the $1^{st}$ tier supplier operations (i.e. wire harnesses manufacturer), in fact, the (s, S) inventory policy according to a base stock approach is adopted for procurement function while the related assessment is based on OEMs demand forecast.

Future work will focus on further investigations of mass customization framework to strengthen the model formulation raised from a real-life problem, besides keeping the abilities of a potential application of the model to another supply chains. For example developing a strengthen backlog's production process to cope with the restricted delivery window exhibited by customized products. Given the different uncertainty levels arisen from the overall dynamic behavior of the network (e.g. market demand, customer preferences…) which compel, for instance, to an inventory management based on forecast, it is worthy to enhance further studies on the most suitable uncertainty approach for such a supply chain, and develop an efficient model to achieve a higher accuracy levels.

# References

1. Eruguz, A.S., Sahin, E., Jemai, Z., Dallery, Y.: A comprehensive survey of guaranteed-service models for multi-echelon inventory optimization. Int. J. Prod. Econ. **172**, 110–125 (2016)
2. Beamon, B.M.: Measuring supply chain performance. Int. J. Oper. Prod. Manag. **19**(3), 275–292 (1999)

3. Brunaud, B., Laínez-Aguirre, J.M., Pinto, J.M., Grossmann, I.E.: Inventory policies and safety stock optimization for supply chain planning. AIChE **65**(1), 99–112 (2019)
4. Da Silveira, G., Borenstein, D., Fogliatto, F.S.: Mass customization: literature review and research directions. Int. J. Prod. Econ. **72**(1), 1–13 (2001)
5. Da Silveira, G., Borenstein, D., Fogliatto, F.S.: The mass customization decade: an updated review of the literature. Int. J. Prod. Econ. **138**(1), 14–25 (2012)
6. Davis, S.M.: From "future perfect": mass customizing. Plan. Rev. **17**(2), 16–21 (1989)
7. Kesen, S.E., Bektaş, T.: Integrated production scheduling and distribution planning with time windows. In: Paksoy, T., Weber, G.W., Huber, S. (eds.) Lean and Green Supply Chain Management. International Series in Operations Research & Management Science, vol. 273. Springer, Cham (2019)
8. Moncayo-Martínez, L.A., Reséndiz-Flores, E.O., Mercado, D., Sánchez-Ramírez, C.: Placing safety stock in logistic networks under guaranteed-service time inventory models: an application to the automotive industry. J. Appl. Res. Technol. **12**(3), 538–550 (2014)
9. Darvish, M., Coelho, L.C.: Sequential versus integrated optimization: production, location, inventory control, and distribution. Eur. J. Oper. Res. **268**(1), 203–214 (2018)
10. Graves, S.C., Schoenmeyr, T.: Strategic safety-stock placement in supply chains with capacity constraints. Manuf. Serv. Oper. Manag. **18**(3), 445–460 (2016)
11. Um, J., Lyons, A., Lam, H.K.S., Cheng, T.C.E., Dominguez-Pery, C.: Product variety management and supply chain performance: a capability perspective on their relationships and competitiveness implications. Int. J. Prod. Econ. **187**, 15–26 (2017)
12. Wei, W., Guimarães, L., Amorim, P., Almada-Lobo, B.: Tactical production and distribution planning with dependency issues on the production process. OMEGA **67**, 99–114 (2016)
13. Zebardast, M., Mapelzi, S., Taish, M.: Mass customization in supply chain level: development of a conceptual framework to manage and assess performance. In: International Conference, APMS, Proceedings, Part II, pp. 81–90 (2013)
14. Zhiqiang, W., Zhang, M., Sun, H., Zhu, G.: Effects of standardization and innovation on mass customization: an empirical investigation. Technovation **48–49**, 79–86 (2016)

# Efficiency Analysis of Performance in Container Terminals, Case Study of Moroccan Ports

Mouhsene Fri[1,2](✉), Kaoutar Douaioui[1], Nabil Lamii[1], Charif Mabrouki[1], and El Alami Semma[1]

[1] Laboratory of Engineering, Industrial Management and Innovation, Faculty of Sciences and Technology, Hassan 1st University, PO Box 577, Settat, Morocco
frimouhsene@gmail.com, kaoutar.douaioui@gmail.com, {n.lamii, charif.mabrouki}@uhp.ac.ma, semmaalam@yahoo.fr

[2] CELOG-Ecole Supérieure des Industries du Textile et Habillement, PO Box 7731 ULFA, Casablanca, Morocco

**Abstract.** This study aims to provide a decision support to port managers in order to evaluate and benchmark performance measurement systems. The decision support can assess the judgment of decision-makers, analyses the sensitivity and the robustness of performance measurement systems and elaborates recommendations. It is composed of two tools: the DELPHI technique and MACBETH method (Measuring Attractiveness by a Categorical Based Evaluation Technique). A case study involving a Moroccan port illustrates the proposed system.

**Keywords:** Decision support · MACBETH · Delphi · Performance measurement systems

## 1 Introduction

Ports occupy a crucial place in the field of international transport, logistics, trade and economic growth [1]. Port terminals are complex environments in which different actors interact at different levels, moreover, influenced by industry 4.0, the port evolve continuously [2, 3]. So, to stay competitive in such a competitive and challenging environment, ports need to improve their performance. On the one hand, ports must be flexible to meet the changing demands of its customers, who are becoming more and more demanding; and on the other hand, the ports must effectively use their facilities and equipment. As a result, ports must optimize the use of their existing infrastructure and equipment in order to increase their productivity and competitiveness. [4], as well as evaluate their performance in order to identify the key factors [5]. This is what motivated the universities to conduct many research to evaluate port performance [6]. But, despite the merits of the mentioned works, there is no standard methodology to assess and analyze performance in ports, which leads us to propose an effective methodology.


A. Saka et al. (Eds.): CPI 2019, LNME, pp. 365–371, 2021.
https://doi.org/10.1007/978-3-030-62199-5_32

To do so, the paper is divided into five sections: the first section aims to introduce the field of research, the second section discuss the performance measurement system and the identification of the convenable key drivers of our systems. However, the development of a performance measurement system is not the end in itself, so in the third section we propose a decision support to evaluate and analyze the performance of organization by using the performance measurement system. In the fourth section we discuss the results by analyzing the sensitivity and robustness of the proposed decision support. Finally, in the five section we draw some conclusions and propose various perspectives in this field.

## 2 Performance Measurement System

Performance measurement system is necessary and vital in each organization. Without PMS we can't measure the system and we can't improve the overall performance. In our case we will use the performance measurement system based on key drivers proposed by [7] in which, authors based their identification of key drivers on the peer-reviewed international journals published between the 1983 and 2014. The Table 1 presents the mentioned key drivers.

**Table 1.** The key drivers of port competitiveness

| Key drivers | Key code | Definition |
|---|---|---|
| Port costs | PC | The cost bearded by port's customers is a function of direct port costs such as port charges, storage and stevedoring, as well as indirect costs incurred during lengthy port stops |
| Hinterland proximity | HP | Hinterland proximity refers to the geographical proximity of the main hinterland markets served by a port (both local/captive markets and others, more distant and contestable) |
| Hinterland connectivity | HC | Hinterland connectivity refers to the efficiency of inland transport networks (e.g. rail and road transport) |
| Port geographical location | PG | Geographical location has an inclusive meaning and refers to the spatial positioning of the port respect to shipping networks, inland market areas, inland transport infrastructure, logistics centers, consuming markets, urban areas, etc. |
| Port infrastructure | PI | Port infrastructure are evaluated on the basis of the number and quality of available infrastructure (e.g. breakwater, quay wall, yard surface, etc.), as well as in relation to their appropriateness respect to the customer's needs and environmental concerns |
| Operational efficiency | OE | Capacity of a port to employ all its resources efficiently to deliver high operational performance (e.g. ship turnaround time, ship waiting times due to congestion, cargo handling productivity, etc.) |

(*continued*)

**Table 1.** (*continued*)

| Key drivers | Key code | Definition |
|---|---|---|
| Port service quality | PSQ | Port service quality refers to the quality of (all) port facilities, and to the capacity of differentiating services supplied from competitors |
| Maritime connectivity | MC | Maritime connectivity refers to the efficiency of shipping transport networks (e.g. number and variety of served destinations, logistics costs, etc.) |
| Nautical accessibility | NA | Nautical accessibility refers to the capacity of a port to accommodate large vessels at any time, regardless of tide and weather conditions. It is affected by natural factors (e.g. depth of inland rivers, tide range, etc.) and the endowment of physical infrastructure (e.g. locks, breakwaters, etc.) |
| Port site | PS | Port site refers to the extension of the entire port area, the quality of terminal layouts and common spaces, as well as its appropriateness respect to the needs of port users |

Decision Support of Performance Measurement Systems
There are lots of ways and methods to evaluate organization but the majority of them are based on expert's judgments, which remains a weakness since if we have a wrong judgment, we will have a mislead conclusion. Thereby, we reinforce this process by adding the DELPHI and MACBETH tools; DELPHI ensures having a consensus between different experts and MACBETH detect possible inconsistencies in judgment.

### 2.1 DELPHI Technique

The Delphi technique was developed by [8] at Rand Corporation. The Delphi process is a structured methodology for obtaining the opinions of several experts who are subjected to a series of intensive questionnaires taking into account the feedback of other experts. Interviewees do not engage in a discussion among themselves, thus they avoid direct discussions. During a Delphi survey, experts participate in an iteration of questionnaires. After each round, they are allowed to consult the intermediate results and have the opportunity to re-examine their judgment on the basis of the judgment of the other experts during the rounds of questionnaires [9]. The number of rounds of the Delphi method depends on the convergence of the judgments of the experts. In our case we called on eight experts in the port field. Who will evaluate and determine the interaction between the different criteria.

### 2.2 MACBETH Tool

The evaluation of the global performance must take into account all criterion and the interaction between different criterion. To achieve that, we chose the MACBETH method [10, 11]. MACBETH is a Multi-criteria decision making (MCDM) approach to assist decision-makers to quantify the attractiveness among the options of each criterion. The strength of this method is that the decision maker is not required to define the weight of the criteria. However, it aims to justify and transform judgments between two elements at a time. After entering the judgments to M-MACBETH software, it assists the experts in the implementation of the whole multi-criteria process and reveal automatically possible inconsistencies in the judgments and generate a weight of each criterion. Furthermore, the M-Macbeth software provides a tool to investigate deeply the sensitivity and the robustness.

## 3 Result

To validate the proposed decision support, we apply this model into the port of Casablanca and the port of Tanger Med, with the fruitful collaboration of eight experts in performance of terminal containers. MACBETH evaluates the global performance on a scale from one to five. Finally, we have the global result presented in Fig. 2 that contains the weight of each key driver of the performance measurement system and shows the level of the overall performance of Tanger med port is 4/5. To have a deeper understanding and validate the model, we conduct a sensitivity, attractiveness and robustness analysis (Fig. 1).

Table of scores

| Options | Overall | PC | HP | HC | PGL | PI | OE | PSQ | MC | NA | PS |
|---|---|---|---|---|---|---|---|---|---|---|---|
| L 1 | 0.00 | 0.00 | 0.00 | 0.00 | 0.00 | 0.00 | 0.00 | 0.00 | 0.00 | 0.00 | 0.00 |
| L 2 | 26.66 | 33.33 | 28.57 | 27.27 | 25.00 | 25.00 | 22.22 | 30.00 | 30.00 | 25.00 | 20.00 |
| L 3 | 67.28 | 66.67 | 85.71 | 63.64 | 50.00 | 50.00 | 77.78 | 80.00 | 60.00 | 50.00 | 70.00 |
| L 4 | 98.39 | 100.00 | 100.00 | 100.00 | 100.00 | 100.00 | 100.00 | 100.00 | 80.00 | 75.00 | 100.00 |
| L 5 | 72.59 | 88.89 | 71.43 | 81.82 | 83.33 | 87.50 | 55.56 | 50.00 | 100.00 | 100.00 | 50.00 |
| [ all upper ] | 100.00 | 100.00 | 100.00 | 100.00 | 100.00 | 100.00 | 100.00 | 100.00 | 100.00 | 100.00 | 100.00 |
| [ all lower ] | 0.00 | 0.00 | 0.00 | 0.00 | 0.00 | 0.00 | 0.00 | 0.00 | 0.00 | 0.00 | 0.00 |
| Weights: | | 0.2857 | 0.0357 | 0.0357 | 0.0357 | 0.1072 | 0.1429 | 0.0714 | 0.0357 | 0.0357 | 0.2143 |

**Fig. 1.** The overall performance of Tanger Med Port

### 3.1 Sensitivity Analysis

Sensitivity analysis aims to calculate the uncertainty in output influenced by the uncertainty applied to the input. MACBETH software provides a tool to analyze the sensitivity; for example, Fig. 3 shows the sensitivity analysis of interaction between Level 5 (L5) and Level 3 (L3). We can conclude that we don't have any level change unless Wpc (weight of criteria pc) from goes from 28.57% to 6.2%, (drops 22.37%) in this case the L3 will be higher than L5. besides that, the other levels remain at their placements. The Table 2 explains the different change if we change the weight of criteria.

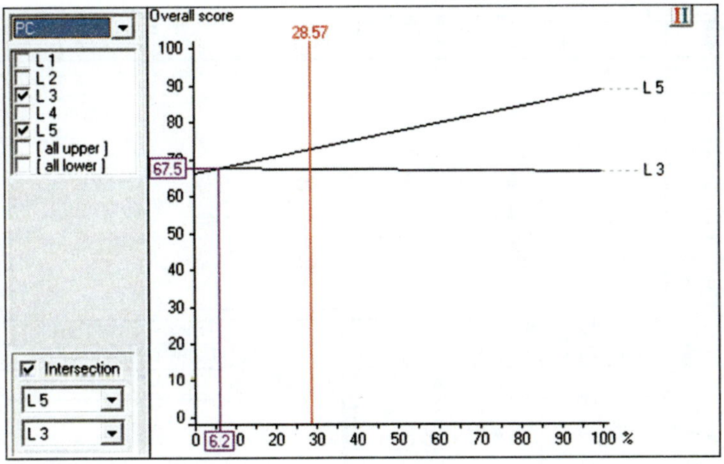


**Fig. 2.** The example sensitivity analysis

**Table 2.** The summary of sensitivity analysis in Tanged Med Port.

| Criteria | Max tolerated change in % | Sensitivity |
|---|---|---|
| PC | 22.37 | If Wpc is below 6.2, L3 outperform L5. otherwise L5 outperform L3 |
| HP | 26.13 | If Whp is below 29.7, L3 outperform L5. otherwise L5 outperform L3 |
| HC | 100 | Any reverse of performance despite the change of weight |
| PG | 100 | Any reverse of performance despite the change of weight |
| PI | 100 | Any reverse of performance despite the change of weight |
| OE | 16.51 | If Woe is below 30.8, L5 outperform L3. otherwise L3 outperform L5 |
| PSQ | 13.96 | If Wpsq is below 21.1, L5 outperform L3. otherwise L3 outperform L5 |
| MC | 54.33 | If Wmc is below 57.9, L4 outperform L5. otherwise L5 outperform L4 |
| NA | 49.3 | If Wna is below 52.6, L4 outperform L5. otherwise L5 outperform L4 |
| PS | 16.56 | If Wna is below 37.9, L5 outperform L3. otherwise L3 outperform L5 |

### 3.2 Robustness Analysis

To evaluate our results, we conduct the robustness analysis using the respective function of the M-MACBETH software. Which shows that a simultaneous change of up to 5 value points across all of the attribute reference levels would not impact the ranking of the alternative levels.

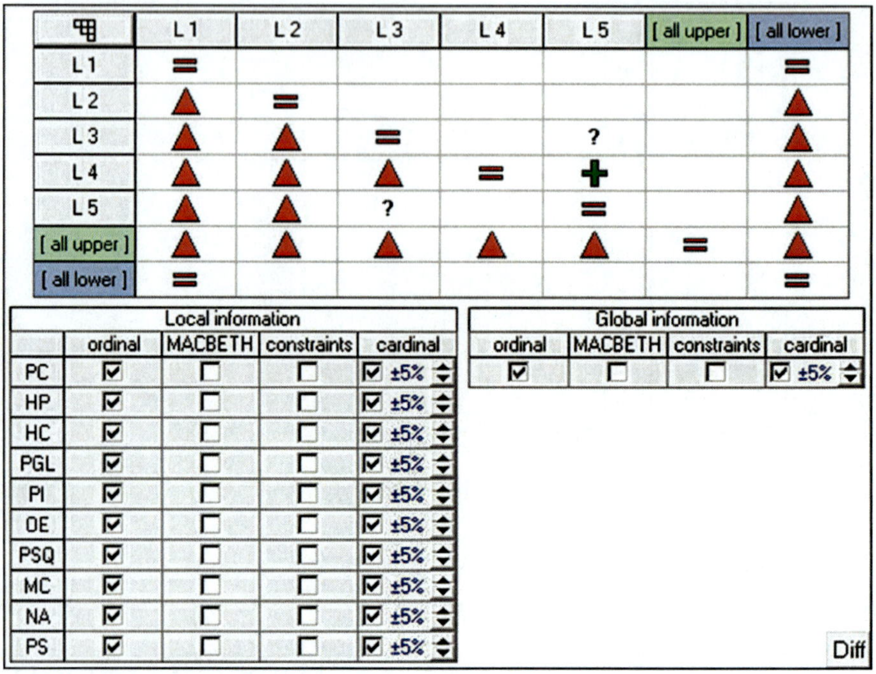

**Fig. 3.** The robustness analysis of Tanger Med Port.

### 3.3 Overall Result

With the same methodology we analyze the sensitivity and robustness of the model applied to Casa Port. We check the global sensitivity of each port by at least 10% variation in the weight of each objective. Furthermore, we have no disturbance in our results which proves the strength of our model. In the same way, applying a simultaneous change up to 5% s over all of the attribute levels would not impact the ranking of the alternative levels, which proves the robustness of our model. The Table 3 shows the overall performance of two Moroccan cases studies.

**Table 3.** Description of global performance of decision support systems.

| Port container terminal | Global performance of 5 | Sensitivity | Robustness |
|---|---|---|---|
| Casa port container terminal | 3 | Stable | Robust |
| Tanger Med container terminal | 4 | Stable | Robust |

## 4 Conclusions and Future Works

This work provides a decision support used to make an evaluation/benchmarking among different port container terminals, to evaluate them, and to detect if the tendency of performance is going higher or lower and to determine the improvements to be made. The strength of our model is the coming of DELPHI method which gives a good judgment, and the MACBETH tool which detects all the inconsistencies during the judgments of the experts and proposes the most consistent judgments.

As a perspective, it is essential to standardize the performance measurement system and the methodology to measure the key drivers. Having different methods of measurement and identification, these indicators became trivial. So, there is still considerable directions for additional research. In addition, all performance measurement systems must take into consideration the new revolution in smart industry and smart logistics.

## References


1. Wang, T., Cullinane, K.: The efficiency of European container terminals and implications for supply chain management. In: Haralambides, H.E. (ed.) Port Management, pp. 253–272. Palgrave Macmillan, London (2015)
2. Douaioui, K., Fri, M., Mabrouki, C., Semma, E.A.: Smart port: design and perspectives, pp. 1–6 (2018)
3. Douaioui, K., Fri, M., Mabroukki, C., Semma, E.A.: The interaction between industry 4.0 and smart logistics: concepts and perspectives. In: 2018 International Colloquium on Logistics and Supply Chain Management (LOGISTIQUA), Tangier, pp. 128–132 (2018). https://doi.org/10.1109/LOGISTIQUA.2018.8428300
4. Benghalia, A., Boukachour, J., Boudebous, D.: Gestion du transfert interne de conteneurs: le cas du port du Havre. Logistique Manag. **24**(1), 57–69 (2016). https://doi.org/10.1080/12507970.2016.1207490
5. Fri, M., Fedouaki, F., Douaioui, K., Mabrouki, C., Semma, E.A.: Supply chain performance evaluation models, state-of-the-art and future directions. Int. J. Eng. Adv. Technol. **9**(1), 6336–6347 (2019). https://doi.org/10.35940/ijeat.a2049.109119
6. Bentaleb, F., Fri, M., Mabrouki, C., Semma, A.: Dry port-seaport system development: application of the product life cycle theory. J. Transp. Logist. **1**(2), 115 (2016). https://doi.org/10.22532/jtl.267840
7. Parola, F., Risitano, M., Ferretti, M., Panetti, E.: The drivers of port competitiveness: a critical review. Transp. Rev. **37**(1), 116–138 (2017). https://doi.org/10.1080/01441647.2016.1231232
8. Dalkey, N., Helmer, O.: An experimental application of the Delphi method to the use of experts. Manag. Sci. **9**, 458–467 (1963)
9. Linstone, H.A., Turoff, M.: Delphi: a brief look backward and forward. Technol. Forecast. Soc. Change **78**(9), 1712–1719 (2011). https://doi.org/10.1016/j.techfore.2010.09.011
10. Costa, C.A.B.e., Chagas, M.P.: A career choice problem: An example of how to use MACBETH to build a quantitative value model based on qualitative value judgments. Eur. J. Oper. Res. **153**(2), 323–331 (2004). https://doi.org/10.1016/s0377-2217(03)00155-3
11. Costa, C.A.B.e., de Corte, J.-M., Vansnick, J.-C.: MACBETH. Int. J. Inf. Technol. Decis. Mak. **11**(02), 359–387 (2012). https://doi.org/10.1142/s0219622012400068

# Supply Chain Planning of Off-Shores Winds Farms Operations: A Review

Mustapha Hrouga[1(✉)] and Nathalie Bostel[2]

[1] Université d'Artois, EA 3926, Laboratoire de Génie Informatique et d'Automatique de l'Artois, department de génie industriel et logistique (LGI2A), 62400 Béthune, France
m.hrouga@gmail.com
[2] Université de Nantes – Faculté des Sciences et Techniques (FST) Bâtiment 34, Laboratoire des Sciences du Numérique de Nantes (LS2N), 2 Chemin de la Houssinière, BP 92208, 44322 Nantes, France

**Abstract.** The development of renewable marine energy at sea is at the mind of the energy transition and blue growth in which France is committed. The domain includes the component manufacturing, assembly, installation and maintenance of the offshore wind farm (OWF). However, the global supply chain of the OWF is very complicated. This complexity is due to the size of wind turbine components. Indeed, this latter is very large: a total height of 150 m and a diameter of 120 m and therefore, require special handling and logistics. To avoid disruption of the global supply chain process of offshore wind, all aspects that can impact the latter must be analyzed. Disturbances can range from capacity problems to quality problems via different weather conditions. Between production and offshore installation phase, the supply chain includes onshore transportation, port handling and shipping. For example, any delay in supply will be fundamentally disrupting the production or installation process of the farm. The produced components from a wind turbine cannot be transported as standard components because they are for the most part very bulky and heavy. To take into account these specificities, transport can be made for example at night when it is possible to use exclusive roads. The purpose of this paper aims to realize a review of logistics planning for OWF operations. A classification according to strategic, tactical and operational decision of logistics planning for the OWF operations was conducted.

**Keywords:** Offshore wind · Installation · Optimization · Maritime supply chain · Supply chain concepts · Logistics strategies

## 1 Introduction

In the last decade, the development of renewable energies has preoccupied the governments around the world. This made these type of energies one of the most important subjects in the energy area. This interest is motivated mainly by the global climate change, the progressive need for energy all while still looking for new market opportunities. In addition, the Energy Transition Law has set an ambitious target to increase total energy consumption by around 32% in 2030. In other words, renewable


A. Saka et al. (Eds.): CPI 2019, LNME, pp. 372–387, 2021.
https://doi.org/10.1007/978-3-030-62199-5_33

energies must therefore represent about 40% relatively to electricity production, 38% as to heat consumption, 15% of fuel consumption and only 10% for gas consumption. More particularly in France, renewable energy represented about 18.7% of electricity consumption in 2015. Despite this progress, there are some challenges to face for the development of renewable energies, such as the preservation of the planet. Indeed, each country involved in the development of renewable energy must guarantee to consumers a sufficient supply of energy, while ensuring also the sustainability of production and supply. The states are also obliged to find alternative sources of renewable energy that must be sustainable and free of greenhouse gas (GHG) emissions.

There are other types of renewable energies such as: hydropower, solar, biomass and geothermal energy. The common point of these various energies is not to produce polluting emissions while also fighting against the greenhouse effect. Despite this fragmentation, wind energy is considered the most efficient and sustainable renewable energy. Indeed, it provides many advantages such as: it does not need fuel, it does not generate greenhouse gases either, as it is 100% natural, renewable and sustainable, and finally it does not produce toxic or radioactive waste. In addition, it helps to fight against climate change. Finally, the wind energy yield increases in winter because the wind is stronger in this season. Regardless of all the advantages mentioned above, the cost of building a wind farm is high since its installation requires rare and expensive means of manufacturing, transport, very special handling, in addition to the components of the wind that require very large storage space. There are two types of wind turbines: the first type is called on-shore wind turbines that we find on earth, also called onshore wind turbines, the second type is called offshore wind that can be found in the sea. We can also find two types of offshore wind turbines: wind putting down and floating wind turbines. The first one is located at medium distances from the beach. As for the floating offshore wind turbine is at sea at distances a little far from the coast. The second one is located in the sea in places with particular weather conditions where the winds are stronger and stable. For this reason the electricity production of an offshore wind farm is significant and exceeds the production of onshore wind farms.

Furthermore, OWFs are often located in the places where the wind is stronger and more regular. The largest offshore wind farm in the world is located in Denmark, it was inaugurated in 2009 under the name of "Horns Rev 2". It is composed of 91 winds with a capacity of 2.3 MW each with a total capacity around of 210 MW. It is located on an area of 35 $km^2$ in the west of the country. The farm could cover at the time of the annual needs of approximately 200 000 inhabitants. Moreover, offshore winds are designed in an identical manner as those of on-shores winds but they have larger sizes. In fact, the construction of wind in the port area and their maritime transport make it possible to build very large winds. For example, General Electric, Areva, Vestas and Simens offer wind with a unit capacity of 5 to 7 MW. They have the particularity of being grouped together in a marine wind farm to facilitate transportation, installation, energy management, transport and maintenance.

Finally, the main reason why renewable energy companies use offshore winds is that the wind resources are more important at sea than on land, we can get more potential energy through higher capacity factors with more stable production. Another common argument is that there is less competition for offshore areas, so potential land use conflicts are avoided [32]. However, these advantages are limited by the installation and operation costs which are much higher than those of onshore winds.

This work focuses mainly on a review for the supply chain planning of offshore wind farm (OWF). In our contribution, we first identify the various issues addressed such as: decision support tools for planning the logistical operations of offshore winds, the proposed models and the methods of resolutions. Then, we identify the different models and resolution methods which will be classified according to the three levels of decisions, strategic, tactical and operational. To the best of our knowledge, we are the first authors to classify the work of supply chain planning of offshore wind farm according to the three levels of decisions: strategic, tactical and operational.

The rest of this paper is organized as follow: an introduction of development and interests of offshore winds have been made in the Sect. 1. A strategic, tactical and operational level of supply chain for offshore wind farm is presented in Sect. 2. Section 3 concludes the paper.

## 1.1 Development of Offshore Wind

Nowadays, the majority of offshore wind farm (OWF) are located in Europe with an installed capacity of 12GW approximately, more specifically in Denmark, Germany and the United Kingdom. These countries have already planned to install double in 2030. Many other countries, including China which is ranked first, have also entered the field of offshore wind energy. The UK offshore wind farm is the first of this kind that has already installed a capacity of 5.15 GW in 2016, this important capacity covers the electricity needs of several million people. However, France is a little behind compared to European countries, although it has the best wind resources in Europe. The figure below represents the megawatt capacity installed around the world in 2016. According to this figure, we note that the United Kingdom is ranked first, unlike France, which has no installed capacity (Fig. 1).

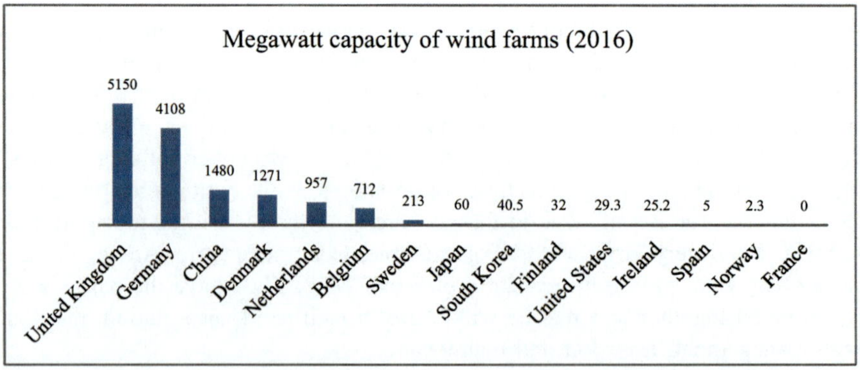


**Fig. 1.** MW capacity of wind farms (2016) (Source: Wind Europe)

According to a study of («Offshore wind in Europe-key trends and statistics 2017», Wind Europe 2018), more than 4149 off-shores winds were connected to the electrical network in Europe at the end of 2017. These winds are spread over 92 wind farms of which 11 countries are involved. The cumulative total installed capacity is approximately 15780 MW for all wind farms. The locations of the wind farms vary according to the technology used in the construction of the offshore winds as well as the geography and winds resources. In fact, almost 99% of this installed capacity in Europe is shared between 6 countries such as: the United Kingdom: 6835 MW with 1753 winds connected to the electrical network, Germany: 5355 MW with 1169 winds, Denmark: 1266 MW with 506 winds, Netherlands: 1118 MW with 365 winds, Belgium: 877 MW with 232 winds and Sweden with a total installed capacity of 202 MW for 86 winds. Table 1 show the different offshore wind farms with the installed capacity of each farm in the European Union.

**Table 1.** Offshore wind Projects with more than 10 years of operation (LORC 2016) [17]

| Country | Project name | Wind farm size (MW) | Winds number |
|---|---|---|---|
| Allemagne | Breitling Demonstration | 2.5 | 1 |
| Allemagne | Ems Emden | 4.5 | 1 |
| Danemark | Frederikshavn | 7.6 | 3 |
| Danemark | HornsRev 1 | 160 | 80 |
| Danemark | Middelgrunden | 40 | 20 |
| Danemark | Nysted 1 | 165.6 | 72 |
| Danemark | Ronland | 17.2 | 8 |
| Danemark | Samso | 23 | 10 |
| Danemark | TunoKnob | 5 | 10 |
| Danemark | Vindeby | 4.95 | 11 |
| Irlande | Arklow Bank 1 | 25.2 | 7 |
| Japon | Sakata | 16 | 8 |
| Japon | Setana | 1.32 | 2 |
| Pays-Bas | IreneVorrink | 16.8 | 28 |
| Pays-Bas | Lely | 2 | 4 |
| Royaume-Uni | Barrow | 90 | 30 |
| Royaume-Uni | Blyth | 4 | 2 |
| Royaume-Uni | Kentish Flats 1 | 90 | 30 |
| Royaume-Uni | North Hoyle | 60 | 30 |
| Royaume-Uni | ScrobySands | 60 | 30 |
| Suède | Bockstigen | 2.5 | 5 |
| Suède | Utgrunden 1 | 10.5 | 7 |
| Suède | YttreStengrund | 10 | 5 |

According to this table, the cumulative installed capacity of OWF increases rapidly. This can be explained by the involvement of various actors in the field of OWF. Several actors have engaged in this area in the last decade. Indeed, this increase is due to the advantages that offer marine renewable energies and the increased needs of inhabitants. However, France has also engaged into the area of offshore wind. Indeed, the first 3 farms are managed by *EDF Energies Nouvelles* since 2016. The three offshore wind farms of Saint-Nazaire (Loire-Atlantique), Fécamp (Seine-Maritime) and Courseulles-sur-mer (Calvados) will have to be operating in 2020.

### 1.2 Actors of Off-Shores Winds

The OWF market is much less fragmented than in onshore wind. The market is dominated by major energy suppliers, such as Vattenfall, Dong Energy or E.ON, which, in partnership with other companies, respond to various national tenders. European companies are the majority, since the Asian markets are still very emerging, and the United States has not yet launched large-scale offshore farms projects. In Europe, the market for offshore winds is shared by Dong Energy (Denmark, 24%), Vattenfall (Swedish company whose offshore wind division is located in Germany with a share of 10.5%), E.ON (Germany, 7.3%) and RWE (Germany, 8.7%).

In the last years, France has engaged in the area of marine renewable energies. Indeed, the two industrial actors General Electric and Siemens have started operating on the French market.

The *Haliade 150* offshore wind will be produced in France by General Electric (GE). The company has finalized the acquisition of Alstom's energy activities in 2015 and has taken over the commitments concerning the creation of factories dedicated to the manufacturing of offshore winds to:

- Saint-Nazaire for the manufacturing of generators and nacelles, the inauguration of the factory took place in December 2014.
- Cherbourg for the manufacturing of the blades, the first stone of the factory was laid in March 2017.
- In Nantes, GE has also created an international research and development center on marine renewable energies (Table 2).

**Table 2.** Actors of offshore winds

| Company | Area | Activities in the wind farm | Position | Onshore | Posed | Floating |
|---|---|---|---|---|---|---|
| Adwen | Offshore Wind manufacturer | Design, manufacture, assembly, installation and maintenance of offshore winds | 3rd actor in the offshore wind market | X | X | X |
| Bouygues-construction | Foundation construction | Construction of supports or foundations for wind turbines:<br>– Gravitation foundations (GBS)<br>– Hulls for floating wind turbines | New entrant, tier 1 supplier for foundation lots (2011) | X | X | X |

(*continued*)

**Table 2.** (*continued*)

| Company | Area | Activities in the wind farm | Position | Onshore | Posed | Floating |
|---|---|---|---|---|---|---|
| DCNS Energies | Shipping, building | – Foundation structures equipped (floats, anchors, dynamic cables) | - Current stage of pilot farm development - Current projects | | X | X |
| GE Renewables (ex Alstom) | Wind turbine manufacturer | Nacelle manufacturer, assembly | Actor has established a stronger presence in Europe with the acquisition of Alstom | X | X | X |
| Ideol | Floating wind turbine technology | Floating foundation design and engineering, supply of key components, offshore installation | Among the 3 world leaders in this new industry, French leader | | | X |
| STX | Ship builder | Design, construction, installation and maintenance. Design and construction of foundations (jackets and floats) | Actor established at European level for foundations. | | X | |
| Vestas | Wind turbine manufacturer | Wind turbine manufacturer | Historical leader | X | X | X |
| Nexans | Cable manufacturer | Modeling and Assistance/Consulting | One of the world's leaders suppliers | X | X | X |
| LM Blades | Blade manufacturer | Design and manufacture of wind turbine blades | One of the world leaders | X | X | X |
| Iméca (groupe REEL) | Offshore lifting | IMECA is involved in lifting applications | Lifting heavy loads | X | X | X |
| Meteoswift | Weather Forecast | Weather forecast and wind production (from a few hours to a few days in advance) | New entrant/emerging market following the energy transition law | X | X | X |

## 2 Classification of Logistics Planning for Offshore Wind Farms Operations

In this section, we define and identify the different phases of the offshore wind farms supply chain and the disturbance factors or constraints of each phase. To complete this study, we will carry out an exploratory study on the logistical operations of the OWFs. In order to succeed this step, we first start with research on websites, press reviews, research projects …, etc. Then, we target the main databases of scientific journals such as Google Scholar, Science Direct, Springer Link or Scopus, as well as the public documents of the main European projects of offshore wind farms installation. Thus, we propose to search for area-specific keywords (and their combination), such as "offshore wind logistics", "decision support, optimization and planning" and "Marine Renewable Energy".

In order to complete this study, we will present a review of the problems of the supply chain and installation of offshore wind farms. The problems will be classified according to the three levels of strategic, tactical and operational decision. As with production planning, offshore wind logistics planning can be broken down into three levels of decisions based on the time horizon over which it is applied: strategic, tactical and operational.

- **Strategic level:** location of offshore wind farms.
- **Tactical level:** planning of operations for offshore wind farms installation.
- **Operational level:** includes daily routing of vessels, weather factors affecting fuel consumption, (local temperature, wind direction and atmospheric pressure), scheduling of operations in the offshore wind farms.

The work on the planning and optimization of offshore wind logistics operations can be classified according to three decision levels as shown in Table 3.

**Table 3.** Decision levels

| Decision level | Description | Supply chain phase | References |
|---|---|---|---|
| Strategic | Location of offshore wind farms | Location of offshore wind farms | [6–10, 12, 14, 15, 20, 23, 30, 31] |
| Tactical | Operations planning for installation of offshore wind farms<br>Vessel tasks planning | Planning of operations for offshore wind farms installation | [1, 3, 15, 19, 25–27, 29] |
| Operational | Logistics network planning for daily vessel routing, Operations scheduling in offshore wind farms | Scheduling of operations in the offshore wind farm | [21, 28] |

## 2.1 Strategic Decision Level

This decision level is generally related to the planning and location of offshore wind farms. The planning aims to identify the appropriate areas for the installation of wind farms at sea. It also aims to identify areas where the establishment of farms is not possible. It is not a question of defining or determining the areas where the project must be carried out, but of defining areas where projects can be reasonably realized. Decisions concerning the location of offshore wind farms are crucial elements of strategic planning in the development of marine renewable energies. The location can be defined as an optimal location of an offshore wind farm taking into account various factors such as: wind speed, water depth, marine infrastructure, coast distance, port accessibility and distance between suppliers (production ports) and assembly points of an offshore wind farm.

In this context, we begin with the work of [20] which aims to locate an offshore wind farm, they look for the most important factors in the location of the farm as well as its entire life cycle. Several parameters are taken into consideration in this location such as wind speed, water depth and distance of appropriate construction ports, size and number of winds to be installed. The first step of the project begins with an analysis of the project life cycle cost. This cost is composed of development, production and acquisition, installation and commissioning, operation and maintenance costs and finally dismantling costs. The authors then used three algorithms: NSGA-I, NSGA-II and PEA2 (Pareto Front). Five main decision variables taken into account in this problem of location which are: production and acquisition costs of components such as turbine, foundation, nacelle, blade and tower, installation cost as vessels, human and material resources, pre-development costs (project management, control, engineering activities), maintenance and disassembly costs). The results showed that the optimal location of offshore wind farms is achieved using the three algorithms. They also showed that the wind speed, the distance between the locations of the suppliers of the main wind components and the installation sites/ assembly ports are important factors to take into account in locating the OWF. [23] have addressed the OWF location problem taking into account its life cycle. They also used multi-objective optimization. They have identified the various constraints or conditions to be respected for the location of an offshore wind farm including the distance to the coast, the water depth, the wind speed and the distance between the winds. To solve this optimization problem, authors have used constraint manipulation (Constraint Handling Techniques) which was later compared with NSGA-II (Pareto Front). The constraints taken into account in the problem are: wind farm boundaries where the wind have to be placed inside the wind farm area, infeasible areas which some areas that are not available to install winds due to human-imposed constraints (e.g. un-exploded ordnances, pre-existing cables, shipwrecks), minimum clearing distance which means a minimum distance between neighboring winds is required to guarantee proper functioning, and finally the number of winds that may be installed. Results show that it is advantageous to use fewer large offshore winds than to use several smaller ones, it depends on the geography of farm and its size. Finally, [12] presented a new approach for the location of offshore wind farms. To solve this problem, the authors proposed two meta-heuristics (the genetic algorithm and optimization by PSO (particle swarms optimization). [8] also addressed the OWF location problem using multi-objective optimization. The elements taken into account in the optimization of this location problem are the wake effects produced by the various winds by optimizing the energy efficiency, the necessary surface and the cable length to connect the winds.

In [15], the authors proposed a mathematical model for the location of ports to optimize the installation of offshore wind farms. For a strategic decision, two models are proposed, the first is based on the analytical hierarchy process (AHP) which aims to select the most appropriate installation port. The second model is a MILP that aims to determine the optimal plan for transporting components from suppliers to the chosen installation port. The feasibility of the proposed models is evaluated for the West Gabbard (UK) offshore wind farm located in the southern part of the North Sea. Their results showed that the total transport costs represent 9% of total supply chain cost. Finally, [30] presented some approaches to design supply chain networks for the installation and maintenance of OWF. The results showed that production, pre-assembly, shipping, construction and maintenance are the key components of the value chain and the most demanding in terms of offshore wind logistics. The authors were interested in the land transport phase, port and maritime transport. With regard to transport, the components produced are transported onshore by specialized heavy vehicles or at sea using vessels. Logistics processes include transportation planning. As part of the planning, we take in consideration: selection, availability check and transport planning of heavy components, transport equipment and fasteners, availability of personnel and logistics service provider. It also includes, route planning, capacity planning, mode optimization and travel operations. The works of the strategic decision on offshore wind farms supply chain can be classified in the table below (Table 4).

**Table 4.** Strategic level

| References | Problem type | Model type | Resolution method | Application case |
|---|---|---|---|---|
| [12, 23]. | Location of offshore wind farms | Mathematical Optimization Model | Metaheuristic (NSGA-II, NSGA-III and PSO) | Real, theoretical |
| [15] and [14] | | MILP | CPLEX, metaheuristic | Theoretical |
| [8] | | Mathematical Optimization Model | Heuristic | Theoretical |
| [20] | | MILP | Metaheuristic (GA) | Theoretical |
| [9, 31] | | | Heuristic | Theoretical |

In a general way, the location of OWFs is the most important step in the development of marine renewable energies. Several factors are required in the location of offshore wind farms: wind speed, distance between component suppliers (manufacturing ports) and assembly ports, coast distance, size or power of winds, and distance between the locations of the winds. It is to be noted that the port must be accessible without tidal constraints i.e. the ease of road access for exceptional convoys. Suitable lifting capacities as well as sufficient storage and assembly space must also be taken into account in the location. It is also important to conduct surveys before locating the farms. In fact, the location of OWFs must take into account the different opinions, especially those of the sea users, elected officials, economic development actors, associations, citizens, etc. Indeed, the investment cost to build an offshore wind farm is important and requires some collaboration. It should also be noted that the location of offshore wind farms is influenced by several constraints: sea routes and access channels to sea ports, because maritime transport is a significant constraint to evaluate. It is therefore necessary, in the location of a wind farm, to establish appropriate measures (safety distance, signaling, circumvention by vessels, etc.).

With regard to the port operations, theses latter must have sufficient surface area for the components of the winds. The hub assembly also requires about 1000 $m^2$. Missing parts at the port will cause significant delays in the transportation process, which will result in significant delays of installation process that can last up to several months. This will also generate additional charter costs about 250,000 Euros per installation vessel per day. The authors point out that port logistics aims to determine the required infrastructure (seaport, storage capacity, quayside capacity, etc.), as much as plan, execute and control pre-assembly processes.

## 2.2 Tactical Decision Level

The tactical decision level relates to the planning of operations for the installation of offshore wind farms. We distinguish two types of planning models: optimization and simulation models.

### 2.2.1 Optimization Models

In this context, we start with the work of [26] who proposed a mathematical model (MILP) to minimize the cost and installation time of an offshore wind farm while taking into account the weather conditions. The proposed model aims to find the optimal plan of installation phase for the offshore wind farm to minimize total vessel operation time. On a short time horizon, the proposed model allows to compute the optimal plan for only one vessel. This model does not consider the possibility of using multiple vessels or probabilistic time data. The authors suggest the extension in a stochastic model. However, in [27], the authors proposed a heuristic that allows solving a larger problem by taking into account longer time horizon and several vessels.

(Irawan et al. 2015) [13] have proposed a mathematical model using integer linear programming (MILP), which aims to minimize the cost and installation time of an offshore wind farm. The model takes into account several constraints such as weather conditions and vessels availability. To solve this problem, the authors have firstly used an exact method where the problem is solved using CPLEX. In the second method, they used a Variable Neighborhood Search (VNS) and Simulated Annealing (SA). The feasible slots are generated using a separate algorithm that takes into account the time required to complete the installation tasks, the order in which all tasks must be performed (sequence), and the weather conditions. [3] have proposed a decision support tool to assist industry to take decisions in the port operations. The proposed tool combines two models of simulation and optimization. The authors assume that the offshore wind can be installed using several scenarios. The components (cables, sub structures and top structures) are prepared at the nearest port of the wind farm site. Then, the components are loaded in the vessel and transported to the wind farm site. The vessel also makes the installation, which can take several days. Another vessel can carry higher structural components (mast, nacelle and blades) for more than six winds. Once the installation process on the site is complete, the vessel can return to the port to pick up others components (Fig. 2).

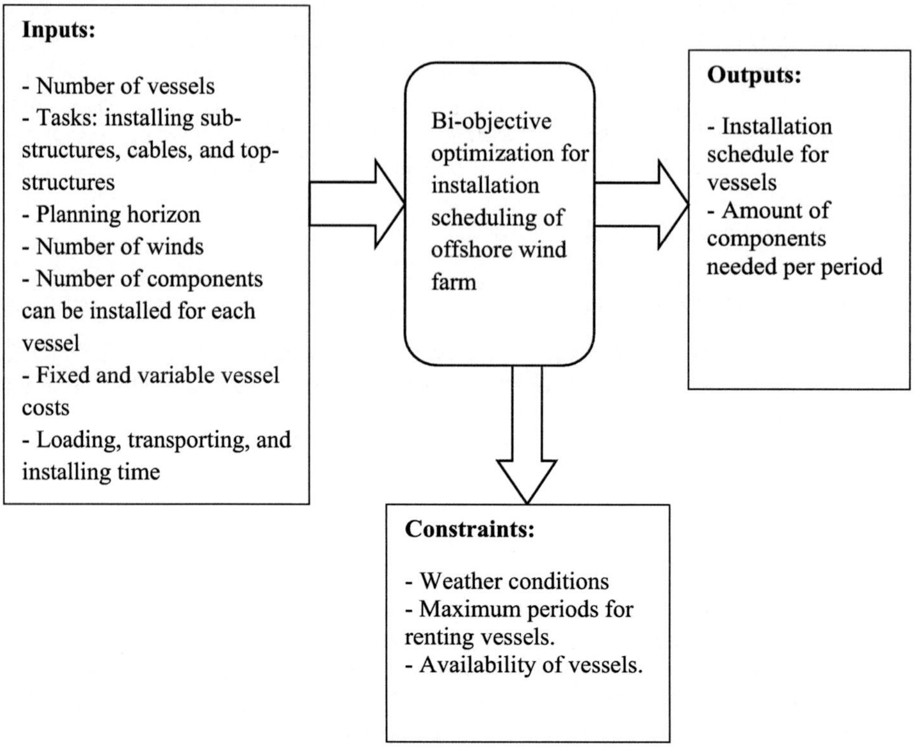


**Fig. 2.** The installation scheduling model [13]

Finally, [25] proposed a MILP to determine the impact of decision variables on the total cost of transportation and installation of offshore wind farm. The objective aims to minimize the total cost of transport and installation. The results showed that the total cost is strongly influenced by the size of the wind and the pre-assembly method. Finally, [2] addressed the problem of installation planning for offshore wind farm using mathematical optimization. Two optimization models have been developed to minimize the cost and farm installation time. Different vessel strategies, ports, time horizons and weather restrictions are considered as input of the problem. Several deterministic test cases with fixed cost parameters and historical meteorological data are implemented in the model.

#### 2.2.2 Simulation Models

There are several publications about simulation of offshore wind farms (OWF). [1] addressed the problem of aggregate planning installation of offshore wind farms. They proposed a model to minimize the cost and time installation of an offshore wind farm. The authors assume that each offshore wind requires four installation operations: the assembly, the laying of foundations, the installation and the connection to the electrical network. Then, each operation belongs to an installation sequence that can be built by a suitable vessel type under specific weather restrictions. They also assume that the components are available at all times in the port. Finally, the components will be built in a predefined scenario named the foundation structure, then the cables and the top structure of the wind. [19] presented a discrete event simulation for transportation and installation of offshore wind farms. They take into account historical wind speeds and actual wave heights as well as supply chain analysis in the offshore wind industry.

(Vis et al. 2016) [29] addressed the consistency between the logistics methods and the performance of the installation of offshore wind farm project while taking into account the impact of the weather. Their study also takes into account the most important factors in the pre-assembly and vessel loading phase as well as the distance from the component suppliers. They then propose a pre-assembly strategy that requires a minimum number of components for installation on the farm and a maximum number of winds to load on the vessel. The result is a decision support tool based on simulation. They tested two assembly methods of the wind: the star mode and the bunny ears mode where the components are assembled at the site. The average pre-assembly process time is approximately 22 h for bunny oriel mode, 16 h for star assembly and 5.5 h to assemble the components in the site.

The works of the tactical decision on offshore wind farm logistic can be classified in the following Table 5.

**Table 5.** Tactical level

| References | Problem type | Supply chain phase | Model type | Resolution method | Application case |
|---|---|---|---|---|---|
| [1] | Operations planning for installation of offshore wind farms | Installation | MILP | CPLEX | Case study |
| [3] | | | MILP/simulation | Discreet event simulation | Case study |
| [26] | | Fabrication, Transport, Installation | MILP | Exact method | German wind farm case study |
| [27] | | | MILP | Exact method | German wind farm case study |
| [2, 25]. | | Installation | Mathematical Optimization Model | Heuristic | Case study |

## 2.3 Operational Decision Level

The operational decision level includes daily vessel routing and setting the tasks to realize.

We start with [21] which developed an optimization-simulation tool that estimates average fuel consumption for weekly vessel schedules. Schedules are built using speed-optimized vessel trips that are then simulated under different weather conditions. The proposed tool takes into account locations, requests and hours of facilities operation, vessel speeds, capacity and meteorological data. The optimization problem has been solved by an algorithm using the Xpress-Optimizer tool, while the simulation problem has been implemented in ARENA.

(Stempinski et al. 2014) [28] considered the scheduling tasks for the installation of tripod foundations. They presented two simulation models: the first one uses a probabilistic evaluation of the downtimes to generate the planning, and the second one is a model of simulation with discrete events. In each case, the weather limits for operation foundation installation are obtained by using a numerical simulation of this process.

In recent decades, information sharing has become increasingly important in supply chain management, mainly because of different aspects such as the quick change of business processes and the diversity of customer demands. These aspects contribute to uncertainties in the management of the supply chain. To address these uncertainties and to improve the performance, supply chain Stakeholders should share their own information throughout the process. However, the majority of research projects focus on technical aspects in the construction, development, installation and operation of offshore wind farms. Regarding the logistics aspect, this phase is often omitted. In fact, very little work has been done on the logistical problem, taking into account the information sharing and resources in the global logistic chain of marine winds [4] (Table 6).

**Table 6.** Operational level

| References | Model type | Supplychain phase | Model type | Resolution method | Application case |
|---|---|---|---|---|---|
| [4, 21] | Scheduling tasks of installation vessels | Installation | Simulation/Optimization | Express Model (ARENA/Xpress-Optimizer) | Real case study |
| [28] | Operations scheduling for installation of offshore wind farm | | Simulation models | Petri nets | Theoretical |

A discrete event simulation model is proposed recently by [5]. The impact of weather restrictions on the loading, transport and installation processes are taken into account in the model. The authors stress that resource sharing in the installation phase of offshore wind farms is essential and should not be neglected [4]. The simulation results show that weather restrictions have a significant effect on installation and resource utilization times. In addition, the resource-sharing approach such as ships and cranes minimizes the time and installation cost of wind farm. [22] studied the impact of information sharing on the offshore wind farm installation process by proposing a simulation model. Finally, [18] proposed a simulation tool to minimize the storage cost of the installation phase for an offshore wind farm. They then presented a heuristic based on the model of [27].

In this work, we have just studied the three levels of decision-making, strategic, tactical and operational applied planning of logistic operations of offshore winds. The study shows that the majority of the works address the problems of installation of offshore wind farms. Very little work was focused on the global supply chain using simulation models. Only three articles focused on the three phases: manufacturing, transportation and installation using simulation models. For the rest of this work, it is interesting to perform the resources sharing in the supply chain of offshore winds, particularly in the installation.

## 3 Conclusion

This work focused on a review of the supply chain planning for offshore wind farms (OWF) operations. We first presented an introduction to offshore wind farms. We have shown the importance of renewable energies, the actors of offshore wind farms, the different classes of offshore winds as well as their characteristics. We also presented the supply chain problem of offshore wind farms. This work is then classified according to the three levels of strategic, tactical and operational decision. We have noted that the work of the scientific literature generally relates to:

- The location of offshore wind farms (OWF),
- The reduction of charter costs for vessels (optimum composition of the fleet),
- Minimization of variable costs of vessel use (optimal deployment of vessels), variable costs are mainly related to fuel consumption,
- Taking into account the two main constraints: weather conditions and the availability of vessels.

The majority of this work focuses on the last phase of installation of a wind farm, they do not take into consideration the global supply chain.

A MILP may be proposed to solve the global supply chain of offshore wind farms. This model must take into account the five phases of supply chain: procurement, transport, storage, handling and installation. Its function will be to minimize the sum of transportation, storage and installation costs.

**Acknowledgements.** Our work is a part of the OPT-EMR project which is co-financed by the CAREN Loire-Atlantique. We also thank the ELSAT 2020 project which authorized us to finalize this work. Finally, we should thank the European Union with the European Regional Development Fund, the French state and the Council of Hauts-de-France Region which funded ELSAT 2020 Project. The support is gratefully acknowledged. The authors also thank the anonymous reviewers for their contribution to improve the quality of this work.

## References

1. Ait-Alla, A., Mortiz, Q., Lutjen, M.: Aggregate installation planning of offshore winds farm. In: Proceedings of the 7th International (CIT 2013), Cambridge (US), pp. 130–35 (2013)
2. Backe, S., Haugland, D.: Strategic optimization of offshore wind farm installation. In: International Conference on Computational Logistics, pp. 285–299. Springer, Cham (2017)

3. Barlow, E., Öztürk, D.T., Revie, M., Akartunalı, K., Day, A.H., Boulougouris, E.: A mixed-method optimisation and simulation framework for supporting logistical decisions during offshore wind farm installations. Eur. J. Oper. Res. **264**(3), 894–906 (2018)
4. Beinke, T., Freitag, M., Zint, H.P.: Ressourcen-Sharing für eine bezahlbare Energiewende. Betrachtung der Produktions-und Errichtungslogistik der Offshore-Wind energie **4**, 7–11 (2015)
5. Beinke, T., Alla, A.A., Freitag, M.: Resource sharing in the logistics of the offshore wind farm installation process based on a simulation study. Int. J. e-Navig. Marit. Econ. **7**, 42–54 (2017)
6. Chen, K., Song, M.X., Zhang, X., Wang, S.F.: Wind turbine layout optimization with multiple hub height wind turbines using greedy algorithm. Renew. Energy **96**, 676–686 (2016)
7. Chen, Y., Li, H., Jin, K., Song, Q.: Wind farm layout optimization using genetic algorithm with different hub height wind turbines. Energy Convers. Manag. **70**, 56–65 (2013)
8. Tran, R., Wu, J., Denison, C., Ackling, T., Wagner, M., Neumann, F.: Fast and effective multi-objective optimisation of wind turbine placement (2013)
9. Feng, J., Shen, W.Z.: Solving the wind farm layout optimization problem using random search algorithm. Renew. Energy **78**, 182–192 (2015)
10. Feng, J., Shen, W.Z.: Design optimization of offshore wind farms with multiple types of wind turbines. Appl. Energy **205**, 1283–1297 (2017)
11. Global Wind Energy Council. Global Wind Energy Outlook (2014)
12. González, J.S., García, Á.L.T., Payán, M.B., Santos, J.R., Rodríguez, Á.G.G.: Optimal wind-turbine micro-siting of offshore wind farms: a grid-like layout approach. Appl. Energy **200**, 28–38 (2017)
13. Irawan, C.A., Jones, D., Ouelhadj, D.: Bi-objective optimization model for installation scheduling in offshore wind farms. Comput. Oper. Res. **78**, 393–407 (2015)
14. Irawan, C.A., Song, X., Jones, D., Akbari, N.: Layout optimization for an installation port of an offshore wind farm. Eur. J. Oper. Res. **259**(1), 67–83 (2017)
15. Irawan, C., Akbari, N., Jones, D., Menachof, D.: A combined supply chain optimization model for the installation phase of offshore wind projects. Int. J. Prod. Res. **56**(3), 1189–1207 (2017)
16. Lange, K., Rinne, A., Haasis, H.D.: Planning maritime logistics concepts for offshore wind farms: a newly developed decision support system. Comput. Logist. 142–158 (2012)
17. LORC (2016). http://www.lorc.dk
18. Lütjen, M., Karimi, H.R.: Approach of a port inventory control system for the offshore installation of wind turbines. In: The Twenty-second International Offshore and Polar Engineering Conference, International Society of Offshore and Polar Engineers (2012)
19. Muhabie, Y.T., Caprace, J.D., Petcu, C., Rigo, P.: Improving the Installation of offshore wind farms by the use of discrete event simulation. In: Improving the Installation of Offshore Wind Farms by the use of Discrete Event Simulation (2015)
20. Mytilinou, V., Kolios, A.J.: A multi-objective optimization approach applied to offshore wind farm location selection. J. Ocean. Eng. Mar. Energy **3**(3), 265–284 (2017)
21. Norlund, E.K., Gribkovskaia, I.: Environmental performance of speed optimization strategies in offshore supply vessel planning under weather uncertainty. Transp. Res. Part D Transp. Environ. **57**, 10–22 (2017)
22. Quandt, M., Beinke, T., Ait-Alla, A., Freitag, M.: Simulation based investigation of the impact of information sharing on the offshore wind farm installation process. J. Renew. Energy (2017)
23. Rodrigues, S., Bauer, P., Bosman, P.A.: Multi-objective optimization of wind farm layouts–complexity, constraint handling and scalability. Renew. Sustain. Energy Rev. **65**, 587–609 (2016)

24. Rodrigues, S., Restrepo, C., Kontos, E., Pinto, R.T., Bauer, P.: Trends of offshore wind projects. Renew. Sustain. Energy Rev. **49**, 1114–1135 (2015)
25. Sarker, B.R., Faiz, T.I.: Minimizing transportation and installation costs for turbines in offshore wind farms. Renew. Energy **101**, 667–679 (2017)
26. Scholz-Reiter, B., Heger, J., Lütjen, M., Schweizer, A.: A MILP for installation scheduling of offshore wind farms. Int. J. Math. Models Methods Appl. Sci. **5**(2), 371–378 (2010)
27. Scholz-Reiter, B., Karimi, H.R., Lütjen, M., Heger, J., Schweizer, A.: Towards a heuristic for scheduling offshore installation processes. In: Proceedings of the 24th International Congress on Condition Monitoring (COMADEM), pp. 999–1008 (2011)
28. Wenzel, F.S., Lüking, J., Martens, L., Hortamani, M.: Modelling installation and construction of offshore wind farms. In: ASME 33rd International Conference on Ocean, Offshore and Arctic Engineering, American Society of Mechanical Engineers (2014)
29. Vis, I.F., Ursavas, E.: Assessment approaches to logistics for offshore wind energy installation. Sustain. Energy Technol. Assess. **14**, 80–91 (2016)
30. Lootz, N.F.: Designing supply chain networks for the offshore wind energy industry. Int. J. Bus. Perform. Supply Chain. Model. **4**(3–4), 271–284 (2012)
31. Zhang, P.Y., Romero, D.A., Beck, J.C., Amon, C.H.: Solving wind farm layout optimization with mixed integer programs and constraint programs. EURO J. Comput. Optim. **2**(3), 195–219 (2014)
32. Breton, S.P., Moe, G.: Status, plans and technologies for offshore wind turbines in Europe and North America. Renew. Energy **34**(3), 646–654 (2009)

# Systematic Review of Macro Parking Models

Hamza Chajae(✉) and Fouad Jawab

MIDLOG Laboratory, Sidi Mohamed Ben Abdellah University Fez - High School of Technology Fez, Fes, Morocco
{Hamza.chajae,Fouad.jawab}@usmba.ac.ma

**Abstract.** The main objective of this paper is to provide synthesized parking solutions and models, compiled by a detailed description of the model, the utility of each model and its main features.

To attain this purpose, we followed a systematic review for a set of articles that we finish by keeping only a reduced number of relevant articles that satisfy our needs, starting by putting the first inclusion criteria which was parking models scale "Macro", afterward to go towards a pertinent analysis we focused more our study by adding more filters and selection criteria to know: the period of models, the editors or the publisher, which most of the exposed studies are published on two database Scopus and Web of Science.

As a result, a synthesized table of the macro model is provided where, four modeling approaches were addressed by the authors: mathematical, probabilistic, economic and analytical. This review will allow practitioners and researchers to easily access a database that includes comprehensive information on parking that can be very useful and will form the basis and background for new modeling proposals.

**Keywords:** Parking · Modeling · Urban logistics · Macro modeling

## 1 Introduction

Nowadays, parking occupies a primordial place in urban logistics and transport in general, it's a structuring element of the functioning of the city (Moufad and Jawab 2018) [1], it contributes considerably to the quality and fluidity of traffic in a dense urban area (Polak and Axhausen 1990), (Feeney 1989) [2] Wong, Tong, Lam, W. C. H. and Fung, R. Y. C. (2000) [3]. However, this finally was seen as a secondary element, always included in general travel costs, according to the classical point of view (Young 1990). [2].

But over time, several researchers have shown that travel and parking costs are independent and must be studied separately, Gillen (1978), Algiers et al. (1975).

Indeed, this traditional vision of parking is no longer valid; nowadays Parking is generally embodied in the management policies and travel strategies used by stakeholders in urban areas.

This is one of the factors that make logistics and transport organization and governance more harmonious, as much as the objective of parking planning and management is the good provision of parking spaces within the city, taking into account the various constraints and variables that may affect this finally. Moreover, we cannot talk


A. Saka et al. (Eds.): CPI 2019, LNME, pp. 388–398, 2021.
https://doi.org/10.1007/978-3-030-62199-5_34

about parking in urban areas without talking about the supply and demand of the parking spaces, which are considered as the input variables of the parking planning, and which help to define the capacity and the absorption rate of vehicles in urban areas.

Rare are the models that treat these primary parking variables as correlated constraints. Some try, attempt to analyze each variable separately and neglect the interaction between the two and don't focus on the reciprocal influence of each variable on the other, otherwise, it's the main problem which the majority of studies should treat and one of the key factors of the parking modeling.

# 2 Background

## 2.1 Parking Epistemology

In the parking modeling, there is no standard or universal model that is capable of studying all the parking system as a whole (Polak et al. 1990; Van der Goot 1982) [2].

Modeling is the way we are representing a system by another more simplified and easier to understand, it could be a logical representation: mathematical, algorithmic…or physical. The mathematical models consist on elaborating a set of mathematical iteration that represent and solve the phenomena, in the objective to attempt the knowledge of, what modifying the primary variables can cause on the modulization, and to predict what can be the consequences on the system.

Moreover, there is a distinction between two categories of models, the mathematical models and the conceptual models Von Bertalanffy (1973) [4] (Fig. 1).

## 2.2 Parking Supply and Demand

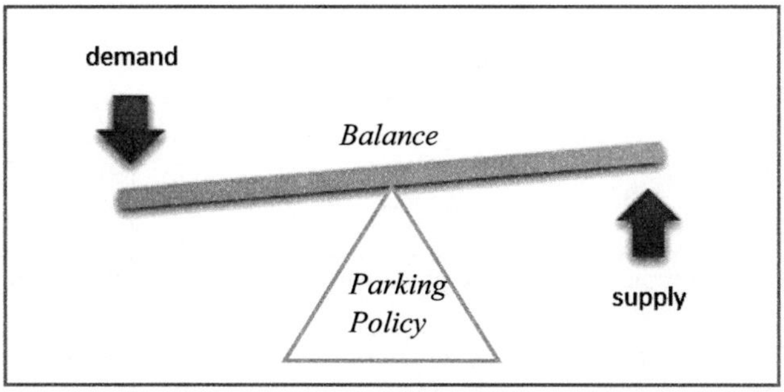


**Fig. 1.** Illustration of the parking supply & demand's balance

Furthermore, it's important to know first what does the offer and parking demand mean, as they are the mean equation's variables:

***Parking offer:*** it refers to the places dedicated for parking in an urban area, taking into consideration all types of parking (the On/Out street and the private, free, illegal, etc.…), it has been defined also as the stock of places available for parking on a discreet interval (Gillen, 1978). [5] this availability of parking spaces is limited and variable in time and depends directly on the uses way and the occupancy rate of conductors.

***Parking demand:*** is always emanated from an activity (purchasing, work, leisure, etc.), whose forecast is most of the time uncontrollable, misunderstood and heterogeneous.

In addition to that, the parking demand refers also the sum of the request of drivers who are seeking parking spaces for a determined period, within an area. (Stephan Lehner, Stefanie Peer 2019) [6].

Studies that deal with parking supply always take into consideration the multiple criteria that impact it, such as location, time and temporary periods per day and seasonality effect, type of parking, driver's, routes traveled, departure times, most frequented destinations, etc.

The analysis of the interaction between parking supply and demand thus remains unsatisfactory and does not achieve the expected results (Florian, 1980; Polak et al. 1990), nevertheless recently efforts have increased on this finally, several studies have focused on parking management and treat it as a complex system with several variables (Young et al. 1990) [2]. Many models have been developed to deal with and respond effectively to this problem and to seek an optimum state taking into account its input variables as well as the other variables that can be qualified as secondary.

## 2.3 Representation of Modeling Scales

The macro models, focus more on purposes with high measurement scales, with large dimension for long term, and who treat high number of elementary action units, furthermore the action units, the mass and the duration of the studied phenomena are among the primordial elements that define the model's scale [7], an illustration of the model's scales organized are organized on Table 1.

**Table 1.** Modeling scales and features.

| | Macro | Meso | Micro |
|---|---|---|---|
| Sample & space studied | High, large | Medium | Specific, restricted |
| The dimension | City, region… | A subdivision | Street, an alleyway |
| Description | Treat a general aspect of matters, for the long term. [strategic] (for example the most abundant problems of parking in a city) | It an intermediate phase between Macro-micro use the medium term [tactical approach] (example: the influencer constraints on parking demand in a specific district) | deals with unitary variables used in the short term [operational] (example the accidents root cause in parking, in 100 m of a specific street) |

## 3 Research Method

In this paper, we provide a systematic literature review of several articles recognized in journals that concern the field of urban logistics and parking policy. As know the systematic literature methodology aims to create a database of articles that can respond to our research question through, a classification, analysis, and interpretation of all Scientifics researches related to the matter [8].

The two main research questions, for which we made the articles selection and analysis are:

- RQ1: What are the main researches presented on the macro parking modeling until 2019?
- RQ2: how the Marco-modeling helps to the understanding of parking issues, and how it can participate in the realization of the coming a modeling?

The choice of these two questions is meticulous, and well-studied because it will allow giving an overall view on different aspects that will considerably help researchers and practitioners to have a clear idea on parking problems and practical insights.

### 3.1 Preliminary Research

several preliminary types of research have been carried out to answer correctly to the article's questions and to achieve the expected results, this preliminary research will allow, on one hand, to collect the relevant, the most frequented and read articles, which will contribute effectively to the achievement of the goal through an object-oriented evaluation and analyze of this articles while taking into account that the articles have passed through filters. the search was carried out through the search of a set of keywords that conduct us to collect all necessary papers, the keywords that we use are: "logistics, parking, model/modeling, Macro".

### 3.2 Article Selection and Study Characteristics

A search strategy has been applied to gather empirical papers on the macro parking models. Starting by collecting a large selection of papers related to the parking model in general, then by reading these papers we were able to classify them according to their scales (Macro, Meso, Micro), as long as our study is only interested in macro modeling we grouped those finally in order to study them in details and to elaborate a state of art.

We obtained a total of 68 articles based on the search that we had realized, afterward, we did a literature screening in order to determine whether the articles, address, firstly an online access to the article, a clear description of the model, a definite feature of each model and finally all articles are treated until the date of 2019, with the following languages: French or English obtained from Scopus and Web of Knowledge, also Elsevier, ACM, Springer, and IEEE. According to filters that we made 51 articles, do not conclude with the pre-established requirements for answering our two research questions, only 17 articles were obtained which we analyze and synthesize intending to respond to our research questions.

# 4 Findings and Discussions

## 4.1 Methodology Used and Models Approach

The research methodology was clearly defined, starting by collecting a large number of scientific articles with French and English language. afterward, several filter and selection criteria were done on the articles and documentation obtained to keep only pertinent articles and those with a value-added. Subsequently, we conducted a literature review on the selected paper, the main goal of this literature review is to define the basic data of macro models left by the author's name, the Model's name, and to provide a synthesized data regrouped on a table (Table 2). In that respect, we will try to give a clear description of each model also its main features and to compare it to the previous models.

From the study and the analysis that we had performed on the papers raised, we reveal that these macro models are divided into four main approaches:

- **Mathematical approach: *1** which uses formulas and mathematical theorems to express through linear program and iteration with a multivariable equation representing and solving the problem, the solution can be recognized in an algorithmic method. Most of the constraints studied in this mathematical approach were the relation between the parking demand and the parking spaces availability also the capacity of parking and the parking pricing…this method is very important and provides brilliant results. However, it should be completed and need to cover the other compartmental and behavioral aspect which is not taken into consideration in this finally.
- **Probabilistic approach: *2** the probabilistic model takes into account the modeling postulate, and based on the collected data we can formulate the percentage of the probability of the realization of the studied phenomena, among the most treated matters in this regard, we find the modeling of the probability of parking space availability taking into account the constraints like parking pricing of parking supply, etc.
- **Economical approach: *3** this method opts for the classical vision and philosophy of economy for treatment of matters, and applies its theorem to represent and study a realistic case.
- **Analytical and rational approach: *4** this kind of approach can combine a multiple and heterogenic method to reach a satisfactory result, it can use for example a mathematical approach and another sociological complementary approach which uses survey and interview to know the behavior of conductors, in that in the aim of addressing a multifield vision of the problematic treated.

All of these approaches are regrouped and well organized, while also adding the used tool for each model, the description and the main features of the models and its source on (Table 2).

**Table 2.** Macro parking models.

| Approach | Model name | Year | Description/Objective | The main feature of the model | Tools/Approach used | Source |
|---|---|---|---|---|---|---|
| Mathematical approach | Trip distribution model *1 | 1980 | Plan the occupation of the relay park according to a mathematical programming approach | Represent the interaction between parking capacity and demand in a spatial context | Mathematical programming approach | [9] |
| | Equilibrium assignment model for the evaluation of parking policies *1 | 1993 | Identify the relationship between road traffic and parking congestio | Segmentation of the demand into classes of users, according to the reason, the duration of the parking and the period of the trip | Multi-user stochastic equilibrium model | [12] |
| | Parking Capacity and Pricing in Park's Ride Trips: A Continuous Equilibrium Network Design Problem *1 | 2002 | The model deals with mathematical problems whose specificities correspond to the properties of the networks. Balance can be formulated through single- or double-level mathematical programming, often as a fixed-point problem or a problem of variational inequation | A tool that helps to solve research and development problems by using linear programming. Then, algorithmic solutions make it possible to solve the problem digitally, through an iterative process that progresses towards a global optimum | Mathematical method | [15] |
| | A time-dependent activity and travel choice model with multiple parking options *1 | 2005 | It is a dynamic deterministic equilibrium model of simultaneous choice of activity (location and duration), a parking location and a route according to a hierarchical structure. The approach is activity-based and studies activity planning behaviors | Model of transport planning tool based on activity type | Variational inequality theory with heuristic algorithmic solutions | [17] |
| | A model for evaluation of transport policies in multimodal networks with road and parking capacity constraints *1 | 2007 | The model focuses on the choice of mode, route, and location of parking | Adding the road capacity as a new variable | Based on a double algorithm of the Lagrangian algorithm | [19] |

(*continued*)

**Table 2.** (*continued*)

| Approach | Model name | Year | Description/Objective | The main feature of the model | Tools/Approach used | Source |
|---|---|---|---|---|---|---|
| | PARKRIDE: Methodology for Parking Modelling *1 | | It is a mathematical methodology proposed to model parking based on the logit model as a model of parking choice. The objective is to find a parking area available for each trip | The utility of the car parks, as well as the characteristics of the drivers, are taken into account | Mathematical methodology | [20] |
| | Dynamic parking charge optimal control *1 | 2017 | This model exposes the commuter's behaviors and trip mode choice regards a dynamic parking offer | The model proposes a dynamic fee parking policy that can impact the commuter's trip mode choice | Mathematical methodology | [23] |
| | The Macroscopic Fundamental Diagram *1 | 2015 | This model treats the parking as a dynamic and a bi-modal approach where a parking pricing strategy is included in the model to measure its impact on the congestion reducing | This model proposes a new concept of dynamic parking pricing in a multi-model urban network | Mathematical methodology | [21] |
| Probabilistic approach | Modeling uncertainty in parking choice behavior *2 | 2003 | The concept was used to represent the probability that a place will be occupied during a given time interval, based on the probability theory | as in previous models, it will contribute to making a probabilistic estimate of obtaining a vacant place in a parking area | Probabilistic method | [16] |
| | Curbside Parking and cruising for parking *2 | 2018 | The model takes into account the curbside parking space and from the occupancy's rate they estimate the road traffic | This model uses the occupancy approximated by parking transaction data, to conclude the cruising 'rate and to measure important performance indicators such as the probability of having a full block-face | Probabilistic methodology | [24] |
| Economical approach | Modeling parking *3 | 1999 | This model aims to study parking in its economic dimension. Their structure meets the standards of the traditional economy insofar as they fully represent the theoretical mechanisms of the market. The balance between supply and demand | Economic modeling is rather less numerous and rare, it helps to define the exploitable gains in parking, departures… | Economic approach | [14] |

(*continued*)

**Table 2.** (*continued*)

| Approach | Model name | Year | Description/Objective | The main feature of the model | Tools/Approach used | Source |
|---|---|---|---|---|---|---|
| Analytical approach | Combined Parking and Traffic Assignment Model *4 | 1981 | Make the parking assignment by considering the interaction between supply and demand of parking by type of parking and by location | Among the first models to use a network extension with a sequence of arcs to connect the parking locations to the nodes of the network | The computable general equilibrium models | [10] |
| | Stated Preference Approach *4 | 1988 | Select a parking type based on disaggregated data collected from traveler responses to parking attribute changes | The model operates in situations where congestion is an important issue and provides a realistic understanding of demand-supply interactions, and parking-transport system interaction | Binomial regression model (model logit) | [11] |
| | Driver Choice of Parking in the City *4 | 1996 | Location of parking lots near to the destination | Present the correlation between walking time and parking location | Multinomial profit model | [13] |
| | Modeling time-dependent travel choice problems in road networks with multiple user classes and multiple parking facilities *4 | 2006 | It is a network equilibrium model that jointly considers the choice of departure time, route, location and parking time with several user classes and different types of parking | Validated by two numerical examples for two different network configurations. The model is dynamic and explicitly recognizes the spatial and temporal interaction between traffic and parking congestion | Based on a super-network approach | [18] |
| | Real-time parking availability prediction *4 | 2015 | The current model presents the work of a real-time parking availability estimation in smart cities | The model focus on two kinds of prediction: the probability of having free spaces and the short-term parking occupancy (data collection is with sensors) | Probabilistic and technological methodology | [22] |
| | Geospatial semantic parking model *4 | 2018 | This model presents how the conductor's behavior changes with they are changing the previous geospatial parking | Model use the standard deviation to mention the behavioral discrepancies between parking on the same and different parking | Geospatial representation, Bigg data | [25] |

## 5 Work Contributions and the Application Field

The models found within this work are with a value-added and with a relevant contribution on all facets, they treated several problematics and provide a guide for a selected database of solution for one of the most troubled problematics in the urban logistics, it offers a set of macro parking models that deal with different logistics aspects, different field of activity and help much the practitioners by providing necessary assistance for the most problems that they can face. The models selected in our paper intervene in several matters, which we can be regrouped as following. First of all, the models that treat the parking lots assignment and localization based on assignment methods and barycenter methods, and also the models that focus on the definition of the parking type and the internal design.

In another hand we have models that deal with the balance of the parking demand & supply, and which are taking into a count the exhaustive constraint faced, (parking offer, availability of places, the parking's type, the traffic jam, and road traffic, capacity of parking, travel time, parking duration, pricing, zoning…), this model can be seen and regrouped by their approach, which is:

- An organizational and managerial approach: that aims to solve all organizational topics by ensuring good cities' accessibility and a good fluidity of flows within cities, smoothed and maintained road traffic whatever the variances…
- Economical approach: that are models that deal with problems in the traditional economic sense and analyze the asset/liability of things.
- Scientific approach: which envisage the parking problems in a scientifically and rational way, using science tools (probability, algorithm, mathematics, logits), that require thereafter an evaluation and testing in the real environment to prove their feasibility, and to complete the other aspects that are missing if necessary.

Further models analyze the definition of the parking choice, which are due to tow elements:

- Human factors: the conductor's behavior and the way of driving.
- A non-human factor: the infrastructure, the road traffic…

## 6 Features Research

As mentioned previously, the parking policy plays a major and determining role in a city, Wong, Tong, Lam (2000) [3], thus this strategic planning involves the implementation of multidimensional scales studies, starting from the macro models going towards Meso and micro models, with aim of brushing all parking problematic aspect. Most of the successful study start necessarily from a macro and generalist context with more or less approximate objectives, than they go to the subordinate's levels, which are the meso and microenvironment that contain more details and specifications, for that, the current article provide a grouped rich macro models about the parking, that we can gather with the other modeling scales (meso & micro) to constitute a useful background to solve some of the parking issues that we are facing and to give an easily accessible database for new propositions for the researchers who want to make a new Macro Parking model's work.

## 7 Conclusion

To conclude, the parking remains an indispensable link in the urban logistics system. several efforts and development stay solicited in this field. Thus, the main objective is to understand deeply the parking problematics into the objective of providing an adequate and customizable solution, taking into account exhaustive constraints, valid and adaptable more or less for any environment. This paper which presents various results of scientific production of Macro parking modeling can be a tool with an added value and a helpful guide for the researchers and stakeholders who want to work in the same field.

## References


1. Moufad, I., Jawab, F.: Performance du transport de marchandises en Ville: Cas de l'implantation des aires de livraison (2018)
2. Polak, J., Axhausen, K., Errington, T.: The application of CLAMP to the analysis of parking policy in Birmingham City Centre. In: PTRC Summer Annual Meeting, 18th, 1990, University of Sussex, United Kingdom (1991)
3. Wong, S.C., Tong, C.O., Lam, W.C.H., Fung, R.Y.C.: Development of parking demand models in Hong Kong. J. Urban Plann. Dev. **126**(2), 55–74 (2000). https://doi.org/10.1061/(asce)0733-9488(2000)126:2(55)
4. Von Bertalanffy, L., Chabrol, J.-B., László, E., et al.: Théorie générale des systèmes. Dunod, Paris (1973)
5. Gillen, D.: Parking policy, parking location decisions and the distribution of congestion. Transportation **7**, 69–85 (1978)
6. Lehner, S., Peer, S.: The price elasticity of parking: a meta-analysis. Department of Socioeconomics, Vienna University of Economics and Business, Welthandelsplatz 1, 1020 Vienna, Austria. https://doi.org/10.1016/j.tra.2019.01.014
7. Michel, G.: Trois échelles d'action et d'analyse. L'abstraction comme opérateur d'échelle. L'Année sociologique, 2006/2, **56**, pp. 285–307. https://doi.org/10.3917/anso.062.0285. https://www.cairn.info/revue-l-annee-sociologique-2006-2-page-285.htm
8. Barbara, K.: Procedures for performing systematic reviews, vol. 33. Keele University, Keele, UK (2004)
9. Florian, M., Los, M.: Determining intermediate origin-destination matrices for the analysis of composite mode trips. Transp. Res. Part B Methodol. **13**(2), 91–103 (1979)
10. Nour Eldin, M.S., El-Reedy, T.Y., Ismail, H.K.: A combined parking and traffic assignment model. Traffic Eng. Control **22**(10), 524–530 (1981)
11. Axhausen, K.W., Polak, J.W.: Choice of parking: stated preference approach. Transportation **18**(1), 59–81 (1991)
12. Bifulco, G.N.: A stochastic user equilibrium assignment model for the evaluation of parking policies. Eur. J. Oper. Res. **71**, 269–287 (1993)
13. Lambe, T.A.: Driver choice of parking in the city. Socio Econ. Plann. Sci. **30**(3), 1996 (1996)

14. Arnott, R., Rowse, J.: Modeling parking. J. Urban Econ. **45**(1), 97–124 (1999). https://doi.org/10.1006/juec.1998.2084
15. García, R., Marín, A.: Parking capacity and pricing in park'n ride trips: a continuous equilibrium network design problem. Ann. Oper. Res. **116**(1–4), 153-178 (2002)
16. Dell'orco, M., Ottomanelli, M., Sassanelli, D.: Modelling uncertainty in parking choice behaviour. In: 82nd Annual Meeting of the Transportation Research Board (2003)
17. Huang, H.-J., Li, Z.-C., Lam, W.H.K., et al.: A time-dependent activity and travel choice model with multiple parking options. Transp. Traffic Theory 717–739 (2005)
18. Lam, W.H.K., Li, Z.C., Huang, H.J., Wong, S.C.: Modeling time-dependent travel choice problems in road networks with multiple user classes and multiple parking facilities. Transp. Res. Part B **40**(5), 368–395 (2006)
19. Li, Z.-C., Huang, H.-J., Lam, W.H.K., et al.: A model for evaluation of transport policies in multimodal networks with road and parking capacity constraints. J. Math. Model. Algorithms **6**(2), 239–257 (2007)
20. Bagloee, S.A., Asadi, M., Richardson, L.: Methodology for parking
21. Zheng, N., Geroliminis, N.: Modeling and optimization of multimodal urban networks with limited parking and dynamic pricing. https://doi.org/10.1016/j.trb.2015.10.008
22. Atzori, L., Iera, A., Morabito, F.: The Internet of Things: a survey. computer networks. **54** (15), 2787–2805 (2010). http://dx.doi.org/10.1016/j.comnet.2010.05.010. [Crossref], [Web of Science®], [Google Scholar]
23. Lin, X., Yuan, P.: A dynamic parking charge optimal control model under perspective of commuter's evolutionary game behavior (2017). https://doi.org/10.1016/j.physa.2017.08.063

# Assessment of a New Model to Optimize Flow in Distribution Networks

Yassine Erraoui(✉), Abdelkabir Charkaoui, and Abdelwahed Echchatbi

Laboratory Industrial Management and Innovation, Faculty of Science and Technology, Settat, Morocco
Erraoui.yassine@gmail.com, charkaoui.a@gmail.com, Echchatbi@gmail.com

**Abstract.** The distribution of goods from suppliers to customers plays an important role in the supply chain. In this paper, the approach of Demand Driven Distribution Resource Planning is proposed in order to optimize the distribution flow in the supply chain. We follow a methodology of presenting the problem of variability, then pointing at the solutions in the literature, among them DRP, and finally giving the proposed model. The purpose is to manage the amplification of the demand Variability, while improving the traditional Distribution Resource Planning. A literature review is presented about the impact of variability on distribution flow, and the solutions proposed in this context. Then, a general study of distribution industries is investigated in order to apply the Demand Driven model; we show the buffer positioning in the distribution network, and the profile and levels for these buffers. After the dynamic adjustment, we present the Demand Driven Planning and the execution based on the net flow equation. The results discuss the approach and the steps of the implementation in the distribution industry.

**Keywords:** DRP · DDDRP · Bullwhip effect

## 1 Introduction

Many researches focus on the optimization of the flow of merchandise through a distribution system. The pressure of the market and the cost of inventories justify the importance of this question. The most widespread method for managing physical flow in the distribution system is Distribution Resource Planning (DRP). In literature, some works focus on the inventory management as one of the keys to optimize the flow (Gümüs and Güneri 2007), other works try to introduce the Just-In-Time concept in the Distribution Resource Planning (DRP) method (Wang et al. 2004). The recent character of the supply chain - where the complexity is growing, the customer tolerance time is short and the product life cycles are restricted - push towards the need for enhancing the conventional methods like DRP. This work focuses on applying the Demand Driven concept to the distribution part of the supply chain. Its originality comes from the method “Demand Driven Material Requirement Planning” (DDMRP), which is a multi-echelon demand and supply planning and execution methodology (Miclo et al. 2016).


A. Saka et al. (Eds.): CPI 2019, LNME, pp. 399–410, 2021.
https://doi.org/10.1007/978-3-030-62199-5_35

Therefore, the purpose of the new approach, which is named "Demand Driven Distribution Resource Planning" (DDDRP) is to manage different forms of variability "demand, operational, supply and management", dealing with various parameters like the lead times, on time-delivery, and to reduce the cost of goods sales. In order to investigate what does the DDMRP includes for good DRP. The structure of this article begins with the identification of Bullwhip effect and in the distribution, with the literature about the causes. Then, it focuses on a multitude of works accomplished in order to enhance flow in distribution network, such as inventory management and naturally the DRP tool. This section finishes with a general overview upon the DDMRP method.

In the last section, we try to introduce the concept of "Demand Driven Distribution Resource Planning", as a method for optimizing flow through the distribution network. We present a general study of the distribution industries (data not included) and try to apply the five steps: positioning buffers in the distribution network, buffering profile and levels, dynamic adjustment, demand driven planning and execution of open supply orders. The results of this work are related to particularities of this approach, and the axes and steps to apply it in a distribution industry. The article finishes with a conclusion and discusses further researches.

# 2 Literature Review

## 2.1 Bullwhip Effect and Variability in Distribution

In the management of supply chain, scientists and industrials notice the fact of Bullwhip effect. It means that the variability process increases as we move from a level to another from the customer to the manufacturer (Lee 2016), presented in (Fig. 1). Some researches discus this phenomenon like in Bagnha and Cohen (1995), Kahn (1987) and Metter (1996), which focuses on analyzing the causes of the Bullwhip effect, and the solutions to reduce its impact. In particular, Lee et al. (1997) identify five main causes to the Bullwhip effect; the use of demand forecasting, supply shortages, lead times, batch ordering, and price variations. Then they establish some suggestions to reduce the impact of this effect, such as the centralization of the demand information (Chen et al. 2000). They also try to quantify the bullwhip effect in order to determining its impact of the demand forecasting.

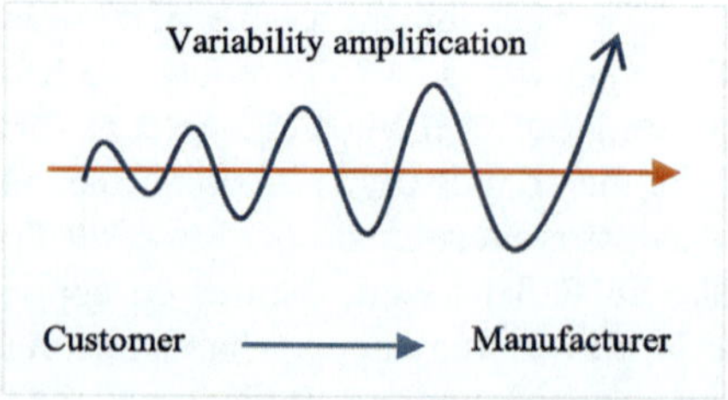


**Fig. 1.** The Bullwhip effect

Thus, the bullwhip effect is manifested on the amplification of variability, which lead to the systematic distortion in demand information. Figure 2 - evoked in Lee (2016) - shows a retail store's sales of a product, alongside the retail's orders issued to the manufacturer which indicates the distortion in demand information.

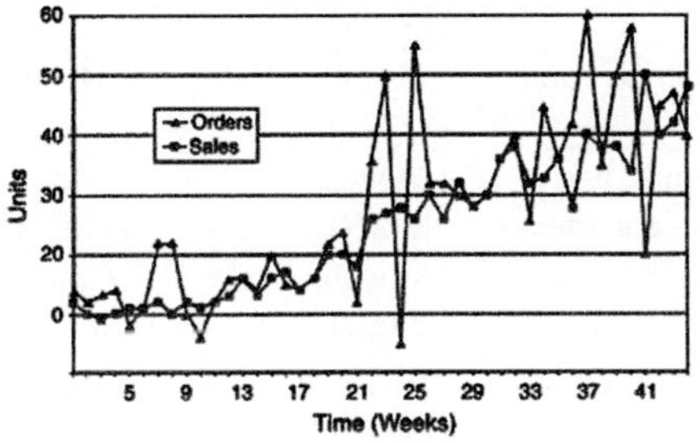


**Fig. 2.** Comparison between orders and sales

Lee et al. (1994) affirm that the bullwhip effect refers to the phenomenon where orders to the supplier tend to have larger variance than sales to the buyer (i.e., demand distortion), and this distortion propagates upstream in an amplified form (i.e., variance amplification). This phenomenon has been recognized in many markets such as Procter & Gamble (Lee et al. 1994). The distortion of demand information implies a number of consequences in the supply chain, leading to serious cost implications. For example, the unplanned purchases of supplies cause excess raw materials, then expensive excess capacity due to the need of space for more inventory. Consequently, this bring additional transportation costs due to inefficient scheduling and premium shipping rates. By a measure published in Fuller (1993), the inefficiencies bear part in responsibility for the $75 billion to $100 billion worth of inventory caught between various members of the $300 billion (annual) grocery industry. Moreover, Trade estimation suggest that these activities can result in excess costs in the range 12.5% to 25% (Kurt Salmon Associates, 1993).

Lee (2016) focuses on the consequence related to the inventory status. They propose that the keys to improve the channel coordination and dampening the bullwhip effect are primordially based on information on inventory status and Sell-through data.

In experimental context for inventory management, Sterman (1989) make in evidence the bullwhip effect in the "Beer Distribution Game", in which the supply chains contain four players who make independent inventory decisions without consultation with other chain members, relying only on orders from the neighboring player as the sole source of communications. The experiment shows that, as far as on moves up in the supply chain, the variances of orders amplify, which increase the bullwhip effect.

Economists like Holt (1960), Blinder (1982), and Blanchard (1983) have concluded that the main role of inventory is like a buffer to smooth production in response to demand fluctuations.

The Table 1 below summarizes the principal causes of the bullwhip effect from the literature.

**Table 1.** Causes of Bullwhip effect

| References | Causes | References | Causes |
|---|---|---|---|
| Lee et al. (1997) | Demand forecasting | Geary et al. (2006) | Multiplier effect |
| Lee et al. (1997) | Order batching | Erkan et al. (2008) | Lack of synchronization |
| Lee et al. (1997) | Price fluctuation | Moyaux et al. (2007) | Misperception of feedback |
| Lee et al. (1997) | Rationing and shortage gaming | Moyaux et al. (2007) | Local optimization without global vision |
| Jafar et al. (2009) | Lead time | Moyaux et al. (2007) | Company processes |
| Chandra and Grabis (2005) | Inventory policy | Alony and Munoz (2007) | Capacity limits |
| Jakšič et al. (2008) | Replenishment policy | Lee et al. (1997) | The strategic interaction of two rational SC members |
| Geary et al. (2006) | Improper control system | Croson and Donohue (2009) | Neglecting time delays in making ordering decisions |
| Lee et al. (1997) | Lack of transparency | Skiadas (1986) | Lack of learning and/or training |
| Alony and Munoz (2007) | Number of echelons | Croson and Donohue (2009) | Fear of empty stock |

## 2.2 Inventory Management in Supply Chain

The inventory management is a major issue in Supply Chain Management, as one of the keys to prevent against the bullwhip effect. Its importance in the SC requires to considering the network of procurement, transformation and delivering firms. This leads to the term, "multi-echelon" network, when the product moves through more than one step before reaching the final customer.

The Table 2 below presents a part of the literature, where the time span is from 2003 to 2016, addressing the multi-echelon inventory management, especially in distribution networks, from an operational research point of view. The purpose is to find the optimal policies and models for inventory management. PS: N (in the table) is presenting a number upper than 2.

**Table 2.** Multi-echelon inventory management in the Supply Chain

| Authors | Number of echelons | Nature of echelon | | | | | | | | Inventory Policy |
|---|---|---|---|---|---|---|---|---|---|---|
| | | Manufacturing site | Warhouse | Retailer | Depot | Stock point | Distribution center | Supplier | Other | |
| So and Zheng (2003) | 2 | – | – | 1 | – | – | – | 1 | – | Order-up-to policy |
| Axsater (2003) | 2 | – | 1 | N | – | – | – | – | – | Continuous review installation stock (R,Q)* policies with given batch quantities |
| Minner (2003) | 2 | – | – | – | 1 | N | – | – | – | Periodic review echelon- order-up-to policies |
| Mitra and Chatterjee (2004) | 2 | – | 1 | 2 | – | – | – | – | – | Periodic-review inventory policy |
| Chiang and Monahan (2005) | 2 | – | 1 | 1 | – | – | – | – | – | One-for-one replenishment inventory control policy |
| Jalbar et al. (2005a) | 2 | – | 1 | N | – | – | – | – | – | Single-cycle policy |
| Jalbar et al. (2005b) | 2 | – | 1 | N | – | – | – | – | – | Nested policy |
| Routroy and Kodali (2005) | 3 | 1 | 1 | 1 | – | – | – | – | – | Continuous review policy (Q, r) |
| Caggiano et al. (2006) | 2 | – | 1 | – | – | N | – | – | – | Integrated real-time model for inventory allocation decisions |
| Caggiano et al. (2007) | N | – | – | – | – | – | – | – | – | Minimizing the system inventory investment by installation base stock levels for each part type at each location |
| Gallego et al. (2007) | 2 | – | 1 | N | – | – | – | – | – | Simple bounds, heuristics, and approximation for distribution systems under local and central control |
| Axsater (2007) | 2 | – | 1 | N | – | – | – | – | – | First come-first served policy |
| Axsater (2008) | 2 | – | 1 | N | – | – | – | – | – | Continuous review policy (R, Q) |
| Berling and Marklund (2009) | 2 | – | 1 | N | – | – | – | – | – | Continuous review installation-stock (R, Q) policy |

(*continued*)

**Table 2.** (*continued*)

| Authors | Number of echelons | Nature of echelon | | | | | | | | Inventory Policy |
|---|---|---|---|---|---|---|---|---|---|---|
| | | Manufacturing site | Warhouse | Retailer | Depot | Stock point | Distribution center | Supplier | Other | |
| Farasyn et al. (2011) | N | – | N | N | – | – | – | N | – | The guaranteed service (GS) model for the multiechelon inventory optimization |
| Eruguz et al. (2013) | N | General study | | | | | | | | (R, S) policy in a multiechelon supply chain |
| Haji et al. (2013) | 2 | – | 2 | – | – | – | – | 1 | – | One-for-one-period ordering policy |
| Fleischhacker et al. (2014) | N | General study | | | | | | | | Provide a new class of multi-echelon inventory models |
| Fattahi et al. (2014) | 2 | 1 | – | 1 | – | – | – | – | – | (s, S) policy is investigated for a supply chain (SC) |
| Berling and Marklund (2014) | 2 | – | 1 | N | – | – | – | – | – | Continuous review (R, nQ) policy |
| Alvarez and Van deir Heijden (2014) | 2 | – | 1 | – | 1 | – | – | – | – | One-for one replenishment policy |
| Dogru and Ozen (2015) | 3 | One central, other local warehouses and Repair Vendor | | | | | | | | Tactical inventory planning tool primarily to support the business tendering process |
| Chen et al. (2015) | N | General study | | | | | | | | (R, Q) policies for serial inventory systems |
| Bertazzi et al. (2016) | 2 | – | – | N | – | – | – | 1 | – | Min-Max exact and heuristic policies |
| Ekanayake et al. (2016) | 1 & N | Comparison between 1 single stage & multi-stage | | | | | | | | Stochastic modelling approach for multi-tier supply chains with multiple inventory items |
| Huh et al. (2016) | N | General study | | | | | | | | Echelon base-stock policies for the system |

### 2.3 Distribution Resource Planning

DRP has become an effective concept for inventory control in the multiproduct, multi-echelon physical distribution environment since 1970 (Watson et al. 2003). The purpose is to gain full visibility of inventory levels in every node of the supply chain, in order to meet the current demand.

There are numerous reports of the benefits that the companies have received through DRP implementation (Forger 1986; Horne 1989; Krepchin 1989; Hammel and Rock 1993; Davis 1994, Frasier Sleyman 1994). Bookbinder and Heath (1988) analyze and compare several lot-sizing policies under DRP assumptions, and Martin (1990) who was Director of Materials Management for Abbott at the time, wrote the seminal work on this subject, Distribution Resource Planning, which explained the logic and the benefits of DRP.

Another contribution for managing DRP and optimizing inventory in Supply chain, is proposed by Wang (2003). It is a Just-In-Time distribution requirements planning system under the limited supply capacity with multi-warehouse and multi-retailer scenarios. The purpose is to minimize the total cost of manufacturing and transportation by establishing an optimal distribution requirement planning. The result show that the integration of the pull SC system is applied to allow the distribution of products is in time, under limited supply capacity.

Some recent studies show that DRP enables the user to establish certain parameters for inventory control and calculate the phase time between inventory requirements (Bhamu et al. 2013), and also the integration of Lean Manufacturing to use the distribution re-source planning (Reyes et al. 2018).

## 3 DDDRP in Distribution Industries: A General Study

The Demand Driven Distribution Resource planning can be explained as a multi-echelon inventory planning and execution solution. The method is based on axes like, DRP, Lean, Six-Sigma, Theory of Constraints and innovation. It will help to execute - in an optimal range of inventory - for a warehouse, the materials prioritized to be purchased, from his supplier, which can be another warehouse or the manufacturing company.

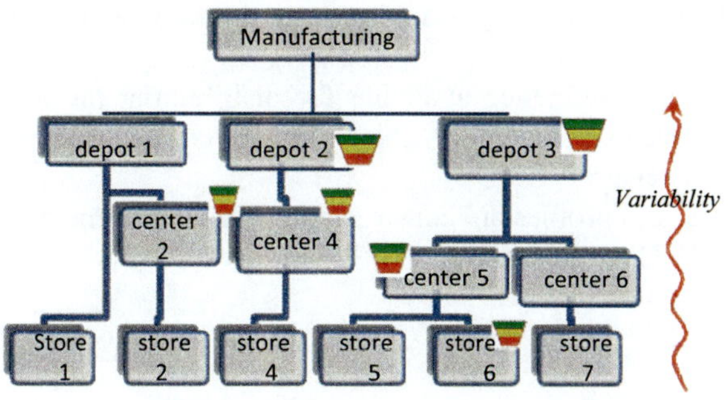


**Fig. 3.** Buffer positioning in distribution network.

The method DDDRP is implemented in 5 main steps.

The first step is "Strategic Inventory Positioning". It consists of positioning a buffer on a warehouse, concerning an article that belongs to the distribution flow, in order to control the amplification of demand variability. Figure 3, shows the positioning of the buffer in the distribution network.

The DDDRP issue is to pull replenishments between strategic buffers, and push planning orders for non-buffered articles. The positioning is done from a financial point of view. The purpose is stop the bullwhip effect by dividing the distribution chain and compressing lead times.

The second step is to define profile and levels; it serves to absorption level of variability in the decoupling points. The buffer contains three zones, the red, yellow and green zone, using some specific methods for calculation, referring to part demand in-formation based in historical and/or forecast and other part settings related to DDDRP, which allow to create values for each zone (Fig. 4).

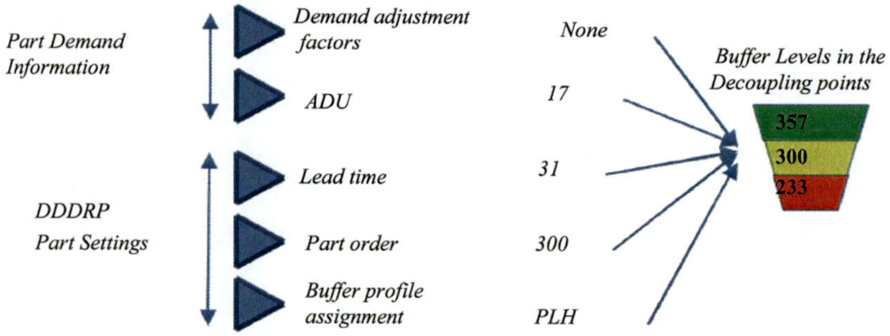


**Fig. 4.** Buffer level in decoupling point

The information needed for buffer level construction concerns multitude of parameters; the lead time of distribution from a warehouse (supplier) to warehouse (receiver), this one is at least the sum of four phases, which constitute the time needed for distribution. It contains the launch and preparation time of the order, the loading, transiting, unloading and stocking. The result is the decoupling lead time (DLT) for each reference between two consecutive warehouses. In the distribution context, it is defined as the longest cumulative coupled lead time chain in a distribution item's product structure.

Moreover, for each reference available for distribution, the average daily usage (ADU), the initial state of inventory, the selling price are primordial parameters to calculate the buffer levels.

The buffer level and profiles are constructed using the parameters above, exploiting the following DDMRP expressions:

$$Red\,Base = ADU * DLT * lead\,time\,factor \tag{1}$$

$$Red\,Safety = Red\,base * variability\,factor \tag{2}$$

$$Total\,red\,zone = Red\,Base + Red\,Safety \tag{3}$$

$$Yellow\,Zone = ADU * DLT \tag{4}$$

$$Green\,Zone = ADU * DLT * Lead\,Time\,factor \tag{5}$$

The buffer sizing lead to the calculation of the average inventory (cost). The issue is to compare and decide between the different ways of positioning buffers.

Moreover, the level of protection flexes up and down based on operating parameters, market changes and planned or future events, this is defined by the Dynamic adjustment of the buffer. The traditional DRP uses the Safety Stock concept, which is adopted in a static point of view. However, the dynamic character of the supply chain is manifested in a variation of the decisive parameters of the buffer levels and profile, especially the average daily usage. Thus, the continuous changes of the parameters push for updating the buffer situation daily. Figure 4 is an example for the lead time compression taken from "Demand Driven Material Requirements Planning (DDMRP)" book (Ptak and Smith 2016).

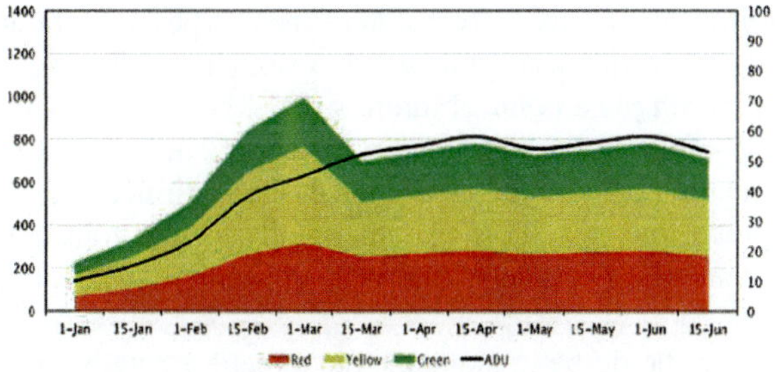


**Fig. 5.** Lead time compression

The demand driven planning is based on the Qualified Sales, without taking forecasts as priority, it is the step of generating supply orders. This generation is done via the equation determining the position of the net flow position. This information is crucial for deciding to issue a supply order or not (Fig. 5).

The net flow equation is defined:

$$H + OS - QS = net\,flow\,position \tag{6}$$

OH = On Hand Quantity, indicating the quantity of stock physically available

OS = Open Supply Quantity, indicating the quantity of stock that has been ordered but not received.

QS = Qualified Sales Order Demand. It is the sum of sales orders past due, sales orders due today, and qualified spikes.

After the planning, the Demand Driven Execution uses - To execute and manage open supply orders - the On-Hand Buffer Status. The lower the On Hand level, the higher the priority to maintain flow, and the execution priority. The DRP systems used to assign the priority by date, which is rejected by this step, where the priority is assigned by buffer Status.

## 4 Conclusion and Future Work

The article discusses the Demand Driven approach in the distribution part for the supply chain. The issue is to remediate to the supply chain variability partially manifested in the demand distortion through warehouses. As one moves away in the distribution network, from the source directly related to the customer, the amplification of variability is noticed forming the bullwhip effect. This study pauses the first keys and steps to adopt the demand driven concept.

Advantages of using this approach are related to shortening the lead times, absorbing variability in buffer positions. Else, there are needs of focus the research on various points evoked in this work:

From variability point of view; the sources of variability are numerous. The study must take in consideration the other forms like the management, the supply and operational variability. We focus on one side of these types, due the importance of the demand variability, as an important factor for customer satisfaction.

The buffer positioning is a point of future work. The purpose is to find the optimal way to implement buffers in the distribution network. Therefore, the subject may use operational research for a number of references, to optimize the objective function related to the average inventory cost. Another study should trait the different factors (variability, lead time …) because of the lack of studies in affecting value to these factors.

Finally, this study should be implemented in a real case study, with all the data and parameters needed to compare with the traditional ways of managing inventory as DRP.

## References


Martin, A.J.: DRP: Distribution Resource Planning. Oliver Wright, Essex Junction (1990)

Bookbinder, J.H., Heath, D.B.: Replenishment analysis in distribution requirements planning. Decis. Sci. **19**, 477–489 (1988)

Ptak, C., Smith, C.: Orlicky's Material Requirements Planning 3/E. McGraw Hill Professional, New York (2011)

Miclo,R., Fontanili, F., Lauras, M., Lamothe, J., Milian, B.: An empirical study of demand-driven MRP. In: Information Systems Logistics and Supply Chain (2016)

Ptak, C., Smith, C.: Demand Driven Material Requirements Planning (DDMRP), Industrial Press, Inc., New York (2016)

Ohno, T.: Toyota Production System. Productivity Press, Portland (1987)

Goldratt, E.M.: What is This Thing Called the Theory of Constraints?. The North River Press, New York (1990)

Martin, A.J.: DRP (Distribution Resource Planning)—can you afford not to have it? Mater. Handling Eng. **40**, 131–139 (1985)

Wang, W., Fung, R.Y., Chai, Y.: Approach of just-in-time distribution requirements planning for supply chain management. Int. J. Prod. Econ. **91**, 101–107 (2004)

Bagnha, M., Cohen, M.: The stabilizing effect of inventory in supply chains. Oper. Res. (1995, forthcoming)

Chen, F., Drezner, Z., Ryan, J.K., Simchi-Levi, D.: Quantifying the Bullwhip effect in a simple supply chain: the impact of forecasting, lead times, and information. Manage. Sci. **46**(3), 436–443 (2000)

Minner, S.: Multiple-supplier inventory models in supply chain management: a review. Int. J. Prod. Econ. **81–82**, 265–279 (2003)

Routroy, S., Kodali, R.: Differential evolution algorithm for supply chain inventory planning. J. Manuf. Technol. Manage. **16**(1), 7–17 (2005)

Ganeshan, R.: Managing supply chain inventories: a multiple retailer, one warehouse, multiple supplier model. Int. J. Prod. Econ. **59**, 341–354 (1999)

Rau, H., Wu, M.-Y., Wee, H.-M.: Integrated inventory model for deteriorating items under a multi-echelon supply chain environment. Int. J. Prod. Econ. **86**, 155–168 (2003)

Chiang, W.K., Monahan, G.E.: Managing inventories in a two-echelon dual-channel supply chain. Eur. J. Oper. Res. **162**, 325–341 (2005)

Kalchschmidt, M., Zotteri, G., Verganti, R.: Inventory management in a multi-echelon spare parts supply chain. Int. J. Prod. Econ. **81–82**, 397–413 (2003)

Tee, Y.S., Rossetti, M.D.: A robustness study of a multi-echelon inventory model via simulation. Int. J. Prod. Econ. **80**, 265–277 (2002)

Dong, L., Lee, H.L.: Optimal policies and approximations for a serial multiechelon inventory system with time-correlated demand. Oper. Res. **51**(6), 969 (2003)

van der Vorst, J.G.A.J., Beulens, A.J.M., van Beek, P.: Modelling and simulating multi-echelon food systems. Eur. J. Oper. Res. **122**, 354–366 (2000)

van der Heijden, M.C.: Multi-echelon inventory control in divergent systems with shipping frequencies. Eur. J. Oper. Res. **116**, 331–351 (1999)

Krepchin, I.P.: PC-based MRP, DRP help Lipton cut inventories. Mod. Mater. Handling **44**, 86–88 (1989)

Horne, R.: Charting a course for integrated logistics. Transp. Distrib. **30**, 45–51 (1989)

Frasier-Sleyman, K.: Forecasting and the continuous replenishment craze of the 1990s. J. Bus. Forecast. **13**, 3–8 (1994)

Forger, G.: How Lotus cut inventory and increased productivity. Mod. Mater. Handling **41**, 70–71 (1986)

Geary, S., Disney, S.M., Towill, D.R.: On bullwhip in supply chains—historical review, present practice and expected future impact. Int. J. Prod. Econ. **101**(1 SPEC. ISS), 2–18 (2006)

Lee, H.L., Padmanabhan, V., Seungjin, W.: The bullwhip effect in supply chains. Sloan Manage. Rev. **38**(3), 93–102 (1997)

Jafar, H., Kazemzadeh, R.B., Chaharsooghi, S.K.: A study of lead time variation impact on supply chain performance. Int. J. Adv. Manuf. Technol. **40**(11–12), 1206–1215 (2009)

Chandra, C., Grabis, J.: Application of multi-steps forecasting for restraining the bullwhip effect and improving inventory performance under autoregressive demand. Eur. J. Oper. Res. **166** (2), 337–350 (2005)

Aharon, B.-T., Boaz, G., Shimrit, S.: Robust multi-echelon multi-period inventory control. Eur. J. Oper. Res. **199**(3), 922–935 (2009)

Lee, H.L., So, K.C., Tang, C.S.: The value of information sharing in a two-level supply chain. Manage. Sci. **46**(5), 626–643 (2000)

Croson, R., Donohue, K.: Impact of POS data sharing on supply chain management: an experimental study. Prod. Oper. Manage. **12**(1), 1–11 (2009)
Alony, I., Munoz, A.: The bullwhip effect in complex supply chains. In: 2007 International Symposium on Communications and Information Technologies Proceedings, Darling Harbour, Sydney, Australia, 17–19 October 2007, pp. 1355–1360 (2007)
Gümüs, A.T., Güneri, A.F.: Multi-echelon inventory management in supply chains with uncertain demand and lead times: literature review from an operational research perspective. Proc. Inst. Mech. Eng. **221**(B10), 1553–1570 (2007)

# Multi-label Classification: New Measure to Remove Cyclical Dependencies

Hamza Lotf(✉) and Mohammed Ramdani

Lab. Informatique de Mohammedia, FSTM, Hassan II University of Casablanca, BP 146, 20650 Mohammedia, Morocco
lotf.hamza@gmail.com, ramdani@fstm.ac.ma

**Abstract.** Classifying data by linking it to a set of labels with a degree of membership is the objective of the multi-labels classification. These labels or classes can have order relationships between them, which can affect the predictive quality of classifiers. Consider these relationships or ignore them, when building the classifier, each has its drawbacks. The first approach facilitates the spread of learning errors and increases complexity of the task, especially if there are cyclical relationships between classes. While the second approach can give inconsistent predictions. There are several approaches intended for solving multi-label classification tasks, some of them take into consideration dependencies between labels and others consider them independent. A new approach called PSI-MC [1] proposes a novel way to teach classifiers the relations between labels without fixing a predefined structure. We propose an approach that uses the same principle as the PSI-MC, and which improves the way to eliminate cycles. Finally, we will test our new approach by using two datasets. According to the Hamming-Loss measure, our proposed approach is slightly better than binary relevance approach that does not take into account the relation between labels.

**Keywords:** Multi-label classification · Cyclical dependencies · Binary classifier · Removing cyclical dependencies · Decision trees

## 1 Introduction

Classification is a topic, which belong to the data-mining field. This technique uses supervised learning, based on training data, to develop a model capable of making predictions on data never before encountered. Different types of classifications exist depending on the nature and number of classes.

**Binary Classification [2]:** This kind of classification is very simple. Prediction is a class that has just two possible values. Each input will have only one value as an output. Binary classification can be used in email filtering to detect spam.

**Multi-class Classification:** Just like binary classification, prediction is a single value but belongs to a predefined set of more than two values.

**Multi-label Classification [3, 4]:** This is a task that aims to predict for each input a vector of values. The size of the output vector is the number of labels in the dataset.


A. Saka et al. (Eds.): CPI 2019, LNME, pp. 411–418, 2021.
https://doi.org/10.1007/978-3-030-62199-5_36

For example, a computer science book can also cover subjects such as mathematics and electronics. The same is true for an image that may contain trees, mountains and cloudy sky.

**Multi-dimensional Classification [5]:** This is a generalization of multi-label classification. Each element of the prediction vector takes the value of a large set of predefined values instead of a single binary value.

**Graduated Multi-label Classification [6]:** It is a generalization of multi-label classification and a special case of multi-dimensional classification. Each input instance can belong to multiple elements of the output vector to a certain degree. This kind of classification can be easily found in the real world. For example, a movie can be classified as (Action: four stars) and (Suspense: two stars).

In this paper, we will present some existing methods known to solve the problems of Multi-Label classification and then we will present the PSI-MC approach [1], which takes into consideration the relationships between labels without imposing a predefined structure.

We outline our approach that proposes an improvement in how to eliminate cycles in Sect. 3. Section 4 discusses testing our approach on real data, and conclusions and perspectives are presented in Sect. 5.

## 2 State of the Art

Learning for multi-label classification is addressed by different approaches in the state of the art. The adaptation of classical methods, the transformation of data, and the use of methods based on a set of classifiers.

This article focuses on methods using data transformation approaches to produce one or more easier tasks. The most commonly used methods are BR (Binary Relevance) [7] and LP (Label Powerset) [2]. BR consists in training k classifiers, each for a label before collecting the results of each classifier to have the prediction vector. This method neglects the existence of dependencies between labels. While LP transforms each label set into a new class. The result is a new multiclass classification problem with the same input number. This method allows you to learn the relationships between labels but only predicts the combinations of labels that already exist in the dataset used for learning.

There are other methods to consider the relationships between labels such as CC [8] (Classifier Chains). This method also involves k classifiers chosen in random order. The first classifier is trained just with the attributes of the original dataset, then this label is added as an attribute for learning the second, and so on until the $k^{th}$ classifier is reached.

The starting order can influence the result of the prediction on the one hand and on the other hand, the fixing of the order does not allow taking into account all the dependencies between the labels. The ECC method (Ensemble of Classifier Chains) allows training a set of CCs with different orders and then use a voting system to decide between the results of each CC.

PSI-MC [1] is a recent approach that allows taking into consideration the relationships between labels. The first principle of the PSI-MC method is to learn a set of classifiers according to the number of classes existing in the initial data. Each classifier will have the objective of predicting the value of a class by using descriptive attributes as an input in addition to the other classes.

This method allows classifiers to learn the relationships between labels. However, cyclical dependencies may occur. The second principle of the PSI-MC method is to remove these cyclical dependencies based on three measures. The first is a pre-selection measure, the second is a selection measure and the third is a chaining interest measure. All the methods mentioned above can be criticized in terms of execution time, which can be very long depending on the size of the data and due to the number of iterations they use.

## 3 Proposed Approach

Our approach proposes a single $S$ measure which represents the score calculated for every class to decide on its degree of importance, instead of the three measures previously mentioned. Let $X = \{x_i\}_{1 \leq i \leq n}$ be the dataset and $C = \{c_l\}_{1 \leq l \leq k}$ the set of labels.

Each element $x_i$ is a vector of p values $(x_{ij})_{1 \leq j \leq p}$ representing descriptive attributes. Value $y_i$ represents the vector of the decision attributes values $(y_{il})_{1 \leq l \leq k}$ associated with the vector $x_i$.

$H : X \rightarrow P(C)$ a classifier that associates each data to a set of labels. We will break down the Multi-Label Classification task into several binary classification subtasks.

$H = \{H_i\}_{1 \leq i \leq k}$ where $H_i$ denote the classifier for the class $c_i$.

Let $D_{out} : X \rightarrow P(H)$ be a function that gives for each classifier all the classifiers that depend on it, and $D_{in} : H \rightarrow P(H)$ a function that provides for each classifier the set of classifiers on which it depends.

The value $p_{i/j}$ represents the weight of the class $c_i$ when predicting the class $c_j$ by the classifier $H_j$. The weight is defined according to the nature of the Binary classifier. In the following example we use decision trees, here is a proposed definition of the weight:

$$\begin{cases} p_{i/j} = 0 & \text{if } H_j \notin D_{out}(H_i) \\ p_{i/j} = -\left(Niv_{i/j} - k\right) & \text{if } H_j \in D_{out}(H_i) \end{cases} \tag{1}$$

Value $Niv_{i/j}$ designates the level where the class $c_i$ appears in the classifier $H_j$. The prediction of the class $c_i$ depends on the value of the class $c_j$ if the value of $Niv_{i/j}$ is greater than zero.

$$P_{out}(H_i) = \sum_{j=1\,\&\, j\neq i}^{k} \mathrm{p}_{i/j} \tag{2}$$

$$P_{in}(H_i) = -\sum_{j=1\,\&\,j\neq i}^{k} \mathrm{p}_{j/i} \tag{3}$$

The measure $S : H \to N$ defined by $S(H_i) = P_{out}(H_i) + P_{in}(H_i)$ which gives each classifier $H_i$ a priority level. If $H_i$ depends on $H_j$ and $S(H_i) > S(H_j)$ we remove the dependency.

Example
Table 1 represents data we are using to explain our approach. The data are composed of eight attributes, two of which are descriptive $\{v_1, v_2\}$ and the others are decision attributes $\{c_1, c_2, c_3, c_4, c_5, c_6\}$. The possible values for a class are {0, 1, 2}.

**Table 1.** Example data.

| | $v_1$ | $v_2$ | $c_1$ | $c_2$ | $c_3$ | $c_4$ | $c_5$ | $c_6$ |
|---|---|---|---|---|---|---|---|---|
| $x_1$ | 30 | 15 | 0 | 2 | 1 | 0 | 2 | 2 |
| $x_2$ | 19 | 15 | 2 | 1 | 2 | 0 | 0 | 1 |
| $x_3$ | 3 | 20 | 0 | 2 | 2 | 1 | 2 | 0 |
| $x_4$ | 51 | 43 | 1 | 0 | 1 | 0 | 2 | 2 |
| $x_5$ | 25 | 30 | 1 | 2 | 2 | 1 | 2 | 0 |
| $x_6$ | 25 | 16 | 0 | 0 | 1 | 1 | 2 | 1 |
| $x_7$ | 29 | 34 | 1 | 1 | 1 | 2 | 0 | 2 |
| $x_8$ | 5 | 16 | 2 | 0 | 1 | 0 | 0 | 2 |
| $x_9$ | 24 | 10 | 2 | 1 | 1 | 2 | 1 | 2 |
| $x_{10}$ | 14 | 33 | 0 | 2 | 2 | 0 | 0 | 0 |
| $x_{11}$ | 35 | 30 | 0 | 0 | 1 | 0 | 2 | 2 |
| $x_{12}$ | 18 | 18 | 2 | 0 | 0 | 0 | 2 | 2 |
| $x_{13}$ | 25 | 60 | 2 | 1 | 0 | 0 | 2 | 2 |
| $x_{14}$ | 19 | 40 | 0 | 2 | 1 | 0 | 0 | 0 |

Figure 1 represents six decision trees, which are the classifiers of $\{c_1, c_2, c_3, c_4, c_5, c_6\}$ classes. For example to generate the $c_1$ classifier, called $H_1$, we consider $\{v_1, v_2, c_2, c_3, c_4, c_5, c_6\}$ as descriptive attributes called also learning inputs.

Table 2 shows the values of the $S$ measure, calculated for all classifiers. A classifier can only depend on others with higher values of the $S$ measure.

Figure 2 present initial dependencies detected between classifiers. We notice the existence of cyclical dependencies between classifiers $(H_2, H_6, H_5, H_4)$.

Figure 3 shows result of removing cycles following our new approach. $H_2, H_4$ and $H_6$ are the only classifiers generated a second time.

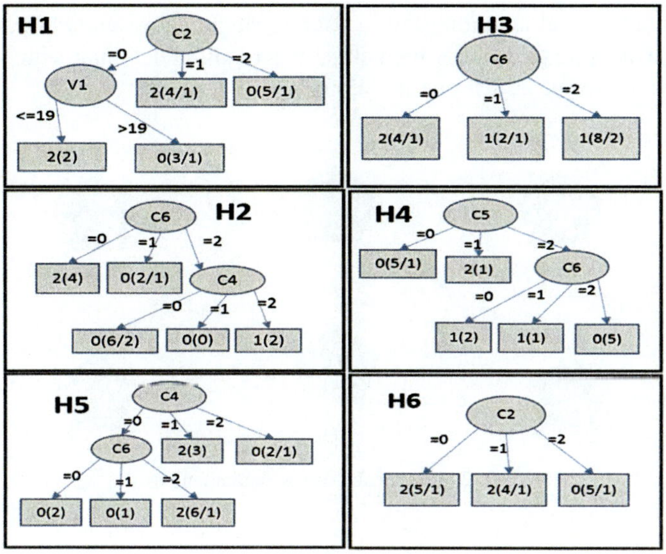


**Fig. 1.** Generated classifiers (Decision trees)

**Table 2.** Values of S measures for the different classifier

| Classifier | $P_{out}$ | $P_{in}$ | $S$ |
|---|---|---|---|
| $H_1$ | 0 | −5 | −5 |
| $H_2$ | 10 | −9 | 1 |
| $H_3$ | 0 | −5 | −5 |
| $H_4$ | 9 | −9 | 0 |
| $H_5$ | 5 | −9 | −4 |
| $H_6$ | 18 | −5 | 13 |

We use the following algorithm to remove cyclical dependencies detected between the classifiers $(H_2, H_6, H_5, H_4)$.

```
For each classifier H_i in H
  For each classifier H_j in H where H_i ≠ H_j
      If  S(H_i) > S(H_j)  then
          Delete class c_j from descriptive attributes of H_i
      End
      End
    Restart learning of H_i
    End
```

This approach helps learning relations between labels and eliminating cyclical dependencies to prevent the risk of over fitting and errors propagation. Existing methods require several iterations and measurements to get result, while our approach uses only one measure to decide which classifiers require relearning with which inputs, and performs the result in just one iteration.

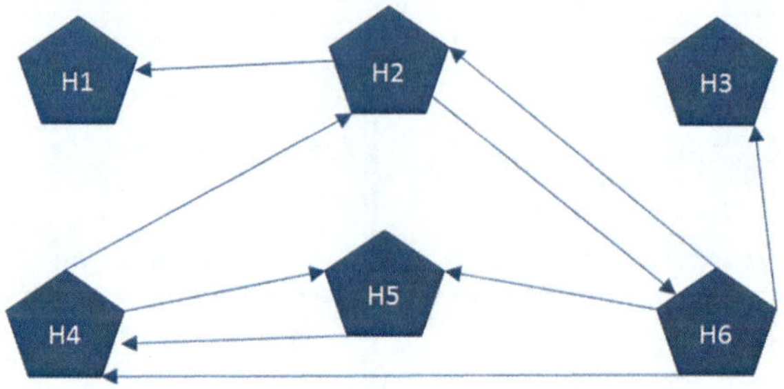


**Fig. 2.** Initial classifiers dependencies

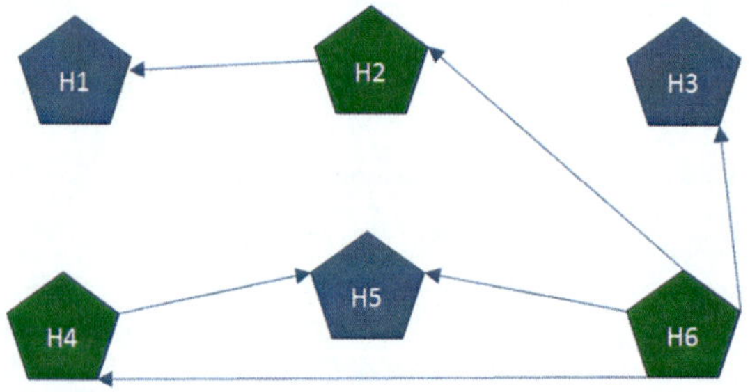


**Fig. 3.** Final classifiers dependencies

## 4 Experimentation

In this section, we describe datasets Table 3 that we have used to make meta-classifier. Yeast is a functional genomics dataset formed by micro-array expression data and phylogenetic profiles. It formed by 103 attributes and 14 labels. The training set contains 1500 genes and 917 genes in testing set. Scene dataset contains 294 attributes and 6 semantic classes (Beach, Sunset, FallFoliage, Field, Mountain, Urban).

To compare our proposed meta-classifier called (B3R) with the BR approach, we used J48 trees as binary classifier, and programs are written with JAVA language using MULAN and WEKA API.

Table 4 shows that according to the Hamming-loss measure on the Yeast dataset, our approach, which takes into consideration relations between labels, performs better than the BR method, which is based on independent classifiers. In the other hand, we notice that both approach are almost equivalent when using the scene dataset. This makes sense because there is no relation between labels in the latter dataset.

**Table 3.** Dataset description

| Dataset | Instances | Attributes | Labels |
|---|---|---|---|
| Yeast | 1500/917 | 103 | 14 |
| Scene | 1211/1196 | 294 | 6 |

**Table 4.** Results according to Hamming-loss measure

| Classifier | Dataset | Hamming-loss |
|---|---|---|
| BR | Yeast | 0.3743 |
| Our approach | Yeast | 0.3354 |
| BR | Scene | 0.2979 |
| Our approach | Scene | 0.2975 |

## 5 Conclusion

In this paper, we proposed a new approach to deal with Multi-Label classification tasks by considering classes dependencies and removing cycles.

Unlike existing approaches, which are based on several measures and iterations, our proposed method use a single measure that introduces the weight of an inter-label dependency to decide on its degree of importance.

Experiments shows that we are slightly better than BR approach according to hamming-loss, even when using datasets with correlated labels or independent ones.

The impact of our approach is not fully studied in this paper, but in future work we plan to use more measures and more datasets with correlated labels. We will compare the performance of our approach with more algorithms and particularly with the PSI-MC.

We plan also to improve this approach by rethinking the proposed measure, by making it applicable with other kind of binary classifiers and implementing it in a Big Data environment.

## References

1. Laghmari, K., Marsala, C., Ramdani, M.: Graded multilabel classification: compromise between handling label relations and limiting error propagation. In: 2016 11th International Conference on Intelligent Systems: Theories and Applications (SITA), December 2016
2. Rivera, A.J., del Jesus, M.J., Herrera, F., Charte, F.: Multilabel Classification Problem Analysis, Metrics and Techniques. Springer, Cham (2006)
3. Zhou, Z.-H., Zhang, M.-L.: A review on multi-label learning algorithms. IEEE Trans. Knowl. Data Eng. **26**(8), 1819–1837 (2014)
4. Freitas, A.A., de Carvalho, A.C.P.L.F.: A Tutorial on Multi-Label Classifcation Techniques. Springer, Heidelberg. pp. 177–195 (2009).
5. Larrañaga, P., Bielza, C., Li., G.: Multi-dimensional classifcation with bayesian networks. Int. J. Approx. Reason. **52**(6), 705–727 (2011)

6. Fürnkranz, J., Brinker, C., Mencía, E.L.: Graded multilabel classification by pairwise comparisons. In: IEEE International Conference on Data Mining (ICDM), December 2014
7. Sarawagi, S., Godbole, S.: Discriminative methods for multi-labeled classification. In: Advances in Knowledge Discovery and Data Mining, vol. 3056, pp. 22–30 (2004)
8. Holmes, G., Frank, E., Jesse, R., Pfahringer, B.: Classifier chains for multi-label classification. Mach. Learn. **85**, 333–359 (2011)

# Determining Learning Styles of Engineering Students and the Impact on Their Academic Achievement

El Haini Jamila(✉)

National School of Applied Sciences, Fez, Morocco
j.elhaini@gmail.com

**Abstract.** The purpose of this research is twofold. Firstly, it aims to identify the learning style preferences of two classes in their fourth year of engineering studying at National School of Applied Sciences of Fez during the 2018–2019 academic year. Secondly, it investigates the relationship between learning styles and the academic achievement of these students. A VARK questionnaire was administrated to the students and the collected data were analyzed by using statistical methods (Pearson correlation).

The findings of this study showed that for the industrial engineering class, the most preferred learning style of students were Kinesthetic with a percentage of 41.17%. This is followed by a multimodal style (23.52%), visual style (20.58%), auditory style (8.82%) and Reading/Writing style (5.88%). For the mechanical and automated systems engineering class, the most preferred learning style of students were visual style and kinesthetic style with the same percentage of 31.57%. This is followed by a multimodal style (18.42%), Reading/Writing style (10.52%) and auditory style (7.89%).

This study will be helpful for instructors in determining appropriate teaching approaches to accommodate the diverse learning styles.

**Keywords:** Learning styles · VARK model · Pearson correlation

## 1 Introduction

Learning styles are various strategies or ways of learning [1]. Their concept owes its evolution in psychology [2]. There are several definitions and models of learning styles. According to [3], learning style as a complex manner in which learners most efficiently and most effectively perceive, process, store, and recall what they are attempting to learn [4]. Another definition of learning style is by Dunn in the paper [5] who described the latter as the way each learner begins to concentrate, process, and retain new and difficult information [1].

The main models of learning styles are: David Kolb's model, Felder's model and Neil Fleming's VARK model.

David Kolb's model classified the learning styles as: convergent (good at problem solving and practical application of ideas); divergent (good imaginative ability and awareness of meaning and values); assimilative (good at inductive reasoning); accommodative (efficient in carrying out plans and like getting involved in new experiences) [2].


A. Saka et al. (Eds.): CPI 2019, LNME, pp. 419–423, 2021.
https://doi.org/10.1007/978-3-030-62199-5_37

The Felder-Silverman Learning Style Model was introduced by Richard Felder and Linda Silverman in 1988 and classified learning styles into four dimensions: active-reflective (processing information), sensing-intuitive (perceiving information), visual-verbal (inputting information) and sequential-global (understanding information). The Index of Learning Styles (ILS) instrument, developed by Felder and Soloman in 1991, comprises 44 questions, 11 for each of the four previously described dimensions [1].

The VARK model was introduced by Neil Fleming in 1987 and has four categories of learners: visual, auditory, reading/writing and kinesthetic [1, 6, 7] and the VARK questionnaire developed by Flemming is used to assess the sensory modalities. Visual learners prefer the use of diagrams and symbolic devices. Read-write learners prefer printed words and texts. Auditory prefer heard information. Kinesthetic learners have to feel or live the experience to learn; they prefer simulations of real practice and experiences [8, 9].

By identifying the learning styles of students, teachers can use different methods and regulate their courses appropriately and according to the conditions in order to match students' learning styles which is the most successful strategy [10–13].

This paper aims to identify the learning style preferences of the industrial engineering class (IE) and the mechanical and automated systems engineering (MASE) class in their fourth year of studying at National School of Applied Sciences of Fez during the 2018–2019 academic year. Also, it investigates the relationship between learning styles and the academic achievement of these students.

## 2 Method

In order to assess the preferred learning styles of the industrial engineering (IE) class and the mechanical and automated systems engineering class (MASE) in their fourth year of studying at National School of Applied Sciences of Fez during the 2018–2019 academic ycar, the Neil Fleming's VARK model was used.

The VARK questionnaire is consisting of The VARK questionnaire of 16 multiple choice questions, each having four choices. All choices correspond to the four sensory modalities which are measured by VARK (visual, aural/auditory, read/write, and kinesthetic). The students can select one or more choices, based on the sensory modalities which are preferred by them, to take in new information [14].

The questionnaire was administrated to the IE class consisting of 34 students and to MASE class consisting of 38 students. This study has paid special attention to assure respondents confidentiality and right to respond or reject the questionnaire.

And in order to determine if there is any relationship between learning styles of students and their academic achievement (final exam's scores), the descriptive method (Pearson correlation) was used using SPSS software.

## 3 Results and Discussion

The learning styles of all 34 students of IE class and of all 38 students of MASE class responses are shown in the following Tables 1 and 2.

**Table 1.** Students' learning styles frequencies of IE class

| | Visual | Auditory | Read/Write | Kinesthetic | Multimodal |
|---|---|---|---|---|---|
| Students number | 7 | 3 | 2 | 14 | 8 |
| Percentage | 20.58 | 8.82 | 5.88 | 41.17 | 23.52 |

**Table 2.** Students' learning styles frequencies of MASE class

| | Visual | Auditory | Read/Write | Kinesthetic | Multimodal |
|---|---|---|---|---|---|
| Students number | 12 | 3 | 4 | 12 | 7 |
| Percentage | 31.57 | 7.89 | 10.52 | 31.57 | 18.42 |

The Table 1 above clarifies that most of students of IE class are kinesthetic learners with a percentage of 41.17%. The kinesthetic learners prefer simulation and hand-on experiences. The table also shows that 23.52% of students prefer a multimodal learning style. This is followed by visual learning style with a frequency of 20.58%. These learners prefer the use of diagrams, symbolic devices and printed information. The least preferred learning styles are Auditory and reading/writing styles with the frequencies 8.82% and 5.88% respectively.

Also, the Table 2 above clarifies that most of students of MASE class are kinesthetic learners with a percentage of 31.57% and visual learners with the same frequency. The table also shows that 18.42% of students prefer a multimodal learning style. This is followed by read/write learning style with a frequency of 10.52%. The least preferred learning style is auditory style with the frequency 7.89%.

So, the best teaching strategy for these students is the use a lot of laboratory applications as possible and to use visuals like PowerPoint slides with images, videos and colours.

In order to investigate the relationship between learning styles and academic achievement of these students based on final exam's scores, we use a descriptive method (Pearson correlation) where the independent variables are learning styles and the dependent variables are academic achievement. Collected data was analyzed using SPSS software. The Tables 3 and 4 below show the results of this analysis.

**Table 3.** Pearson correlation of different learning styles for IE class

| | Academic achievement | |
|---|---|---|
| *Learning styles* | | |
| Visual style | Pearson correlation | 0.83 |
| | Signification | 0.021 |
| Auditory style | Pearson correlation | 0.023 |
| | Signification | 0.39 |
| Reading/Writing style | Pearson correlation | −0.36 |
| | Signification | 0.43 |
| Kinesthetic style | Pearson correlation | 0.87 |
| | Signification | 0.012 |

**Table 4.** Pearson correlation of different learning styles for MASE class

| | Academic achievement | |
|---|---|---|
| *Learning style* | | |
| Visual style | Pearson correlation<br>Signification | 0.91<br>0.049 |
| Auditory style | Pearson correlation<br>Signification | −0.54<br>0.39 |
| Reading/Writing style | Pearson correlation<br>Signification | 0.18<br>0.031 |
| Kinesthetic style | Pearson correlation<br>Signification | 0.61<br>0.032 |

From the results of the Table 3 above, the visual learning style shows a strong correlation with the academic achievement of the IE students because the Pearson correlation is equal to 0.83 which is higher than 0.8 and we can also conclude that this correlation is highly significant (because the coefficient signification is less than 0.05). The same results for the Kinesthetic learning style (Pearson correlation is equal to 0.87 which is higher than 0.8 and the signification is equal to 0.012). In the other hand, the auditory learning style shows a very weak correlation equal to 0.023. Reading/Writing style shows a negative and weak correlation. We can conclude that visual and kinesthetic style influence the academic achievement for IE class.

For the MASE class as shown in the Table 4, we can conclude that the visual learning style shows a very strong correlation with the academic achievement of the MASE students because the Pearson correlation is equal to 0.91 (very near of 1) which is higher than 0.8 and this correlation is highly significant (because the coefficient signification is less than 0.05). Also, we can admit that there is a correlation between kinesthetic learning style (Pearson correlation is equal to 0.61 and the signification is equal to 0.032). In the other hand, the Reading/Writing learning style shows a very weak correlation equal to 0.18 which is highly significant (signification equal to 0.031). Auditory style shows a negative correlation which is not significant. We can conclude that visual and kinesthetic style influence the academic achievement for MASE class.

So, we can conclude that the same learning styles influence the academic achievement for both classes. Therefore, including practice and visual aids in the teaching strategy is very necessary.

## 4 Conclusion

This research is carried out to determine the learning styles of two classes at National school of applied sciences of Fez, Morocco and to investigate if the learning style Influences students Academic Performance. Based on the VARK questionnaire, the findings show that the most preferred learning styles for both alos, by using the statistical method, the findings showed that only visual learning and kinesthetic learning are the only styles that have a significant impact on the academic performance of the students of both classes while the other two styles are not significant drivers in impacting the academic performance.

So, in order to find appropriate teaching methods and to achieve educational goals, the teaching strategy must integrate practice and use visual aids.

## References


1. Asiah, N., Ab, G., Nik, R.: Learning styles of business students at a Malaysian polytechnic. Int. J. Educ. Res. **3**(10) (2015)
2. Gaikwad, H.V.: Analysis of learning styles of engineering students for improving engineering education. J. Eng. Educ. Transform. (2017). eISSN 2394-1707. https://doi.org/10.16920/jeet/2017/v0i0/111788
3. James, W.B., Gardner, D.L.: Learning styles: implications for distance learning. In: New Directions for Adult and Continuing Education, vol. 67, pp. 19–31 (1995)
4. Rahmani, J.: Learning styles and academic achievement: a case study of Iranian high school girls' students. Procedia Soc. Behav. Sci. **51**, 1030–1034 (2012)
5. Dunn, R.: Understanding the Dunn and Dunn learning styles model and the need for individual diagnosis and prescription. Read. Writ. Learn. Disabil. **6**, 223–247 (1990)
6. Mlambo, V.: An analysis of some factors affecting student academic performance in an introductory biochemistry course at the University of the West Indies. Carib. Teach. Sch. **1** (2), 79–92 (2012)
7. Montemayor, E., Aplaten, M.C., Mendoza, G.C., Perey, G.M. Learning styles of high and low academic achieving freshman teacher education students: an application of the Dunn and Dunn's learning style model (2011). http://www.eisrjc.com/documents/Learning_Styles_Of_High_And_Low_Academic_Achieving_Freshman_1325667415.pdf. Accessed 7 Mar 2015
8. Baykan, Z., Nacar, M.: Learning styles of first-year medical students attending Erciyes University in Kayseri, Turkey. Adv. Physiol. Educ. **31**, 158–160 (2007)
9. Ictenbas, B.D., Eryilmaz, H.: Determining learning styles of engineering students to improve the design of a service course. Procedia Soc. Behav. Sci. **28**, 342–346 (2011)
10. Langlois, J., Thach, S.: Teaching and learning styles in the clinical setting. Fam. Med. **33**, 344–346 (2001)
11. Miller, P.: Learning styles: the multimedia of the mind. Education Resources Information Center, 451: 140 (2001)
12. Tanner, K., Allen, D.: Approaches to biology teaching and learning: learning styles and the problem of instructional selection-engaging all students in science courses. Cell Biol. Educ. **3**, 197–201 (2004)
13. Dinakar, C., Adams, C., Brimer, A., Silva, M.D.: Learning preferences of car givers of asthmatic children. J. Asthma **42**, 683–687 (2005)
14. Hadi, P., Jamil, S., Javaher, K., Masood, Y., et al.: Using VARK approach for assessing preferred learning styles of first year medical sciences students: a survey from Iran. J. Clin. Diagn. Res. **8**(8), GC01-GC04 (2014). https://doi.org/10.7860/jcdr/2014/8089.4667

# WSN's Life-Time Improvement Passing from Hierarchical to Hybrid Routing Techniques: A Comparative Study

Hicham Qabouche(✉), Aïcha Sahel, and Abdelmajid Badri

LEEA & TI, FSTM, Hassan II University of Casablanca, Casablanca, Morocco
qabouche.hicham@gmail.com

**Abstract.** Wireless sensor networks consist of a large number of nodes deployed randomly in an area of interest. Theses nodes have sensing, computation, and wireless communications capabilities. In another hand nodes have energy constraints since they are equipped by non-rechargeable batteries. To handle this issue, routing protocols consist of designing the network in order to collect and transmit data with less energy consumption. In this paper we present a life-time and throughput comparative study passing from classical clustering protocols (LEACH) to a hybrid routing protocol, combines between clustering and Multi-hop Routing techniques (SMR). SMR introduces another type of nodes, called Independent Nodes (IN), whose role is to help Cluster Heads (CHs) to route their collected data to the Base Station (BS). Finally, a comparative study is done, in the first place, between LEACH and SMR in a homogeneous network. And in a second place between LEACH, DEEC and SMR in heterogeneous environment. Those simulations show a great improvement in network's life-time and throughput using SMR technique in both homogeneous and heterogeneous environment when the base station is far from the center of the network, and less performance when the SB is placed in the middle of the network.

**Keywords:** WSN · Routing protocols · LEACH · SMR · Life-time

## 1 Introduction

Wireless Sensor Networks (WSNs) are attracting the attention of many researchers thanks to their participation in a variety of practical applications [1]. The main problem of WSNs is energy harvesting, because nodes are equipped with a non-rechargeable battery. Many routing, power management, and data dissemination protocols have been specially designed for WSNs where energy awareness is an essential design issue [2]. In another hand, routing protocols plays a very important role in routing the data collected by the sensors to the base station with the minimum cost of energy. Although, routing protocols defines the structure of the network and data transmission strategy. For this reason, LEACH


A. Saka et al. (Eds.): CPI 2019, LNME, pp. 424–437, 2021.
https://doi.org/10.1007/978-3-030-62199-5_38

[3], provided a clustering-based algorithm consisting of creating an optimal number of clusters determined by a probability $p_{opt}$. The election process is based on a random probabilistic method. Elected CHs are responsible of collecting data from cluster members and transmit it in a direct way to the BS. Furthermore, LEACH present a great improvement compared to direct schemes [3], but the main problems of LEACH are:

- ◇ not considering nodes residual energy in election process.
- ◇ CHs transmit directly to the BS, so knowing that the energy is related to distance to BS, the energy consumed by far CHs is high than energy consumed by close CHs.

In another hand the main idea of SMR [1] is to dived the network into levels, counted from the BS, as in MHT-LEACH [4] and IMHT-LEACH [5], and introduces another type of nodes called independent nodes NIs, created by limiting the number of cluster members to No, as in EDMHT-LEACH [6]. In addition, the cluster is divided into two levels, for that reason SMR consider inter and intra-cluster communication.

## 2 Related Work

A lot of researches are done in order to handle energy harvesting problem for WSN. Thus most of the current clustering algorithms are: **LEACH** [3], selects CHs periodically and drains energy uniformly by role rotation. Each node decides itself whether or not a CHs distributed by a probability. Under the homogeneous network, LEACH performs well, but its performance become badly in the heterogeneous network [7]. In **PEGASIS** [8], nodes organizes theirs self to form a chain. This method is difficult to implement due to the requirement of global knowledge of the network topology. In a heterogeneous environment, there is, **SEP** [10] considers that the network is consist of two types of nodes, *normal nodes* and *advanced nodes* whom have more energy than normal nodes. SEP is based on the same principle of LEACH except that each type of nodes has a specified election probability based on the proportion of each type in the network. **DEEC** [7], is an election energy aware protocol. It uses the same technique of LEACH, it introduce nodes residual energy in CHs election process so that nodes with more energy have more chance to be CHs than nodes with less energy. This process distribute consumption of energy in the network. Analysis of this protocol is not considered on this paper rather it's used on comparative study.

## 3 Radio Model

In this paper we use the common radio model of a node to represent consumed energy in different transmissions of collected data through the network to the BS (see Fig. 1). To compute the consumed energy, we use the energy dissipation model proposed in [9]. The required energy to transmit and receive a k-bits message is calculated using the following equations [9]:

$$E_{Tx}(d) = LE_{ele} + \begin{cases} L\epsilon_{fs}d^2 & \text{si} \quad d < d_0 \\ L\epsilon_{amp}d^4 & \text{si} \quad d \geq d_0 \end{cases} \tag{1}$$

$$E_{Rx} = lE_{ele} \tag{2}$$

With d is the distance between the two nodes, $\mathbf{E_{TX}}$ is the energy dissipated during transmission and $\mathbf{E_{RX}}$ is the consumed energy in reception. $\epsilon_{fs}$ is the transmission power in the free space transmission model, $\epsilon_{amp}$ is the transmission power in the multi-hop transmission model. $\mathbf{d_0}$ is the limit value between free space and multi-hop transmission models, it's expression is giving in [9,11] as follow:

$$d_o = \sqrt{\frac{\epsilon_{fs}}{\epsilon_{amp}}} \tag{3}$$

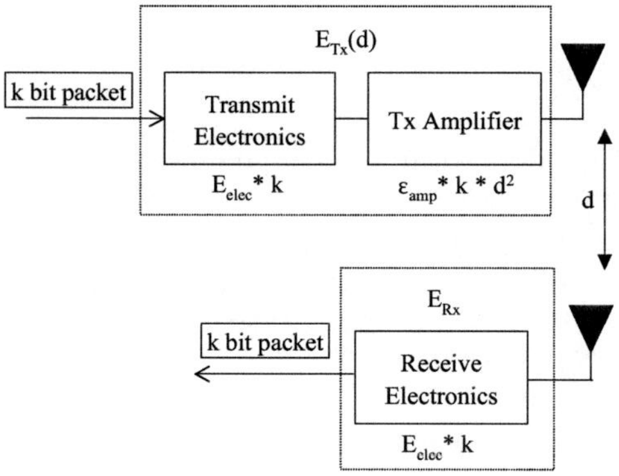


**Fig. 1.** Radio model of a node

## 4 LEACH: Low Efficent Adaptative Cluster Hierarchy

LEACH is a self-organizing, adaptive clustering protocol that uses randomization to distribute the energy load evenly among the sensors in the network [3]. It consists in electing an optimal number $k_{opt} = p_{opt}N$ of CHs. The latter are responsible for creating their clusters and receiving data collected by the cluster member nodes, after that, they are responsible of routing the data to the base station. Knowing that the energy consumption in the CHs is large compared to the normal nodes and in order to balance the energy consumption between the different nodes, LEACH is divided into rounds and each round starts with a set-up phase in which the clusters are formed, followed by a communication phase during which data is routed to the base station (see Fig. 2).

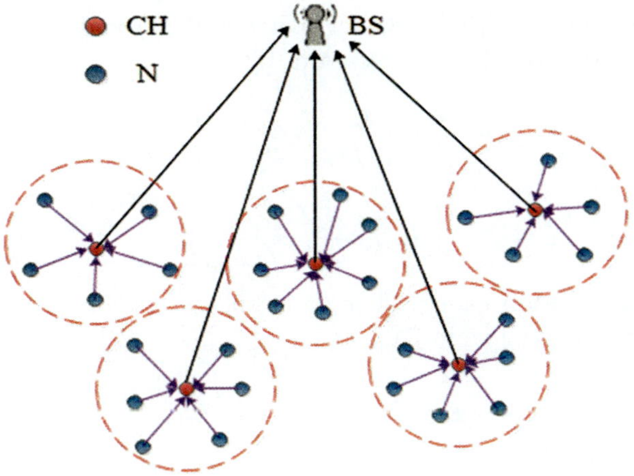


**Fig. 2.** Example of a hierarchical architecture model.

This protocol is applied on a homogeneous network whose initial energy of each node is $\mathbf{E_0}$, which gives that the total initial energy of the network is given by the Eq. (5).

$$E_{initot} = \sum_{i=1}^{N} E_i = NE_0 \tag{4}$$

LEACH is divided into rounds, each round contains a **set-up** phase, and a **communication** phase [1,3]. In the **Set-up phase** each node decides whether it will be a CH or not, based on the following procedure: *each node chooses a random number between 0 and 1, if the chosen number is less than the threshold value* $\mathbf{T(n)}$, *the node becomes a CH.* With $\mathbf{T(n)}$ is given by the following expression as in [3]:

$$T(n) = \begin{cases} \frac{p_{opt}}{1-p_{opt}\left(r mod \frac{1}{p_{opt}}\right)} & \text{if } \{n\} \in G \\ 0 & \text{otherwise.} \end{cases} \tag{5}$$

With $p_{opt}$ is the optimal probability of a node to be a CH and G is the set of legitimate nodes to be CHs for the round r. After the election, each CH for the round r, broadcasts an advertising message to all the nodes using the CSMA MAC protocol, to avoid collision with other CHs messages. Then each normal node decides which cluster will belong by sending a JOIN-REQ to the corresponding CH. At this time, each CH generates a time schedule using the TDMA which manages the transmission periods of each cluster members.

In the **Communication phase**, each cluster member collect data and send is to his CH in the time slot reserved by the TDMA. The CH aggregate data received and send it to the BS using the node radio model discussed earlier.

The energy consumed by a cluster member (CM) and a CH are expressed respectively in (6) and (7) as expressed in [11]:

$$E_{cm} = L.E_{elec} + \begin{cases} L\epsilon_{fs}d_{toCH}^2 & \text{si} \quad d_{toCH} < d_0 \\ L\epsilon_{amp}d_{toCH}^4 & \text{si} \quad d_{toCH} \geq d_0 \end{cases} \tag{6}$$

$$E_{CH} = \frac{N}{k}L\left(E_{data} + E_{elec}\right) + \begin{cases} L\epsilon_{fs}d_{toBS}^2 & \text{si} \quad d_{toBS} < d_0 \\ L\epsilon_{amp}d_{toBS}^4 & \text{si} \quad d_{toBS} \geq d_0 \end{cases} \tag{7}$$

$E_{data}$ is the aggregation energy. The total energy consumed by a cluster in a single round is expressed as in [10].

$$E_{clus} = E_{CH} + \frac{N}{k_{opt}}E_{cm} \tag{8}$$

With N is the total number of nodes in the network and $k_{opt}$ is the optimal number of clusters.

Let $e = \frac{1}{p_{opt}}$, the epoch, defined as the number of rounds necessary so that all nodes are being legitimate again to be elected as CHs [7]. Since the network is homogeneous, in terms of energy, all nodes have the same possibility of being a CH once per epoch. Figure 3 shows the dynamic process of creating clusters in two different rounds. *All nodes marked with the same symbol belongs to the same cluster.*

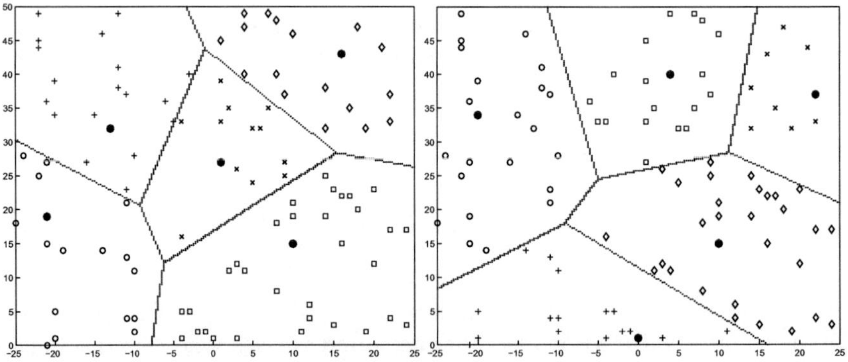

**Fig. 3.** Network's state in two different rounds

LEACH protocol has gained much popularity in the WSN research field, due to its outstanding success [1], prolonging network's life-time and throughput compared to direct schemes. But it still have a lot of limitations, we mention, not considering node's residual energy on election process. and the fact that CHs transmit theirs collected data directly to the BS. Knowing that the election process is totally random, far CHs will consume more energy than close CHs, this results that far nodes dies quickly due to the imbalanced energy consumption.

# 5 SMR: Static Multi-hop Routing

**SMR** is a hybrid routing protocol, that combines between Hierarchical and flat routing schemes in order to prolong network's life-time and throughput. The main idea of this protocol is to divide the network into many levels, starting from the BS, the length of each level is $\frac{d_o}{2}$. Each node calculate it's level according to the distance to the BS. For example, the CH located at a distance less than $\frac{d_o}{2}$ from the BS will belong to the first level, whereas the CH with a distance equal or larger than $\frac{d_o}{2}$ but less than $d_o$ will belong to the second level and so on [1] (see Fig. 4).

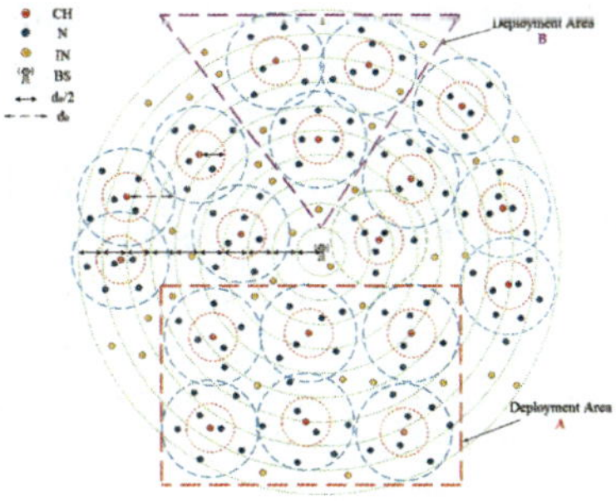


**Fig. 4.** The general topology of a WSN using the proposed approach

SMR is composed of four phases as in [6].

## 5.1 Initial Phase

In this phase, the election of CH's is done. The process is the same as in LEACH, but the difference is in the threshold expression. The new election threshold is expressed as follow [6,12]:

$$T(n_i) = \begin{cases} max\left(\frac{p_i}{1-p_i.\left(rmod\frac{1}{p_i}\right)} \times \frac{E_{res}}{E_{init}}, T_{min}\right) & \text{si } \{n_i\} \in G \\ 0 & \text{Otherwise.} \end{cases} \tag{9}$$

Here, the $\mathbf{E_{res}}$ is the amount of energy, which the node still has after a period of working.$\mathbf{E_{init}}$ denotes the amount of energy incorporated in a node before starting its activity. $\mathbf{T_{min}}$ refers to the minimum value of the threshold, which can be used if the $\mathbf{E_{res}}$ value has descended to low values [1]. As fixed in [6], each cluster have a limited, permitted, number of members fixed to $N_o = \frac{N}{k_{opt}}$. Moreover, the proposed approach assumes that the sensor node is only able to join the cluster is the distance between them is less than $d_o$.

The limited number of CMs and distance approach results the apparition of another types of nodes that's not belong to any cluster. Those nodes are called

**Independent Nodes** (INs) whose used to help CHs route data to the BS. The new approach add by this protocol is clusters leveling. It assume that each cluster is also divided into two levels, and each one is a $\frac{d_o}{2}$ length.

## 5.2 Announcement Phase

After election process each node, CHs, CMs and INs, declare it self in the network broadcasting a message containing it's ID, localization (x, y) and level.

## 5.3 Selection Phase

Diffused messages are helping nodes far from the BS to create theirs **Routing tables** (RTs), containing the hops to passe through to attend the BS. We can examine two cases:

**Intra-communication:** CMs existing in cluster's first level (FCM) send their data directly to the CH. CMs existing in cluster's second level, create a RT containing the closest FCMs to route their data. The process is that each node **Ns** send a request message to the FCMs, this one checks its TDMA schedule, if there is an empty slot it sends back an acceptance message and it's add to CM's RT. If not it passes to the other closest FCM.

**Inter-communication:** each node **Ns** send a request message to the **New Hop Node** (NHN), NHN can be a CH or an IN existing on lower levels than the sending node. the NHN checks its TDMA schedule, if there is an empty slot it sends back an acceptance message and it's add to it's RT. If there is no empty slot, the NHN send a refusal message.

## 5.4 Routing Phase

SMR consist of creating a static road to route data collected by nodes to the BS every round. We have two types of communications, intra and inter cluster communication.

**Intra-cluster Routing:** As mentioned above, each cluster is divided into two levels. The nodes of the 1st level communicate directly with the CH and the CMs of 2nd level uses their RTs to select the closest node of 1st level which forward then the data towards the CH. The chosen route will be used during all the present round.

**Inter-cluster Routing:** Routing is done by the same way, the difference is that the communication in this case is between CHs and INs. CHs/INs existing on the lowest level send directly to the BS, and those far frome the SB send to others in lower levels, and so on until data is routed to the SB.

The energy consumed in transmitting or receiving data is calculated using the same Eq. (1) and (2).

## 6 Simulation Results

In this section we will evaluate the performance of the hybrid protocol SMR performing three experiments. the three cases of simulations, type of the network, and protocols analyzed per experiment are resumed on Table 1. A MATLAB program is used to evaluate the performance and to make the comparative study in a network of 200 nodes randomly deployed in an area of 200 × 200 m.

**Table 1.** Experiments specifications

| Experiment | Nbr. nodes | SB localisation | Network type | Compared protocols |
|---|---|---|---|---|
| 1 | 200 | 100 × 300 | Homogeneous | LEACH-SMR |
| 2 | 200 | 100 × 300 | Heterogeneous | LEACH-DEEC-SMR |
| 3 | 200 | 100 × 100 | Heterogeneous | LEACH-DEEC-SMR |

Analysis will be divided into three sections: in the first experiment, we will discuss the performance of SMR compared to LEACH in a homogeneous network (nodes have the same initial energy) and the SB is located far from the network. After that, SMR performance is evaluated in case of a heterogeneous network (nodes have different initial energy) for a SB located far from the network (second experiment) and for a SB located in the center of the network (third experiment). Common parameters used in the simulations, are mentioned in Table 2.

**Table 2.** The essential simulation parameters

| Parameters | Value |
|---|---|
| a | 1 |
| Énergie initiale ($E_0$) | 0.5 J |
| p | 0.2% |
| $E_{elec}$ | 50 nJ/bit |
| $E_{fs}$ | 10 pJ/bit/m$^2$ |
| $E_{amp}$ | 0.0013 pJ/bit/m$^4$ |
| $E_{data}$ | 5 nJ/bit |
| $T_{min}$ | 0.03 |
| L | 6400 bit |

### 6.1 1st Experiment

Figure 5 represents the network life time of SMR and LEACH. Based on the obtained results, we can denote that SMR, have a great improvement in network life time compared to LEACH. This result is achieved because the SMR suppose that the network is divided into levels around the BS and the CHs. Particularly, the CHs and the INs determine their levels around the BS, whereas the CMs will identify their levels based on the distances to its CH. Thus, all sensor nodes of the network use multi-hop routing to deliver the data to the destinations either the BS or the CHs.

Due to network leveling, the distance of transmission is reduced to be less or equal to $\frac{d_o}{2}$, so we use the free space model that consumes less energy. Thus, it explains prolonging stability zone and life-time occurred by SMR.

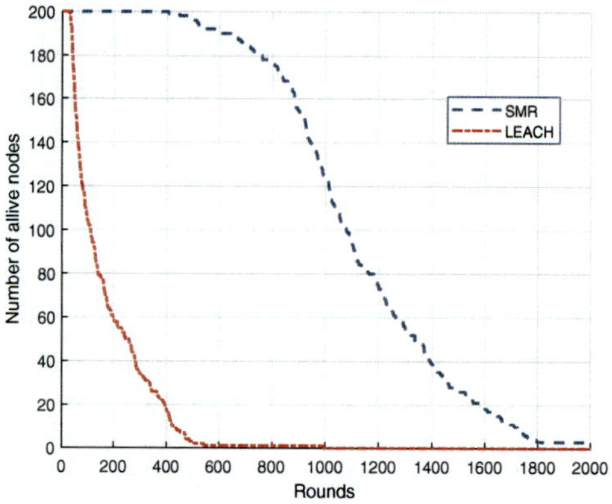


**Fig. 5.** The comparison of the number of alive nodes in the SMR and LEACH for experiment 1.

In Fig. 6, we represent the number of packets sent to the BS and to CHs. We observe that due to the increasing of network's life time, the number of packets sent using SMR is very large than the number of packets sent using LEACH.

### 6.2 Experiments 2 and 3

From Table 1, we can denote that the difference between the second and the third experiments is the localization of the SB. For this reason we will discuss results in the same section. To create an heterogeneous network, each node will have a specific energy equal to $E_0(1 + a_i)$, with $0 < a_i \leq 1$ [7]. In this case each node will have a specific exceed of energy that makes it unique in the network. Since SMR have done well in the exp. 1, in exp. 2, SMR is evaluated, for the first time, in a heterogeneous network and compared with LEACH and DEEC, knowing that the SB is localized, far from the network, in (100m, 300m).

The results of simulation of the second and the third experiments are presented on Figs. 7 and 8.

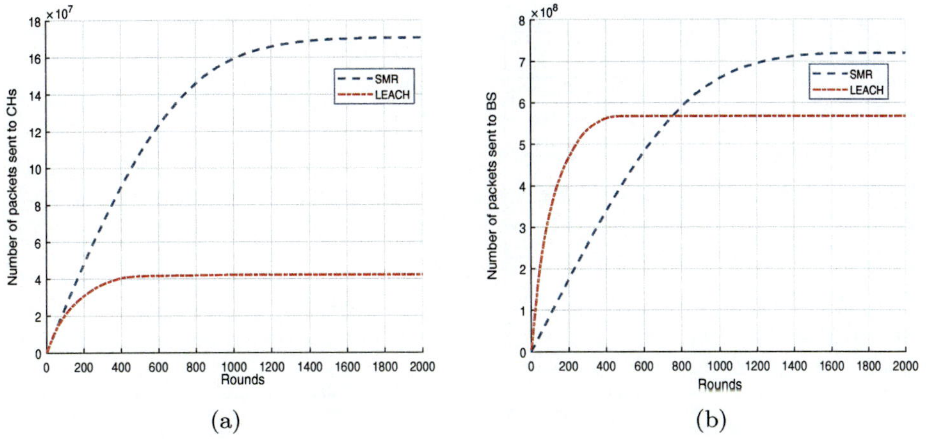


**Fig. 6.** Comparison of the number of packets sent to CHs (a) and to BS (b) using LEACH and SMR techniques in experiment 1.

Figure 7, represent network's life time of SMR, DEEC and LEACH in an heterogeneous network with the SB is located respectively at (100m, 300m) (a), and (100 m,100 m) (b).

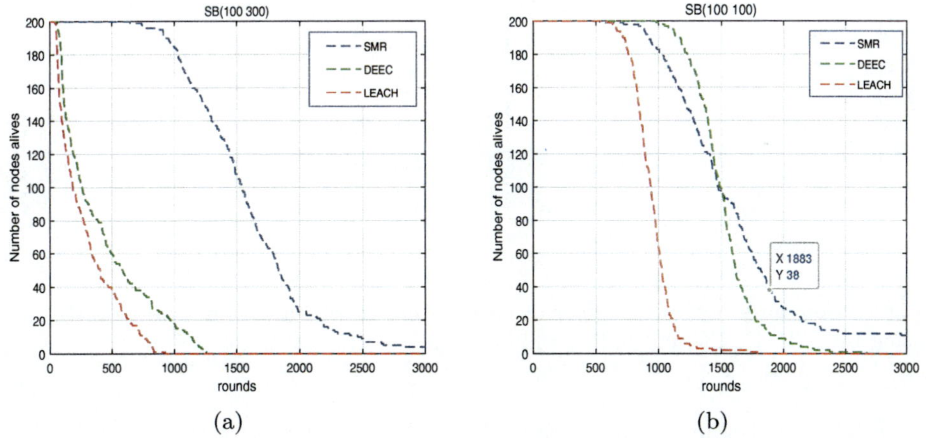


**Fig. 7.** The comparison of the number of alive nodes in the SMR, DEEC and LEACH, in experiments 2 & 3.

We can clearly observe that SMR perform very well in the case where the SB is located far from the network (Fig. 7(a)). As in homogeneous network, stability zone is larger and network's life time is prolonged and the number of packets sent to the SB (resp. CHs) Fig. 8(a) (resp. Fig. 8(b)) is very large using SMR compared to DEEC and LEACH.

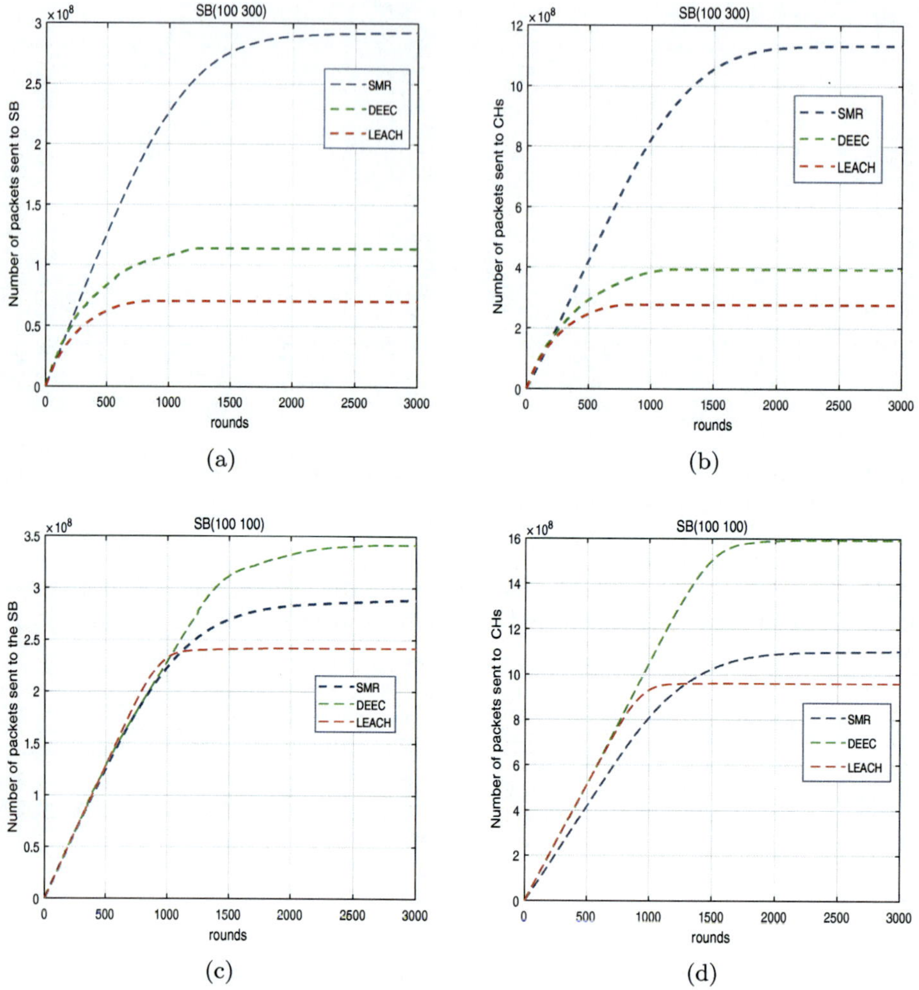


**Fig. 8.** Comparison of the number of packets sent to BS (a), (c) and to CHs (b), (d) using LEACH, DEEC and SMR techniques for experiments 2 & 3.

In the other side, when the SB is located in the center of the network (Fig. 7(b)), SMR have an improvement compared to LEACH but not compared to DEEC.

Life-time is prolonged, but the stability zone is still shorter, than stability zone obtained using DEEC. Furthermore, the number of packets sent to SB and CHs, using SMR, is greater than number obtained using LEACH, but less than number obtained using DEEC, Fig. 8(c) and (d).

### 6.3 Results Discussion

Analyzing obtained results, SMR showed a remarkable improvement, in network's life time and throughput, in both homogeneous and heterogeneous networks, when the SB is located far from the network, compared to LEACH and DEEC. This can be explained by the fact that in LEACH and DEEC, CHs transmit their data directly to the BS, no matter what the distance is. This results a raise in energy consumption specially for far CHs and create an unbalanced consumption of energy. In the other hand SMR limited the distance of transmission to $\frac{d_0}{2}$, for both inter and intra communications, using multi-hop transmission and introducing the INs. This solution reduced the energy consumed by nodes during transmission and results prolonging network life time and improving throughput. In thc third simulation, the little improvement in stability zone using SMR, compared to LEACH can be explained by the fact that the number of nodes per clusters in SMR is limited to $N_o$, this reduce load on CHs and save more energy aggregating collected data and explain the slow raise in the number of packets sent to CHs using SMR in the third experiment, Fig. 8(d). Rather that the stability zone is not improved compared to DEEC, that is explained by the fact that SMR is based on the same election process of LEACH, not using an election process based on the global and residual energy of nodes in each round.

Despite the stability zone is not prolonged, Network's life-time using SMR is extended compared to LEACH and DEEC, this is, as explained before, due to the multi-hop communication schemes.

As regards throughput. In experiments 1 and 2, SMR showed a great raise in the number of packets sent to the BS and the CHs. That is clearly related to the great improvement in network's life-time and stability zone extension.

### 6.4 Statistics

More statistics presenting information about rounds of the First Nodes Dead (FND), Half Node dead (HND) and All Nodes Dead (AND) of compared protocols in the three experiment are resumed in Table 3.

From this statistics we can see that passing from a homogeneous network to a heterogeneous network (exp. 1 to exp. 2) have increased performance of SMR and LEACH. This is explained by the random value of energy ($a(i)$) add to each sensor node to create a heterogeneity in the network.

**Table 3.** FND, HND and AND rounds of SMR, DEEC and LEACH in all experiments

| Exp. | Technique | First Node D | Half Node D | All Node D |
|---|---|---|---|---|
| 1 | LEACH | 26 | 109 | 933 |
| | SMR | 404 | 1078 | >2000 |
| 2 | LEACH | 41 | 183 | 926 |
| | DEEC | 56 | 258 | 1256 |
| | SMR | 647 | 1539 | >3000 |
| 3 | LEACH | 593 | 937 | 1786 |
| | DEEC | 980 | 145 | 2674 |
| | SMR | 534 | 1472 | >3000 |

# 7 Conclusion

In this paper, we have analyzed two techniques for routing data in WSNs: the LEACH and SMR techniques. LEACH is a hierarchical based routing protocol consisting of a random process of election of CHs. Those elected nodes (CHs) are responsible of creating clusters, collecting data and send it to the BS. Furthermore, LEACH presented a great success comparably to direct schemes but still have limitations. The second technique is a hybridization between multi-hop and clustering concepts. The principal idea of this protocol is assuming that all nodes could be in one of three situations during the network rounds: CH, IN or CM. Using network leveling strategy, SMR introduces two types of data routing: the first type is the intra-cluster data routing and the second type is the inter-cluster data routing. The SMR technique have presented a new approach for disseminating the data through the network levels.

Three simulations were done to evaluate the performance of the SMR. For the first one SMR was compared to LEACH on an homogeneous network with the SB is located far from the network. The Second was on an heterogeneous network, where the SB is located far from the network and the third simulation is the same as the second but, this time, the SB is located in the center of the network.

Simulations results demonstrated that SMR's routing technique have prolonged network's life-time and increased throughput, compared to LEACH, in the first experiment, and compared to LEACH and DEEC in the second experiment. The reduction in transmission distances, which is obtained by applying network leveling approach, has reduced the energy dissipation, and has increased the stability of the network. In the third simulation, SMR showed an improvement in stability compared to LEACH, but not a considerable performance compared to DEEC when the SB is located at the center of the network.

Finally, basing on simulations results, we can conclude that SMR performs very well when the SB is far from the network, in both homogeneous and heterogeneous networks, compared to LEACH and DEEC. In the other hand SMR didn't show a significant improvement when the SB is in the center of the network.

## References


1. Alnawafa, E., Marghescu, I.: New energy efficient multi-hop routing techniques for wireless sensor networks: static and dynamic techniques. Sens. J. Basel. (2018). https://doi.org/10.3390/s18061863
2. Al-Karaki, J.N., Kamal, A.E.: Routing techniques in wireless sensor networks: a survey. IEEE Wirel. Commun. **11**(6), 6–28 (2004). https://doi.org/10.1109/MWC.2004.1368893
3. Heinzelman, W.R., Chandrakasan, A., Balakrishnan, H.: Energy-efficient communication protocol for wireless microsensor networks. In: Proceedings of the 33rd Annual Hawaii International Conference on System Sciences, Maui, HI, USA, vol. 2, p. 10 (2000). https://doi.org/10.1109/HICSS.2000.926982
4. Alnawafa, E., Marghescu, I.: MHT: multi-hop technique for the improvement of leach protocol. In: 15th RoEduNet Conference: Networking in Education and Research, Bucharest, pp. 1-5 (2016)
5. Alnawafa, E., Marghescu, I.: IMHT: improved MHT-LEACH protocol for wireless sensor networks. In: 8th International Conference on Information and Communication Systems (ICICS), Irbid, pp. 246–251 (2017)
6. Alnawafa, E., Marghescu, I.: EDMHT-LEACH: enhancing the performance of the DMHT-LEACH protocol for wireless sensor networks. In: 16th RoEduNet Conference: Networking in Education and Research (RoEduNet), Targu Mures, pp. 1-6 (2017)
7. Qing, L., Zhu, Q., Wang, M.: Design of a distributed energy-efficient clustering algorithm for heterogeneous wireless sensor networks. Comput. Commun. **29**, 2230–2237 (2006)
8. Lindsey, S., Raghavendra, C.S.: PEGASIS: power-efficient gathering in sensor information systems. In: Proceedings, IEEE Aerospace Conference, Big Sky, MT, USA, pp. 3-3 (2002)
9. Heinzelman, W., Chandrakasan, A., Balakrishnan, H.: An application-specific protocol architecture for wireless micro sensor networks. IEEE Trans. Wireless Commun. **1**(4), 660–670 (2002)
10. Smaragdakis, G., Matta, I., Bestavros, A.: SEP: a stable election protocol for clustered heterogeneous wireless sensor networks (2004)
11. Zhao, Z., Xu, K., Hui, G., Hu, L.: An energy-efficient clustering routing protocol for wireless sensor networks based on AGNES with balanced energy consumption optimization. Sensors (Basel) **18**(11), 3938 (2018). https://doi.org/10.3390/s18113938
12. Selvakennedy, S., Sinnappan Sinnappan, S.: A configurable time-controlled clustering algorithm for wireless sensor networks. In: 11th International Conference on Parallel and Distributed Systems (ICPADS 2005), Fukuoka, pp. 368–372 (2005)

# Defect Modes in One-Dimensional Periodic Closed Resonators

Ilyas Antraoui(✉) and Ali Khettabi

Laboratoire de Dynamique et d'optique des Matériaux,
Département de Physique, Faculté des Sciences, Université Mohamed I,
60000 Oujda, Morocco
{antraoui_ilyas1718, a.khettabi}@ump.ac.ma

**Abstract.** In this paper, the study of the presence of defects in an acoustic periodic structure is used in an original way. This structure is composed by closed acoustic resonators. The defect resonator is located in the center of the structure. The Sylvester's theorem is applied to examine the properties and the effects of these defect modes. We consider 1D geometry and in this case, it's known that only the plane mode is propagated. The higher-order modes and visco-thermic effects are ignored. We consider the linear acoustic model and the fluid in the waveguides is air. Two kinds of defects state are studied: the first case is corresponding of a waveguide without branched resonator and the second case is corresponding of a closed resonator with a great height. The introduction of a defect in the structure leads to create a resonance peak in the acoustic band gap. To our knowledge this study is original in the field of acoustic wave propagation. We show that the position and the frequency of the defect peak in the band gap are well be controlled by increasing the length of the defect. Finally, it can be said that the interest of this study is reflected in the fact that the band gap is exploited in an original way. Also, this study allows us to detect the presence or not of defects in a periodic structure.

**Keywords:** Acoustic periodic structure · Defect mode · Closed resonators · Sylvester's theorem · Band gap

## 1 Introduction

Wave propagation in one-dimensional periodic structures possessed a distinct and interesting range of physical properties that opened surprisingly new and successful avenues of research. For example, 1D photonic crystals are artificial materials that have several electromagnetic properties due to the existence of prohibited frequency intervals called photonic band gaps (PBGs) [1–6]. By analogy with these materials, 1D periodic acoustic structures such as side-branch and Helmholtz resonator [7–9], periodic expansion chambers [10, 11] and other systems, are also capable of creating acoustic band gaps (ABGs) in which the acoustic wave cannot propagate [12]. This is of interest for applications, such as elastic/acoustic filters and improving acoustic attenuation performance to reduce noise [7, 10]. The problem of a tube branched mounted perpendicular to a waveguide has been studied by many authors to suppress


A. Saka et al. (Eds.): CPI 2019, LNME, pp. 438–445, 2021.
https://doi.org/10.1007/978-3-030-62199-5_39

the noise. In 2005, Ji. use a numerical approach based on the three-dimensional boundary element method (BEM) to determine the acoustic length correction of closed cylindrical side-branched tube mounted perpendicular to a cylindrical main pipe [13]. Chenzhi et al. focused on improving noise attenuation performance of the Helmholtz resonator (HR) at low frequencies with limited space [14]. Meng Xiao et al. present a theoretical framework and two experimental methods for determining the Zak phase in a periodic acoustic system [15]. The objective of the present work is to use Sylvester's theorem [16, 17] to examine the properties and the effects of defect modes in a periodic acoustic structure. This structure is composed by closed acoustic resonators. The defect resonator is located in the center of the structure. We consider 1D geometry and in this case, it's known that only the plane mode is propagated. The higher-order modes and visco-thermic effects are ignored. We consider the linear acoustic model and the fluid in the waveguides is air. In this paper, two kinds of defects state are studied: the first case is corresponding of a waveguide without branched resonator and the second case is corresponding of a closed resonator with a big height. The numerical results show that the presence of a defect in the structure leads to create a resonance peak in the acoustic band gap. It is, therefore possible to create very narrow bandwidths within the acoustic band gap. We have shown that the increasing the length of the defect is a very useful technique to control the position and frequency of the defect peak in the acoustic band gap. The interest of this study results in the fact of exploiting the band gap for the reduction of noise. Also, this study allows us to detect the presence or not of defects in a periodic structure.

## 2 The Structure Used and Theory

We consider a one-dimensional periodic structure of closed resonators consisting of finite waveguides of height $d_2$ and cross-sectional area $s_2$ grafted perpendicularly to the main guide of length $d_1$ and section $s_1$. Figures 1 (a) and (b) show respectively two

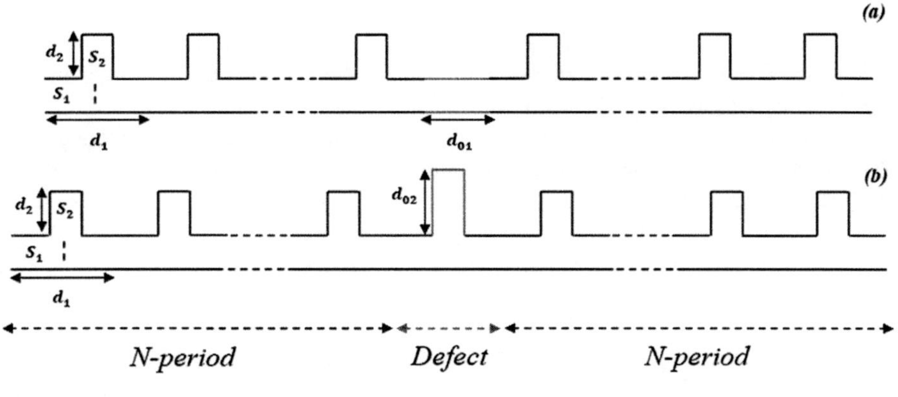


**Fig. 1.** 1D periodic structure of closed resonators with a defect located in the center of the structure. (a) is a waveguide defect of variable length $d_{01}$ and (b) a closed resonator defect of heights $d_{02} > d_2$.

types of defects that are introduced one by one into the center of our regular structure: the first is a waveguide of variable length $d_{01}$ (Fig. 1.a), and the second is a closed resonator of heights $d_{02} > d_2$ (Fig. 1.b).

In the first study, the transfer matrix method (*TMM*) is used to calculate the transmission coefficient and dispersion relation of our system. In the absence of the defect, the transfer matrix of a period or an elementary cell is given by:

$$T_c = \begin{pmatrix} A & B \\ C & D \end{pmatrix} \begin{pmatrix} 1 & 0 \\ y_f & 1 \end{pmatrix} \begin{pmatrix} A & B \\ C & D \end{pmatrix} \quad (1)$$

$$T_c = \begin{pmatrix} A_c & B_c \\ C_c & D_c \end{pmatrix} \quad (2)$$

With $A = cos\left(k\frac{d_1}{2}\right)$, $B = jz_{c1}\, sin\left(k\frac{d_1}{2}\right)$, $C = j\frac{sin\left(k\frac{d_1}{2}\right)}{z_{c1}}$ *and* $D = A$.

$A$, $B$, $C$ and $D$ are the elements of the transfer matrix of the main waveguide of length $d_1$ and characteristic acoustic impedance $z_{c1} = 1/y_{c1} = \rho c/s_1$. $\rho$ and $c$ are respectively the density and the speed of sound in the air, $k$ is the wave number and $y_f$ is the acoustic admittance of the closed resonator.

In this type of resonator, the acoustic flow rate is zero at the end of his cavity. Therefore, the acoustic admittance of a closed resonator is given by the following expression:

$$y_f = jy_{c2}\, tan(kd_2) \quad (3)$$

Where $d_2$ and $y_{c2}$ are respectively the length and the acoustic admittance of the closed resonator.

From the matrix Eq. (2), the dispersion relationship of an elementary cell is calculated using Bloch's theorem [18], this relation is:

$$cos(Kd) = cos(kd_1) - \frac{M}{2} sin(kd_1) tan(kd_2) \quad (4)$$

Where $M$ is the ratio of cross-sections, $K$ is the Bloch wave number and $d = d_1 + d_2$ is the length of a cell or a period.

The matrix equation corresponding to the periodic structure is calculated using Sylvester's theorem applied at the unimodular acoustic transfer matrix (Eq. 2), this equation is given in the form:

$$T_c^N = \begin{pmatrix} A_cU_N - U_{N-1} & B_cU_N \\ C_cU_N & D_cU_N - U_{N-1} \end{pmatrix} \quad (5)$$

Where $U_N = \frac{sin(NKd)}{sin(Kd)}$ is the Chebycheff polynomial of second kind of order $N$ [7].

In the presence of one defect, placed on the center of the periodic structure, the corresponding final matrix of this structure $T_{cd}$ is given by:

$$T_{cd} = T_c^N \, T_{def} \, T_c^N = \begin{pmatrix} T_{11} & T_{12} \\ T_{21} & T_{22} \end{pmatrix} \tag{6}$$

Where $T_{def}$ is the transfer matrix of each type of defects studied here. $T_{11}$, $T_{12}$, $T_{21}$ and $T_{22}$ are the elements of the final matrix of this system which contains a defect placed at the center of the periodic structure.

And therefore, the intensity transmission coefficient $T$ is obtained by the following relation:

$$T = \left| \frac{2}{T_{11} + y_{c1} T_{12} + z_{c1} T_{21} + T_{22}} \right|^2 \tag{7}$$

## 3 Results and Discussions

In this section, we analyzed the effects of the presence of defects in a periodic structure with closed resonator. In the first party, we focused on the acoustic properties of the transmission and the dispersion relation of a perfect periodic structure. Figure 2 shows the band structure (Fig. 2.a) and the transmission spectrum (Fig. 2.b) for the perfect closed tubes system. The geometric parameters used are: $d_1$ = 0.6 m, $d_2$ = 0.15 m, $M$ = 0.75 is the ratio of cross-sections ($M = s_2/s_1$) and the number of repeated cells $N$ = 20 cells. We note that the results obtained are in perfect agreement with the Elmalki and Khettabi [7] report, which used a periodic network of lateral resonators closed by of the Green function method. In the band structure, the opening of the acoustic band gaps is well justified by the transmission spectrum. This shows that these bands occur in the same frequency range with the band gap of the transmission is formed. In these regions, the number of Bloch waves ($K$) is complex, leading to an evanescent wave and therefore to a zero transmission for the structure [12]. Furthermore, we see that this closed tube system works as a low-pass filter due to the first band that occurs is the pass band.

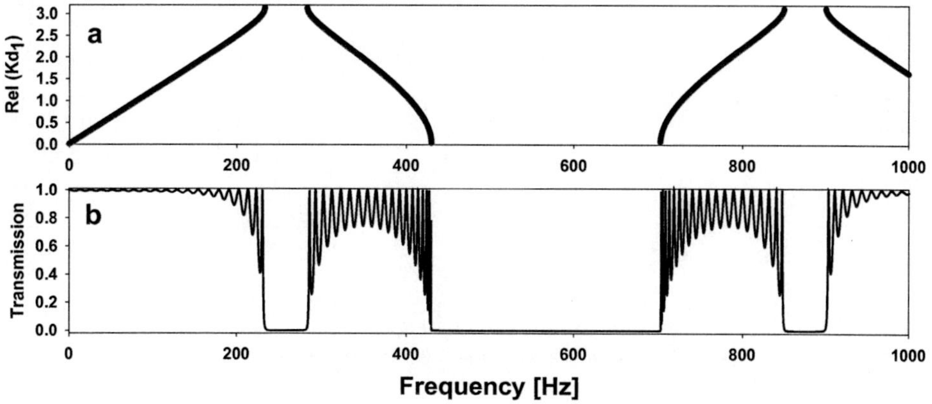


**Fig. 2.** (a) The band structure, (b) The transmission spectrum for the perfect closed tube system.

Then, we developed the study of Elmalki and Khettabi by introducing geometric defects into our periodic structure acoustic. We specifically analyze the properties and the behavior of defect mode inside the first acoustic band gap of limited frequency range between 228.96 Hz and 288.52 Hz for the two kinds of defects studied here. Figure 3 represents the transmission spectra as a function of the frequency at normal incidence for four different values of $d_{01}$. The parameters used here are $d_1 = 0.6$ m, $d_2 = 0.15$ m, $M = 0.75$, and $N = 10$. The insertion of a defect guide at the center of the structure results in the creation of a defect mode that appears inside the band gap with a good transmission rate. More precisely, it is demonstrated that the defect peak is shifted to the left inside the band gap when $d_{01}$ increases. Furthermore, when $d_{01} = d_1/2 = 0.3$ m, this resonance peak is located in the center of the acoustic band gap at a frequency $f = 258.68$ Hz by a good transmission $T = 98\%$. It is therefore interesting to note that the behavior of the defect mode depends strongly on the variation in the length $d_{01}$. Thus, we are able to control the position of the defect peak in the band gap by increasing the length of the defect guide $d_{01}$. These results are very effective for manufacturing of narrow transmission bands in the acoustic band gap, to design frequency selective filters [19].

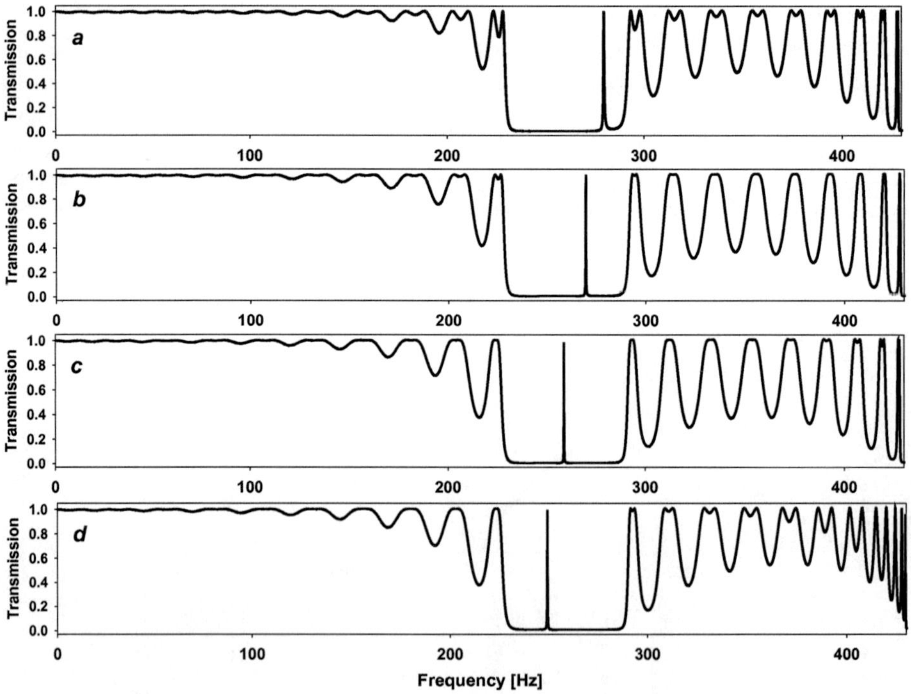


**Fig. 3.** Variation of transmission as a function of frequency for different values of $d_{01}$. (a) For $d_{01} = 0.1$ m, (b) $d_{01} = 0.2$ m, (c) $d_{01} = 0.3$ m and (d) $d_{01} = 0.4$ m.

Then, we discuss the numerical results for the second kind of defect studied here, which is a closed lateral chimney from a height $d_{02} > d_2$ integrated into the center of the periodic structure. The transmissions spectra as a function of the frequency for different values of $d_{02}$ are illustrated in Fig. 4. When $d_{02}$ is greater than $d_2 (d_{02} > d_2)$, a defect mode appears in the acoustic band gap. The appearance of this mode is essentially due to a disturbance of the spatial periodicity of the structure. As shown in the figure, it is clear that the position and the frequency of the defect peak are regularly moved within the band gap when the value of $d_{02}$ is increased. More precisely, for $d_{02} = 2.2\,d_2$, a defect mode appears closer to the pass band at the frequency $f = 274.16\,\text{Hz}$, but if the value of $d_{02}$ is higher, we see that this defect peak moves from right to left in the band gap for three values $d_{02}$ of different frequencies which are respectively: $f = 266.44\,\text{Hz}$ for $d_{02} = 2.5\,d_2$, $f = 252.8\,\text{Hz}$ for $d_{02} = 3\,d_2$ and $f = 244.76\,\text{Hz}$ for $d_{02} = 3.5\,d_2$. This means that the control of the defect peak in the acoustic band gap depends mainly on the increase for the height $d_{02}$. Moreover, it is interesting to deduce that when $d_{02} = 3\,d_2$ the defect mode is centered in the middle of the acoustic band gap.

Besides, a very important remark shows that the transmission deteriorated in the first pass band due to an increase in the $d_{02}$ value. However, in the case where the defect is a waveguide of length $d_{01}$ (see Fig. 3), this band remains unchanged,

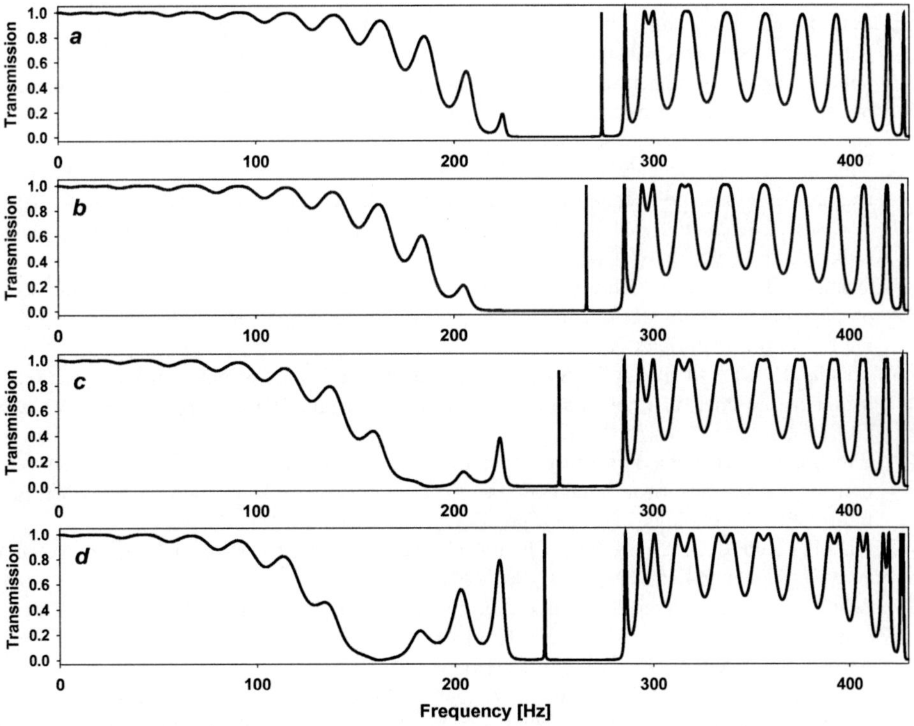


**Fig. 4.** Variation of transmission as a function of frequency for different values of $d_{02}$. (a) For $d_{02} = 2.2\ d_2$, (b) $d_{02} = 2.5\ d_2$, (c) $d_{02} = 3\ d_2$ and (d) $d_{02} = 3.5\ d_2$.

regardless of the value of the $d_{01}$ defect guide. This is helpful to choose the second case ($d_{02} > d_2$) if we need to create a narrow band gap of zero transmission in the pass band.

## 4 Conclusion

This study showed that the use of Sylvester's method is very effective and also very simple. The effect of the defect modes has been carefully examined. These modes appear in the forbidden band and the position of the corresponding peaks in this band is controlled by the dimensions of the defect. These dimensions to be varied are both the length and height of the connected guide representing the defect in the periodic acoustic structure. We also show that these variations have allowed peaks with a high transmission rate. Finally, we conclude that the interest of this study is reflected in the fact that the band gap is exploited in an original way. Also, this study allows us to detect the presence or not of defects in a periodic structure. These results can be used in the development of non-destructive testing systems for the detection of geometric defects in a periodic structure.

## References

1. Lou, J., He, L., Yang, J., Kitipornchai, S., Wu, H.: Wave propagation in viscoelastic phononic crystal rods with internal resonators. Appl. Acoust. **141**, 382–392 (2018)
2. Villa-Arango, S., Betancur-Sánchez, D., Torres, R., Kyriacou, P., Lucklum, R.: Differential phononic crystal sensor: towards a temperature compensation mechanism for field applications development. Sensors **17**(9), 1960 (2017)
3. Villa-Arango, S., Betancur-Sánchez, D., Torres, R., Kyriacou, P.: Use of transient time response as a measure to characterize phononic crystal sensors. Sensors **18**(11), 3618 (2018)
4. Aly, A.H., Mehaney, A.: Phononic crystals with one-dimensional defect as sensor materials. Indian J. Phys. **91**(9), 1021–1028 (2017)
5. Wang, Z.G., Lee, S.H., Kim, C.K., Park, C.M., Nahm, K., Nikitov, S.A.: Acoustic wave propagation in one-dimensional phononic crystals containing Helmholtz resonators. J. Appl. Phys. **103**(6), 064907 (2008)
6. Mandal, B., Chowdhury, A.R.: Spatial soliton scattering in a quasi phase matched quadratic media in presence of cubic nonlinearity. J. Electromagn. Waves Appl. **21**(1), 123–135 (2007)
7. Elmalki, M., Khettabi, A.: Study of various periodic acoustic lattices by two methods: transfer matrix and Green's method. In: 2017 International Conference on Electrical and Information Technologies (ICEIT), pp. 1–3. IEEE (2017)
8. Lu, J.H., Kuo, C.C., Hsiao, F.L., Chen, C.C.: Acoustic filter based on Helmholtz resonator array. Appl. Phys. Lett. **101**(5), 051907 (2012)
9. Zhaoa, H., Lu, Z., Guan, Y., Liu, Z., Li, G., Liu, J., Ji, C.Z.: Effect of extended necks on transmission loss performances of Helmholtz resonators in presence of a grazing flow. Aerosp. Sci. Technol. **77**, 228–234 (2018)
10. Khettabi, A., Antraoui, I.: Study of various periodic study of a finite network of one-dimensional periodic expansion chambers by the transfer matrix method and Sylvester theorem. In: AIP Conference Proceedings, vol. 2074, no. 1, p. 020003 (2019)

11. Liu, J., Yu, D., Wen, J., Zhang, Z.: Analysis of an ultra-low frequency and ultra-broadband phononic crystals silencer with small size. J. Theor. Comput. Acoust. **27**(02), 1850026 (2019)
12. King, P.D.C., Cox, T.J.: Acoustic band gaps in periodically and gausiperiodically modulated Waveguide. J. Appl. Phys. **102**(1), 014902 (2007)
13. Ji, Z.L.: Acoustic length correction of closed cylindrical side-branched tube. J. Sound Vib. **283**(3–5), 1180–1186 (2005)
14. Cai, C., Mak, C.M., Shi, X.: An extended neck versus a spiral neck of the Helmholtz resonator. Appl. Acoust. **115**, 74–80 (2017)
15. Xiao, M., Guancong, M., Yang, Z., Sheng, P., Zhang, Z.Q., Chan, C.T.: Geometric phase and band inversion in periodic acoustic systems. Nat. Phys. **11**(3), 240 (2015)
16. Khettabi, A.: Application of the Sylverster theorem to study 1D acoustic periodic expansion chamber. Unpublished
17. Wu, C.J., Chung, Y.H., Syu, B.J.: band gap extension in a one-dimensional ternary metal-dielectric photonic crystal. Prog. Electromagn. Res. **102**, 81–93 (2010)
18. Zhang, X., Forrest, S.R.: Generalized phase matching condition for lossy periodic photonic structures. Opt. Express **18**(2), 1151–1158 (2010)
19. Wu, C.J., Wang, Z.H.: Properties of defect modes in one dimensional photonic crystals. Prog. Electromagn. Res. **103**, 169–184 (2010)

# Integration of a Prognosis Model of a Rotating Microwave Oven Guidance System Subject to Linear Degradation

Imad El Adraoui[1,2(✉)], Hassan Gziri[1,2], and Ahmed Mousrij[1,2]

[1] IMMII Laboratory, FST, Hassan I University, Settat, Morocco
aimadeladraoui@gmail.com, hgziri@gmail.com, mousrij@gmail.com

[2] FST-Settat, Hassan I University, Km 3, B.P.: 577 Casablanca Settat Road, 26000 Settat, Morocco

**Abstract.** In this article, we propose a prognosis model for estimating the remaining useful life (RUL) before the failure of a wear bearing to guide rotating trees in microwaves, in the company called (COSUMAR - Morocco).

It is a matter of establishing a model of the degradation envisaged (deterioration by phenomenon of wear) during the favorable mission to the decrease of the reliability over time. The proposed approach is based mainly on the model of behavior of the system under predefined working conditions (Evolution of the degradation).

This work consists in establishing an adaptive model for this phenomenon, an empirical law envisaged, a degradation based on the computation of the rate of wear by the formula of ARCHARD. This model helps us predict the remaining useful life (RUL) of the wear bearing before failure, the main objective of this law is to help the maintenance manager to take the action before the failure.

**Keywords:** Integration · Model · Maintenance · Degradation · Prognosis · Complex systems

## 1 Introduction

During use, all systems and components during operation evolve physically over time, with behaviors degraded according to reliability experts.

The degraded behavior results in a type of strategic and predictive maintenance, which varies according to different criteria (operational safety, cost, working environment,…).

Therefore, it is essential for the maintenance manager to establish a predictive system based on interpolation data. This predictive assessment will allow him to choose the action plan to be implemented before the system failure.

This prediction action is essentially based on the prognostic process, which aims to predict the future state of a system before failures occur. So, the estimate of the remaining useful life is considered a central point in the scenario.

The method adopted consists of generating a model of the system's behavior, based on equations and empirical laws of tribology (HOLM and ARCHARD). It deals with the case of the wear bearings monitored during the operating time. The objective is to


A. Saka et al. (Eds.): CPI 2019, LNME, pp. 446–458, 2021.
https://doi.org/10.1007/978-3-030-62199-5_40

estimate the remaining useful life (RUL) before the failure of the entire guidance system is not satisfied.

# 2 Modeling of the Microwave Oven

## 2.1 Introduction

The microwave oven (Fig. 1) is an essential system in the sugar conditioning process in the COSUMAR sugar factory, this oven is used to solidify incoming sugar loaves by creating a magnetic field inside for a period of time, called the solidification time.

**Fig. 1.** Microwave oven in problematic situations

## 2.2 Modeling the Microwave Oven with a Use Case Diagram (Fig. 2)

The purpose of this diagram is to show the functionalities offered by a system by identifying the services it provides: it therefore makes it possible to model requirements from a complementary point of view to that presented by the requirements diagram.

The statement of a use case must be non-technological, since it is defined in terms of expected results [1].

The controller (technician) of the microwave oven has the role of following with the aid of a PLC, three essential steps for a better quality of the product (solidified sugar):

- Sugar heating: the function of the microwave oven is to solidify sugar at a temperature of 160 °C based on the creation of a magnetic field inside.
- Moving the bar: five trees rolling to guide the load of sugar bread in translation animate the movement of the bar.
- Loading the bar: The sugar loaf is loaded by an automatic loader outside the microwave oven.

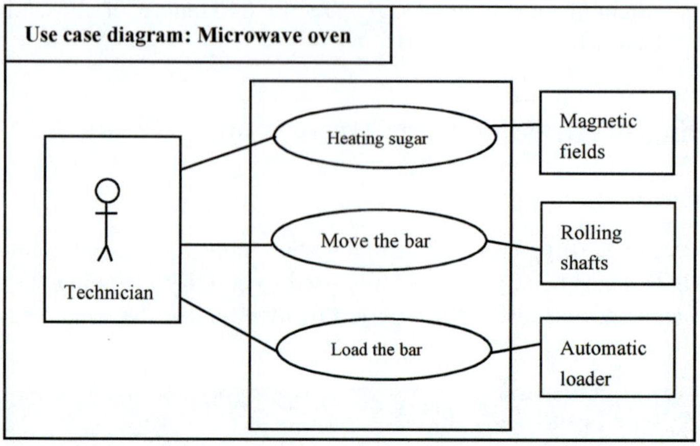


**Fig. 2.** Microwave oven use case diagram

The bar loaded by the sugar loaves (Fig. 3) is guided in translation by five rolling shafts, the latter are guided in rotation by Aluminum wear bearings in order to reduce the coefficient of friction and to guarantee the principle of interchangeability of the sacrificial element, in our case it is the bearing.

The rotational guidance is considered with a dry friction mode (without lubrication), this mode is characterized by a life time insufficient to carry out a large production series because of the frequent changes of the bearings.

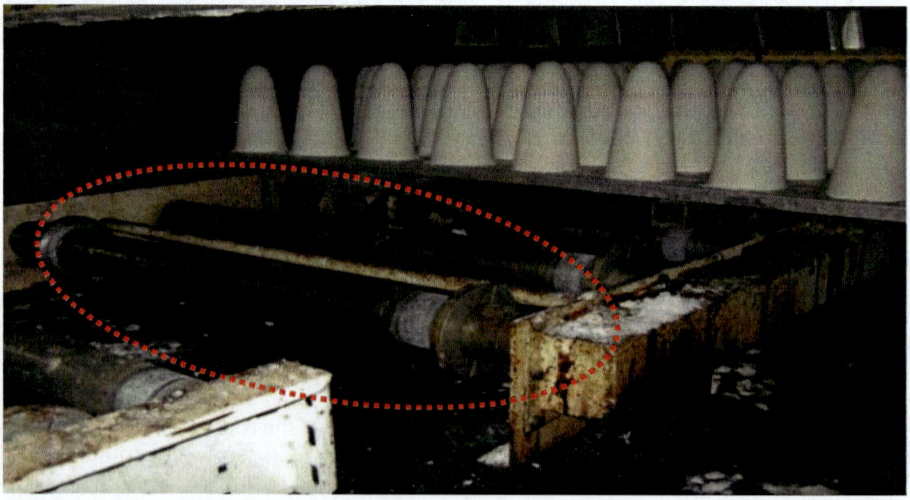

**Fig. 3.** Sugar bar guided in translation

## 2.3 Modeling of the Microwave Oven by Means of a State Diagram (Fig. 4)

The status diagram is attached to a block that can be the system, a subsystem or a component. The behavior described by this type of diagram is used to show the different states taken by the block according to the events that happen to it. A report represents a situation of a finite duration during which a system performs an activity, satisfies a certain condition or is waiting for an event. The transition from one state to another is done by crossing a transition [1].

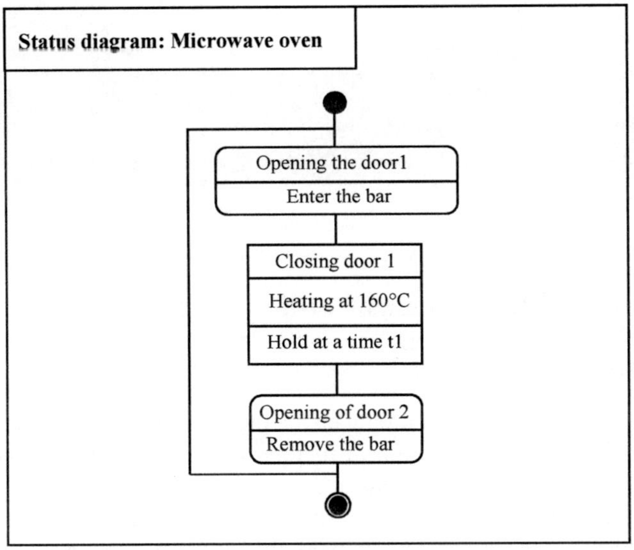


**Fig. 4.** Microwave oven status diagram

The working cycle of the microwave oven is problematic as follows:

- Supply of the oven by an alternating electric current, the latter is converted into direct current by means of a rectifier, a diode and a capacitor.
- Opening of the door 1 (at the entrance of the oven) by an automatic action programmed by a detector of presence of the loaded bar and entered the bar of the loaves of sugar.
- Closing the door 1.
- Automatic closing of the door 1 and heating of the sugar at 160° C for a duration t1 to solidify the sugar loaves.
- Opening of the door 2 (at the exit of the oven) by an automatic action programmed after the time t1 consumed by the solidification and exit of the bar of the loaves of sugar.
- Closing the door 2.

As a result, the cycle resumes the same scenario (Fig. 5).

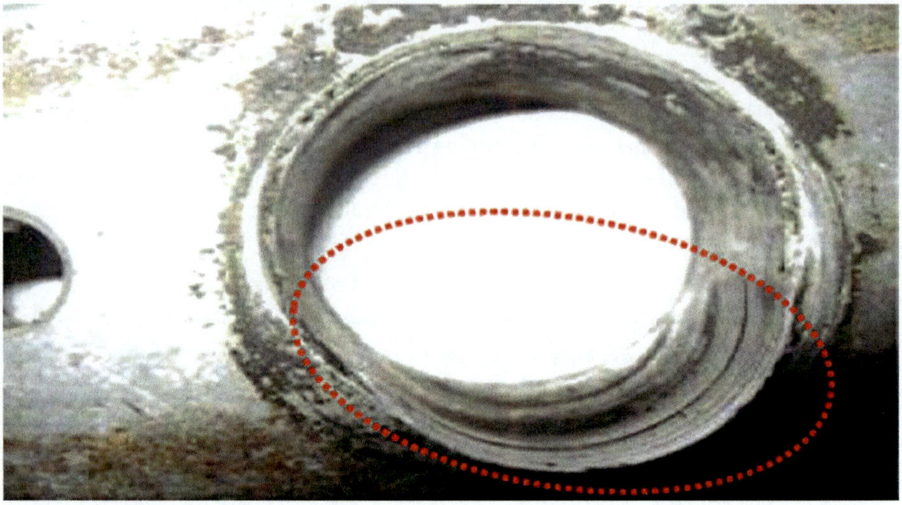

**Fig. 5.** Faulty wear bearing

# 3 Application of the Prognosis on the Wear Bearing

## 3.1 Introduction

In this chapter, we will develop a prognosis method for the wear degradation problem described above. The implementation of method is mainly based on a model derived from physical laws or simulation models.

The use of the method involves very specific knowledge related to the failure and mode of operation of the system. The method obtained is used to visualize and monitor the behavior of the system and its degradation during its use [2].

As an example of the application of this method, the prognosis based on a real model applied for the suspension of a half vehicle for different road profiles (Luo et al. 2008). The method is based on a complete knowledge of the model.

The degradation representing the length of the crack propagating over the suspension spring was not measurable. It was estimated through an indicator of its effect (Fig. 6) by analogy with the degradation model shown in (Fig. 7), which has the same quasi-affine behavior [2], which represents a wear degradation model of a gear according to the CETIM document. In order to predict the state of degradation and its variance according to the road profile.

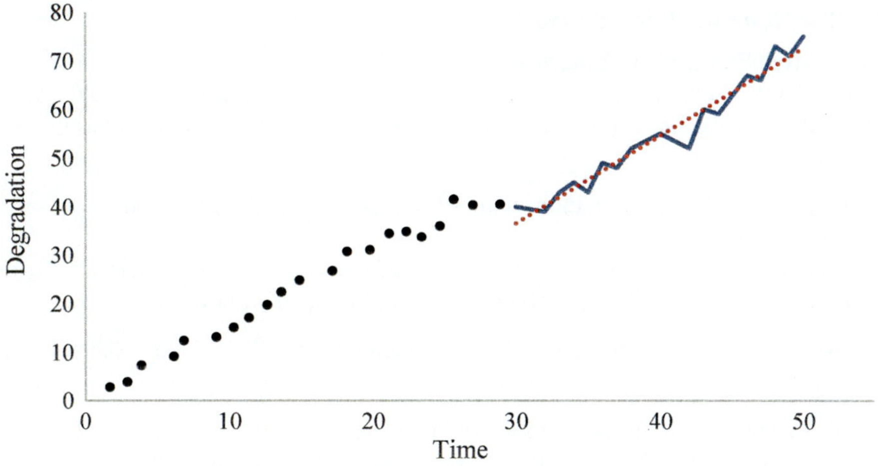


**Fig. 6.** Degradation representing the length of the crack on a suspension spring (ANFIS system)

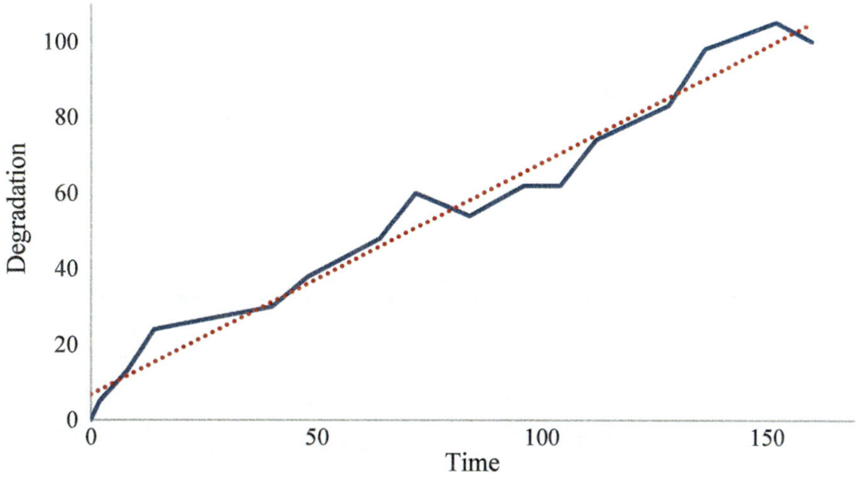


**Fig. 7.** Evolution of the wear degradation of a gear (Doc.CETIM) [3]

We note that the evolution model of the degradation is almost linear to a smaller square. So in this sense, we will consider our behavior of degraded wear and tear of the bearings in problematic is almost refined according to the two referential approaches in Luo et al. (2008) and (Doc.CETIM).

## 3.2 Modeling the Degradation Law

### 3.2.1 Modeling and Assumptions

To model wear behavior, we can use the empirical model developed by HOLM and ARCHARD for the linear phase of wear (Luo et al. 2008). This model is based on the following points:

- Contact is made on a minimum number of asperities randomly distributed over the contact area;
- These roughnesses must withstand the applied stress and maintain their deformation in the elastic zone of the materials according to HOOK's law.

For the implementation of our wear degradation model, we adopt the following assumptions:

- The temperature is uniform in the contact area;
- The influence of temperature evolution on the wear rate is negligible;
- The sliding length L *(perimeter)* is almost constant during wear;
- The contact is made with continuous sliding;
- The wear rate is continuous and linear;
- The associated wear is adhesive.

### 3.2.2 Calculation of the Degradation Rate

The notations and associated numerical values required to calculate the wear rate are summarized in the following table [2]:

**Table 1.** Associated numerical values required to calculate the wear rate

| Parameters | Definitions | Values |
|---|---|---|
| $F_z$ | Model of the normal contact force, assumed constant | 400 N |
| L | Sliding length | 157 mm |
| H | Hardness of the softest material | 245 MPa |
| K | Wear coefficient in the case of dry mode | $15.10^{-6}$ |
| $V_t$ | Total volume used | 1500 $mm^3$ |

The value of the wear coefficient in the case of dry mode is justified for the value of the total volume and calculated empirically for a sample of a bearing in failed condition (Table 1), in an approximate way at the end of which we validated the model.

The ARCHARD's law that we will introduce into our calculation, its initial shape expresses, for an adhesive wear configuration, in slip, a relationship between the used volume of the quantities characteristic of the contact [5]:

$$V_a = K\frac{F_z}{H}L \tag{1}$$

According to Eq. (1) we find:

$$V_a = 3,8.10^{-3}\,\mathrm{mm}^3$$

According to the calculated wear degradation rate, a law of evolution of the wear volume as a function of the time parameter $t$ is associated (Fig. 8) since the sliding length is proportional to time.

The degradation behavior varies linearly along a straight line refining towards the indicated threshold (1500 $mm^3$) as shown.

The intersection of the line refines which models the evolution of the wear degradation and the indicated threshold is located at the abscissa 21000 min, this value for which the change of the bearing is predicted before avoiding any kind of failure of the system as a whole.

The drop in wear degradation schematized by the vertical segment is the event of the maintenance manager's intervention, i.e. the change of the defective wear bearing by another one who will follow the same behavior as the one before it.

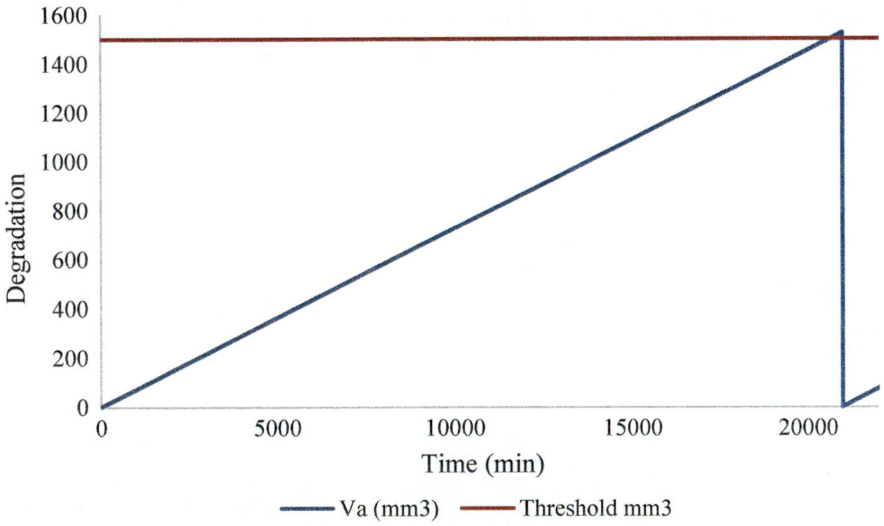


**Fig. 8.** Evolution of the wear volume as a function of time for dry mode

It is found that in the case of use of a wear bearing in dry mode (favorable friction), the service life is no longer sufficient to achieve the production output in a specified time, which requires the search for another solution to reduce the downtime caused by the frequent change of worn bearings.

Suppose that failure times are available for identical systems that have worked under the same conditions (Fig. 9) [2].

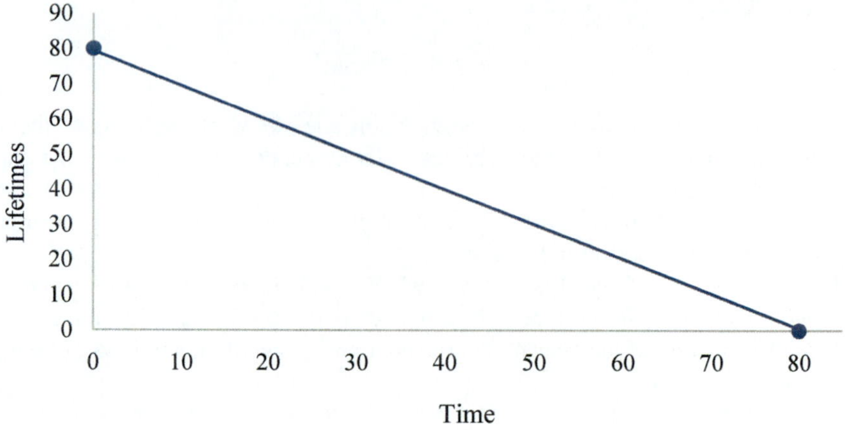


**Fig. 9.** Systems lifetimes

The line refines contained in (Fig. 10), shows the evolution of the difference between the threshold and the volume torn off by wear as a function of the time (min) of the bearing during its operation with a degradation of the material.

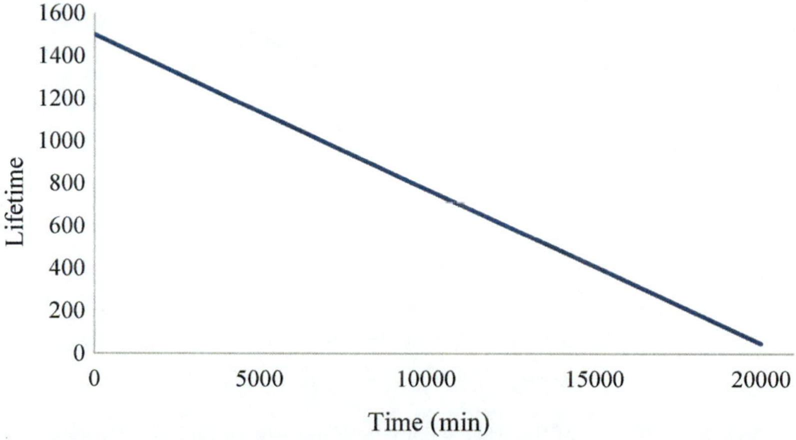


**Fig. 10.** Trajectory of the difference between the threshold and the volume used with the dry mode

The slope of the straight line is negative, which justifies the decrease in the volume allocated for degradation before the intervention until total failure *(the intersection with the abscissa axis)*. This last one is possibly the date of the maintenance which will be taken into account (21000 min).

Therefore, the service life decreases according to an affine law until the total failure, the latter is due to the anticipation of several factors, which are included namely: the temperature, the pressure, the friction, the wear…

### 3.3 Remediation of the Problem

To minimize friction and wear between two surfaces in relative motion, it is preferable to separate them with a lubricating fluid film (Fig. 11) which can be a liquid or a gas. The load supported by the mechanism is created either by the movement of the surfaces (hydrodynamic) or by an external pressure source (hydrostatic). The thickness of the film must be greater than the height of the surface roughness (roughness), otherwise we are in mixed lubrication or boundary lubrication [6].

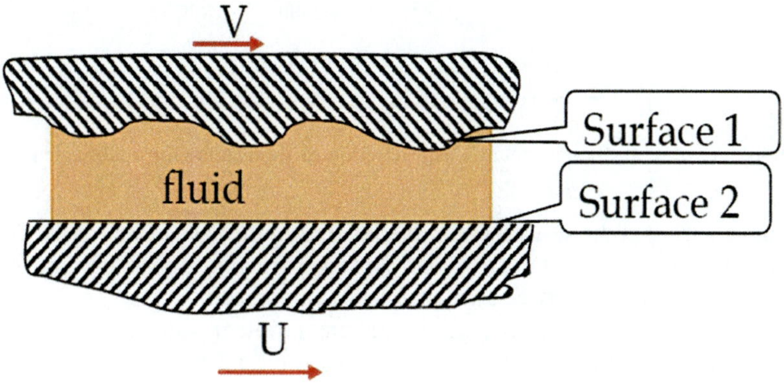


**Fig. 11.** Lubricated contact model

In order to remedy the problem of degradation, conditional maintenance will be adopted initially, the objective of which is to reduce friction and, consequently, to minimize the rate of degradation, which is based on the lubrication.

The lubrication mode, when may be associated with our problem, is a grease lubrication, namely synthetic grease to meet the high temperature requirement (160 °C) in the microwave oven.

***Justification of the Choice***

Synthetic grease is more effective than conventional fat. It is designed for, bearings, chains, joints and other lubricants. This grease is the result of cutting-edge technologies and is specially designed for maximum efficiency. Exceptional resistance to pressure, temperature and oxidation. Very good adhesion on all types of substrates. Insoluble in boiling water [7].

In the second step consists in predicting the remaining useful life (RUL) to intervene with an action to change the bearing into a failed state, this approach is called predictive maintenance.

Let's take the same model with another adaptive value for wear with lubricant, i.e., with a wear coefficient in the case of the lubricated mode of order $3.10^{-6}$, we obtain the result illustrated in (Fig. 12).

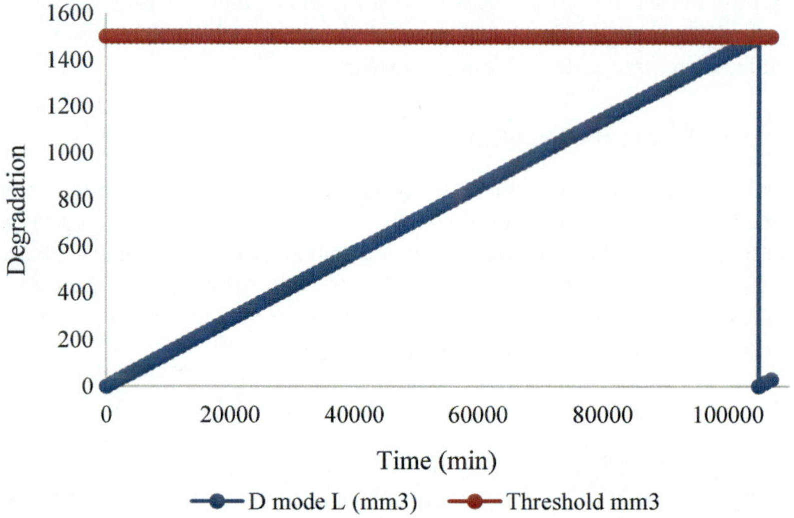


**Fig. 12.** Evolution of wear volume as a function of time (min) for a lubricated mode

It can be seen that the service life of the bearing in the case of lubricated contact is increased compared to that of the dry contact mode, which justifies the improvement of the coefficient of friction between the two solids in contact.

So, we are of course remedying the failure problem with a lifetime of about 105.000 min, which reflects an improvement in the ratio 5 times.

The result found gave us a precise idea on a factor which can improve our solution namely lubrication, this last is an operation which serves to separate the two surfaces in contact by a film of lubricant in order to minimize the coefficient of friction (Fig. 13).

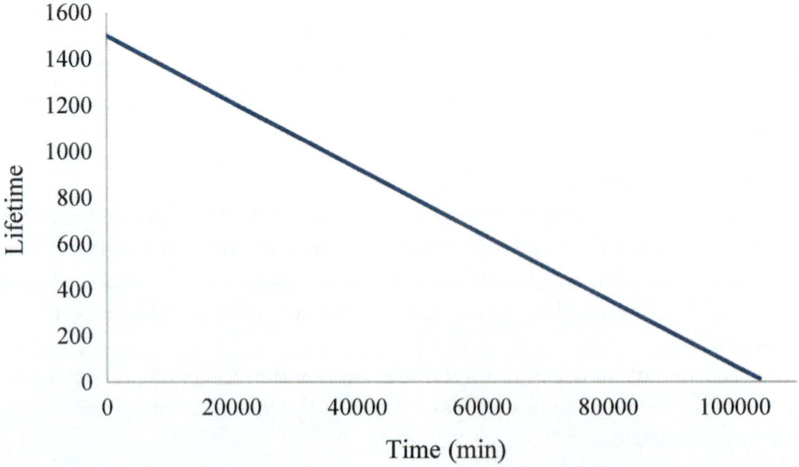


**Fig. 13.** Trajectory of the difference between the threshold and the volume used with the lubricated mode ($mm^3$)

According to the law of reliability degradation, a refine line of negative slope was found, which justifies the decrease in the volume allocated for degradation before the intervention until total failure *(the intersection with the abscissa axis).* The latter is possibly the date of maintenance in the case of lubricated contact mode, which will be taken into account (105.000 min).

***Discussion:***

In this chapter, we have developed a model for predicting the state of degradation of a wear bearing. We began our case study with an introduction to the linear model of wear degradation using methods experimentally based on research centers, namely CETIM and (Luo et al. 2008).

Consequently, a modeling of wear degradation is based on assumptions and empirical laws that explain the evolution of the phenomenon under consideration. The key law introduced in this chapter is the HOLM and ARCHARD law.

The preliminary calculation found by the empirical law of HOLM and ARCHARD, allowed us to develop a model of linear evolution of degradation and predict the remaining useful life (RUL) of the bearing in dry contact mode.

The effective forecasting policy followed in our problematic is led to a minimization of the downtimes caused by the linear and continuous degradation of wear levels. Its efficiency is related to its ability to predict the future state of the component failure; therefore, we will have the ability to provide some adequate solution to increase the service life, meeting the conditions of use: Pressure, friction and temperature.

Our prognosis and the results obtained will be used to estimate the residual life. However, in the case where it is unfeasible to implement the model-based approach and as we have shown in our paper.

In other words, the prognosis is only realized after a certain evolution in the time of the linear degradation and the transition to a remarkable degraded level of functioning. This will allow us to ensure a horizon of analysis time that will be sufficient for the proper deployment of a prediction.

## 4 Conclusion

In this paper, we have presented a model for the prognosis of wear degradation for a rotating guidance system (wear bearing). The objective of the methodology adopted is to be:

- Able to estimate the useful life in relation to the performance and requirements desired by the production system.
- Able to target the intervention horizon before the failure has arrived.
- Propose adequate solutions in the future to minimize downtime and increase production.

The prognosis model is based on an empirical HOLM and ARCHARD model based on real data whose objective is to generate adaptive behavior to the law of bearing degradation and predict useful life.

Prediction by the physical model requires more or less accurate and misused data in the field to give a remaining useful life span (RUL) close to reality. This allows us to decide on the action of maintenance intervention before the failure.

The integrated model is a tool that can therefore enable, the person in charge of the packaging area (COSUMAR), to control the optimization of his forecast maintenance policy.

This study is the track of a field work which will be interested primarily in the improvement of the production within the zone of conditioning while minimizing the downtime.

The optimal time requirement is used to think of a constructive solution to internal realization (without having another external partner) that responds to the constraints of temperature (160 °C), friction, pressure, wear, etc. …

## References


1. Caignot, A., et al.: Industrial Sciences of the Engineer, Vuibert Edition, pp. 47–53 (2013)
2. Aggab, T.: Prognosis of complex systems by the joint use of hidden Markov model (MMCs) and observer, thesis, pp. 40–41 (2016)
3. Zambelli, G., Vincent, L.: Materials and contacts: a tribological approach. published under the direction of Ms. Polytechnic and University Press Romandes, Lausanne, p. 75 (1998)
4. Downing, A., Mccabe, J.F., Gordon, P.H.: The effect of artificial forces between orthodontic brackets and archwires. British J. Orthod. **22**(1), 41–46 (1995)
5. Archard, J.F.: Contact and rubbing of flat surfaces. J. Appl. Phys. **24**, 24 (1953)
6. Caubet, J.J., Cartier, M.: Analysis of stresses resulting from the contact of two friction bodies. Intrinsic curves of friction. The technical memories of the C.E.T.I.M., no. 1, pp. 6–44 (1969)
7. http://www.zoneindustrie.com. Accessed 10 Juin 2019

# State of the Art of Bone Regeneration

Fatima Haddani(✉) and Anas El Maliki

Applied Polymer Research Team, ERAP ENSEM, Hassan II University, Casablanca, Morocco
fatimahaddani7@gmail.com

**Abstract.** Scaffold design for bone tissue engineering try to mimic the function of the natural extracellular matrix, to promote the regeneration of damage tissue.

Improving their bioactivity requires a better understanding of the structural effects and mechanical behavior on the bone reconstruction process. This article reviews the relevant literature on mechanobiology of bone reconstruction at different length scales. In the first part the reader is introduced to structure and bone composition. Then the remodeling process associated with cellular mechanotransduction is presented. Indeed, bone cells are extremely sensitive to mechanical loads, so bone has the ability to optimize its architecture according to the mechanical stresses it undergoes. The interest and specificity of the piezoelectric effects of bone compared to a conventional piezoelectric material are analyzed. Finally, we suggest that interdisciplinary approaches, combining mechanobiology and Materiobiology will inspire innovative ideas to satisfy the design requirements of scaffolds for bone tissue regeneration.

**Keywords:** Bone regeneration · Mechanobiology · Bone remodeling · Piezoelectric

## 1 Introduction

Biomineralization is an important process in the remodeling and regeneration of bone tissue [1]. It consists of a co-precipitation of inorganic ions (calcium, magnesium, phosphate, etc.) dissolved in body fluids on an organic substrate to form a compact and anisotropic hybrid nanocomposite consisting of hydroxyapatite nanocrystals complexly bound to collagen [2].

Although there are clear connections between fundamental biological concepts and the biomechanical aspects of growth (mass change), remodeling (property change) and morphogenesis (shape change) in living systems, Comprehensive discussion of these concepts can be found in Taber's excellent review [3, 4]. It is well known that the interactions between mechanics and biology are crucial to interpret and describe this phenomenon correctly. Indeed, changes induced by mechanical forces (pressure, shear or elongation) are considered as one of the determining factors for modifying the biological behavior of cells by triggering cascades of reactions through adaptation processes (mechanotransduction) and thus stimulating in particular the remodeling and repair of bone tissue [2]. This explains the increasing amount of research into the mechanical characteristics of bone.


A. Saka et al. (Eds.): CPI 2019, LNME, pp. 459–479, 2021.
https://doi.org/10.1007/978-3-030-62199-5_41

This Review highlights that the Mechanobiology study of bone reconstruction and its relationship with the microstructure and piezoelectric effects of bone, are very important to improve the design and development of scaffolds for bone tissue engineering.

We focus primarily on mechanical behavior of bone tissues at different scales. Several theoretical models and mainly experimental results are examined.

## 2 Mechanobiology of Bone Reconstruction

### 2.1 Multi-scale Bone Structure

Bone is a dynamic multi-scale structure; it is optimized to offer maximum performance with a minimum of material (to be both strong and relatively lightweight) [5, 6]. The inside of the bone consists of bone marrow surrounded by two main types of tissue as shown in Fig. 1: The cortical bone that makes up the rigid outer wall of the bone. It is relatively dense and resistant, with a low porosity of 5% to 10%. It surrounds trabecular bone tissue that has a porous open-celled honeycomb structure made trabecular mesh surrounded by bone marrow [1]. Since the morphology of the trabeculae is controlled by applied mechanical forces (pressure, shear or stretching), the trabeculae are mainly oriented in the direction of the main loading constraints [7].

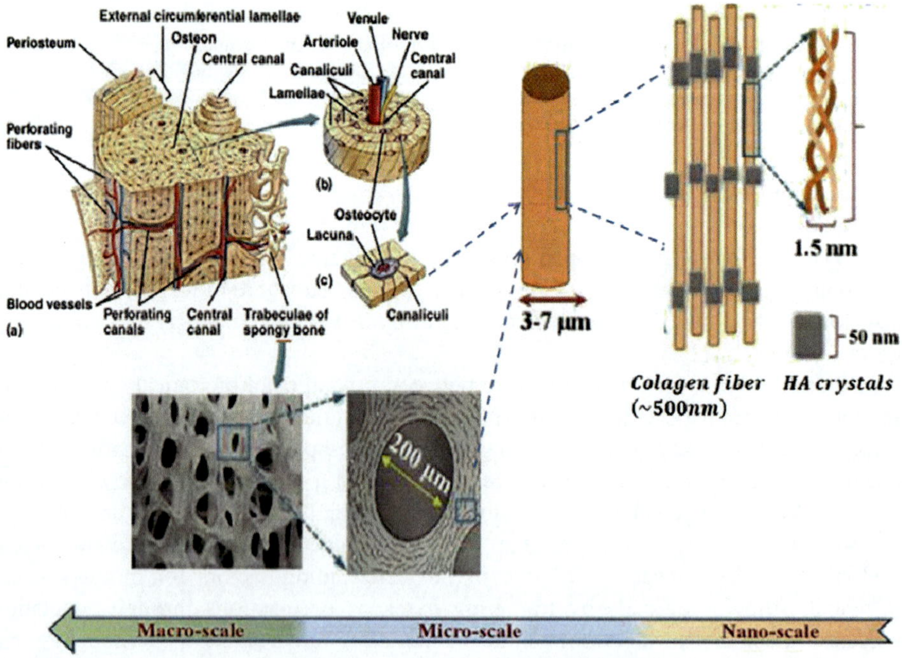


**Fig. 1.** Multiscale aspects of bone structure [1, 6, 8]

## 2.2 Bone Composition

Bone tissue is a connective tissue consisting of an extracellular organo-mineral matrix and bone cells (osteoblasts, osteocytes and osteoclasts). The bone mineral idealized as hydroxyapatite (Hap), $Ca_{10}(PO_4)_6(OH)_2$, is a carbonate-hydroxyapatite, approximated by the formula: $(Ca, X)_{10}(PO_4, HPO_4, CO_3)_6(OH, Y)_2$ [9], where X are cations (magnesium, sodium, strontium ions…) that can substitute for the calcium ions, and Y are anions (chloride or fluoride ions) that can substitute for the hydroxyl group [1]. These ionic substitutions also affect the formation as well as the overall properties of apatite. Thus, for example, the greater solubility and lower crystallinity (smaller crystal size) of bone or dentin compared to enamel (Table 1) can be explained by the higher concentrations of $Mg^{2+}$ and $CO_3^{2-}$ in bone or dentin compared to enamelled apatite, Also, strontium, even present in trace amounts, has a proven role in bone strengthening since it leads to a decrease in bone resorption. Inorganic phase is chemically linked to collagen type I fibers, which have specific sites for biomineralization, to form an anisotropic hybrid nano-composite with a hierarchical morphology.

## 2.3 Biomechanics of Bone

Since the Second World War, several research studies have been carried out on the mechanical characteristics of bone [12]. Indeed, the study of the mechanical behavior of bone, at different scale, is the main step towards to assess the risk of fracture, understand bone consolidation and select bone biomaterials [1, 13]. At the nanoscale, bone derives its tensile strength from its collagen component and its compressive strength from its mineral component. The arrangement of apatite crystals in small units (Fig. 1) protects the bone from crack propagation. The microscopic scale represents an arrangement of collagen fibers, reinforced with hydroxyapatite crystals, in concentric lamellae forming packed Haversian systems for cortical bone and trabeculae for trabecular bone. The fibers are oriented in different ways for two adjacent lamellae, which explains the mechanical performance and anisotropy of the bone. Micromechanical studies carried out by Ascenzi and Bonucci [14, 15] on osteons isolated from cortical bone have revealed that the stress deformation curve in osteons is very strongly dependent on the orientation of collagen fibers bundles. At macroscopic scale, by reviewing the literature, it can be concluded that the mechanical characteristics of bone, as revealed by the tests, vary greatly depending on the conditions under which bone samples are prepared [16], also in vivo bone and muscle combine to form a composite beam that is much stronger than isolated bones [17]. Measurements on different bones of the four essential mechanical properties (Young's modulus, tenacity, hardness and tensile strength) have been the subject of numerous publications. Examples of orders of magnitude of these properties are presented in Table 2.

**Table 1.** Comparative composition and physical properties of inorganic phases of adult human enamel, dentine, and bone. W. Suchanek et al. [10] (after LeGeros [11]).

| | | Enamel | Dentine | Bone |
|---|---|---|---|---|
| Chemical composition | Calcium, $Ca^{2+}$ | 36,5 | 35,1 | 34,8 |
| | Phosphorus, as P | 17,7 | 16,9 | 15,2 |
| | Ca/P molar | 1,63 | 1,61 | 1,71 |
| | Carbonate, as $CO_3^{2-}$ | 3,5 | 5,6 | 7,4 |
| | Magnésium $Mg^{2+}$ | 0,44 | 1,23 | 0,72 |
| | Potassium $K^{+}$ | 0,08 | 0,05 | 0,03 |
| | Sodium $Na^{+}$ | 0,5 | 0,6 | 0,9 |
| | Fluoride $F^{-}$ | 0,01 | 0,06 | 0,03 |
| | Chloride $Cl^{-}$ | 0,3 | 0,01 | 0,13 |
| | Pyrophosphate $P_2O_7^{4-}$ | 0,022 | 0,10 | 0,07 |
| | Trace elements Sr2+, Pb2+, Zn2+, Cu2+, Fe3+, etc. | | | |
| Total inorganic (mineral) | | 97,0 | 70,0 | 65,0 |
| Total organic | | 1,5 | 20 | 25 |
| Absorbed $H_2O$ | | 1,5 | 10 | 10 |
| Crystallographic properties | a-axis (±0.003 A°) | 9.441 | 9.42 | 9.41 |
| | c-axis (±0.003 A°) | 6.880 | 6.88 | 6.89 |
| | Crystallinity index | 70–75 | 33–37 | 33–37 |
| | Crystallite size (aver.), | 1,30 × 300 | 200 × 40 | 250 × 30 |
| Ignition products (800 °C) | | β – TCP + HAp | β – TCP + HAp | HAp + CaO |

**Table 2.** Mechanical Properties of human bones [18–20]

| | Cancellous bone | Cortical bone | Cartilage |
|---|---|---|---|
| Compressive strength (MPa) | 0,1–16 | 130–200 | N/a |
| Tensile strength (MPa) | 80–150 | 50–160 | 3,7–10,5 |
| Young's modulus (GPa) | 0,02–0.5 | 7–30 | 0,7–15,3 (MPa) |
| Fracture toughness $\left(MPa.m^{1/2}\right)$ | N/a | 2–12 | N/a |

However, many questions remain to understand specific mechanobiology, which explains why a coupled experimental and numerical approach [21, 22] is often crucial to improve the prediction of behavior and the evolution of bone tissue and to understand mechanical-biological couplings [3, 23] .

***Trabecular Bone***

Trabecular bone is a hierarchical material; its mechanical properties are studied by various experiments and model-calculations [24]. It is formed by anastomosed bone trabeculae (adjacent bone lamellae). Each of these behaves like a rigid body and communicates with the others to form a united structure. In order to incorporate information related to microstructure, trabecular bone can be modelled by a micropolar material (Cosserat's theories) [25].

***Single Trabeculae***

Lorenzetti and co-workers [7] presented a new method for determining the elastic modulus of single trabeculae in the natural network and for studying its influence at the macroscopic level. They compared the experimental results of the 3-point bending tests and the FE analysis (Fig. 2) for estimated Young's modulus for two trabeculae from a sample of the femoral head of an adult healthy sheep; they found 15.0 and 16.8 GPa respectively. They proved the elastic behavior of the trabeculae as well as the influence of their volume on the mechanical properties. Mechanical competences of single trabeculae in the post-yield region of both tensile and bending tests have also been Investigated [26]. Due to the small amplitude of the displacements that occur in the sample, an optical acquisition system was implemented to exclude errors associated with the relatively high and non-reproducible compliance with the sample boundaries. Droplets of fluorescent microspheres were deposited on the surface of the samples (Fig. 3). The Voce exponential non- linear model (Voce, 1948), shown in Eq. (1), was used to describe the experimental data, Table 3 summarizes the significant difference between the young donor and the old donor, for the tensile and bending tests separately [26].

$$\begin{cases} if\ \varepsilon \le \varepsilon_y : & \sigma = E.\varepsilon \\ if\ \varepsilon > \varepsilon_y : & \sigma = \sigma_y + R.\left(1 - e^{-B\varepsilon_{pl}}\right) \end{cases} \tag{1}$$

With the corresponding parameters: E (Elastic modulus, GPa), $\sigma_y$ (yield stress, MPa), $\varepsilon_y$ (yield strain, -), $\varepsilon_{pl}$ (plastic strain, -), B (exponential hardening coefficient, -), R (post-yield hardening stress, MPa).

**Table 3.** The evaluated mechanical parameters calculated for tensile and bending test mode, reported for the young and old donor separately [26].

| | Tensile | Bending |
|---|---|---|
| Young donor | $\begin{cases} if\ \varepsilon \le 0.68 : & \sigma = 11.84.\varepsilon \\ if\ \varepsilon > 0.68 : & \sigma = 78.01 + 49.4.\left(1 - e^{-154.6\varepsilon_{pl}}\right) \end{cases}$ | $\begin{cases} if\ \varepsilon \le 2.02 : & \sigma = 4.43.\varepsilon \\ if\ \varepsilon > 2.02 : & \sigma = 77 + 63.6.(1 - e^{-42.4\varepsilon_{pl}}) \end{cases}$ |
| Old donor | $\begin{cases} if\ \varepsilon \le 0.75 : & \sigma = 15.56. \\ if\ \varepsilon > 0.75 : & \sigma = 115.28 + 62.52\,.(1 - e^{-256.6\varepsilon_p} \end{cases}$ | $\begin{cases} if\ \varepsilon \le 1.57 : & \sigma = 10,16.\varepsilon \\ if\ \varepsilon > 1.57 : & \sigma = 157.1 + 95.19.(1 - e^{-24.86\,\varepsilon_{pl}}) \end{cases}$ |

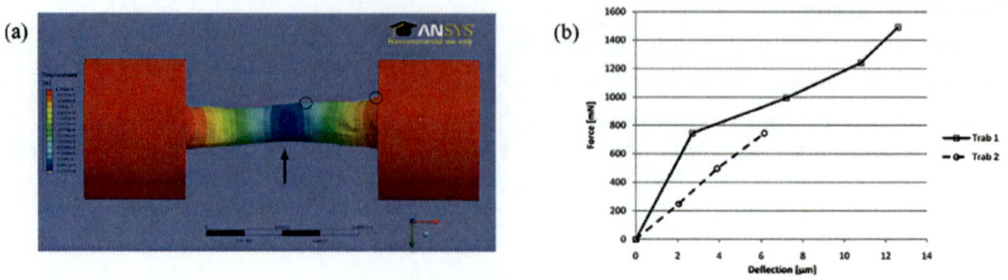


**Fig. 2.** (a) Relationship between applied force and measured deformation show that both the trabeculae are loaded within the linear-elastic range [7]. (b) FE analysis of the deflection of the trabeculae under a load of 0.745 N. Scale indicates the displacement in [m] [7].

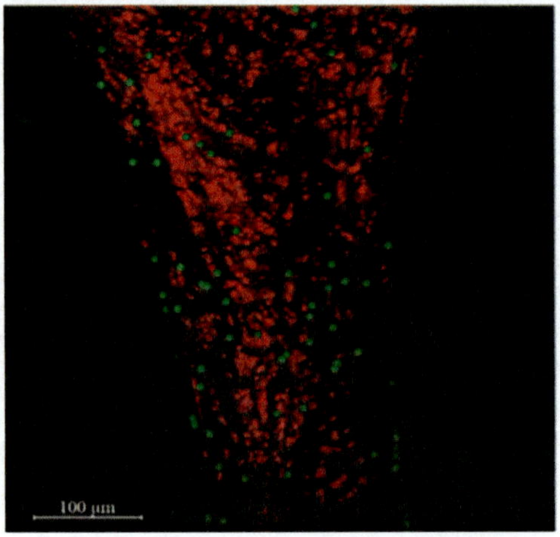


**Fig. 3.** Single trabeculae (red) covered with fluorescent Microspheres (green), as viewed with a confocal microscope [26]

### *Micropolar Continuum Model*

Generalized continuum mechanics theories aim to explain the effects of microstructure on the macroscopic properties of bone and to introduce an intrinsic characteristic length of the material into the formulation of the constituent equations. They follow three possible main strategies, namely integral non-local models [27], higher-order gradient models [3], and micropolar [13, 16, 28] (otherwise coined Cosserat) theories [29, 30]. For the assessment of effective mechanical properties of vertebral trabecular bone and their relationship to microstructural parameters, Goda et al. [25] constructed a 3D anisotropic micropolar continuum model. To incorporate information related to microstructure by taking into account the influence of neighboring points in the formulation of constitutive equations, vertebral trabecular bone is modeled as a cellular material with an idealized periodic structure made of open 3D hexagonal cells, which is effectively orthotropic (Fig. 4). In order to reflect the impact of transverse shear deformations on the effective properties of vertebral trabecular structure, the scaling behavior of the equivalent elastic is determined (Fig. 5), considering the following geometrical parameters: $L_v = 0.5L$; $h = 0.5L$; $\theta = 30°$. The obtained results show that the absence of transverse shear leads to overestimate Young's moduli ($E_x^*$ and $E_y^*$) and there is no such effect of transverse shear on $E_z^*$.

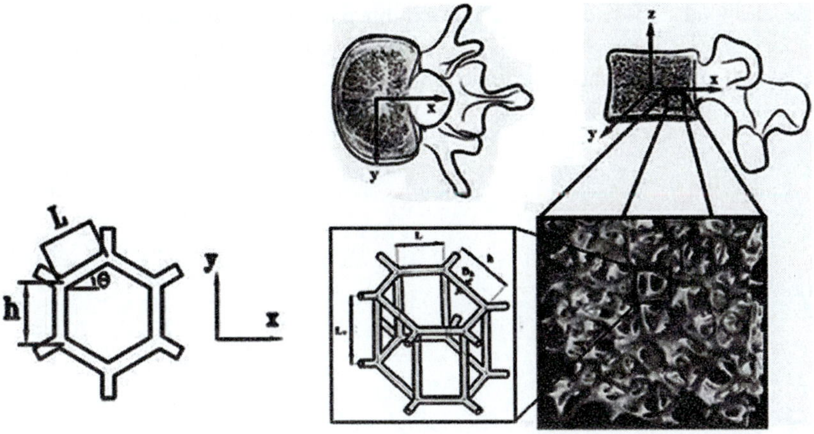


**Fig. 4.** General anisotropic unit cell for 3D hexagonal models. The geometric parameters include the cell angle θ, the in-plane strut lengths (L, h), and the vertical length $L_v$ [6, 20, 25].

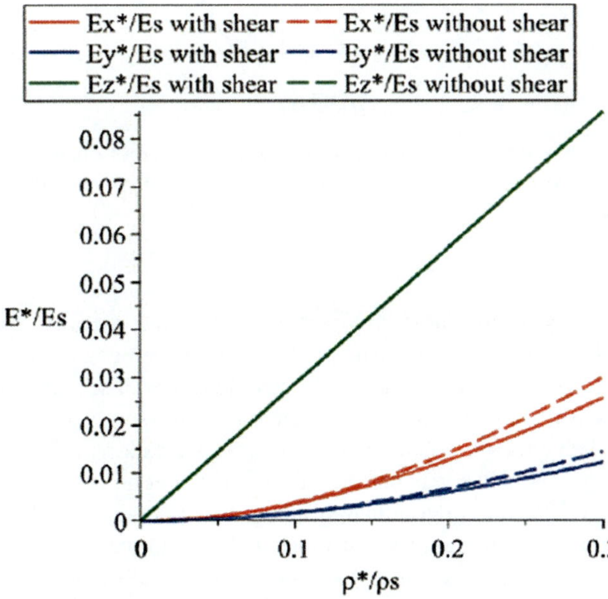


**Fig. 5.** Effective elastic moduli $E_x^*$, $E_y^*$, and $E_z^*$ of vertebral trabecular bone versus the relative density ρ*/ρs and impact of transverse shear [6, 25]

### *Compact Bone*

#### *Loading/Unloading Cycles*

In 1978, Bonfield and O'Connor [31] performed measurements of elastic and non-elastic tensile deformation behavior of longitudinal compact bone samples from bovine and rabbit bones using the micro deformation technique. They found that for high stresses, the load and discharge curves exhibit persistent residual deformation and an open loop appearance in hysteresis (Fig. 6 curve C). The inelastic deformation disappeared slowly after discharge, provided that it waited long enough (up to 40 min) [12].

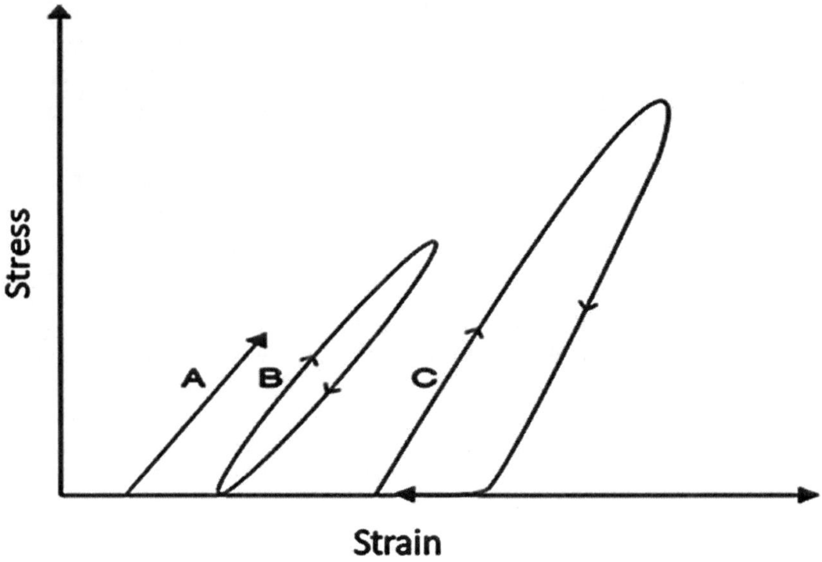


**Fig. 6.** The three types of bone loading/unloading cycles (based on Bonfield and O'Connor). A: elastic phase; B: closed hysteresis loop; C: hysteresis loop initially open and which closes with time at zero stress [12].

#### *Single Osteons*

After carefully studying the lamellar structure of isolated osteons, Ascenzi and his collaborators [14, 15], studied the tensile properties of osteonic lamellar samples - those at the initial stage of calcification, and fully calcified ones. They have shown that the tensile load increases only slightly as calcification progresses. A comparative analysis of the torsional behavior of osteons with different orientations of collagen and crystallites in successive lamellae was performed by ascenzi et al. [14]. High shear moduli were reported, 23 GPa for longitudinal structure osteons and 17 GPa for alternating orientation lamellae were more than four times higher than the modules reported in the literature on macroscopic bone tissue specimens. Lakes [28] elucidated the structural and continuum reasons for the difference in stiffness between osteons and bone whole, as shown in Fig. 7. The ability of the bone to redistribute stress around these defects (before any remodeling) is associated with the presence of couple-stress effects involving a reduction in the stress concentration factor around holes [13].

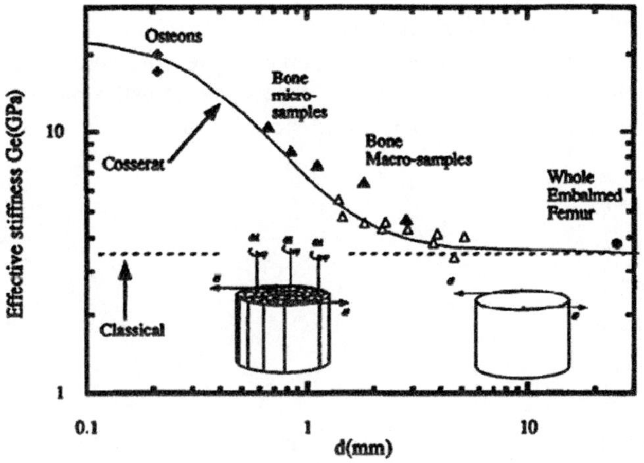


**Fig. 7.** Effective torsional stiffness vs. diameter., ♦ Osteons, fresh, wet, after Ascenzi et al. (1994); Δ, bone macro-samples, fresh, wet, after Yang and Lakes (1981); A, bone micro-samples, fresh, wet, after Lakes and Yang (1983); •, whole embalmed femur, after Huiskes et al. (1981); -, Solid curve, Cosserat elasticity, torsion, assuming technical elastic constants to be: G = 3.5 GPa, $l_l$ = 0.22 mm, $l_b$ = 0.44 mm, ψ = 1.5, N = 0.62. Apparent shear modulus seems to vary with diameter. Inset drawings: right, classical elastic continuum with stress due to force increments dF; left, bone as a Cosserat elastic continuum with both force increments dF and torque increments dM upon osteons. (- - - -): classical elasticity, torsion. Apparent shear modulus is independent of diameter. The solid line represents a Cosserat continuum view which incorporates distributed moments (couple stress) as well as distributed forces (stress) [28].

***Flexural Rigidity***

Yang et al. [16] have shown that size effects in quasistatic bending of compact bone are consistent with micropolar theory. They estimated an average characteristic bending length of 0.45 mm. as for the magnitude of the ratio in stiffness of osteons to that of whole bone, a factor of 3.5 stiffening effect was observed [28]. From an analytical point of view, of linear isotropic material in the micropolar theory of Eringen (1968) [16], flexural rigidity $J$ can be expressed as:

$$J = J_1\Omega = \frac{\pi d^4 E}{64} \times \left[ 1 + \frac{8N^2}{\nu+1} \left( \frac{1-(\beta/\gamma)^2}{\delta_n^2} + \frac{\left(\frac{\beta}{\gamma}+\nu\right)^2}{\zeta(\delta_a)+8N^2(1-\nu)} \right) \right] \quad (2)$$

In which:

$J_1$ is the classical flexural rigidity, $\Omega$ is the ratio $J/J_1$, $N = k/[2(\mu+k)]$ is the coupling number, E is Young's modulus, $\zeta$ Bessel function and $\delta^{-1} = \left[\frac{\gamma(\mu+k)}{k(2\mu+k)}\right]^{1/2}$ is a micropolar length parameter for bending.

Figure 8 show that the behavior of compact bone contradicts the predictions of classical elasticity theory. The experimental data are adjusted more precisely by an exact solution of the bending problem in micropolar theory.

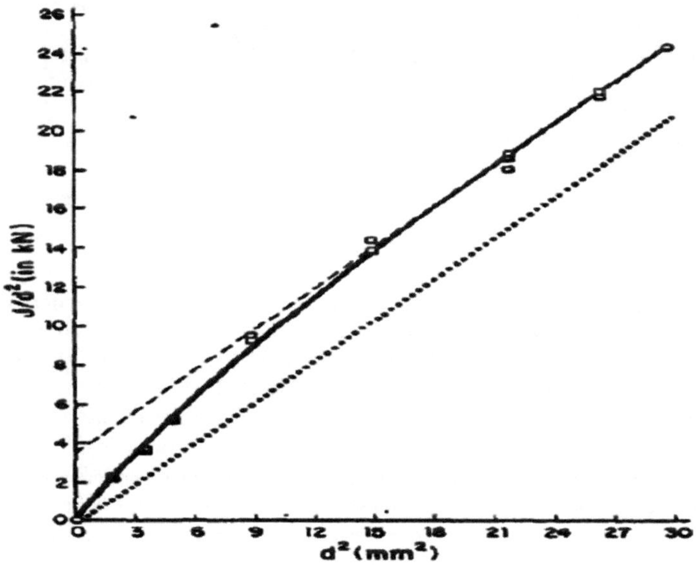


**Fig. 8.** Behavior of a human compact bone sample in bending, curve fitting based on an exact solution in micropolar theory. Solid curve is an optimal theoretical curve (Eq. 2) for $E = 12.367\,\text{GN/m}^2$, N = 1.0, $\frac{\beta}{\gamma} = 1.0$, and $\delta^{-1} = 0,638\,\text{mm}$ . Squares represent experimental data. Dotted line represents the predicted behavior of a classically elastic material. The dashed line represents the expected behavior of a micropolar solid for which $\frac{\beta}{\gamma} = -\nu$ [16].

The micropolar Young's modulus is defined by Gauthier and Jahsman in 1975:

$$E = \frac{(2\mu+k)(3\lambda+2\mu+k)}{2\lambda+2\mu+k} \tag{3}$$

This E is what is measured in flexion only if the sample is much thicker than the characteristic length [16].

## 2.4 Bone Remodeling and Reconstruction

Bone remodeling is a biological process [32] that allows bone to optimize its internal structure to adapt to mechanical loading [33] and to regulate phosphate metabolism and calcium availability [34]. This dynamic process is essentially based on the cells' ability to interact with their environment. Indeed, this adaptive renewal involves three types of cells: (i) Osteoblasts responsible for bone mineralization on the surface of a pre-existing organic matrix, (ii) Osteoclasts, their function is the resorption of bone tissue, (iii) and Osteocytes, which are mechanoreceptor cells, are the orchestrator of bone remodeling through regulation of osteoclast and osteoblast activity [35].

***Wolff's Laws***

In 1866 Culmann, an engineer and mathematician, proposed to biologists a law to be confirmed: the skeleton is designed to support the maximum load with the minimum material [12]. The idea was accepted and in 1892 Wolff, a German anatomist and surgeon, laid down his famous laws which made it possible to better understand the reorganization of bone architecture [33] (increase in bone mass observed in athletes and loss of bone tissue during immobilization) and which read as follows [5, 36]:

- Bone is formed and resorbed according to the mechanical stresses it undergoes: optimizations of strength with respect to weight.
- Its resistance varies according to the direction of the loads incurred: alignment of trabecular structure with stress directions.
- Muscular activity leads to a change in stress and therefore bone activity: self-regulation of bone structure by cells after mechanical stimulation.

***Optimal Density Distribution***

Mechanical constraints imposed by the environment can lead to tissue remodeling by triggering a cascade of reactions on the cells (mechanotransduction) [3]. Understanding of these phenomena is important for the prevision of bone remodeling and bone reconstruction in many orthopedic and surgical applications [37]. The optimal distribution of bone density was studied by topological optimization methods based on homogenization [38]. The high and intermediate density distributions obtained (Fig. 9 (a)) are in satisfactory agreement with the cortical and trabecular bone tissue models observed in the real proximal femur (Fig. 9(b)).

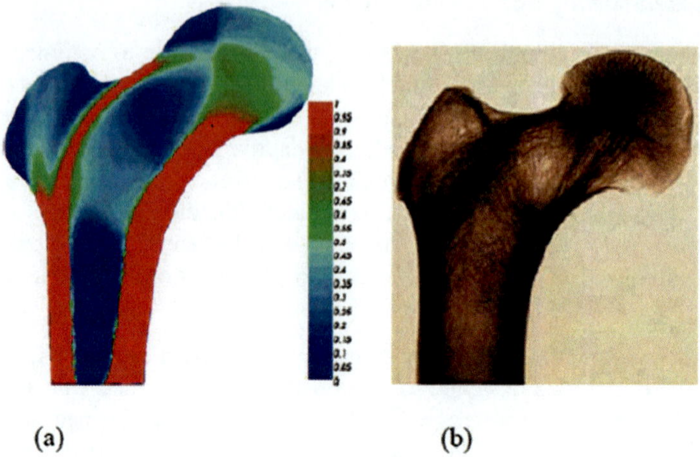


**Fig. 9.** (a) Optimal density distribution of femoral bone based on homogenization approach. (b) Roentgenogram of the internal structure of the femur [38]

To describe the mechanobiological interactions that occur in bone tissue remodeling, D. George et al. [21] considered the activities and location of osteoclasts and osteoblasts by modelling the biological stimulus as an energy function, based on applied mechanical loads and depending on the strain developed, and the material parameters E (Young's modulus) and ν (Poisson's ratio). They studied the effect of the different intensities and directions of mechanical loading applied to characterize the quality, rate and density of bone remodeling

$$U(\varepsilon, \rho_b) = \frac{1}{2} \frac{E(\rho_b).\upsilon}{(1+\upsilon)(1+2\upsilon)} \varepsilon_{ii}^2 + \frac{1}{2} \frac{E(\rho_b)}{(1+\upsilon)} \varepsilon_{ij}\varepsilon_{ij} \tag{4}$$

Where E and ν are the Young's modulus and the Poisson coefficient of the mixture, respectively. The Young's modulus is assumed to be dependent on the densities $\rho_b$ of bone and $\varepsilon = \frac{1}{2}(\nabla u + \nabla^T u)$ is the deformation tensor.

Hasuike et al. [39] confirmed, with in vivo microfocus computerized tomography (micro-CT), that Low Intensity Pulse Ultrasound (LIPUS) accelerates bone regeneration of non-critical calvary defects in rats (Fig. 10). Scala et al. [18] used the experimental results of [19] and a continuum mixing model [19] to simulate bone growth inside a cavity. They assumed that cell activation is mainly due to the intensity of the charge detected by the osteocytes rather than the frequency of the applied signal, so they studied the effects of an "average" static load to observe the evolution of the system on a weekly basis to monitor the evolution of bone mass density. From the Fig. 12 we can deduce that bone growth is always triggered in areas where the stimulus is much higher. This is reasonable if we think that a high stimulus induces osteoblasts to synthesize new bone tissue more easily.

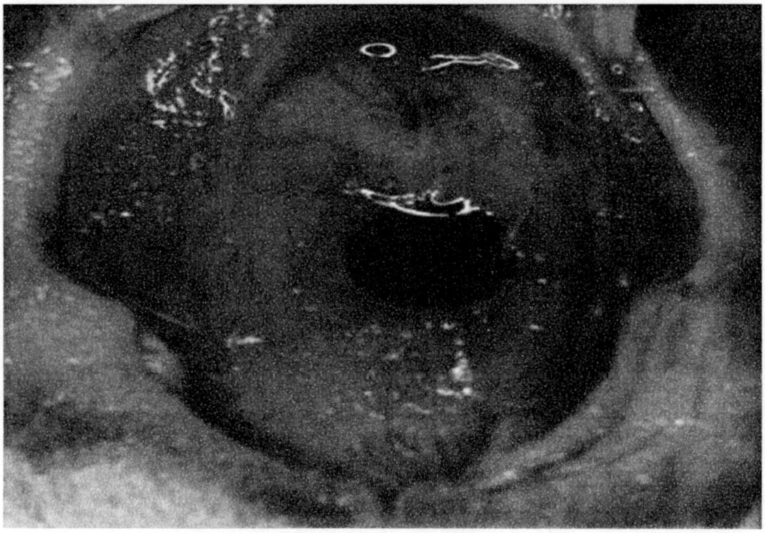

**Fig. 10.** 2.7 mm calvarial defects trephined on the midsagittal suture [39].

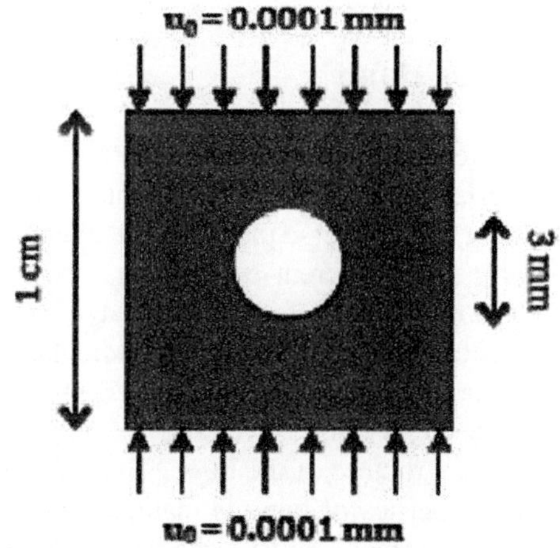


**Fig. 11.** Geometry of the considered bone defect and applied boundary displacement [23].

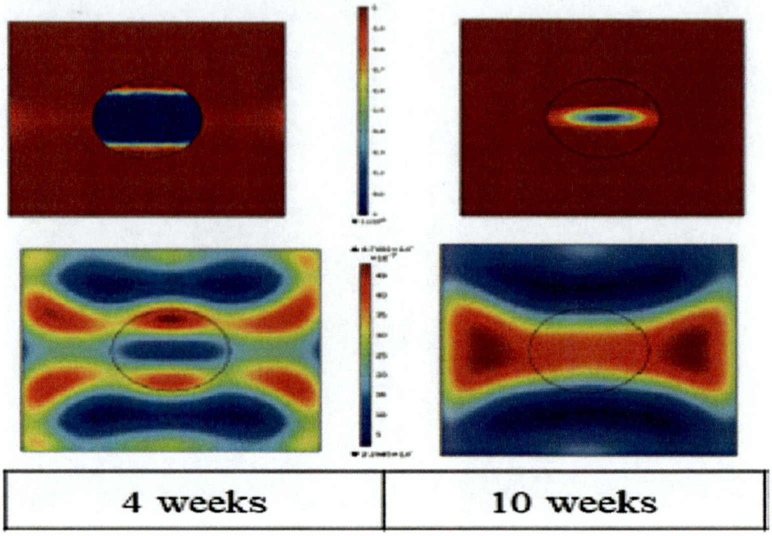


**Fig. 12.** Time evolution of tissue growth and biological stimulus under the application of the boundary conditions shown in Fig. 11 [23].

## 3 Piezoelectric Effect of Bone

Bone has a piezoelectric effect similar to that of crystalline materials: mechanical stresses cause electrical polarization and vice versa. In late 1950, Fukada et al. [40], the first to be confirmed the direct and converse piezoelectric effect in bone, have measured the piezoelectric constant by three different experiments, that is, measurement of the static direct effect, the dynamic direct effect and the dynamic converse effect. The measured values of piezoelectric constant in two perpendicular directions (Fig. 13) are represented in Table 3. The authors have revealed the dependence of the piezoelectric constant and the angle between the bone axis and the pressure direction, which is consistent with the anisotropic structure of the bone. The influence of the piezoelectric effect on the physiological properties of bone is manifested in the ability of bones to detect stresses [41] in order to protect against large deformations, growth, remodeling and fracture healing [2, 42]. However, the constitutive equation of a classic piezoelectric material may be inadequate to describe the electromechanical behavior of bone, and polarization may result from the gradient-stress as well as the stress itself [43]. The stress gradient effects in piezoelectricity can be explained by microstructural considerations. Halperin et al. [42] studied the piezoelectric response, with nanoscale resolution, of a human cortical bon (tibial shaft)e near the Haversian Channel using a piezoelectric force microscope (PFM). Quasi homogeneous piezoresponse nanometer scale imaging of the studied bone samples evidence that piezoelectric properties are uniform for the bones of a specific human individual [42] (Fig. 14 and Table 4).

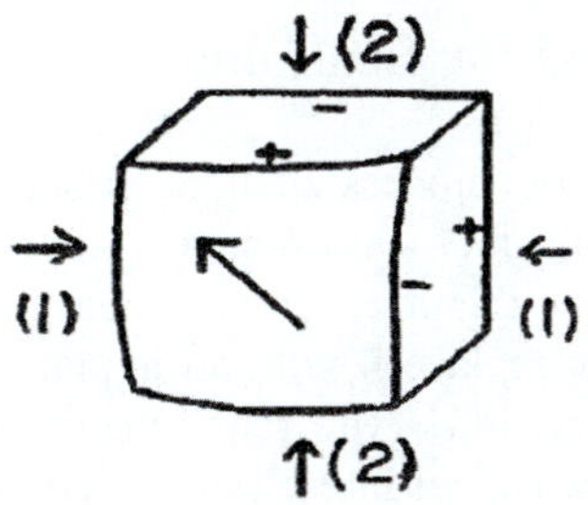


**Fig. 13.** Specimens of bone. Signs represent those of electric polarization which appear in square planes when pressure is applied on each side plane [40]

**Table 4.** Piezoelectric constant of bone of man (CGS-UES) [39]

| Direct effect | | Converse effect |
|---|---|---|
| Static | Dynamic | Dynamic |
| (1) $2.0 \times 10^{-9}$ | $2.9 \times 10^{-9}$ | $2.9 \times 10^{-9}$ |
| (2) $2.2 \times 10^{-9}$ | $3.6 \times 10^{-9}$ | $3.5 \times 10^{-9}$ |

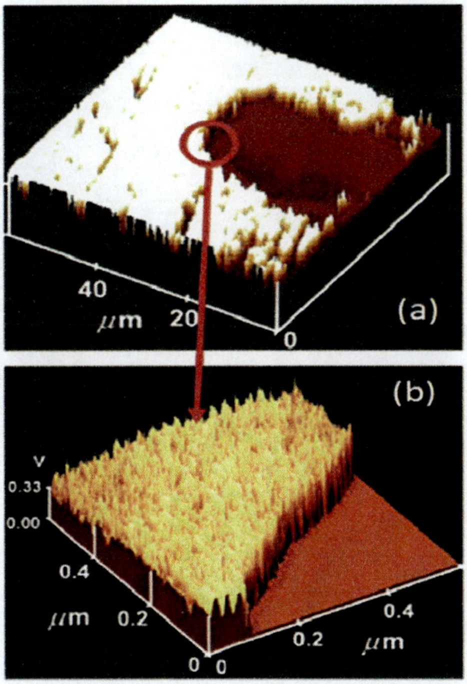


**Fig. 14.** PFM image of cross-section of a sample from mature human cortical bone (tibial shaft): (a) piezoresponse image near the Haversian canal; (b) nanoscale resolution piezoresponse image in the vicinity of the Haversian canal [41]

## 4 Design Requirements for Scaffolds

During the natural regeneration process of bone tissue, cells receive many signals, transmitted by the surrounding microenvironment. Goal of bone-tissue engineering is to provide scaffolds that mimic this microenvironment in order to regenerate diseased or damaged tissue rather than replace it with an implant when the bone deficit is too important (several centimeters) [44]. The requirements of scaffolds for bone- tissue engineering are sufficient mechanical integrity [45], controlled pore size [46, 47] and Cell-Scaffold Interactions that can regulate cell behaviors (e.g., cell adhesion, spreading, proliferation, cell alignment, and the differentiation or self-maintenance of stem cells) [48, 49]. the interdependencies of these factors have been demonstrated.

- **Sufficient mechanical integrity:** the mechanical properties of bone scaffolds are considered to be one of the most important selection criteria [1] the mechanical properties of bone scaffolding are considered as one of the most important selection criteria since they act as a physical support structuring and regulating cellular activity in fact bone cells are sensitive to mechanical signals. Harley et al. [45] performed a quantitative mechanical analysis on a series of Collagen–glycosaminoglycan (CG) scaffolds that exhibited a stress and strain behavior characteristic of open-cell low-density foams with distinct regimes of linear elasticity, collapse plateau and densification regimes (Fig. 15).

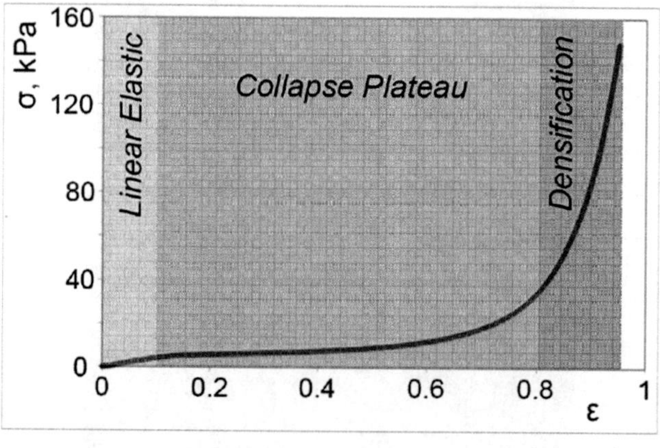


**Fig. 15.** Characteristic stress–strain curves observed for the dry CG scaffold [45]

- **Pore size:** Pore size must be optimized to achieve a compromise between a large specific surface area [47] (Fig. 16(a)), necessary for the initial adhesion of cells, and the advantage of larger interconnected pores, which ensure cell migration, nutrient flow and vascularization (Fig. 16(b)) [46]. Various techniques have been applied to control porosity, pore size, pore shape and interconnectivity such as dissolvable water sacrificial 3D printed moulds [50], electrospinning [51], robocasting [52] and Thermally Induced Phase Separation [53], etc.

(a)

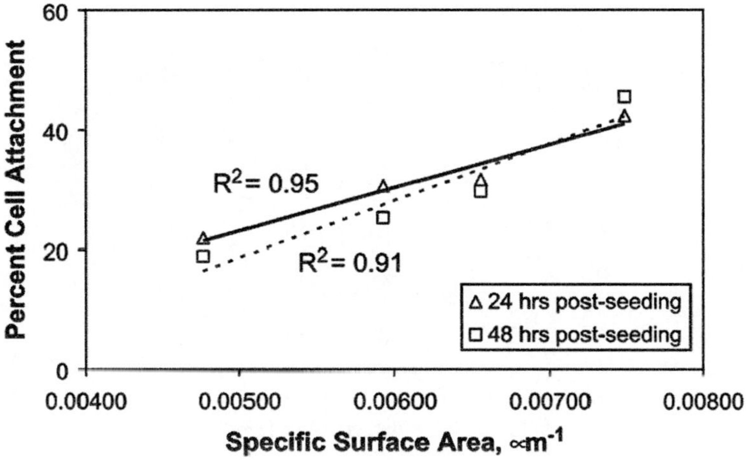


(b)

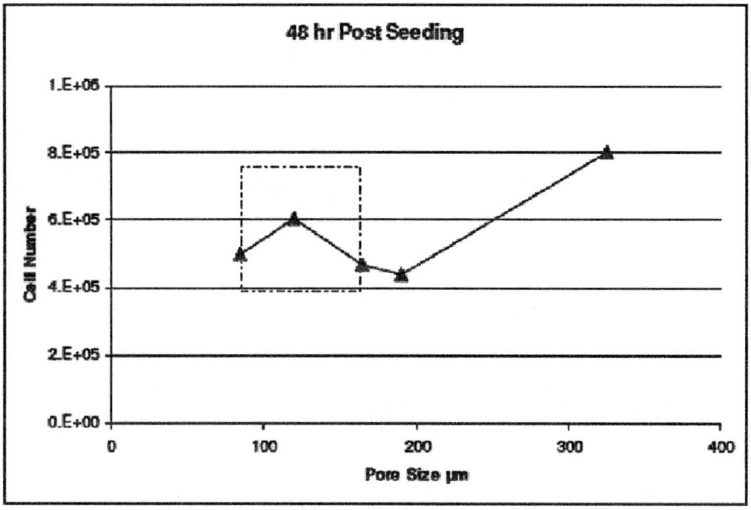


**Fig. 16.** (a) Attached cell number plotted against specific surface area showing a strong linear relationship at 24 (solid line) and 48 h post seeding (dashed line) [47]. (b) Effect of mean pore size on cell number [46].

- **Cell-Scaffold Interactions:** Cell migration behavior is known to be governed by scaffold properties, specifically the pore size and Young's modulus [48] and it has been proven that the use of piezoelectric stimuli, mechanically induced variations in surface charge, makes it possible to restore the physiological electric microenvironment, which promotes the modulation of bone marrow mesenchymal stem cells (BM-MSC) and stimulates their proliferation and differentiation [53] (Fig. 17). It is also interesting to note that biological cells can contract a scaffold, Freyman et al. [54] have shown that biological cells can contract a scaffold and that the contractile forces they generate are limited in strength and not in displacement (Fig. 18).

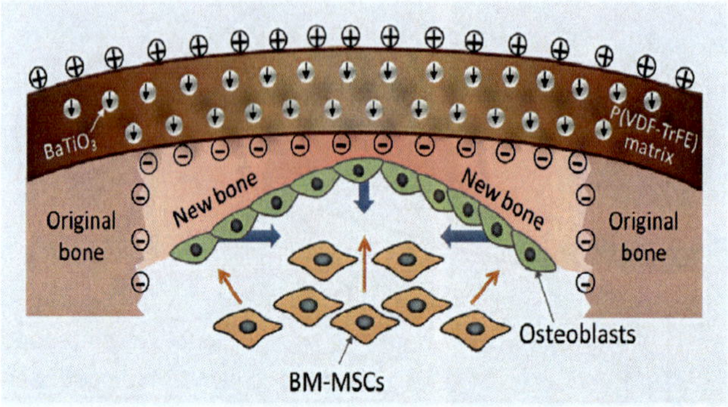


**Fig. 17.** Illustration of biomimetic electric microenvironment created by BTONP/P(VDF-TrFE) composite membranes encouraging bone defect repair [53].

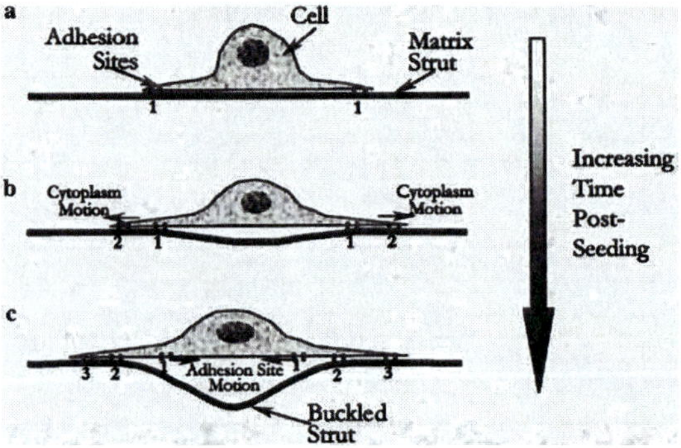


**Fig. 18.** Schematic of cell elongation and matrix contraction [54].

## 5 Conclusion

Although bone tissue regeneration is currently the subject of extensive and multidisciplinary research, the complexity of bone architecture and the variability of its properties pose challenges in the development of scaffolding for bone tissue engineering. Therefore, it is necessary to fully understand the processes of natural regeneration to improve the interaction between scaffolding and living tissue by optimizing their composition and structure. Our research is based on the results of the mechanobiology of bone reconstruction and its relationship with the microstructure and piezoelectric effects of bone, to improve interactions between bone tissue and poly (vinylidene fluoride) (PVDF) scaffolds by incorporating hydroxyapatite nanoparticles, which have a chemical structure similar to that of a living bone. The impact of hydroxyapatite nanoparticles and the various parameters of the elaboration processes on the structural and morphological properties of scaffolding will be studied by X-ray diffraction; scanning electron microscopy and Fourier transform infrared spectroscopy.

The thermal properties of the sample will be studied by differential thermal analysis, and the mechanical tests will be carried out with a mechanical test machine coupled with electrical detection equipment.

## References


1. Gao, C., Peng, S., Feng, P., Shuai, C.: Bone biomaterials and interactions with stem cells. Bone Res. **5**, 1–33 (2017). 17059
2. Rodriguez, R., Rangel, D., Fonseca, G., Gonzalez, M., Vargas, S.: Results in Physics Piezoelectric properties of synthetic hydroxyapatite-based organic-inorganic hydrated materials. Results Phys. **6**, 925–932 (2016)
3. Louna, Z., Goda, I., Ganghoffer, J.-F.: Identification of a constitutive law for trabecular bone samples under remodeling in the framework of irreversible thermodynamics. Continuum Mech. Thermodyn. **30**(3), 529–551 (2018). https://doi.org/10.1007/s00161-018-0619-9
4. Taber, L.A.: Biomechanics of growth, remodeling, and morphogenesis Appl. Mech. Rev. **48** (8). 487–545 (1995)
5. Bert, M.: The Wolff's laws clinical consequences Part I Spécial implant. Actual Odonto Stomatol. **288**(2), 1–5 (2018)
6. Ganghoffer, G.F., Goda, I.: Micropolar Models of Trabecular Bone, 1st edn. Multiscale Biomechanics Elsevier-ISTE Press Ltd, London (2018). ISBN 9781785482083 263-316
7. Lorenzetti, S., Carretta, R., Müller, R., Stüssi, E.: A new device and method for measuring the elastic modulus of single trabeculae. Med. Eng. Phys. **33**(8), 993–1000 (2011)
8. LeGeros, R.Z.: Properties of Osteoconductive Biomaterials-calcium phosphates. Clin. Orthop. Relat. Res. **395**, 81–98 (2002)
9. Suchanek, W., Yoshimura, M.: Processing and properties of HAp based biomaterials for use as hard tissue. J. Mater. Res. **13**(1), 94–117 (1998)
10. Legros, R.Z., Balmain, N., Bonel, G.: Structure and composition of the mineral phase of periosteal bone. J. Chem. Res. **1**, 8–9 (1986)
11. Meyrueis, P.: Biomechanics of bones and treatment of fractures. EMC Rhumatol. **1**(1), 64–93 (2004)
12. Ripamonti, C., Lisi, L., Buffa, A., Gnudi, S., Caudarella, R.: The trabecular bone score predicts spine fragility fractures in postmenopausal caucasian women without osteoporosis independently of bone mineral density. Med. Arch. **72**(1), 46–50 (2018)
13. Lakes, R.: The torsional properties of single selected osteons. J. Biomech. **28**(11), 1409–1410 (1995)
14. Ascenzi, A., Benvenijti, A., Bonucci, E.: The tensile properties of single osteonic lamela technical problem and preliminary results. J. Biomechanics **15**(1), 29–37 (1982)
15. Yang, J.F.C., Lakes, R.J.: Experimental Study of Micropolar and couple stress elasticity in compact bone bending. J. Biomech. **15**(2), 91–98 (1982)
16. Van den Abbeele, M., et al.: A subject-specific biomechanical control model for the prediction of cervical spine muscle forces. Clin. Biomech. **51**, 58–66 (2018)
17. Yang, S., Leong, K.F.: The design of scaffolds for use in tissue engineering. Part I. Traditional factors. Tissue Eng. **7**(6), 679–689 (2002)
18. Hutmacher, D.W., Schantz, J.T., Xu, C., Lam, F., Tan, K.C., Lim, T.C.: State of the art and future directions of scaffold-based bone engineering from a biomaterials perspective. J. Tissue Eng. Regen. Med. **1**, 245–260 (2007)
19. Engine, C.: The machanical behavior of cancellous bone. J. Biomech. **18**(5), 317–328 (1985)

20. George, D., Spingarn, C., Dissaux, C., Nierenberger, M., Rahman, R.A., Rémond, Y.: Examples of multiscale and multiphysics numerical modeling of biological tissues. Biomed. Mater. Eng. **28**(1), S15–S27 (2017)
21. Carretta, R., Stüssi, E., Müller, R., Lorenzetti, S.: Prediction of local ultimate strain and toughness of trabecular bone tissue by raman material composition analysis. Biomed. Res. Int. **2015**, 1–9 (2015)
22. Scala, I., Spingarn, C., Rémond, Y., Madeo, A., George, D.: Mechanically-driven bone remodeling simulation: application to LIPUS treated rat calvarial defects. Math. Mech. Solids **22**(10), 1976–1988 (2017)
23. Madeo, A., George, D., Lekszycki, T., Nierenberger, M., Rémond, Y.: A second gradient continuum model accounting for some effects of micro-structure on reconstructed bone remodelling Comptes Rendus Mecanique A second gradient continuum model accounting for some effects of micro-structure on reconstructed bone remodelling. Comptes Rendus Mec. **340**(8), 575–589 (2012)
24. Goda, I., Assidi, M., Ganghoffer, J.F.: A 3D elastic micropolar model of vertebral trabecular bone from lattice homogenization of the bone microstructure. Biomech. Model. Mechanobiol. **13**(1), 53–83 (2014)
25. Carretta, R., Luisier, B., Bernoulli, D., Stu, E., Mu, R., Lorenzetti, S.: Novel method to analyze post-yield mechanical properties at trabecular bone tissue level. J. Mech. Behav. Biomed. Mater. **20**, 6–18 (2013)
26. Charlebois, M., Jirásek, M., Zysset, P.K.: A nonlocal constitutive model for trabecular bone softening in compression. Biomech. Model. Mechanobiol. **9**, 597–611 (2010)
27. Lakes, R.S.: Dynamical study of couple stress effects in human compact bone. J. Biomech. Eng. **104**, 6–11 (1982)
28. Cosserat, E., Cosserat, F.: On the theory of elasticity. Annales de la faculté des sciences de Toulouse **10**(3,4), 1–116 (1896)
29. Vardoulakis, I.: Cosserat Continuum Mechanics. vol. 87, pp. 99–119. Springer, Cham (2019). https://doi.org/10.1007/978-3-319-95156-0
30. Bonfield, W., O'Connor, P.: Anelastic deformation and the friction stress of bone. J. Mater. Sci. **13**(1), 202–207 (1978)
31. George, D., Allena, R., Rémond, Y.: Cell nutriments and motility for mechanobiological bone remodeling in the context of orthodontic periodontal ligament deformation. J. Cell. Immunother. **4**(1), 26–29 (2018)
32. Stoltz, J.F., et al.: Influence of mechanical forces on bone: introduction to mechanobiology and mechanical adaptation concept. J. Cell. Immunother. **4**(1), 10–12 (2018)
33. Thi, M.M., et al.: Mechanosensory responses of osteocytes to physiological forces occur along processes and not cell body and require αVβ3 integrin engineering cell biology. Proc. Nat. Acad. Sci. **110**(52), 21012–21017 (2013)
34. Bonewald, L.F.: The amazing osteocyte. J. Bone Min. Res. **26**(2), 229–238 (2011)
35. Cowin, S.C.: Wolff's law of trabecular architecture at remodeling equilibrium. J. Biomech. Eng. **108**, 84–88 (1986)
36. Gross, T.S., et al.: Strain gradients correlate with sites of periosteal bone formation. J. Bone Min. Res. **12**(6), 982–988 (1997)
37. Goda, I., et al.: Optimal internal architectures of femoral bone based on relaxation by homogenization and isotropic material design. Mech. Res. Commun. **76**, 64–71 (2016)
38. Hasuike, A., Sato, S., Udagawa, A.: In vivo bone regenerative effect of low-intensity pulsed ultrasound in rat calvarial defects. YMOE **111**(1), 12–20 (2011)
39. Fukada, B.E., Yasuda, I.: Piezoelectric of bone. J. Phys. Soc. Jpn. **12**(10), 1158–1162 (1957)
40. Fernández, J.R., García-Aznar, J.M., Martínez, R.: Piezoelectricity could predict sites of formation/resorption in bone remodelling and modelling. J. Theor. Biol. **292**, 86–92 (2012)

41. Halperin, C., et al.: Piezoelectric effect in human bones studied in nanometer scale. Nano Lett. **4**(7), 1253–1256 (2004)
42. Lakes, R.S.: The role of gradient effects on in the piezoelectricity of bone. IEEE Trans. Biomed. Eng. **27**(5), 282–283 (1980)
43. Brie, J. et al.: A new custom made bioceramic implant for the repair of large and complex craniofacial bone defects. J. Cranio-Maxillofacial Surg. **41**, 403–407 (2012)
44. Harley, B.A., et al.: Mechanical characterization of collagen – glycosaminoglycan scaffolds. Acta Biomaterialia **3**, 463–474 (2007)
45. Murphy, C.M., Haugh, M.G.: The effect of mean pore size on cell attachment, proliferation and migration in collagen glycosaminoglycan scaffolds for tissue engineering. Biomaterials **31**(3), 461–466 (2010)
46. O'Brien, F.J., et al.: The effect of pore size on cell adhesion in collagen-GAG scaffolds. Biomaterials **26**, 433–441 (2005)
47. Harley, B.A., et al.: Microarchitecture of three-dimensional scaffolds influences cell migration behavior via junction interactions. Biophys. J. **95**(8), 4013–4024 (2008)
48. Li, Y., Xiao, Y., Liu, C.: The horizon of materiobiology: a perspective on material-guided cell behaviors and tissue engineering. Chem. Rev. **117**(5), 4376–4421 (2017)
49. Mohanty, S., et al.: Fabrication of scalable and structured tissue engineering scaffolds using water dissolvable sacrificial 3D printed moulds. Mater. Sci. Eng., C **55**, 569–578 (2015)
50. Haider, A., Haider, S., Kang, I.: A comprehensive review summarizing the effect of electrospinning parameters and potential applications of nanofibers in biomedical and biotechnology. Arab. J. Chem. **11**(8), 1165–1188 (2018)
51. Houmard, M., et al.: On the structural, mechanical, and biodegradation properties of HA/β-TCP robocast scaffolds. J. Biomed. Mater. Res. Part B Appl. Biomater. **101**(7), 1–37 (2013)
52. Akbarzadeh, R., Yousefi, A.M.: Effects of processing parameters in thermally induced phase separation technique on porous architecture of scaffolds for bone tissue engineering. J. Biomed. Mater. Res. Part B Appl. Biomater. **102**(6), 1304–1315 (2016)
53. Zhang, X., et al.: Nanocomposite membranes enhance bone regeneration through restoring physiological electric microenvironment. ACS Nano **10**(8), 7279–7286 (2016)
54. Freyman, T.M., Yannas, I.V., Yokoo, R., Gibson, L.J.: Fibroblast contractile force is independent of the stiffness which resists the contraction. Exp. Cell Res. **272**, 153–162 (2002)

# A Good Practice of IoT Protocols

Sakina Elhadi(✉), Abdelaziz Marzak, and Nawal Sael

Laboratory of Modeling and Information Technology,
Faculty of Sciences Ben M'SIK, University Hassan II, Casablanca, Morocco
elhadi.sakinal993@gmail.com, marzak@hotmail.com,
saelnawal@hotmail.com

**Abstract.** Internet of thing (IoT) is a modern concept that appeared in the 90s. IoT is actually one of the future challenges and opens up prospects for technological and scientific developments. The success of the Internet of Things depends greatly on the choice of protocol to use. Nevertheless, one of the major problems of IoT is the lack of suitable user manual that respect the characteristics, features and requirements of IoT systems compared to classic applications. In earlier paper, we presented a comparative study of the existing IoT protocols according to various criteria, as well as evaluated their capabilities and compared their main characteristics and behaviors in terms of various metrics. In this conceptual paper, we propose a model of use of IoT protocols based on critics already mentioned in the comparative study. The purpose of these models is to enable researchers and developers to choose the appropriate protocols for an IoT application by allowing these models before the application is released. This article aims to provide first an IoT characteristics, then the importance of protocols for the IoT Application. In addition, we present in these models of choice of application and network layer protocols.

**Keywords:** Internet of things · Networks protocols · Applications protocols · Models

## 1 Introduction

The Internet of Things (IoT) is a key technology for digital and virtual technologies. All objects are able to exchange information and communicate with each other also to interact with their users using the Internet. According to Gartner IoT is growing rapidly, there are approximately 8.4 billion things connected in 2017, and this figure should reach 20.4 billion by 2020. As much to tell us that we are witnessing a real digital revolution that will radically change our lifestyles. Owing to these advances in IoT, connected and smart solutions will enable to use software to monitor a real-world experience and respond across the full range of consumer the IoT applications. To manage 'things' heterogeneity and data streams over large scale and secured deployments, IoT and data platforms are becoming a central. part of the IoT. To respond to this fast-growing demand, we see more and more platforms being developed, acquiring the right choice of protocols for each type of IoT application. The complexity of IoT applications, always connected, with high load peaks at certain times, is a challenge for the researchers and the developers. Due to the architecture, IoT systems are constructed


A. Saka et al. (Eds.): CPI 2019, LNME, pp. 480–489, 2021.
https://doi.org/10.1007/978-3-030-62199-5_42

in a rather different way from traditional software systems. The remainder of this document is organized as follows: Sect. 2 presents the different IoT characteristic. Section 3, presents the importance of protocols IoT. In Sect. 4 we expose the related work. Next in Sect. 5 we propose some models of choice application and network protocols. We conclude study Sect. 6.

## 2 Internet of Things

The concept of the Internet of Things was invented in 1999 by Kevin Ash-ton, executive director of the Auto-IDCentre, MIT. Since 2005, various newspapers have started publishing articles on IoT such as The Guardian, Scientific American and Boston Globe. In 2011, IoT experienced growth and massive inter-est. Later, IT giants such as Cisco, IBM and Ericson took many educational and commercial initiatives with the IoT [1].

According to our research there is several definitions, but there is no current standard definition for the Internet of Things. the International Telecommunication Union defines the Internet of Things as a "Global infrastructure for the information society, enabling advanced services by interconnecting objects (physical or virtual) through information technology and existing or evolving interoperable communication" [2]. In the book "Internet of Things" by Pierre-Jean Benghozi, Sylvain Bureau and Françoise MassitFolléa, the internet of things is defined as a network of networks which allows via standardized and unified electronic identification systems, and wireless mobile devices, to directly and unambiguously identify digital entities and physical objects and thus be able to recover, store, transfer and process, without discontinuity between the physical and virtual worlds, the related data [3].

These last years, here are a variety of application domains for the Internet of Things sectors, whether it is in Smart home, Smart City The smart city, Smart mobility, Health Medical and health, Retail industry or Smart factor [4].

## 3 IoT Protocols

Currently, the internet of thing passed the myth stage and take an increasingly important in our life of all activity sectors. The IoT applications have unique characteristics and requirements of the IoT systems.

Protocols are competing to become the main choice for connected objects, but for each application, the most appropriate protocol will be different. the choice is not obvious, because each technology has its own advantages and disadvantages.

According to a state-of-the-art study, we have classified almost all protocols according to four main layers and common between all architectures proposed for IoT [5]. At this time, there is no standard architecture for the Internet of things. There are several articles where researchers propose their own architecture according to their needs [6].

The protocols are classified into four layers which are: device layer, network layer, middleware layer and application layer [7].

This classification simplifies application programmers' and service providers' jobs (Fig. 1).

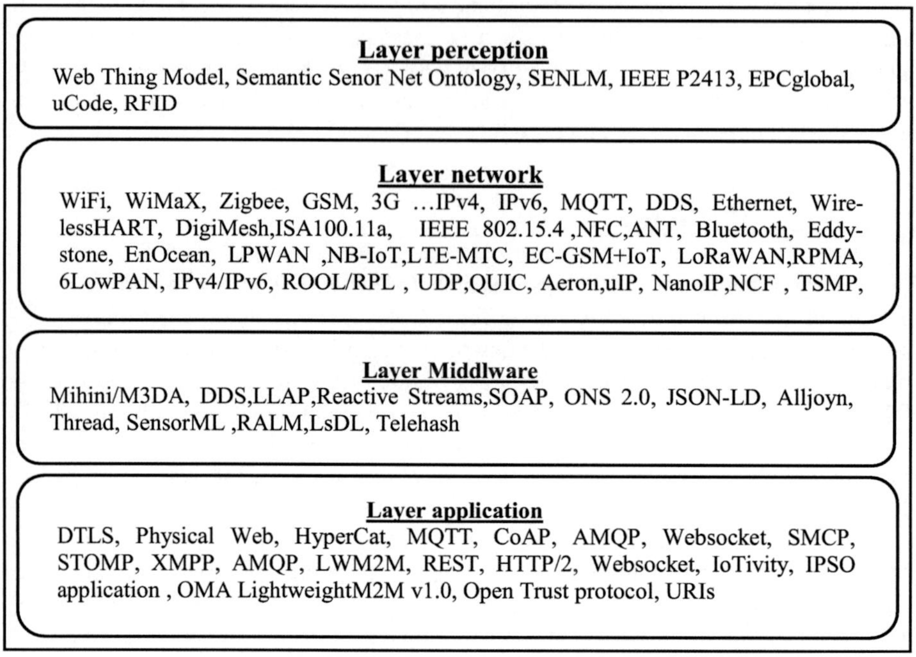


**Fig. 1.** Protocols IoT

## 4 Related Work

The protocols are a must inevitable for the development of Internet of Things applications. However, selecting a protocol is a daunting task and discouraging for all because it depends on the nature of the IoT system and its needs. So, it is important to know the widely accepted and emerging protocols for some IoT systems.

The work presented by Nitin Naik, it presents an evaluation of the four established messaging protocols, MQTT, CoAP, AMQP and HTTP for IoT systems. It conducts a thorough and relative analysis based on certain interrelated criteria in order to better understand their strengths and limitations. Thus, based on this detailed evaluation, the user can decide on his choice [8].

In this article [9], reports the observed performance of the protocols. they quantitatively compare the performances of the IoT protocols, namely MQTT, CoAP, DDS and a UDP-based custom protocol in a medical context. The performance of the protocols was evaluated using an emulator network.

In [10], researchers provide quantitative performance analysis of CoAP and MQTT protocols on various conditions of offered traffic, probability of packet loss and delay (Table 1).

For the paper [11], they are analyzed the performance of communication protocols for Internet of Things platforms such as MQTT, AMQP, CoAP and XMPP. First of all, they present a comparative study of the main protocols that allows the implementation of an IoT platform, to determine their computational load, overhead and network bandwidth. Then they tested this protocol to determine CPU processing, transferred bytes and sent packets for different communication scenarios.

This paper [12], review only the application layer protocols of IoT. Specifically, MQTT, MQTTSN, AMQP, CoAP, XMPP, and DDS data protocols of IoT. They compared these protocols with the challenging issues such as security, caching, resource discovery, support to QoS etc. Then, they analyzed the performance of these protocols with various metrics such as network packet loss rate, message size, bandwidth consumption and latency.

These previous studies are not yet complete and does not provide a good practice of the IoT protocols that will allow the developers to choose the right protocols for such an IoT application. There are several criteria that need to be evaluate.

# 5 A Good Practice of IoT Protocols

## 5.1 Comparative Study of Protocols

In a previous work we presented a comparative study of the different network and application protocols. This research analyzes the main services of the IoT protocols. In particular, it focuses on network and application layer protocols. This study evaluates the capabilities of their main characteristics and behaviors in terms of various metrics of these protocols. The comparison presented in this article would be beneficial for researchers and developers in selecting an appropriate protocol for IoT applications. This study also allows us to have a thorough knowledge of these technologies and protocols and presents a choice of help that is suitable for the implementation of an IOT solution. To simplify the use of the comparative study and the good practice of network and application protocols we have proposed a model thereafter. First of all, we classify the charters of the protocols in the form presented in the following's tables, then we realize automata according to our classification.

a) **Networks Protocols**

We classified the criteria in three categories General Information, Network Services, Data Management Services. Our studies are focused on ten protocols namely: Wifi, Bluetooth, LoRa, Zigbee, Z-Wave, Cellular, NFC, Sigfox, Neul, 6LowPan.

**Table 1.** The criteria of the network protocols

| Service | Description | Criteria |
| --- | --- | --- |
| General Information | General description | Specification, Application, Market Adoption, Cost, Energy needed |
| Network Services | Management of the address format<br>Correspondence of addresses<br>Acknowledgment of receipt<br>The direction of the information flow<br>Sequence control<br>Flow management<br>Routing | Network type, Topology, Power, Network size |
| Data Management Services | The management of the data format<br>Detection of transmission errors<br>Information loss management<br>The direction of the information flow | Technical Modulation (MT), Spread Spectrum (SP), Range (R), Security (S), Risk of data collision (RoC), Data Rate (DT) |

b) **Application Protocols**

We classified the criteria in three categories General Information, Messaging and data transfer services, Error Management Services and Information Loss, Data Management Services. Our studies are focused on ten protocols namely: CoAP, XMPP, RESTful HTTP, WebSocket, AMQP, DDS (Table 2).

**Table 2.** The criteria of the application protocols

| Service | Criteria |
| --- | --- |
| General Information | Standard, Application, Benefits, Disadvantages |
| Messaging and data transfer services | Transport, 2G, 3G, 4G Adequacy (nodes 1000 s) |
| Error Management Services and Information Loss | Security and QoS, Quality factor |
| Data Management Services | Aptitude LLN (nodes 1000 s), Calculates Resources, Architecture, Technologies |

## 5.2 Proposed Models

### a) Proposed Models of Network Protocols

The proposed models allow the good practice of the use of the protocol's networks. We have realized four models according to the services proposed, there are two for long-distance protocols and the others for the small-distance protocols (Fig. 2).

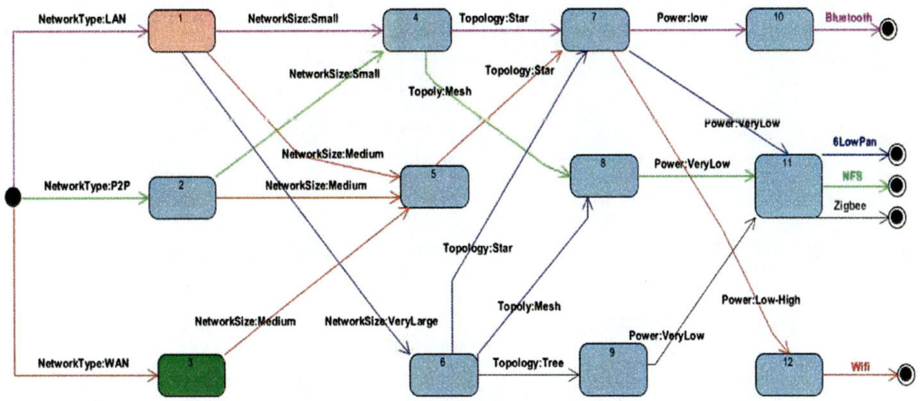


**Fig. 2.** The small-distance protocols: Network Services

This model for the small distance protocols which are: WIFI, Bluetooth, Zigbee, NFC, 6LowPan. The model allows the choice between the protocols based on the exact measurements of the network criteria. the network types are: LAN, WAN and P2P. There are two types of the network size, which are: Medium and small. The types of topology are: Mesh, Tree, Star. The power enters Low, Very Low, Low- High. Bluetooth, NFS and Zigbee have only one path, so only one choice for each criterion.

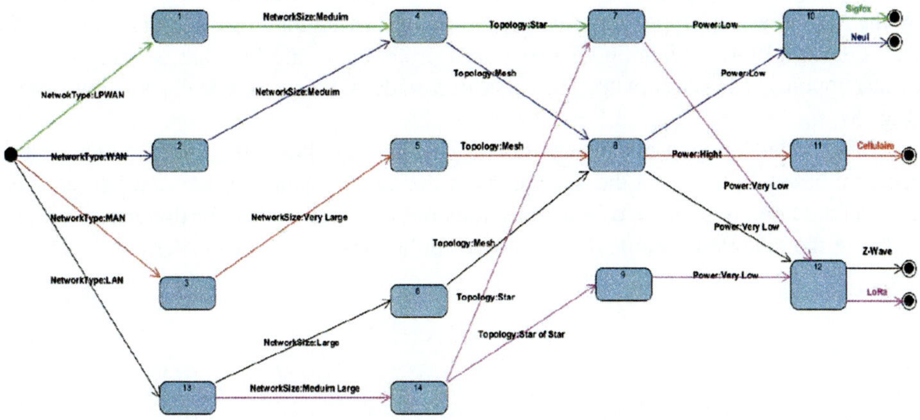


**Fig. 3.** The small-distance protocols: Data Management Services

However, the WIFI, there is the choice between the network types, consequently the WIFI is adaptable with the three networks type. Zigbee is adaptable to all types of topology. To deduce that the WIFI remains the most used protocol in the IoT according to the criteria mentioned (Fig. 3).

This model for small distance protocols. the model allows the choice between the protocols based on the exact measurements of the data management service criteria. The criteria are already mentioned in the previous table. NFS, 6lowPan, Bluetooth, WIFI has only one way to give a single choice of measurements of the criteria, but Zigbee there is the choice of technical moderation between QPSK and BPSK (Fig. 4).

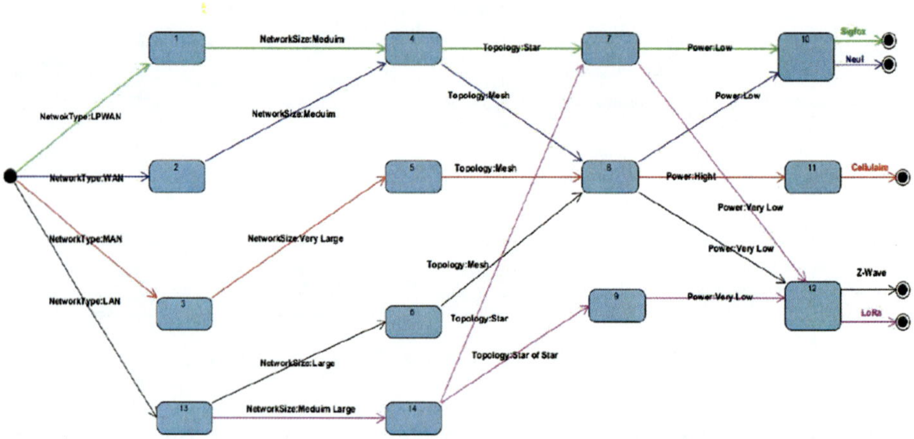


**Fig. 4.** The long-distance protocols: Network Service

This model for protocols at a distance long which are: Sigfox, Neul, Cellular, z-Wave, LoRa. Also, the model allows the choice between the protocols based on the exact measurements of the criteria of the network. For example, the cellular protocol is the best known and used in the world for long distance IoT applications, characterized by the MAN network type, very large network size, mesh topology and high power. As well the Lora Protocol among the most used protocols in the IoT that has star and star of star topology, LAN network size, medium wide network size and very low power (Fig. 5).

This model for long-distance protocols, the model allows the choice between the protocol based on the exact measurements of the data management service criteria. The Criteria are already explained in the previous models, same thing for this model allows to know the adaptable protocol for a basic application of a set of criteria.

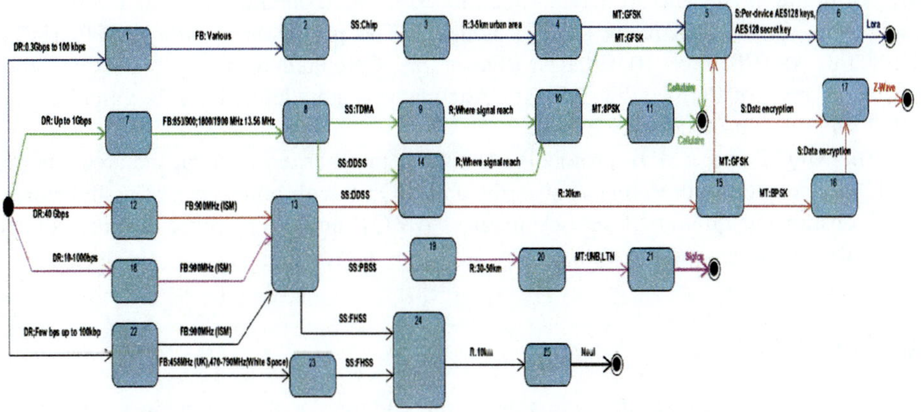


**Fig. 5.** The long-distance protocols: Data Management Services

b) **Proposed Models of Application Protocols**

The proposed models allow the good practice of using protocol applications. This model toward seven the application protocols that are: CoAP, XMPP, Rest-full HTTP, MQTTT, AMQP, DDS and WebSocket (Fig. 6).

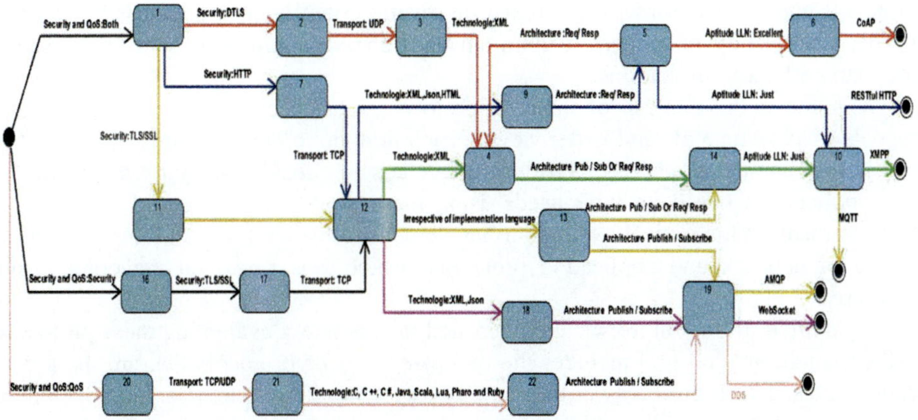


**Fig. 6.** The application protocols

The model for the application protocols that are: CoAP, XMPP, Restful HTTP, MQTTT. This model allows to choose the adaptable protocol for an application. It is based on the criteria that are already listed in the previous table.

This model allows developers and researchers to easily choose the adaptable protocol for an IoT application. It is based on measures of the criteria that are already mentioned in the table preceding of application protocols. There are some criteria that are not in the model and are already mentioned in the table are: "Calculate resources" and "2G, 3G, 4G Adequacy (1000 s nodes)". The protocols CoAP, XMPP, HTTP RESTful, MQTT have 10 Ks RAM/Flash for "Calculate resources", but the other protocols are not defined.2G, 3G, 4G Adequacy is not defined for WebSocket and DDS, the other protocols are excellent.

The MQTT and AMQP protocols are similar in all criteria except the architecture, AMQP is adapted with Publish/Subscribe and MQTT with both types of architectures. DDS ensure the quality of service quality, AMQT and WebSocket ensure security while the others ensure the both.

## 6 Conclusion

This paper has presented an evaluation of the widely and the emerging accepted the application and network protocols for IoT systems. We focused on the application protocols on CoAP, XMPP, Restful HTTP and MQTTT. While we focused on the network protocols on WIFI, Bluetooth, LoRa, Zigbee, Z-Wave, Cellular, NFC, Sigfox, Neul and 6LowPan.

First, we presented the concept of the Internet of Things and its importance in several areas. Subsequently, we presented a classification of all protocols used in each of the IoT architecture. Then there is the related work where we have illustrated the latest studies of researchers who are focused on the assessment of application protocols in most articles. Thus, we presented the results of our comparative study of network protocols and application which is based on several criteria which are classified by the service. This study guarantees a thorough and relative analysis to better understand their strengths and limitations.

To facilitate this analysis of choice of protocols, it has been illustrated using custom models to make an agile and wider view of each protocol relative to other protocols for researchers and developers. As a result, researchers can decide on their relevant use in IoT applications based on their needs using these models that is based on accurate measurements. This critical evaluation has demonstrated a greater and a comparative model of network and application protocols; which was based on the criteria and measures.

In addition, In the future, we are interested in concretely evaluating these protocols using a single model who includes the two categories of protocols that are the application and network protocols in the same IoT system; the researchers and developers can choose at the same time the application and network protocol that are adapted to their needs for the IoT system.

## References

1. Suresh, P., Daniel, J.V., Parthasarathy, V., Aswathy, R.H.: A state of the art review on the Internet of Things (IoT) history, technology and fields of deployment. In: 2014 International Conference on Science Engineering and Management Research (ICSEMR), Chennai, India, pp. 1–8 (2014)

2. Présentation générale de l'Internet des objets [archive] (rapport ITU-T Y.2060), ITU-T Study Group 20, ITU, juin 2012
3. Benghozi, P.-J., Bureau, S., Massit-Folléa, F.: Définir l'internet des ob-jets. In: L'Internet des objets : Quels enjeux pour l'Europe, Éditions de la Maison des sciences de l'homme, Paris, pp. 15–23 (2012)
4. Bujari, M.F., Mandreoli, F., Martoglia, R., Montangero, M., Ronzani, D.: Standards, Security and Business Mod-els Key Chall.pdf, al. (2018)
5. Elhadi, S., Marzak, A., Sael, N., Merzouk, S.: Comparative Study of IoT protocols. SSRN Electron. J. (2018)
6. Darwish, D.G., Square, E.: Improved Layered Architecture for Internet of Things **4**(4), 10 (2015)
7. Bozdogan, Z., Kara, R.: Layered model architecture for internet of things. J. Eng. Res. Appl. Sci. **4**, 5 (2015)
8. Naik, N.: Choice of effective messaging protocols for IoT systems: MQTT, CoAP, AMQP and HTTP. In: 2017 IEEE International Systems Engineering Symposium (ISSE), Vienna, Austria, pp. 1–7 (2017)
9. Chen, Y., Kunz, T.: Performance evaluation of IoT protocols under a constrained wireless access network. In: 2016 International Conference on Selected Topics in Mobile & Wireless Networking (MoWNeT), Cairo, Egypt, pp. 1–7 (2016)
10. Collina, M., Bartolucci, M., Vanelli-Coralli, A., Corazza, G.E.: Internet of Things application layer protocol analysis over error and delay prone links. In: 2014 7th Advanced Satellite Multimedia Systems Conference and the 13th Signal Processing for Space Communications Workshop (ASMS/SPSC), Livorno, Italy, pp. 398–404 (2014)
11. Rizzo, L.: Dummynet: a simple approach to the evaluation of network protocols. ACM SIGCOMM Comput. Commun. Rev. **27**(1), 31–41 (1997)
12. Anusha, M., Babu, E.S., Reddy, L.M., Krishna, A.V., Bhagyasree, B.: Performance analysis of data protocols of internet of things. Int. J. Pure Appl. Math. **115**(6), 37–47 (2017). ISSN 1311-8080 (printed version); ISSN 1314-3395. http://www.ijpam.eu

# A Latency and Energy Trade-Off for Computation Offloading Within a Mobile Edge Computing Server

Youssef Hmimz(✉), Tarik Chanyour, Mohamed El Ghmary, and Mohammed Ouçamah Cherkaoui Malki

LIIAN Laboratory, FSDM, SMBA University, Fez, Morocco
youssef.hmimz@usmba.ac.ma

**Abstract.** The Mobile Edge Computing (MEC) provides leading-edge services to multiple smart mobile devices (SMDs). Besides, computation offloading is a promising service in the 5G networks: it reduces battery drain and applications' execution time. These SMDs generally possess limited battery power and processing capacity. In addition, the local CPU frequency allocated to processing has a huge impact on SMDs energy consumption. In this paper, we consider a multi-user MEC system, where multiple SMDs demand computation offloading to a MEC server. The weighted sum of the overall energy consumptions and latencies represent the optimization problem's objective. In this problem, we jointly optimize offloading decisions, radio resource allocation and local computational resources allocation. The results obtained using our heuristic scheme show that it achieves good performance in terms of energy and latency. Accordingly, its achievement is encouraging compared to both cases where we perform local execution only or complete tasks offloading only.

**Keywords:** Mobile edge computing · Computation offloading · Resource allocation · Meaningful trade-off

## 1 Introduction

In recent years, the digital revolution hugely impacted applications that use mobile Internet. Among these recent applications we find interactive gaming, virtual reality, Internet of Things (IoT), natural language processing, etc. [1]. The applications diversity led smart mobile devices (SMDs) to gradually become the main equipment of people's daily lives. In addition, a large number of IoT terminal devices become practical in various industrial applications. These applications become more and more gourmand in computing resources and energy. However, these SMDs have limited computing capabilities and battery power life. Mobile Edge Computing concept has been recognized as the next generation computing infrastructure that can allocate nearby customized services with requiring of good transmission bandwidth, additional data storage and their processing [2]. It is the future technology that can be the more preferred for latency constrained and computation-intensive SMDs. Moreover, many state of the art studies considered the decision making to offloading computation with


A. Saka et al. (Eds.): CPI 2019, LNME, pp. 490–499, 2021.
https://doi.org/10.1007/978-3-030-62199-5_43

multiple SMDs. They studied resource allocation while they optimize energy [3–7]. In [5], the authors have studied partitioning both radio and computational resources, without taking into account the radio resources insufficiency. On the other hand, the authors of [6, 7] dealt with this problem while they fixed computational resources in advance. However, when radio resources are critical, they penalized some devices by obliging them to execute locally their tasks with expending their latency constraints.

In this paper, we study the computation-offloading problem of several SMDs within one MEC server, and we propose an offloading scheme to investigate the trade-off between energy consumption of SMDs and latency of their tasks. Hence, we formulate an optimization problem and establish the objective function, i.e. the minimum value of weighted sum of mobile energy and time consumption, where we jointly optimize the tasks' offloading selection, and the allocation of both radio and local processing resources. Actually, the radio resources partitioning sub-problem only is NP-hard (see [6]). Accordingly, we propose and evaluate a heuristic scheme named Weighted Minimization of Energy and Latency by Resource Partitioning (WMELRP).

The remainder of this paper is organized as follows. In Sect. 2, we describe the system's model and present the problem's formulation. In Sect. 3, we present our proposed solution. Then, evaluation and result are presented in Sect. 4. Finally, Sect. 5 concludes the paper.

## 2 System Model and Problem Formulation

### 2.1 System Model

Figure 1 shows the system model where it is present N SMDs connected to one base station that is connected with a local MEC server. The set of all SMDs is denoted $\mathbb{G} = \{e_1, e_2, \ldots, e_N\} \triangleq \{1, 2, \ldots, N\}$ and we have $|\mathbb{G}| = N$ . Through a cellular communication system, the base station (BS) is also connected to the Internet. Each SMD $i$ has a unique offloadable task $\tau_i$ that it is described in three terms as: $\tau_i \triangleq \langle d_i, \lambda_i, T_i^{max} \rangle$. The first one denoted $d_i$[bits] and it identifies the amount of the input parameters and program codes to transfer from this SMD to the MEC server. The second one denoted $\lambda_i$ [cycles] specifies the workload referring to the computation amount needed to accomplish the processing of SMD's task. The third one denoted $T_i^{max}$ [seconds] is the maximum latency required by the computation task. In addition, we can use the technics presented at [8, 9] to obtain the desired values of $d_i$ and $\lambda_i$.

In our work, we consider an OFDMA system where the spectrum is divided into $K^T$ equal subchannels of size W [Hz], which are denoted $\mathbb{K} = \{1, 2, \ldots, K\}$, and we consider $K$ as the number of subchannels available at the time decision ($K \leq K^T$). We assume that every SMDs accesses the BS through the OFDMA system where each subchannel in this system is orthogonal to the others. In addition, the achievable uplink rate for SMD $i$ in any assigned subchannel can be written as:

$$r_i = W \log_2\left(1 + \frac{p_i g_i}{W N_0}\right) [5], \tag{1}$$

**Fig. 1.** The system model.

where $W$ is the bandwidth of a subchannel, $p_i$ is the fixed transmit power of SMD $i$ in any allocated subchannel, $g_i$ is its channel gain in one subchannel, and $N_0$ is the noise power spectral density. Moreover, knowing the total allocated subchannels $y_i$ for an SMD $i$, the achievable uplink rate for this SMD can be written as:

$$R_i = y_i r_i = y_i W \log_2\left(1 + \frac{p_i g_i}{WN_0}\right). \tag{2}$$

The execution nature decision for a task $\tau_i$ either locally or by offloading to the edge MEC server is denoted $x_i$ where $x_i \in \{0; 1\}$. $x_i = 1$ that the decision is the offloading of task $\tau_i$ $\tau_i$ to the MEC server, and $x_i = 0$ indicates that $\tau_i$ is executed locally. Additionally, for subchannels allocation, we note the total number of allocated subchannels for SMD $i$ by $y_i$ where $y_i \in [\![0; K]\!]$. $y_i = 0$ indicates that the decision is the local execution of task $\tau_i$, and $y_i \neq 0$ indicates that the decision is the offloading of this task.

Considering to the offloading decision, if the SMD $i$ executes locally its task, the processing time is $t_i^L = \frac{\lambda_i}{f_{i,L}}$ [seconds], where $f_{i,L}$ is a located CPU frequency of SMD $i$. Additionally, its energy consumption is given by: $e_i^L = k_{i,L}.f_{i,L}^2.\lambda_i$[J] [10], where $k_{i,L}$ is the energy coefficient corresponding to the specific chip architecture of SMD $i$. However, if task $\tau_i$ is offloaded to the MEC server, its processing time is expressed by: $t_i^O = t_i^T + t_i^S + t_i^R$, where $t_i^T$ is the transmitting time of this task to the BS, and it is given by $t_i^T = \frac{\mathrm{d}_i}{y_i r_i}$ [seconds]. $t_i^S$ is the time to execute the task $\tau_i$ at the MEC server, and it can be formulated as $t_i^S = \frac{\lambda_i}{f_{i,S}}$ [seconds], where $f_{i,S}$ is the assigned CPU frequency to compute task $\tau_i$ on the MEC server. $t_i^R$ is the time to receive the result out from the BS. Because the data size of the result is usually more ignored compared to the input data size, we ignore this relay time and its energy consumption as adopted by [6]. Hence, for the task $\tau_i$: $t_i^O = \left(\frac{d_i}{y_i r_i} + \frac{\lambda_i}{f_{i,S}}\right)$[seconds]. So, the energy consumption of the communication process can be obtained by multiplying the resulting transmission period by the transmission undertaken power $p_i$. Thus, we obtain: $e_i^T = \frac{p_i d_i}{y_i r_i}[\mathrm{J}]$. Similarly, energy consumption at the MEC server while processing the task $\tau_i$ is given by: $e_i^S = k_{i,S}.f_{i,S}^2.\lambda_i[\mathrm{J}]$, where $k_{i,S}$ is the energy coefficient depending on the specific chip architecture of the MEC Server which is dedicated to SMD $i$.

Now, the energy consumption to offload the task $\tau_i$ is $e_i^O = e_i^T + e_i^S$, and consequently the energy consumption to process this task is given by:

$$E_i = (1 - x_i)e_i^L + x_i\left(e_i^T + e_i^S\right), \tag{3}$$

and the time to execute the task $\tau_i$ is given by:

$$T_i = (1 - x_i)t_i^L + x_i t_i^O. \tag{4}$$

## 2.2 Problem Formulation

In this section, we present the formulation of our optimization problem. Our objective is to minimize the weighted sum energy and latency of all the SMDs. In [9], the authors studied a naive trade-off between energy consumption and latency by using a pair of weight $(\alpha, \beta) \in [0;1]^2$ which balances the energy and latency, where $\alpha$ is the weight associate to the overall energy consumption, $\beta$ is the weight associate to the overall latency, and $\alpha + \beta = 1$. However, neglecting the units of measurement, the values of the energies and the latencies generally belong to disjoint intervals. Hence, a naive trade-off is less meaningful. To obtain a trade-off more meaningful that normalize the overall energy and overall latency, we use specific weights. In view of that, for all devices, we define the offloading decisions' set $x$ given by $x = (x_1, \ldots, x_N)$, the local frequency decisions' set $f_L$ given by $f_L = \left(f_{1,L}, \ldots, f_{N,L}\right)$, and the uplink subchannels allocation decisions' set $y = (y_1, \ldots, y_N)$. The obtained problem is formulated as:

$$\begin{aligned}
& min_{\{x,y,f_L\}} \sum\nolimits_{i \in \mathbb{G}} \{\alpha \frac{E_i}{E^{max}} + \beta \frac{T_i}{T^{max}}\}, \qquad \textbf{(P)}\\
& \text{s.t.}(C_1)\, x_i \in \{0;1\}, i \in \mathbb{G};\\
& (C_2)\, y_i \in [\![0;K]\!], i \in \mathbb{G};\\
& (C_3) \sum\nolimits_{i \in \mathbb{G}} y_i \le K;\\
& (C_4)\, x_i \le y_i \le x_i y_i, i \in \mathbb{G};\\
& (C_5)\, F_{i,L}^{min} \le f_{i,L} \le F_{i,L}^{max}, i \in \mathbb{G};\\
& (C_6)\, (1 - x_i)t_i^L + x_i t_i^O \le T_i^{max}, i \in \mathbb{G};\\
& (C_7)\, (1 - x_i)e_i^L + x_i e_i^T \le E_i^{max}, i \in \mathbb{G}.
\end{aligned}$$

The role of $E^{max} \triangleq \sum_{i \in \mathbb{G}} E_i^{max}$ and $T^{max} \triangleq \sum_{i \in \mathbb{G}} T_i^{max}$ is to normalize the energy and latency for the objective function, and to eliminate their units. Thus, our proposed weights are $\frac{\alpha}{E^{max}}$ and $\frac{\beta}{T^{max}}$.

The sum of all allocated subchannels given by $\sum_{i\in\mathbb{G}} y_i$ in constraint $(C_3)$ must not exceed $K$ that is the total available subchannels. The constraint $(C_4)$ is the mathematical formulation of an important offloading decision property that is $y_i = 0 \Leftrightarrow x_i = 0$, which it must hold for any SMD $i$. Constraint $(C_5)$ indicates that the allocated variable local frequency $f_{i,L}$ for SMD $i$ belongs necessary to a priori fix interval given by $\left[F_{i,L}^{min}, F_{i,L}^{max}\right]$. $(C_6)$ shows that the processing time of the task $\tau_i$ must satisfy latency requirement $T_i^{max}$. The final constraint $(C_7)$ is important especially for battery-critical SMDs. For an SMD $i$, it imposes that the effective consumed energy to execute its task must not exceed the tolerated amount $E_i^{max}$.

## 3 Problem Resolution

In this section, we will introduce how we derive our solution from the obtained optimization problem. We propose our solution called Weighted Minimization of Energy and Latency by Resource Partitioning (WMELRP). In addition, to simplify the description of our solution, we denote $\mathbb{W}_i(x_i; y_i) = \alpha \frac{E_i}{E^{max}} + \beta \frac{T_i}{T^{max}}$.

If the task $\tau_i$ must be executed locally ($x_i = 0$), constraint $(C_6)$ is reformulated as: $f_{i,L} \geq \frac{\lambda_i}{T_i^{max}} \triangleq f_{i,L}^-$, and $(C_7)$ is reformulated as follows: $f_{i,L} \leq \sqrt{\frac{E_i^{max}}{k_{i,L}\lambda_i}} \triangleq f_{i,L}^+$. Besides, taking into account the constraint $(C_5)$, we have:

$$\left(f_{i,L} \geq f_{i,L}^- \; and \; f_{i,L} \geq F_{i,L}^{min}\right) \Rightarrow f_{i,L} \geq \max\left\{f_{i,L}^-; F_{i,L}^{min}\right\};$$

$$\left(f_{i,L} \leq f_{i,L}^+ \; and \; f_{i,L} \leq F_{i,L}^{max}\right) \Rightarrow f_{i,L} \leq \min\left\{f_{i,L}^+; F_{i,L}^{max}\right\}.$$

Knowing that $x_i = 0 \Leftrightarrow y_i = 0$, we have $\mathbb{W}_i(0;0) = \frac{\alpha k_{i,L}\lambda_i}{E^{max}} f_{i,L}^2 + \frac{\beta\lambda_i}{T^{max}} \frac{1}{f_{i,L}}$, and we obtain $\frac{\partial \mathbb{W}_i(0;0)}{\partial f_{i,L}} = \frac{\lambda_i\left(2\alpha k_{i,L} T^{max} f_{i,L}^3 - \beta E^{max}\right)}{E^{max} T^{max} f_{i,L}^2}$. Moreover, we note that $2\alpha k_{i,L} T^{max} f_{i,L}^3 - \beta E^{max} = 0 \Rightarrow f_{i,L} = \sqrt[3]{\frac{\beta E^{max}}{2\alpha k_{i,L} T^{max}}}$, and $\frac{\partial\left(2\alpha k_{i,L} T^{max} f_{i,L}^3 - \beta E^{max}\right)}{\partial f_{i,L}} = 6\alpha k_{i,L} T^{max} f_{i,L}^2 > 0$.

Hence, $\mathbb{W}_i(0;0)$ is strictly decreasing w.r.t. the variable $f_{i,L}$ over the $\left]0; \sqrt[3]{\frac{\beta E^{max}}{2\alpha k_{i,L} T^{max}}}\right]$ interval, and it is strictly increasing over the $\left[\sqrt[3]{\frac{\beta E^{max}}{2\alpha k_{i,L} T^{max}}}; +\infty\right[$ interval. Therefore, we can derive the following optimum value $f_{i,L}^*$ given by:

$$f_{i,L}^* = \begin{cases} \emptyset \quad if \quad f_{i,L}^- > F_{i,L}^{max} \; \mathbf{or} \; f_{i,L}^+ < F_{i,L}^{min} \; \mathbf{or} \; f_{i,L}^- > f_{i,L}^+ \\ \min\left\{f_{i,L}^+; F_{i,L}^{max}\right\} \; if \; \sqrt[3]{\frac{\beta E^{max}}{2\alpha k_{i,L} T^{max}}} \geq \min\left\{f_{i,L}^+; F_{i,L}^{max}\right\} \\ \max\left\{f_{i,L}^-; F_{i,L}^{min}\right\} \; if \; \sqrt[3]{\frac{\beta E^{max}}{2\alpha k_{i,L} T^{max}}} \leq \max\left\{f_{i,L}^-; F_{i,L}^{min}\right\} \\ \sqrt[3]{\frac{\beta E^{max}}{2\alpha k_{i,L} T^{max}}} \quad otherwise \end{cases} \tag{5}$$

If task $\tau_i$ is offloaded ($x_i = 1$), we reformulate $(C_6)$ as follows: $y_i \geq \frac{d_i}{r_i\left(T_i^{max} - \frac{\lambda_i}{f_{i,S}}\right)}$. Thus, to satisfy $(C_6)$, the minimum number of subchannels to allocate to SMD $i$ is: $y_i^{m1} \triangleq \left\lceil \frac{d_i}{r_i\left(T_i^{max} - \frac{\lambda_i}{f_{i,S}}\right)} \right\rceil$, where $\lceil . \rceil$ is the ceil function. In addition, $(C_7)$ can be reformulated as follows: $y_i \geq \frac{p_i d_i}{r_i E_i^{max}}$. So, to satisfy $(C_7)$, the minimum allocated subchannels number that SMD $i$ must have is: $y_i^{m2} \triangleq \left\lceil \frac{p_i d_i}{r_i E_i^{max}} \right\rceil$. As a result, the minimum attributable subchannels for SMD $i$ to offload its task is:

$$\underline{y_i} \triangleq \max\left\{y_i^{m1}; y_i^{m2}\right\}. \tag{6}$$

In this work, we suppose that $K$ represent the number of available subchannels is sufficient to satisfy all SMDs, and we set $K = \sum_{i \in \mathbb{G}} \underline{y_i}$.

For each offloadable task $\tau_i (x_i = 1 \Leftrightarrow y_i > 0)$, we have $\mathbb{W}_i(1; y_i) = \frac{1}{y_i}\left(\frac{\alpha p_i d_i}{r_i E^{max}} + \frac{\beta d_i}{r_i T^{max}}\right) + \frac{k_{i,S} \cdot f_{i,S}^2 \cdot \lambda_i}{E^{max}} + \frac{\beta \lambda_i}{f_{i,S} T^{max}}$, and $\mathbb{W}_i(1; y_i)$ is strictly decreasing w.r.t. the variable $y_i$. Moreover, for a large value of $\frac{\alpha p_i d_i}{r_i E^{max}} + \frac{\beta d_i}{r_i T^{max}}$, to decrease the value of $\mathbb{W}_i(1; y_i)$, the value of $y_i$ must be large enough (because the second part $\frac{k_{i,S} \cdot f_{i,S}^2 \cdot \lambda_i}{E^{max}} + \frac{\beta \lambda_i}{f_{i,S} T^{max}}$ is independent of $y_i$). So, in order to minimize our objective function, for each offloadable task $\tau_i$, $\frac{\alpha p_i d_i}{r_i E^{max}} + \frac{\beta d_i}{r_i T^{max}}$ represents an influential weight to estimate the optimum value of $y_i$. Thus, an approximate value of $y_i$ that can minimize our objective function is given by:

$$\tilde{y}_i = \frac{\frac{\alpha p_i d_i}{r_i E^{max}} + \frac{\beta d_i}{r_i T^{max}}}{\sum_{\substack{i \in G \\ x_i = 1}} \left(\frac{\alpha p_i d_i}{r_i E^{max}} + \frac{\beta d_i}{r_i T^{max}}\right)} K. \tag{7}$$

From the above results, we designed the following Algorithm 1 to classify tasks and distribute subchannels.

In this algorithm, to ensure that constraints $(C_6)$ and $(C_7)$ are satisfied and to minimize our objective function value, we try to allocate $\max\left\{\underline{y_i}; [\tilde{y}_i]\right\}$ subchannels to an SMD $i$ to which they ensure that the offloading of its task is favored over the local execution (the other SMDs must be execute their tasks locally). If some tasks must be executed locally, then, for an offloadable task $\tau_i$, the $\tilde{y}_i$'s recalculated value is greater than the $\tilde{y}_i$'s initial value (because the denominator of the $\tilde{y}_i$'s recalculated value is less than the denominator of the $\tilde{y}_i$'s initial value). So, the $z_i$ value is greater than the $[\tilde{y}_i]$'s initial value. Thus, the value of $\max\left\{\underline{y_i}; z_i\right\}$ is greater than the $\max\left\{\underline{y_i}; [\tilde{y}_i]\right\}$'s initial value. Hence, the offloading of task $\tau_i$ is more favored over its local execution.

According to (7) we have $\sum_{\substack{i \in G \\ x_i = 1}} \tilde{y}_i = K$, generally $\sum_{\substack{i \in G \\ x_i = 1}} \lfloor \tilde{y}_i \rfloor < K$, and with final values of $z_i$ (after line 17: $z_i = \lfloor \tilde{y}_i \rfloor$ or $z_i = \lfloor \tilde{y}_i \rfloor + 1$) we have $\sum_{\substack{i \in G \\ x_i = 1}} z_i = K$. But, we can have $\sum_{\substack{i \in G \\ x_i = 1}} \max\left\{\underline{y_i}; z_i\right\} > K$, i.e., $\sum_{\substack{i \in G \\ x_i = 1}} y_i > K y_i > K$ and the constraint $(C_3)$ is not satisfied. In this case, from the group of SMDs that can execute their tasks locally ($f^*_{i,L} \neq \varnothing$) and they have $\underline{y_i} > z_i$, we private some of them of their subchannels and they execute their tasks locally. If the constraint $(C_3)$ is still not satisfied, for some SMDs with $f^*_{i,L} = \varnothing$, we reduce their allocated subchannels $y_i$ to $z_i$.

---

**Algorithm 1**: Weighted Minimization of Energy and Latency by Resource Partitioning (WMELRP)

---

**Initialization:** $x_i = 1, \forall i \in \mathbb{G}; Y = 0; K_I = K;$
1: **for** each $i \in \mathbb{G}$ **do**
2: Calculate $f^*_{i,L}$ according to (5), $\underline{y_i}$ according to (6), and $\tilde{y}_i$ according to (7);
3: **end for**
4: **for** each $i \in \mathbb{G}$ **do**
5: **if** $f^*_{i,L} \neq \emptyset$ **then**
6: $f_{i,L} = f^*_{i,L};$
7: **if** $\mathbb{w}_i(0;0) < \mathbb{w}_i\left(1; \max\left\{\underline{y_i}; \lfloor \tilde{y}_i \rfloor\right\}\right)$ **then**
8: $x_i = 0; y_i = 0;$
9: **end if**
10: **end if**
11: **end for**
12: **for** each $i \in \mathbb{G}$ with $x_i = 1$ **do**
13: Recalculate $\tilde{y}_i$ according to (7); $z_i = \lfloor \tilde{y}_i \rfloor$; $K = K - z_i$;
14: **end for**
15: **while** $K > 0$ **do**
16: $i_0 = \operatorname{argmax}_{i \in \mathbb{G} / x_i = 1}\{\tilde{y}_i - z_i\}$; $z_{i_0} = z_{i_0} + 1$; $K = K - 1$;
17: **end while**
18: **for** each $i \in \mathbb{G}$ with $x_i = 1$ **do**
19: $y_i = \max\left\{\underline{y_i}; z_i\right\}$; $Y = Y + y_i$;
20: **end for**
21: **while** $Y > K_I$ **do**
22: $i_1 = \operatorname{argmax}_{i \in \mathbb{G} / x_i = 1 \,\&\, f^*_{i,L} \neq \emptyset}\left\{\underline{y_i} - z_i\right\}$; $Y = Y - y_{i_1}$; $x_{i_1} = 0$; $y_{i_1} = 0$;
23: **end while**
24: **while** $Y > K_I$ **do**
25: $i_2 = \operatorname{argmax}_{i \in \mathbb{G} / x_i = 1 \,\&\, f^*_{i,L} = \emptyset}\left\{\underline{y_i} - z_i\right\}$; $Y = Y - \left(\underline{y_{i_2}} - z_{i_2}\right)$; $y_{i_2} = z_{i_2}$;
26: **end while**

---

## 4 Evaluation

The presented results in this work are averaged for 100 time executions. We implement our WMELRP solution using C++ language. For each SMD $i$, the number of CPU cycles $\lambda_i$ required by the computing task is randomly selected between the values 500 and 2500 Megacycles. The size of the data $d_i$ required by the task is randomly selected between the values 1000 and 3000 KB. The deadline $T_i^{max}$ is uniformly defined from 0.5 s to 3 s. The achievable uplink rate $r_i$ takes values between the values 700 KB/s and 800 KB/s. The local CPU frequency $f_{i,L}$ will be optimized between the values $F_{i,L}^{min} = 1\text{MHz}$ and $F_{i,L}^{max} = 60\text{MHz}$. The CPU frequency of the MEC server set to be $F_{i,S} = 6\,\text{GHz}$. The transmission power is set to be $p_i = 0.1$ Watt. For the energy efficiency coefficients, we set $k_{i,L} = 10^{-26}$ and $k_{i,S} = 10^{-29}$. We choose that 20% of the SMDs have the threshold energy $E_i^{max}$ is uniformly chosen in $[0.5, 0.95] * \lambda_i.k_{i,L}.\left(F_{i,L}^{max}\right)^2$, and the others SMDs are $E_i^{max}$ is uniformly chosen in $[1.05, 5] * \lambda_i.k_{i,L}.\left(F_{i,L}^{max}\right)^2$. Finally, we take $(\alpha, \beta) = (0.3, 0.7)$, and we vary the number of SMDs $N$ between 2 and 20.

As the performance study of the proposed solution in [7, 11] with three situations (in MEC context), Fig. 2 shows a comparison between the weighted sum cost of energy and latency in the following three situations: 1) all tasks are offloaded to the MEC server. 2) the offloading decision is done using our WMELRP solution. 3) all tasks are executed locally, in this case we ignore the local constraints and we set $f_{i,L} = \sqrt[3]{\frac{\beta E^{max}}{2\alpha k_{i,L} T^{max}}}$.

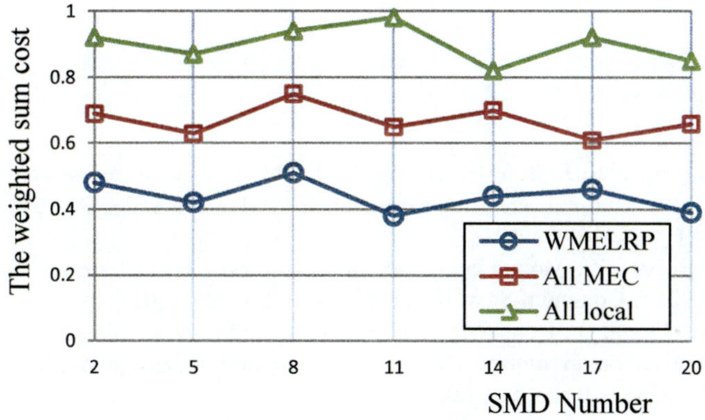


**Fig. 2.** The weighted sum cost of latencies and energy consumptions under different number of SMDs and with $(\alpha, \beta) = (0.3, 0.7)$.

From Fig. 2, we first notice that, for the three execution situations with N varies from 2 to 20, the values of the weighted sum cost do not exceed 1. This comes back to the divisions by $E^{max}$ and $T^{max}$ in the expression of the objective function. Then, we notice that for the local execution of all the tasks, the weighted sum cost varies between 0.82 and 0.94. For the offloading of all the tasks, the weighted sum cost varies between 0.61 and 0.75, and when the offloading decision is done using our WMELRP solution, the weighted sum cost varies between 0.38 and 0.51. In addition, our proposed solution saves between 45% and 61% of weighted sum cost compared to the local execution of all tasks, and it saves on average 51%. Also and compared to offloading all tasks, it saves between 24% and 41% weighted sum cost, and it saves on average 34%. Therefore, we show that our proposed heuristic scheme efficiently manages the offloading decisions.

## 5 Conclusion

In this paper, we considered a one-cell Mobile Edge Computing system serving several smart mobile devices (SMDs). Each SMD has a unique task that can be offloaded for processing to a MEC server. For this system, we studied computation offloading and resources allocation with a Latency/Energy Trade-off. We formulated an optimization problem to minimize the weighted sum of energy consumption and tasks' processing time. To solve this problem more effectively, we proposed a heuristic solution called Weighted Minimization of Energy and Latency by Resource Partitioning (WMELRP). Numerical results showed that our WMELRP solution provides good performance compared to both local execution and complete tasks offloading. As a future work, we plan to generalize our study to the multi-sever MEC case with considering networks state, the availability of radio resources, devices mobility, and wireless communication interference.

## References

1. Fu, Z., Ren, K., Shu, J., Sun, X., Huang, F.: Enabling personalized search over encrypted outsourced data with efficiency improvement. IEEE Trans Parallel Dist Syst. **27**(9), 2546–2559 (2016)
2. Mach, P., Becvar, Z.: Mobile edge computing: a survey on architecture and computation offloading. IEEE Commun. Surv. Tutor. **19**(3), 1628–1656 (2017)
3. Chen, M.H., Liang, B., Dong, M.: Joint offloading and resource allocation for computation and communication in mobile cloud with computing access point. In: Proceedings of IEEE INFOCOM, Atlanta, GA (2017)
4. Chen, M.H., Liang, B., Dong, M.: Joint offloading decision and resource allocation for multi-user multi-task mobile cloud. In: Proceedings of the IEEE International Conference on Communications (ICC), pp. 1–6 (2016)
5. Zhao, P., Tian, H., Qin, C., Nie, G.: Energy-saving offloading by jointly allocating radio and computational resources for mobile edge computing. IEEE Access **5**, 11255–11268 (2017)
6. Zhang, K., et al.: Energy-efficient offloading for mobile edge computing in 5g heterogeneous networks. IEEE Access **4**, 5896–5907 (2016)

7. Hmimz, Y., Chanyour, T., El Ghmary, M., Malki, M.O.C.: Energy efficient and devices priority aware computation offloading to a mobile edge computing server. In: 5th International Conference on Optimization and Applications (ICOA), pp. 1–6. IEEE (2019)
8. Chun, B., Ihm, S., Maniatis, P., Naik, M., Patti, A.: Clonecloud: elastic execution between mobile device and cloud. In: Proceedings of the ACM Sixth Conference on Computer Systems, Salzburg, Austria, pp. 301–314 (2011)
9. Yang, L., Cao, J., Yuan, Y., Li, T., Han, A., Chan, A.: A framework for partitioning and execution of data stream applications in mobile cloud computing. ACM SIGMETRICS Perform. Eval. Rev. **40**(4), 23–32 (2013)
10. Chen, X., Jiao, L., Li, W., Fu, X.: Efficient multi-user computation offloading for mobile-edge cloud computing. IEEE/ACM Trans. Netw. **24**(5), 2795–2808 (2016)
11. Ning, Z., Dong, P., Kong, X., Xia, F.: A cooperative partial computation offloading scheme for mobile edge computing enabled internet of things. IEEE Internet Things J. **6**(3), 4804–4814 (2018)

# Monitoring of Production Systems Using Artificial Intelligence Tools

Wafi Morad(✉) and Gziri Hassan

IMMII Laboratory, FST, Hassan I University, Settat, Morocco
morad.wafi@gmail.com, hgziri@gmail.com

**Abstract.** Monitoring of industrial systems is necessary to ensure reliability and long-term availability of equipments.

The overall purpose of this study is to identify and evaluate the use of the artificial intelligence (AI) tools in monitoring complex industrial systems and more precisely the artificial neural networks (ANN) or connectionist systems that stand out from the rest by their capacity of learning and solving problems in the same way that a human brain would.

This paper is a state of the art of monitoring techniques and neural architectures the most used in static and dynamic monitoring.

This work is divided into two main parts: in the first one we will present a comparative study of the different architectures of neuron networks that are most used in static monitoring such as RBF and MLP. This comparison led to the choice of RBF.

Since dynamic monitoring (dynamic detection, prognosis) requires taking into account the temporal aspect; we devoted the second part to dynamic neural networks (temporal). The tests done using a dynamic detection benchmark and a prognostic benchmark allowed us to choose RRBF as the optimal neural configuration for dynamic monitoring.

**Keywords:** ANN · Detection · Prognostic · RBF · RRBF

## 1 Introduction

Today in an industrial environment increasingly characterized by a fierce competitiveness, now the company should satisfy the requirements: Quality, cost, time. To maintain this balance, it must ensure the reliability, availability and safety of the facilities. Hence the need to put a very good maintenance policy.

Manufacturers tend to replace traditional maintenance based on equipment redundancy with predictive maintenance that has become wide-spread. The goal of predictive maintenance is to reduce downtime and cost of maintenance under the premise of zero failure manufacturing through monitoring the working condition of equipment and predicting when equipment failure might occur. Prediction for future potential fault allows maintenance to be planned before the fault happens. Ideally, maintenance schedule can be optimized to minimize the cost of maintenance and achieve zero failure manufacturing [1].


A. Saka et al. (Eds.): CPI 2019, LNME, pp. 500–515, 2021.
https://doi.org/10.1007/978-3-030-62199-5_44

This methodology is mainly based on monitoring, to check the operating status of the system. It consists of two parts that are detection and diagnosis.

In the literature there are two methods of monitoring: with model and without model. But the complexity of industrial processes makes obtaining a model very difficult or impossible, so we will rely on monitoring by artificial intelligence (AI).

Artificial neural networks (ANNs) are mathematical tools with intensive utilization in the resolution of many real-world complex problems, especially in classification and prediction ones. ANNs pretend to emulate biological human neural networks learning from the experience and generalizing previous behaviors [2].

In industry ANN can be used in conventional monitoring (detection, static diagnostic), and dynamic monitoring

This work will be organized as follows: We start with some definition and a state of the art of monitoring techniques. The second part is devoted to the operating principles of Neural Networks (a strong AI tool that differs from other tools in its ability to learn), and to the main ANN architectures used for classical monitoring. In the third part we address the temporal aspect in neuron networks, and we present an application by which we will compare the neural architectures used in predictive detection and dynamic monitoring.

## 2 Industrial Monitoring

### 2.1 Classical Monitoring

Monitoring includes detecting and classifying faults by observing the evolution of the system, then diagnosing them by locating failures and identifying root causes [3]. Monitoring has two main functions: detection and diagnostic (see Fig. 1).

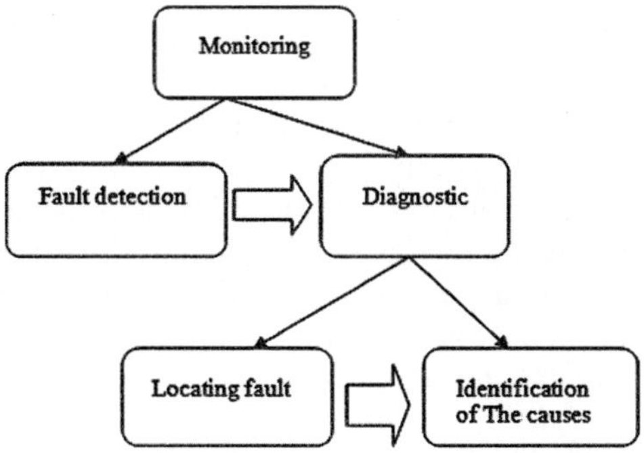


**Fig. 1.** Components of industrial monitoring.

**Detection:** Detection of system failures requires the ability to classify observable situations as normal or unusual.

**Diagnostic:** In order to continuously monitor the evolution of a system and determine the causes of failures, it would therefore seem interesting to make the best use of all available information, whether digital (sensor data) or symbolic (historical, environments, repairs made, etc.).

The purpose of the diagnostic function is to investigate the causes and locate the organs that have led to a particular observation.

From the observation of a state of failure, the diagnostic function is responsible for finding the fault that is at the origin [3]. This problem is difficult to solve. If, for a given fault, it is easy to predict the resulting fault, but the reverse approach of identifying the fault from its effects is complicated (see Fig. 2).

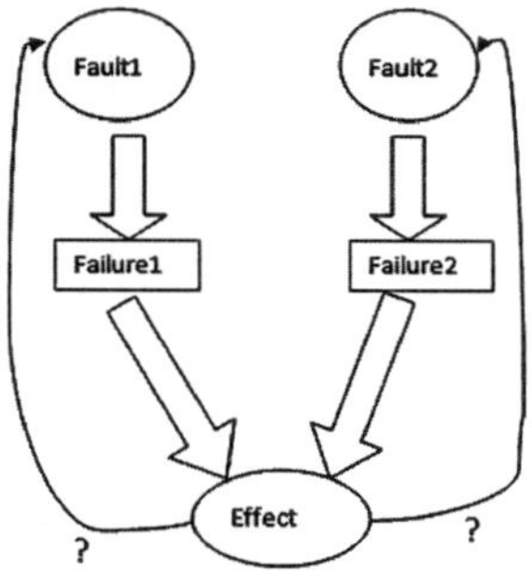


**Fig. 2.** Difficulty of diagnostic: two faults led to the same effect.

### 2.2 Dynamic Monitoring

In recent years, with the increasing demand for machining quality and manufacturing complexity, the complexity and integration of industrial equipment has been increasing drastically. On the one hand, an unexpected failure can result in a devastating accident and financial losses for the company owing to the interaction behaviors among industrial equipments. On the other hand, early detection and prediction of a fault can prevent it from growing and eventually turning into critical problems. Hence, increasing attention has been paid to prognosis in modern industry [4].

As with conventional monitoring, predictive monitoring is a passive, informational device that analyses the present and past state of the system and provides indicators of future trends in the system's evolution. Predictive monitoring consists of: predictive (dynamic) detection and predictive diagnostic, also known as prognostic [3].

**Predictive detection** is the prediction of future failure. In other words, the purpose of predictive detection is to detect degradation instead of failure [5]. That enables the planning of maintenance before the fault happens.

**Predictive diagnosis** identifies the causes and the location of the organs responsible for a particular degradation. If the performances are below the threshold defined in the equipment's functional specification, there is no further degradation but dysfunction. This operation is based either on a priori knowledge of the laws of the evolution of aging phenomena, or on a learning phase. According to ISO 13381 (2004), the prognosis is the estimation of the operating time before failure and the risk of existence or appearance of one or many dysfunctions [6].

In terms of reliability, the lifetime of a system is called remaining useful life (RUL). This approach addresses the issue of predictive maintenance of systems and components in order to reduce costs [7].

The Fig. 3 summarizes the difference between classic and dynamic monitoring.

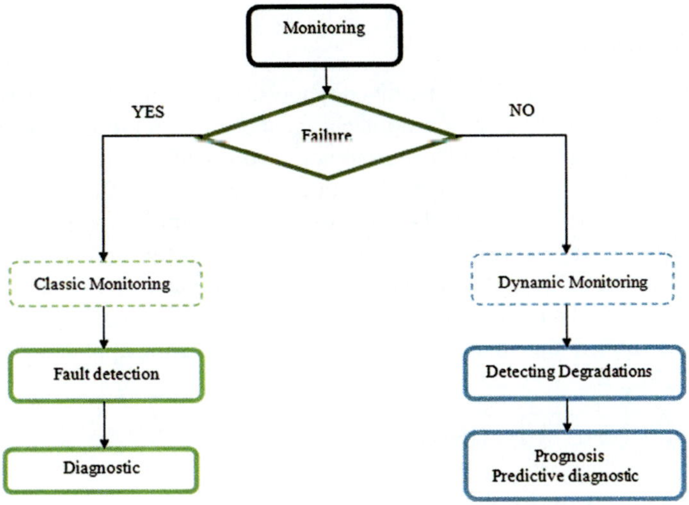


**Fig. 3.** Classic and dynamic monitoring.

After presenting the basic elements of monitoring covered by our contributions, we present an overview of the main techniques related to industrial monitoring.

## 2.3 Monitoring Techniques

The monitoring methods can be classified in two categories: model-based monitoring methodologies and without any model monitoring. The first class contains essentially control system techniques based on the difference between the system model's outputs and the equipment's output [20]. But to obtain a formal model especially for complex equipments is very difficult or impossible that present the major disadvantage of these techniques. The second class of monitoring techniques is more interesting and not sensitive to this problem. These techniques can be subdivided into subclasses: the probabilistic ones and the Artificial Intelligence ones.

For the AI techniques, a database covering some system parameters is required. This provides the adaptability learning process in order to give certain performance to the monitoring system [21].

### 2.4 Monitoring by IA

The use of artificial intelligence makes it possible to overcome the complexity of the diagnosed systems.

AI can be characterized by the ability to treat:

- a lot of information,
- non-homogeneous data (digital/symbolic),
- context-dependent data,
- Incomplete data [8].

In the case of using artificial intelligence tools, monitoring function is often considered only as a form recognition application; the forms represent the input vector composed of the equipment's data (measurable and qualifiable data), and the classes represent the different modes of operation and/or dysfunction [9].

Artificial neural networks (ANNs) seem to be very promising prognostic tools: they draw examples and try to capture the subtle relationship between data. These are efficient calculation techniques that are perfectly suited to practical problems, where it is easier to collect data than to formalize the behavior of the system being studied. Current developments confirm the value of using ANNs in forecasting applications [10].

## 3 Artificial Neural Network (ANN)

### 3.1 Artificial Neuron Model

The artificial neural networks can be characterized as computational models, with particular properties, whose main characteristics can be defined as intrinsic ability for parallel operation; no need for knowledge of possible mathematical models that describe the behavior of a given application; systems inspired by the human brain; and ability to learn from experience [11].

Basic building block of every ANN is artificial neuron (see Fig. 4).

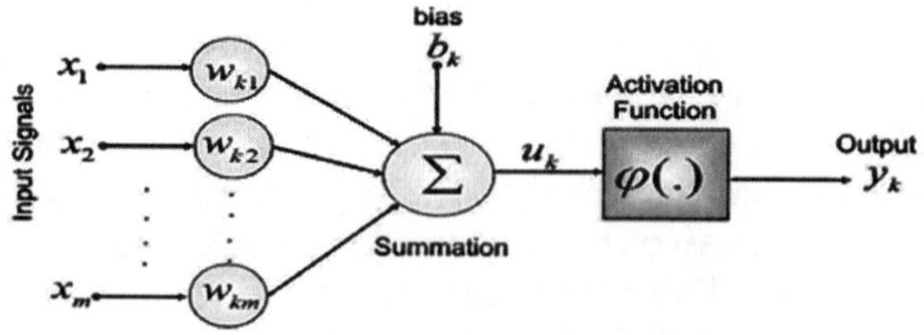


**Fig. 4.** Artificial neuron model (Perceptron).

Each node in a neural network is a processing unit which contains a weight ($w_{ki}$) and summing function followed by a transfert function ($\varphi$) (see Fig. 5) [12].

The summer output, referred to as the total synaptic input of the neuron, is given by the inner product of the input and weight vectors [12]. The output value of the neuron is given by:

$$y_k = \varphi\left(\sum x_i * w_{ki} + b_k\right) \tag{1}$$

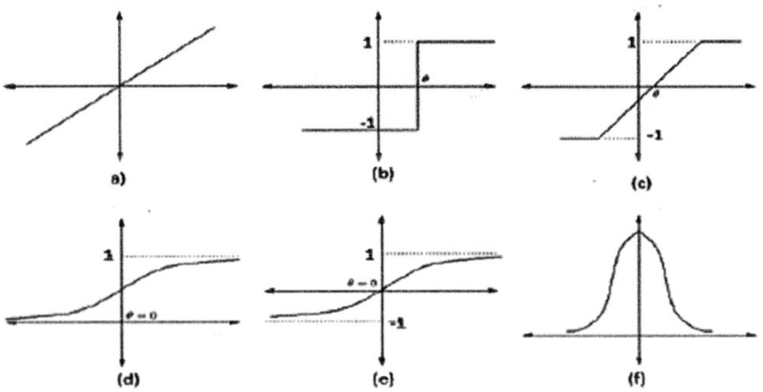

**Fig. 5.** Most activation functions used are: a) The Linear function, b) The signum function, c) The Piece-wise linear function, d) The sigmoid function, e) The Hyperbolic tengent function, f) The Gaussian function.

ANN can be classified into two main categories:

- **Static networks** (feed-forward networks): Time is not a significant parameter and the modification of the input only results in a stable modification of the output and does not lead to feedback to this input. These architectures are the most commonly used in diagnostics.
- **Dynamic networks** (or loops): These networks are characterized but the presence of at least one feedback loop at the neurons or between layers, and consider the temporal aspect of the phenomenon.

### 3.2 ANN Learning

Learning is one of the key features of intelligence that distinguishes ANN from other tools.

Before an ANN can be used to perform any desired task, it must be trained to do so. Basically, training is the process of determining the arc weights which are the key elements of an ANN. The knowledge learned by a network is stored in the arcs and nodes in the form of arc weights and node biases [13].

There are two main types of learning: supervised learning and unsupervised learning.

**Supervised Learning:** The Supervised learning is where you have input variables (x) and an output variable (Y) and you use an algorithm to learn the mapping function from the input to the output.

The goal is to approximate the mapping function so well that when you have new input data (x) that you can predict the output variables (Y) for that data.

It is called supervised learning because the process of algorithm learning from the training dataset can be thought of as a teacher supervising the learning process (see Fig. 6). We know the correct answers; the algorithm iteratively makes predictions on the training data and is corrected by the teacher. Learning stops when the algorithm achieves an acceptable level of performance [14].

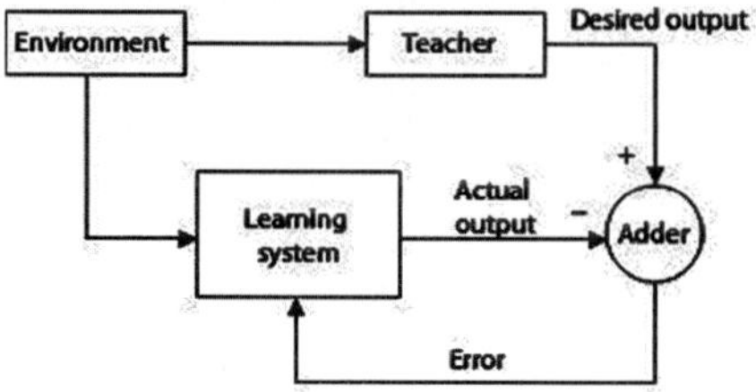


**Fig. 6.** Supervised learning.

**Unsupervised Learning:** Self-Organizing neural networks learn using unsupervised learning algorithm to identify hidden patterns in unlabelled input data. This unsupervised refers to the ability to learn and organize information without providing an error signal to evaluate the potential solution (see Fig. 7). The lack of direction for the learning algorithm in unsupervised learning can sometime be advantageous, since it lets the algorithm to look back for patterns that have not been previously considered [15].

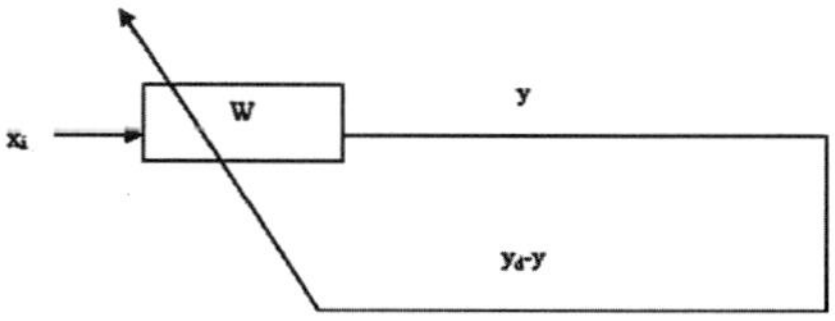


**Fig. 7.** Unsupervised learning.

**Hybrid Learning:** It is a fusion between the two previous models where the network is divided into two sections: a section where the weights are determined by the supervised learning and a part by the unsupervised learning.

## 4 Static Monitoring by ANN

IN synthesis of [5] the most used static neural architectures in the static monitoring are:

- Multi-layer perceptron (MLP).
- Network with radial basic function (RBF).

### 4.1 Multilayer Perceptron (MLP)

MLP is a feed-forward artificial neural network. It consists of at least three layers of nodes (see Fig. 8).

An MLP is typically composed of several layers of nodes. The first or the lowest layer is an input layer where external information is received. The last or problem solution is obtained. The input layer and output layer are separated by one or more inter- mediate layers called the hidden layers. The nodes in adjacent layers are usually fully connected by acyclic arcs from a lower layer to a higher layer. Figure 8 gives an example of a fully connected MLP with one hidden layer [13].

The activation function of each neuron is the sigmoid function [16].

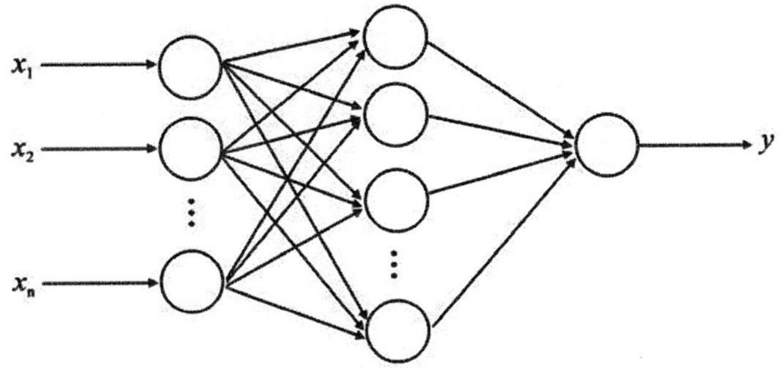


**Fig. 8.** Structure of a MLP.

### 4.2 Radial Function Based Networks (RBF)

The RBF neural network is a classical three-layered feed-forward neural network, with a cascade structure including input, hidden, and output layers. RBFs are used as the hidden neurons, producing nonlinear mappings of the input vector. The outputs of the hidden layer are linearly weighted in order to calculate the final outputs [17]. Each cell in the hidden layer uses a radial-based hi-function (see Fig. 9).

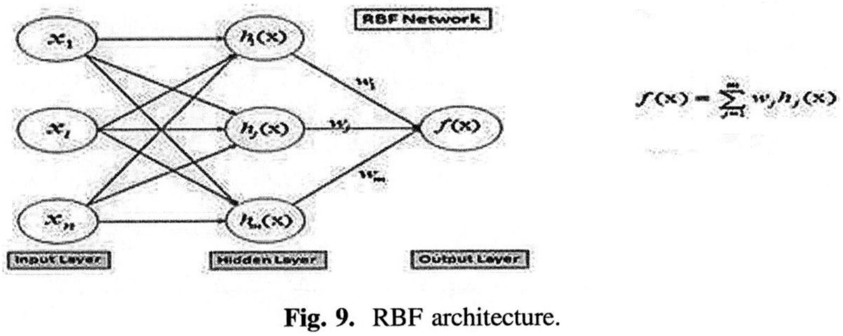


**Fig. 9.** RBF architecture.

These basis functions are normally chosen to be Gaussians and the output activity of the jth hidden unit is then given by:

$$h_j(x) = \exp\left(-\frac{\|x_i - c_j\|^2}{2\sigma_j^2}\right) \quad (2)$$

where $x_i$ is the input pattern, $c_j$ is the prototype pattern or centre for the jth unit and $\sigma_j$ is a scalar width parameter for that unit (giving a radially symmetric response for each Gaussian kernel).

### 4.3 Comparison Between RBF and MLP

Based on the work [3, 5, 16] which compared the different architectures of neural networks, we have summarized in the following table (see Table 1) the advantages and disadvantages of the two neural architectures most used in static monitoring.

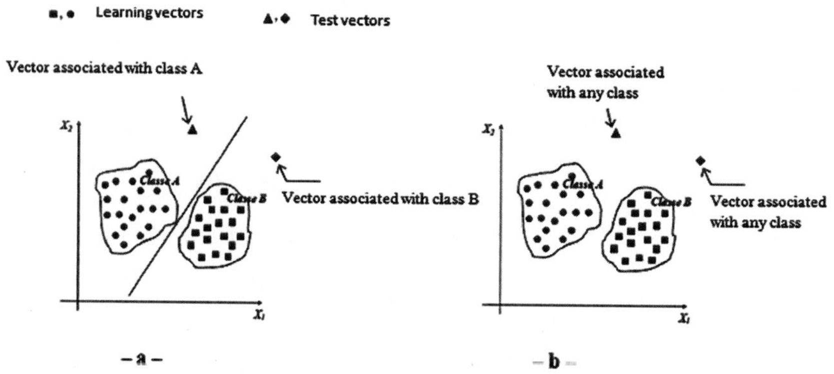


**Fig. 10.** Difference in generalisation capabilities between MLP (a) and RBF (b).

**Table 1.** Comparaison between MLP AND RBF.

| ANN | Description | Learning | Advantage | Weakness |
|---|---|---|---|---|
| MLP | – At least 3 layers<br>– Activation function in each layer: Sigmoid | Supervised | – Accepts muted data and non-linear classification<br>– Global representation of space (All neurons in the network participate in the output of the network)<br>– Simple Architecture | – Classifies the elements that do not belong to any class in the nearest class (see Fig. 10)<br>– The number of hidden layers and neurons in the hidden layers is indefinite |

(*continued*)

**Table 1.** (*continued*)

| ANN | Description | Learning | Advantage | Weakness |
|---|---|---|---|---|
| RBF | – 3 layers<br>– hidden layer uses a radial-based hi-function | Hybrid | – Accepts muted data and non-linear classification<br>– Ability to say «I don't know»<br>– Local representation of space (some of the neurons participate in to the network output)<br>– Great precision | – Complex learning<br>– Requires large computing capacity |

One of the advantages is that unlike the MLP, the RBF are able to say "I don't know". This RBF characteristic is very important in industrial monitoring.

RBF attract the attention of the human expert on a new mode encountered while the MLP tends to give a wrong answer and the expert will never know that a new mode has been encountered.

## 5 Dynamics Monitoring by ANN

Predictive detection and prognosis are the essential functions of dynamic monitoring. These two functions require taking into account the notion of time.

In order to explicitly take the time the recurrent networks have been developed.

The authors of [18] extracted three recurrent neural networks for dynamic surveillance and prognostic applications:

- Recurrent Radial Basis Function (RRBF)
- R2BF
- Dynamic General Neural Network (DGNN).

### 5.1 Recurrent Radial Basis Function (RRBF)

The RRBF is a hierarchical neural network inspired by the conventional radial basis function networks (RBF) [19], it has two types of memories: a dynamic memory materialized by looped neurons in the input layer (with a sigmoid as activation function), and a static memory (hidden layer) to memorize the prototypes. The output layer represents the decision layer (see Fig. 11).

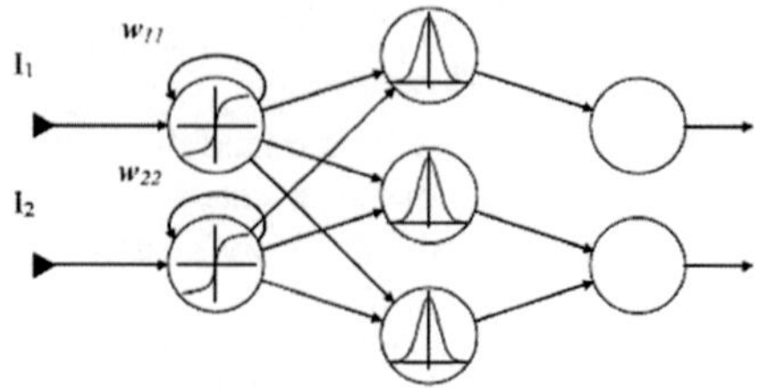


**Fig. 11.** RRBF architecture.

### 5.2 R2BF

The R2BF network is a recurrent neural network which consists of 4 layers (see Fig. 12). It is an hybrid architecture between Gaussian neurons (local representation) and sigmoid neurons (global representation) [3].

The input vector of each Gaussian neuron is then:

$$x(t) = (s_1(t), \ldots, s_n(t), y_1(t-1), \ldots, y_m(t-1)) \quad (3)$$

with $s_i(t)/_{i:1\ldots n}$ representing the inputs of the network at time t, and $y_i(t-1)/_{i:1\ldots m}$ (i: 1 … m) the outputs of sigmoidal neurons the second hidden layer at time $t-1$.

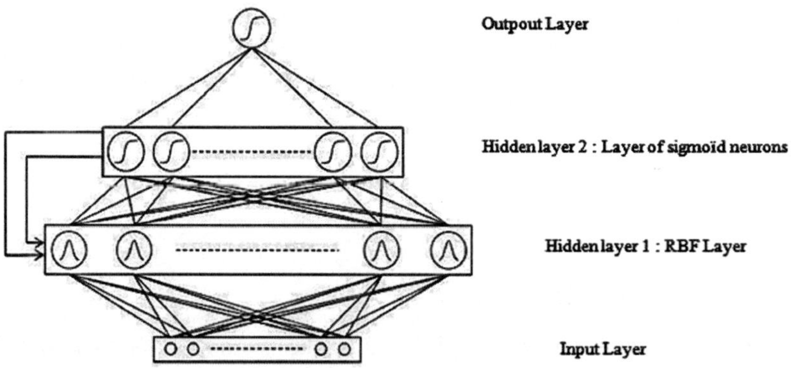


**Fig. 12.** Structure of R2BF.

### 5.3 Dynamic General Neural Network (DGNN)

The DGNN networks are based both on the structure of the MLP and on the RBF, thus making it possible to use the advantages of each of them: the generalization capacities of the MLPs and the computational efficiency of the RBFs. The main purpose of these networks is to solve identification problems in non-linear systems [18] (Fig. 13).

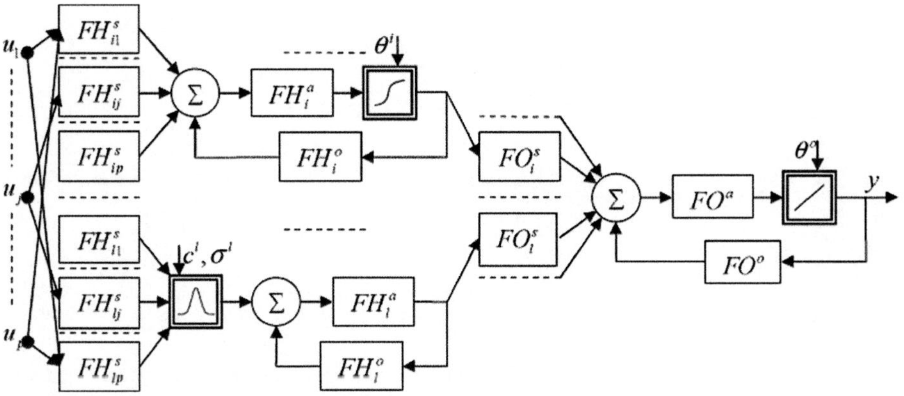

**Fig. 13.** The topology of DGNN with p inputs, one hidden layer containing n hidden neurons and one output. The hidden neurons are characterised by sigmoid or Gaussian activity functions. Here $FHij^s$, i = 1,…,n, j = 1,…,p represents the synaptic hidden filter corresponding to the connection considered from input j to hidden neuron i; $FHi^a$ and $FHi^o$ denote the activation filter and the output filter of the ith hidden neuron, respectively; $FOi^s$ specifies the synaptic filter corresponding to the connection considered from the ith hidden neuron to the output neuron; $FO^a$ and $FO^o$ denote the activation filter and the output filter of the output neuron, respectively; $\theta$ specifies the bias of the sigmoid hidden neurons and the bias of the output neuron; $c,\sigma$ indicate the centres and the standard deviations of the Gaussian hidden neurons [22].

## 5.4 Comparison Between the 3 Recurrent Neural Network

To compare these three recurrent neural networks we will base on the work of [3, 5, 18], where the authors proposed two tests by these networks on a prognostic benchmark (gas oven) and a monitoring dynamic benchmark (monitoring of a robot arm).

**Prediction by Recurrent Neural Networks**
In this first test the incoming gas flow is considered as the input series u(t) and the CO2 concentration as the output y(t).

From the input and output at the preceding moments the recurrent neural network must determine the output at the moment t + 1 (see Fig. 14).

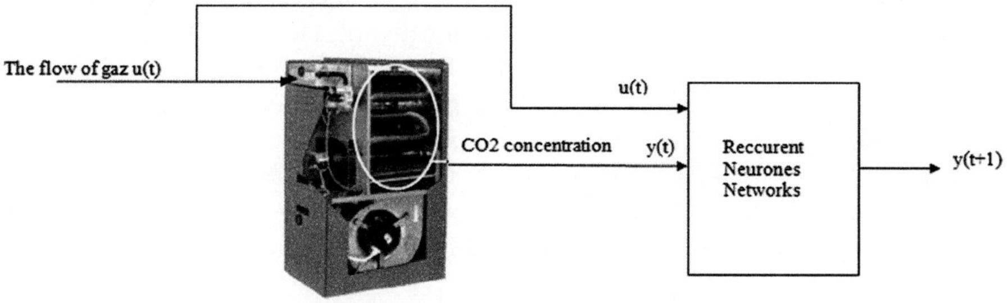


**Fig. 14.** Prediction of the CO2 concentration.

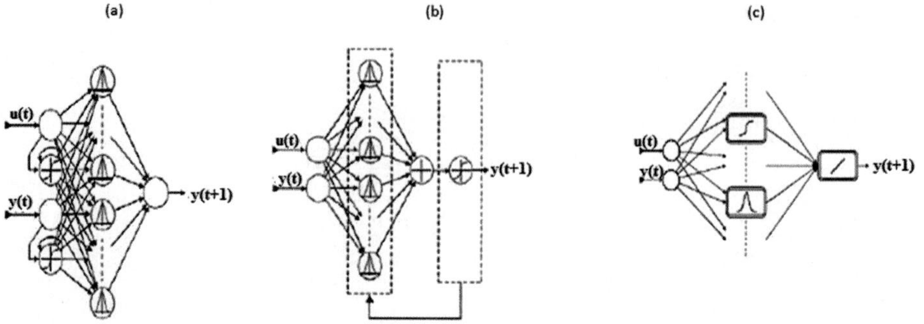


**Fig. 15.** Prediction of CO2 concentration with: (a) RRBF network, (b) R2BF network, (c) DGNN network

The result of this simulation is summarized in the following table (see Table 2) (Fig. 15).

**Table 2.** Simulation results from prediction of CO2 concentration with recurrent neural network.

| ANN | Mean error | Learning time | Recognition time |
|---|---|---|---|
| RRBF | 0.86% | 0.26 s | 0.2 ms |
| R2BF | 1.25% | 0.81 s | 0.3 ms |
| DGNN | 2.37% | 260000 s (3 days) | 2.1 ms |

In terms of mean error and learning time (see Table 2) the RRBF network appears to be the one with the lowest prediction error. However, the R2BF network is also particularly interesting in these results even if its learning time is 3 times greater than the RRBF but remains acceptable since it does not exceed the second, on the other hand the network DGNN takes a huge time in the learning phase.

So RRBF and R2BF are very good tools for prediction.

**Dynamic Diagnostic by Recurrent Neural Networks**

The purpose of this application is to recognize the type of collision of a robot arm from a sensor signal acquisition using recurrent neural networks (see Fig. 16). The robot arm is equipped with three force sensors (Fx, Fy, Fz). The answers given by these three sensors will inform us about the existence or not of sudden contact of the robot arm with an obstacle.

Four types of collisions are likely to occur (see Fig. 17): head-on (C Av), back (C Ar), left (C Ag), and right (C Ad). Each type of collision is characterized by a temporal evolution of the three measurement signals [18] (Fig. 18).

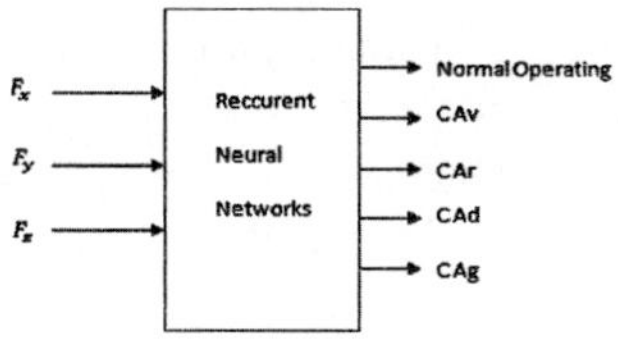


**Fig. 16.** Robot arm monitoring application.

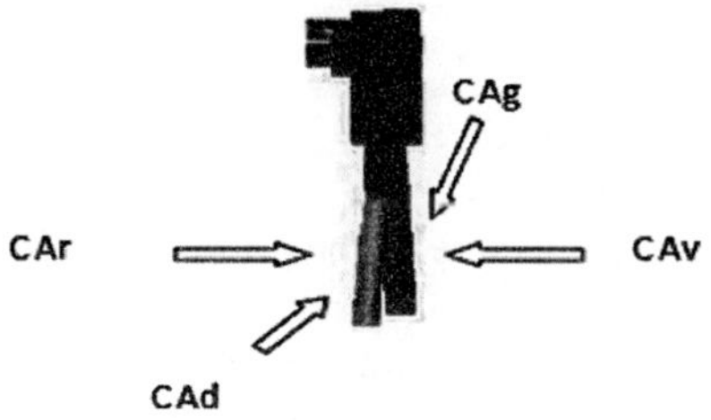


**Fig. 17.** Different types of collisions.

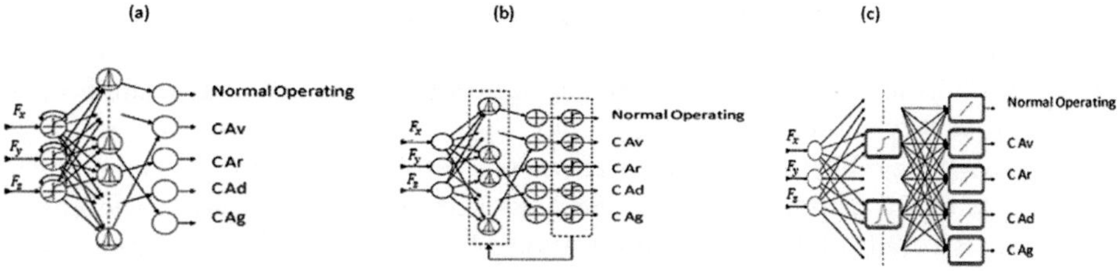


**Fig. 18.** Prediction of collision types by: (a) RRBF network (b) R2BF network (c) DGNN

**Table 3.** Recognition and learning time of recurrent neuron network applied for dynamic monitoring.

| ANN | Learning time | Recognition time |
|---|---|---|
| RRBF | 1.03 s | 2 ms |
| R2BF | 14 s | 1 ms |
| DGNN | 13423 s (3.73 h) | 20 ms |

**Table 4.** Results of detection and classification types' collision with recurrent neural network.

| | Detection% | | | Classification% | | |
|---|---|---|---|---|---|---|
| | RRBF | R2BF | DGNN | RRBF | R2BF | DGNN |
| FN | | | | 73,68 | 54,32 | 0 |
| C Av | 100 | 62,33 | 80,00 | 50 | 7 | 80 |
| C Ar | 100 | 50,40 | 66,67 | 60 | 8,8 | 0 |
| C Ad | 100 | 78,00 | 50 | 50 | 12 | 0 |
| C Ag | 100 | 68,50 | 100 | 75 | 21 | 0 |

Based on the results obtained on Table 3 and 4, it can be seen that the learning time at the DNGG level is too long.

The only network that has produced regular results with short learning time is the RRBF network. While these results are excellent in error detection, there is still a fair degree of error classification. This is partly due to the weakness of the learning base.

The results of the two test sequences showed that the RRBF network was the most efficient of the three networks, both by its results and by its learning times. For dynamic monitoring and prognostic applications, our choice will be the RRBF network.

## 6 Conclusion and Perspectives

In this paper we have presented a state of the art of the most used neural technique in monitoring.

This comparative study of this neural technique led us to the choice of RBF network for static monitoring. For dynamic monitoring where the time plays an important role we choose RRBF. Thanks to it dynamic memory obtained by a self-connection of the input neurons, RRBF is able to learn temporal sequences.

In a lot of works by using ANN, diagnosis is seen as a classification problem. It means that ANN identifies a process mode that reflects failure state. In this sense, neural networks are good tool for intelligent detection rather than diagnostic because they can't identify causes of failures.

To solve this problem, we can associate neural networks with methods that provide a representation of causal analysis of the links between failures, causes, and effects. Several tools of AI allow such representation. We can mention causal graphs, contextual graphs, fuzzy logic, Petri net. These tools, through their modeling and expression capabilities, provide explanatory models for the casual analysis failures.

## References


1. Li, Z., Wang, K.S., He, Y.: Industry 4.0 – potentials for predictive maintenance. In: International Workshop of Advanced Manufacturing and Automation (IWAMA), pp. 44–46. Atlantis Press, Manchester (2016)
2. Olivencia Polo, F.A., Ferrero Bermejo, J., Gómez Fernández, J.F., Crespo Márquez, A.: Failure mode prediction and energy forecasting of PV plants to assist dynamic maintenance tasks by ANN based models. Renew. Energy **81**, 227–238 (2015)
3. Tel-00011708. https://tel.archives-ouvertes.fr/tel-00011708, last accessed 2019/05/20
4. Li, Z., Wang, Y., Wang, K.-S.: Intelligent predictive maintenance for fault diagnosis and prognosis in machine centers: Industry 4.0 scenario. Adv. Manuf. **5**(4), 377–387 (2017). https://doi.org/10.1007/s40436-017-0203-8
5. Tel-00006003. https://tel.archives-ouvertes.fr/tel-00006003. Accessed 10 Jan 2019
6. Rafaël Gouriveau, R., El Koujok, M., Zerhouni, N.: Spécification d'un système neuro-flou de prédiction de défaillances à moyen terme. In: Rencontres Francophones sur la Logique Floue et ses applications, LFA 2007, Nîmes, pp. 65–72 (2007)

7. Debbah, Y., Cherfia, A., Saadi, A.: Application de la méthodes des reseaux de neuronne pour la prédiction des vibrations induites par défauts combinés (DESALIGNEMENT ET BALOURD). Sciences Technologie B **43**, 73–78 (2016)
8. Handle/123456789/1697. http://dlibrary.univ-boumerdes.dz:8080/handle/123456789/1697. Accessed 21 Nov 2018
9. Mahdaoui, R., Mouss, H., Chouhal, O., Kadri, O., Houassi, H.: La Surveillance Industriel Dynamique par les Systèmes Neuro-Flous Temporels: Application à un système de Production. In: 5th International Conference: Sciences of Electronic, Technologies of Information and Telecommunications (2009)
10. Zemouri, R., Zerhouni, N.: Autonomous and adaptive procedure for cumulative failure prediction. Neural Comput. Appl. **21**, 313–331(2012)
11. Monteiro, N.A.B., da Silva, J.J., da Rocha Neto, J.S.: Soft sensors to monitoring a multivariate nonlinear process using neural networks. J. Control Autom. Electr. Syst. **30**(1), 54–62 (2018). https://doi.org/10.1007/s40313-018-00426-x
12. El Farissi, O., Moudden, A., Benkacha, S.: Using artificial neural networks for recognition of control chart pattern. Int. J. Comput. Appl. IJCA (0975 – 8887) **116**(3), 46–50 (2015)
13. Zhang, G., Eddy Patuwo, B., Hu, M.Y.: Forecasting with artificial neural networks: Int. J. Forecast. **14**(1), 35–62 (1998)
14. https://machinelearningmastery.com/supervised-and-unsupervised-machine-learningalgorithms/. Accessed 14 July 2019
15. Sathya, R., Abraham, A.: Comparison of supervised and unsupervised learning algorithms for pattern classification. Int. J. Adv. Res. Artif. Intell. (IJARAI) **2**(2), 34–38(2013)
16. hal-01260830. https://hal.archives-ouvertes.fr/hal-01260830. Accessed 10 Apr 2019
17. Zhang, W., Kou, J., Wang, Z.: Nonlinear aerodynamic reduced-order model for limit-cycle oscillation and flutter. AIAA J. **54**(10), 3304–3311 (2016)
18. Palluat, N., Racoceanu, D., Zerhouni, N.: Utilisation des réseaux de neurones temporels pour le pronostic et la surveillance dynamique. Article in Revue d intelligence artificielle **19**(6), 911–948 (2005)
19. Hartono, P.: Classification and dimensional reduction using restricted radial basis function networks. Neural Comput. Appl. **30**(3), 905–915 (2016). https://doi.org/10.1007/s00521-016-2726-5
20. Zemouri, R., Racoceanu, D., Zerhouni, N.: Recurrent radial basis function network for time-series prediction. Eng. Appl. Artif. Intell. **16**(5–6), 453–463 (2003)
21. Telmoudi, A.J., Tlijani, H., Nabli, L., Ali, M., M'Hiri, R.: A new RBF neural network for prediction in industrial control. Int. J. Inf. Technol. Decis. Making **11**(04), 749–775 (2012)
22. Ferariu, L., Marcu, T.: Evolutionary design of dynamic neural networks applied to system identification, In: 15th Triennal World Congress, IFAC, Barcelona (2002)

# Integrating Artificial Intelligence in Knowledge Management: A Primer

Hayat El Asri[1(✉)], Laila Benhlima[1], and Abderrahim Agnaou[2]

[1] Mohammadia School of Engineering, Mohammed V University, Rabat, Morocco
hayatelasri@research.emi.ac, benhlima@emi.ac.ma
[2] Al Akhawayn University in Ifrane, Ifrane, Morocco
a.agnaou@aui.ma

**Abstract.** Over the years, the recurring and systematic use of data, information, and knowledge in managing organizational expertise has given rise to Knowledge Management (KM) as a distinct field of research and application. With the continuous growth of institutional and corporate knowledge, it has become imperative to integrate Artificial Intelligence (AI) in the process of managing the growing body of knowledge as a strategic asset. An AI-based KM aims at improving decision making, promoting innovation, and predicting future trends in organizations. Combining AI and KM is referred to as "cognitive computing" because they both need, use, and are about data. Thus, context-specific applications of AI in KM are in order, namely agent technology, expert systems, building information modelling, ontologies, analytics, and knowledge bases. Organizations have realized that the way they go about approaching KM was wanting. That is why emphasis shifted from technology to people, behaviors, and ways of working. This paper presents a primer for integrating AI in KM. It investigates the current AI trends and techniques that are suitable for integration in KM.

**Keywords:** Knowledge Management · Artificial Intelligence · Organizational performance

## 1 Introduction

With the ever-increasing amount of data and information that organizations around the world produce on a daily basis, Knowledge Management (KM) has proven to be of extreme importance. Value creation, information storing, and knowledge capturing and sharing all contribute tremendously in carving an organization's competitive advantage. Over the past two decades, a lot of research has been conducted in the field of KM. Nonetheless, the integration of Artificial Intelligence (AI) techniques is still recent and has mainly been applied in the e-health domain. The objective of this primer paper is to explore potential ways of integrating AI in KM with the aim to drive institutional effectiveness.


A. Saka et al. (Eds.): CPI 2019, LNME, pp. 516–526, 2021.
https://doi.org/10.1007/978-3-030-62199-5_45

## 2 General Background

In the literature about KM, there is a popular graphical representation that shows the relationships between data, information, and knowledge in the form of a pyramid. Sometimes, there is an additional layer of wisdom added on top to form what is referred to as the DIKW (Data-Information-Knowledge-Wisdom) Pyramid. In some rare cases, there are additional layers added to the hierarchy models, a "facts" layer and a "measurement" layer at the bottom and an "enlightenment" layer at the top.

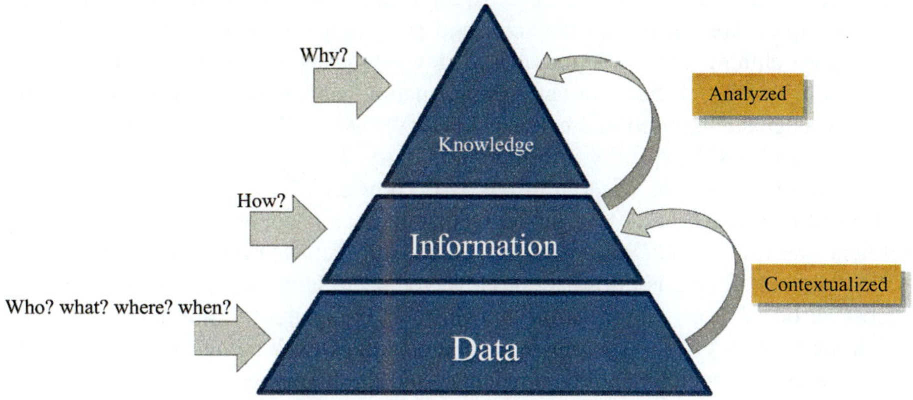


**Fig. 1.** Data, information, and knowledge

Data refer to "facts of the world", a collection of structured and unstructured raw facts and figures. While unstructured raw facts and figures represent any data sets that do not have a recognizable structure, structured raw facts and figures represent data with a loosely recognizable structure. These raw facts and figures are usually meaningless without relevant relational structure. "Data are the things given to the analyst, investigator, or problem-solver; they may be numbers, words, sentences, records, assumptions – just anything given, no matter what form and of what origin" [1].

Information is generated when these facts and figures are processed (categorized and contextualized) to figure out the relevance among pieces of data. In other words, information is a set of meaningfully processed data. This processing occurs when linking data items with the aim of finding relevance. Linking the data from various tables (databases) generates information, i.e. meaningful relations between data. IT may help a lot in the process of transformation of data into information. However, Floridi (2004) considers information as "a polymorphic phenomenon and a polysemantic concept" [2].

Knowledge is the abstract interpretation of information, and it refers to the know-how, the implied understanding. For Chaffey & Wood (2005), knowledge is "the combination of data and information, to which is added expert opinion, skills and experience, to result in a valuable asset which can be used to aid decision making" [3]. It is "an organized body of facts, principles, procedures and information acquired over time" [4]. It is "the cumulative stock of information and skills derived from use of information by the recipient" [5]. It is the "justified true belief" [6]. Knowledge is at the highest level of the classical pyramid because it is richer, deeper, and the most valuable.

Wisdom, on the other hand, refers to the combined understanding of data, information, and knowledge. It is exclusively a human capability, and it exists beyond any IT system. While wisdom increases individual efficiency and institutional effectiveness, it requires good judgment and ethics [7]. Ackoff (1989) states that wisdom "is the ability to increase effectiveness." He further adds that wisdom "adds value, which requires the mental function that we call judgement. The ethical and aesthetic values that this implies are inherent to the actor and are unique and personal" [8].

In approaching the DIKW Pyramid, Rowley [9] makes the implicit assumption that "data can be used to create information; information can be used to create knowledge, and knowledge can be used to create wisdom" [9]. She further sows the four concepts together as follows: "Typically information is defined in terms of data, knowledge in terms of information, and wisdom in terms of knowledge" [9].

Be that as it may, the DIKW model has lately been challenged. Frické [10] asks a pertinent question: "What would be the relationship between data and information?" [10]. For him, "all data is information. However, there is information that is not data. Almost all of science is information, but, in most contexts, it is not data." Bernstein [11], in turn, posits that "it is wrong to assume that the DIKW model accurately reflects the stages of the development of knowledge, and the hierarchy itself seems due for a fresh reappraisal if not necessarily banishment from the canon of information science" [11].

## 3 Types of Knowledge

Davenport and Prusak (1998) define knowledge in a way that fits the purposes of this work as "a fluid mix of framed experience, values, contextual information, and expert insight that provides a framework for evaluating and incorporating new experiences and information. It originates and is applied in the minds of knowers. In organizations, it often becomes embedded not only in documents or repositories but also in organizational routines, processes, practices and norms" [12].

Two types of knowledge can be recognized: tacit and explicit. Tacit knowledge is the "knowledge gained from personal experiences" [13]. It is personal knowledge that is hard to transmit and express. Furthermore, tacit knowledge is at the ontological dimension in which its clarification necessitates an extensive process of socialization, but whose sharing has become possible through communities of practice [14].

Explicit knowledge, on the other hand, refers to the knowledge that is codified, formal, shared, contextualized, documented, and transformed [15]. It is knowledge that is non-personal, easy to collect and to transfer using human or digital means (Fig. 1).

## 4 Knowledge Hierarchy

Knowledge hierarchy is important to visualize the flow from data to wisdom, although wisdom is frequently left off the chart. Figure 2 represents the knowledge hierarchy.

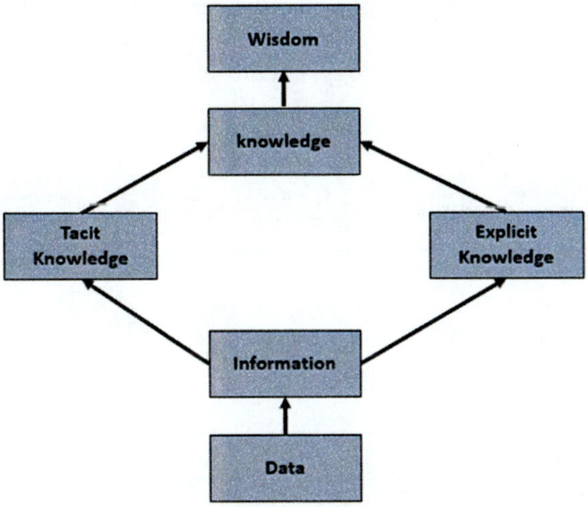


**Fig. 2.** Knowledge hierarchy

As presented in Fig. 2, data is the lowest level in the hierarchy because raw data must be contextualized in order to be useful. The second layer is information, followed by both tacit and explicit knowledge. The highest level of the hierarchy is wisdom. This latter requires judgement, critical thinking, ethics, and values.

## 5 Knowledge Growth

The humanity's development relies on knowledge. People, organizations, and institutions generate, use, and share knowledge on a regular basis. In fact, knowledge is the most important intellectual capital, and plays a vital role in corporate growth [16]. Thus, one may say that knowledge is what affects, positively or negatively, the growth of any institution, as it is considered a "strategic asset". For that reason, KM has become a leading trend in organizations.

## 6 Knowledge Management

Nonaka and Takeuchi [6] define KM as "The process of applying a systematic approach to the capture, structuring, management, and dissemination of knowledge throughout an organization to work faster, reuse best practices, and reduce costly rework" [6]. Two decades later, the definition evolved. In 2018, the Knowledge Management Institute (KMI) defined it as "a multidisciplinary field that encompasses psychology, epistemology, and cognitive science, whose goals are to enable people and organizations to collaborate, share, create, use and reuse knowledg" [17].

KM touches on several fields, to name but a few, informatics, computer science, cognitive science, business, psychology, management, and marketing. Nonetheless, for it to become well established within a firm/an organization, it needs a number of enablers such as: information technology, culture, and strategy. Figure 3 represents the KM enablers and processes.

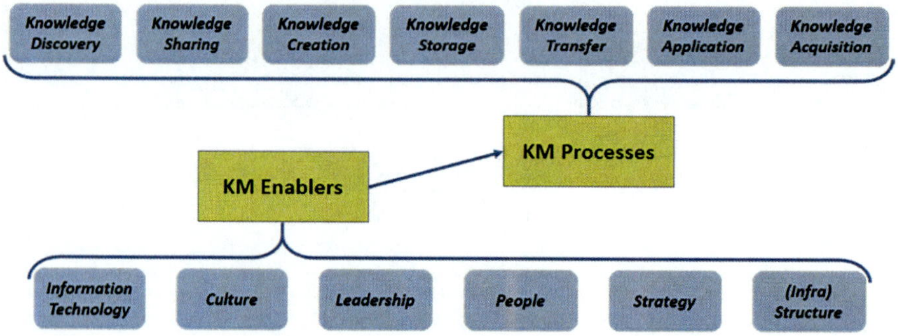


**Fig. 3.** KM enablers and processes

The main two processes are Knowledge Discovery (KD) and Knowledge Sharing (KS). KD refers to the development of new tacit and/or explicit knowledge from data, information, or prior knowledge, while KS refers to "process where individuals mutually exchange their implicit (tacit) and explicit knowledge to create new knowledge" [18]. The most effective way to share knowledge is through systematic transfer. To achieve that, organizations need a supportive environment with four enablers: culture, technology, infrastructure, and measurement [19].

The benefits of integrating KM in organizations are countless. KM has been proven to improve decision making, work efficiency, and access to information. It also promotes knowledge sharing, foster KD, and encourages innovation. Therefore, adopting it provides a great competitive advantage. Most organizations nowadays under-stand how important is KM in capturing the most "valuable" knowledge, storing it, for the purpose of using it when needed (Table 1 and Fig. 4).

## 7 Integrating AI in KM

Artificial Intelligence, or AI for short, "is a technology where knowledge is captured, shared, carefully developed and transformed into the right format in organizations" [20]. AI and KM, at their core, are both about knowledge, and they both aim at improving decision making in organizations; however, according to Suss (2018), there are some AI techniques that are more suitable for integrating with KM than others [21]. These techniques are: cognitive insight, cognitive engagement, and process automation. Cognitive insight combines machine learning and human-data interaction, and is used for automation and categorization purposes. Process automation uses robotics technologies to automate digital and physical tasks. Lastly, cognitive engagement uses intelligent agents that can offer a service.

Because both AI and KM need, use, and are about data, combining the two together is referred to as cognitive computing. Schizas (2017) defines cognitive computing as "the simulation of human thought processes in a computerized model… [It] involves self-learning systems that use data mining, pattern recognition and natural language processing to mimic the way the human brain works" [22].

If one simplest definition of leaning is turning content into knowledge, then cognitive computing is the simulation of this complex process by way of computers. Hurwitz et al. [23] state that "to create a cognitive system, there needs to be organizational structures for the content (…), [which] provide meaning to unstructured content" [23]. Other subsets of AI have also been used with KM, namely machine learning and deep learning. Figure 3 provides an overview of the three areas.

**Table 1.** Artificial Intelligence, Machine Learning, and Deep Learning (adapted from edureka.co [24]).

| Artificial Intelligence | Machine Learning | Deep Learning |
|---|---|---|
| Emerged as a field in the 1950s | Emerged as a field in the 1960s | Emerged as a field in the 1970s |
| Simulation of human intelligence | Teaching machines to make decisions by themselves | Using Artificial Neural Networks to solve complex problems |
| An umbrella research field | A technique of AI | A subfield of machine learning |
| Needs techniques and/or processes to solve problems | Interprets, processes, and analyzes data | Implement neural networks of high dimensional data |

In AI-based KM, the current trends at the moment is the use of multi-agent systems, expert systems, analytics, and business intelligence. Several applications, mainly in the e-health domain, have used one or more, or a combination of several techniques in the development of their proposed solution. Nevertheless, other application domains are yet to be further researched and investigated.

The benefits of integrating AI in KM are, but not limited to, simplifying KD, boosting information accessibility, improving real time decision making, promoting research and innovation, and providing more predictive search capabilities. A KM system with built-in AI capability is expected not only to bring contextual knowledge and find effective solutions for a given business problem, but also help organizations identify the missing link in knowledge flow. Among the benefits of combining AI and KM, there is smart search capabilities that help users locate data/information more quickly, business intelligence that correlates and predicts industry trends, and access to old and new data to support decision-making.

# 8 Applications of AI in KM

## 8.1 Agent Technology

Agent technology has been wildly used for over a decade as a solution to real world issues. In combination with KM, agent technology has been used in the e-health application domain, but is still considered underused.

## 8.2 Expert Systems

Expert systems are intelligent software programs that are able to reproduce human knowledge and "expertise", and to emulate human decision-making. Moreover, expert systems are said to be an idyllic solution for converting both explicit and tacit knowledge into a form that is available to several users, which is a key process in KM, indeed [25]. Therefore, these systems can be of great added value to KM systems, mainly in the following KM phases: knowledge creation, knowledge transfer, and knowledge use/reuse.

## 8.3 Building Information Modelling

Building information modelling is an emerging industrial and procedural technology for managing the building design and the data in construction. However, it may be seen simply as a solution for capturing and sharing knowledge in construction projects.

## 8.4 Ontologies

Ontologies represent a description of knowledge as concepts within a specific domain of application, and the relationship between them. Ontologies are explicit knowledge-based specifications of conceptualizations that "typically describe a taxonomy of the tasks that define the knowledge" [26]. Further-more, ontologies provide common understanding of information; therefore, their integration in KM can be extremely beneficial for several reasons. First, they provide a good alternative to hard coding, a useful method for capturing data, and an easy technique for capturing relationships.

## 8.5 Analytics

Analytics refers to the set of AI-enabled techniques used to analyze data, extracted and categorized from various sources, to identify recurring behavioral patterns with the aim of enhancing work performance. Moreover, it uses quantitative and predictive models to facilitate better decision making. Business data analytics provides a range of tools for statistical analysis, forecast, optimization, and modelling. The purpose of analytics in KM is to help stakeholders understand the impact of KM approaches on business processes and make informed decisions. More straightforwardly, analytics refers to AI-enabled processes of transforming data/information into meaningful knowledge patterns for organization stakeholders, especially decision-makers.

### 8.6 Knowledge Bases

A knowledge base is a centralized database for managing and sharing structured and unstructured information. It is a library of information whose main purpose is to provide self-help. A knowledge base could be the first step towards integrating AI techniques in KM. In fact, most knowledge bases use basic AI techniques for search and retrieval.

All these technologies provide a good competitive edge to the organizations using them. However, because of the limitations that each of these technique might have, combining several techniques can increase the effectiveness of the system. Nonetheless, understanding the requirements and the main objective behind this integration is key. Furthermore, taking external factors into consideration is just as important.

## 9 Rationale for AI-Based KM

AI-based KM provides several benefits that KM alone cannot. KM is a good start for companies to gain a competitive edge, access structured information easily, manage overload, cope with change, share best practices, identify the gap in knowledge, save institutional memory, and harness institutional effectiveness. However, when AI techniques, such as analytics, ontologies, agents, and expert systems are integrated, it provides so much more. On top of what KM provides, AI-based KM:

- improves decision making;
- enhances work efficiency;
- promotes innovation;
- fosters KD;
- predicts market trends;
- identifies recurrent behavioral patterns.

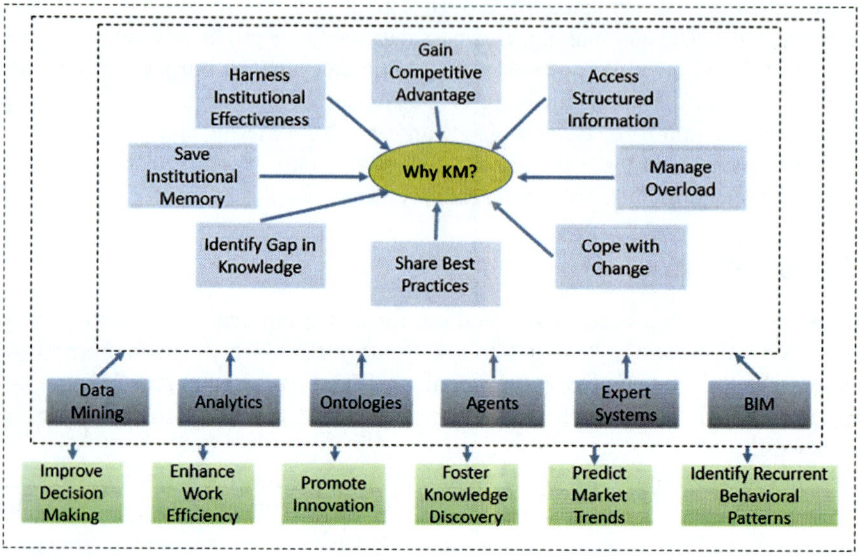


**Fig. 4.** AI-based KM

## 10 KM Caveats

Organizations have recognized that the way they approach KM was deficient; therefore, they shifted the focus from technology to people. According to [27], the limited success of KM was due to:

- Focus on technology rather than business activity and its people.
- KM was not linked to business processes and ways of working.
- Too much hype and impression.
- Companies spent too much money with little or no return on investments.
- Most KM literature was very conceptual and lacking in practical advice.
- Inability to translate KM theory into practice.
- Lack of incentives for employees who rightly asked the question: "what's in it for me?"
- Insufficient top management buy-in.

## 11 Conclusion

Managing knowledge is one of the main challenges that organizations face. Because of the overwhelming, yet needed, amount of knowledge that organizations produce, need, use, and share on a daily basis, managing it using modern technologies has proven to be vital. Proper and efficient management of organizational knowledge is key to ensuring the sustainability of staying ahead of competition. This article presented some AI techniques that can be integrated in KM for better management. Conceptualizing an integrated KM system will be the next step in this research.

## References


1. Machlup, F., Mansfield, U., (eds.): Semantic quirks in studies of information. In the Study of Information: Inter- Disciplinary Message. Wiley, New York (1983)
2. Floridi, L., (ed.), Information. In the Blackwell Guide to the Philosophy of Computing and Information. Blackwell, Oxford (2004)
3. Chaffey, D., Wood, S.: Business Information Management: Improving Performance using Information Systems. FT Prentice Hall, Harlow (2005)
4. Blanchard, P., Thacker, J.: Effective training: systems strategies and practices. In Learning, Motivation, and Performance, pp. 58–90. Prentice Hall, New Jersey (2009)
5. Burton-Jones, A.: Knowledge Capitalism: Business, Work and Learning in the New Economy. Oxford University Press Oxford (1999)
6. Nonaka, I., Takeuchi, H.: The Knowledge-Creating Company: How Japanese Companies Create the Dynamics of Innovation. Oxford University Press, New York (1995)
7. Jashapara, A.: Knowledge Management: An Integrated Approach. FT Prentice Hall, Harlow (2005)
8. Ackoff, R.L.: From data to wisdom. J. Appl. Syst. Anal. **16**, 3–9 (1989)
9. Rowley, J.E.: The wisdom hierarchy: representations of the DIKW hierarchy. J. Inform. Sci. **33**, 80–163 (2007)
10. Frické, M.: The knowledge pyramid: a critique of the DIKW hierarchy. J. Inform. Sci. **35**, 131–142 (2008)
11. Bernstein, J.H.: The data-information-knowledge-wisdom hierarchy and its antithesis. In: Proceedings North American Symposium on Knowledge Organization, pp. 68–75 NY, Syracuse (2009)
12. Davenport, T.H., Prusak, L.: Working Knowledge: How Organizations Manage What They Know. Harvard Business School Press, Boston (1998)
13. Alexander, R.: Implicit, Tacit, or Explicit: All Knowledge Is Valuable. Bloomfire (2018)
14. Omotayo, F.: Knowledge Management as an Important Tool in Organisational Management: A Review of Literature, Library Philosophy and Practice, University of Nebraska-Lincoln (2015)
15. Magalhães, I.: Tacit and Explicit Knowledge. Linkedin.com (2015)
16. Yang, J.: Turning Knowledge Management Strategy into Corporate Growth, pp. 85 –94 (2001)
17. KMI: The Connection Between Artificial Intelligence and Knowledge Management. KM Institute (2017)
18. Van den Hooff, B., De Ridder, J.: Knowledge sharing in context: the influence of organizational commitment, communication climate and CMC use on knowledge sharing. J. Knowl. Manage. **86**(6), 117-130 (2004)
19. O'Dell, C., Grayson, J.: Knowledge transfer: Discover your value proposition. Strateg. Leadersh. **27**(2), 10–15 (1999)
20. Curatti. The Bond Between Artificial Intelligence and Knowledge Management (2018)
21. Suss, T.: How Is Artificial Intelligence Changing Knowledge Management? – Unika (2018)
22. Schizas, C.N.: Cognitive computing for supporting eHealth. Health and Technol. **7**(1), 11–12 (2016). https://doi.org/10.1007/s12553-016-0162-2

23. Hurwitz, J.S., Kaufman, M., Bowles, A.: Cognitive Computing and Big Data Analytics. Wiley, Indianapolis (2015)
24. Lateef, Z.: Most Frequently Asked Artificial Intelligence Interview Questions. Edureka.co (2019)
25. Lamont, J.: Expert Systems and KM Are a Natural Team. [online] Kmworld.com (2000)
26. O'Leary, D.E.: Using AI in knowledge management: Knowledge bases and ontologies. In: IEEE Intelligent Systems (1998)
27. Servin, G., de Brun, C.: ABC of Knowledge Management, NHS National Library for Health: Knowledge Management Specialist Library (2005)

# Trends and Applications of Cooperative Intelligent Transport Systems (C-ITS)

Hanae Lahmiss(✉) and Abdellah Khatory

Laboratory of Energy Production and Sustainable Development, CED: Sciences and Engineering Technology of the Faculty of Science and Technology, Sidi Mohamed Ben Abdellah University, Fez, Morocco
hanae.lahmiss@gmail.com, abdellah.khatory@usmba.ac.ma

**Abstract.** Intelligent transport system (ITS) is a controlled system which uses sophisticated technologies to increase road safety and enhance management of traffic flow as well as to support travelers of all classes, drivers, passengers and pedestrians. With the development of internet and wireless communication technologies, current studies have moved towards cooperation on the road and between transport systems, a new trend was born: the cooperative intelligent transport systems (C-ITS), these systems based on availability of the information and interaction of the vehicle with its environment and other vehicles. In this paper the C-ITS trends and their applications in transport systems are reviewed and discussed. Our work aims to present a comparative study of wireless communication technologies used in transport systems, two types of wireless communication technology are indicate: (1) Radio frequency (RF) technology like Dedicated Short-Range Communications (DSRC), Radio-frequency identification (RFID) and WIFI, (2) Optical wireless communication (OWC) based on visible light technology (VLC). Results show that the DSRC technology is more used and applied to make the cooperation between road entities. A case study of using wireless communication technology in Moroccan highways based on DSRC technology is presented.

**Keywords:** Intelligent transport systems (ITS) · Cooperative intelligent transport systems (C-ITS) · DSRC · Vehicle to vehicle (V2V) communication · Vehicle to infrastructure (V2I) communication

## 1 Introduction

An intelligent transport system can be defined as a system based on the integration of the information and communication technologies in transport systems (vehicles, infrastructure) for improving road safety and the efficiency of transport networks as well as reducing the losses of life and property.

The intelligent transport system (ITS) is a system has services to support users of transport systems (drivers, passengers, pedestrians) and to assist the performance of the road network by using information and communication technologies in the urban and rural surfaces including intermodal and multimodal aspects [1]. ISO TC204 defined


A. Saka et al. (Eds.): CPI 2019, LNME, pp. 527–536, 2021.
https://doi.org/10.1007/978-3-030-62199-5_46

ITS as information, communication, and control systems including traffic management, public transport, commercial transport and emergency services [2].

Intelligent transport system is characterized by a powerful capacity in terms of energy, localization and computation then the time sensitivity which allows the optimization of the time needed to deliver a safety information to nodes, finally the security and privacy which the most important characteristic that allows the quick and safe exchange of information by the implementation of security protocols in ITS [3].

Actually traditional transport solutions are insufficient to optimize the road traffic network so the development of the ITS systems calls for smart synergy of stakeholders of transport systems with the objective is to reduce the number of deaths and injuries, the technologies used in intelligent transport systems are: **computer technologies** to solve problems in transport systems, **floating car data** concept is used in ITS to determine vehicle's information for example the speed and the location of the vehicle, **detection technologies** is used to obtain information about activities carried in a transport system and the use of **wireless communication** which allows vehicles to communicate with each other and with their environment. The application of wireless communication in transport systems allows the appearance of cooperative intelligent transport systems (C-ITS) these systems are current trend developed with the development of wireless communication technologies.

In this study a review is made on cooperative intelligent transport systems (C-ITS) trends with the purpose of to study the C-ITS applications on transport systems and a presentation of the utilization of electronic toll (Jawaz) in Morocco is made.

This paper is structured as follows: in Sect. 2 we present the cooperative intelligent systems and the principle of interconnection between Road entities. Section 3 Reviews the C-ITS applications and wireless communication technologies used in transport systems. In Sect. 4 the utilization of electronic toll in Morocco (Jawaz) is studied. Section 5 presents the conclusion and perspectives for future researches.

## 2 Cooperative Intelligent Transport Systems

Cooperative Intelligent Transport System (C-ITS) is a key enabler of future road traffic management systems [4], this system allows the real time connectivity between transport entities for improving road safety and optimize the flow of traffic, it is based on the connectivity and collaboration to exchange data and information for increase the road safety levels. The C-ITS is composed of On-Board Units (OBU), that is fixed inside the vehicle and a Road Side Unit (RSU), which is fixed along the road. The OBU and RSU interact with each other using dedicated short-range communication (DSRC). The communication can be vehicle-to-vehicle (V2V), vehicle-to-infrastructure (V2I), or infrastructure-to vehicle (I2V) [5].

Following figure shows the principle of interconnection between Road entities (Fig. 1).

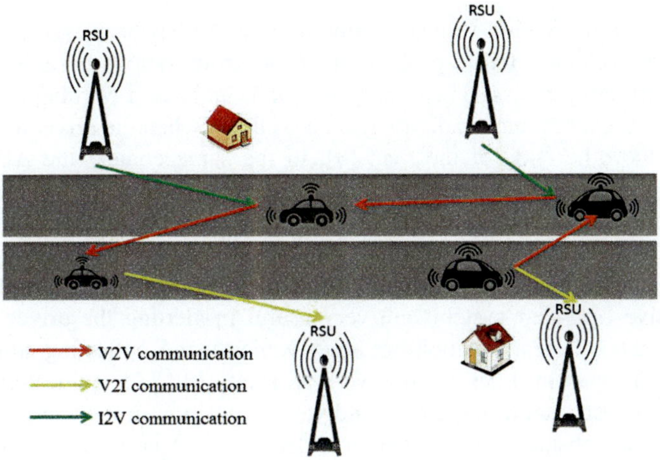


**Fig. 1.** The principle of interconnection between road entities

# 3 A Review of C-ITS Applications

## 3.1 Data Basis

In order to ensure a high quality of the publications included in our review, an automatic search was performed on the databases of the most prestigious publishers with additional criteria illustrated in Table 1.

**Table 1.** Data bases

| | |
|---|---|
| Electronic databases | Science Direct |
| | IEEE Xplorer |
| | Elsevier |
| | Web of science |
| Language | English |
| Publication period | 2008–2018 |

In this Data bases references [2, 6, 7] were excluded.

## 3.2 C-ITS Applications

The increase of congestion phenomenon and the number of accidents on the road makes appear intelligent transport systems that have applied for several objectives, but road safety is the main purpose of these systems, for example to alert the driver to the dangers present at the intersections or turns, detecting and alerting the non-respect of the safety distance between the vehicle and other nearby vehicles and to reduce the number of deaths and injuries on the road.

Wireless communication vehicle to anything (V2X) has been applied to minimize the number of accidents and to protect road users from various hazards, for example pedestrian alert, left turn assistance function which can issue a warning to the driver or may apply the brakes automatically to avoid a collision when the driver turns to the left and crosses the other side of the road [3]. In the same context the AEVW system (Approaching Emergency Vehicle Warning) is a V2V (vehicle-to-vehicle) communication system based on the DSRC (Dedicated range short communication) technology which aims to disseminate the warning messages of safety between an emergency vehicle and other vehicles in the vicinity to facilitate and secure road traffic [8].

Cooperative risk alert systems are very useful in alerting the driver to accidents occurring in the different positions of road vehicles and collecting information on incidents on the road in order to improve road safety [9], ITS application in vehicles allows direction estimation in the road and vehicle position in its lane, so the driver is warned when an obstacle is detected in order to avoid collisions [10], as well as improving driver comfort by providing meteorological information, an internet service and information on the journey to be taken [3].

ITS are applied in road infrastructure for public transit management systems to improve the efficiency and safety of data management and public operations, it allows automated information retrieval, storage for planning and overall transportation management [10]. The commercial vehicle administration center is responsible for administrative tasks related to the regulation of powers, taxes and the security of commercial vehicles (trucks, buses, etc.), it issues identification information, as well as collects costs and taxes. For example, it communicates with motor carriers to process certification applications, collect fuel taxes, weight/distance taxes, and other taxes and fees associated with commercial vehicles [10].

In environmental field, the architecture and the principle of a model emission charging using a cooperative intelligent transport system (C-ITS) has been developed, that the communication between vehicles and infrastructure is essential to reduce traffic-related gas emissions in order to solve environmental problems related to transport [11]. This goal is also presented in [9], which use floating car data to collect traffic information, including congestion obtained using ITS, and calculate fuel consumption to estimate $CO_2$ emissions. These systems help to reduce $CO_2$ emissions and create a society which respects environment.

Intelligent Transportation Systems (ITS) fall under the category of smart mobility in the context of the Smart City. They have been applied to optimize passenger transport and logistics as well as to provide efficient, safe and environmentally friendly passenger and cargo services in the Smart City [12]. Electronic Vehicle Identification (EVI) is an important ITS technology that can be implemented in both urban and interurban scenarios in a smart city, the ITS is used to combat terrorist threats, identify and track vehicles, for traffic management and crime prevention, for resistance to fraud, and for the dynamic regulation of traffic in urban and interurban areas [13]. Intelligent Transportation Systems (ITS) are widely used to assess traffic conditions by calculating the number of vehicles on the road in order to estimate traffic flow using cameras installed in parking areas and at the same time along the road side [14].

ITS is also applied in strategic state projects, the Slovak Republic has implemented Intelligent Transport Systems (ITS) for traffic management and maximized protection of critical road and rail transport infrastructure to make them more resilient and safer infrastructure [15]. In Russia, digital technologies are applied to eliminate or minimize transport losses, for example when calculating passenger traffic for an urban transport route, it is assumed that users will always choose the most optimal route [16]. In Italy, intelligent transport systems are applied to evaluate and analyze the health of Italian local public transport using on-board video supervision, in order to monitor potential security threats [17], as well as emergency management, parking in case incident and automatic payment are taken into account for the management of public transport in Italy [18].

### 3.3 Wireless Communication Technologies Used in Transport Systems

Actually this is the era of connectivity, wireless communication technologies was developed very fast, current studies have moved towards cooperation on the road, to obtain the cooperation and communication between vehicles with each other and with infrastructure, the radio frequency(RF) technology and optical wireless communication (OWC) technology were applied on transport systems.

In Table 2 we specify types of wireless communication technology discussed or applied. This table and Fig. 2 show that more works have been done on the integration of the radio frequency technology on transport systems which the Dedicated Short Range Communications (DSRC) technology is used more than GPS, RFID, WIFI and VLC technologies to provide and facilitate the cooperation and connectivity on road. DSRC is opened on 5.9 GHz of the radio frequency spectrum, DSRC has low latency and high reliability and more secure and efficient than WIFI [19, 20] demonstrate that DSRC is more suitable for safety related V2V traffic application, and [21] proved that the DSRC performance can persist through air density changes, which helps to make up for lost human visibility on roads during foggy times. As well as the vehicle to vehicle (V2V), vehicle to infrastructure (V2I) and vehicle to anything (V2X) communication systems are covered by numerous researchers in comparison with vehicle to pedestrian (V2P) or infrastructure to vehicle (I2V) communication systems because the incidence of accidents can be due to lack of the V2V, V2I and V2X communication.

**Table 2.** Wireless communication technologies used in Transport systems

| Reference | Wireless communication technology | Type of communication | Journal/Scientific event |
|---|---|---|---|
| [11] H. Vojvodic, S. Mandzuka, M. Vujic | – DSRC<br>– Cellular technologies | V2I; I2V | IEEE, 24th Telecommunications Forum (TELFOR), 2016 |
| [12] A. Sumalee and H. W. Ho | – RFID: Radio-frequency identification<br>– GPS: Global positioning satellite<br>– WIFI, 3G/4G | V2V; V2P; V2I; P2I | IATSS Research, 2018 |
| [13] O. Ghim, Y. Gary | – RFID; GPS; DSRC;<br>– GSM: Global System for Mobile | V2I | Not cited, 2008 |

(*continued*)

**Table 2.** (*continued*)

| Reference | Wireless communication technology | Type of communication | Journal/Scientific event |
|---|---|---|---|
| [19] A. Fitah, A. Badri, M. Moughit, et A. Sahel | – DSRC: Dedicated Short Range Communications (IEEE 802.11p)<br>– WIFI (IEEE 802.11a)<br>– VANET: vehicular-ad-hoc network | V2V; V2I | Procedia Computer Science, 2018 |
| [22] Z. Kljaic, E. Briski, H. Vojvodic, N. Amin | – GPS<br>– GNSS: global navigation satellite system | Not cited | 41st International Convention on Information and Communication Technology, 2018 |
| [23, 24] I. Kabashkin | – DSRC<br>– VANET | V2X; V2V; V2I | IEEE, Advances in Wireless and Optical Communications (RTUWO), 2015, 2017 |
| [25] K. Siddiqi, A. D. Raza, and S. S. Muhammad | – VLC: Visible light communication | V2V | International Conference on Broadband Communications for Next Generation Networks and Multimedia Applications (CoBCom), 2016 |
| [26] N. Sharma and V. Saini | – OWC: optical wireless communication links | V2V; V2I | International Conference on Soft Computing Techniques and Implementations (ICSCTI), 2015 |
| [27] P. Fang, Y. Bao, J. Shen, and J. Chen | – VLC | I2V | IEEE, International Conference on Connected Vehicles and Expo (ICCVE), 2015 |
| [28] A. Cailean, B. Cagneau, L. Chassagne, S. Topsu, Y. Alayli, and J.-M. Blosseville | – VLC | V2V; V2I | IEEE Intelligent Vehicles Symposium, 2012 |
| [29] A. E. Coronado Mondragon, E. S. Coronado Mondragon, C. E. Coronado Mondragon, and F. Mung'au | – DSRC<br>– Cellular network<br>– Satellite | V2I | Expert Systems with Applications, 2012 |
| [30] A. Ghosal, F. Bai, R. Debouk, and H. Zeng | – DSRC | V2V; V2I | SAE International, 2011 |
| [31] B. Drilo, D. Saric, et R. Filjar | – GPS; GNSS; Galileo; Glonass<br>– Cell-ID | Not cited | IEEE, 1st International Conference on Wireless Communication, Vehicular Technology, 2009 |

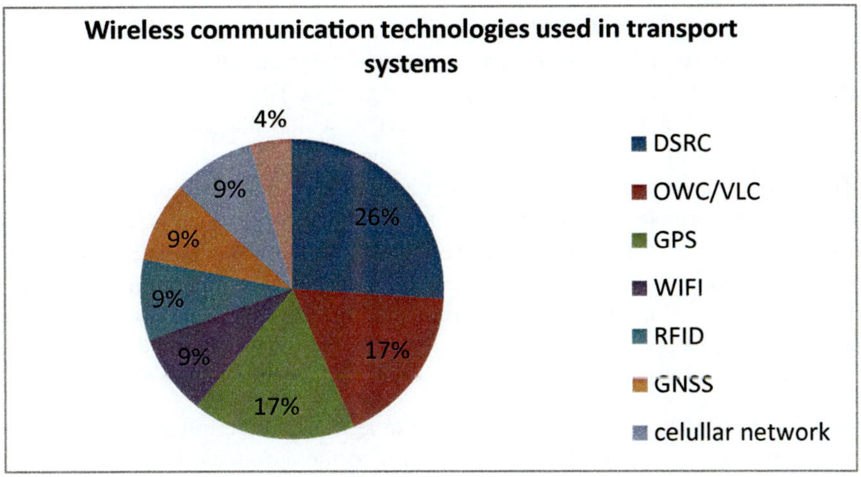


**Fig. 2.** Wireless communication technologies used in transport systems

## 4 The Electronic Toll Collection in Morocco

In recent years, investment in infrastructure has been accelerated, including highway of Morocco (ADM) has expanded the toll stations that are experiencing a great influx and launched the Jawaz electronic toll collection to increase the performance of Moroccan highways and the comfort of drivers.

### 4.1 The General Operation of an Electronic Toll Collection

The electronic toll collection is a fast motorway payment system using DSRC technology (Dedicated Short Range Communication). The DSRC works in a more efficient way than WIFI, allowing drivers to cope with the dangers on the road by using radio waves to transmit information across a network. The badge equipped with a DSRC chip and attached the windshield of the vehicle captured by the receiver installed at the entrance and exit of a paid section. This system allows the interoperable exchange of data between the badge (on-board Unit) and the road Side Unit (RSU).

The electronic toll collection allows the passage without stopping at a speed with a speed 30 and 50 km/h. Now in Norway the electronic tolling are at fast speed allows the passage to a speed of 100 km/h [7].

### 4.2 Jawaz Morocco

The electronic toll collection in Morocco is launched in June 2014 this solution facilitates the payment of the toll and streamlines the traffic. It works by means of a case (badge) stuck to the windshield of the vehicle which allows the users do not to mark the stop: the transaction is carried out automatically at the approach of the toll, the barrier rises instantly so that the driver passes constantly at a speed of 20 km/h and the count of the user is debited from the customer's balance, subscribers have for their passage the reserved lanes (indicated by an orange "j"). The year 2016 was marked by the progressive generalization of this toll service, which has more than 50,000 customers and covers 80% of the toll stations of the motorway network in Morocco [6].

Highways are not just infrastructure, but a set of services that support, facilitate and make more comfortable the movement of millions of people. This is why highway of Morocco (ADM) has been working to diversify the toll collection offer throughout the motorway network in order to consolidate the concept of cooperation and interconnection between road entities (vehicles, infrastructure) to save time, user comfort, and traffic flow and protect the environment.

The benefits of the Jawaz Morocco electronic toll collection are:

**Saving Time:** Saving a lot of time, avoiding time waste at the toll stations, the user of the highway simply passes without waiting line to the automatic toll system.

**Traffic Fluidity:** Jawaz avoids the problem of congestion on Moroccan motorways which leads to loss of productivity because of the delays of people going to the place of work, and late deliveries of supplies or services rendered with people, significant delays or even canceled.

**User Comfort:** Jawaz's goal is to ensure that users travel on the motorway network in the best possible conditions, in particular by avoiding traffic congestion that cause stress problems and increase the risk of accidents.

**Protect Environment:** The ETC system minimizes CO2 emissions and fuel consumption on waiting line at the toll sections, which increases environmental pollution.

## 5 Conclusion and Perspectives

This paper has presented a brief review of cooperative Intelligent transport systems (C-ITS) applications in transport systems and the wireless communication technologies used to provide the inter-connection between road entities, that the DSRC system is more used than other technologies. A presentation of Moroccan electronic toll collection (Jawaz) is made to discuss the benefits of this system for highway users.

In Morocco we need to consolidate the concept of cooperation and the connection on the road between the drivers in real time, the use of Jawaz is already a considerable step towards the connection between the entities of the road.

For future research, we aim to study ITS applications in urban Logistic in Morocco that ITS can have a significant impact on the performance of logistics operations and supply chain, and optimize freight transport using intelligent transport systems and wireless communication technologies, as well as we aim to study the application of DSRC technology in freight transport and their advantages in this field.

## References


1. Williams, B.: Intelligent Transport Systems Standards. Artech House, Boston (2008)
2. ISO/TC 204 - Intelligent transport systems. http://www.iso.org/cms/render/live/en/sites/isoorg/contents/data/committee/05/47/54706.html
3. Hamida, E., Noura, H., Znaidi, W.: Security of cooperative intelligent transport systems: standards, threats analysis and cryptographic countermeasures. Electronics **4**, 380–423 (2015)
4. Javed, M.A., Zeadally, S., Hamida, E.B.: Data analytics for cooperative intelligent transport systems. Veh. Commun. **15**, 63–72 (2019)
5. Khalil, U., Javid, T., Nasir, A.: Automatic road accident detection techniques: a brief survey. In: 2017 International Symposium on Wireless Systems and Networks (ISWSN), Lahore, pp. 1–6. IEEE (2017)
6. Daoudi, L.: Les infrastructures au Maroc, 13
7. Télépéage (2019). https://fr.wikipedia.org/w/index.php?title=T%C3%A9l%C3%A9p%C3%A9age&oldid=157325911
8. Mostaed, M., Aldabas, K., Olaverri-Monreal, C.: Approaching emergency vehicle warning (AEVW) (2017)
9. Sun, L., Li, Y., Gao, J.: Architecture and application research of cooperative intelligent transport systems. Procedia Eng. **137**, 747–753 (2016)
10. Nkoro, A.B., Vershinin, Y.A.: Current and future trends in applications of intelligent transport systems on cars and infrastructure. In: 17th International IEEE Conference on Intelligent Transportation Systems (ITSC), Qingdao, China, pp. 514–519. IEEE (2014)
11. Vojvodic, H., Mandzuka, S., Vujic, M.: The use of cooperative intelligent transport systems for vehicle emission charging. In: 2016 24th Telecommunications Forum (TELFOR), Belgrade, Serbia, pp. 1–4. IEEE (2016)
12. Sumalee, A., Ho, H.W.: Smarter and more connected: future intelligent transportation system. IATSS Res. **42**, 67–71 (2018)
13. Gary, O.G.Y.: Electronic vehicle identification in the intelligent city. In: IET Road Transport Information and Control Conference and the ITS United Kingdom Members' Conference (RTIC 2008), Institution of Engineering and Technology, Manchester, UK, p. 41 (2008)
14. Yang, Z., Pun-Cheng, L.S.C.: Vehicle detection in intelligent transportation systems and its applications under varying environments: a review. Image Vis. Comput. **69**, 143–154 (2018)
15. Janušová, L., Čičmancová, S.: Improving safety of transportation by using intelligent transport systems. Procedia Eng. **134**, 14–22 (2016)
16. Efimova, O.V., Dmitrieva, E.I., Tereshina, N.P., et al.: Eliminating transport losses with the aid of intelligent digital environment. In: 2018 IEEE International Conference on Quality Management, Transport and Information Security, Information Technologies (IT&QM&IS), St. Petersburg, pp. 103–105. IEEE (2018)
17. Papola, A., Tinessa, F., Marzano, V., et al.: Quantitative overview of efficiency and effectiveness of public transport in Italy: the importance of using ITS. In: 2017 5th IEEE International Conference on Models and Technologies for Intelligent Transportation Systems (MT-ITS), Naples, Italy, pp. 895–900. IEEE (2017)
18. Yatskiv, I., Savrasovs, M., Udre, D., et al.: Review of intelligent transport solutions in Latvia. Transp. Res. Procedia **24**, 33–40 (2017)
19. Fitah, A., Badri, A., Moughit, M., et al.: Performance of DSRC and WIFI for intelligent transport systems in VANET. Procedia Comput. Sci. **127**, 360–368 (2018)

20. Xu, Z., Li, X., Zhao, X., et al.: DSRC versus 4G-LTE for connected vehicle applications: a study on field experiments of vehicular communication performance. J. Adv. Transp. **2017**, 1–10 (2017)
21. El-Said, M., Bhuse, V., Arendsen, A.: An empirical study to investigate the effect of air density changes on the DSRC performance. Procedia Comput. Sci. **114**, 523–530 (2017)
22. Kljaic, Z., Briski, E., Vojvodic, H., et al.: Benefits of utilisation of GPS error mitigation models for intelligent transport systems. In: 2018 41st International Convention on Information and Communication Technology, Electronics and Microelectronics (MIPRO), Opatija, pp. 1121–1125. IEEE (2018)
23. Kabashkin, I.: Reliability of bidirectional V2X communications in the intelligent transport systems. In: 2015 Advances in Wireless and Optical Communications (RTUWO), Riga, Latvia, pp. 159–163. IEEE (2015)
24. Kabashkin, I.: Reliable v2x communications for safety-critical intelligent transport systems. In: 2017 Advances in Wireless and Optical Communications (RTUWO), Riga, pp. 251–255. IEEE (2017)
25. Siddiqi, K., Raza, A.D., Muhammad, S.S.: Visible light communication for V2V intelligent transport system. In: 2016 International Conference on Broadband Communications for Next Generation Networks and Multimedia Applications (CoBCom), Graz, Austria, pp. 1–4. IEEE (2016)
26. Sharma, N., Saini, V.: Analysis of intelligent transport system with optical vehicle-to-vehicle communication. In: 2015 International Conference on Soft Computing Techniques and Implementations (ICSCTI), Faridabad, India, pp. 155–158. IEEE (2015)
27. Fang, P., Bao, Y., Shen, J., et al.: A visible light communication based infra-to-vehicle intelligent transport demo system. In: 2015 International Conference on Connected Vehicles and Expo (ICCVE), Shenzhen, China, pp. 140–141. IEEE (2015)
28. Cailean, A., Cagneau, B., Chassagne, L., et al.: Visible light communications: application to cooperation between vehicles and road infrastructures. In: 2012 IEEE Intelligent Vehicles Symposium, Alcal de Henares, Madrid, Spain, pp. 1055–1059 IEEE (2012)
29. Coronado Mondragon, A.E., Coronado Mondragon, E.S., Coronado Mondragon, C.E., et al.: Estimating the performance of intelligent transport systems wireless services for multimodal logistics applications. Expert Syst. Appl. **39**, 3939–3949 (2012)
30. Ghosal, A., Bai, F., Debouk, R., et al.: Reliability and Safety/Integrity Analysis for Vehicle-to-Vehicle Wireless Communication. SAE Int. J. Passeng. Cars Electron. Electric. Syst. **4**, 156–165 (2011)
31. Drilo, B., Saric, D., Filjar, R.: The role of telecommunications in development of new-generation intelligent transport systems. In: 2009 1st International Conference on Wireless Communication, Vehicular Technology, Information Theory and Aerospace & Electronic Systems Technology, Aalborg, Denmark, pp. 125–127. IEEE (2009)

# Automated Detection of Craniofacial Landmarks on a 3D Facial Mesh

El Rhazi Manal(✉), Zarghili Arsalane, and Majda Aicha

Faculty of Science and Technology, Intelligent System and Application Laboratory, 2202 Fez, Morocco
{manal.elrhazi,arsalane.zarghili, aicha.majda}@usmba.ac.ma

**Abstract.** 3D faces analysis has always been an active research field in computer vision and virtual reality, more specifically, detecting 3D facial landmarks automatically is of high importance step that can be used for various purposes such as maxillofacial discipline, corrective, plastic and aesthetic surgery, in which a high number of landmarks are needed to achieve more accurate detection results. The problem of accurately detecting the 3D facial landmarks has received a lot of attention, and several research studies were conducted but few of them have detected a great number automatically and precisely. In this article, we introduce a novel technique to automatically localize 30 3D facial landmarks using the geometric information extracted from the facial 3D shape. This method was tested on two 3D face databases; the Basel Face Model dataset and the sample data set from the CranioGUI dataset. The experimental results prove that the presented methodology can detect the 30 landmarks for 75% of the 3D meshes on the two databases and the detection accuracy is comparable to the results obtained by manually landmarking the facial meshes.

**Keywords:** 3D face analysis · Landmarks detection · Geometric information · 3D shape

## 1 Introduction

Lately, with the expanded availability of three-dimensional scans, using 3D faces has become a good deal to overcome the traditional limitations mostly those related to 2D images, such as lighting conditions and viewpoint. This is an advantage also for various domains such as the area of craniofacial research where manually manipulation of faces has been the basic technique for the study of craniofacial dysmorphology, and surgical planning as well as the outcome assessment [1].

Therefore the use of 3D faces has become widely expanded in different disciplines, which first tend to localize the landmarks as a key step for further analysis. The 3D facial landmarks have a wide application in several domains as face recognition, face registration, facial attractiveness analysis, facial expression analysis, motion capture, facial mesh reconstruction, face relighting and face animation, which makes the task of automatically detecting them with accuracy the main challenge especially when a considerable number of 3D facial landmarks are needed. In such a case, avoiding time


A. Saka et al. (Eds.): CPI 2019, LNME, pp. 537–548, 2021.
https://doi.org/10.1007/978-3-030-62199-5_47

consuming and the made efforts during the manual labeling process as well as the errors that can be occurred during the landmarking are of great importance.

The existing methods for 3D facial landmarks detection can be classified into two categories: those that are based on the geometric information and those that are based on the trained statistical feature models [2]. Regarding the methods that are dependent on the trained statistical feature models, authors in [3–5] presented examples of this use. And of those based geometric information and shape descriptors, a detailed survey has been presented in our previous paper [6].

The geometric information-based methods may also be classified into four types: (1) methods based on curvature analysis, like in [7–9], (2) methods based on combining 2D texture with the 3D shape information as in [10–12], (3) methods based on matching the 3D query face with a manually landmarked one and establishing correspondences as in [2, 13, 14], and (4) methods based generic image descriptors [15–17] (see Related works section).

Geometry based techniques have attracted a good deal of attention, and have shown good results in 3D facial landmarks localization as they have the following strengths: discriminant capacity, quick to compute, concise to store, pose independent and efficient to match.

In the presented study, we provide a novel methodology to automatically detect 30 landmarks on a 3D face mesh, using geometric information provided by the meshes. A part of this work is actually inspired by [1], where the authors used as first step a geometric technique to detect 10 landmarks on 3D face mesh, then to detect additional 10 landmarks, they established a dense correspondence between the template of the 3D mesh and a complete set of 20 landmarks using a matching procedure.

This method has been chosen for the following reasons: the simplicity of the algorithm, the reliability of the obtained results, the number of the localized landmarks and the medical information basis on which they were based on as well. In this article, we aim to extend the geometric technique used to localize a wide number of 3D facial landmarks.

To assess the performances of the proposed method, we calculate the distance between the manually located and the automatically detected landmarks.

The major contributions of this paper are:

- Applying the geometric information based technique to localize a high number of 3D facial landmarks automatically; this for the best of our knowledge the first work using the geometric based technique to localize 30 landmarks.
- Performing a classification of geometric methods into four themes using a systematic mapping study of the already existing geometric methods for 3D landmarks detection.
- The accuracy of the localized landmarks is comparable to the manually detected ones on 3D meshes and the other proposed methods in the literature.
- Applying our method to different 3D face meshes datasets, and the obtained results show that the suggested method is good as the powerful existing method for 3D face landmarks detection.

The rest of this paper is as follows: In Sect. 2, we present some related works reported in the literature for detecting 3D facial landmarks using geometric information. The technique used to detect 3D facial landmarks is outlined in Sect. 3. Details regarding experimental results are discussed in Sect. 4, a discussion is presented in Sect. 5 and a conclusion in the last section.

## 2 Related Works

In this section, we introduce a classification of the geometric method for 3D landmarks detection by performing a systematic mapping study [18]. In the literature several papers have been suggested for landmarks detection from a 3D face using geometric methods, the systematic mapping is a technique to generate categories of approaches sharing the same techniques using previous works of the same research context. By searching for the articles that employ geometric methods for 3D facial landmarks detection in the period from 2010 to 2018, 30 papers were found after excluding papers that are not in the field. We characterize then the resulted papers based on the approaches used for landmarks detection.

Methods based geometric information used to detect 3D facial landmarks can be classified into four types:

- Methods based on curvature analysis:

A high number of works based curvature analysis have been published, for example, in [19] the authors detected 9 landmarks using differential analysis: the coefficients of the Fundamental Forms, the first, the second, and the mixed derivates, Curvedness and Shape Indexes, the curvatures, and the Tangent Map. In another paper [9] they extended this method to detect 17 landmarks. In [7] the same authors presented 105 new descriptors to detect 6 landmarks by deriving and composing primary ones. Newly generated descriptors are holistic or global and they have shown to have some advantages compared to local ones. Similarly in [20], to localize 14 landmarks, authors combined different surface curvature analysis descriptors and then they detected the points located on the face profile using the symmetric profile extraction. This extraction is based on valleys and peaks of a profile curve representing points with maximal and minimal values of a curvature. In [8] authors presented various combinations of the fundamental forms coefficients, the first, second and mixed derivatives, the curvatures, shape, and curvedness indexes as descriptors to detect 13 landmarks on fetal images suffering from cleft lip pathology. Moreover, authors in [21] combined the surface curvature analysis results and the obtained relief curve from the depth data in order to localize 5 facial landmarks.

- Methods based on combining 2D texture with the 3D shape information:

In [22] authors have first detected 3 landmarks using curvature analysis, then 7 additional points were localized using a 2D + 3D EBGM algorithm. In another work [23], authors proposed a new method to localize 17 landmarks based on efficient dense feature extraction (SURF-like descriptors) and a comparison of the 2D local landmark classification outcome with (projected) 3D shape vectors using the multiclass Hausdorff distance. In [24], the authors tried to map 3D facial models to 2D geometry-images using conformal mapping to localize 7 landmarks on the 2D face using the best method for that purpose. At last, a point to point mapping is used to map back all the detected landmarks in the 2D images to the 3D models.

- Methods based on matching the 3D query face with a manually landmarked one and then establishing correspondences:

An example of this approach is presented in [25] where authors aimed to localize 14 landmarks using a graph matcher trained with manually detected landmarks, this approach uses a query graph and a model graph as input and returns, for each node of the query, a set of probable candidate labels in the model with associated scores, then the finest labels are identified using a scale-adapted rigid registration. Additionally in [26], After aligning the face to a frontal pose, a set of 25 facial meshes were manually landmarked and used to generate a facial atlas of landmarks, as well as a curvature map for further meshes to be landmarked. Authors in [27] have detected 21 landmarks on the edge of the face by integrating the cascaded coupled-regressor approach with a 3D point distribution model a to predict the 3D facial landmarks learned from manually landmarked 3D scans.

- Approaches based on a generic image descriptors:

In [17], authors detected 6 facial landmarks using the Scale Invariant Feature Transform (SIFT) [28] by using windows around each feature and then identify the points with the largest scale as landmarks. Also in [15], 3D Constrained Local Model framework based on the 3D facial geometry features histogram is proposed to detect 3D facial landmarks, the geometric features used are: Histogram of Oriented Normal Vectors (HONV) [29] and Local Normal Binary Patterns (LNBP) [30] using 3D Point Distribution Model.

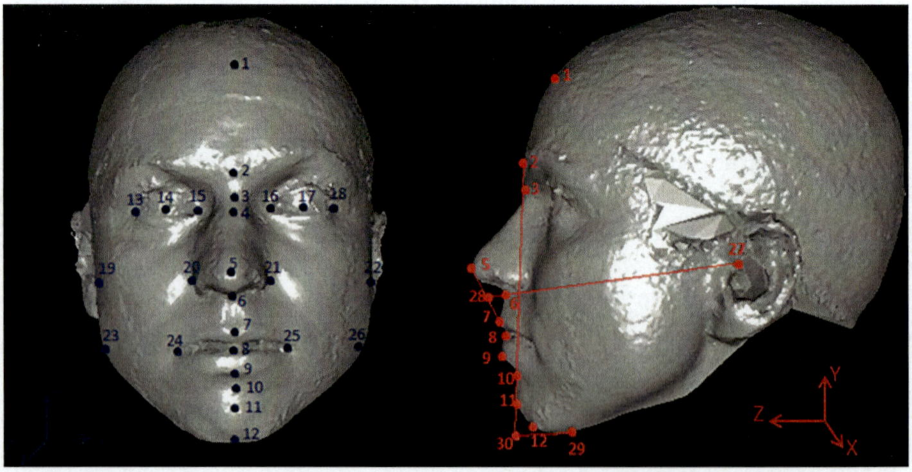


**Fig. 1.** 3D facial landmarks on a frontal face (left) and a profile side (right).

## 3 Method

This section presents our suggested method to localize 30 3D facial landmarks using geometric based technique as shown in Fig. 1. The detected landmarks are some of those needed in the craniofacial anthropometry, which is largely used in several fields to analyze and quantify the facial anthropometry. To achieve this goal, first the input 3D face must be preprocessed by normalizing its orientation and pose using the method presented in [31], that aims to correct the angular rotations of roll and yaw by minimizing the difference between the right and the left sides of the 3D face, then the pitch is processed by minimizing the difference between the height of the forehead and the height of the chin. Some 3D face needed a manual correction of the pose and orientation.

To detect the 3D facial landmarks, we followed the technique used in [1] to localize the 10 needed established landmarks which are: the pronasale (5), the subnasale (6), the left and the right alares (20), (21), the left and the right chelion (24), (25), and the exocanthion and the endocanthion on the left and the right eye (13), (15), (16), (18). And then we extend this approach to localize 19 additional landmarks on the 3D face, as shown in Fig. 1. In the extended method before detecting a landmark, the research zone is restricted using previously detected landmarks, which makes the step of 3D face pose normalization a crucial and necessary step on the detection process.

On the frontal side, to localize the pupils of the two eyes; (14) and (17) in Fig. 1, the research region is limited to $x(13) < x < x(15)$ and $y(15) \leq y \leq y(3)$ for the right eye and $x(16) < x < x(18)$ and $y(16) \leq y \leq y(3)$ for the left eye as the pupils are located in these regions, and then we find for each region the local maxima.

To detect the point in the middle of the two eyes (4) in Fig. 1, we find the intersection of the line having the x-value similar to the pronasale (5) in Fig. 1 and the line with the same y-value as the pupils. We add that the look of the eyes must be forward so that the pupils can be localized on the same line.

To locate the stomion (8) in Fig. 1, we use the results obtained for the two chelion, and we find the intersection of the line that has the x-value similar to the pronasale (5) and the one that goes along the chelion (24) and (25).

To find the ganthion (12) in Fig. 1, the region is limited to below the stomion $y < y(8)$, and then we find the point with the smallest y-value on the pronasale (5) line.

The sublabiale (10) is located by restriction of the region to $y(12) < y < y(8)$, and then we detect the local minima on the line that has the x-value similar to the pronasale (5).

To detect the pogonion (11) in Fig. 1, we restricted the region to $y(12) < y < y(10)$. and then we locate the local maxima in this area.

To locate the labiale superius (7) in Fig. 1, we restrict the region to $y(8) < y < y(6)$, and then on the line with the x-value as the pronasale, we locate the local maxima. The same thing for the labiale inferius (9), the local maxima is found on the restricted research region $y(10) < y < y(8)$.

The two upper cheek extremities (19) and (22) are detected by restricting the regions to $x < x(20)$ for the left point and $x > x(21)$ for the right one, and then on the line that has the same y-value as the pronasale (5), we find the local minima.

The same thing for the two lower cheek extremities (23) and (26) are detected by restricting the regions to $x < x(24)$ for the left point and $x > x(25)$ for the right one, and then on the line that has the same y-value as the stomion (8), we detect the smallest x-value on this line.

To detect the forehead point (1) in Fig. 1, we restrict the research region to $y > y(3)$, and then on the line whit the x-value as the pronasale (5), we detect the local minima.

The last detected point on the frontal side is the nasale (2) in Fig. 1, this point is detected by restricting the research region to $y(3) < y < y(1)$, and starting from the sellion (3), we locate the first local maxima on the line that has the x-value similar to the pronasale (5).

On the profile side, we change the orientation of the head by rotation of 90° then four landmarks are detected. First, the tragion (27) is localized by restricting the research zone to $z < z(18)$ and $y(6) < y < y(18)$, and then we detect the local minima.

To find the subnasale angle (28) in Fig. 1, we detect the intersection of the line that connects the tragion (27) and the subnasale (6), and the one that connects the pronasale (5) and the labiale superius (7).

To locate the point (29) in Fig. 1, the research region is limited to $y < y(11)$ and z $< z(12)$, and then we locate the first point having the tangent to the chin tip.

To localize the point (30) in Fig. 1 that forms the mentocervical angle, we detect the intersection of the tangent that goes along the chin tip (29) and the line that connects the nasale (2) and the pogonion (11).

## 4 Experimental Results

In this section, we assess the reliability and the efficiency of the proposed approach to localize 30 3D facial landmarks on two 3D face databases, and then we compare our results to those achieved by [1], and to other recent works in the literature.

The accuracy is measured by calculating the average distance using the Euclidian distance and the standard deviation of the localized landmarks to those detected manually.

The databases used in this experiment are the BaselFaceModel database [32] and the sample data set from the CranioGUI dataset [33].

For both databases, we manually locate the 30 landmarks on the 3D faces, then to perform our extended approach. For a given face we calculate the average error (mm) of the 3D landmarks detection as well as the standard deviation using the Euclidean distance as shown in Table 1.

For each automatically detected facial landmark using our method (first column in Table 1), we have calculated the average distance to the corresponding manually annotated one as well as the standard deviation for each face in the tested 3D face databases. The average distance of the automatically detected 30 landmarks to the manually labeled points for the sample data set from the CranioGUI dataset is 3.59 mm and the average standard deviation for the same dataset is 2.32 mm. For the BaselFaceModel database, the average distance of the 30 detected landmarks to their manually annotated points is 4.01 mm and the average standard deviation is 2.54 mm. both of these results prove the significant accuracy of the presented method.

**Table 1.** Average distances and the standard deviations of automatically detected landmarks to the manually located ones for the two 3D face databases.

| | Facial landmarks | Average distance and standard deviation for the sample data set from the CranioGUI dataset | Average distance and standard deviation for BaselFaceModel |
|---|---|---|---|
| 1 | Forehead | 3.56 ± 3.16 mm | 4.27 ± 2.41 mm |
| 2 | Glabella | 2.57 ± 1.19 mm | 2.58 ± 2.2 mm |
| 3 | Nasion | 4.83 ± 5.17 mm | 3.41 ± 2.15 mm |
| 4 | Eyes centre | 1.46 ± 2.68 mm | 2.43 ± 1.93 mm |
| 5 | Pronasale | 2.49 ± 1.65 mm | 2.56 ± 2.03 mm |
| 6 | Subnasale | 2.06 ± 1.48 mm | 2.16 ± 1.66 mm |
| 7 | Labliale superius | 5.47 ± 3.89 mm | 3.96 ± 2.86 mm |
| 8 | Stomion | 2.38 ± 1.54 mm | 3.09 ± 3.93 mm |
| 9 | Labiale inferius | 5.88 ± 4.92 mm | 3.34 ± 3.78 mm |
| 10 | Sublabiale | 3.35 ± 2.91 mm | 4.03 ± 2.11 mm |
| 11 | Pogonion | 1.34 ± 1.89 mm | 2.87 ± 2.24 mm |
| 12 | Ganthion | 3.21 ± 2.75 mm | 3.4 ± 3.05 mm |
| 13 | Endocanthion left | 4.88 ± 2.85 mm | 4.36 ± 1.29 mm |
| 14 | Lefts eye pupil | 2.96 ± 2.43 mm | 2.47 ± 2.06 mm |
| 15 | Exocanthion left | 4.05 ± 1.97 mm | 3.33 ± 2.51 mm |
| 16 | Exocanthion right | 2.98 ± 2.75 mm | 3.05 ± 2.28 mm |
| 17 | Right eye pupil | 2.69 ± 1.68 mm | 2.78 ± 1.94 mm |
| 18 | Endocanthion right | 2.7 ± 1.47 mm | 3.37 ± 2.61 mm |
| 19 | Cheek's superius point left | 4.59 ± 2.5 mm | 5.32 ± 1.18 mm |
| 20 | Alare left | 2.62 ± 1.52 mm | 2.77 ± 1.88 mm |
| 21 | Alare right | 1.87 ± 1.52 mm | 3.97 ± 1.94 mm |
| 22 | Cheek's superius point right | 6.3 ± 3.67 mm | 3.79 ± 2.11 mm |
| 23 | Cheek's inferius point left | 7.6 ± 2.01 mm | 8.73 ± 3.5 mm |
| 24 | Chelion left | 3.43 ± 2.18 mm | 4.29 ± 2.31 mm |
| 25 | Chelion right | 3.93 ± 3.88 mm | 4.35 ± 2.25 mm |

(*continued*)

**Table 1.** (*continued*)

| | Facial landmarks | Average distance and standard deviation for the sample data set from the CranioGUI dataset | Average distance and standard deviation for BaselFaceModel |
|---|---|---|---|
| 26 | Cheek's inferius point right | 6.61 ± 3.41 mm | 6.24 ± 4.09 mm |
| 27 | Ear point (Tragion) | 8.73 ± 6.04 mm | 8.69 ± 5.32 mm |
| 28 | Subnasale angle | 1.54 ± 0.95 mm | 3.06 ± 2.4 mm |
| 29 | Menton | 6.72 ± 3.43 mm | 6.45 ± 2.72 mm |
| 30 | Ganthion 2 | 6.98 ± 4.19 mm | 5.15 ± 3.4 mm |

By comparing our results to those obtained by [1] and other recent works in the literature that aim to localize 3D landmarks on neutral and frontal faces using similar databases (Table 2), we confirm that the achieved results by the extended approach are as good as those reported by the original work [1] and some other competing methods. Moreover, the presented approach can detect a large number of 30 3D facial landmarks unlike the presented approaches in the literature. As presented in Table 2, by using our approach, the average distance of the automatically detected 30 landmarks to the manually labeled points for the sample data set from the CranioGUI dataset is 3.59 mm and the average standard deviation is 2.32 mm.

Authors in [10] have detected 8 landmarks with an average distance of 4.26 mm to the manually located landmarks. Authors claimed that the proposed approach can locate 3D facial landmarks on faces even in the presence of occlusions and facial expression variations. Similarly, in a competing work, authors have located 17 landmarks [11] and claimed that the average distance and the average standard deviation to the manually labeled 3D landmarks are 1.39 mm and 0.98 mm respectively. This method has proved to be effective and robust in locating the 17 facial landmarks even for 3D faces from different ethnicities comparing to other works in the literature. In [34], authors have detected 9 facial landmarks with an average distance of 2.67 mm and a standard deviation of 10.12 mm to the manually located landmarks. This method has shown some drawbacks concerning the accuracy of the detection as well as the localization process as some landmarks were reported missing. In [9], 8 facial landmarks have been located with an average distance of 2.6 mm and a standard deviation of 2.7 mm to the manually located landmarks. This method is an extension of the previously discussed method; the obtained results show an improvement of the detection accuracy due to the partitioning of the research zones. In [13], authors have located 18 facial landmarks with an average distance of 3.38 mm and a standard deviation of 0.94 mm to the manually detected landmarks. The proposed method is one of the methods that can localize a high number of 3D facial landmarks with a significant accuracy; this number can be extended by using a dense correspondence technique. In another work [14], 14 facial landmarks were localized with an average distance of 3.34 mm and a standard deviation of 1.93 mm to the manually detected landmarks. The authors claimed that the presented approach can handle missing data as well as facial occlusions. Authors in [24] have detected 7 facial landmarks with an average distance of 2.6 mm and a standard deviation of 1.95 mm to the manually detected landmarks. In [35], authors have detected 13 landmarks with an average distance of 4.21 mm to the manually detected 3D facial landmarks, the suggested method can localize an important number of facial landmarks on faces and can also be applied for faces in the presence of facial expression variations and occlusions, the reported detection mean error can be due to the thresholding technique used in the paper. In the original work, 10 facial landmarks have been detected using a geometric-based information technique with an average distance of 3.12 mm and a standard deviation of 1.98 mm to the manually detected 3D facial landmarks.

In fact, the high number of 3D facial landmarks are often needed for various purposes such as the craniofacial anthropometry field that include a large set of face analysis studies, these studies require a wide number of facial landmarks to be detected accurately in order to get a great and relevant results of analysis, especially in the field the facial aesthetic quality analysis where the more landmarks are detected the more the analysis is reliable.

**Table 2.** Comparison of the average distances and the standard deviations of the presented method against some existing methods in the literature.

| | Our approach | [1] | [10] | [11] | [34] | [9] | [13] | [14] | [24] | [35] |
|---|---|---|---|---|---|---|---|---|---|---|
| Average distance | 3.59 | 3.12 | 4.26 | 1.39 | 2.67 | 2.6 | 3.38 | 3.34 | 2.6 | 4.21 |
| Standard deviation | 2.32 | 1.98 | – | 0.98 | 10.12 | 2.7 | 0.94 | 1.93 | 1.95 | – |
| Number of the detected landmarks | 30 | 10 | 8 | 17 | 9 | 8 | 18 | 14 | 7 | 13 |

## 5 Discussion

3D facial landmarks detection has been discussed in many research studies over the past decades, and a number of research studies have been suggested to achieve good and accurate landmarks detection. Particularly, the localization of the 3D facial landmarks using the geometric information provided by the mesh has proved to be relevant as reported by many research studies in which authors have compared the efficiency of the methods based on the geometric information to other proposed methods in the literature such as the statistical-based methods for 3D facial landmarks detection.

In the presented paper, we introduced an extended approach based on the analysis of the 3D geometric information provided by a mesh of the original work in [1]. In [1], authors have detected 10 3D facial landmarks using a geometric information-based method and then they applied a template matching approach to localize an additional 10 landmarks. In our work, we aimed to localize a larger set of 3D facial landmarks to satisfy the needs of certain craniofacial studies such as the analysis of 3D faces aesthetic quality, as the domain experts need a high number of landmarks to get more reliable results of analysis [36, 37].

In the presented paper, two 3D face databases were used; the sample data set from the CranioGUI dataset and the BaselFaceModel database. Experimental results showed that the presented method is as effective as the first study and the other works based geometric information presented previously in the literature.

As Table 2 reported, the obtained results highlight the accuracy of our approach in detecting a high number of 3D facial landmarks.

Concerning the computational efficiency of the presented approach, a PC with: Intel Core 2 Duo, 2.20 GHz with 3 GB RAM is used. The required time for the detection and the labeling for each 3D scan is 4.42 (s) as the average which makes it applicable in many face analysis applications in the real world. While the required time for the original study in [1] to detect 10 landmarks is 4 (s) as average. Table 3 illustrates the computational cost of the proposed approach for both the sample data set from the CranioGUI dataset and the Basel Face Model database.

**Table 3.** Computational cost of the proposed approach

| 3D database | Average computational cost for each scan |
|---|---|
| A sample data set from the CranioGUI dataset | 4.42 (s) |
| Basel Face Model database | 6.86 (s) |

## 6 Conclusion

In this paper, we introduced a new approach to detect 30 3D facial landmarks using the geometric information provided by the 3D mesh, this approach is extended from the original study of Liang [1] who has detected 10 landmarks using a similar method. The experimental results prove that our method is as great as the original study and many other presented works in the literature for detecting the 3D facial landmarks using geometric information. Particularly, our approach can localize a high number of 30 facial landmarks accurately with a significant speed of detection.

## References

1. Liang, S., Wu, J., Weinberg, S. M., Shapiro, L.G.: Improved detection of landmarks on 3D human face data. In: 35th Annual International Conference of the IEEE, Engineering in Medicine and Biology Society (EMBC), Osaka, Japan, pp. 6482–6485 (2013)
2. Perakis, P., Passalis, G., Theoharis, T., Kakadiaris, I.A.: 3D facial landmark detection and face registration. Technical report, University of Athens (2010)
3. Xu, C., Wang, Y., Tan, T., Quan, L.: Robust nose detection in 3D facial data using local characteristics. In: the International Conference on Image Processing, Singapore, pp. 1995–1998 (2014)
4. Zhao, X., Dellandrea, E., Chen, L., Kakadiaris, I.A.: Accurate landmarking of three-dimensional facial data in the presence of facial expressions and occlusions using a three-dimensional statistical facial feature model. IEEE Trans. Syst. Man, Cybern. Part B **41**(5), 1417–1428 (2011)

5. Yu, T.-H., Moon, Y.-S.: A novel genetic algorithm for 3D facial landmark localization. In: 2nd IEEE International Conference on Biometrics: Theory, Applications and Systems, Arlington, VA, USA, pp. 1–6 (2008)
6. Elrhazi, M., Zarghili, A., Majda, A.: Survey on the approaches based geometric information for 3D face landmarks detection. IET Image Process **13**(8), 1225–1231 (2019)
7. Marcolin, F., Vezzetti, E.: Novel descriptors for geometrical 3D face analysis. Multimedia Tools Appl. **76**(12), 13805–13834 (2016)
8. Moos, S., et al.: Cleft lip pathology diagnosis and foetal landmark extraction via 3D geometrical analysis. Int. J. Interact. Des. Manuf. (IJIDeM) **11**(1), 1–18 (2014)
9. Vezzetti, E., Marcolin, F.: 3D landmarking in multiexpression face analysis: a preliminary study on eyebrows and mouth. Aesthetic Plast. Surg. **38**(4), 796–811 (2014)
10. Perakis, P., Theoharis, T., Kakadiaris, I.A.: Feature fusion for facial landmark detection. Pattern Recognit. **47**(9), 2783–2793 (2014)
11. Guo, J., Mei, X., Tang, K.: Automatic landmark annotation and dense correspondence registration for 3D human facial images. BMC Bioinform. **14**(1), 232 (2013)
12. Fan, X., Wang, H., Luo, Z., Li, Y., Hu, W., Luo, D.: Fiducial facial point extraction using a novel projective invariant. IEEE Trans. Image Process. **24**(3), 1164–1177 (2015)
13. Gilani, S.Z., Shafait, F., Mian, A.: Shape-based automatic detection of a large number of 3D facial landmarks. In: IEEE Conference on Computer Vision and Pattern Recognition, Boston, MA, USA, pp. 4639–4648 (2015)
14. Sukno, F.M., Waddington, J.L., Whelan, P.F.: 3-D facial landmark localization with asymmetry patterns and shape regression from incomplete local features. IEEE Trans. Cybern. **45**(9), 1717–1730 (2015)
15. Cheng, S., Zafeiriou, S., Asthana, A., Pantic, M.: 3D facial geometric features for constrained local model. In: IEEE International Conference on Image Processing, Paris, France, pp. 1425–1429 (2014)
16. Bennamoun, M., Guo, Y., Sohel, F.: Feature selection for 2D and 3D face recognition. In: Encyclopedia of Electrical and Electronics Engineering, Hoboken, NJ (2015)
17. Berretti, S., Ben Amor, B., Daoudi, M., Del Bimbo, A.: 3D facial expression recognition using SIFT descriptors of automatically detected keypoints. Vis. Comput. **27**(11), 1021 (2011)
18. Petersen, K., Feldt, R., Mujtaba, S., Mattsson, M.: Systematic mapping studies in software engineering. In 12th International Conference on Evaluation and Assessment in Software Engineering, Italy, pp. 68–77 (2008)
19. Vezzetti, E., Marcolin, F., Stola, V.: 3D human face soft tissues landmarking method: an advanced approach. Comput. Ind. **64**(9), 1326–1354 (2013)
20. Galvánek, M., Furmanová, K., Chalás, I., Sochor, J.: Automated facial landmark detection, comparison and visualization. In: Proceedings of the 31st Spring Conference on Computer Graphics, Smolenice, Slovakia, pp. 7–14 (2015)
21. Segundo, M.P., Silva, L., Bellon, O.R.P., Queirolo, C.C.: Automatic face segmentation and facial landmark detection in range images. IEEE Trans. Syst. Man, Cybern. Part B **40**(5), 1319–1330 (2010)
22. Gupta, S., Markey, M.K., Bovik, A.C.: Anthropometric 3D face recognition. Int. J. Comput. Vis. **90**(3), 331–349 (2010)
23. Sangineto, E.: Pose and expression independent facial landmark localization using dense-SURF and the Hausdorff distance. IEEE Trans. Pattern Anal. Mach. Intell. **35**(3), 624–638 (2013)
24. Fan, X., Jia, Q., Huyan, K., Gu, X., Luo, Z.: 3D facial landmark localization using texture regression via conformal mapping. Pattern Recognit. Lett. **83**, 395–402 (2016)

25. Creusot, C., Pears, N., Austin, J.: 3D face landmark labelling. In: Proceedings of the ACM Workshop on 3D Object Retrieval, Firenze, Italy, pp. 27–32 (2010)
26. Li, M., et al.: Rapid automated landmarking for morphometric analysis of three dimensional facial scans. J. Anat. **230**(4), 607–618 (2017)
27. Jourabloo, A., Liu, X.: Pose-invariant 3D face alignment. In: Proceedings of the IEEE International Conference on Computer Vision, Santiago, Chile, pp. 3694–3702 (2015)
28. Lowe, D.G.: Distinctive image features from scale-invariant keypoint. Int. J. Comput. Vis. **60**(2), 91–110 (2004)
29. Tang, S. et al.: Histogram of oriented normal vectors for object recognition with a depth sensor. In Asian Conference on Computer Vision, pp. 525–538 (2012)
30. Sandbach, G., Zafeiriou, S., Pantic, M.: Local normal binary patterns for 3D facial action unit detection. In: 19th IEEE International Conference on Image Processing, Orlando, FL, USA, pp. 1813–1816 (2012)
31. Wilamowska, K., Shapiro, L., and Heike, C.: Classification of 3D face shape in 22q11. 2 deletion syndrome. In: IEEE International Symposium on Biomedical Imaging: From Nano to Macro, Boston, MA, USA, pp. 534–537 (2009)
32. Paysan, P., Knothe, R., Amberg, B., Romdhani, S., Vetter, T.: A 3D face model for pose and illumination invariant face recognition. In: Sixth IEEE International Conference on Advanced Video and Signal Based Surveillance, Genova, Italy, pp. 296–301(2009)
33. Brinkley, J.F., Fisher, S., Harris, M.P., Holmes, G., Hooper, J.E., Jabs, E.W., Jones, K.L., Kesselman, C., Klein, O.D., Maas, R.L., Marazita, M.L., Selleri, L., Spritz, R.A., van Bakel, H., Visel, A., Williams, T.J., Wysocka, J., FaceBase Consortium, Chai, Y.: The FaceBase Consortium: a comprehensive resource for craniofacial researchers [WWW Document] (n. d.). https://doi.org/10.1242/dev.135434. Epub 10 June 2016
34. Vezzetti, E., Marcolin, F.: Geometry-based 3D face morphology analysis: soft-tissue landmark formalization. Multimedia Tools Appl. **68**(3), 895–929 (2012)
35. Vezzetti, E., Marcolin, F., Tornincasa, S., Ulrich, L., Dagnes, N.: 3D geometry-based automatic landmark localization in presence of facial occlusions. Multimedia Tools Appl. **77** (11), 1–29 (2017)
36. Park, H.S., Rhee, S.C., Kang, S.R., Lee, J.H.: Harmonized profiloplasty using balanced angular profile analysis. Aesthetic Plast. Surg. **28**(2), 89–97 (2004)
37. Schmid, K., Marx, D., Samal, A.: Computation of a face attractiveness index based on neoclassical canons, symmetry, and golden ratios. Pattern Recognit. **41**(8), 2710–2717 (2008)

# Design of a Learner Model for Integration into an Adaptive Hypermedia System

Mehdi Tmimi[1(✉)], Mohamed Benslimane[1], Mohammed Berrada[2], and Kamar Ouazzani[2]

[1] Laboratory of Transmission and Processing of Information, EST, USMBA, Fez, Morocco
mehdi.tmimi@usmba.ac.ma, benslimane_mohamed@live.fr
[2] Computer and Interdisciplinary Physics Laboratory, ENS, USMBA, Fez, Morocco
mohammed.berrada@gmail.com, kamar_ouazzani@yahoo.fr

**Abstract.** This article is part of our research on adaptive hypermedia systems and specifically the development of the learner model.

In our previous work, we proposed a new conception of the learner model based on six facets describing all the information and data of the learner.

Certainly after later research, we have thought to redesign the learner model by making some changes in order to make it more efficient. so in this article we will introduce first our designed learner model. then its improved version while discussing the new modifications and update we have made. and finally we will conclude by an overview of the life cycle of the learner model within the adaptive hypermedia system.

**Keywords:** E-learning · Learner model · Adaptive hypermedia system · Emotions · Cognitive abilities · Knowledge · Competency

## 1 Introduction

The learner model is a very promising solution for the representation and description of learner information within adaptive hypermedia system. The adaptive hypermedia systems are E-learning systems that personalize the learning experience for the learners based on their preferences and current knowledge [1].

As our main goal is to design and develop our own adaptive hypermedia system, we are currently working on designing its three fundamental models. Which are: learner model, domain model and adaptation model.

So, this article presents our work done on the learner model that we published in two papers [2, 3].

In the first paper we proposed the basic elements of our six-faceted learner model. namely: (personal data, competency and knowledge, history, learning style, cognitive ability and emotional state). these six facets represent the different dimensions that the learner can have during the learning phases and have been deduced from a set of existing approaches and ontologies. Also our proposal model is founded on three different abstraction levels of the LMPA1234 approach [4]


A. Saka et al. (Eds.): CPI 2019, LNME, pp. 549–558, 2021.
https://doi.org/10.1007/978-3-030-62199-5_48

While in the second paper, we presented our in-depth research on these six facets that we used to propose a complete conception of our learner model.

However, with the progress of our research we have thought to improve our proposal. In fact, the two facets ‘historic’ and ‘competency and knowledge’ have been the subject of our improvements, of which we have first introduced the recommendations of the Experience API [5] and CMI5 [6] and then delegate the facet of competency and knowledge to the domain model.

In the following sections, we will first detail the six facets of our proposed learner model while discussing all the work that contributed to the realization of this model. then we introduce the improved learner model and the motivations behind these improvements, and finally we will conclude with an overview of the life cycle of the learner model within the adaptive hypermedia system.

## 2 Detailed Studies on the Learner Model

First, we have studied the existing learner models found in the literature in order to collect as much information as possible to use them for its development. In fact, we have deduced 6 facets or dimensions of learner data that we illustrated in Fig. 1.

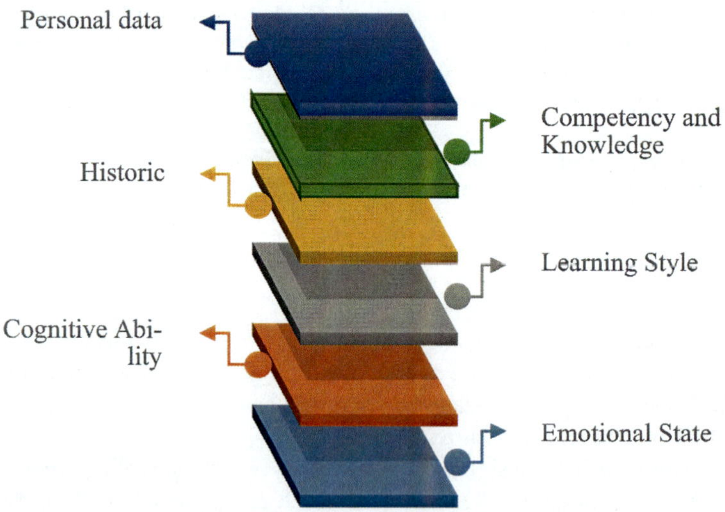


**Fig. 1.** The six facets of the learner model

### 2.1 Personal Data Facet

This facet contains all the personal data of the learner. such as: name, identity, gender, date of birth, etc. The particular point of this facet is its stability over time. In fact, the personal data do not change overtime, but it can have an influence on the learning activities, especially on the adaptation process. For example: The researcher Collier has proved that the successful acquisition of a language depends on the age of the learner [7].

### 2.2 Competency and Knowledge Facet

This facet contains any type of information regarding the knowledge and skills acquired by a learner. Such as: diplomas, certificates, projects, etc.

In the literature, we have a lot of work on the concept of competency. The main challenge was always to correctly define a competency and differentiate it from other terms such as performance, knowledge and skills [8].

In our first Learner Model proposal, we relied on a new approach called the Competency Approach [9]. This approach encompasses not only the learning model, but also the domain model.

### 2.3 Historic Facet

Historic facet is responsible for logging and reporting any type of actions performed by the learner during its learning activities within the adaptive hypermedia system.

Certainly this facet seems to be easy to implement. However we anticipate two major problems that are: the large amount of data to be stored and the diversity of action to be reported. Hence logging and archiving processes are inevitable to develop.

### 2.4 Learning Style/Preferences Facet

The learning style facet will manage all the preferences of the learner in terms of: type of learning, the nature and size of learning activities, the type of support desired, etc.

All these preferences have been the subject of many theories of the psychological profile. We cite the Felder-Silverman learning style model (FSLSM); that we chose to implement; combines many existing psychological theories and describes in more detail the behaviour of learners in their learning phase [10].

Finally, the mastery of languages and the physical limitations (handicap) of the learner were also introduced in this facet.

### 2.5 Cognitive Ability Facet

The cognitive ability translates the capacity to perform all the functions involved in cognition. Cognition can be defined as the mental process of knowing, including aspects such as: awareness, perception, reasoning, and judgment.

In the context of learning, this facet will describe the level of cognitive abilities of the learners whatever the domain of knowledge to acquire.

In the Table 1, we present some characteristics of the cognitive abilities of a learner.

**Table 1.** Example of learner's cognitive abilities

| Cognitive ability | Description |
|---|---|
| Memory capacity | Keeps a precise amount of information in the active state for a short time. |
| Reasoning ability | The ability to build concepts from examples |
| Information processing speed | How fast the learner is in acquiring information |
| Associative learning skills | It's the skill of linking new knowledge to old ones. |

To implement this facet, we used the referential proposed in the work of CAFOC Nantes/CFA Pays de la Loire [11].

This referential distinguishes 6 cognitive processes. namely: concentrate, develop his memory, know his learning style, communicate, organize, search and process information.

Each one of these cognitive process contains several levels on different elements. In the Tables 2 and 3, we show a description of the cognitive process “organize”:

**Table 2.** Descriptive of the cognitive process: ‘organize’

| | Level 1 | Level 2 | Level 3 |
|---|---|---|---|
| Define the act of learning | Identify the main aspects of the act of learning | Identify his own abilities | Develop his learning abilities |
| Characterize learning styles | Being able to identify his or her learning methods | Evaluate the one that suits him best | Implement learning methods adapted to different situations |

**Table 3.** Descriptive of the different learning styles in the context of learning activities

| Characterize learning styles | Descriptive |
|---|---|
| Visual | Apprehend objects or work to be done by visual representations |
| Auditory | Apprehends the objects or the work to be performed by auditory representations |
| kinaesthetic | Apprehend objects or work to be done by going directly to the object, appropriating it by manipulation |
| Deductive | From definitions, principles, to deductions |
| Inductive | Starting from a concrete situation to conceive an abstraction |
| Analytic | Apprehend each element separately and then make groupings to build a set |
| Global | First, understand the global structure then locate particular cases |
| Speak | Must verbalize before writing |
| Write | Must write before verbalizing |
| Reflexive | Progresses step by step, ensuring the accuracy of the results at different stages. Tend to postpone his answer until he is sure of his accuracy |
| Impulsive | Make a first gesture that he criticizes, makes mistakes that he rectifies later. Answer before thinking, the important thing being to be fixed right away |

In short, this referential groups cognitive abilities into dimensions, each dimension contains several elements containing several levels. This has been implemented in a similar way in our proposed model.

## 2.6 Emotional State Facet

This facet represents the emotional state of the learner. in the literature, we found many works that modelled emotions in relation to learning phases. In Fig. 2, We show the Kort's four quadrant model [12], which classifies the learner's emotions in two dimensions according to the learning phases in which he/she might be involved [13].

In our proposed model we took this model and added to it a third dimension called the dominance dimension [14], which represent the ability to control the situation that caused the emotion.

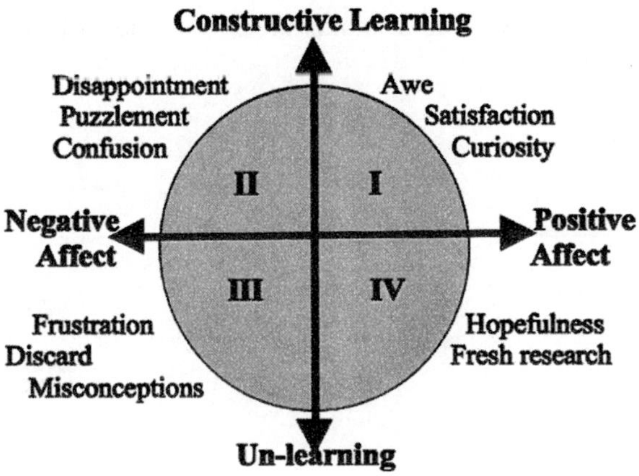


**Fig. 2.** The four-quadrant model of Kort, Reilly and Picard linking emotions in the learning phases [12]

For the identification of these states, there are three types of methods to detect emotions of the learner. In Table 4 we present the three types and their methods:

**Table 4.** Identification Methods of the emotional state of the learner

| Type | Method |
|---|---|
| Human observation | - Facial expression<br>- Movement of the head<br>- Gestures<br>- Speech<br>- Etc. |
| Sensor data | - Analysis of the face<br>- Analysis of the voice<br>- Physiological signs<br>- Field of text<br>- Etc. |
| Log DATA | - Correlation<br>- Classification<br>- Machine learning techniques<br>- Etc. |

# 3 Our Contribution

In this section we will respectively introduce our first and published proposal of the learner model, then we will present the motivation and the recommendations that we introduced into our model, and finally we will present our updated and improved proposal.

## 3.1 Our First Proposal

The above were a summary of our research on the six facets. in fact, we have done extensive research that allowed us to model the learner model using the UML2 class diagram shown in Fig. 3.

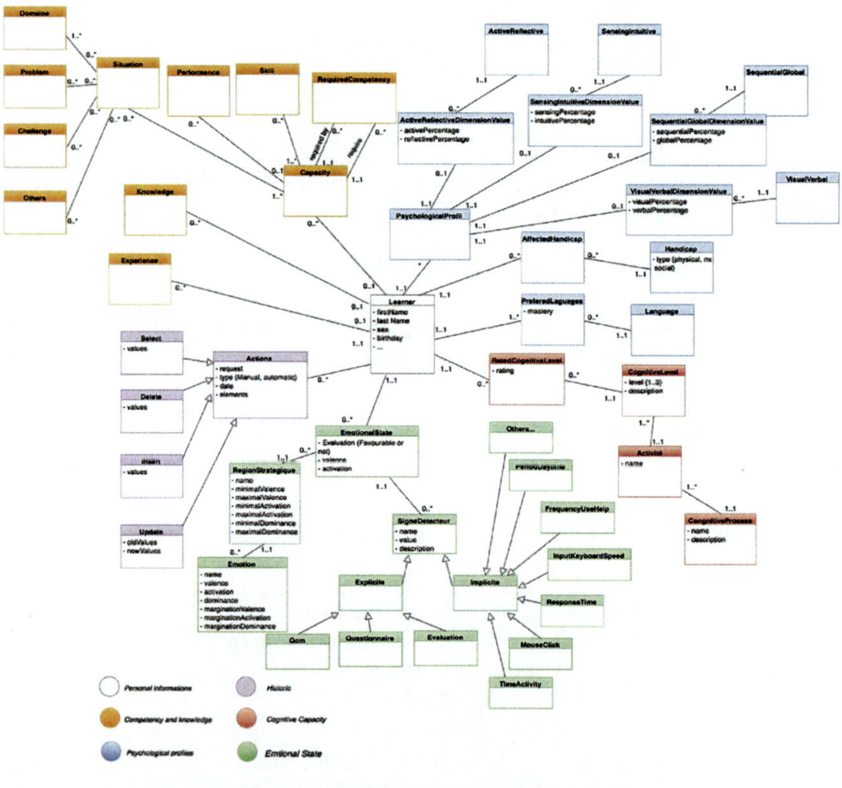


**Fig. 3.** Class diagram of our learner model

## 3.2 Modifications and Adjustments

As cited previously the two facets 'historic' and 'competency and knowledge' are the subject of the modifications that we made.

Regarding the historic facet, we wanted to introduce in our model the recommendations of xApi and CMI5. These specifications allow online learning software to record and track all types of learning experiences. So instead of recording the historic of learners in a set of transactions composed of multiple Insert/Delete/Update/Select statements, we will use the xAPI and CMI5 Statement Data Model shown in Fig. 4.

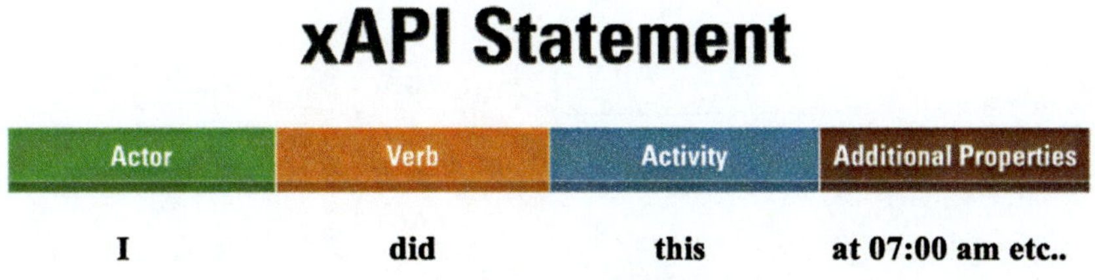


**Fig. 4.** Structure of the xAPI statement

Regarding the competency and knowledge facet, we used to have a very complicated conception shown with yellow colour in Fig. 3. In fact, the real information about competency and knowledge and all its relations and dependencies should be in the domain model.

In short, we will delegate the facet of competency and knowledge to the domain model while removing all the previous work we have done in this facet and linking its components to the learner as achieved objectives.

### 3.3 Our Improved Proposal

According to the different recommendations and updates mentioned above, we have developed the improved design of the learner model shown in Fig. 5.

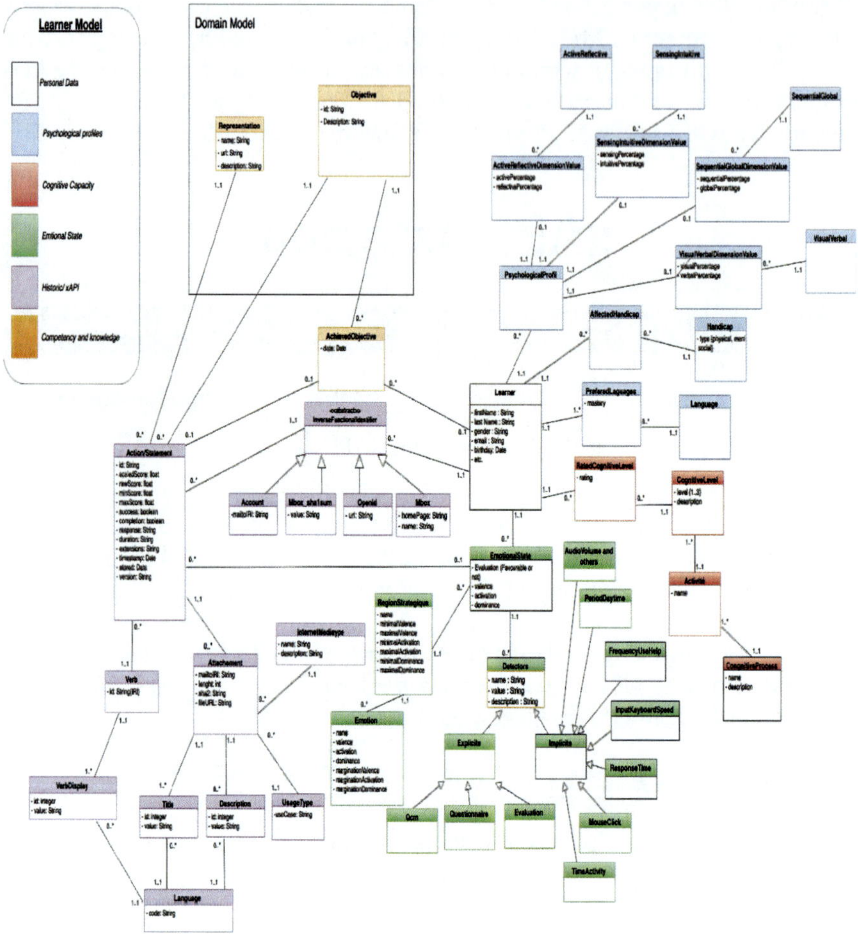

**Fig. 5.** Class diagram of our Improved Proposal of Learner Model

## 4 Life Cycle of the Learner Model

In addition to these 6 facets, our model will be based on the three levels of abstraction of the LMPA1234 approach. The goal is to make it more flexible and adaptable whatever the target area.

Indeed, the LMPA1234 approach defines the learner model on three levels. which are:

- Level 3: learner model design language.
- Level 2: learner model templates.
- Level 1: Learner model.

At Level 3 we will have all the elements of our learning model. And then at Level 2, instructional designers can make a selection and restriction of the elements of our model in order to adapt them to their strategic and pedagogic needs. In Table 5, we show an example of using such an approach where the instructional designer does not want to include age and gender in the adaptation process.

**Table 5.** Example of using the three levels of abstraction of LMPA123 on the personal data facet

| Level | Elements |
|---|---|
| Level 1 | - Last name<br>- First name<br>- Age<br>- Gender |
| Level 2 | Last name<br>First name |
| Level 3 | - Last name: Stevenson<br>- First name: TOM |

Finally, regarding the lifecycle of the learner model, we shown in Fig. 6 that first the learner model will be initialized and assigned to stereotype profiles that are supposed to partially represent the learner profile. And following a continuous assessment of information collected directly or indirectly from the learner, his learning profile will change and tends to get closer to his exact profile.

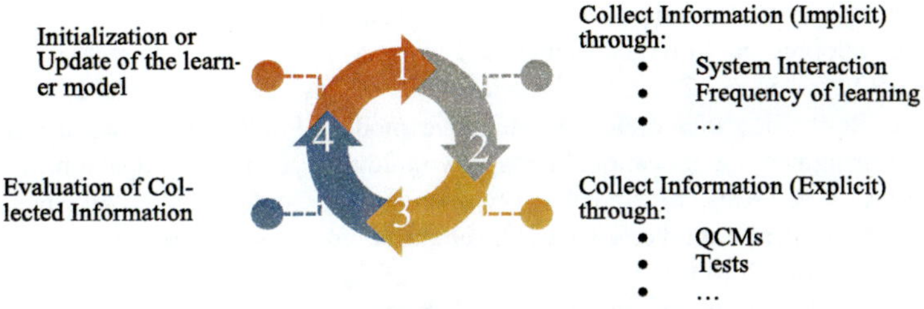


**Fig. 6.** Lifecycle of the Learner Model

## 5 Conclusion and Perspectives

In this article, we have presented first all the work that we have done on the learner model, from its abstract model to a detailed and improved design. Then we briefly explained the LMPA1234 approach as well as the life cycle of the learner model within adaptive hypermedia systems.

Nowadays, we are still looking to improve our proposal while working on the two remaining models (adaptation model and domain model) in order to build the complete architecture of our adaptive hypermedia system.

## References


1. De Bra, P., Aroyo, L., Cristea, A.: Adaptive web-based educational hypermedia. In: Mark, L., Alexandra, P. (eds.) Web Dynamics, Adaptive to Change in Content, Size, Topology and Use, pp. 387–410. Springer (2004)
2. Tmimi, M., Benslimane, M., Berrada, M., Ouazzani, K.: Elaboration d'une ontologie apprenant pour les hypermédias adaptatifs, Conférence Internationale sur les Approches Pédagogiques & E-Learning, Morocco, Appel'2016. (2016)
3. Tmimi, M., Benslimane, M., Berrada, M., Ouazzani, K.: A proposed conception of the learner model for adaptive hypermedia. Int. J. Appl. Eng. Res. **12**(24), 16008–16016 (2017)
4. Stéphanie, J., Thi-Thu-Hong, P.: Différents niveaux de modélisation pour des profils d'apprenants. EIAH 2011 (2011)
5. Chew, L.K.: Tracking Learning Experiences Using the Experience API. eLearning Forum Asia 2015 (ELFA) (2015)
6. Cohen, E., et al.: AICC/ CMI5_Spec_Current cmi5 Specification Profile for xAPI. Aviation Industry CBT Committee (AICC). (2016). https://github.com/AICC/CMI-5_Spec_Current/blob/quartz/cmi5_spec.md. Accessed 19 June 2019
7. Collier, V.P.: The Effect of Age on Acquisition of a Second Language for School. New Focus. The National Clearinghouse for Bilingual Education. No: 2, Winter: 1987–1988 (1988)
8. Miller, G.E.: The assessment of clinical skills/competence/performance. Acad. Med. **65**(9) (1990) (September supplement 1990)
9. Hachmoud, A., Khartoch, A., Oughdir, L., Alami, S.K.: Bridging the gap between competency based approach and intelligent tutoring systems. Int. J. Appl. Eng. Res. ISSN 0973-4562 **12**(19), 9099–9111 (2017)
10. Graf, S., Viola, S.R., Kinshuk, Leo, T., Representative Characteristics of Felder-Silverman Learning Styles: An Empirical Model. IADIS Press (2006)
11. Anon: Developing cognitive capacities and working methods of apprentices'-March 2008 GIP FCIP Experience - CAFOC of Nantes/ CFA of Pays de la Loire the Loire (2008)
12. Kort, B., Reilly, R., Picard, R.: An affective model of interplay between emotions and learning: reengineering educational pedagogy-building a learning companion. In: Okamoto, T., Hartley, K.R., Klus, J.P. (eds.) IEEE International Conference on Advanced Learning Technology: Issues, Achievements and Challenges, pp. 43–48. Wisconsin: IEEE Computer Society, Madison (2001)
13. Jraidi, I.: Modélisation des émotions de l'apprenant et interventions implicites pour les Systèmes Tutoriels Intelligents, University of Montreal (Canada), ProQuest Dissertations Publishing, 2014. NS28321 (2013)
14. Mehrabian, A.: Pleasure-arousal-dominance: a general framework for describing and measuring individual differences in temperament. Curr. Psychol. **14**(4), 261–292 (1996). https://doi.org/10.1007/bf02686918

# Reconfiguration of Flexible Manufacturing Systems Considering Product Morpho-Dimensional Characteristics and Modular Design

Chaimae Abadi[1](✉), Imad Manssouri[1], and Asmae Abadi[2]

[1] Laboratory of Mechanics, Mechatronics and Command, Team of Electrical Energy, Maintenance and Innovation, ENSAM-Meknes, Moulay Ismail University of Meknes, Meknes, Morocco
chaimae.abadi@gmail.com

[2] Mechanical and Industrial Engineering Department, INSA, Euro-Mediterranean University of Fez, B.P. 30 070 UEMF Campus, Fez, Morocco

**Abstract.** Recently, manufacturing companies are required to be more flexible and more reconfigurable in order to follow the continuous market changes and the high competition between trademarks. Actually, reconfiguration of Flexible Manufacturing Systems (FMS) has become one of the most widely used solutions by factories even if it stills hard to be implemented. Therefore, we propose in this paper a structured methodology based on modular design in order to reconfigure the FMS easily and efficiently. Our proposed approach takes into account product morpho-dimensional characteristics. To validate it, we apply it on a case study concerning the flexible machining cell of our high engineering school "ENSAM". Finally, we discuss the expected contributions of new artificial intelligence technologies in the context of Flexible Manufacturing Systems reconfiguration.

**Keywords:** Flexible Manufacturing Systems · Reconfiguration · Modular design · Product morpho-dimensional characteristics · MES · Artificial Intelligence

## 1 Introduction

Nowadays, rapid reconfiguration of Flexible Manufacturing Systems (FMS) has become one of the main interests of factories because of its ability to resolve problems of competitiveness and continuous market changes. In fact, the rapid advancement of technologies will continue to generate complexity and uncertainty [1]. So, flexibility and reconfigurability are required even if they aren't evident to implement.

Thus, in our paper, we present a structured approach in order to assist companies in the reconfiguration of their FMS basing on modular design and product morpho-dimensional characteristics. The integration of these two basic elements plays a key role in the quick identification of the changes and modifications location. Artificial Intelligence tools are explored for the possible automation of the reconfiguration approach.


A. Saka et al. (Eds.): CPI 2019, LNME, pp. 559–565, 2021.
https://doi.org/10.1007/978-3-030-62199-5_49

## 2 Flexible Manufacturing Systems (FMS)

### 2.1 Definition of FMS and Their Characteristics

As its name indicates, the Flexible Manufacturing System is a manufacturing system with a high flexibility. So, to better define it, it is indispensable to define firstly "the flexibility" concept. In fact, the flexibility is the ability of a manufacturing system to be quickly and automatically adapted to changing product life cycle factors: product and process parameters, loads, machine failure and instability of environment [2–4]. Thus, we conclude that flexible manufacturing system is a computer controlled programmable machining center configuration. It contains integrated software that manages the product life cycle: production planning, part programs, production tools and work orders [5, 6].

### 2.2 Flexible Manufacturing Systems and Reconfigurability

Manufacturing flexibility and reconfigurability are two keys of success in the marketplace for any factory. They allow the company to follow the continuous changes of its environment and to achieve a competitive advantage by adapting their products to customers need quickly and with a minimum cost and effort.

Concerning reconfiguration, it is done according to a specific scale composed of three levels [7]. The first one touches on the entire system. Actually, in this level, the majority of resources in the factory are reconfigured to be adjusted to a new manufacturing process. As for the second level, it implies the implementation of few new workstations in the factory. The third level involves minor updates of the control software, grippers, jig and fixtures, and other auxiliary devices. It is in this context that the reconfiguration of FMS is included. Actually, the reconfiguration of FMS doesn't need radical changes in the factory. It consists only in doing some updates of existing workstations programs in order to adapt the FMS to the new specifications of our product.

### 2.3 Flexible Manufacturing Systems Reconfiguration and Modular Design

Today, industrial companies face many challenges. Indeed, customer needs are numerous and renewable, product life cycles are too short and competition between companies, in terms of quality, of product life cycle time and of costs, is too high. So, industrials should adapt continually their processes in order to respond to these needs. In this context the modular design has appeared and it has proven its efficiency. Actually, its approaches allow changeability by sharing common components and a platform across a product family [8].

Thus, we conclude that the integration of modular design in a methodology of flexible manufacturing system reconfiguration will facilitate its implementation and will make it more effective. This is the major contribution of our paper.

## 3 Proposed Approach

Our aim in this paper is to develop a structured methodology to assist manufacturing engineers in the rapid reconfiguration of their flexible manufacturing systems. We present in Fig. 1 (see Fig. 1) the flowchart of the proposed approach.

We are placed in the industrialization phase of a new product. In our case, it is not a question of designing a new manufacturing process, but only of reconfiguring the existing manufacturing system. After the development of the product architecture, the reconfiguration process must be initiated.

The first step in our approach is to decompose the product into a set of modules. To do so, we will gather, in the same module, the parts that contribute to the insurance of a specific product function. Then, we will analyze each module j and check in the enterprise memory if similar modules, that can ensure the same functions with some minor modifications, have been previously developed by the design team.

If a similar module does exist in the data base, then a deductive approach will be applied in order to adapt the ancient programs and configurations to meet the new specificities of the module. We propose in this case to use the different techniques of Group Technology (GT). In fact, we will list all the parts of the considered module. GT propose to identify and group together into families the different parts in order to take advantage of their similarities in design and production [9]. Dedicated algorithms, using Jaccard similarity coefficient for example, for grouping products for FMS based on their operational similarities could be used in this step [10]. Then, we will assign a morpho-dimensional code to each part in order to specify to which family of parts it belongs exactly. Once it is specified, we will analyze the typical Process Plan of the parts' family and develop the new one by integrating complementary operations in the typical Process Plan. All we have to do now is to program the new operations to be performed on our monitoring information system (adjusting the teaching positions in the handling robot, doing some minor modifications in the G-code introduced to the CNC machine and in the Process Plans defined in the SCADA and MES software …).

If no similar module has been ever developed previously by the design team, then a generative approach will be adopted to develop the Process Plan of the new module. To do so, we will start by analyzing the definition drawing of each new part i (elementary surfaces, geometric shapes, cotations…) and we will classically identify the different machining entities, group the operations into phases and determinate the positioning of the part on the corresponding machine for each machining phase.

The same steps will be performed for all the new parts that are manufactured by machining in the FMS. Finally, the assembly sequence will be developed for this new module and the new configurations will be added in the SCADA and MES software. The details of these steps will be described in the case study developed in the next section of this paper.

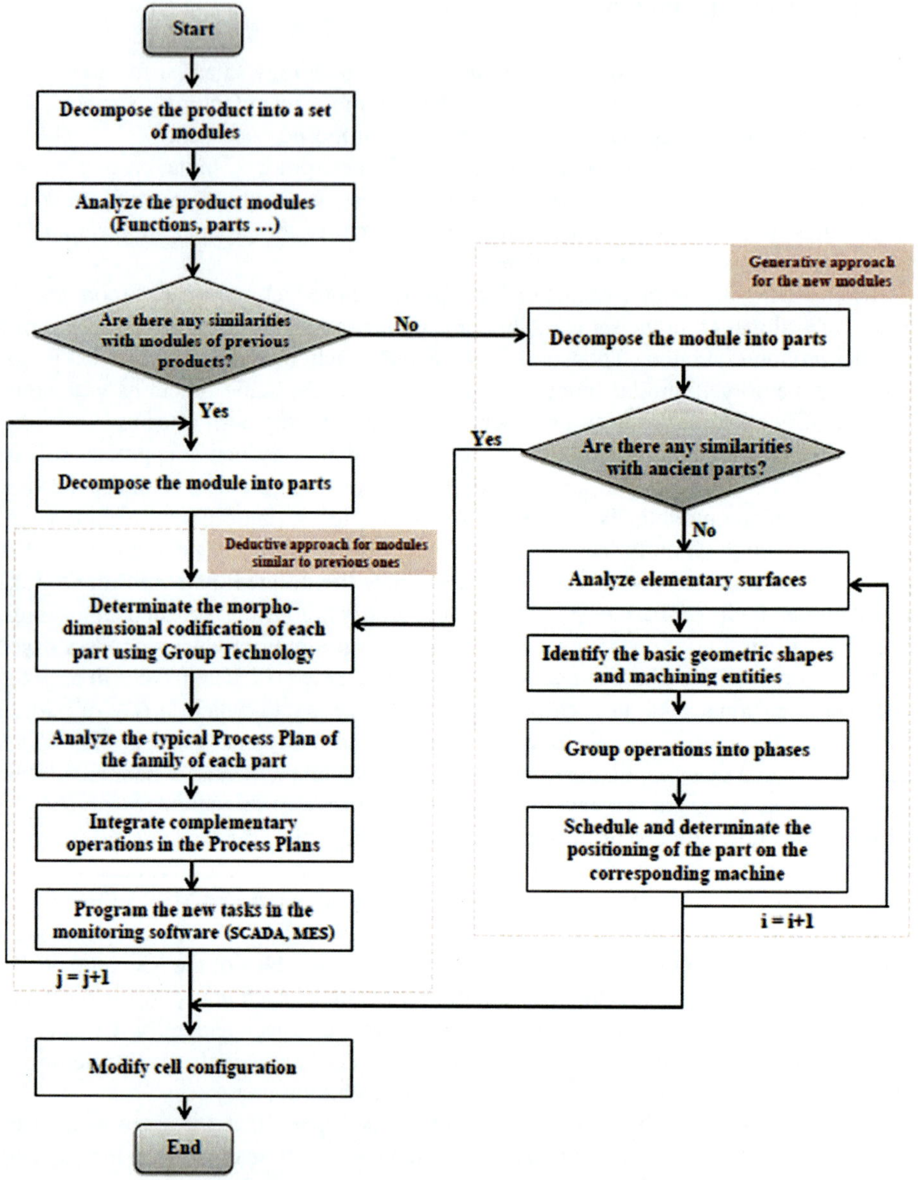


**Fig. 1.** The proposed approach considering product characteristics in the Reconfiguration of Flexible Manufacturing Systems.

## 4 Case of Study

In this section, we will develop a case study concerning the rapid reconfiguration of the flexible machining cell of ENSAM. To do so, we will follow the steps of our approach. The existing configuration of ENSAM's cell allows the production of different intermediate products for desk sets (baseplates and penholders) and variants of assembled Desk sets. The hardware and monitoring information system of the studied FMS is represented in Fig. 2 (see Fig. 2).

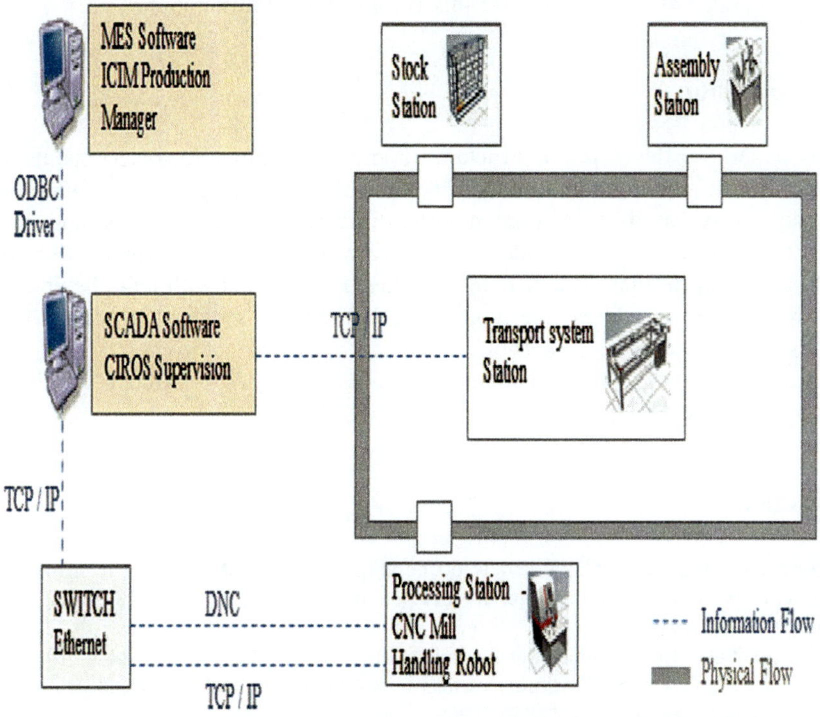


**Fig. 2.** The monitoring information system of ENSAM flexible cell.

The new product that we have to produce is a new variant of Desk sets with a third instrument (a watch) and with a new base plate (which has four holes instead of three). In our case study, the product is composed of one single module and similar ones have already been configured and produced in the flexible cell so we have to implement a deductive approach in order to only adjust the existing Process Plans and configurations.

We have started by defining the morpho-dimensional codification of each part of the module (for the base plate for example, it is 42 14X). We then introduced the new base plate information to the MES software ICIM Production Manager (an aluminum one with four holes). Then we modified the Typical Process Plan of the Base plates' family by adding a new operation to it. It consists in the drilling of the fourth hole. This reconfiguration has been done on the SCADA software CIROS Production [11] and on the milling program on the CNC machine.

The “Pick and Place” program of the handling robot has been also reconfigured in order to take into account the new dimension of the base plate. For the pen holder, the reconfiguration is the same. Predefined programs already exist in ICIM Manager Database and allow its production. In [12], the possibility of developing a PLM module within ENSAM cell in order to automate its reconfiguration process has been discussed. In the next section, we will discuss the different artificial intelligence technologies that can allow us to realize this automated reconfiguration.

## 5 Conclusion

In this paper, we developed a structured approach in order to assist manufacturing engineers in the reconfiguration of their Flexible Manufacturing Systems. Our proposed methodology is based on product morpho-dimensional characteristics and modular design and it takes into account the two main cases of deductive and generative product development approaches. The methodology has been applied within the flexible cell of ENSAM in order to address the monitoring information system (SCADA, MES …) reconfiguration issues.

As a perspective, we propose to use the new technologies of artificial intelligence in order to automate our approach.

## References

1. Prasad, D., Jayswal, S.C.: A review on flexibility and reconfigurability in manufacturing system. In: Innovation in Materials Science and Engineering, pp. 187–200. Springer, Singapore (2019)
2. Mascarenhas, M.B.: Planning for flexibility. Long Range Plan. **14**(5), 78–82 (1981)
3. Gupta, Y.P., Goyal, S.: Flexibility of manufacturing systems: concepts and measurements. Eur. J. Oper. Res. **43**(2), 119–135 (1989)
4. Nagarur, N.: Some performance measures of flexible manufacturing systems. Int. J. Prod. Res. **30**(4), 799–809 (1992)
5. Mehrabi, M.G., Ulsoy, A.G., Koren, Y., Heytler, P.: Trends and perspectives in flexible and reconfigurable manufacturing systems. J. Intell. Manuf. **13**(2), 135–146 (2002)
6. Browne, J., Dubois, D., Rathmill, K., Sethi, S.P., Stecke, K.E., et al.: Classification of flexible manufacturing systems. FMS Magazine **2**(2), 114–117 (1984)
7. Kim, D.Y., Park, J.W., et al.: A modular factory testbed for the rapid reconfiguration of manufacturing systems. J. Intell. Manuf., 1–20 (2019)
8. Agrawal, T., Sao, A., Fernandes, K.J., Tiwari, M.K., Kim, D.Y.: A hybrid model of component sharing and platform modularity for optimal product family design. Int. J. Prod. Res. **51**(2), 614–625 (2013)
9. Kashkoush, M., ElMaraghy, H.: Product family formation for reconfigurable assembly systems. Procedia CIRP **17**, 302–307 (2014)

10. Abdi, M.R., Labib, A.W.: Grouping and selecting products: the design key of reconfigurable manufacturing systems (RMSs). Int. J. Prod. Res. **42**, 521–546 (2004)
11. Festo: Festo Didactic, CIROS Supervisions, User Manual (2010)
12. Asmae, A., Souhail, S., El Moukhtar, Z., Ben-Azza, H.: Using ontologies for the integration of information systems dedicated to product (CFAO, PLM…) and those of systems monitoring (ERP, MES…). In: The International Colloquium on Logistics and Supply Chain Management, pp. 59–64. IEEE, Rabat (2017)

# Crack Propagation Modeling Using the Extended Isogeometric Analysis Technique

Soufiane Montassir[1,4(✉)], Abdeslam El Elakkad[2], H. Moustabchir[3], and Ahmed Elkhalfi[1,4]

[1] Mechanical Engineering Laboratory, Faculty of Sciences and Techniques, B.P 2202 Route Imouzzer, Fes, Morocco
Soufiane.montassir@usmba.ac.ma

[2] Department of Mathematics, Regional Centre for Professions of Education and Training, Fes, B.P: 243, Sefrou, Morocco

[3] Laboratory of Systems Engineering and Applications (LISA), National School of Applied Sciences of Fez, Fez, Morocco

[4] Department of Mechanical Engineering, Faculty of Science and Technology, Sidi Mohamed Ben Abdellah University Fez, Fez 30000, Morocco

**Abstract.** In this work, we implemented the extended isogeometric method for cracked structures in 2D. In order to approximate, the displacement fields we have worked with interpolation functions that are based on Non-Uniform Rational B-spline (NURBS). The traditional approximations used in the isogeometric method are extended by the insertion of enrichment functions, which are capable of capturing discontinuities and singularities in the crack tip. This method allows modeling the cracks without being in conformity with a given mesh. For the purpose of showing the ability to model cracks by this method, the result obtained by the extended isogeometric method is compared with the XFEM method.

**Keywords:** XIGA · XFEM · CRACK · B-spline

## 1 Introduction

Nowadays, the failures of many engineering structures caused mainly from crack-like surface defects. For this reason, the evaluation of different failure modes includes cracking is of major importance to ensure the exploitation of these structures. As a result, the treatment and prediction of cracks is a challenge for scientific researchers and finite element specialists. Even if the various works are done by the MEF [1] for the calculation of structures, it remains unable to simulate the problems having discontinuities caused by cracks, holes, other bi-material interfaces and modeling the propagation of discrete cracks. This last requires a remeshing on each increment and a mesh compliant, which is difficult and very expensive in terms of time.

To solve the numerical difficulties corresponds to the problems produced by the cracks, several techniques were introduced such as, The element free Galerkin approach [2], Boundary element method [3], Extended Finite element method [4], peridynamic models [5], modeling with phase-field [6]. Among all the various processing techniques


A. Saka et al. (Eds.): CPI 2019, LNME, pp. 566–576, 2021.
https://doi.org/10.1007/978-3-030-62199-5_50

for modeling structures, the most used method so far is the XFEM, this method based on the unit partition [7], and it allowed meshing the structure without taking into account the crack to describe the opening of the crack and the singularity at its tip, special shapes functions are inserted. The approximation of the geometry and the solution performed by various basic functions during the use of such calculation technique as FEM, XFEM. Consequently, they present discretization errors [8]. To eliminate these errors, Hugues et al. [9] developed a new computation tool called Iso- Geometric Analysis (IGA). This method created a relationship between computer-aided design CAD and finite element method FEM. The idea is to apply the basic functions of CAD in the geometric representation and to build the finite approximations. While the basic Lagrangian function with the finite element method is the best known in the CAE, the most basic functions widespread use of CAD are non-uniform and rational B-spline functions (NURBS). The use of the IGA approach offers us the ability to make a simple refinement as well as an exact description of the engineering structures, so the accuracy and robustness make distinguish the method from the conventional method.

In recent years the IGA has appeared successfully in various engineering problems [10] particularly in the mechanics of fracture [11]. Concerning cracking problems, to describe the phenomenon of discontinuity on the long lips of the crack and the singularity at the crack tip, the enrichment functions through the unit partition method integrated into IGA. There are many works like De Luycker et al. [12]; he applied the method XFEM and IGA to study the mechanics of linear rupture. In the following Gorashiet et al. [13] was extended this method to an enriched method (under the XIGA name) to capture the behavior of cracked structures in 2D. XIGA has been a real success in the modeling of fracture mechanics [14], cracking of a thin shell [15].

This work is done to study a stationary crack in a 2D structure using the implementation of the XIGA. We will present the concept of the XFEM and a generalization on the isogeometric method then a description of the XIGA method. Finally, simple examples to compare the new approach with the XFEM approach.

## 2 The Extended Finite Element Method

### 2.1 Representation of the Crack

The XFEM method has appeared as an alternative to the classical finite element method, it based on the principle of unit partition [7], it is able to follow the crack without being in conformity with the mesh thus it doesn't require to do this remeshing operation.

At first, the shape of the crack meshed with triangles independently of the structural mesh.

Second, the level set functions called ($\psi$, $\varphi$) are calculated at nodes of the geometry around to the crack surface as despite in Fig. 1.

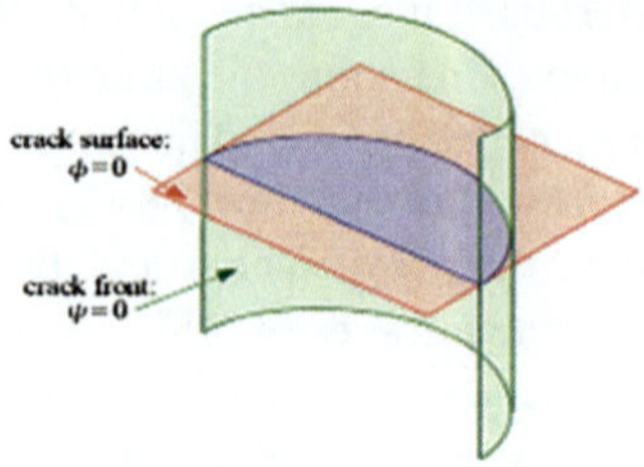


**Fig. 1.** Level set function

## 2.2 Enrichment Approach

The essential idea of XFEM is to add enrichment function to the classical finite element method [16]. Therefore, after the crack is presented by level set function, it is achievable to enrich the displacement field by inserting enrichment shape functions and related degrees of freedom to the discretization. In fact, the mesh of the structure is created without any regard of the crack and XFEM could take it into account only by appending specific enrichment function. According to the partition of unity, discontinuous shape function (called Heaviside) presented by Eq. (1) is added at the nodes corresponds to the elements completely cut by the crack, thus the singular shape functions described by Eq. (2) are added at the nodes of the elements including the crack front.

$$\mathrm{H(x)} = \begin{cases} +1 \text{ if } \varphi(\mathrm{x}) > 0 \text{ above the crack} \\ -1 \text{ if } \varphi(\mathrm{x}) < 0 \text{ Below the crack} \end{cases} \tag{1}$$

$$\begin{gathered} F_1(\mathrm{x}) = \sqrt{r}\sin(\theta/2),\ F_2(\mathrm{x}) = \sqrt{r}\sin(\theta/2)\sin(\theta), \\ F_3(\mathrm{x}) = \sqrt{r}\cos(\theta/2),\ F_4(\mathrm{x}) = \sqrt{r}\cos(\theta/2)\sin(\theta). \end{gathered} \tag{2}$$

With r and $\theta$ are polar coordinates attached to the crack tip.

Let $\Omega$ represent the solution domain, $\Omega_H$ the nodes of the elements cut by the crack and $\Omega_F$ the nodes of the elements which crack tip placed. So the displacement field with the XFEM discretization can be represented as follows:

$$U(x) = \sum\nolimits_{i\in\Omega} N_i(x)u_i + \sum\nolimits_{i\in\Omega_H} N_i(x)H(x)a_i + \sum\nolimits_{i\in\Omega_F} N_i(x)\left[\sum\nolimits_{j=1,..4} F_J(x)b_{J,i}\right] \tag{3}$$

Where $N_i$ is the finite element shape function linked to the node i, then $u_i$, $a_i$, $b_{J,i}$ represent the classical, discontinuous and singular degrees of freedom [16].

The isotropic enrichment functions represented in Eq. (2) are the basic and the most used type of enrichment functions [17]. Since the structure has an elastic behavior, the use of this type of enrichment function increases the precision of the approximation.

In order to obtain an optimal discretization, that is to say, to have a minimum number of degrees of freedom, Fig. 2 Shown the topology of enrichment followed.

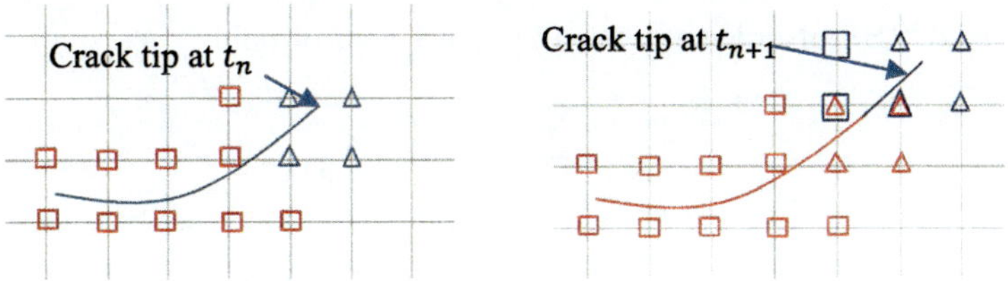


**Fig. 2.** Enrichment technique

Only the elements that have cut through the crack enriched. There are two strategies of enrichment; geometry enrichment which based on the implementation of a fixed zone around the crack tip. This method improves the convergence of the solution [18], for the other approach that might have been examined is the degrees of freedom, collecting that has been presented in [19].

During the propagation of the crack, novel singular and discontinuous degrees of freedom are included in the approximation (see Fig. 2).

# 3 A Brief Presentation of the Isogeometric Analysis IGA

The most used basic functions of IGA are the NURBS; they commonly used to approximate the fields of displacement and geometry in a given domain. IGA has made it possible to eliminate all discretization errors caused by the geometric approximation.

The vector of knots $\xi$ can be ordered in the following way;

$$\xi = \{\xi_1, \xi_2, \xi_3, \ldots \ldots \xi_{n+p+1}\} \tag{4}$$

With $\xi_i \in \Re$ is the i-th knot, i is the index of the node, i ∈ {1,2,3……n + p + 1} with p is the polynomial degree of the b-spline and n is the number of the associated function.

## 3.1 Shape Function

For a given order p, the B-spline basis functions are defined recursively from the node vector by applying the cox-of-boor formula, starting with the constant functions (P = 0):

$$N_{i,0} = \begin{cases} 1 & \xi_i \leq \xi < \xi_{i+1} \\ 0 & \text{Otherwise} \end{cases} \tag{5}$$

Then, we build for p > 0:

$$N_{i,p}(\xi) = \frac{\xi - \xi_i}{\xi_{i+p} - \xi_i} N_{i,p-1}(\xi) + \frac{\xi_{i+p+1} - \xi}{\xi_{i+p+1} - \xi_{i+1}} N_{i+1,p-1}(\xi) \tag{6}$$

A curve NURBS of order p defined by n + 1 control point:

$$P(\xi)=\sum_{i=0}^{n} R_{i,p}(\xi)X_i \tag{7}$$

Where:

$$R_{i,p} = \frac{N_{i,p}(\xi)w_i}{\sum_{i=0}^{n} N_{i,p}(\xi)w_i} \tag{8}$$

We take note that:
$\{R_{i,p}\}$: Function NURBS
$\{X_i\}$: $\{X_{i_1}, X_{i_2}\}$ the coordinates of the control point set.

### 3.2 The Discretization of a Cracked Structure with IGA

In the context of elastic linear mechanics, consider a domain with the following boundary conditions (see Fig. 3):

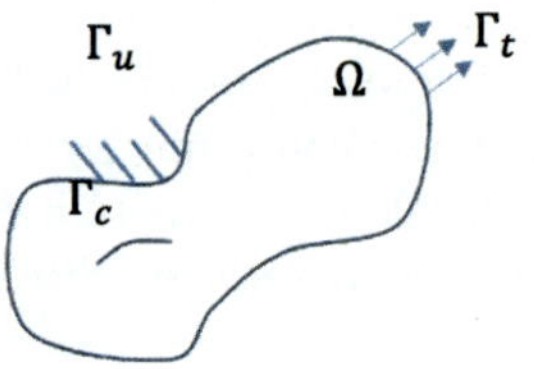


**Fig. 3.** Mechanical problem

$\Gamma_u$: Dirichlet boundary, $\Gamma_t$: Neumann boundary, $\Gamma_c$: crack surface
Therefore, the equilibrium equation and the boundary conditions of this problem can be represented in the following form:

$$\nabla.\sigma + b = 0 \text{ on } \Omega \tag{9}$$

$$\sigma \cdot \mathrm{n} = \bar{\mathrm{t}} \text{ on } \Gamma_t \tag{10}$$

$$\sigma \cdot \mathrm{n} = 0 \text{ on } \Gamma_c \tag{11}$$

$$\mathrm{u} = \bar{u} \text{ on } \Gamma_u \tag{12}$$

Where, σ, b, u corresponds respectively to Cauchy stress tensor, body force, displacement.

The law hook gives the behavior law of a linear elastic problem:

$$\sigma = D\varepsilon \tag{13}$$

With the deformation, a tensor can be written in the form:

$$\varepsilon = \nabla_s \mathrm{u} \text{ on } \Omega \text{ with } \nabla_s = \begin{bmatrix} \partial/\partial x & 0 \\ 0 & \partial/\partial y \\ \partial/\partial y & \partial/\partial x \end{bmatrix} \tag{14}$$

And D represents the elastic matrix.

Any deformable-body under external stress has the weak formulation of the following equilibrium equation:

$$\int_{\Omega} \sigma : \mathrm{u}\, d\Omega - \int_{\Omega} \mathrm{b} : \mathrm{u}\, d\Omega - \int_{\Gamma_t} \bar{t} : \mathrm{u}\, d\Gamma_t = 0 \tag{15}$$

Through this equation, we can obtain the following discrete equation system:

$$[\mathrm{K}]\{\mathrm{U}\} = \{\mathrm{f}\} \tag{16}$$

With, K represents the global stiffness matrix.

U the vector of nodal unknowns

f the vector of forces

In this work, the displacement field approximation and domain geometry are done by integrating the NURBS of the IGA technique.

$$U^h(\xi) = \sum\nolimits_{i=1}^{n_{en}} R_i(\xi) u_i \tag{17}$$

$$X(\xi) = \sum\nolimits_{i=1}^{n_{en}} R_i(\xi) X_i \tag{18}$$

Hence, R ($\xi$) represents the basic function NURBS and $n_{en}$ = (p + 1) × (q + 1) defines the number of control points in $\xi_1$ and $\xi_2$ directions, then p and q are the orders of the curve in the direction $\xi_1$ and $\xi_2$ respectively.

$\xi$ = ($\xi_1$, $\xi_2$) represents the parametric space

X = $(X_1, X_2)$ Represents physical coordinates

$U^h$ Represents the displacement approximation

## 4 Extended Isogeometric Analysis (X-IGA)

This new tool (XIGA) has been improved to allow the modeling of cracks without taking into account the mesh chosen for the calculation. In this method, using the unit partitioning principle in enriching the conventional displacement approximations by suitable enrichment functions. To model the discontinuities, various types of enrichment exist. XIGA removes all unchangeable compatible mesh. To extract local discontinuous fields and singular fields using the concept of the XFEM method, the formula for the enriched displacement approximation is represented as follows:

$$U^h(\xi) = \sum_{i=1}^{n_{en}} R_i(\xi)u + \sum_{j=1}^{n_s} R_i(\xi)\{H(\xi) - H(\xi_j)\}a_j + \sum_{k=1}^{n_t} R_k(\xi)\sum_{\alpha=1}^{4} [\beta_\alpha(\xi) - \beta_\alpha \xi_k)]b_k^\alpha \tag{19}$$

In this formulation above, the polynomials of Lagrange have been replaced by basic functions NURBS $R_i$ to have a new numerical method. $u_i$ represents the traditional degree of freedom and $a_j$ are the enriched degree of freedom corresponds to the crack lip, $b_k^\alpha$ introducing the enriched degree of freedom corresponds to the crack tip. $n_s$ and $n_t$ respectively represent the number of basis functions having the crack lips in their support and the number of basic functions corresponds to the crack tip in the support as shown in Fig. 4.

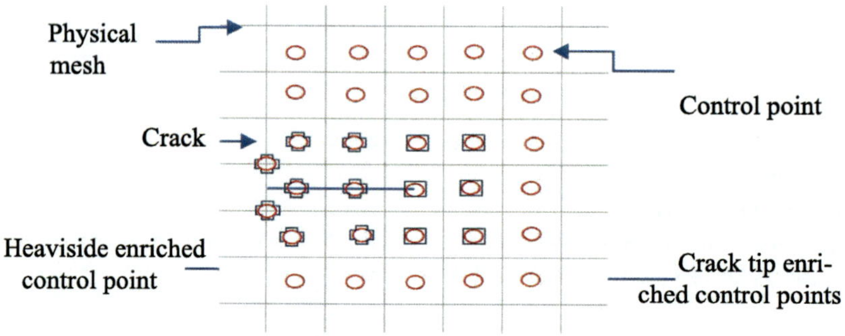


**Fig. 4.** A quadrature NURBS mesh with the enrichment approach

In the formula below, H($\xi$) represents the Heaviside function [16] it receives the value +1 when it is in the upper side of the crack and −1 on the opposite side. $\beta_\alpha(\xi)$ represents the functions of enrichment of the crack tip so the Eq. (20) represents these functions in local polar coordinates (r,$\theta$):

$$\beta_\alpha(\xi) = \sqrt{r}\,(\sin(\theta/2), \sin(\theta/2)\sin(\theta), \cos(\theta/2)\cos(\theta)) \tag{20}$$

The various matrices constituting the XIGA model can be represented in the following way:

$$[K_{ij}] = \begin{bmatrix} K_{ij}^{uu} & K_{ij}^{ua} & K_{ij}^{ub} \\ K_{ij}^{au} & K_{ij}^{aa} & K_{ij}^{ab} \\ K_{ij}^{bu} & K_{ij}^{ba} & K_{ij}^{bb} \end{bmatrix} \tag{21}$$

$$\{\mathrm{f}\} = \left\{f_i^u, f_j^a, f_k^{b1}, f_k^{b2}, f_k^{b3}, f_k^{b4}\right\}^T \tag{22}$$

$$f_i^u = \int_{\Omega} R_i^T \cdot b\, d\Omega + \int_{\Gamma} R_i^T \cdot \bar{t}\, d\Gamma \tag{23}$$

$$f_j^a = \int_{\Omega} R_j^T \{\mathrm{H}(\xi) - \mathrm{H}(\xi_j)\}\ \mathrm{b}\, d\Omega + \int_{\Gamma} R_j^T \{\mathrm{H}(\xi) - \mathrm{H}(\xi_j)\}\ \mathrm{t}\, d\Gamma \tag{24}$$

$$f_k^{b\alpha} = \int_{\Omega} R_k^T \{\beta_\alpha(\xi) - \beta_\alpha(\xi_k)\}\, \mathrm{b}\, d\Omega + \int_{\Gamma} R_k^T \{\beta_\alpha(\xi) - \beta_\alpha(\xi_k)\}\, \mathrm{t}\, d\Gamma \tag{25}$$

$$K_{ij}^{r,s} = \int_{\Omega} B_i^{rT} = \mathrm{D} B_j^s d\Omega \text{ with } \mathrm{r, s = u, a, b} \tag{26}$$

$$B_i^a = \begin{bmatrix} (R_i)_{,x1} H & 0 \\ 0 & (R_i)_{,x2} H \\ (R_i)_{,x2} H & (R_i)_{,x1} H \end{bmatrix} \tag{27}$$

$$B_i^{b_\alpha} = \begin{bmatrix} (R_i \beta_\alpha)_{,x1} & 0 \\ 0 & (R_i \beta_\alpha)_{,x2} \\ (R_i \beta_\alpha)_{,x2} H & (R_i \beta_\alpha)_{,x1} \end{bmatrix} \quad (\alpha = 1, 2, 3, 4) \tag{28}$$

$$B_i^b = [B_i^{b1} B_i^{b2} B_i^{b3} B_i^{b4}] \tag{29}$$

# 5 Numerical Result

To evaluate the performance and accuracy of the XIGA, we will take a simple problem of elastic linear mechanics; an edge and center crack problems. We used the simulation code developed by [20] and the result of the simulation will be compared with the XFEM method. We have chosen an order 3 for basic functions NURBS in both parametric directions. To do the simulation we took the properties of the following material: E = $10^7$ MPA, $\nu$ = 0,3 for edge crack and E = 3 $10^7$ MPa, $\nu$ = 0,3 for center crack. We assumed the plane strain condition.

## 5.1 Example of a Plate with an Edge Crack

The geometry of the plate is of dimension W × L as the Fig. 5 indicate. In the top edge, uniform stress is applied $\sigma$ = 1. a is the crack length with a = 0,45. In the bottom right corner of the domain U1 = U2 = UR3 = 0 and in the top right corner U1 = UR3 = 0. The result of the displacement Uy for an edge crack is represented in the Fig. 6.

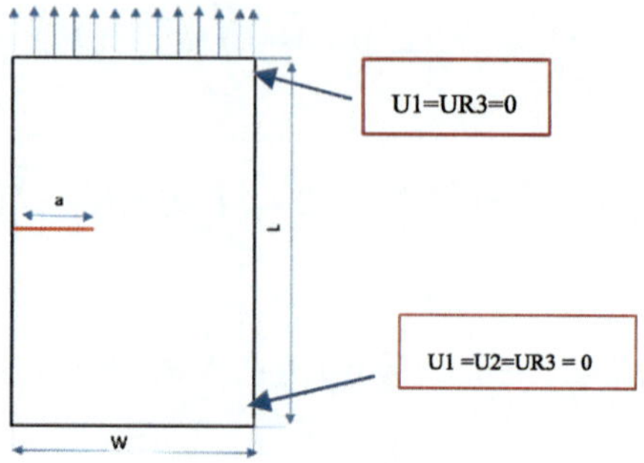


**Fig. 5.** Edge crack in tension

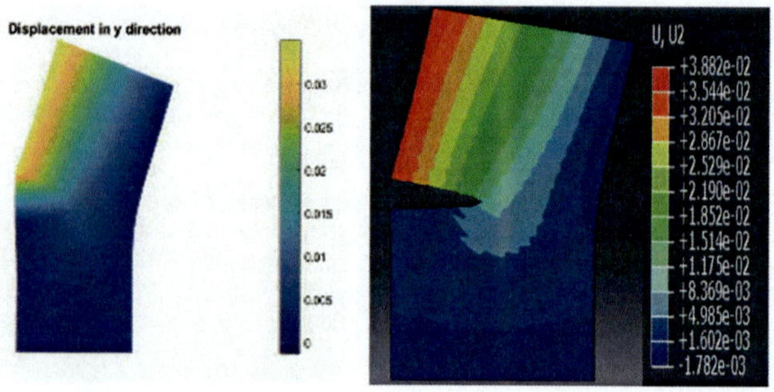


(a) Illustration with XIGA (b) illustration with XFEM

**Fig. 6.** Displacement contour plot Uy for an edge crack

## 5.2 Example of a Plate with a Center Crack

The geometry of the plate is of dimension W × L as the Fig. 7 indicate. In the top edge, uniform stress is applied $\sigma = 1$. a is the crack length with a = 0,25. The result of the displacement Uy for a center crack is represented in the Fig. 8.

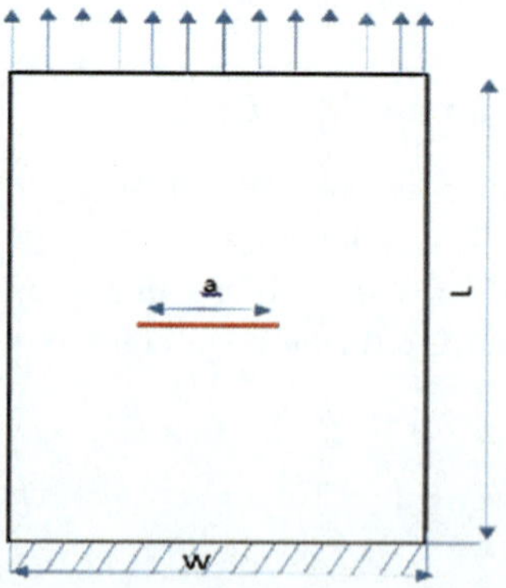


**Fig. 7.** Plate with a center crack

(b) Illustration with XIGA (b) illustration with XFEM

**Fig. 8.** Displacement contour plot Uy for a center crack

The results obtained by the XFEM method and the XIGA method are identical so we can use this novel method to study the various defects produced by the crack and to follow the crack propagation.

## 6 Conclusion

In this work, the simulation of a cracked structure is done by the implementation of the extended isogeometric analysis X-IGA method, which is the extended of the isogeometric method. By applying the concept of the XFEM method, the enrichment functions are integrated to extract the discontinuous and singular local fields. The integration of this method in the finite element calculation makes it possible to make a connection between the CAD and FEM to minimize the errors caused by the discretization thus obtaining the results more efficient. For future researches, we decide to apply this method to follow the propagations of the cracks on the shell structures, which are subjected to internal pressures like pipelines.

## References

1. Barsoum, R.S.: On the use of isoparametric finite elements in linear fracture mechanics. Int. J. Numer. Meth. Eng. **10**(1), 25–37 (1976)
2. Belytschko, T., Lu, Y., Gu, L.: Element-free Galerkin methods. Int. J. Numer. Meth. Eng. **37** (2), 229–256 (1994)
3. Yan, A., Nguyen Dang, H.: Multiple-cracked fatigue crack growth by BEM. Comput. Mech. **16**(5), 273–280 (1995)
4. Dolbow, J., Belytschko, T.: A finite element method for crack growth without remeshing. Int. J. Numer. Methods Eng. **46**(1), 131–150 (1999)
5. Silling, S.A.: Linearized theory of peridynamic states. J. Elast. **99**(1), 85–111 (2010)

6. Kim, J.: A generalized continuous surface tension force formulation for phase-field models for multi-component immiscible fluid flows. Comput. Methods Appl. Mech. Eng. **198**(37–40), 3105–3112 (2009)
7. Melenk, J.M., Babuška, I.: The partition of unity finite element method: basic theory and applications. Comput. Methods Appl. Mech. Eng. **139**(1–4), 289–314 (1996)
8. Bhardwaj, G., Singh, I., Mishra, B., Bui, T.: Numerical simulation of functionally graded cracked plates using NURBS based XIGA under different loads and boundary conditions. Compos. Struct. **126**, 347–359 (2015)
9. Hughes, T.J.R., Cottrell, J.A., Bazilevs, Y.: Isogeometric analysis: CAD, finite elements, NURBS, exact geometry and mesh refinement. Comput. Methods Appl. Mech. Eng. **194** (39–41), 4135–4195 (2005)
10. Cottrell, J.A., Hughes, T.J.R., Bazilevs, Y.: Isogeometric Analysis: Towards Integration of CAD and FEA, 1st edn. Wiley, Chichester (2009)
11. Verhoosel, C.V., Scott, M.A., de Borst, R., Hughes, T.J.R.: An isogeometric approach to cohesive zone modeling. Int. J. Numer. Methods Eng. **87**(1–5), 336–360 (2011)
12. De Luycker, E., Benson, D., Belytschko, T., Bazilevs, Y., Hsu, M.: X-FEM in isogeometric analysis for linear fracture mechanics. Int. J. Numer. Meth. Eng. **87**(6), 541–565 (2011)
13. Ghorashi, S.S., Valizadeh, N., Mohammadi, S.: Extended isogeometric analysis for simulation of stationary and propagating cracks. Int. J. Numer. Meth. Eng. **89**(9), 1069–1101 (2012)
14. Singh, I.V., Bhardwaj, G., Mishra, B.K.: A new criterion for modeling multiple discontinuities passing through an element using XIGA. J. Mech. Sci. Technol. **29**(3), 1131–1143 (2015)
15. Nguyen-Thanh, N., Valizadeh, N., Nguyen, M., Nguyen-Xuan, H., Zhuang, X., Areias, P., et al.: An extended isogeometric thin shell analysis based on Kirchhoff–Love theory. Comput. Methods Appl. Mech. Eng. **284**, 265–291 (2015)
16. Moës, N., Dolbow, J., Belytschko, T.: A finite element method for crack growth without remeshing. Int. J. Numer. Meth. Eng. **46**(1), 131–150 (1999)
17. Moës, N., Belytschko, T.: Extended fnite element method for cohesive crack growth. Eng. Fract. Mech. **79**(7), 813–833 (2002)
18. Béchet, E., Minnebo, H., Moës, N., Burgardt, B.: Improved implementation and robustness study of the X FEM for stress analysis around cracks. Int. J. Numer. Methods Eng. **64**(8), 1033–1056 (2005)
19. Laborde, P., Pommier, J., Renard, Y., Salaün, M.: High-order extended finite element method for cracked domains. Int. J. Numer. Meth. Eng. **64**(3), 1033–1056 (2005)
20. Nguyen, V.P., Anitescu, C., Bordas, S., Rabczuk, T.: Isogeometric analysis: an overview and computer implementation aspects. Math. Comput. Simul. **117**, 89–116 (2015)

# Towards a Scheduling Optimization Support Tool for a Perfume Manufacturing Process

Adam Souabni[1,2], Khalil Tliba[1,2], Thierno M. L. Diallo[2], Romdhane Ben Khalifa[1], Olivia Penas[2(✉)], Noureddine Ben Yahia[1], and Jean-Yves Choley[2]

[1] ENSIT Laboratoire de Mécanique, Productique et Energétique (LR18ES01), Tunis University, Tunis, Tunisia
adam.souabni@gmail.com, {Khalil.tliba, Romdhane.khalifa, Noureddine.yahia}@ensit.rnu.tn

[2] Supméca - Institut Supérieur de Mécanique de Paris, Laboratoire QUARTZ (EA 7393), Saint-Ouen, France
{Thierno.diallo, Olivia.penas, Jean-Yves.choley}@supmeca.fr

**Abstract.** The satisfaction of an increasingly personalized and unpredictable demand that certain companies are experiencing demand flexible production systems. The large number and complexity of the parameters to be taken into account most often require a scheduling support system to meet delivery times and optimize the use of resources. We present the industrial and research issues then a synthesis of the state of the art on the different types of scheduling problems. We chose a simulation approach for our scheduling tool. After defining the specifications of the scheduling tool and how the corresponding industrial data have been collected, the RFLP (Requirements, Functional, Logical and Physical) models of the perfume manufacturing case study on which the scheduling support will be based, are then presented.

**Keywords:** Scheduling · Simulation · Perfume manufacturing · Industry 4.0

## 1 Introduction

A large number of companies are experiencing an evolution of its market with the multiplication of their direct customers, an increasingly personalized and unpredictable demand. Indeed, even if most of distribution is done through wholesalers, the development of electronic commerce is becoming increasingly important and can no longer be neglected. Splitting the volume of orders and multiplying distribution points result in ever smaller batch sizes, frequent series changes and finally require a high degree of flexibility and adaptability of production systems.

The objective of the work presented in this article is to develop a scheduling tool to help companies face these increasingly personalized demands applied to a perfume manufacturing workshop.

The paper is organized as follows: industrial context and state of the art related to the research issue addressed and the approach selected are first presented; needs


A. Saka et al. (Eds.): CPI 2019, LNME, pp. 577–586, 2021.
https://doi.org/10.1007/978-3-030-62199-5_51

analysis regarding the data gathering and scheduling tool specifications are highlighted in Sect. 2; compliant modeling of the case study manufacturing workshop is described in Sect. 3 followed by the conclusion and the announcement of future work.

### 1.1 Industrial Context

The evolution of personalized customer's request, combined with continuous changes in product formats, requires a high degree of flexibility to increase the ability of production system to adapt to these changes. In parallel, the increasing introduction of information technologies into manufacturing processes, which appears to be the 4th industrial revolution (Industry 4.0), is reflected in the use of intelligent and connected production systems. Thus, the control of industrial automated production systems is evolving today by integrating embedded intelligence. This intelligence makes it possible to extract as much elaborate information as possible and to increase the reactivity of the control system. At the same time, the management of the production tool requires taking into account many parameters (product-process characteristics, supply, subcontracting, distribution, installation capacity, etc.) that complicate the decision-making process.

Besides, production lines are generally adapted to a product, so it is useful to take into account the interdependence of product and process data [1], to improve their design and thus product quality, and to allow a high degree of operational exploitation. Existing tools and approaches make it possible to model and simulate production lines at different scales for the control and management of production lines: process analysis in the context of the "digital factory" [2]; analysis of the behavior of mechatronic systems in production lines [3].

### 1.2 Research Issue

The improvement of flexibility can be achieved at the process control level (optimization of manufacturing processes and human resources scheduling …), but also in the implementation of the manufacturing system (choice of equipment and the definition of their relating architecture and configuration). In this paper, we focus on the first approach tackling the scheduling optimization issue.

### 1.3 State-of-the-Art and Proposal

Production scheduling is defined by S.C. Graves [4] as "*the allocation of available production resources over time to best meet certain criteria*". In a context of tasks or operations to be performed under time and resource constraints, a scheduling problem consists in giving which process occupies which machine and the times to start the tasks, in order to optimize a certain criterion.

Scheduling approaches depend on the of the scheduling problem typology they address. Different criteria can be consider to classify scheduling problems such as 1) Source of requirements, 2) Manufacturing process structure, 3) Parameters knowledge confidence and 4) The production system environment. When considering the first criteria, two kinds of situations can be distinguished: open shops for which requirements are based on the customers' needs and closed shops for which requirements are based on stock levels. Considering the structure of the manufacturing process, i.e. the number of stages and associated processors we can distinguish between processes with a unique stage and processes with several stages. The level of confidence in the knowledge of process parameters is another factor in the selection of a scheduling method. Indeed, manufacturing processes with constant and determinist parameters will be easier to schedule than dynamic and stochastic ones. Finally, the production system environment can be as static or dynamic. As a result, we can classify existing scheduling approaches according to these different points of view (see Table 1).

**Table 1.** Scheduling methods categories and related existing literature

| | Point of view | Example of existing literature |
|---|---|---|
| Source of requirements | Open shop | [5] |
| | Closed shop | [6] |
| Manufacturing process structure | One stage/One processor | [7] |
| | Multistage flow/Job shop | [8] |
| Parameters knowledge confidence | Determinist | [9] |
| | Stochastic | [10] |
| Production system environment | Static | [11] |
| | Dynamic | [10] |

Scheduling problems are generally NP-difficult [5, 12], so it is mainly methods of approached resolution that are developed for practical cases. The majority of published resolution models and techniques apply to simple, deterministic and static scheduling problems [10]. However, most of the real scheduling problems are stochastic (e.g. uncertain operating time) and dynamic (customer requests, equipment or operator unavailability, etc.) and depend on many parameters that are not always known. To that respect, we propose to use simulation to address our scheduling problems. Nevertheless, we must pay particular attention to its limitations highlighted by [10]. Simulation models do not always have a theoretical meaning, in our case, this point is not critical since the proposed model will only represent the necessary and sufficient view to address the scheduling problem, without attempting to represent the system in an exhaustive way. Additionally, the interpretation of the simulation results is limited because it depends on the scenarios tested (quantity, perimeter, hypothesis). Finally, as the value, accuracy and credibility of simulation results are largely dependent of the quality of the simulation models and then the judgment and skills of the programmer, we will particularly watch out on the model and scenario data adequacy regarding the simulation objective.

## 2 Scheduling Tool Specification Definition

The decision support tool to be developed has to optimize the manufacturing process based on a strategic production plan including dynamic change of the market behavior.

### 2.1 Functional Analysis

Performing the modeling of the manufacturing system that will be used as support to the scheduling tool, requires the knowledge of the relevant parameters, in order to adapt the model perimeter and refinement to the objective of the simulation. For that respect, we realized a processes mapping with a "Suppliers, Inputs, Processes, Outputs and Customers" (SIPOC) in order to identify the different process steps from the supplier (inputs) to the customer (outputs) and their operational parameters. The manufacturing process include different stages through which all production batches pass. Each stage consists of one or more workstations that can be different. Then the implementation of a scheduling tool that, for a set of batches over a given period, will define the order of completion and the assignment of batches to each workstation.

Then regarding the objective of the analysis (here scheduling optimization), the second step consists in selecting, among all the parameters identified, those that are relevant to the objective fixed. These relevant parameters will be those that have to be both collected for the selected industrial scenarios (strategic production plan) and provided in the system modeling to be developed.

Finally, the tool to be developed (see Fig. 1) will have to take into account the current situation of the production system (current production in progress and corresponding availability of production equipment and downstream systems), the current demand (manufacturing order with earliest starting date and latest ending date). Based on this information, the scheduling tool will have to provide an "optimal" scheduling according to one or more performance criteria: maximum completion time, average time to complete all of the jobs, cost, etc.

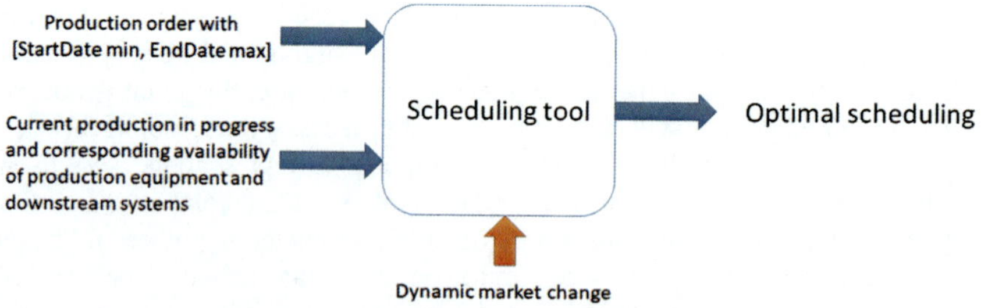


**Fig. 1.** Inputs and output data of the scheduling tool

### 2.2 Data Gathering

The simulation scenarios will be built on a strategic production plan and some assumptions based on production history, observed and future market trends. Consequently, our approach requires the collection of useful data to feed the modeling of the manufacturing system in order to ensure a reliable scheduling. To meet this need, we conducted questionnaires to the staff of the company. These questionnaires later made it possible to build a database to feed our model with real input data in order to have realistic simulations. The questionnaires were written from the specifications of the industrial company that is the subject of our case study and field observations of the perfume manufacturing process.

Indeed, the collection of information and documents is a key step in our approach. We conducted our questionnaires in the form of individual interviews with employees attached to the perfume juice manufacturing department and who have good experience in their positions. In parallel, we have identified company-specific documents that contain important data for our study.

Based on the SIPOC representation of the case study (Fig. 4), these interviews have made it possible to build a database with the following information: Production plan at X months: annual and monthly quantity (seasonality and integrating peaks of activities), reference number of products, prioritization of references, manufacturing process stages time, batch size, opening time of the workshop, human resources availability, parameters of consumption of the lines of conditioning, quality level.

## 3 System Modeling

### 3.1 Presentation of the Manufacturing Process

The process of the perfume manufacturing is generally accomplished by two main phases: the production of perfume juice and the packaging on line. Our research work focuses on the first process of perfume juice manufacturing as shown the Fig. 2.

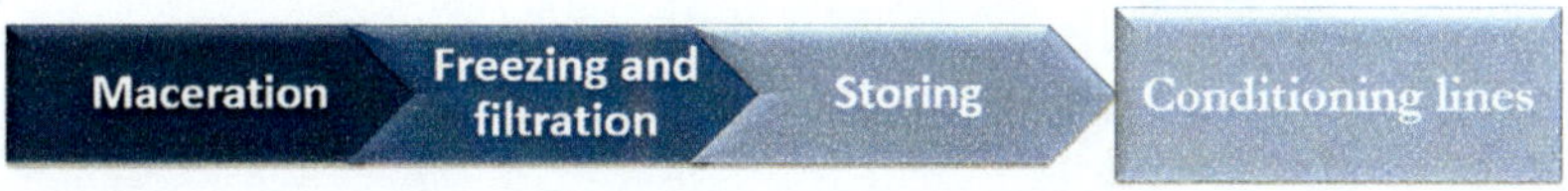


**Fig. 2.** Perfume manufacturing process

This process begins with the maceration phase by mixing the alcohol and the perfume concentrate and then letting it macerate for a specific period of time according to the product specifications. Once the maceration of the mixture is done, the second phase consists in the ice-filtration by freezing and filtering the macerated perfume juice in order to eliminate impurities and residues of the juice concentrate after the mixture saturation. Finally, the process is completed by the third phase of storage. The stored perfume juice will be then conveyed to the second phase of production: packaging on line. Each process step needs to fulfill specific requirements (such as: availability of material and human resources, cleaning of tanks, etc.) and to satisfy input parameters of product.

### 3.2 Modeling Tool and Case Study Description

Two modeling tools have been chosen, in order to describe the case study: perfume manufacturing process.

The free Aris Express software [13] provides a static description of the process (activities, inputs/outputs, resources, product flows, location, equipment and consumables). The 3DEXPERIENCE Platform is a business experience platform that integrates a design framework named RFLP as unique data referential to support a Systems Engineering process with requirements (R), functions (F) and logical product definition (L) such as components including 0-1D models, physical product definition (P) including multidimensional models [14]. Even if this platform is mainly dedicated to products design, we have chosen it, since it provides an integrated modeling platform supporting functional analysis and dependency links traceability to facilitate verification process regarding the scheduling optimization objective. The perfume manufacturing process consists of 4 main activities: cleaning, maceration, freezing and filtration and storing. The cleaning process is shown Fig. 3 as an example.

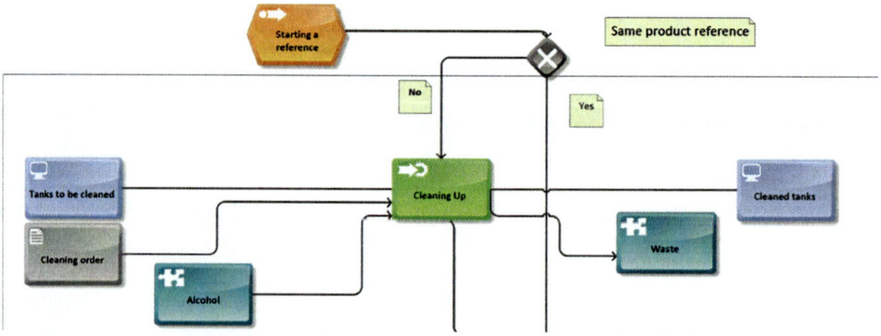


**Fig. 3.** Cleaning stage description

### 3.3 SIPOC Analysis

The relevant parameters for each activities to be taken into consideration for the simulation have been identified from the SIPOC view. An extract of the SIPOC of the perfume manufacturing process is given in Fig. 4.

| Process | Supplier | INPUT | Resources | Parameters | OUTPUT | Requirements | Product features | Customer |
|---|---|---|---|---|---|---|---|---|
| Cleaning | Manufacturing service agent (hot water)<br>Raw material supplier (Alcohol)<br>Manufacturing Manager (Cleaning Order) | Hot water<br>Alcohol<br>Cleaning order | Manufacturing service agent<br>Movable pump<br>Hot water production unit | Quantities of alcohol and hot water according to the size of the tank<br>Hot water temperature | Cleaned tank | Cleaning quality | Juice composition | Maceration<br>Freezing<br>Storing |

**Fig. 4.** Extract of the case study SIPOC

An extract of relevant parameters and variables are presented in Table 2.

**Table 2.** Extract of relevant parameters and variables for the system modeling

| Scope | Parameters | Variables |
|---|---|---|
| Product | Juice reference<br>Juice volume<br>Juice composition | |
| Infrastructure | Tanks number<br>Tank volume | |
| | Tank stirring system<br>Freezing/Filtration flow rate | |
| Process | Maceration time | Storing time |
| | Freezing/Filtration time | Human resource need |
| Scheduling | Production plan (delay) | Cleaning need |
| | Manufacturing orders | Storing need |

## 3.4 Requirements

The requirements of the perfume manufacturing process (see Fig. 5) have been categorized in 4 types: the requirements related to (1) the organization notably to the human resources and business hours, to (2) the rules and regulations (Quality, Safety, Good Manufacturing Practice), to (3) the infrastructure (tanks and freezing/filtration equipment) and finally to (4) the product (composition, stirring, volume) and the scheduling process (cleaning, filter change and storing activities).

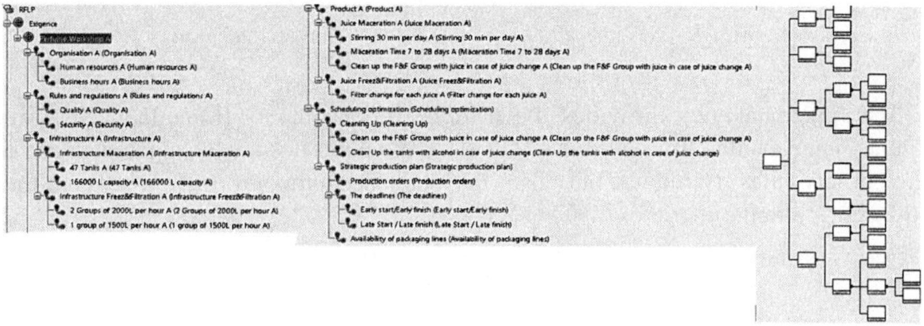


**Fig. 5.** Case study requirements

## 3.5 Functional Architecture

The functional architecture of the perfume manufacturing process, composed of three main functions (Maceration including dosing, cleaning and quality sampling, Freezing/filtration and Storing) is presented in Fig. 6 and Fig. 7.

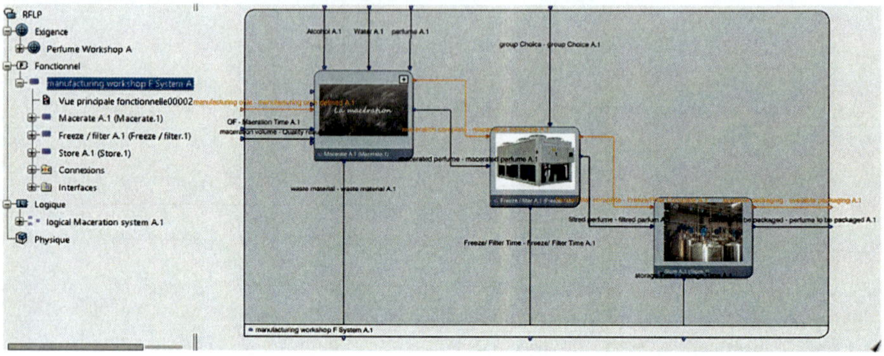


**Fig. 6.** Case study functional architecture

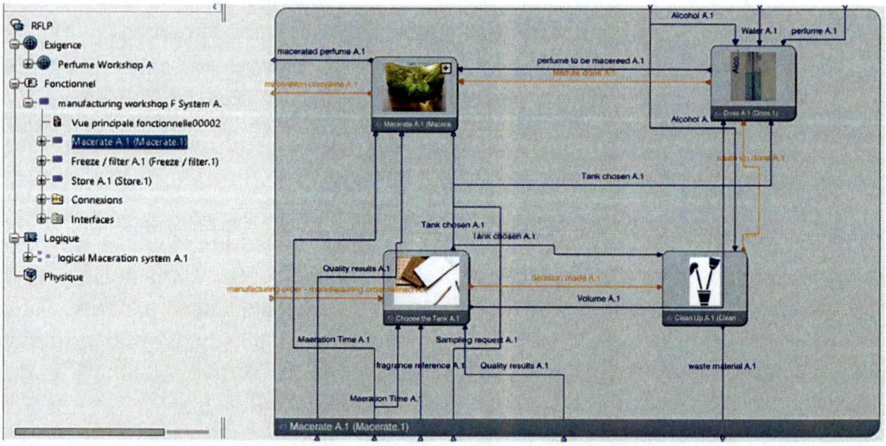


**Fig. 7.** Maceration stage detailed functional architecture.

The functional view provides the external inputs/outputs (Manufacturing Order features, juice availability for packaging, etc.), the main flows exchanges between the different activities (in blue) but also the sequence/temporal logic between these activities (in orange).

### 3.6 Logical Architecture

The logical architecture integrates the system behavior to be simulated to meet the scheduling optimization objective. An example is given in Fig. 8 for the dosing activity, this model provides the process time, by measuring the delay between the start of the dosing (tank volume becomes different from zero, and the end when the tank volume equal the juice volume desired).

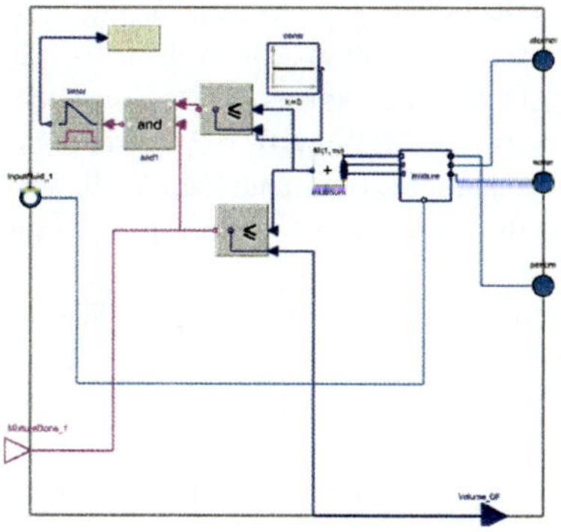


**Fig. 8.** Dosing activity logical modeling

### 3.7 Physical Architecture

The physical architecture represent the 3D model of the infrastructure and of the products to dynamically visualize the online process, it will be particularly interesting when we will optimize the manufacturing process through its equipment architecture reconfiguration or modification. For the scheduling optimization, we did not need this representation of the system.

## 4 Conclusion and Future Work

This research work focuses on the development of a scheduling support tool. The case study was provided by one of our industrial partners in the cosmetics sector. The first phase of the work was to define the industrial problem and to realize a literature review. The results of this first phase (industrial problems definition and literature review) led us to a simulation approach for the implementation of the scheduling aid tool. Indeed, our scheduling problem like most real scheduling problems are complex, stochastic and dynamic problems. After the specification of the support tool to put in place, we realized the RFLP model of the production system on which the scheduling support tool will be based. Future work will address the development and implementation of the scheduling support tool.

**Acknowledgements.** The LMPE and Quartz laboratories sincerely thank the financial support of the PHC UTIQUE project SA2QPI No. 19G1127. The authors would specifically thank our industrial partner for the provision of the case study data.

## References

1. Mhenni, F., Penas, O., Hammadi, M., et al.: Systems engineering approach for the conjoint design of mechatronic products and their manufacturing systems. In: Systems Conference (SysCon), 2018 Annual IEEE International, pp 1–8. IEEE (2018)
2. Cheutet, V., Lamouri, S., Paviot, T., Derroisne, R.: Consistency management of simulation information in digital factory. In: MOSIM10, p. 9 (2010)
3. Penas, O., Plateaux, R., Patalano, S., Hammadi, M.: Multi-scale approach from mechatronic to cyber-physical systems for the design of manufacturing systems. Comput. Ind. **86**, 52–69 (2017)
4. Graves, S.C.: A review of production scheduling. Oper. Res. (1981). https://doi.org/10.1287/opre.29.4.646
5. Gonzalez, T.: Open shop scheduling to minimize finish time. JACM **23**, 665–679 (1976)
6. Tavakkoli-Moghaddam, R., Jolai, F., Vaziri, F., et al.: A hybrid method for solving stochastic job shop scheduling problems. Appl. Math. Comput. **170**, 185–206 (2005)
7. Ferreira, D., Morabito, R., Rangel, S.: Relax and fix heuristics to solve one-stage one-machine lot-scheduling models for small-scale soft drink plants. Comput. Oper. Res. **37**, 684–691 (2010). https://doi.org/10.1016/j.cor.2009.06.007
8. Linn, R., Zhang, W.: Hybrid flow shop scheduling: a survey. Comput. Ind. Eng. **37**, 57–61 (1999). https://doi.org/10.1016/S0360-8352(99)00023-6
9. Jain, A.S., Meeran, S.: Deterministic job-shop scheduling: Past, present and future. Eur. J. Oper. Res. **113**, 390–434 (1999)
10. Mohan, J., Lanka, K., Rao, A.N.: A review of dynamic job shop scheduling techniques. Procedia Manuf. **30**, 34–39 (2019). https://doi.org/10.1016/j.promfg.2019.02.006
11. Sen, T., Sulek, J.M., Dileepan, P.: Static scheduling research to minimize weighted and unweighted tardiness: a state-of-the-art survey. Int. J. Prod. Econ. **83**, 1–12 (2003)
12. Johnson, S.M.: Optimal two- and three-stage production schedules with setup times included. Naval Res. Logistics Q. **1**, 61–68 (1954). https://doi.org/10.1002/nav.3800010110
13. Software AG ARIS Express. In: Aris community. https://www.ariscommunity.com/aris-express. Accessed 13 July 2019
14. Chauvin, F., Fanmuy, G., Systèmes, D.: System Engineering on 3DEXPERIENCE platform-UAS use case. In: CSDM (Posters). Citeseer, Paris, France, pp 113–126 (2014)

# Product Lifecycle Management Effect on New Product Development Performance

Ghita Chaouni Benabdellah(✉) and Karim Bennis

L.I.R.E.M (Laboratoire Interdisciplinaire de Recherche En économie et Management), FSJES-Sidi Mohamed Ben Abdellah University, Fes, Morocco
ghitabenabdellahl@gmail.com

**Abstract.** Today's hyper-competitive worldwide markets, turbulent environment, demanding customers, and diverse technological advancements force any companies developing new products to look into all the possible areas of improvement in the entire new product development (NPD) process. As a key determinant of firm's competitiveness, efficient NPD is hard to imagine out of Product Lifecycle Management (PLM) strategy. In fact, PLM provides a dynamic and integrated information management platform that enables firms to effectively manage the creation, modification, and exchange with product information on the collaborative NPD. Moreover, by adopting PLM systems, firms can derive numerous benefits such as increasing the speed to market of new products, improving the response to market demands, delivering more new products in a shorter time, reducing product development project costs, and reducing material and energy consumption costs. Hence, this paper presents an overview that mediates the effect of PLM system on NPD performance. Therefore, it is expected that the results will contribute to the literature and practice by increasing the understanding of how PLM system influences NPD performance.

**Keywords:** New product development · Product lifecycle management systems · Performance

## 1 Introduction

In order to outperform their competitors, manufacturing companies needs to improve the development of new products to meet the challenges of reducing the product lifecycle and the heterogeneity of customer preferences.

New Product Development (NPD) is a complex, uncertain but controlled process that must meet defined criteria of quality, cost and time [1]. This process includes several interrelated activities such as the creation and evaluation of new product ideas, the integration of product requirements into final design specifications… [2].

To increase NDP performance, companies need to collaborate effectively with external partners. For this collaboration to be powerful, the company must develop a mutual understanding between them and their partners through the implementation of a competent inter-organizational process [3]. This process creates an increased demand for information processing the success of which depends on the information technology (IT) solutions implemented by the company. The contribution of the information


A. Saka et al. (Eds.): CPI 2019, LNME, pp. 587–596, 2021.
https://doi.org/10.1007/978-3-030-62199-5_52

technology solution (IT) to the performance of product development has been demonstrated, whether in an intra organizational framework [4] or inter-organizational [5].

There are in the literature several technological solutions implemented by the company to enhance product development, but in this work, we will focus on Product Lifecycle Management (PLM). PLM is a key driver of innovation and success [6], it is a strategic solution for the integrated management of product- related information throughout a product's life cycle.

This work is organized as follows. Section 2 presents a conceptual background for each Product Lifecycle Management (PLM) and New Product Development (NPD). First, we present a literature review of the PLM systems in the technological point of view, in the managerial point of view and finally as a strategic business approach. Second, a literature review of NPD performance and its various activities. Section 3 presents the effect of PLM systems in NPD performance. First, we explained the value of implementing PLM systems in NPD. After that, we illustrated the key factors of PLM systems that affect the NPD performance. In this sub-section, we analyzed two models of how PLM systems affect NPD performance. Finally, in the fourth section, we conclude with some further researches that might complete this work.

## 2 Background

### 2.1 PLM Systems

A new product always requires a tool to manage the collaboration of its many resources that are often dispersed in different geographical areas. What is important to know is that inappropriate management of these complex processes can result in a waste of resources. The most effective way for the management of these processes by integrating people, data, processes and business systems into the new product development is by adopting PLM.

The term PLM has been widely adopted and defined, in the early twenty-first century, by different communities with different interpretations. In fact, it seems to be hard to recognize a definition especially when the visions changes in depending to whom you are talking to (engineers, managers, suppliers, customer, etc.).

PLM is an integrated technology [7] for grouping into single application information about the project, the product, and the development process. It provides a virtual representation of a physical product and allows virtual management of collaboration throughout the development process by storing, coordinating and controlling all development information [8].

Stark [9], author of two books about PLM and founder of a PLM consulting firm, considers only managerial features. He considers the Product Lifecycle Management (PLM) as a holistic term and define it as the business activity of managing, in the most effective way, a company's products all the way across their lifecycles; from the very first idea for a product, all the way through until it is retired and disposed of. [9] Does not consider the technological and collaborative features that have instead a primary role in this study. [10]. Terzi [11] considers the PLM solution as a business strategy for creating and sustaining such a product- oriented knowledge environment. Product data are shared among actors, processes, and organizations in the different phases of the product life cycle for achieving desired performances and sustainability for the product and related services. As a technology solution, PLM is an integrator of tools and technologies that streamlines the flow of information through the various stages of the product life cycle and seeks to provide the right information at the right time and in the right context. CIMdata [12], proposes a definition of PLM based on a research firm "PLM is a strategic business approach that applies a consistent set of business solutions in support of the collaborative creation, management, dissemination, and use of product definition information across the extended enterprise, and spanning from product concept to end of life-integrating people, processes, business systems, and information. PLM forms the product information backbone for a company and its extended enterprise." Angelo Coralo[10] consider that is classical and outdated to consider PLM as a set of technologies. The new way to conceive PLM is as "a strategic business approach". CIMdata [12] uses the word "approach" to capture all the various elements composing PLM, and it underlines the ability to integrate "people, processes, business systems, and information".

The adjectives "strategic business" used for "approach" underlines the need to manage the entire product lifecycle to obtain business efficiency and effectiveness. In fact, correct management of product lifecycle enables the enterprise to obtain competitive advantages creating better products in less time, at a lower cost, and with fewer defects; in other words, it creates value for the company.

Therefore, the definition that sums up all the previous is the one of Co.PLM community of practice that defined PLM as follows: "PLM Product lifecycle management—is a strategic business approach that supports all the phases of the product lifecycle, from concept to disposal, providing a unique and timed product data source. Integrating people, processes, and technologies and assuring information consistency, traceability, and long-term archiving, PLM enables organizations to collaborate within and across the extended enterprise."

The proposed definition of PLM focalized on management, technological, and collaboration aspects. The specified terms are:

(1) Strategic business approach;
(2) Phases of product lifecycle;
(3) A unique and timed product data source;
(4) Consistency, traceability, and long-term archiving;
(5) Integrating people, processes, and technologies;
(6) Collaborating within and across the extended enterprise.

## 2.2 NPD Performance

New product development (NPD) refers to an entire process from idea generation, through product design and manufacturing, to bringing a new product to the market. NPD is much important to a firm's profitability and competitiveness; it is the key driver of a firm's overall success [13]. Yet, in today's marketplace, developing successful new products can be particularly difficult: competitive pressures are building, consumer tastes are rapidly changing, technological changes are accelerating and product life cycles are becoming shorter [14]. In other words, customers want a product with better performance, better quality, lower prices, and instant availability.

NPD is a strategic weapon for a firm to compete and differentiate itself and outperform its competitors [15]. In this context, [16] indicated that NPD is one of the most important determinants of sustained company performance and represents a key challenge for firms. It generally involves various complex and interdependent activities that can be divided into three macro-categories: "discovery", "Development" and "Industrialization":

- **Discovery:** involve the specifications reception defined by marketing, the validation of the budget assigned to the project and the presentation of the Business Case, in which the preventative spending items are detailed. This means Generating and assessing new product opportunities and ideas.
- **Development:** including the creation of prototypes, the tests development, tests, and certifications as well as the preparation of the all necessary documents for the industrialization phase.
- **Industrialization:** or commercialization, involves implementing marketing plans to launch products on the market. The industrialization phase ends when the production starts [5, 17].

Improving NPD performance is a major driver for developing, achieving and sustaining the competitive advantage of a firm.

We can't talk about NPD's performance without mentioning effectiveness and efficiency. Measures of effectiveness are concerned with the quality of the final product while measures of efficiency are concerned with the number of resources required to complete the project [18]. In other words, effectiveness is for quality (i.e. overall technical performance and meeting customer requirements) and efficiency for time to market, R&D budget and unit manufacturing costs.

Intuitively, an NPD process that is effective and efficient should result in a product that is successful in the marketplace but that is not always the case. NPD is still one of the "riskiest endeavors of the modern corporation" [19].

# 3 PLM Effect on NPD Performance

## 3.1 Benefits of Implementing PLM in NPD

Due to the increasing number of environmental and regulatory rules and requirements, companies have to manage increasing product and manufacturing complexities. At the same time, companies have to be able to collaborate closely within their own organization and with partners and suppliers located in various parts of the world to meet the customer's demand: products with more attractive designs, better performance, better quality, lower prices, and instant availability. (PLM Technology Guide, 2008).

To meet these needs, companies must adopt PLM.

PLM systems form the backbone of product information for firms and their NPD partners [20–22] thereby assisting the firms in managing the creation, modification, and exchange of product information with NPD partners at different levels of their respective organizational hierarchies, from the strategic to operational levels. Because PLM systems are designed to fulfill the requirements of managing large quantities of information and knowledge and exchanging information and knowledge among product development participants[23, 24]. This strategic tool increases additional value by its characteristics that enable converting manually managed processes into automated processes.

Moreover, implementing PLM systems enables firms to reduce the time and cost of developing new products by providing a virtual product design platform that enables firms to expedite feasibility analysis and reduce the costs of product testing and validation [25, 26].

## 3.2 Key Drivers of NPD Performance

### 3.2.1 Management Capabilities for NPD Performance

To improve NPD performance, a PLM system is implemented to support NPD activities. So how can PLM systems affect NPD performance?

According to [27], PLM systems can enhance the firm's performance only by the integration of PLM systems capability into management capabilities thereby improving NPD management. By developing the firm's management capabilities, we mean the firm's process management capability, coordination capability, and absorptive capability (Fig. 1).

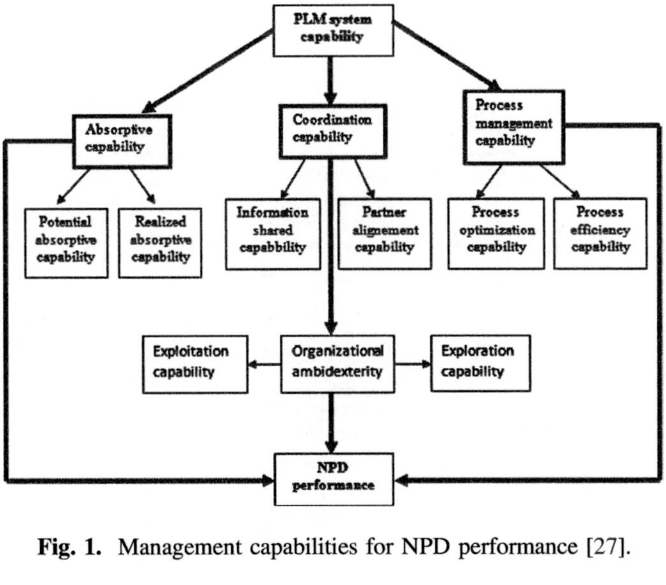


**Fig. 1.** Management capabilities for NPD performance [27].

**Coordination Capability:** Appropriate information sharing mechanism is important to be established in firms [2, 28] so as the alignment of the NPD partners' production strategies with its own NPD strategies [29, 30]. In other words, an appropriate information sharing system enhances NPD activity performance by creating a mutual understanding between the firm's inter- organization and NPD partners [31, 32] The organizational ambidexterity: helps firms to improve competencies to make with the discontinuous environment changes. It is the capacity to simultaneously exploit existing capabilities and the exploration of new possibilities to value creation.

**Absorptive Capability:** The ability of firms to identify and exploit knowledge received from external sources. The potential absorptive capability is the capacity to acquire and integrate this knowledge and the realized absorptive capability is the ability to modify and exploit them.

**Process Management Capability:** A firm's capacity to control and improve NPD management processes. The Process efficiency capability: A firm's ability to regulate the duration and cost of NPD, the Process optimization capability is the capacity to optimize NPD processes by increasing resource utilization and reducing waste generation.

### 3.2.2 Reliability and Productivity for Product Development

Shortening the development time by reducing the complexity of the project is the main concern for product development [33–35]. Also, it is necessary to have finished products while respecting the good conditions of quality and that implies high reliability of the process of developing the product.

While quality applies to the developed object, to the finished product, reliability applies to the development process product (Fig. 2).

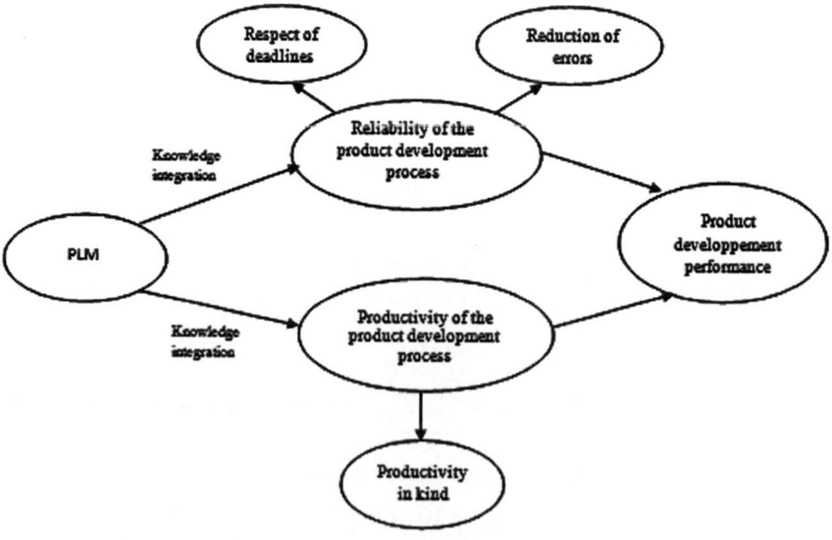


**Fig. 2.** Reliability and productivity for NPD performance.

Reliability is the ability to produce collectively and repeatedly with finished product quality and timely product release [36]. The reliability of the development process results from the ability to quickly identify development problems in order to implement the necessary adjustments as quickly as possible in order to meet the project implementation commitments in accordance with the cost, timeframe and specific characteristics.

Although respect of deadlines is a key dimension of product development performance [37], the method of calculating average delays is more difficult. for the [38] approach, the average project delay results from an analysis of the gap between the planned and the realized at the end of the industrialization process.

In product development, the detection and correction of problems must be done upstream, during the design to avoid the complexity of the project. We retain glich's approach [39] of avoiding problems of communication between actors.

Accelerating the product development process by improving productivity is essential for organizations to remain competitive. Productivity in-kind compares the quantities produced with the given workforce. Through the use of PLM, productivity can be measured through time realization of task, development time... In other words, the use of PLM makes it possible to analyze productivity gains in kind, whether through the ability to develop new products or by reducing cycle time.

## 4 Discussion and Conclusion

In this article, we have combined definitions of Product lifecycle management systems (PLM) and New Product Development (NPD) performance and examined the relationships among these components in NPD contexts.

In other words, we have analyzed the effect of the implementation of such technology to support NPD that results to be an effective method for achieving superior NPD performance by a different method.

Time effort and resources are invested by firms for implementing PLM systems to improve NPD performance. However, according to [27] such investment may not lead to NPD performance if the PLM systems implemented cannot be used to develop the improved management capabilities required in NPD contexts. This model suggests that PLM system capability influence NPD performance through three management capabilities: process management capability, coordination capability, and absorptive capability.

Coordination capability operates as a platform that facilitates interaction between firms and their NPD partners. This platform facilitates sharing the information of product updates between themselves and their partners and aligns NPD partners' product development activities.

PLM system capability influence NPD performance through absorptive capability. Absorptive capability is the ability for firms, using PLM systems, to define, standardize and modify workflows and information generated during product development and from external sources.

For the process management capability, it is the ability for firms to control the cost and duration of the NPD process. Also, it is the capacity to increase the utilization of resources to optimize the NPD process.

The first model adopts developing management capabilities in NPD context by the PLM system is the key to enhance NPD performance.

This second model adopted by [8] states that the complexity of the project must be reduced by reducing the time of the process development and having a finished product with the best quality. This cannot be executed without a performed PLM system that provides a knowledge integration platform to increase the reliability of the product development process and its productivity.

Improving the reliability of the product development process is done by the implementation of a PLM system to modify the organizational routines and the evolvement of the actors. The development problems should be quickly identified and adjust to meet the project characteristics to have a reliable product development process.

Improving the productivity of the product development process is done by accelerating the product development process.

This work has certain conceptual and methodological limits that make it possible to envisage future research.

First, this study focuses only on three management capabilities. Further researches could consider other management capabilities that might influence NPD performance. Second, a distinction between the specificities of the projects would probably enrich this work. In particular, we could introduce the complexity of projects at the technological, organizational or market level. Finally, a case study is recommended in order to test the two models of how PLM systems affect NPD performance.

## References

1. Le Masson, P., Weil, B., Hatchuel, A.: Les processus d'innovation: Conception innovante et croissance des entreprises. Lavoisier, Paris (2006)
2. Thomas, E.: Supplier integration in new product development: computer mediated communication, knowledge exchange and buyer performance. Industr. Market. Manag. **42** (6), 890–899 (2013)
3. Mishra, A.A., Shah, R.: In union lies strength: collaborative competence in new product develop- ment and its performance effects. J. Oper. Manage. **27**(4), 324–338 (2009)
4. Iansiti, M., MacCormack, A.: Developing products on Internet time. Harvard Bus. Rev. **75** (5), 108–117 (1997)
5. Pavlou, P.A., El Sawy, O.A.: From IT leveraging competence to competitive advantage in turbulent environments: the case of new product development. Inf. Syst. Res. **17**(3), 198–227 (2006)
6. D'Avolio, E., Pinna, C., Bandinelli, R., Terzi, S.: Analysing product development process and PLM features in the food & fashion industries. In: IFIP 14th International Conference on Production Lifecycle Manag. Springer International Publishing, Seville, Spain (2017)
7. Mostefai, S., Batouche, M.: Data integration in product lifecycle management: an ontology-based approach. In: PLM 2005: International Conference on Product Life Cycle Management, Lyon, France (2005)

8. Merminod, V., Mothe, C., Rowe, F.: Effets de Product Lifecycle Management sur la fiabilité et la productivité: une comparaison entre deux contextes de développement produit. M@n@gement **12**(4), 294–331 (2009)
9. Stark, J.: Product Lifecycle Management: 21st Century Paradigm for Product Realisation. Spinger, London (2011)
10. Corallo, A., Latino, M.E., Lazoi, M., Lettera, S., Marra, M., Verardi, S.: Defining product lifecycle management: a journey across features, definitions, and concepts. ISRN Industrial Engineering (2013)
11. Terzi, S., Bouras, A., Dutta, D., Garetti, M., Kiritsis, D.: Product lifecycle management-from its history to its new role. Int. J. Prod. Lifecycle Manage. **4**(4), 360–389 (2010)
12. CIMdata: Product lifecycle management—empowering the future of business. Technical report (2002)
13. Loch, C., Stein, L., Terwiesch, C.: Measuring development performance in the electronics industry. J. Prod. Innovation Manage. Int. Publ. Prod. Dev. Manage. Assoc. **13**(1), 3–20 (1996)
14. Menon, A., Chowdhury, J., Lukas, B.A.: Antecedents and outcomes of new product development speed: an interdisciplinary conceptual framework. Industr. Market. Manage. **31** (4), 317–328 (2002)
15. Chan, S.L., Ip, W.H.: A dynamic decision support system to predict the value of customer for new product development. Decis. Support Syst. **52**(1), 178–188 (2011)
16. Graner, M., Mißler-Behr, M.: Key determinants of the successful adoption of new product development methods. Eur. J. Innovation Manage. **16**(3), 301–316 (2013)
17. Hilletofth, P., Eriksson, D.: Coordinating new product development with supply chain management. Industr. Manage. Data Syst. **111**(2), 264–281 (2011)
18. Olson, E.M., Walker Jr, O.C., Ruekerf, R.W., Bonnerd, J.M.: Patterns of cooperation during new product development among marketing, operations and R&D: implications for project performance. J. Prod. Innovation Manage. Int. Publ. Prod. Dev. Manage. Assoc. **18**, 258–271 (2001)
19. Cooper, R.G., Edgett, S.J., Kleinschmidt, E.J.: Benchmarking best NPD practices—I. Res. Technol. Manage. **47**(1), 31–43 (2004)
20. Pol, G., Merlo, C., Legardeur, J., Jared, G.: Implementation of collaborative design processes into PLM systems. Int. J. Prod. Lifecycle Manage. **3**(4), 279 (2008)
21. Cantamessa, M., Montagna, F., Neirotti, P.: Understanding the organizational impact of PLM systems: evidence from an aerospace company. Int. J. Oper. Prod. Manage. **32**(2), 191–215 (2012)
22. Stark, J.: Product lifecycle management. In: Product Lifecycle Management, vol. 1, pp. 1–29. Springer, Cham
23. Fielding, E.A., McCardle, J.R., Eynard, B., Hartman, N., Fraser, A.: Product lifecycle management in design and engineering education: international perspectives. Concurr. Eng. **22**(2), 123–134 (2014)
24. Segonds, F., Mantelet, F., Nelson, J., Gaillard, S.: Proposition of a PLM tool to support textile design: a case study applied to the definition of the early stages of design requirements. Comput. Ind. **66**, 21–30 (2015)
25. Gecevska, V., Chiabert, P., Anisic, Z., Lombardi, F., Cus, F.: Product lifecycle management through innovative and competitive business environment. J. Industr. Eng. Manage. **3**(2), 323–336 (2010)
26. Hadaya, P., Marchildon, P.: Understanding product lifecycle management and supporting systems. Industr. Manage. Data Syst. **112**(4), 559–583 (2012)
27. Tai, Y.M.: Effects of product lifecycle management systems on new product development performance. J. Eng. Technol. Manage. **46**, 67–83 (2017)

28. Ma, C., Yang, Z., Yao, Z., Fisher, G., Fang, E.E.: The effect of strategic alliance resource accumulation and process characteristics on new product success: exploration of international high-tech strategic alliances in China. Ind. Mark. Manage. **41**(3), 469–480 (2012)
29. Acur, N., Kandemir, D., Boer, H.: Strategic alignment and new product development: drivers and performance effects. J. Prod. Innov. Manage **29**(2), 304–318 (2012)
30. Tavani, N., Sharifi, H., Ismail, H.S.: A study of contingency relationships between supplier involvement, absorptive capacity and agile product innovation. Int. J. Oper. Prod. Manage. **34**(1), 65–92 (2014)
31. Hsu, T.T., Tsai, K.H., Hsieh, M.H., Wang, W.Y.: Strategic orientation and new product performance: the roles of technological capability. Can. J. Admin. Sci. **31**(1), 44–58 (2014)
32. Schilke, O.: On the contingent value of dynamic capabilities for competitive advantage: the nonlinear moderating effect of environmental dynamism. Strateg. Manage. J. **35**(2), 179–203 (2014)
33. Clark, K.B., Fujimoto, T.: The power of product integrity. Harvard Bus. Rev. **68**(6), 107–118 (1990)
34. Calvi, R., Blanco, É., Koike, T.: Coopérer en conception pour améliorer les supply chains de demain. Revue française de gestion **3**, 187–202 (2005)
35. Malhotra, A., Gosain, S., El Sawy, O.A.: Absorptive capacity configurations in supply chains: Gearing for partner-enabled market knowledge creation. MIS Q. **29**(1), 145–187 (2005)
36. Hannan, M.T., Freeman, J.: Structural inertia and organizational change. Am. Sociol. Rev. **49**, 149–164 (1984)
37. Marciniak, R., Pagerie, M.: Gestion de projet: guide pratique de la réussite de tous vos projets et produits industriels. Weka, Paris (1999)
38. Cooper, R.G., Kleinschmidt, E.J.: Stage gate systems for new product success. Market. Manage. **1**(4), 20–29 (1993)
39. Hoopes, D.G., Postrel, S.: Shared knowledge, "glitches", and product development performance. Strateg. Manage. J. **20**(9), 837–865 (1999)

# Social Responsibility Performance: A Case Study of a Multinational Electrical Company Located in Casablanca, Kingdom of Morocco

Mohammed Hadini(✉), Said Rifai, Mohamed Ben Ali, Otmane Bouksour, and Ahmed Adri

CIM and Industrial Engineering Laboratory (LMPGI), University Hassan II Casablanca High School of Technology, Casablanca, Morocco
Mohammed.hadinil@gmail.coml,
mohamed-dob12@hotmail.fr, dptgmp@gmail.com,
bouksour2@gmail.com, ahmedadri@gmail.com

**Abstract.** This article aims to present a statistical approach involving latent and manifest variables applied in order to measure the organization's social responsibility performance of a worldwide enterprise (electrical sector) located in Casablanca, Kingdom of Morocco. The main idea is to measure the impact of Social Responsibility's practices (which are divided into five segments: Leadership, Strategy & Planning, Human Resources, Partnerships & Resources, Process Management, and Organizational governance) on the total performance (which is divided into six perspectives: (Human Rights Results perspectives, Labour practices Results perspectives, the environment Results perspectives, Consumer issues Results perspectives, Fair operating practices Results perspectives, community involvement and development Results perspectives)), enabling the company to characterize her performance regarding to the ISO 26000 standard's. For this, we use the SEM's resolution based on the Partial Least squares (PLS) method via the XL-STAT software on a sample of 120 questionnaires administered face-to-face with managers, technicians, engineers and directors of this company. The obtained results could be examined in order to plan the improvements and develop an action plan. It is necessary to control social responsibility performance to ensure that it is either good enough, or that something is being done to improve it.

**Keywords:** Corporate social responsibility · Partial least squares · Modeling approach · Company · Performance

## 1 Introduction

We live in a world in which the resources available are not sufficient to meet all of our desires. We have to make the most efficient use of those scare resources. Levels of performance, of individuals departments and organizations are therefore tied to standards which determine what counts as inadequate, satisfactory or good. Organizations are concerned with performance in the pursuit of goals. The performance of an organization as a whole determines its survival.


A. Saka et al. (Eds.): CPI 2019, LNME, pp. 597–612, 2021.
https://doi.org/10.1007/978-3-030-62199-5_53

Given this context, Social Responsibility is the best way to stimulate an organization.

Social Responsibility (SR) is a perspective that expects organizations to act ethically, and in ways that consider, contribute to and benefit economic development, quality of working life, local communities and society at large [1].

The goal of this article is to measure the impact of the social responsibility performance of this worldwide enterprise (electrical sector) located in Casablanca, Kingdom of Morocco, through the use of the structural equation method according to PLS approach (Partial Least Square) and via the Xl-stat software. This is an empirical study and it relies on questionnaire-based surveys administered via a direct contact with the managers, technicians, engineers and directors of this worldwide enterprise (electrical sector).

## 2 Conceptual Framework and Research Methodology

As mentioned before, this article deals with a worldwide enterprise based in Casablanca, Kingdom of Morocco. A worldwide enterprise could be defined as a large company that operates abroad thanks to its subsidiaries and following a globally-scaled strategy and organization. A worldwide enterprise operates in several geographical areas, but with one unique centre or one main decision-making centre. The decisions of a worldwide enterprise are mainly based on criteria of economies of scale, tax policies and repatriation of profits.

## 3 Definition of the Structures of the Conceptual Model

### 3.1 Equations Structural Modeling

The Structural Equation Modelling (SEM) is a method to define complex interacting systems [2] and it allows studying the causal connections between multiple latent variables. These variables represent a concept but we can only measure them with manifest variables (MV) [3]. MES is used for the generalization of many classic models such as principal components analysis, factor analysis, and canonical analysis. These statistical models are used in several research fields [4, 5] especially in the marketing field to construct satisfaction indicators [6]. This type of modeling is thus important to test the hypotheses of our conceptual model. There are two methods of modeling via (MES) for estimating the existing relationships between the constructs: partial least square (PLS), based on PLS variances, and linear structural relationship (LISREL), based on maximum likelihood the LISREL method and the PLS method [7, 8].

### 3.2 ISO 26000 Standard: A Quick Survey

Social Responsibility (SR) is a perspective that expects organizations to act ethically, and in ways that consider, contribute to and benefit economic development, quality of working life, local communities and society at large [1]. Social Responsibility provides many benefits for companies such as boosting investors' confidence, attracting the best talent, maximizing brand value, attracting new clients, giving to the customers a reason to be loyal, and maintaining a good relationship with the local communities.

The ISO 26000 guideline [9] is considered an international framework developed to help organizations to address the Social Responsibility relevant to the impact of their decisions on the society and environment and the stakeholders expectations.

The ISO26000 guidelines are conceptualized around the following social responsibility principles:

- The Accountability principle: an organization should be accountable for its impacts on society, the economy and the environment. This principle suggests that an organization should accept appropriate scrutiny and also accept a duty to respond to this scrutiny [9].
- Transparency: Organization should be transparent in its decisions and activities that impact society and the environment [9].
- Ethical Behaviour: Organization should behave ethically [9].
- Respect Stakeholder: Organization should respect, consider and respond to the interests of its stakeholders [9].
- Respect Rule of Law: Organization should accept that respect for the rule of law is mandatory [9].
- Respect International Norms: Organization should respect international norms of behaviour, while adhering to the principle of respect for the rule of law [9].
- Respect Human Rights: Organization should respect human rights and recognize both their importance and their universality [9].

According to the ISO 26000 organization should address the following seven principles of social responsibility to identify relevant issues and set priorities to their stakeholders [9].

- **Organizational governance**: Is the system by which an organization makes and implements decisions in pursuit of its objectives [9].
- **Human rights:** All humans have the right to fair treatment and the elimination of discrimination, torture, and exploitation [9].
- **Labor practices:** Those working on behalf of the organization are not a commodity. The goal is to prevent unfair competition based on exploitation and abuse [9].
- **Environment:** The organization has a responsibility to reduce and eliminate unsustainable volumes and patterns of production and consumption and to ensure that resource consumption per person becomes sustainable [9].
- **Fair operating practices:** Building systems of fair competition, preventing corruption, encouraging fair competition, and promoting the reliability of fair business practices help to build sustainable social systems [9].
- **Consumer issues:** The promotion of just, sustainable, and equitable economic and social development with respect to consumer health, safety, and access is the organization's responsibility [9].
- **Community involvement and development:** The organization should be involved with creating sustainable social structures where increasing levels of education and well-being can exist [9] (Fig. 1).

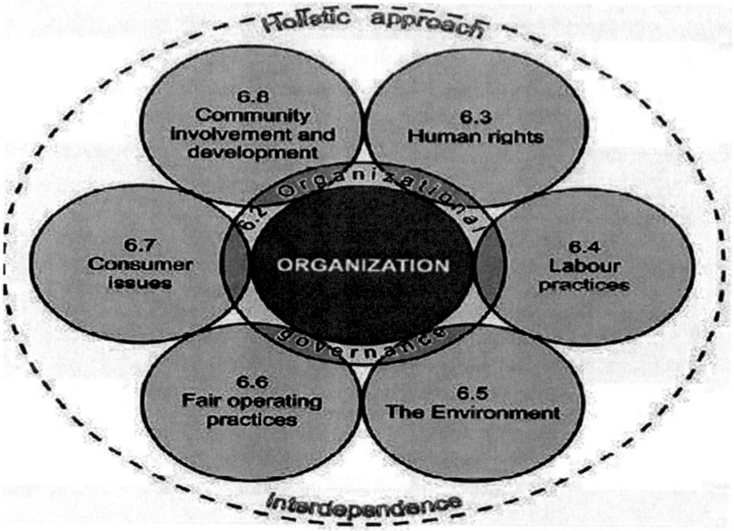


**Fig. 1.** The seven core subjects according ISO26000 (Source: ISO 26000:2010)

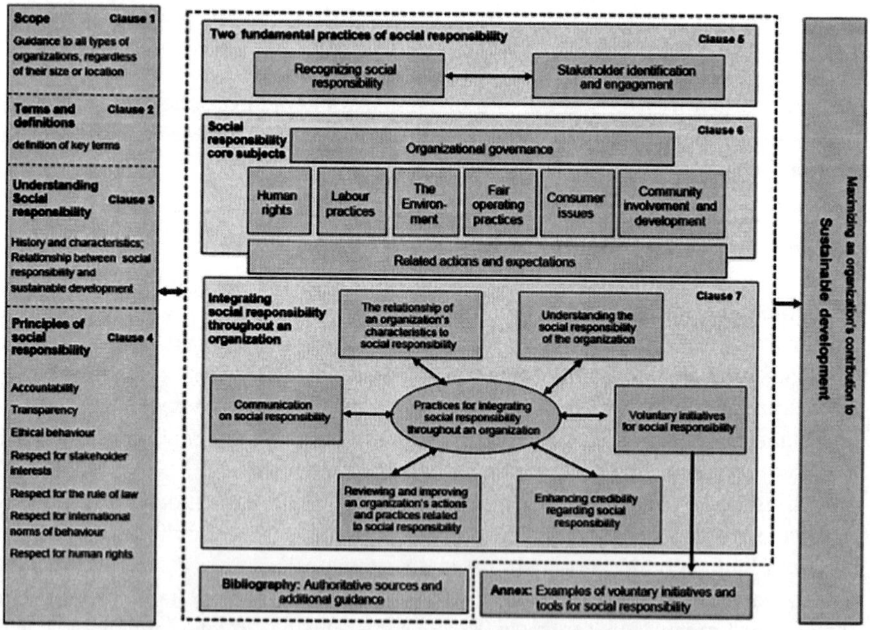


**Fig. 2.** Overview of ISO 26000 (Source: ISO 26000:2010)

# 4 Conceptual Framework and Research Methodology

## 4.1 Choice of Criteria of Means (Social Responsibility Practices)

The preparation and development of our questionnaire was conducted after a theoretical and empirical exploration stage. The empirical exploration consisted in meetings with the managers of this multinational electrical company located in Casablanca, Morocco.

The theoretical exploration was conducted following a review of the literature [8, 10–12] and of the major quality awards (Malcolm Baldrige National Quality Award (MBNQA), Deming, European Foundation for Quality Management (EFQM)). Ultimately, we have selected six criteria of means (social responsibility practices): 1. Leadership, 2. Strategy & Planning, 3. Human Resources, 4. Partnerships & Resources, 5. Process Management and 6. Organizational governance. To assess the significance of the implementation of and presence of social responsibility practices, we used a Likert-type scale of five points. It starts with "Very Low" (1) and ends with "Very high" (5).

## 4.2 Choice of Criteria Results: Social Responsibility Performance Perspectives

Our choice of performance perspectives is based on the 6 well-known perspectives exactly as mentioned in the ISO 26000:

1. Human Rights Result Perspective, 2. Labour practices Result Perspective, 3.the environment Result Perspective, 4. Consumer issues Result Perspective, 5. Fair operating practices Result Perspective, 6. Community involvement and development Result Perspective.

# 5 Presentation of the Research Model

Our model relies on 12 criteria which are divided into 2 families: (Fig. 4).

- 6 criteria refer to the means (Leadership, policy and Strategy, Human Resources, Partnerships & Resources, Process Management and Organizational governance).
- 6 criteria refer to results (Human Rights Results, Labour practices Results, the environment Results, Consumer issues Results, Fair operating practices Results, community involvement and development Results).

There are causal relations between the criteria of means and the criteria of results. In other words, the means in place are the causes of the given results (Table 1).

**Table 1.** Codes used in the causal model

| Constructs of the proposed model | Code | Title |
|---|---|---|
| Social Responsibility practices | LSHP | Leadership |
| | STP | Strategy and planning |
| | HRS | Human resources |
| | PRS | Partnerships and resources |
| | PMAN | Process management |
| | OG | Organizational governance |
| Criteria results | HU.R | Human rights result |
| | LP.R | Labour practices result |
| | ENV.R | The environment result |
| | CI.R | Consumer issues |
| | FOP.R | Fair operating practices |
| | CID.R | Community involvement and development result |

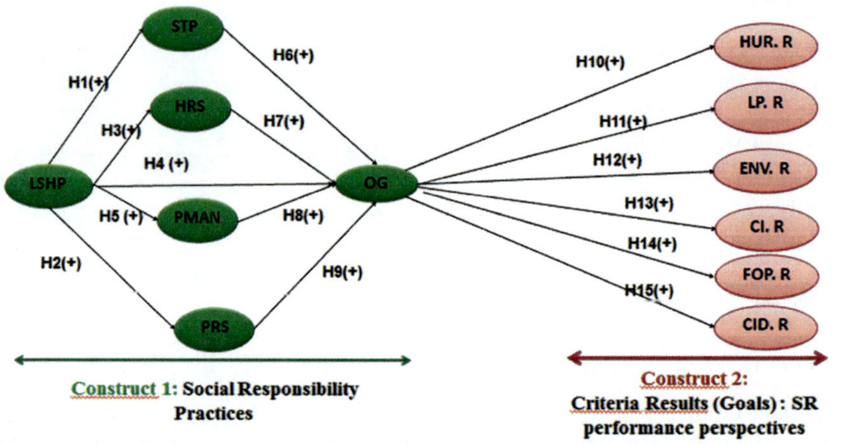


**Fig. 3.** Proposed model

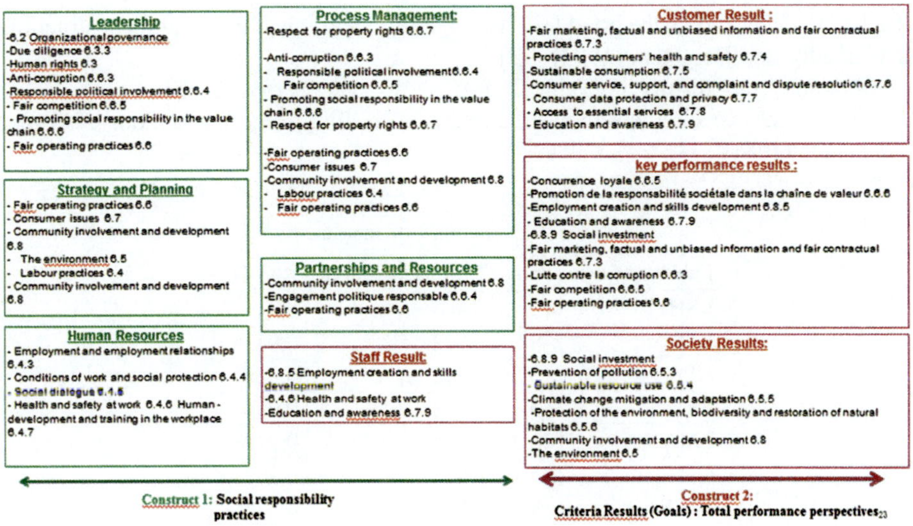


**Fig. 4.** Summary of the ISO 26000 requirements according to the EFQM model

## 6 Research Methodological Framework

Our study is an empirical type administered via face-to-face questionnaires from managers, technicians, engineers and directors of a worldwide enterprise located in Casablanca, Morocco (the survey response rate reached 97%, out of a total population of 130 employees).

### 6.1 Sampling

This research is conducted on a worldwide enterprise that operates in the Electrical sector and which is located in Casablanca, Morocco

### 6.2 Evaluation of the Proposed Model

The variables of the research model are 12 which are divided into 2 families:

- 6 variables that are related to social responsibility practices.
- The other 6 variables are related to the social responsibility performance.

The latent variables of our model are operationalized via the use of different items. These items are collected and evaluated by means of Likerts scales of five-degree ranging from the Very low to the Very high.

### 6.3 Reliability of the Measurements and Unidimensionality of the Blocks

To examine the reliability of measurements we use "Cronbach Alpha" and "Rho D.G." indexes. For each dimension, the value of this internal consistency is greater than 0.8

(Table 2), which shows the good level of reliability, and this, following the instructions of Nunnally and Bernstein (1994) [13].

Similarly, the results in Table 3 also show satisfactory results where Rhô.D. G > 0.8 for all measurement models, and this according to the instructions of Fornell and Larker (1981) [14].

**Table 2.** Reliability of measures

| Latent variable | Items | Cronbach's alpha | Rho D.G. |
|---|---|---|---|
| LSHPP | 12 | 0,904 | 0,911 |
| STP | 9 | 0,972 | 0,983 |
| HRS | 13 | 0,915 | 0,922 |
| PRS | 9 | 0,918 | 0,934 |
| PMAN | 13 | 0,975 | 0,979 |
| OG | 5 | 0,935 | 0,969 |
| HU.R | 8 | 0,800 | 0,875 |
| LP.R | 5 | 0,851 | 0,886 |
| ENV.R | 4 | 0,889 | 0,890 |
| CI.R | 5 | 0,839 | 0,864 |
| FOP.R | 7 | 0,876 | 0,916 |
| CID.R | 7 | 0,890 | 0,891 |

**Table 3.** Eigen values latent variables causal model

| LSHPP | STP | HR | PRS | PMAN | OG | HU.R | LP.R | ENV.R | CI.R | FOP.R | CID.R |
|---|---|---|---|---|---|---|---|---|---|---|---|
| 7,134 | **4,631** | **6.228** | **4,944** | **8,089** | **4,944** | **5,243** | **5,082** | **3,255** | **2,716** | **2,721** | **2,286** |
| 0,880 | 2,124 | 0.732 | 0,909 | 0,566 | 0,909 | 0,839 | 1,011 | 0,425 | 0,566 | 0,993 | 0,466 |
| 0,437 | 1,108 | 0.487 | 0,465 | 0,271 | 0,465 | 0,630 | 0,483 | 0,179 | 0,436 | 0,766 | 0,302 |
| 0,418 | 0,648 | 0.414 | 0,279 | 0,215 | 0,279 | 0,378 | 0,426 | 0,141 | 0,282 | 0,384 | 0,248 |
| 0,347 | 0,244 | 0.335 | 0,186 | 0,169 | 0,001 | 0,312 | 0,384 | | 0,189 | 0,136 | 0,112 |
| 0,288 | 0,061 | 0.304 | 0,175 | 0,148 | | 0,209 | | | | 0,186 | 0,052 |
| 0,184 | 0,036 | 0.267 | 0,042 | 0,129 | | 0,205 | | | | 0,175 | 0,007 |
| 0,147 | 0,002 | 0.153 | 0,000 | 0,112 | | 0,182 | | | | 0,042 | |
| 0,103 | 0,000 | 0,013 | | 0,101 | | | | | | | |
| 0,000 | | 0,000 | | 0,079 | | | | | | | |
| 0,000 | | 0,000 | | 0,000 | | | | | | | |
| | | | | 0,000 | | | | | | | |

### 6.4 Evaluation of the Measurement Model

There are three ways of linking the manifest variables to the latent variables whose scheme can be MIMIC, formative, or reflective. The external evaluation of the measurement models depends on the nature of the selected schema or pattern (MIMIC, formative, reflective) [8, 15].

The same author has confirmed that the reflective type schema (Fig. 2) is the most suitable in most structural equation models and that such a choice is mainly based on the subjectivity of the researcher's subjectivity. Each manifest variable is linked to its latent variable by a simple regression (Fig. 5):

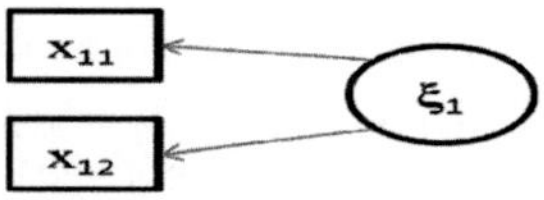


**Fig. 5.** Reflective diagram

The relation between the latent variable and the group of manifest variables related to it is expressed as follows: $x_{kj} = \pi_{kj} * \xi_k + \varepsilon_{kj}$

$\mathbf{x_{kj}}$: Vector related to the N$^{th}$ manifest variable of the latent variable $\xi_k$
$\xi$: Latent variable
**k**: Index of latent variables
$k_j$: Index of block manifest variables
$\pi$: Loading related to $\mathbf{x_{kj}}$
$\varepsilon_{kj}$: Term of errors (measurement errors of latent variables)

### 6.4.1 Convergent Validity

We can talk about a good convergent validity if AVE > 0.5 and this following Fornell and Larcker guidelines (1981) [14]. The results in Table 2 clearly confirm the good convergent validity of our measurement model. The rhô.D.G > 0.7 coefficient appears also significant for all the measurement model [13] (Table 4).

**Table 4.** Quality index of measurement models

| Variable latente | Average Variance Extracted (AVE) | D. G. Rhô |
|---|---|---|
| LSHPP | 0,713 | 0,911 |
| STP | 0,901 | 0,983 |
| HRS | 0,697 | 0,922 |
| PRS | 0,706 | 0,934 |
| PMAN | 0,819 | 0,979 |
| OG | 0,737 | 0,969 |
| HU.R | 0,651 | 0,875 |
| LP.R | 0,655 | 0,886 |
| ENV.R | 0,635 | 0,890 |
| CI.R | 0,813 | 0,864 |
| FOP.R | 0,679 | 0,916 |
| CID.R | 0,542 | 0,891 |

### 6.4.2 Divergent Validity (Descriminant)

According to the recommendations of Chin (2010) [16] the divergent validity is retained only if the items belonging to a single construct do not contribute significantly with the others constructs To assess the said validity, the square root of the average extracted variance (AVE) for each factor (Latent Variable) shall be compared with the correlation between the two-by-two factors. The results of Table 5 show that the square root of the AVE is greater than the correlations between the different dimensions of our model, the divergent validity is ensured.

This indicator is calculated as follows [14]:

$$\text{AVE} = \frac{\sum\left[\gamma_i^2\right]\text{var(VL)}}{\sum\left[\gamma_i^2\right]\text{var(VL)} + \sum[\text{var}(\varepsilon_i)]}$$

With $\gamma_i$: factorial contributions of manifest variables associated with a Latent Variable (VL); var: Variance; $\varepsilon_i$: Terms of errors related to each manifest variable; VL: Latent Variable.

**Table 5.** The discriminating validity

| | STP | HRS | PRS | PMAN | OG | HU.R | LP.R | ENV.R | CI.R | FOP.R | CID.R | (AVE) |
|---|---|---|---|---|---|---|---|---|---|---|---|---|
| STP | **0,947*** | | | | | | | | | | | **0,901** |
| RH | 0,823 | **0,825** | | | | | | | | | | **0,697** |
| PRS | 0,836 | 0,805 | **0,844** | | | | | | | | | **0,706** |
| MP | 0,837 | 0,780 | 0,829 | **0,915** | | | | | | | | **0,819** |
| OG | 0,591 | 0,516 | 0,609 | 0,621 | **0,869** | | | | | | | **0,737** |
| HU.R | 0,585 | 0,564 | 0,604 | 0,675 | 0,683 | **0,817** | | | | | | **0,651** |
| LP.R | 0,701 | 0,656 | 0,729 | 0,819 | 0,663 | 0,704 | **0,819** | | | | | **0,655** |
| ENV.R | 0,771 | 0,734 | 0,734 | 0,816 | 0,558 | 0,651 | 0,802 | **0,787** | | | | **0,635** |
| CI.R | 0,499 | 0,476 | 0,519 | 0,444 | 0,364 | 0,321 | 0,500 | 0,445 | **0,912** | | | **0,813** |
| FOP.R | 0,510 | 0,456 | 0,503 | 0,528 | 0,408 | 0,373 | 0,498 | 0,461 | 0,585 | **0,814** | | **0,679** |
| CID.R | 0,368 | 0,317 | 0,394 | 0,424 | 0,305 | 0,309 | 0,434 | 0,398 | 0,357 | 0,547 | **0,726** | **0,542** |

* The square root

## 6.5 Validation of the Structural Model

The structural model defines the nature of the relationships existing between its latent variables. To test the validity of adjusting our internal model via the PLS algorithm, we will resort the following:

### 6.5.1 Goodness of Fit Index (GoF)

This index takes into account both the performance of the structural model and that of the measurement model [8, 17]. It is defined by the geometric mean of the communities mean (or AVE) on the set of latent variables ($\overline{H^2}$) and the mean of the coefficients of determinations ($R^2$) associated with endogenous latent variables: ($\overline{R^2}$): GoF = ($\sqrt{\overline{H^2} \times \overline{R^2}}$).

According to Wetzels et al. (2009) [18], the usual values of this index are 0.1, 0.25 and 0.36. They refer respectively to a small, medium and large adequacy of the model. Therefore, and following instructions of Wetzels et al. [18], and according to the obtained results (Table 6), the search model can be retained based on the threshold (GoF > 0.5), and this, according to the instructions of Wetzels et al. (2009) [18].

This confirms the validity of the structural model.

**Table 6.** The discriminating validity

| | GOF | GoF (Bootstrap) | Standard error | Critical ratio (CR) |
|---|---|---|---|---|
| Absolute | 0,705 | 0,702 | 0,032 | 22,710 |
| Relative | 0,931 | 0,915 | 0,029 | 33,326 |
| External model | 0,992 | 0,989 | 0,021 | 41,381 |
| Internal model | 0,933 | 0,921 | 0,009 | 107,060 |

### 6.5.2 The Coefficient of Determination ($R^2$)

On the one hand, the usual values of $R^2$ are 0.67, 0.33 and 0.19. These are respectively considered as substantial, moderate and low [19]. On the other hand, the structural model is retained when $R^2 > 0.67$ [19]. From what preceded, the results of $R^2$ and $R^2$-adjusted (Table 7) are substantial to moderate. In order to validate the structural model (internal), we will remove the link relative to the growth factor (EC) ($R^2 \lll 0.67$).

**Table 7.** $R^2$ and $R^2$ Adjusted Results

| | LSHPP | STP | HRS | PRS | PMAN | OG | HU.R | LP.R | ENV.R | CI.R | FOP.R | CID.R |
|---|---|---|---|---|---|---|---|---|---|---|---|---|
| $R^2$ | – | 0,849 | **0,871** | **0,886** | **0,938** | **0,921** | **0,973** | **0,852** | **0,856** | **0,716** | **0,675** | **0,746** |
| $R^2$-Ajusted | – | 0,849 | **0,871** | **0,886** | **0,938** | **0,921** | **0,973** | **0,852** | **0,856** | **0,716** | **0,675** | **0,746** |

### 6.5.3 Effect Size $f^2$

This index allows us to check the validity and magnitude of structural coefficients.

$$f^2 = \frac{R^2 \text{ incl} - R^2 \text{ excl}}{1 - R^2 \text{ incl}}$$

According to (Cohen, 1988) [20], the usual values of this index are of Effect size $f^2$ are 0.02, 0.15, and 0.35. They refer respectively to a weak, medium and strong effect. The results of Table 8 our survey clearly show the validity of the measurement model (external) and the validity of the structural model (internal).

### 6.5.4 Structural Equations of the Conceptual Model

Our model holds a single exogenous variable which is the "Leadership" one and it contains eleven endogenous variables. Each endogenous variable is explained by one or more variables in addition to an error term.

The internal model is defined by linear equations connecting the latent variables with each other. For all endogenous $\xi k$, we have $\xi_k = \sum_{i:\xi i \rightarrow \xi k} \beta_{kj} +_k$.

With $\beta_{kj}$ representing the coefficient associated with the relation between the variables ξk and ξi. ζk is the error term and $\xi_i \rightarrow \xi_k$: ξi explains ζk in the model.

Our model is constituted of 12 equations that have been tested based on the PLS approach through the XL-STAT software. The structural equations of the conceptual model are as follows:

1. STP = 0,93801*LSHPP
2. HRS = 0,89913*LSHPP
3. PRS = 0,90824*LSHPP
4. PMAN = 0,92157*LSHPP
5. OG = 0,94107*LSHPP + 0,85916*STP + 0,49923*HRS + 0,18154*PRS + 0,67168*PMAN
6. HU.R = 0,80717*OG
7. LP.R = 0,65826*OG
8. ENV.R = 0,37754*OG
9. CI.R = 0,62605*OG
10. FOP.R = 0,28960*OG
11. CID.R = 0,21729*OG

## 7 Hypothesis Testing

A global hypothesis will be tested: HG: "social responsibility practices positively impact the SR performance (goals) of this worldwide enterprise". Thus, for each causal relationship, we have formulated a derived hypothesis and since we have 15 causal relationships, we have set up 15 hypotheses derived. These hypotheses will also be subjected to tests of confirmation.

**Table 8.** Research Hypothesis Tests

| Causal relationship | Path Coefficient (β) | T* student | Effect size ($f^2$) | Pr > \|t\| (P-Value) | Validation the hypothesis |
|---|---|---|---|---|---|
| H1:LSHPP → STP | 0.938 | 29.599 | 7.301 | 0.000 | Valid |
| H2:LSHPP → PRS | 0.908 | 24.429 | 4.973 | 0.000 | Valid |
| H3:LSHPP → HRS | 0.899 | 22.508 | 4.222 | 0.000 | Valid |
| H4: LSHPP → PMAN | 0,921 | 20,511 | 5,617 | 0.000 | Valid |
| H5:LSHPP → OG | 0,941 | 20,247 | 7,874 | 0.000 | Valid |
| H6: STP → OG | 0,859 | 13,986 | 0,291 | 0.000 | Valid |
| H7: PRS → OG | 0,181 | 5,039 | 0,093 | 0.000 | Valid |
| H8: HRS → OG | 0,499 | 5,185 | 0,142 | 0.000 | Valid |

(*continued*)

**Table 8.** *(continued)*

| | | | | | |
|---|---|---|---|---|---|
| H9: PMAN → OG | 0,671 | 6,055 | 0,327 | 0.000 | Valid |
| H10:OG → HU.R | 0,807 | 7,249 | 0,562 | 0.000 | Valid |
| H11:OG → LP.R | 0,658 | 5,303 | 0,174 | 0.000 | Valid |
| H12:OG → ENV.R | 0,377 | 4,562 | 0,163 | 0.000 | Valid |
| H13:OG → CI.R | 0,626 | 6,962 | 0,378 | 0.000 | Valid |
| H14:OG → FOP.R | 0,289 | 4,976 | 0,152 | 0.000 | Valid |
| H15:OG → CID.R | 0,217 | 2,953 | 0,145 | 0.000 | Valid |

According to Table 8, we can confirm the validity of all assumptions about social responsibility practices and total performance (T > 1.96).

The final model can be represented as follows (Fig. 6):

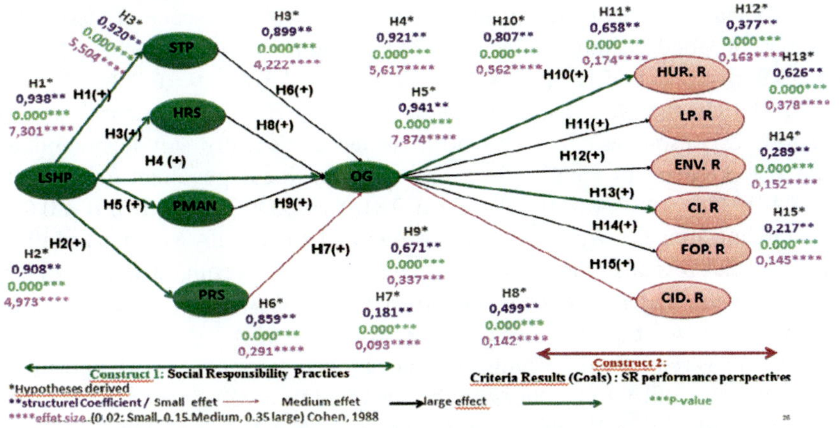


**Fig. 6.** Final model estimated by the PLS

**Table 9.** Indirect effects between latent variables

| | LSHPP | STP | HRS | PRS | PMAN | OG | HU.R | LP.R | ENV.R | CI.R | FOP.R | CID.R |
|---|---|---|---|---|---|---|---|---|---|---|---|---|
| LSHPP | 0,000 | | | | | | | | | | | |
| STP | 0,000 | | | | | | | | | | | |
| HRS | 0,000 | 0,000 | | | | | | | | | | |
| PRS | 0,000 | 0,000 | 0,000 | | | | | | | | | |
| PMAN | 0,000 | 0,000 | 0,000 | 0,000 | | | | | | | | |
| OG | 0,000 | 0,000 | 0,000 | 0,000 | 0,000 | | | | | | | |
| HU.R | **0,946** | **0,334** | **0,040** | **0,0102** | **0,246** | 0,000 | | | | | | |
| LP.R | **0,923** | **0,362** | **0,035** | **0,038** | **0,223** | 0,000 | 0,000 | | | | | |
| ENV.R | **0,826** | **0,348** | **0,030** | **0,025** | **0,276** | 0,000 | 0,000 | 0,000 | | | | |
| CI.R | **0,922** | **0,367** | **0,028** | **0,032** | **0,622** | 0,000 | 0,000 | 0,000 | 0,000 | | | |
| FOP.R | **0.783** | **0,214** | **0,016** | **0,071** | **0.283** | 0,000 | 0,000 | 0,000 | 0,000 | 0,000 | | |
| CID.R | **0.888** | **0,232** | **0,022** | **0,018** | **0.088** | 0,000 | 0,000 | 0,000 | 0,000 | 0,000 | 0,000 | |

### 7.1 Results Interpretation

The main objective of this empirical study is to test, on the one hand, the impact of some practices of the quality on the social responsibility performance factors. The results obtained via the structural equations and the XL-STAT software (Fig. 3 and Tables 8 and 9) presented above, help us come up with the following recommendations:

- Leadership, which is an exogenous latent variable, has statistically significant positive influence(s) hence demonstrating a very large impact on the latent variables "Strategy and Planning", "Human Resources", and "Partnerships & Resources", and "Process Management" and "Organizational governance": (($\beta = 0.938$ $t = 29.599$, $f^2 = 7.301$, $p < 0.01$); ($\beta = 0.908$, $t = 24.429$, $f^2 = 4.973$, $p < 0.01$); ($\beta = 0.899$, $t = 22.508$, $f^2 = 4.222$, $p < 0.01$)).; ($\beta = 0{,}921$, $t = 20{,}511$, $f^2 = 5{,}617$, $p < 0.01$)).; ($\beta = 941$, $t = 20{,}247$, $f^2 = 7{,}874$, $p < 0.01$)).
- The latent variables "Strategy & Planning" and "Process Management" demonstrate positive and statistically significant influences, with the presence of a considerable number of good medium effects on "Organizational governance" (($\beta = 0.859$, $t = 13.986$, $f^2 = 0.291$, $p < 0.01$), ($\beta$ = (($\beta = 0.671$, $t = 6.055$, $f^2 = 0.327$, $p < 0.01$). On the other hand, the latent variable "Human Resources" and "Partnerships & Resources" indicates positive and statistically significant influences, with the presence of small effects on "Organizational governance", ($\beta = 0.499$, $t = 5.185$, $f^2 = 0.142$, $p < 0.01$).; ($\beta = 0{,}181$, $t = 5{,}039$, $f^2 = 0{,}093$, $p < 0.01$)).
- The Organizational governance has positive and statistically significant influences on the social responsibility factors of "Human Rights Result" and the "Consumer issues" with a large effects importance ($\beta = 0.807$, $t = 7.249$, $f^2 = 0.562$, $p < 0.01$).; ($\beta = 0{,}626$, $t = 6{,}962$, $f^2 = 0{,}378$, $p < 0.01$)).
- The latent variables concerning "Organizational governance" has positive and statistically significant influences on the"community involvement and development Result" with a small effects importance ($\beta = 0{,}217$, $t = 2{,}953$, $f^2 = 0{,}145$, $p < 0.01$)).
- In the end, we notice that The latent variable Organizational governance has positive and statistically significant influences on "Labour practices Result", "the environment Result", "Fair operating practices", and the "community involvement and development Result" with a medium effects importance ($\beta = 0.658$, $t = 5.303$, $f^2 = 0.174$, $p < 0.01$).; ($\beta = 0{,}377$, $t = 4{,}562$, $f^2 = 0{,}163$, $p < 0.01$)). ($\beta = 0.289$, $t = 4.976$, $f^2 = 0.152$, $p < 0.01$);

## 8 Discussion and Conclusion

Our article tackles the issue of measuring the impact of social responsibility practices (divided into 6 practices) on the SR performance (divided into 6 perspectives) of a worldwide Electrical enterprise located in Casablanca, Kingdom of Morocco. The practices that are related to the used means show how the organization works, and whether the means help attain the desired results. For that, this kind of enterprises should seek to constantly improve, the social responsibility practices which exhibit the low and medium direct and indirect influences: “Strategy & planning”, “human Resources”, “Partnerships & resources” and “Organizational governance”.

Also one must become aware of the fact that leadership and commitment of leaders vis-à-vis social responsibility must be visible, perpetual, and proactive, and should take place at all management levels.

Research has shown that excellent results are not achieved by result-based criteria, but by improved performance in criteria of means. This is why organizations are more aware of criteria of means than criteria of results.

## References

1. Huczynski, A.A., Buchanan, D.A.: Organizational Behaviour, 6th edn., pp. 44 (2007
2. Fernandes, V.: En Quoi L’approche PLS Est-Elle Une Méthode a (Re)-Découvrir Pour Les Chercheurs En Management? Management **15**(1), 102 (2012)
3. Elhasbi, A., Barkaoui, M., Bouksour, O., Kamach, O.: The territorial attractiveness for foreign investments of Mediterranean cities: the case of city of Tangier. Int. J. Euro-Mediterr. Stud. Table Contents **9**(2), 3–26 (2016)
4. Roussel, P., Durrieu, F., Campoy, E., El Akremi, A.: Méthodes D’équations Structurelles: Recherches et Applications En Gestion. Economica, Paris (2002)
5. Jacobowicz, E.: Contributions aux modeles d’equations structurelles a variables Latentes, These de Doctorat, Conservatoire national des arts et metiers de Paris (2007)
6. Clémence, D.: Une Méthode Alternative L’approche PLS: Comparaison et Application Aux Modèles Conceptuels Marketings.“ Statistique Aplliquée **52**(3), 37–72 (2004)
7. Lacroux, A.: L’analyse Des Modèles de Relations Structurelles Par La Méthode PLS: Une Approche Émergente Dans La Recherche Quantitative En GRH. Paper presented on XXème Congres (AGRH), Toulouse, 17–21 November (2009)
8. Benali, M., Rifai, S., Bouksour, O., Barrijal, S.: Young manufacturing enterprises in the growth phase located in the Wilaya of Tetouan, Morocco: a model for measurement of the impact of quality on their growth. Asian J. Manage. Res. **6**(1), 208–222 (2015)
9. ISO 26000: Guidelines Corporate Social Responsibility. ISO, Geneva, p. 127 (2010)
10. Benali, M., Rifai, S., Bouksour, O., Barrijal, S.: Proposal of a causal model measuring the links between quality practices and corporate social responsibility (CSR): case of the manufacturing companies located in Tangier Free Zone and in the Gueznnaya industrial area in Tangier (Morocco). Int. J. Innovation Appl. Stud. **11**(3), 650–662 (2015)
11. Fethallah, W., Chraibi, L., Sefiani, N.: Assessment tool for social responsibility performance according to the ISO 26000. World Acad. Sci. Eng. Technol. Int. J. Industr. Manuf. Eng. **10**(10), 3533–3537 (2016)
12. Chraibi, L., Sefiani, N.: A new performance measurement corporate social responsibility model for small and medium-sized industrial enterprises. In: Proceedings of the International Conference on Industrial Engineering and Operations Management Rabat, Morocco, 11–13 April 2017
13. Gefenet, D.W., Straub, D., Boudreau, M.-C.: Structural equation modelling and regression: guidelines for research practice. Commun. Assoc. Inf. Syst. **4**(7), 1–79 (2000)
14. Hulland, J.: Use of partial least squares (PLS) in strategic management research: a review of four recent studies. Strateg. Manag. J. **20**(2), 195–204 (1999)
15. Chenhall, R.H.: Integrative strategic performance measurement systems, strategic alignment of manufacturing, learning and strategic outcomes: an exploratory study. Acc. Organ. Soc. **30**(5), 395–422 (2005)

16. Nunnally, J.C., Bernstein, I.H.: Psychometric Theory, 3eme edn. McGraw-Hill, New York (1994)
17. Fornell, D.F., Larcker, C.: Evaluating structural equation models with unobservable variables and measurement error. J. Mark. Res. **18**(1), 39–50 (1981)
18. Chin, W.: The partial least squares approach for structural equation modeling. In: Marcoulides, G.A. (ed.) Modern Methods for Business Research, pp. 295–236. Lawrence Erlbaum Associates, London (1998)
19. Wetzels, M., Odekerken-Schroder, G., Vanopen, C.: Using PLS path modeling for assessing hierarchical construct models: guidelines and empirical illustration. MIS Q. **33**(1), 177–195 (2009)
20. Cohen, J.: Statistical power analysis for the Behavioral sciences, 2e edn. Lawrence Erlbaum Associates, Hillsdale (1988)

# Industry 4.0 and Lean Six Sigma: Results from a Pilot Study

Cherrafi Anass[1(✉)], Belhadi Amine[2], El Hassani Ibtissam[1], Imane Bouhaddou[1], and Said Elfezazi[2]

[1] Moulay Ismail University, Meknes, Morocco
a.cherrafi@ensam.umi.ac.ma
[2] Cadi Ayyad University, Marrakesh, Morocco

**Abstract.** In the recent years, the two important initiatives, namely, Lean Six Sigma and Industry 4.0 have attracted the attention of many professionals and researchers. However, the relationship between the two concepts needs to be investigated. Motivated by this gap, this paper aims to study the connection between Lean Six Sigma and Industry 4.0 in the Moroccan context based on a descriptive survey questionnaire. The findings indicate that Lean Six Sigma and Industry 4.0 are synergetic and compatible. More studies are required in order to validate this result.

**Keywords:** Lean six sigma · Industry 4.0 · Survey · Morocco

## 1 Introduction

In today's global competition environment, organizations have to be efficient, agile, responsive and cost-effective by reducing the different operational costs [1], In addition, organizations need to satisfy the strong consumer demand for customized products [2]. To respond to these challenges, changes and innovations are inevitable. This has encouraged many firms to explore new ways to achieve the operational excellence. In this context, many initiatives have emerged including Lean Six Sigma. It is based on the integration of two powerful approaches Lean Manufacturing and Six Sigma. Lean Six Sigma aims to improve operational performance by eliminating waste and reducing variability [3]. It provides the key elements to achieve operational excellence. However, given the increasing complexity of process, many organizations have found that Lean Six Sigma is not sufficient to survive in the era of the globalization. Recently, an important advance in digital technologies known as Industry 4.0 has emerged to help companies in the different industrial sectors to improve manufacturing productivity and cost efficiency by using an arsenal of technologies including artificial intelligence, cloud computing, internet of the things, additive manufacturing, and robotics [4]. By implementing the adequate technologies, organizations can boost their efficiency, and increase throughput and quality.

Recently, some evidences from industry show that organizations are able to integrate Lean and some technologies to take advantage of both initiatives [5]. However, to the best of our knowledge, there is no study to investigate the relationship between


A. Saka et al. (Eds.): CPI 2019, LNME, pp. 613–619, 2021.
https://doi.org/10.1007/978-3-030-62199-5_54

Lean Six Sigma and Industry 4.0. It has not been examined how the introduction of industry 4.0 will influence the continuous improvement imitative such as Lean Six Sigma and how this approach will influence the deployment of Industry 4.0 in the manufacturing sectors. In light of the shortcomings of existing literature on the Lean Six Sigma and Industry 4.0, this paper aims to address this gap and provides evidences to assist managers in achieving the operational excellence. It aims to explore the connection between Lean Six Sigma and Industry 4.0 using a survey questionnaire instrument.

The rest of the paper is structured as follows: Sect. 2 presents a literature review. Section 3 presents the research methodology followed by the presentation of findings in Sect. 4 and the conclusion of the paper is presented in the final section.

## 2 Literature Review

### 2.1 Lean Six Sigma

Lean Six Sigma is a business improvement approach that integrates two powerful continuous improvement methodologies; Lean Manufacturing and Six Sigma [6]. The term 'Lean Six Sigma' was used in the first time in the late 1990s and early 2000s to describe this integration [7]. Lean Six Sigma aims to increase shareholder value by improving quality and speed resulting in enhanced customer satisfaction and cost reduction [8]. It uses techniques and tools from both strategies in order to eliminate wastes, defects and reduce variation in the different processes to improve bottom line results [9]. Lean Six Sigma uses a structured problem-solving methodology based on five stages called DMAIC (Define-Measure-Analyse-Improve-Control) [10]. It has been successfully implemented in different organizations around the globe [3]. However, in many firms, the potential of Lean Six Sigma transformation is exhausted. Recent advances in digital technology offer the opportunity for organizations to move towards a new industrial era where production systems are more autonomous, intelligent, automated and integrated [11].

### 2.2 Industry 4.0

Industry 4.0 -or the fourth industrial revolution- is a concept that has emerged from many developed countries and it was originate from "High-Tech Strategy 2020" developed by German government in 2011 [12]. According to [13], industry 4.0 is the physical digitalization of different manufacturing processes to develop a connected and smart factory in order to develop e-value chains. Prause and Weigand define Industry 4.0 as the "combination of cyber-physical systems with automated systems" in order to achieve the efficiency and effectiveness of the manufacturing operation with minimum human intervention [14]. Industry 4.0 is characterized by an important transformation in the manufacturing systems connectivity due to the integration of Information and Communication Technologies (ICT) – especially the Internet of Things (IoT) – forming the so-called Cyber-Physical Systems (CPS) [15]. According to many studies, industry 4.0 can lead organizations to develop more flexible operations and intelligent products

[16]. It will help also companies to capture and analyze significant amounts of data in real-time for better decision making and strategic development [17]. In addition, industry 4.0 can contribute to the business growth by producing highly customized products with high productivity, efficiency and quality [18]. It is based on many technological concepts such as the cyber-physical systems, the internet of things, big data, autonomous robots, cloud computing, additive manufacturing, augmented reality and artificial intelligence [18].

## 3 Methodology

This study aims to identify the relationship between Lean Six Sigma and Industry 4.0. The study attempts to shed some light on the following question:

- What is the future of Lean Six Sigma: it will become obsolete or will offer a compatibility and synergy when implemented with industry 4.0 and their new technologies?

In order to answer this research question, a survey-based study appeared appropriate. The survey questionnaire was developed based on a brainstorming session with three academics and two professionals in the field of industrial management and new technologies. It consists of two sections, the first section aimed to capture general information about the participating companies. The second section aimed to identify the relationship between Lean Six Sigma and Industry 4.0 (See Appendix).

A set of elements was determined in order to select the appropriate sample of companies. Consequently, the sample selection was based on two conditions:

1. Company is located in Morocco;
2. Implementing Lean Six Sigma or Industry 4.0 or both for at least 12 months;

The questionnaire was mailed to 94 organizations in Morocco. After sending two reminders to sample organizations, 28 questionnaires were returned with only 23 completed and valid responses. This represented a response rate of 24.5% that have been considered as acceptable by many researchers in this field [19]. In addition, this survey was covered only Moroccan organizations that have deployed Lean Six Sigma/Industry 4.0 as an approach to improve their operational performance; a low response rate was excepted as these techniques are recent and developed.

## 4 Results and Analysis

The analysis of the survey is very crucial to develop a clear picture on the relationship between Lean Six Sigma and Industry 4.0. Some of the major findings of the questionnaire are presented below.

- The majority of the participated organizations had between 50 and 150 workers (86%) and 14% of the organizations replied had more than 150 workers. No organizations (less than 50 employees) responded to the survey (See Fig. 1).

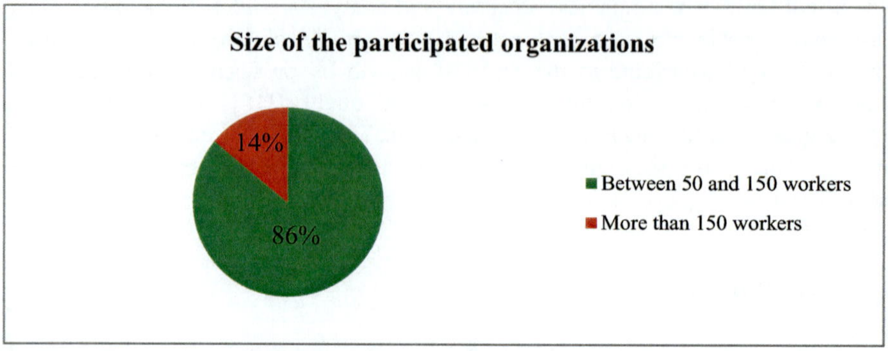


**Fig. 1.** Size of the participated organization.

- The respondents to the survey included employees from different functions. Details of the participant's positions are shown in Fig. 2. In addition, the types of manufacturing sectors that have participated in this study are indicated in Fig. 3.

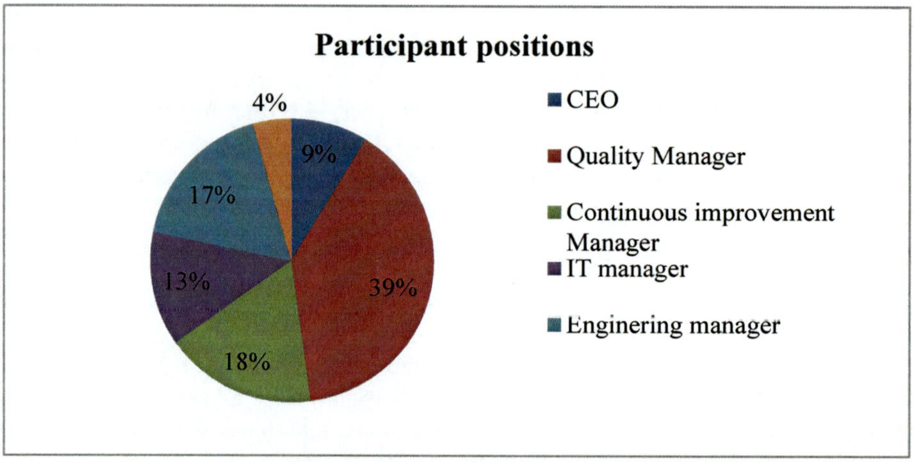


**Fig. 2.** Participant positions.

- More than 83% of the respondents believe that organizations aiming to improve their performance need to understand the interplay between Lean Six Sigma and Industry 4.0.
- 92% of the respondents believe that Industry 4.0 can benefit from Lean Six Sigma and that Industry 4.0 can boost Lean Six Sigma projects.
- More than 98% think that although Industry 4.0 and the new technologies will change business dramatically, Lean Six Sigma will stay relevant.
- 74% of the respondents indicate that Industry 4.0 and Lean Six Sigma need an important alignment when implemented together.

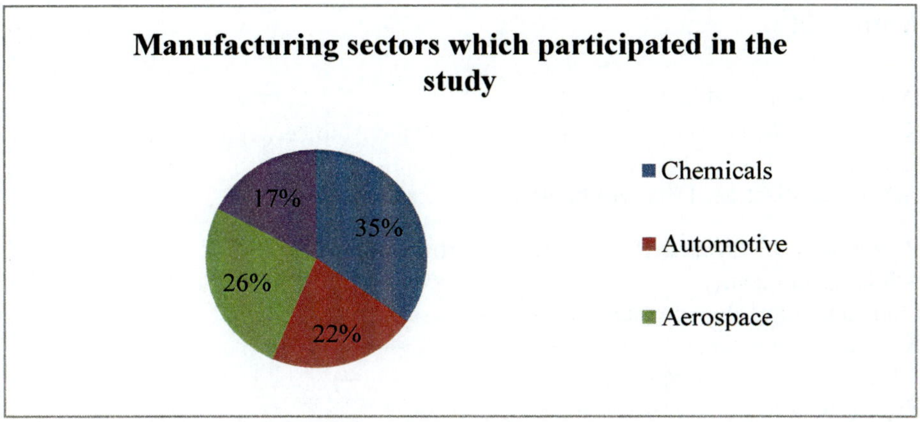


**Fig. 3.** Manufacturing sectors which participated in the study.

It is evident from the finding that there is an acknowledgement among professionals that integration of Lean Six Sigma and industry 4.0 can offer many opportunities for organizations to achieve operational excellence. By aligning Lean Six Sigma and Industry 4.0 technologies, companies can drive continuous improvement and maximize efficiency in order to reach the highest level of operational excellence. Organizations cannot base only on either Lean Six Sigma or Industry 4.0 or deploy one initiative without the other. Lean Six Sigma is important for industry 4.0 in order to avoid the automation of waste or instable process. New technologies developed by industry 4.0 are important for achieving a higher level of impact from Lean Six Sigma projects.

## 5 Conclusion

The aims of this paper were to study the relationship between Lean Six Sigma and Industry 4.0. A survey was developed to collect data from Moroccan organizations that have implemented Lean Six Sigma/Industry 4.0 initiatives. The results indicate that Lean Six Sigma and Industry 4.0 are compatible and synergetic. Therefore, organizations can implement the two strategies simultaneously. These results can encourage practitioners to integrate the two approaches and researchers to conduct more studies to validate statistically the relationship between Lean Six Sigma and Industry 4.0 in different contexts. This study has certain limitations. It was conducted in Morocco and the analysis was descriptive. Further research will be performed to develop a practical roadmap for the integration of Lean Six Sigma and Industry 4.0.

## Appendix

Dear expert/professional,
Kindly provide us your perception about the relationship between Lean Six Sigma and Industry 4.0. This makes part of a research about the future of Lean Six Sigma in the era of digitalization. Data will be used only academically.

**Personal data: Experts/Professionals identification**
Affiliation/company: ——————————
Main activity of the company: ——————————
Number of employees: ——————————
Job title/areas of expertise: ——————————
Experience (years): ——————————
Primary product (s) (if applicable): ——————————
Lean Six Sigma implementation: Yes: No:
Industry 4.0 implementation: Yes: No:

**Personal data: Perception about the relationship between Lean Six Sigma and Industry 4.0**
Please indicate how much you agree with a few following statements:

| Statement | Strongly agree | Agree | Neither agree nor disagree | Disagree | Strongly disagree |
|---|---|---|---|---|---|
| S1: Industry 4.0 can help Lean Six Sigma | | | | | |
| S2: Industry 4.0 can profit from Lean Six Sigma | | | | | |
| S3: Industry 4.0 should be integrated in Lean Six Sigma projects | | | | | |
| S4: Industry 4.0 and Lean Six Sigma can be integrated and implemented simultaneously | | | | | |
| S5: Lean Six Sigma will stay relevant and useful in the era of Industry 4.0 | | | | | |

## References

1. Fatorachian, H., Kazemi, H.: A critical investigation of industry 4.0 in manufacturing: theoretical operationalisation framework. Prod. Plann. Control **29**(8), 633–644 (2018)

2. Souza das Neves, J., Silva Marins, F., Akabane, G., Kanaane, R.: Deployment the MES (Manufacturing Execution System) aiming to improve competitive priorities of manufacturing. Independent J. Manag. Prod. **6**(2), 449–463 (2015)
3. Albliwi, S., Antony, J., Abdul Halim Lim, S., van der Wiele, T.: Critical failure factors of Lean Six Sigma: a systematic literature review. Int. J. Qual. Reliab. Manag. **31**(9), 1012–1030 (2014)
4. Wu, D., Weiss, B.A., Kurfess, T., Wang, L., Davis, J.: Introduction to the special issue on smart manufacturing. J. Manufact. Syst. **48**, 1–2 (2018)
5. Buer, S.V., Strandhagen, J.O., Chan, F.T.: The link between Industry 4.0 and lean manufacturing: mapping current research and establishing a research agenda. Int. J. Prod. Res. **56**(8), 2924–2940 (2018)
6. Cherrafi, A., Elfezazi, S., Govindan, K., Garza-Reyes, J.A., Benhida, K., Mokhlis, A.: A framework for the integration of green and lean six sigma for superior sustainability performance. Int. J. Prod. Res. **55**(15), 4481–4515 (2017)
7. Thomas, A.J., Francis, M., Fisher, R., Byard, P.: Implementing Lean Six Sigma to overcome the production challenges in an aerospace company. Prod. Plann. Control **27**(7–8), 591–603 (2016)
8. Ben Ruben, R., Vinodh, S., Asokan, P.: Implementation of Lean Six Sigma framework with environmental considerations in an Indian automotive component manufacturing firm: a case study. Prod. Plann. Control **28**(15), 1193–1211 (2017)
9. Moya, C.A., Galvez, D., Muller, L., Camargo, M.: A new framework to support Lean Six Sigma deployment in SMEs. Int. J. Lean Six Sigma **10**(1), 58–80 (2019)
10. Sunder, M.V.: Rejects reduction in a retail bank using Lean Six Sigma. Prod. Plann. Control. **27**(14), 1131–1142 (2016)
11. Schumacher, A., Erol, S., Sihn, W.: A maturity model for assessing Industry 4.0 readiness and maturity of manufacturing enterprises. Procedia Cirp **52**, 161–166 (2016)
12. Bibby, L., Dehe, B.: Defining and assessing industry 4.0 maturity levels–case of the defence sector. Prod. Plann. Control **29**(12), 1030–1043 (2018)
13. Castelo-Branco, I., Cruz-Jesus, F., Oliveira, T.: Assessing Industry 4.0 readiness in manufacturing: Evidence for the European Union. Comput. Ind. **107**, 22–32 (2019)
14. Prause, M., Weigand, J.: Industry 4.0 and object-oriented development: incremental and architectural change. J. Technol. Manag. Innovation **11**(2), 104–110 (2016)
15. Reischauer, G.: Industry 4.0 as policy-driven discourse to institutionalize innovation systems in manufacturing. Technol. Forecast. Soc. Chang **132**, 26–33 (2018)
16. Hahn, G.J.: Industry 4.0: a supply chain innovation perspective. Int. J. Prod. Res. 1–17 (2019)
17. Schwab, K.: The Fourth Industrial Revolution. Penguin Random House, U.K. Tec. News: 26. Harting Technology Group (2016)
18. Ghobakhloo, M.: The future of manufacturing industry: a strategic roadmap toward Industry 4.0. J. Manufact. Technol. Manag. **29**(6), 910–936 (2018)
19. Shah, R., Chandrasekaran, A., Linderman, K.: In pursuit of implementation patterns: the context of Lean and Six Sigma. Int. J. Prod. Res. **46**(23), 6679–6699 (2008)

# Experimentation of MASK Applied to Formalize the Design Technique of Ornamental Patterns

Imane El Amrani[1(✉)], Abdelmjid Saka[1], Nada Matta[2], and Taoufik Ouazzani Chahdi[3]

[1] National School of Applied Sciences of Fez, Avenue My Abdallah KM 5 Route D'Imouzzer, BP72 Fes, Morocco
{imane.elamrani2,abdelmjid.saka}@usmba.ac.ma

[2] University of Technology of Troyes, 12 Rue Marie Curie, 10010 Troyes, France
nada.matta@utt.fr

[3] Euro-Mediterranean University of Fes, Fes, Morocco
t.ouazzani@ueuromed.org

**Abstract.** The ornamental art found in Moroccan architecture is revealing ingenuity and a very valuable know-how that has been passed from generation to generation in the secrecy of the workshops and in the absence of documentation. The technique widely used for geometric drawing of patterns is called "Hasba" and its description remains ambiguous hence the interest to formalize the method construction rules through a formal modeling of the knowledge involved. The method used for this purpose is MASK, which has been chosen for its many advantages (ease of appropriation, flexibility, generic aspect, etc.). Its application to the field of Moroccan craftsmanship is the subject of this paper. It has been realized through a case study with expert craftsmen. However, this experiment revealed several challenges that are largely due to the informal nature of the craftsmanship domain and the artisans' profile. Thus, an adaptation of the methodological framework was necessary and allowed us to validate our approach. The result is a set of interleaved and interconnected models constituting a small "knowledge book" on the geometric drawing technique called "Hasba".

**Keywords:** MASK · Knowledge capitalization · Knowledge modeling · Hasba · Knowledge management

## 1 Introduction

Crafts trades have been always considered an ancestral heritage very rich in terms of know-how and ingenuity. A know-how that has been passed from generation to generation in the secrecy of workshops and the absence of documents. A knowledge, which is largely tacit and implicit. For instance, the art of geometric drawing of the ornamental patterns. It is about ambiguous and rather complex rules which make it possible to draw the patterns in an extreme precision. The main objective of this paper


A. Saka et al. (Eds.): CPI 2019, LNME, pp. 620–632, 2021.
https://doi.org/10.1007/978-3-030-62199-5_55

is to formalize the knowledge related to this technique through a modeling approach. To do so, no document references has been found except the work in [11], a research paper that describes in free text the "Hasba" method. The modeling is first based on the knowledge addressed in this paper and then enriched and completed through the capitalization process with artisans. The method that will be used for capitalization and then modeling comes from the field of knowledge engineering and knowledge management. It is about the MASK method (Method for Analyzing and Structuring Knowledge) [1], which seems to be the best suited for our case because it is based on an easy-to-use graphical programming language, which interferes with the knowledge (Knowledge level) instead of the system (symbolic level) [2]. Other methods, such as KOD or CommonKADS [9] are more specifically dedicated to the development of knowledge-based systems (previously called "expert systems"). Besides, the models building is very formal and their concrete applications in different business contexts are very few [2].

Thus, the main idea of this article is to experiment the MASK method to formalize the Moroccan craftsmen know-how. The experimentation is carried out through a case study about the geometric drawing of Moroccan ornamental patterns that artisans call in arabic "the Hasba". To do this, the article is organized as follows: Sect. 2 is a short presentation of knowledge capitalization methods and the reasons for choosing MASK as a methodological framework. Section 3 is a presentation of the MASK method and its different models presented and described one by one. In Sect. 4 the authors discuss the difficulties encountered in the use of MASK [1, 5] within Moroccan craftsmanship, thus exposing the constraints related to the nature of the highly informal field and the rules transgressed accordingly. In Sect. 4, the application of MASK on the case study is described, an approach has been described for this purpose and the result is knowledge formalization in the form of models. Finally, results are discussed and the perspectives are presented in the conclusion.

## 2 Methods for Knowledge Capitalization

There are several methods in the literature for knowledge capitalization that can be classified into two broad categories [5, 6]: A range of methods for the development of knowledge-based systems (KBS), formerly known as "expert systems" and which are rather located in the field of artificial intelligence such as CommonKADS and KOD and a second range of methods for the knowledge formalization in order to build a business memory or an organizational memory, among them MASK and REX methodologies.

CommonKADS and KOD are used for developing KBS. The modeling result is located at a symbolic level and requires a thorough mastery of the modeling language, which makes their appropriation a little complicated. REX is a methodology used to formalize an experience feedback acquired within a project or an activity [5]. It is mainly used to constitute a project memory through a process of the experience formalization [2]. MASK method is based on previous work of Ermine (1993) and was used at the beginning to capitalize the knowledge of researchers approaching retirement in three particular areas: the nuclear tests, the fast neutrons and the laser enrichment of uranium [12]. It has been then successfully evaluated on various organizations, for instance: the self-motivated sector within PSA Peugeot Citroen in France, Thales (defense systems), INRS (research institutes for security), Decathlon (sports goods) [1, 2, 5] and others. The MASK models representation is the result of long studies carried out in the field of knowledge engineering [4]. Represented in the form of graphical diagrams, they have the advantage of being simple to understand and generic enough to be applied in various domains [3]; its rather generic character allows an easy application to the very specific field of the craft industry. The set of models once built make up the book of knowledge that represents the ultimate goal of the MASK method. Thus, among all the methodologies cited above, MASK method appeared the best adapted to our case study. For that, the authors preferred to devote the following part (part 3 of the paper) to well define the method through its different knowledge models before engaging in the case study.

## 3 MASK Methodological Framework

MASK (Method of Acquisition and Structuring of Knowledge) [1, 5], which was originally only a knowledge capitalization method has largely evolved and gave birth to three components: MASK II, MASK I and MASK III that constitute in this order a global approach for knowledge management. Starting from knowledge mapping and strategic alignment (MASK II), to knowledge sharing and transfer (MASK III), including the Knowledge formalization (MASK I). In this paper, the authors are interested in particular in "MASK I" which is in fact the origin of the MASK method. Besides, the name "MASK" often refers to "MASK I".

According to MASK, the knowledge is perceived as information that makes sense in a given context [3]. This decomposition of knowledge into "Information, meaning and context" according to the "Semiotic" hypothesis is coupled with a second decomposition, which defines knowledge according to three other points of view, namely: structure, function and evolution according to the systemic hypothesis [3]. Cognitive genius is at the basis of the formalisms advocated for modeling the sense that those advocated for modeling context are based on the systemic [4]. From this follows 6 models of knowledge [3]:

- The so-called "phenomenon" and "activity" models that describe the activity in which knowledge is implemented and the phenomena that this activity seeks to control. These two models correspond to the analysis of the contextual point of view [3].
- The so-called "task" (or "know-how") and "concept" models which are intended to describe the knowledge and skills that are found in the knowledge system. These two models correspond to the semantic point of view of the "sense" [3].
- The so-called "historical" and "line" models that make it possible to describe the evolution of knowledge in a given context and the guiding lines that brought knowledge to the current state by giving it meaning [3].

The knowledge representation is based on a graphic modeling that the authors define in the following. It is about six different models (from Figs. 1, 2, 3, 4, 5 and 6) [6, 7].

### 3.1 Overview of MASK Knowledge Models

**The Phenomenon Model**

In a given system of knowledge, phenomena are generally characterized by the physical, chemical or social sciences, and are therefore described by mathematical, numerical or other models, most often complicated to understand and therefore to take into account in a knowledge management activity. However, in a knowledge management project, these phenomena are approached in a completely different way [5]. In the case of MASK, a phenomenon is modeled by starting with its initiator or triggering event, and at the end, the consequences of the phenomenon on the activity or system in question are defined. Until then, we are only telling facts. It is indeed "the elements of influence" which conceal know-how and a real expertise to find the parameters, which influence the whole phenomenon, and to find the answers to the following questions; what makes the consequences worse? how can the expert avoid this phenomenon? Thus, we will focus the capitalization effort on the level of "Elements of influence".

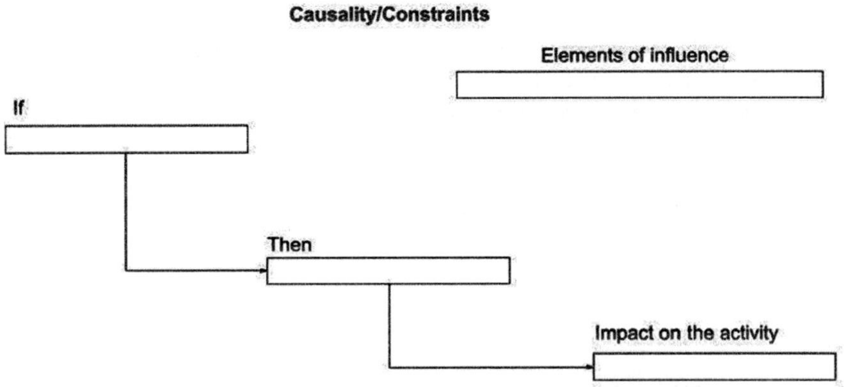


**Fig. 1.** "The Phenomenon Model" or "The Constraint Model"

**The Process Model**

It is a model that describes an activity through its inputs, outputs, resources and actors [3]. Flows upstream of the activity that may be streams of material or information are the inputs, those downstream outputs are the results provided by the activity. For the resources needed for the activity accomplishment, on the one hand we find: the means used to accomplish the activity in terms of materials, information and procedures etc. which are called resources on the diagram and in the other hand: the knowledge which is also necessary to carry out an activity, it is the knowledge resources. The graphic representation of all these elements is shown in Fig. 2. Graphically, the activity is represented by a rectangle symbolizing the black box, and arrows represent the other elements. Thus we can model a process by breaking it down into a succession of activities. When an activity is represented by a shaded rectangle, it means that it will be decomposed in another diagram, thus allowing a hierarchical description [5].

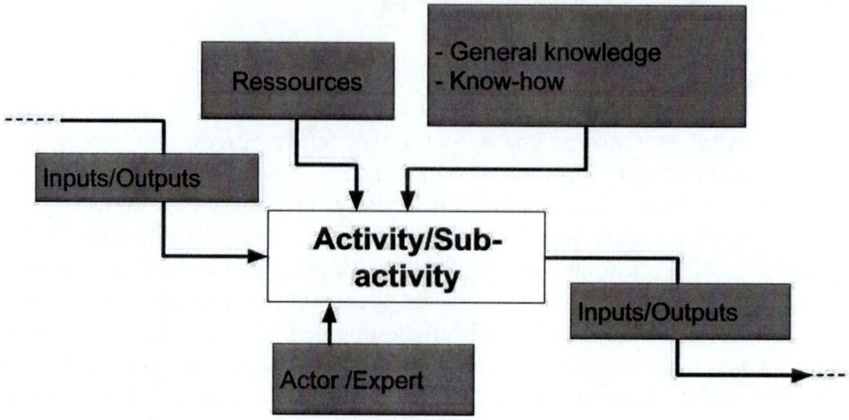


**Fig. 2.** "The Process Model" or "The Activity Model"

**The Task/Problem Solving Model** The task model describes the strategy implemented to solve a problem related to the knowledge system considered [5]. However, the traditional logic of the description of the reasoning in actions or successive tasks is not that advocated in MASK. The main objective is to highlight the objectives set through the actions taken to solve the problem. It is decomposition on several levels. A first level to describe the overall objective then in lower one the sub objectives (sub-goals) and in final level the actions implemented to achieve these objectives [6, 7]. The goal behind this representation is to offer a rather generic resolution strategy without involving any method of proceeding that can sometimes differ from one expert to another.

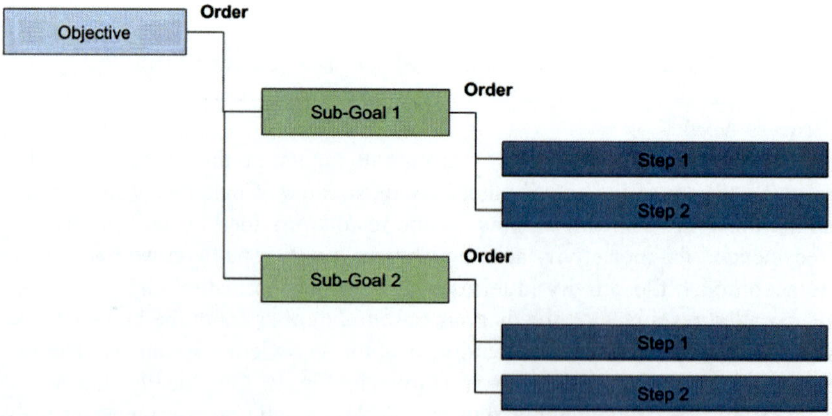


**Fig. 3.** The "Task" model or the "Problem Solving" model

### The Concept Model

While the task model describes a dynamic knowledge, the concept model represents the static aspect of knowledge [5]. Representation is a kind of fine taxonomy of domain inspired by semantic networks and object-oriented models [6, 7].

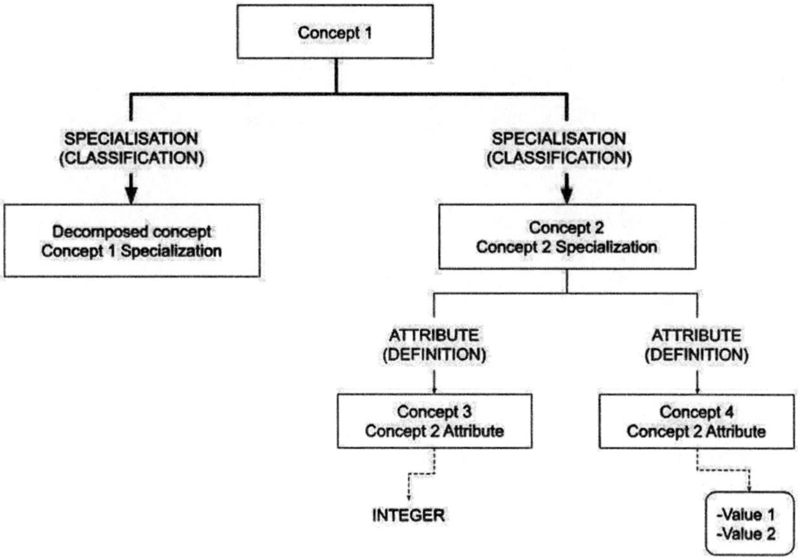


**Fig. 4.** The "Concepts" model

### The Historical Model

The historical model describes the evolution of knowledge and the path that has led to innovations, discoveries and improvements in order to comprehensively apprehend the guidelines that led to knowledge evolution [5].

During the historical analysis, a small number of seemingly relevant elements are used to describe the historical context and the evolution context. These elements are identified separately, and their arrangement over time is described succinctly, naming the large characteristic classes over time for these elements. These classes can be attached to goals that may have evolved over time. The time references are given by milestones, which are significant dated events for the considered element [5–7].

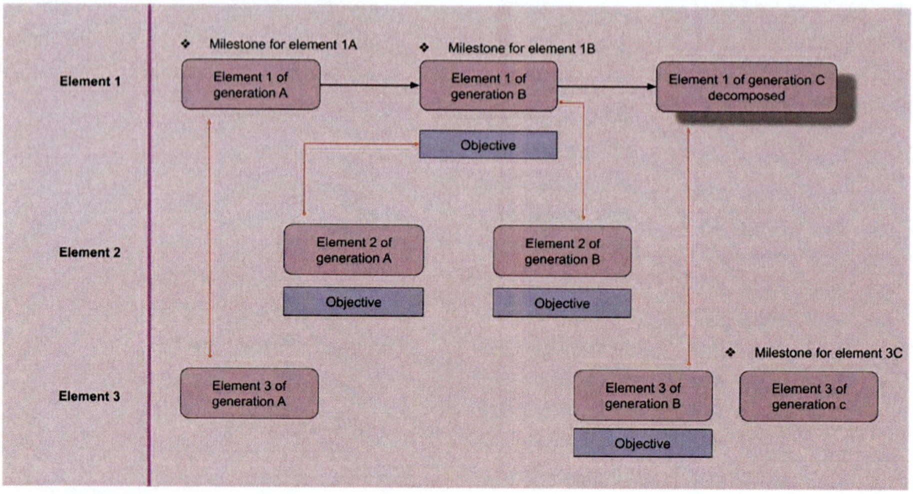


**Fig. 5.** The "Historical" model

## The Lineage Model

This is a tool for understanding afterwards. The lineage model proposes a general temporal reading, which aims at giving a relevant image of the evolution of the main system objects/concepts and which is reconstructed from a posteriori analysis and reasoned argumentation [3, 5].

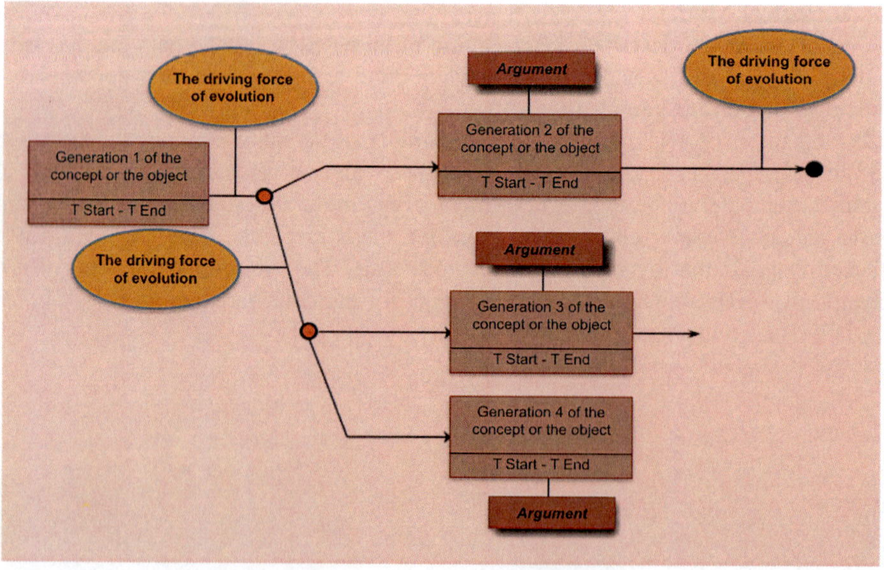


**Fig. 6.** The "Lineage" model

## 4 Issues of Applying MASK Within the Handicraft Sector

The MASK method was born in the CEA (Commissariat à l'Energie Atomique) in 1993 [5] and was subsequently applied in other fields, most of which are industrial fields [2] characterized by their formal aspect and whose actors (the knowledge holders/experts) are able to be part of the very formal loops of knowledge capitalization. Unfortunately, this doesn't fit the very specific field of craftsmanship. It is rather an informal environment [2] where the acquisition of expertise is done through practice in the absolute absence of abstract structuring. A completely oral transfer without any rules formalization. This on the ground reality does not allow a systematic application of the MASK method. In particular, two fundamental principles that characterize this method. It is about:

- First principle: The models Co-construction Principle with experts. Indeed, during the modeling activity, the "cognitician" engineer presents the models to the expert and invites them to be part of the models construction activity; they can fill the models by themselves, suggest modifications and give comments [1]. In this way, the expert sees himself as the master of his knowledge, becomes more motivated and seeks to highlight what he knows. This is a key success factor in MASK.
- Second principle: The models Validation with experts. Once the models have been built, experts must revise them for validation. This is a preliminary step to the publication of the knowledge book [1].

It is obvious that these two principles are difficult to apply to our case study. Most craftsmen do not have a sufficient education level to be able to learn the models and do the modeling on their own. Thus the knowledge extraction will be mainly based on direct observation of the craftsman activity using video recording and the interviewing using concrete examples and real situations.

## 5 Experimentation of MASK Within the Handicraft Sector

The geometric drawing of ornamental patterns or "the Hasba" technique is the subject of the MASK method experiment. The goal is to formalize this technique in the form of knowledge models. For our case study, we chose a wood artisan, which sculpts by hand the result of their drawing on wooden supports.

### 5.1 MASK Models Elicitation

The elicitation process of MASK models is summarized in the diagram of Fig. 8. A first draft of modeling will be carried out based on information resources, mainly the research lead by Thalal [11], which attempt to describe the geometric drawing method "the Hasba method" in free text approach. Afterwards, the models thus constructed will be enriched and completed through a series of interviews with Maâlem craftsmen. Video recording is the main tool for knowledge gathering. To help us to ask the right questions, we will always refer to a reliable guide, these are the generic models of CommonKads standard [8, 10] (Commonkads is a knowledge elicitation method for the development of knowledge-based systems). In our case, the "Hasba" method is indeed a technique of designing geometric patterns, for this reason the "Design" generic model (Fig. 7) was chosen to guide the process of knowledge elicitation. It describes the stages that make up a design process from the idea "Intention" to the "artifact" through various stages. Among them, three represent real expertise and need the most attention. Indeed, the most critical expert know-how is revealed within them, it is about "specification", "assessment" and definition of "Rules and Laws". It is therefore recommended to focus the elicitation effort on these three elements.

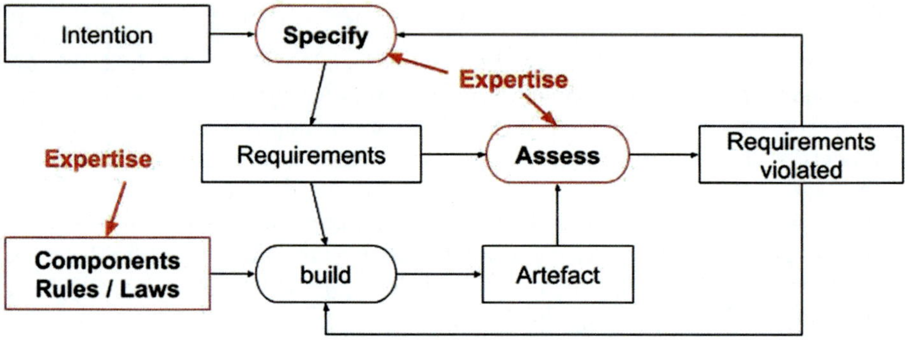


**Fig. 7.** The "Design" generic model of CommonKADS

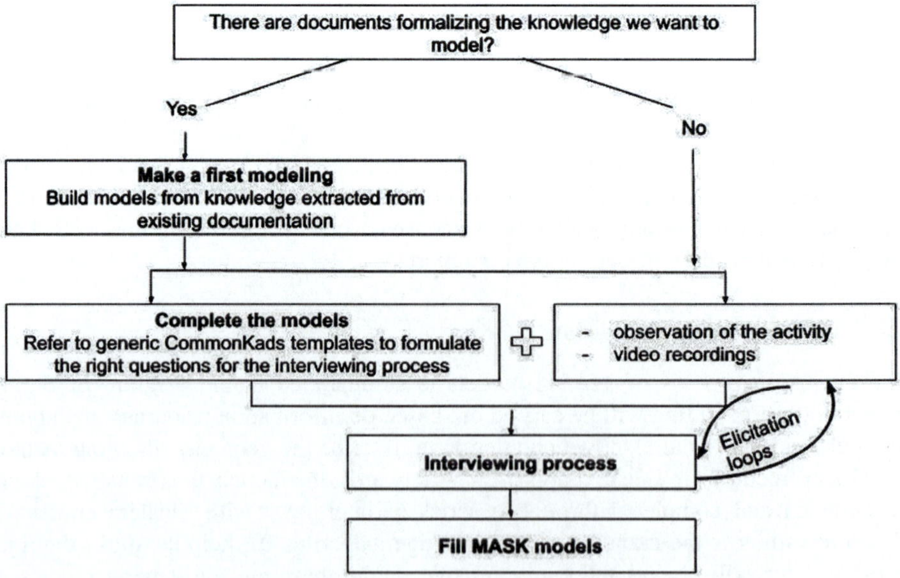


**Fig. 8.** The knowledge elicitation approach to build MASK models

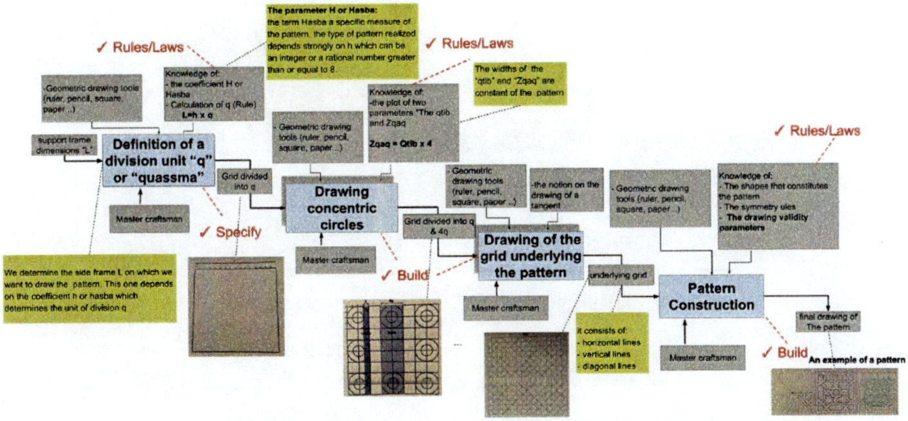


**Fig. 9.** The process model of the geometric drawing method "Hasba"

### 5.2 Case Study: MASK to Formalize the "Hasba" Method

We will proceed by following the procedure presented earlier in Fig. 8 for the knowledge elicitation related to the "Hasba" method. To do this, a single documentary reference will serve as a basis to build the first models. It is about is the work included in [11]. Since this is a method that can be broken down into steps, the process model is the first one to be built (Fig. 9 and Fig. 10) from which the other models will emerge.

We notice that the three critical knowledge elements suggested in the CommonKADS generics model (Fig. 7) appear well in the MASK process model (Fig. 9), namely "Specify", "Assess" and "Rules/Laws". In order to further extract the tacit knowledge that has escaped formalization in the documentation text, we have resorted to interviews with wood artisans whom we have selected for this study. Thus, videos were filmed and questions were asked to the maâlems. To illustrate this, we have chosen to present an example in relation to the "assess" component; this is the elicitation of a rule for evaluating the validity of the constructed pattern that appears in the zoom on the process model (Fig. 11). It checks whether the work performed is valid and will be accepted by peers.

The results of the interviewing process enabled us to model these validation rules in the form of a "Constraint" model which is shown in Fig. 12.

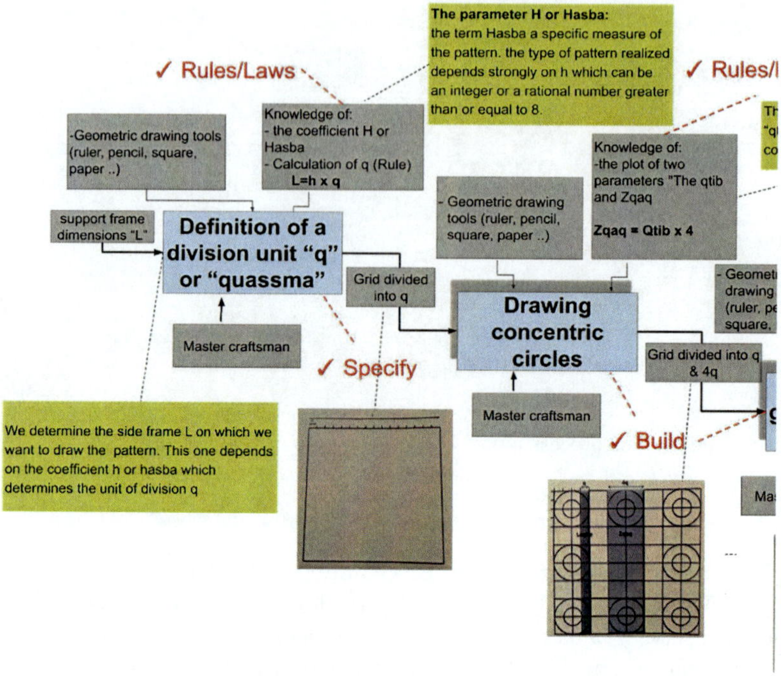


**Fig. 10.** A zoom on the process model of the "Hasba" method

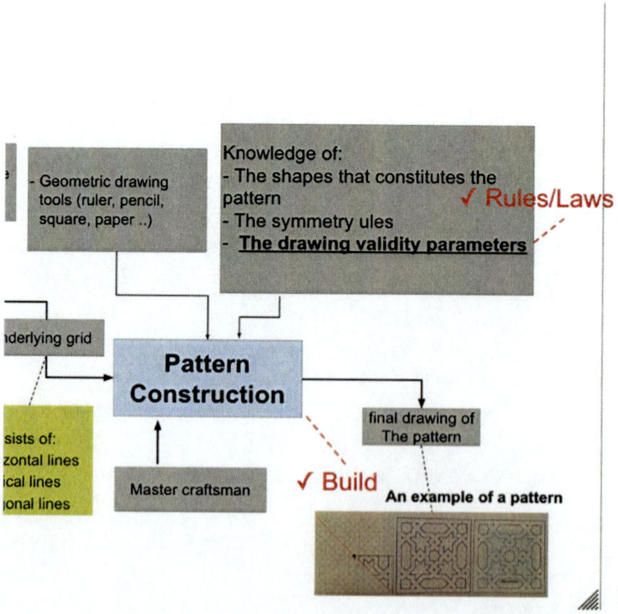


**Fig. 11.** A zoom on the process model to highlight the "Validity Rule" to be modelled

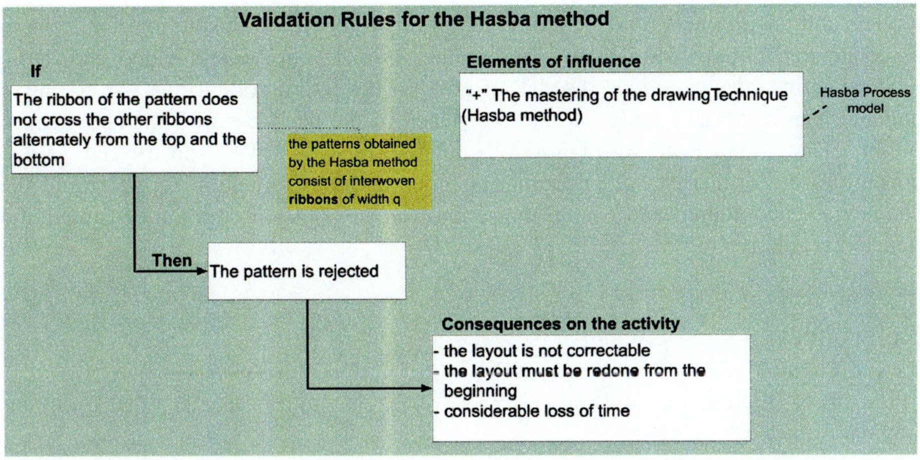


**Fig. 12.** Constraint model about the validation rule of the Hasba method

# 6 Results and Discussion

The models validation by experts remains a difficult requirement to meet, only if one may chance upon an expert craftsman with a fairly high level of education. It would then be possible to explain the models to the expert and ensure that he is able to read and understand them well. This was possible in our case thanks to the training centre of Moroccan crafts of Fez that opened its doors to us and allowed us to observe closely the craftsmen activity and to interview the most talented experts artisans who act as trainers.

In any knowledge capitalization project, it is important to have thought ahead to the exploitation of the deliverable and its re-use, namely, who can benefit and how. For our case, our deliverable corresponds to the formalisation of a design technique of geometrical patterns within the Moroccan craft industry that can also be called procedure, but which is represented in a graphic form called models with links between them. This procedure will feed a library of documents on behalf of apprentices to assist them in their learning but also for the trainers who can use it as a course support.

# 7 Conclusion

This paper enabled us to test and validate the application of MASK method to capitalize the know-how of Moroccan craftsman. This has been done through a very specific case study about the geometric drawing technique of ornamental patterns called "the Hasba". A procedure for the MASK models elicitation is proposed (Fig. 8) and knowledge models are constructed (Fig. 9 and 12). Through this experiment, we were able to highlight the failure to comply with certain principles of MASK method, namely: the models co-construction principle as well as their validation with the experts, this is due to the craftsmen profile who do not have the ability to apprehend the models and take ownership of it. To remedy this problem, a first solution would be to associate the models with illustrations in the form of photos and video and audio recordings in order to reduce the risk of error in the interpretation that the cognitive engineer can commit. Finally, we are convinced of the usefulness of this approach to preserve and enhance the know-how of craftsmen that otherwise run the risk of disappearing. Nevertheless, we encourage a more in-depth reflection to overcome the challenges encountered and ensure that a knowledge book on crafts is consulted and updated regularly.

## References


1. Matta, N., Ermine, J.L., Aubertin, G., Trivin, J.Y.: How to capitalize knowledge with the MASK method? (2001)
2. El Amrani, I., Saka, A., Matta, N., Chahdi, T.O.: A methodology for building knowledge memory within the handicraft sector. Int. J. Knowl. Manage. (IJKM) **15**(3), 45–65 (2019)
3. Castillo, O., Matta, N., Ermine, J.L.: Une méthode pour l'appropriation de savoir-faire, capitalisé avec MASK. In: EGC 2004: Extraction et gestion des connaissances, 4èmes journées, pp. 1–15, January 2004
4. Chaillot, M., Ermine, J.-L.: Le livre de connaissances électronique. Revue des Sciences et Technologies de l'Information - Série Document Numérique, Lavoisier **1**(1), 75–98 (1997)
5. Aries, S., Le Blanc, B., Ermine, J.L.: MASK: une méthode d'ingénierie des connaissances pour l'analyse et la structuration des connaissances. Management et ingénierie des connaissances, modèles et méthodes, Hermes sciences, pp. 208, Traité IC2, Série Management et Gestion des STIC (2008)
6. Ermine, J.L.: Les systèmes de connaissances (2000)
7. Ermine, J.L.: La gestion des connaissances (2003)
8. Leblanc, B., Dagorn, H., Bard, D., Ermine, J.L: Knowledge management in human radioprotection. In: International Symposium on Management of Industrial and Corporate Knowledge, Compiègne, pp. 211–220, October 1994
9. Guaglianone, M.T., Matta, N.: MNEMO (methodology for knowledge acquisition and modelling): definition of a global knowledge management approach combining knowledge modelling techniques. Adv. Inf. Sci. Serv. Sci. **7**(6), 161–162 (2012)
10. Breuker, J., Van de Velde, W. (eds.): CommonKADS Library for Expertise Modelling, Reusable Problem Solving Components. IOS press, Amsterdam (1994)
11. Thalal, A., Aboufadil, Y., Elidrissi Raghni, M.A., Benatia, J., Jali, A., Oueriagli, A.: Conception des motifs ornementaux par la méthode Hasba dans l'art géométrique marocain. Fondation de la Mosquée Hassan II de Casablanca (eds.). Colloque International 2013, LNCS, vol. 9999, pp. 125–148
12. Prax, J.Y.: Le manuel du Knowledge Management: une approche de 2e génération. Dunod, Paris (2003)

# Individual and Collective Competencies Modeling in Industrial Engineering

Bensouna Ikram(✉), Fikri Benbrahim Chahinaze, Sefiani Naoufal, and Azzouzi Hamid

Research Laboratory in Engineering, Innovation and Management of Industrial Systems, FST Tangier (AEU), Old Road of the Airport, Km 10 Ziaten. BP: 416, 90060 Tangier, Morocco
ikram.bensouna48@gmail.com

**Abstract.** Companies are leaning more and more towards a competency strategy. Competences represent a very important capital for the company both at the operational level and at the strategic level. However, companies find it difficult to find a framework or a model that will allow them to properly model the competency and reach a consensus on this notion. The purpose of this article is to propose an extensive model that takes into account all the factors surrounding the competency, individual or collective, which will enable the company to properly map its competency capital to facilitate its assessment and to achieve maximum performance. We start with a literature review on the notion of competence, its different typologies and components, as well as a return to the already existing models that were judicious in proposing a global model (MUCAC) that meets companies' expectations especially in the field of industrial engineering. The MUCAC model represents a real opportunity for companies wanting to immerse themselves in better management of their human capital.

**Keywords:** Competency modeling · Performance · MUCAC model

## 1 Introduction

Nowadays, competency management has become an indispensable factor for the success of strategic choices within companies. Competence management allows companies, regardless of their size, to make the most efficient choices possible for their human resources. that is to say, to assign the right person for the right position and in the right situation, this makes it possible to exploit the human capital well and to extract the best contributions from it. Individual and collective competencies are becoming increasingly important in the strategies developed by companies, whether in the short or the long term. In industrial engineering, few works deal with this theme of competency management. This has led us to find adequate answers to the following questions: how can a company visualize or map all the competencies that are present? How can companies create competencies? or even how can they evaluate them?

Through this article we have dealt with competencies modeling (individual and collective), we started with a literature review on competencies notion, it turned out that


A. Saka et al. (Eds.): CPI 2019, LNME, pp. 633–647, 2021.
https://doi.org/10.1007/978-3-030-62199-5_56

this notion is not easy enough and avoidable to approach, on the contrary it takes several forms in function of several factors. Then, we made a passage on some already existing models that treat the competency modeling from different angles in various application domains, and in the end we proposed a global competency management model that breaks down two loops, one of competencies creation and another one for its use or its mobilization to face a given situation by answering a predefined objective. This model allows companies to have a consensus on the competence concept and allows all employees to have a clear vision about the competency modeling approach adopted by the organization.

## 2 Around the Competence Concept

After having carried out several researches on 'competence', it turned out that this word takes several forms: term, notion, or concept and its definition have always known progress. The competence notion emerges at different times and in different scientific fields: linguistics, ergonomics, psychology, educational sciences, sociology, human resources management and business strategy. Given the diversity of definitions and fields of competence, it is more appropriate to begin by highlighting the main typologies of the competence and its components.

### 2.1 Individual Competence

As the name suggests, individual competence is primarily related to the individual entity. An individual is competent when he knows what he is doing and when he is facing a given situation with a certain relevance that sets him apart from others. Individual competence is based on several attributes:

**Knowledge:** reflects the knowledge acquired from professional experiences or learning through the educational achievements. It is composed of:

*Theoretical knowledge:* (declarative knowledge/knowing how to understand, knowing how to interpret) represented by scientific knowledge (Le Boterf 1995).

*Procedural knowledge*: refers to teaching and learning strategies.

**Skills:** the ability of an actor (individual or a work group) to perform a task based on acquired or learned knowledge. It is composed of:

*Procedural skills*: the approaches, methods, procedures applied and respected by the actor in a given situation.

*Social skills*: how to act within a group or organizational structure.

*Cognitive skills:* the intellectual operations allowing to approach, to describe a problem, to reason by analogy, to formulate hypotheses, to generalize until the creation of a new information from an already existing one.

*Experiential skills*: is an empirical knowledge related to an experience given in a professional or educational context.

**Attitudes:** the ability of an actor to produce actions and reactions respecting the elements that surround him. They are professional qualities related to the personality of the individual.

**The knowledge of how to do it**: Guy Le Boterf defines the knowledge of how to do it as an implicit knowledge acquired in the heat of the moment, with experience, by the repeated practice of the "in the heat of the moment" handling of professional problems, by approximations addressed on a case-by-case basis, by the recurrence of similar situations, by the systematic treatment of experience feedback, by working in a lining or a companionship (Le Boterf 2008).

## 2.2 Collective Competence

Collective competence is different from the sum of individual competences (Pemartin 1999). Indeed, collective competence rests on the coordination between the members of the team-work and the association of several attributes.

**Collective memory:** (Girod-Séville 1995) A set of knowledge acquired or created during an interaction with another individual or a group of people. It is a confrontation of knowledge via several individuals.

**Common standards:** (Montmollin 1984) and (Veltz and Zarifian 1994) A common representation allowing groups of individuals to perpetuate all the work done in coordination in the form of a heritage.

**Shared language:** (Le Boterf 1994) is a language specific to the collective labor; it allows it to distinguish itself from other collective labor groups. It is a group identity that allows it to save time in useless explanations by having abbreviations and terms shared between the labor groups.

**Relational factor:** is a factor that allows the identification of agreement and understanding degree between the members of the group. Indeed the collective competence is different from the sum of the individual competences, we can have a competent person X in his field Y but not competent in the working group. This justifies the relationship aspect between the members of the group that can lead to conflict which negatively impacts the group or even the collective competence. Such factor is articulated around 3 axes, the first one is Trust, it is the sense of security allowing the person X to express themselves and to share their ideas without fear of judgments or others opinions. In general, this step concerns a new member joining the team-work and if it is poorly exploited it generates a gap between the members of the group or a poor integration of this new member. The second axe is Debate, it's co-constructing solutions from each idea presented separately, it's creating a space for discussion, analysis and comparison between a set of solutions proposed in order to settle an arrangement and it's the third axe constituting the relational factor which makes it possible to keep and agree on the proposed solution, it can be a relevant idea presented separately or a merger of several ideas, the important thing is that at this stage the solution of the problem must be defined in a common way.

**Collective Autonomy:** is the ability of the team-work to progress and move forward without having to give permissions and validations request. The group presents a power within that should not be misused. This autonomy will allow the group to strengthen its confidence and develop more and more relevant solutions beyond what the superior will think. Because managers are more interested to results and performances than the obstacles that face the group. The team-wok is supposed to present final results that will allow reaching collective performance beyond the circumstances or the blocking points.

## 3 Competency Models

There are several studies that took the lead to model competences. The competency modeling in a company allows it to explain and map the principle of the competence concept and to make it easily exploitable by each individual and team-work (Draganidisand Mentzas 2006), so it constitutes a reference to provide training (Naquin and Holton 2006). Thus, it makes it possible, among other things, to facilitate the competence evaluation thanks to a clear visibility of the components that form it.

Over time, competency management will become more and more important, competence will be important for the achievement of the objectives of the company (Hamel and Prahalad 1994). In fact, competency modeling is a tool that allows companies to agree on a common language. Competency models reflect the strategies, objectives and vision of the company in regards of its employees. A competency model can also be defined as a descriptive tool that aims to identify competencies needed to meet the expectations of the company (Sampson and Fytros 2008). The model should provide a clear and comprehensive definition for each competency and the elements for its creation and evaluation, to allow evaluators to have a common basis of review and HR managers to develop training and recruitment plans for where necessary and according to the results obtained from the evaluation made. Competency models can act as a checklist or audit to identify existing competencies. The final list of competencies expected in the competency model are sometimes amazing for the company. Because the company can identify competencies that they did not take into account beforehand. And thus having a competence model will allow a better anticipation of the hazards, and a better adjustment of position in case of need. The structure of the model must therefore support the use of competencies in human resource functions (Kubes et al. 2004). Indeed, the competency model is considered as a linkage tool between the development of human resources and the strategy elaborated by the company. Therefore competency modeling aims to recruit staff, create training programs, benchmark performance, collect information about learning and work experiences, plan and manage structural changes within companies (Overton 2013). On the paragraphs bellow we will try to go through some competency models.

### 3.1 The Competence Model of Miranda et al. (2016)

The Miranda model is based on the idea that competence has three different key points: the skills that are the elements of KSA (Knowledge, Skills, Attitude), the context in which the skill itself has been acquired and the evaluation in terms of how the skill itself was acquired. Thus, the model includes two notions, the static notion and the dynamic notion. For this model "Competence, associated skills and context are static and are always within some kind of taxonomy, whereas the relationship between competence and context is not static but it depends on where the person has acquired the skill itself" (Miranda et al. 2016).

### 3.2 The (MADDEC) Model of Coulet (2011)

Coulet's model (MADDEC) takes into account the individual and collective competences and processes that contribute to their mobilization and construction. According to Coulet (2011), "any activity, whether individual or collective, is structured by a scheme". On the other hand (Vergnaud 2007), defines the scheme as "an invariant form of organization of activity and conduct for a given class of situations". The four components of the scheme:

**Operative invariants:** that are conceptual foundations of the activity, according to Vergnaud (2007). "Their main function is to collect and select relevant information and infer useful consequences for action, control and use of subsequent information".

**Inferences:** these are adjustments according to the circumstances which, according to Vergnaud (1990), concern the mobilized scheme (valid for a class of situations), according to the specificities of the situation and the task. They are based on past experiences in order to choose from all the available task rules, the most adequate ones.

**Task rules:** these are modalities of the implementation of the desired result. They also represent schemes subordinated to the higher scheme.

**Expectations:** they are the expected results of the activity undertaken, they are mediated by artifacts. Coulet also uses a notion of regulation through two loops:

**Short loop:** Matching task rules with special circumstances. The interest at this level is the elaboration of the results.

**Long Loop:** This loop is based on an understanding of the reasons of success or failure. We talk about an interest in success.

### 3.3 The CKIM Model of Vergnaud et al. (2004)

The CKIM model of Vergnaud et al. (2004) assumes that competence has four main characteristics: 1) a competence may be required by a domain or acquired by the individuals of a domain, 2) the competence has resources structured into categories 3) competence is realized in a context and 4) competence is related to the accomplishment of one or more tasks or missions (Harzallah 2000a, b). Indeed, the CKIM model is a competences management tool that seeks to determine who can accomplish a mission, to better assign human resources to tasks without looking at methods and ways to perform them. In the medium and long term, this model seeks accounting the competences acquired in a system, to pilot processes or to define a strategy. It has four main parts: Competence & Resources, Domain, Knowledge and Individuals. Knowledge is defined as information taking on certain meanings in a given context, and allowing tasks to be performed (Ermine 2000; Collin 2001; Gardoni 1999).

### 3.4 The Global Model of the Competence «Collective, Individual, Organizational» (MGCCIO) of Fikri Benbrahim et al. (2018b)

The MGCCIO global competence model (Fikri Benbrahim et al. 2018b) is used to represent professional competences. It details the concept of competence in order to have a global and coherent vision taking into account: the hierarchy of skills, individual competence and its typologies (knowledge, skills, attitudes and the way of being),

Collective competence and its typologies (knowing how to cooperate, having a common reference, having a collective memory, engaging collectively, learning collectively and having a common language), organizational competence and its typologies (cultural aspects, routines and organizational learning), the repository of individual competences and the repository of collective competences, the process and/or the projects described through the referential of the collective competences, the reference system of employment relating to the frame of reference of the individual competences, the resources and its typologies, assessment of individual competences and the level of collective competence through performance indicators.

### 3.5 The Project Ecosystem Competency Model of Park (2016)

The "Project ecosystem competency model" (Park 2016), is a model that aims to create a sustainable performance within companies. This model is based on three main entities:

**Design entity and maturity of the PECM**: via the levels: portfolio, program, project and operations.

**Entity to update success factors**: (multi-project management) through the process of identifying success factors and performance index.

**Engineering Project Management Maturity Entity**: via competences upgrades. Three types of competencies were discussed: engineering competencies, system engineering competencies and project management competencies. The essential point for improving the EPM competence is the development of a project data warehouse system. So basically the PECM model is about creating sustainable performance in companies as well as valuing project stakeholders for creating sustainability based on a data warehouse system from classified competences.

### 3.6 The Competency Model for Reading Energy Efficiency in Production Engineering of Müller-Frommeyer et al. (2017)

The competency model (Müller-Frommeyer et al. 2017) aims to list the competencies set for students working with a factory in the learning setting. The model is based on a total of 12 skills, and is part of 3 large families: professional/methodological skills, social skills and personal skills. The results of this model provide an opportunity to establish and adapt competency-based teaching concepts in higher education and organizations. This model is a real means of identifying and naming a set of competencies representing human capital for the company.

### 3.7 The Systemic-Interactionist Model to Design a Competency-Based Curriculum of Andronache et al. (2015)

The competency-based interactionist-system model illustrates the role of the interaction of knowledge, skills and attitudes with competencies training. This mobilization is the result of a definite objective which then generates an evaluation of the results obtained (feedback). The model uses several notions:

**Cognitive content**: which is considered content related to knowledge and leading to its formation and development. It is also defined as a knowledge system that facilitates the learning process.

**Actional content**: are complete cognitive contents in a dialectical relationship with them, fulfill a role in training and competences development. The role of this action content is to facilitate application and transfer in specific situations (theoretical or practical) while facilitating problem solving.

**Attitudinal content**: as a knowledge system designed to guide motivations, train and develop students in the exercise of a profession. The model also assumes the existence of sub-competences that represent the description of each component of sub-competences (knowledge, skills, attitudes), the sum of all sub-competences descriptors representing a general description of the quality of the competence.

### 3.8 The Graphic Model of Competence Management Ontology of Fikri Benbrahim et al. (2018b)

The ontological model of Fikri Benbrahim et al. (2018b) allows managers to identify, roll out and develop individual and collective competences based on an ontological analysis. It also ensures the reuse of information and the processing of data concerning recruitment. The model is operational and adjustable according to the company's expectations. This model uses the following concepts: company, employee, strategy, competency framework, individual and collective competency, competency assessment, action plan, skill building, attribution, recruitment and development. In addition, it is dissect in five steps:

**Step 1:** This is the step of understanding the company, its expectations, strengths and weaknesses.
**Step 2:** Develop an action plan (Plan) to ensure the success of the company's strategic plan which is: skill improvement, attribution and recruitment.
**Step 3:** The implementation of the action plan (Do) based on the results of the first step.
**Step 4:** The evaluation (verification) based on the strategy developed by the company through a framework competency.
**Step 5**: The company conducts another competences assessment and competence strategy audit and develops skills to elaborate corrective action plans to address identified gaps between required and available competences and also to improve the weaknesses of the competence management system (Act).

### 3.9 Competency Mapping and Job Performance of Kalita and Singha (2018)

It is an approach that connects the human behaviors with the skill and its application in a company. The goal is to strengthen the competitive advantages of the company over others. The approach considers employee competence as the sum of Skills, Attitude and knowledge.

The study is based on the following objectives:

1. Identify the influence of selected competency dimensions on job performance.
2. To measure the overall competencies level of employees in relation to work performance,
3. Evaluate the relationship between competency level and job performance.
4. Suggest corrective measures to improve the skills and job performance of the organization's employees.

So, this model has proven to be an effective HR tool to improve the performance level of the organization.

### 3.10 Competency Model for Logistics Employees in Smart Factories of Kohl et al. (2019)

The logistics-specific competency model consists of four different categories: self, social, methodological and professional competencies. These categories are divided into 24 sub-competencies in order to detail the main groups without a prioritization or weighting between the sub-competencies. Summing up, the competency model is applicable in different processes and fully depicts the logistical activities. Competences were sufficient to map future scenarios in the processes. The ability to determine logistics-specific competences, clarify and specify upcoming changes, and induce early engagement with future challenges. In addition, the model can serve as a basis for strategic staff planning during digitization. Several indicators of new technologies have been defined as indicator of important process changes. In addition, organization-specific adjustments may occur in the enterprise, which may also be an indicator of revision of the competency model. The purpose is to keep the logistics competency model up-to-date. Regular exams can ensure that the competences required for future changes are systematically identified.

## 4 Comparative Analysis of the Models Studied and Contribution to the Competency Modeling

### 4.1 Synthesis of the Studied Models

The competency modeling as we have seen through the models above, has been exploited in different ways. Each model brings back a new idea in order to achieve a representation as significant as possible. As you can see, we tried to summarize some models that we have studied from 2004 to 2019 especially because the competency modeling is a theme that has been developed and approached according to several domains. Starting with the CKIM model which links between the domain the resources and the context in which the competence is mobilized in associated mission. For the MADDEC model it is based on the scheme notion, Coulet affirms that each activity is structured by a scheme that plays the role of an organizational actions process, then the three models of Daniel, Lena and Markus which share approximatively the same concept in the sense of the competence categories: social, personal, professional and methodological competencies. For the Miranda model, it relies on three part the part KSA (Knowledge, skills, attitudes), the context and the part of competencies assessment. In regards on the Rijumany model it connects the skills to the behavior. For the park model, he put the point on the concept of sustainable performance in industrial companies. According to Fikri she proposed two models in her first model the competences was presents in a global way, and in her second model she calls and takes into account the both notions of individual and collective competence. Therefore, as a synthesis, we have tried to list the set of factors surrounding this thematic which are purpose, situation, knowledge, performance, resources and competence. Competences could be divided into individual (skills, attitudes and knowledge) and collective (shared language, common standards and collective memory).The analysis of the ten models is summarized in Table 1.

**Table 1.** Comparative analysis of studied models

| Models / Components | | Miranda's model 2016 | Coulet's MADDEC Model 2011 | Nathalie's CKIM model 2004 | Chahinaze's MGCCIO model 2018 | Chang-woo's PECM model 2016 | Lena's Competency model for EEPE 2017 | Daniel's systemic-interactionist Model 2015 | Chahinaze's and al. ontological model 2018 | Rijumani Kalita1 and Dr. Seema S Singha Approach 2018 | Markus kohl and al. competency model |
|---|---|---|---|---|---|---|---|---|---|---|---|
| Situation | | + | + | + | | | | | | | |
| Purpose | | + | + | + | | + | | + | | | + |
| Results | | | + | | | + | | + | + | | + |
| Knowledge | | | | + | | | + | + | | + | + |
| Individual competence | Skills | + | | + | + | | + | + | | + | |
| | Knowledge | + | | + | + | | + | + | | + | |
| | Attitude | + | | + | + | | | + | | + | |
| | Knowledge of how to do it | | + | | | + | Application of knowledge | | + | | + |
| Collective competence | Collective memory | | | | + | | | | | | |
| | Common standards | | | | + | | | | | | |
| | Common language | | + | | + | + | (Capacity for Teamwork) | | + | | + |
| | Relational factor | | | | | | | | | | |
| | Collective autonomy | | | | | | | | | | |
| Knowledge of how to act in common | | | | | | | | | | | |
| Resources | | + | + | + | + | + | | | | | |
| Performance | | | | | + | + | | | + | + | |
| Collective performance | | | | | | | | | | | |
| Scheme | | | + | | | | | | | | |

## 4.2 Contribution to Competency Modeling

Based on the literature review discussed above we proposed a competency model called Model of Use, Creation and Assessment of Competences (MUCAC). It takes into account all the elements by turning competence from the reception of source material to the development of the competences with its typologies and components.

**Actor:** An input element can be a company, a department manager, a manager, a simple employee who is facing a situation.

**Situation:** it's an activity to carry out, an event to be faced, a problem to solve, a project to achieve … etc. (Le Boterf 2008).

**Purpose:** it's an expected result, a desired goal.

**Resources:** it's a necessary condition to achieve the objective defined, they are means, tools, environmental links …

**Competences:** they can be individual (knowledge, skills, attitudes, knowledge of how to do it) or collective (collective memory, common standards, common language, collective autonomy, relational factor).

**Results:** these are the output elements of resource mobilization in coordination with the competences.

**Performance:** it's a tool to measure and evaluate the effectiveness of the result obtained according to the programmed or well-defined objective.

**Knowledge:** according to Vergnaud et al. (2004) it's the information that takes on a certain meaning in a given context, and allows the performing of tasks.

**Scheme:** it's a way of identifying, sorting, combining, interpreting, extrapolating, differentiating and mobilizing knowledge to face a single situation (Perrenoud 1994). The scheme is composed of operative invariants, inferences, rule of actions and anticipations (Vergnaud 2007).

**Collective performance:** represents means of evaluating the collective competence in the framework of collaboration between several individuals through the knowledge of how to act in common.

**Knowledge of how to act in common:** is a commitment to problem solving in cooperation, it's the solidarity of the working group. It represents the ability of the group to advance in common to accept others by remaining open to each idea presented, it is the force that makes the difference between an individual working alone and a working group with collective responsibilities. Knowing how to act in common represents the link between individual competence and collective competence; it is a transition between the individual entity and the group entity.

The proposed model relies on two processes, one of competence-building and the other one for the use of it, with the measurement of its relevance through a performance indicator that links the defined objective and the results obtained.

**Competences creation process**: this process starts with the reception of knowledge, it is the result of a set of analyzed data processed and stored in the memory of the individual. Individuals during their lives, whether professional or personal, acquire experience and memorize the results obtained during the various problems encountered in the form of knowledge. The knowledge, if it is guided by a scheme generates a competence. The competence is not only having baggage which means having the different components of the competence (to have competences) but the real competence lies in the result of an intelligent mobilization of the different knowledge acquired (to be competent). The scheme plays the role of an organizational structure and manager of knowledge; it allows individuals to choose among a set of knowledge that is best suited to the situation encountered, to draw the attractive results that can be used to solving the problem or dealing with the project concerned. To be competent is therefore to use with relevance and intelligence the knowledge acquired. The process of achieving this skill is called "scheme".

**Competences using/mobilization process:** an individual is confronted to a situation, it can be the leader of the company, the manager of service etc., for example a methods department responsible is faced to an industrialization of a new product. It starts by defining the goal, for example industrialize the product X in a well-defined period with a budget to respect. To achieve this goal, it is necessary to use two elements: resources (machines, tools, applicable documents … etc.) and competences (a staff to mobilize) that can be individual (knowledge, skills, attitudes, and knowledge of how to do it) or collective (collective memory, common standards, common language, relational factor, collective autonomy). Indeed, the transition from an individual competence to a collective one is guided by the knowledge of how to act in common and it allows individuals to work in a group. We will mention an evaluation factor or a measure of relevance known as collective performance. Moreover, more and more companies rely on the collective competency that represent strength for the company or even a very powerful capital of expertise. The mobilization of resources and competencies generates a result that can be positive or negative; this judgment can be done through the performance of the result obtained. The performance therefore allows positioning the result obtained according to the objective set at the beginning. The process is therefore said to be efficient if the result obtained is as close as possible to the objective defined while respecting the required conditions, for example (deadlines, budget, etc.).

The MUCAC Fig. 1 model is based on the following definition of competence:

> Competence is a process mobilized to face a defined goal in a particular situation. It uses a set of knowledge guided and structured by a scheme. The result obtained from this mobilization can be judged by its performance in relation to the defined objective, or by the collective performance when we speak about collective competences.

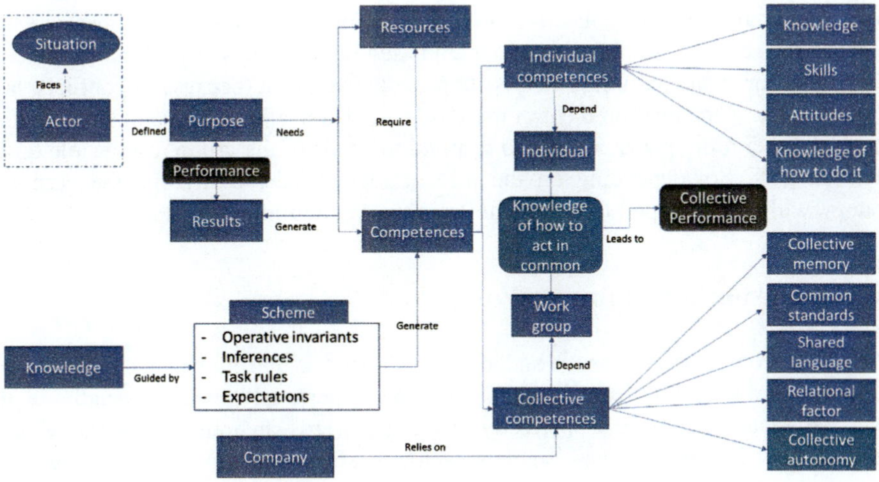


**Fig. 1.** Competences Model (MUCAC) proposed by Ikram Bensouna et al. (2019)

## 5 Proposal Applications of the MUCAC Model

The proposed model is intended to provide a clear consensus on the competency concept and to facilitate its management within the company. Indeed, we cannot talk about competency management if we do not have the same definition. This model represents the ultimate basis for creating tools or methods for competency management. In addition, the MUCAC model allows HR, managers and employees to have an apparent mapping in the sense of:

**Competencies definition**: for example, in an organization, when we speak about a competent element (individual or work group), we do not have the same definition in mind. That is to say, how is this person competent? what are his assets? what are the parameters that make it possible to decipher the competency term? Therefore, these are unanswered rotating questions. As a result, having the same definition or the same mapping within the company will create trust and credibility among members of the entire company and will also facilitate the competency assessment.

**Competencies creation**: for example in a need of a competency X, the manager knows very well that he must offer a training around the knowledge forming the desired or requested competence, or he must select individuals having knowledge within the framework of the required competence. For instance, in the case of a mechanical design project for a Japanese customer, the manager must select from the members of his team the one who has knowledge on design and Japanese language. The validation of an individual's competency is based on the way which they use their knowledge and actions implemented. Having competences does not necessarily mean that the person is competent, but only if he guides or calls his knowledge through a well-structured process known as 'scheme'.

**Competencies assessment**: thanks to two essential indicators, performance as a global indicator that allows to place the results obtained in relation to the defined objectives, and the collective performance within the framework of collective competence. That is the measure of the performance of a working group based on an acting on common. Indeed, the collective performance makes it possible to measure and visualize the strength or weakness of a team based on the issues or projects handled by this group. For example, if the result is supposedly irrelevant (because of conflicts, lack of skills … etc.), the measurement of these two indicators allows the company to define training or recruiting plans in order to respond to needs of the group responsible or the HR managers. So, depending on the measurement of the indicators, the company strategy will be developed (recruitment, training, teambuilding … etc.)

## 6 Conclusion and Future Work

Nowadays, companies are more and more interested in the human factor. They are constantly growing both professionally and on a personal level. Especially in the industrial sector, Morocco is perfectly moving towards an industrial strategy: new projects, new industrial zones, so new factories are setting up and therefore more competition for existing companies. Therefore, how can companies cope with this abrupt change in order to mobilize human resources? and how can the company easily choose the right people for the right job and for the right projects? how can they map their human capital? and how can they measure the performance of mobilized human resources?

The proposed MUCAC model takes place in this cloud of questions, to face the hazards encountered by companies and to best meet the requirements and expectations of the company whether at the strategic or operational level. This model will allow small, medium and large companies to better understand their human capital, to better visualize weaknesses in choosing the best action plan.

The analysis of the notion of competence has led us to have several divergences about this concept, which is due to the different links that can be established with this concept. In this model, we have taken into account all the factors surrounding this notion within the company. We insisted in particular on four links:

1) Link between competence and knowledge: The conditioning of knowledge through a mobilized scheme will lead to competence, here we have the support on how a competence can be created.
2) Link between individual and collective competence: through a common act knowledge that allows one or more individuals to work collectively.
3) Link between competence and performance: via the comparison of the results achieved according to the objectives defined, we talk about an efficient system in the case where the final results coincide as much as possible with the defined objective.
4) Link between collective competence and collective performance: we speak about a collective performance when the collective work to achieve the goal defined is coordinated between all the members of the group. The more knowledge of how to act in common is relevant the more the collective performance is maximal.

These links show that the MUCAC model is a global model that takes into account a large number of factors influencing competence.

In perspective, we intend to use this proposed model to establish means to measure and evaluate competence. Wouldn't it be very interesting to measure competences? to say that a person X is more competent than a person Y in the domain Z? Would it not be relevant to say that Working Group A is more efficient than Working Group B? All of this, automatically, according to criteria that the company itself will fix. This approach will allow the company to strengthen its credibility towards its employees, and to create a certain confidence and comfort for the future of these employees. Not only that, but the company will be able to base itself on this approach and be able to make evolutions of posts, plans of formation or more to face the brutal hazards met.

## References


Le Boterf, G.: De la compétence, essai sur un attracteur étrange. N.49, Editions d'organisation, Paris (1995)

Le Boterf, G.: Construire les compétences individuelles et collectives, Agir et réussir avec compétence, les réponses à 100 question. Editions Eyrolles, Paris (2008)

Pemartin, D.: Gérer par les compétences ou comment réussir autrement. Management et Société (EMS), Caen (1999)
Girod-Séville, M.: La mémoire organisationnelle. Harmattan Montéreal, Paris (1995)
Montmollin, M.: L'intelligence de la tâche- Eléments d'ergonomie cognitive. Peter Lang, New York (1984)
Veltz, P., Zarifian, P.: Travail collectif et modèles d'organisation de la production. Le Travail Humain **57**(3), 239–249 (1994)
Le Boterf, G.: De la compétence, essai sur un attracteur étrange. N.67, Editions d'Organisation, Paris (1994)
Draganidis, F., Mentzas, G.: Competency based management: a review of systems and approaches. Inf. Manag. Comput. Secur. **14**(l), 51–64 (2006)
Naquin, S.S., Holton, E.F.: Leadership and managerial competency models: a simplified process and resulting model. Adv. Dev. Hum. Resour. **8**(2), 144–165 (2006)
Hamel, G., Prahalad, C.K.: Competing For the Future. Harvard Business School Press, Boston (1994)
Sampson, D., Fytros, D.: Competence models in technology-enhanced competence-based learning. In: Adelsberger, H.H., Kinshuk, Pawlowski, J.M., Sampson, D.G. (eds.) Handbook on information Technologies for Education and Training, pp. 155–177. Springer, Heidelberg (2008)
Kubes, M., Spillerova, D., Kurnicky, R.: Manazerske kompetence – Zpusobilosti vyjimecnych manazeru, Grada publishing, Praha (2004)
Overton, W.F.: Reasoning, Necessity, and Logic: Developmental Perspectives. Jean Piaget Symposia Series. Taylor & Francis, New Jersey (2013)
Miranda, S., Orciuoli, F., Loia, V., Sampson, D.: An ontology-based model for competence management. Data Knowl. Eng. **107**, 51–66 (2016)
Coulet, J.C.: La notion de compétence : un modèle pour décrire, évaluer et développer les compétences, Le Travail Humain, 1 vol. 74, pp. 1–30. Presses Universitaires de France (2011)
Coulet, J.C.: Le concept de schème dans la description et l'analyse des compétences professionnelles: formalisation des pratiques, variabilité des conduites et régulation de l'activité. In: Merri, M. (ed.) Activité humaine et conceptualisation. Questions à Gérard Vergnaud, Toulouse, pp. 297–306. PUM (2007)
Vergnaud, G.: La théorie des champs conceptuels, Recherches en Didactique des Mathématiques, vol. 10, no. 2–3, pp. 133–170. La Pensée Sauvage, Grenoble (1990)
Vergnaud, N., Harzallah, M., Briand, H.: Modèle de Gestion Intégrée des Compétences et Connaissances. In: Conférence: Extraction et gestion des connaissances (EGC 2004), Actes des quatrièmes journées Extraction et Gestion des Connaissances, Clermont Ferrand, France (2004)
Harzallah, M.: Modélisation des aspects organisationnels et des compétences pour la réorganisation d'entreprises industrielles, Thèse de l'université de Metz (2000a)
Harzallah, M.: Guide d'utilisation de la base de compétences du service maintenance de Trémery, Rapport de recherche, LGIPM 2000–1, Université de Metz (2000b)
Ermine, J.-L.: Les systèmes de connaissances, p. 144. Hermes Science Publication, France (2000)
Collin, R.: Gestion des connaissances et aide à la décision. Séminaire «Connaissances, compétences et technologies. Pourquoi l'homme est la mesure de toute information?» (2001)
Gardoni, M.: Maîtrise de l'information non structurée et capitalisation de savoir et savoir-faire en Ingénierie Intégrée. Cas d'étude Aérospatiale. Thèse de l'Université de Metz (1999)
Fikri Benbrahim, C.: Thèse: Proposition des démarches de pilotage des compétences collectives et modélisation d'un système d'aide à la décision pour le développement des compétences au niveau de la supplychain, Maroc (2018a)

Park, C.: Project ecosystem competency model. Procedia Soc. Behav. Sci. (226), 116–123 (2016)
Müller-Frommeyer, L.C., Aymans, S.C., Bargmann, C., Kauffeld, S., Herrmann, C.: Introducing competency models as a tool for holistic competency development in learning factories: challenges, example and future application. Procedia Manuf. (9), 307–314 (2017)
Andronache, D., Bocoş, M., Neculau, B.C.: A systemic-interactionist model to design a competency-based curriculum. Procedia Soc. Behav. Sci. (180), 715–721 (2015)
Fikri Benbrahim, C., Jaber, E., Hajira, B., Cheikhrouhou, N., Sefiani, N., Reklaoui, K.: Towards an ontological approach to company compenteces management. In: 2018 International Colloquium on Logistics and Supply Chain Management (LOGISTIQUA), France (2018b)
Kalita, R., Singha, S.S.: Competency mapping and job performance: a case study of big bazaar. Int. J. Commer. Manag. Res. 98–102 (2018)
Kohl, M., Heimeldinger, C., Brieke, M., Fottner, J.: Competency model for logistics employees in smart factories. In: Karwowski, W., Trzcielinski, S., Mrugalska, B. (eds.) Advances in Manufacturing, Production Management and Process Control. AHFE 2019. Advances in Intelligent Systems and Computing, vol. 971. Springer, Cham (2020)
Perrenoud, Ph.: L'ambiguïté des savoirs et du rapport au savoir dans le métier d'enseignant, Université de Genève, Faculté de psychologie et des sciences de l'éducation (re-pris dans Perrenoud, Ph., Enseigner : agir dans l'urgence, décider dans l'incertitude. Savoirs et compétences dans un métier complexe 1996, ESF chapitre 6, Paris, pp. 129–159 (1994)

# Author Index





A. Saka et al. (Eds.): CPI 2019, LNME, pp. 649–651, 2021.
https://doi.org/10.1007/978-3-030-62199-5